Practice Problems *for the*
Mechanical Engineering PE Exam

A Companion to the
Mechanical Engineering Reference Manual

Twelfth Edition

Michael R. Lindeburg, PE

The Power to Pass
www.ppi2pass.com

Professional Publications, Inc. • Belmont, California

How to Locate and Report Errata for This Book

At PPI, we do our best to bring you error-free books. But when errors do occur, we want to make sure you can view corrections and report any potential errors you find, so the errors cause as little confusion as possible.

A current list of known errata and other updates for this book is available on the PPI website at **www.ppi2pass.com/errata**. We update the errata page as often as necessary, so check in regularly. You will also find instructions for submitting suspected errata. We are grateful to every reader who takes the time to help us improve the quality of our books by pointing out an error.

PRACTICE PROBLEMS FOR THE MECHANICAL ENGINEERING PE EXAM
Twelfth Edition

Current printing of this edition: 3

Printing History

edition number	printing number	update
12	1	New edition. Revised thermodynamic properties of steam incorporated.
12	2	Minor corrections.
12	3	Minor corrections.

Printed in the United States of America

PPI
1250 Fifth Avenue, Belmont, CA 94002
(650) 593-9119
www.ppi2pass.com

Library of Congress Cataloging-in-Publication Data
Lindeburg, Michael R.
    Practice problems for the mechanical engineering PE exam: a companion to the mechanical engineering reference manual / Michael R. Lindeburg. -- 12th ed.
        p. cm.
    Includes index.
    ISBN: 978-1-59126-050-9
    1. Mechanical engineering--Problems, exercises, etc. 2. Mechanical engineering--Examinations--Study guides. I. Lindeburg, Michael R. Mechanical engineering reference manual. II. Title.

TJ159.L5234   2006
621.3076--dc22
                                                    2006045126

Topics

Topic I: Mathematics

Topic II: Fluids

Topic III: Thermodynamics

Topic IV: Power Cycles

Topic V: Heat Transfer

Topic VI: HVAC

Topic VII: Statics

Topic VIII: Materials

Topic IX: Machine Design

Topic X: Dynamics and Vibrations

Topic XI: Control Systems

Topic XII: Plant Engineering

Topic XIII: Economics

Topic XIV: Law and Ethics

Mathematics

Machine Design

Fluids

Dynamics and Vibrations

Thermodynamics

Control Systems

Power Cycles

Plant Engineering

Heat Transfer

Economics

HVAC

Law and Ethics

Statics

Materials

Where do I find help
solving these Practice Problems?

Practice Problems for the Mechanical Engineering PE Exam presents complete, step-by-step solutions for more than 500 problems like those found on the Mechanical PE exam. You can find all the background information, including charts and tables of data, that you need to solve these problems in the *Mechanical Engineering Reference Manual*.

The *Mechanical Engineering Reference Manual* may be purchased from Professional Publications at **www.ppi2pass.com** or from your favorite bookstore.

Table of Contents

Preface and Acknowledgments vii

How to Use This Book viii

Topic I: Mathematics

Systems of Units 1-1
Engineering Drawing Practice 2-1
Algebra 3-1
Linear Algebra 4-1
Vectors 5-1
Trigonometry 6-1
Analytic Geometry 7-1
Differential Calculus 8-1
Integral Calculus 9-1
Differential Equations 10-1
Probability and Statistical Analysis 11-1
Numerical Systems 12-1
Numerical Analysis 13-1

Topic II: Fluids

Fluid Properties 14-1
Fluid Statics 15-1
Fluid Flow Parameters 16-1
Fluid Dynamics 17-1
Hydraulic Machines 18-1
Fluid Power 19-1
Fans and Ductwork 20-1

Topic III: Thermodynamics

Inorganic Chemistry 21-1
Fuels and Combustion 22-1
Energy, Work, and Power 23-1
Thermodynamic Properties
   of Substances 24-1
Changes in Thermodynamic Properties 25-1
Compressible Fluid Dynamics 26-1

Topic IV: Power Cycles

Vapor Power Equipment 27-1
Vapor Power Cycles 28-1
Combustion Power Cycles 29-1
Nuclear Power Cycles 30-1
Advanced and Alternative
   Power-Generating Systems 31-1
Gas Compression Cycles 32-1
Refrigeration Cycles 33-1

Topic V: Heat Transfer

Fundamental Heat Transfer 34-1
Natural Convection, Evaporation,
   and Condensation 35-1

Forced Convection and Heat Exchangers 36-1
Radiation and Combined Heat Transfer 37-1

Topic VI: HVAC

Psychrometrics 38-1
Ventilation 39-1
Heating Load 40-1
Cooling Load 41-1
Air Conditioning Systems and Controls 42-1

Topic VII: Statics

Determinate Statics 43-1
Indeterminate Statics 44-1

Topic VIII: Materials

Engineering Materials 45-1
Material Properties and Testing 46-1
Thermal Treatment of Metals 47-1
Properties of Areas 48-1
Strength of Materials 49-1
Failure Theories 50-1

Topic IX: Machine Design

Basic Machine Design 51-1
Advanced Machine Design 52-1
Pressure Vessels 53-1

Topic X: Dynamics and Vibrations

Properties of Solid Bodies 54-1
Kinematics 55-1
Kinetics 56-1
Mechanisms and Power Transmission
   Systems 57-1
Vibrating Systems 58-1

Topic XI: Control Systems

Modeling of Engineering Systems 59-1
Analysis of Engineering Systems 60-1

Topic XII: Plant Engineering

Management Science 61-1
Instrumentation and Measurements *(no problems)*
Manufacturing Processes 63-1
Materials Handling and Processing *(no problems)*
Fire Protection Systems *(no problems)*
Environmental Engineering *(no problems)*
Electricity and Electrical Equipment 67-1
Illumination and Sound 68-1

Topic XIII: Economics

Engineering Economic Analysis 69-1

Topic XIV: Law and Ethics

Engineering Law 70-1
Engineering Ethics 71-1
Engineering Registration in the
  United States *(no problems)*

Preface and Acknowledgments

Putting out a practice problems manual for one of my reference manuals has always been a major undertaking. For one thing, it's a lot of work. For another, "good enough" just won't do. Over the years, I have learned what amount of detail will help you learn from a solution, as opposed to just presenting the numbers. (I have also learned that if I leave out certain steps or items, then I get a lot of inquisitive letters, phone calls, and email!)

The style standards set by Professional Publications' production department are as strict as my content standards. There is a proper way to edit, typeset, illustrate, and proofread a book. The production department won't do it any other way.

Most textbook authors see their problem sets or solutions manuals as a necessary evil: something required by their contract, something to bang out as quickly as possible, an afterthought stuck between two editions. The finished product seems to say, "Here are the numbers and the answer. Sure, we've been a little sloppy with units, and maybe we've omitted a few steps. But a little struggling is good for you. You can figure it all out. Somehow."

But not Professional Publications. All that struggling with vague content and sloppy production wastes your time. And before an exam, time is one thing that an examinee doesn't have much of. There was no way we were going to cut corners on this book.

There are also several major changes that drove the development of this new edition.

- All steam tables (both English and SI versions of saturated pressure, saturated temperature, and superheated and compressed water tables) have been "modernernized." Values from the old Keenan and Keyes tables, dating back to 1936, have been replaced with values calculated from the 1997 ASME/NIST correlations. The tables provided in this book's "sister" volume, *Mechanical Engineering Reference Manual* (MERM), have been changed to match.

- End-of-chapter problems are no longer included in MERM. The difficulty in keeping MERM and *Practice Problems for the Mechanical Engineering PE Exam* (MEPP) synchronized over the years was a considerable source of confusion to readers around the world. Now, the problem statements will only appear in MEPP, and there is no way that MERM problem statements and MEPP solutions can diverge.

All told, however, this is very much a maintenance edition. There is little new material. However, some pages are renumbered, and some chapter references have changed because some of MERM's appendices have changed. An instructor trying to use this book in a classroom environment with some students having one printing and other students having the new printing would experience chaos and confusion. Hence, the need for a new edition.

The team that produced this edition is substantial in number and talent. I'll begin by mentioning the staff at Professional Publications. Cathy Schrott, production department manager, monitored the typesetting, illustrating, and printing. Kate Hayes typeset the ninth, tenth, and eleventh editions, and now she's had the chance to update it again. The illustrations, originally converted from pen-and-ink by Yvonne Sartain in a previous lifetime, were revised by Tom Bergstrom.

Sarah Hubbard, editorial department manager, oversaw the editing, proofreading, and administrative aspects of bring out a new edition. She also supervised the student interns who integrated the new steam table values into this edition, particularly Nate Ginzton, now a professional engineer himself, but at the time of his contribution a student at Stanford.

For me, each new edition of each book has become another opportunity to be humbled. Each new edition requires a retreat from the security of pages whose kinks have supposedly been ironed out in a previous edition, and a charge into the critical, hot spotlight of a new edition. You would think that after all these years of writing problems and solutions I would know virtually all of the ways you and I make mistakes. But even knowing the ways mistakes are made doesn't mean we can avoid them all.

If you think you've found something questionable in a solution, or if you think there is a better way to solve a problem, visit our website at **www.ppi2pass.com/errata**. I personally read and review all comments for this book. I like to learn new things, too.

Michael R. Lindeburg, PE

How to Use This Book

This book is a companion to *Mechanical Engineering Reference Manual*. Since it is a practice problems book, there are a few, but not many, ways to use it.

Since most of the problems in the book can be solved in either customary U.S. or SI units, your first decision will be which set of units you will work in. Don't get me wrong: The exam doesn't give you such a choice. Exam problems are either in customary U.S. or SI units, not both. So, you have to be proficient with both. I recommend that you solve half of the problems in customary U.S. units and half in SI. Then, if you have time, go back to solve all of the problems a second time, using the alternate units.

The big decision you have to make is whether you really work the practice problems or not. Some people think they can read a problem statement, think about it for about ten seconds, read the solution, and then say "Yes, that's what I was thinking of, and that's what I would have done." Sadly, these people find out too late that the human brain doesn't learn very efficiently that way. Under pressure, they find they know and remember little. For real learning, you have to spend some time with a stubby pencil.

There are so many places where you can get messed up solving a problem. Maybe it's in the use of your calculator, like pushing log instead of ln, or forgetting to set the angle to radians instead of degrees, and so on. Maybe it's rusty math. What is $\ln(e^x)$, anyway?

How do you factor a polynomial? Maybe it's in finding the data needed or the proper unit conversion. Maybe it's just trying to find out if that funky code equation expects L to be in feet or inches. These things take time. And you have to make the mistakes once so that you don't make them again.

If you do decide to get your hands dirty and actually work these problems, you'll have to decide how much reliance you place on the practice problems book. It's tempting to turn to a solution when you get slowed down by details or stumped by the subject material. You'll probably want to maximize the number of problems you solve by spending as little time as possible. I want you to struggle a little bit more than that.

Studying a new subject is analogous to using a machete to cut a path through a dense jungle. By doing the work, you develop pathways that weren't there before. It's a lot different than just looking at the route on a map. You actually get nowhere by looking at a map. But cut that path once, and you're in business until the jungle overgrowth closes in again.

So, do the problems. All of them. Do them in both sets of units. Don't look at the answers until you've sweated a little. And, let's not have any whining. Please.

1 Systems of Units

PRACTICE PROBLEMS

1. Convert 250°F to degrees Celsius.

  (A) 115°C
  (B) 121°C
  (C) 124°C
  (D) 420°C

2. Convert the Stefan-Boltzmann constant (0.1713×10^{-8} Btu/ft^2-hr-°R^4) from English to SI units.

  (A) 5.14×10^{-10} W/m^2·K^4
  (B) 0.95×10^{-8} W/m^2·K^4
  (C) 5.67×10^{-8} W/m^2·K^4
  (D) 7.33×10^{-6} W/m^2·K^4

3. How many U.S. tons (2000 lbm per ton) of coal with a heating value of 13,000 Btu/lbm must be burned to provide as much energy as a complete nuclear conversion of 1 g of its mass? (Hint: Use Einstein's equation: $E = mc^2$.)

  (A) 1.7 tons
  (B) 14 tons
  (C) 779 tons
  (D) 3300 tons

SOLUTIONS

1. The conversion to degrees Celsius is given by

$$
\begin{aligned}
°C &= \left(\tfrac{5}{9}\right)(°F - 32°F) \\
&= \left(\tfrac{5}{9}\right)(250°F - 32°F) \\
&= \left(\tfrac{5}{9}\right)(218°F) \\
&= \boxed{121.1°C \; (121°C)}
\end{aligned}
$$

The answer is (B).

2. In customary U.S. units, the Stefan-Boltzmann constant, σ, is 0.1713×10^{-8} Btu/hr-ft^2-°R^4.

Use the following conversion factors.

$$
\begin{aligned}
1 \text{ Btu/hr} &= 0.2931 \text{ W} \\
1 \text{ ft} &= 0.3048 \text{ m} \\
T_{°R} &= \frac{9}{5} T_K
\end{aligned}
$$

Performing the conversion gives

$$
\sigma = \left(0.1713 \times 10^{-8} \frac{\text{Btu}}{\text{hr-ft}^2\text{-°R}^4}\right)\left(0.2931 \frac{\text{W}}{\frac{\text{Btu}}{\text{hr}}}\right)
$$

$$
\times \left(\frac{1 \text{ ft}}{0.3048 \text{ m}}\right)^2 \left(\frac{1°R}{\frac{5}{9}K}\right)^4
$$

$$
= \boxed{5.67 \times 10^{-8} \text{ W/m}^2\text{·K}^4}
$$

The answer is (C).

3. The energy produced from the nuclear conversion of any quantity of mass is given as

$$
E = mc^2
$$

The speed of light, c, is 3×10^8 m/s.

For a mass of 1 g (0.001 kg),

$$
\begin{aligned}
E &= mc^2 \\
&= (0.001 \text{ kg})\left(3 \times 10^8 \, \frac{\text{m}}{\text{s}}\right)^2 \\
&= 9 \times 10^{13} \text{ J}
\end{aligned}
$$

Convert to customary U.S. units with the conversion
1 Btu = 1055 J.

$$E = (9 \times 10^{13} \text{ J}) \left(\frac{1 \text{ Btu}}{1055 \text{ J}} \right)$$
$$= 8.53 \times 10^{10} \text{ Btu}$$

The number of tons of 13,000 Btu/lbm coal is

$$\frac{8.53 \times 10^{10} \text{ Btu}}{\left(13,000 \, \dfrac{\text{Btu}}{\text{lbm}} \right) \left(2000 \, \dfrac{\text{lbm}}{\text{ton}} \right)} = \boxed{3281 \text{ tons (3300 tons)}}$$

The answer is (D).

2 Engineering Drawing Practice

PRACTICE PROBLEMS

1. Two views of an object are shown. Prepare a free-hand drawing of the missing third view.

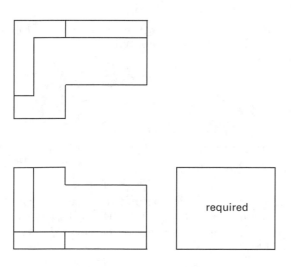

3. Two views of an object are shown. Prepare a free-hand drawing of the missing third view.

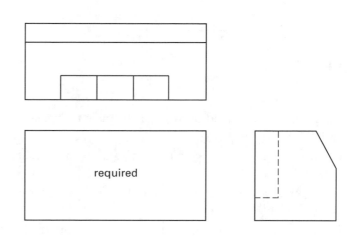

2. Two views of an object are shown. Prepare a free-hand drawing of the missing third view.

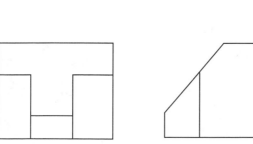

4. A pictorial sketch of an object is shown. Prepare a freehand three-view orthographic drawing set.

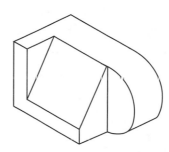

5. Plan and elevation views of an object are shown.

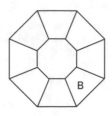

plan view

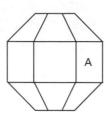

elevation view

(a) Prepare a freehand drawing of the view that shows surface A in true shape.

(b) Referring to part (a), what is this view of surface A called?

(c) Prepare a freehand drawing of the view that shows surface B in true shape.

(d) Referring to part (c), what is the view of surface B called?

6. Prepare a freehand isometric drawing of the object shown. Hint: Lay off horizontal and vertical (i.e., isometric) lines along the isometric axes shown.

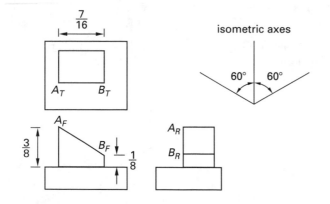

7. Two lines in an oblique position are shown. Prepare a freehand right auxiliary normal view.

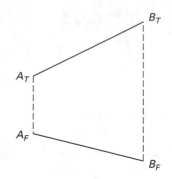

8. Two lines in an oblique position are shown. Prepare a freehand rear auxiliary normal view.

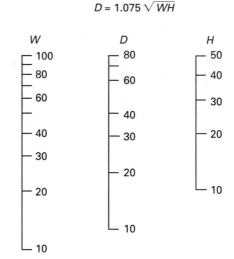

9. A parallel-scale nomograph has been prepared to solve the equation $D = 1.075\sqrt{WH}$. Use the nomograph to estimate the value of D when $H = 10$ and $W = 40$.

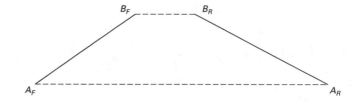

SOLUTIONS

1.

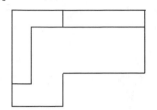

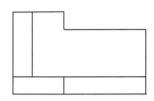

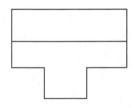

2.

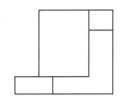

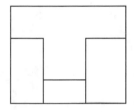

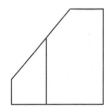

3.

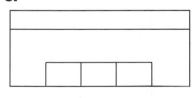

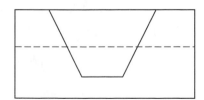

4.

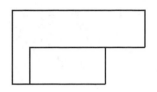

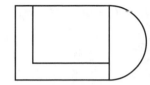

5. (a)

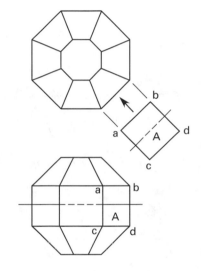

(b) An ⸢elevation auxiliary⸣ view shows surface A in true shape.

(c)

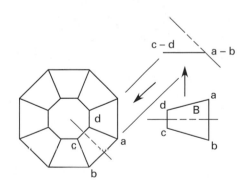

(d) An $\boxed{\text{oblique}}$ view shows surface B in true shape.

6.

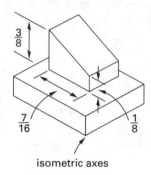

isometric axes

7.

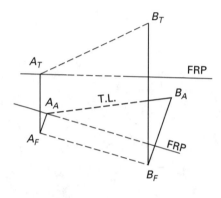

8.

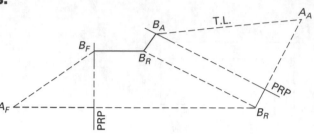

9. A straight line through the points $H = 10$ and $W = 40$ intersects the D-scale at approximately $\boxed{21.5.}$

3 Algebra

PRACTICE PROBLEMS

Series

1. Calculate the following sum.

$$\sum_{j=1}^{5}[(j+1)^2 - 1]$$

(A) 15
(B) 24
(C) 35
(D) 85

Logarithms

2. If every 0.1 sec a quantity increases by 0.1% of its current value, calculate the doubling time.

(A)   14 sec
(B)   70 sec
(C)   690 sec
(D)   69,000 sec

SOLUTIONS

1. Let $S_n = (j+1)^2 - 1$.

For $j = 1$,
$$S_1 = (1+1)^2 - 1 = 3$$

For $j = 2$,
$$S_2 = (2+1)^2 - 1 = 8$$

For $j = 3$,
$$S_3 = (3+1)^2 - 1 = 15$$

For $j = 4$,
$$S_4 = (4+1)^2 - 1 = 24$$

For $j = 5$,
$$S_5 = (5+1)^2 - 1 = 35$$

Substituting the above expressions gives

$$\sum_{j=1}^{5}\left((j+1)^2 - 1\right) = \sum_{j=1}^{5} S_j$$
$$= S_1 + S_2 + S_3 + S_4 + S_5$$
$$= 3 + 8 + 15 + 24 + 35$$
$$= \boxed{85}$$

The answer is (D).

2. Let n represent the number of elapsed periods of 0.1 sec, and let y_n represent the amount present after n periods.

y_0 represents the initial quantity.

$$y_1 = 1.001 y_0$$
$$y_2 = 1.001 y_1 = (1.001)\left((1.001)y_0\right) = (1.001)^2 y_0$$

Therefore, by deduction,

$$y_n = (1.001)^n y_0$$

The expression for a doubling of the original quantity is

$$2y_0 = y_n$$

Substitute for y_n.

$$2y_0 = (1.001)^n y_0$$
$$2 = (1.001)^n$$

Take the logarithm of both sides.

$$\log(2) = \log(1.001)^n$$
$$= n \log(1.001)$$

Solve for n.

$$n = \frac{\log(2)}{\log(1.001)} = 693.5$$

Since each period is 0.1 sec, the time is given by

$$t = n(0.1 \text{ sec})$$
$$= (693.5)(0.1 \text{ sec}) = \boxed{69.35 \text{ sec}}$$

The answer is (B).

4 Linear Algebra

PRACTICE PROBLEMS

Determinants

1. What is the determinant of matrix $\mathbf{A}$?

$$\mathbf{A} = \begin{bmatrix} 8 & 2 & 0 & 0 \\ 2 & 8 & 2 & 0 \\ 0 & 2 & 8 & 2 \\ 0 & 0 & 2 & 4 \end{bmatrix}$$

(A) 459

(B) 832

(C) 1552

(D) 1776

Simultaneous Linear Equations

2. Use Cramer's rule to solve for the values of $x, y,$ and z that simultaneously satisfy the following equations.

$$x + y = -4$$
$$x + z - 1 = 0$$
$$2z - y + 3x = 4$$

(A) $(x, y, z) = (3, 2, 1)$

(B) $(x, y, z) = (-3, -1, 2)$

(C) $(x, y, z) = (3, -1, -3)$

(D) $(x, y, z) = (-1, -3, 2)$

SOLUTIONS

1. Let D be the determinant. Expand by cofactors of the first row since there are two zeros in that row. (The first column can also be used.)

$$D = 8 \begin{vmatrix} 8 & 2 & 0 \\ 2 & 8 & 2 \\ 0 & 2 & 4 \end{vmatrix} - 2 \begin{vmatrix} 2 & 0 & 0 \\ 2 & 8 & 2 \\ 0 & 2 & 4 \end{vmatrix} + 0 - 0$$

by first row:

$$\begin{vmatrix} 8 & 2 & 0 \\ 2 & 8 & 2 \\ 0 & 2 & 4 \end{vmatrix} = (8)[(8)(4) - (2)(2)] - (2)[(2)(4) - (2)(0)]$$
$$= (8)(28) - (2)(8) = 208$$

by first row:

$$\begin{vmatrix} 2 & 0 & 0 \\ 2 & 8 & 2 \\ 0 & 2 & 4 \end{vmatrix} = (2)[(8)(4) - (2)(2)]$$
$$= 56$$

$$D = (8)(208) - (2)(56) = \boxed{1552}$$

The answer is (C).

2. Rearrange the equations.

$$\begin{aligned} x &+ y && = -4 \\ x && + z &= 1 \\ 3x &- y &+ 2z &= 4 \end{aligned}$$

Write the set of equations in matrix form: $\mathbf{AX} = \mathbf{B}$.

$$\begin{bmatrix} 1 & 1 & 0 \\ 1 & 0 & 1 \\ 3 & -1 & 2 \end{bmatrix} \begin{bmatrix} x \\ y \\ z \end{bmatrix} = \begin{bmatrix} -4 \\ 1 \\ 4 \end{bmatrix}$$

Find the determinant of the matrix $\mathbf{A}$.

$$|\mathbf{A}| = \begin{vmatrix} 1 & 1 & 0 \\ 1 & 0 & 1 \\ 3 & -1 & 2 \end{vmatrix}$$

$$= 1 \begin{vmatrix} 0 & 1 \\ -1 & 2 \end{vmatrix} - 1 \begin{vmatrix} 1 & 0 \\ -1 & 2 \end{vmatrix} + 3 \begin{vmatrix} 1 & 0 \\ 0 & 1 \end{vmatrix}$$

$$= (1)\big((0)(2) - (1)(-1)\big)$$
$$\quad - (1)\big((1)(2) - (-1)(0)\big)$$
$$\quad + (3)\big((1)(1) - (0)(0)\big)$$
$$= (1)(1) - (1)(2) + (3)(1)$$
$$= 1 - 2 + 3$$
$$= 2$$

Find the determinant of the substitutional matrix $\mathbf{A}_1$.

$$|\mathbf{A}_1| = \begin{vmatrix} -4 & 1 & 0 \\ 1 & 0 & 1 \\ 4 & -1 & 2 \end{vmatrix}$$

$$= -4 \begin{vmatrix} 0 & 1 \\ -1 & 2 \end{vmatrix} - 1 \begin{vmatrix} 1 & 0 \\ -1 & 2 \end{vmatrix} + 4 \begin{vmatrix} 1 & 0 \\ 0 & 1 \end{vmatrix}$$

$$= (-4)\big((0)(2) - (1)(-1)\big)$$
$$\quad - (1)\big((1)(2) - (-1)(0)\big)$$
$$\quad + (4)\big((1)(1) - (0)(0)\big)$$

$$= (-4)(1) - (1)(2) + (4)(1)$$
$$= -4 - 2 + 4$$
$$= -2$$

Find the determinant of the substitutional matrix $\mathbf{A}_2$.

$$|\mathbf{A}_2| = \begin{vmatrix} 1 & -4 & 0 \\ 1 & 1 & 1 \\ 3 & 4 & 2 \end{vmatrix}$$

$$= 1 \begin{vmatrix} 1 & 1 \\ 4 & 2 \end{vmatrix} - 1 \begin{vmatrix} -4 & 0 \\ 4 & 2 \end{vmatrix} + 3 \begin{vmatrix} -4 & 0 \\ 1 & 1 \end{vmatrix}$$

$$= (1)\big((1)(2) - (4)(1)\big)$$
$$\quad - (1)\big((-4)(2) - (4)(0)\big)$$
$$\quad + (3)\big((-4)(1) - (1)(0)\big)$$

$$= (1)(-2) - (1)(-8) + (3)(-4)$$
$$= -2 + 8 - 12$$
$$= -6$$

Find the determinant of the substitutional matrix $\mathbf{A}_3$.

$$|\mathbf{A}_3| = \begin{vmatrix} 1 & 1 & -4 \\ 1 & 0 & 1 \\ 3 & -1 & 4 \end{vmatrix}$$

$$= 1 \begin{vmatrix} 0 & 1 \\ -1 & 4 \end{vmatrix} - 1 \begin{vmatrix} 1 & -4 \\ -1 & 4 \end{vmatrix} + 3 \begin{vmatrix} 1 & -4 \\ 0 & 0 \end{vmatrix}$$

$$= (1)\big((0)(4) - (-1)(1)\big)$$
$$\quad - (1)\big((1)(4) - (-1)(-4)\big)$$
$$\quad + (3)\big((1)(1) - (0)(-4)\big)$$

$$= (1)(1) - (1)(0) + (3)(1)$$
$$= 1 - 0 + 3$$
$$= 4$$

Use Cramer's rule.

$$x = \frac{|\mathbf{A}_1|}{|\mathbf{A}|} = \frac{-2}{2} = \boxed{-1}$$

$$y = \frac{|\mathbf{A}_2|}{|\mathbf{A}|} = \frac{-6}{2} = \boxed{-3}$$

$$z = \frac{|\mathbf{A}_3|}{|\mathbf{A}|} = \frac{4}{2} = \boxed{2}$$

The answer is (D).

5 Vrectors

PRACTICE PROBLEMS

Dot Products

1. Calculate the dot products for the following vector pairs.

(a) $\mathbf{V}_1 = 2\mathbf{i} + 3\mathbf{j}; \mathbf{V}_2 = 5\mathbf{i} - 2\mathbf{j}$

(b) $\mathbf{V}_1 = 1\mathbf{i} + 4\mathbf{j}; \mathbf{V}_2 = 9\mathbf{i} - 3\mathbf{j}$

(c) $\mathbf{V}_1 = 7\mathbf{i} - 3\mathbf{j}; \mathbf{V}_2 = 3\mathbf{i} + 4\mathbf{j}$

(d) $\mathbf{V}_1 = 2\mathbf{i} - 3\mathbf{j} + 6\mathbf{k}; \mathbf{V}_2 = 8\mathbf{i} + 2\mathbf{j} - 3\mathbf{k}$

(e) $\mathbf{V}_1 = 6\mathbf{i} + 2\mathbf{j} + 3\mathbf{k}; \mathbf{V}_2 = \mathbf{i} + \mathbf{k}$

2. What is the angle between the vectors in Probs. 1(a), 1(b), and 1(c)?

Cross Products

3. Calculate the cross products for each of the five vector pairs in Prob. 1.

SOLUTIONS

1. (a) $\mathbf{V}_1 \cdot \mathbf{V}_2 = \mathbf{V}_{1x}\mathbf{V}_{2x} + \mathbf{V}_{1y}\mathbf{V}_{2y}$

$\qquad = (2)(5) + (3)(-2)$

$\qquad = \boxed{4}$

(b) $\quad \mathbf{V}_1 \cdot \mathbf{V}_2 = (1)(9) + (4)(-3)$

$\qquad = \boxed{-3}$

(c) $\quad \mathbf{V}_1 \cdot \mathbf{V}_2 = (7)(3) + (-3)(4)$

$\qquad = \boxed{9}$

(d) $\quad \mathbf{V}_1 \cdot \mathbf{V}_2 = \mathbf{V}_{1x}\mathbf{V}_{2x} + \mathbf{V}_{1y}\mathbf{V}_{2y} + \mathbf{V}_{1z}\mathbf{V}_{2z}$

$\qquad = (2)(8) + (-3)(2) + (6)(-3)$

$\qquad = \boxed{-8}$

(e) $\quad \mathbf{V}_1 \cdot \mathbf{V}_2 = (6)(1) + (2)(0) + (3)(1)$

$\qquad = \boxed{9}$

2. (a) $\quad \cos\phi = \dfrac{\mathbf{V}_1 \cdot \mathbf{V}_2}{|\mathbf{V}_1||\mathbf{V}_2|}$

$\qquad = \dfrac{4}{\sqrt{(2)^2 + (3)^2}\sqrt{(5)^2 + (-2)^2}}$

$\qquad = 0.206$

$\qquad \phi = \cos^{-1}(0.206) = \boxed{78.1°}$

(b) $\quad \cos\phi = \dfrac{\mathbf{V}_1 \cdot \mathbf{V}_2}{|\mathbf{V}_1||\mathbf{V}_2|}$

$\qquad = \dfrac{-3}{\sqrt{(1)^2 + (4)^2}\sqrt{(9)^2 + (-3)^2}}$

$\qquad = -0.077$

$\qquad \phi = \boxed{94.4°}$

(c) $\quad \cos\phi = \dfrac{\mathbf{V}_1 \cdot \mathbf{V}_2}{|\mathbf{V}_1||\mathbf{V}_2|}$

$\qquad = \dfrac{9}{\sqrt{(7)^2 + (-3)^2}\sqrt{(3)^2 + (4)^2}}$

$\qquad = 0.236$

$\qquad \phi = \boxed{76.3°}$

3. (a)
$$\mathbf{V}_1 \times \mathbf{V}_2 = \begin{vmatrix} \mathbf{i} & \mathbf{V}_{1x} & \mathbf{V}_{2x} \\ \mathbf{j} & \mathbf{V}_{1y} & \mathbf{V}_{2y} \\ \mathbf{k} & \mathbf{V}_{1z} & \mathbf{V}_{2z} \end{vmatrix}$$

$$= \begin{vmatrix} \mathbf{i} & 2 & 5 \\ \mathbf{j} & 3 & -2 \\ \mathbf{k} & 0 & 0 \end{vmatrix}$$

Expand by the third row.

$$= \mathbf{k} \begin{vmatrix} 2 & 5 \\ 3 & -2 \end{vmatrix} = \boxed{-19\mathbf{k}}$$

(b)
$$\mathbf{V}_1 \times \mathbf{V}_2 = \begin{vmatrix} \mathbf{i} & 1 & 9 \\ \mathbf{j} & 4 & -3 \\ \mathbf{k} & 0 & 0 \end{vmatrix}$$

Expand by the third row.

$$= \mathbf{k} \begin{vmatrix} 1 & 9 \\ 4 & -3 \end{vmatrix} = \boxed{-39\mathbf{k}}$$

(c)
$$\mathbf{V}_1 \times \mathbf{V}_2 = \begin{vmatrix} \mathbf{i} & 7 & 3 \\ \mathbf{j} & -3 & 4 \\ \mathbf{k} & 0 & 0 \end{vmatrix}$$

Expand by the third row.

$$\mathbf{k} \begin{vmatrix} 7 & 3 \\ -3 & 4 \end{vmatrix} = \boxed{37\mathbf{k}}$$

(d)
$$\mathbf{V}_1 \times \mathbf{V}_2 = \begin{vmatrix} \mathbf{i} & 2 & 8 \\ \mathbf{j} & -3 & 2 \\ \mathbf{k} & 6 & -3 \end{vmatrix}$$

Expand by the first column.

$$= \mathbf{i} \begin{vmatrix} -3 & 2 \\ 6 & -3 \end{vmatrix} - \mathbf{j} \begin{vmatrix} 2 & 8 \\ 6 & -3 \end{vmatrix}$$

$$+ \mathbf{k} \begin{vmatrix} 2 & 8 \\ -3 & 2 \end{vmatrix}$$

$$= \boxed{-3\mathbf{i} + 54\mathbf{j} + 28\mathbf{k}}$$

(e)
$$\mathbf{V}_1 \times \mathbf{V}_2 = \begin{vmatrix} \mathbf{i} & 6 & 1 \\ \mathbf{j} & 2 & 0 \\ \mathbf{k} & 3 & 1 \end{vmatrix}$$

Expand by the second row.

$$= -\mathbf{j} \begin{vmatrix} 6 & 1 \\ 3 & 1 \end{vmatrix} + (2) \begin{vmatrix} \mathbf{i} & 1 \\ \mathbf{k} & 1 \end{vmatrix}$$

$$= \boxed{2\mathbf{i} - 3\mathbf{j} - 2\mathbf{k}}$$

6 Trigonometry

PRACTICE PROBLEMS

1. A 5 lbm (5 kg) block sits on a 20° incline without slipping.

(a) Draw the freebody with respect to axes parallel and perpendicular to the surface of the incline.

(b) Determine the magnitude of the frictional force on the block.

    (A)   1.71 lbf (16.8 N)
    (B)   3.35 lbf (32.9 N)
    (C)   4.70 lbf (46.1 N)
    (D)   5.00 lbf (49.1 N)

SOLUTIONS

1. (a)

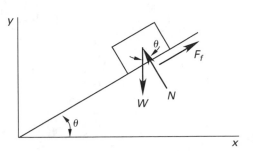

Customary U.S. Solution

(b) The mass of the block is $m = 5$ lbm.

The angle of inclination is $\theta = 20°$. The weight is

$$W = \frac{mg}{g_c}$$

$$= \frac{(5 \text{ lbm}) \left(32.2 \dfrac{\text{ft}}{\text{sec}^2} \right)}{32.2 \dfrac{\text{lbm-ft}}{\text{lbf-sec}^2}}$$

$$= 5 \text{ lbf}$$

The frictional force is

$$F_f = W \sin \theta$$
$$= (5 \text{ lbf})(\sin 20°)$$
$$= \boxed{1.71 \text{ lbf}}$$

The answer is (A).

SI Solution

(b) The mass of the block is $m = 5$ kg.

The angle of inclination is $\theta = 20°$. The gravitational force is

$$W = mg$$
$$= (5 \text{ kg}) \left(9.81 \dfrac{\text{m}}{\text{s}^2} \right)$$
$$= 49.1 \text{ N}$$

The frictional force is

$$F_f = W \sin \theta$$
$$= (49.1 \text{ N})(\sin 20°)$$
$$= \boxed{16.8 \text{ N}}$$

The answer is (A).

7 Analytic Geometry

PRACTICE PROBLEMS

1. The diameter of a sphere and the base of a cone are equal. What percentage of that diameter must the cone's height be so that both volumes are equal?

- (A) 133%
- (B) 150%
- (C) 166%
- (D) 200%

SOLUTIONS

1. Let d be the diameter of the sphere and the base of the cone.

The volume of the sphere is given by

$$V_{\text{sphere}} = \left(\tfrac{4}{3}\right)\pi r^3 = \left(\tfrac{4}{3}\right)\pi\left(\frac{d}{2}\right)^3$$
$$= \left(\frac{\pi}{6}\right)d^3$$

The volume of the circular cone is given by

$$V_{\text{cone}} = \left(\tfrac{1}{3}\right)\pi r^2 h = \left(\tfrac{1}{3}\right)\pi\left(\frac{d}{2}\right)^2 h$$
$$= \left(\frac{\pi}{12}\right)d^2 h$$

Since the volume of the sphere and cone are equal,

$$V_{\text{cone}} = V_{\text{sphere}}$$
$$\left(\frac{\pi}{12}\right)d^2 h = \left(\frac{\pi}{6}\right)d^3$$
$$h = 2d$$

The height of the cone must be 200% of the diameter.

The answer is (D).

8 Differential Calculus

PRACTICE PROBLEMS

1. What are the values of a, b, and c in the following expression such that $n(\infty) = 100$, $n(0) = 10$, and $dn(0)/dt = 0.5$?

$$n(t) = \frac{a}{1 + be^{ct}}$$

(A) $a = 10$, $b = 9$, $c = 1$
(B) $a = 100$, $b = 10$, $c = 1.5$
(C) $a = 100$, $b = 9$, $c = -0.056$
(D) $a = 1000$, $b = 10$, $c = 0.056$

2. Find all minima, maxima, and inflection points for

$$y = x^3 - 9x^2 - 3$$

(A) maximum at $x = 0$
inflection at $x = 3$
minimum at $x = 6$
(B) maximum at $x = 0$
inflection at $x = -3$
minimum at $x = -6$
(C) minimum at $x = 0$
inflection at $x = 3$
maximum at $x = 6$
(D) minimum at $x = 3$
inflection at $x = 0$
maximum at $x = -3$

SOLUTIONS

1. If c is positive, then $n(\infty) = 0$, which is contrary to the data given. Therefore, $c \leq 0$. If $c = 0$, then $n(\infty) = a/(1 + b) = 100$, which is possible depending on a and b. However, $n(0) = a/(1 + b)$ would also equal 100, which is contrary to the given data. Therefore, $c \neq 0$.

c must be less than 0.

Since $c < 0$, then $n(\infty) = a$, so $\boxed{a = 100.}$

Applying the condition $t = 0$ gives

$$n(0) = \frac{a}{1 + b} = 10$$

Since $a = 100$,

$$n(0) = \frac{100}{1 + b}$$
$$100 = (10)(1 + b)$$
$$10 = 1 + b$$
$$\boxed{b = 9}$$

Substitute the results for a and b into the expression.

$$n(t) = \frac{100}{1 + 9e^{ct}}$$

Take the first derivative.

$$\frac{d}{dt}n(t) = \left(\frac{100}{(1 + 9e^{ct})^2}\right)(-9ce^{ct})$$

Apply the initial condition.

$$\frac{d}{dt}n(0) = \left(\frac{100}{(1 + 9e^{c(0)})^2}\right)\left(-9ce^{c(0)}\right) = 0.5$$
$$\left(\frac{100}{(1 + 9)^2}\right)(-9c) = 0.5$$
$$(1)(-9c) = 0.5$$
$$c = \frac{-0.5}{9}$$
$$= \boxed{-0.0556}$$

The answer is (C).

Substitute the terms a, b, and c into the expression.

$$n(t) = \frac{100}{1 + 9e^{-0.0556t}}$$

2. Determine the critical points by taking the first derivative of the function and setting it equal to zero.

$$\frac{dy}{dx} = 3x^2 - 18x = 3x(x - 6)$$
$$3x(x - 6) = 0$$
$$x(x - 6) = 0$$

The critical points are located at $x = 0$ and $x = 6$.

Determine the inflection points by setting the second derivative equal to zero. Take the second derivative.

$$\frac{d^2y}{dx^2} = \left(\frac{d}{dx}\right)\left(\frac{dy}{dx}\right) = \frac{d}{dx}(3x^2 - 18x)$$
$$= 6x - 18$$

Set the second derivative equal to zero.

$$\frac{d^2y}{dx^2} = 0 = 6x - 18 = (6)(x - 3)$$
$$(6)(x - 3) = 0$$
$$x - 3 = 0$$
$$x = 3$$

This inflection point is at $x = 3$.

Determine the local maximum and minimum by substituting the critical points into the expression for the second derivative.

At the critical point $x = 0$,

$$\left.\frac{d^2y}{dx^2}\right|_{x=0} = (6)(x - 3) = (6)(0 - 3)$$
$$= -18$$

Since $-18 < 0$, $x = 0$ is a local maximum.

At the critical point $x = 6$,

$$\left.\frac{d^2y}{dx^2}\right|_{x=6} = (6)(x - 3) = (6)(6 - 3)$$
$$= 18$$

Since $18 > 0$, $x = 6$ is a local minimum.

The answer is (A).

9 Integral Calculus

PRACTICE PROBLEMS

Elementary Operations

1. Find the integrals.

(a)
$$\int \sqrt{1-x}\, dx$$

(b)
$$\int \frac{x}{x^2+1}\, dx$$

(c)
$$\int \frac{x^2}{x^2+x-6}\, dx$$

Definite Integrals

2. Calculate the definite integrals.

(a)
$$\int_1^3 (x^2+4x)\, dx$$

(b)
$$\int_{-2}^2 (x^3+1)\, dx$$

(c)
$$\int_1^2 (4x^3-3x^2)\, dx$$

Areas by Integration

3. Find the area bounded by $x=1$, $x=3$, $y+x+1=0$, and $y=6x-x^2$.

Fourier Series

4. Find a_0 for the two waveforms shown.

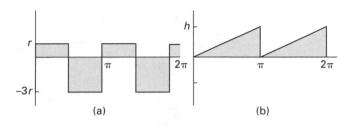

(a)   (b)

5. For the two waveforms shown, determine if their Fourier series is of type A, B, or C.

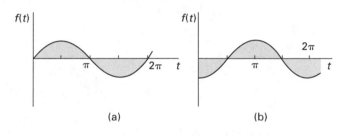

(a)   (b)

type A: $f(t) = a_0 + a_2\cos 2t + b_2\sin 2t$
$\qquad + a_4\cos 4t + b_4\sin 4t + \cdots$

type B: $f(t) = a_0 + b_1\sin t + b_2\sin 2t + b_3\sin 3t + \cdots$

type C: $f(t) = a_0 + a_1\cos t + a_2\cos 2t + a_3\cos 3t + \cdots$

SOLUTIONS

1. (a) $\int \sqrt{1-x}\,dx = \int (1-x)^{\frac{1}{2}}\,dx$

$$= \boxed{\left(-\frac{2}{3}\right)(1-x)^{\frac{3}{2}} + C}$$

(b) $\int \dfrac{x}{x^2+1}\,dx = \dfrac{1}{2}\int \dfrac{2x}{x^2+1}\,dx$

$$= \boxed{\frac{1}{2}\ln\left|(x^2+1)\right| + C}$$

(c) $\dfrac{x^2}{x^2+x-6} = 1 - \dfrac{x-6}{x^2+x-6}$

$$= 1 - \frac{x-6}{(x+3)(x-2)}$$

$$= 1 - \frac{\frac{9}{5}}{x+3} + \frac{\frac{4}{5}}{x-2}$$

$\int \dfrac{x^2}{x^2+x-6}\,dx = \int \left(1 - \dfrac{\frac{9}{5}}{x+3} + \dfrac{\frac{4}{5}}{x-2}\right)dx$

$$= \int dx - \int \frac{\frac{9}{5}}{x+3}\,dx + \int \frac{\frac{4}{5}}{x-2}\,dx$$

$$= \boxed{x - \frac{9}{5}\ln\left|(x+3)\right| + \frac{4}{5}\ln\left|(x-2)\right| + C}$$

2. (a) $\int_1^3 (x^2+4x)\,dx = \left[\dfrac{x^3}{3} + 2x^2\right]_1^3 = \boxed{24\tfrac{2}{3}}$

(b) $\int_{-2}^2 (x^3+1)\,dx = \left[\dfrac{x^4}{4} + x\right]_{-2}^2 = \boxed{4}$

(c) $\int_1^2 (4x^3 - 3x^2)\,dx = \left[x^4 - x^3\right]_1^2 = \boxed{8}$

3.

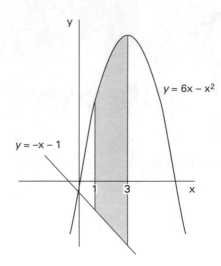

$$\text{area} = \int_1^3 \left((6x-x^2)-(-x-1)\right)dx$$

$$= \int_1^3 (-x^2 + 7x + 1)\,dx$$

$$= \left[-\frac{x^3}{3} + \frac{7}{2}x^2 + x\right]_1^3 = \boxed{21\tfrac{1}{3}}$$

4. For waveform (a):

$$a_0 = \frac{1}{2\pi}\int_0^{2\pi} f(t)\,dt$$

$$= \frac{1}{\pi}\int_0^{\pi} f(t)\,dt$$

$$= \frac{1}{\pi}\left((r)\left(\frac{\pi}{2}\right) + (-3r)\left(\frac{\pi}{2}\right)\right) = \boxed{-r}$$

For waveform (b):

$$a_0 = \frac{1}{2\pi}\int_0^{2\pi} f(t)\,dt$$

$$= \frac{1}{\pi}\int_0^{\pi} f(t)\,dt$$

$$= \left(\frac{1}{\pi}\right)\left(\frac{1}{2}\pi h\right)$$

$$= \boxed{h/2}$$

5. For waveform (a): Since $f(t) = -f(-t)$,
$\boxed{\text{it is type B}}$

For waveform (b): Since $f(t) = f(-t)$, $\boxed{\text{it is type C}}$

10 Differential Equations

PRACTICE PROBLEMS

1. Solve the following differential equation for y.

$$y'' - 4y' - 12y = 0$$

(A) $A_1 e^{6x} + A_2 e^{-2x}$
(B) $A_1 e^{-6x} + A_2 e^{2x}$
(C) $A_1 e^{6x} + A_2 e^{2x}$
(D) $A_1 e^{-6x} + A_2 e^{-2x}$

2. Solve the following differential equation for y.

$$y' - y = 2xe^{2x} \qquad y(0) = 1$$

(A) $y = 2e^{-2x}(x-1) + 3e^{-x}$
(B) $y = 2e^{2x}(x-1) + 3e^x$
(C) $y = -2e^{-2x}(x-1) + 3e^{-x}$
(D) $y = 2e^{2x}(x-1) + 3e^{-x}$

3. The oscillation exhibited by the top story of a certain building in free motion is given by the following differential equation.

$$x'' + 2x' + 2x = 0 \qquad x(0) = 0 \qquad x'(0) = 1$$

(a) What is x as a function of time?
   (A) $e^{-2t} \sin t$
   (B) $e^t \sin t$
   (C) $e^{-t} \sin t$
   (D) $e^{-t} \sin t + e^{-t} \cos t$

(b) What is the building's fundamental natural frequency of vibration?
   (A) $1/2$
   (B) 1
   (C) $\sqrt{2}$
   (D) 2

(c) What is the amplitude of oscillation?
   (A) 0.32
   (B) 0.54
   (C) 1.7
   (D) 6.6

(c) What is x as a function of time if a lateral wind load is applied with a form of $\sin t$?
   (A) $\frac{6}{5}e^{-t}\sin t + \frac{2}{5}e^{-t}\cos t$
   (B) $\frac{6}{5}e^t \sin t + \frac{2}{5}e^t \cos t$
   (C) $\frac{2}{5}e^{-t}\sin t + \frac{6}{5}e^{-t}\cos t + \frac{2}{5}\sin t$
        $- \frac{1}{5}\cos t$
   (D) $\frac{6}{5}e^{-t}\sin t + \frac{2}{5}e^{-t}\cos t + \frac{1}{5}\sin t$
        $- \frac{2}{5}\cos t$

4. (*Time limit: one hour*) A 90 lbm (40 kg) bag of a chemical is accidentally dropped in an aerating lagoon. The chemical is water soluble and nonreacting. The lagoon is 120 ft (35 m) in diameter and filled to a depth of 10 ft (3 m). The aerators circulate and distribute the chemical evenly throughout the lagoon.

Water enters the lagoon at a rate of 30 gal/min (115 L/min). Fully mixed water is pumped into a reservoir at a rate of 30 gal/min (115 L/min).

The established safe concentration of this chemical is 1 ppb (part per billion). How many days will it take for the concentration of the discharge water to reach this level?

   (A) 25 days
   (B) 50 days
   (C) 100 days
   (D) 200 days

5. A tank contains 100 gal (100 L) of brine made by dissolving 60 lbm (60 kg) of salt in pure water. Salt water with a concentration of 1 lbm/gal (1 kg/L) enters the tank at a rate of 2 gal/min (2 L/min). A well-stirred mixture is drawn from the tank at a rate of 3 gal/min (3 L/min). Find the mass of salt in the tank after 1 hr.

   (A) 13 lbm (13 kg)
   (B) 37 lbm (37 kg)
   (C) 43 lbm (43 kg)
   (D) 51 lbm (51 kg)

SOLUTIONS

1. Obtain the characteristic equation by replacing each derivative with a polynomial term of equal degree.

$$r^2 - 4r - 12 = 0$$

Factor the characteristic equation.

$$(r - 6)(r + 2) = 0$$

The roots are $r_1 = 6$ and $r_2 = -2$.

Since the roots are real and distinct, the solution is

$$y = A_1 e^{r_1 x} + A_2 e^{r_2 x}$$

$$= \boxed{A_1 e^{6x} + A_2 e^{-2x}}$$

The answer is (A).

2. The equation is a first-order linear differential equation of the form

$$y' + p(x)y = g(x)$$
$$p(x) = -1$$
$$g(x) = 2xe^{2x}$$

The integration factor $u(x)$ is given by

$$u(x) = \exp\left(\int p(x)dx\right)$$

$$= \exp\left(\int (-1)dx\right)$$

$$= e^{-x}$$

The closed form of the solution is given by

$$y = \left(\frac{1}{u(x)}\right)\left(\int u(x)g(x)dx + C\right)$$

$$= \left(\frac{1}{e^{-x}}\right)\left(\int (e^{-x})(2xe^{2x})dx + C\right)$$

$$= e^x\left(2(xe^x - e^x) + C\right)$$

$$= e^x\left(2e^x(x - 1) + C\right)$$

$$= 2e^{2x}(x - 1) + Ce^x$$

Apply the initial condition $y(0) = 1$ to obtain the integration constant C.

$$y(0) = 2e^{(2)(0)}(0 - 1) + Ce^0 = 1$$
$$(2)(1)(-1) + C(1) = 1$$
$$-2 + C = 1$$
$$C = 3$$

Substituting in the value for the integration constant C, the solution is

$$y = \boxed{2e^{2x}(x - 1) + 3e^x}$$

The answer is (B).

3. (a) The differential equation is homogeneous, second-order, and linear, with constant coefficients. Write the characteristic equation.

$$r^2 + 2r + 2 = 0$$

This is a quadratic equation of the form $ar^2 + br + c = 0$ where $a = 1$, $b = 2$, and $c = 2$.

Solve for r.

$$r = \frac{-b \pm \sqrt{b^2 - 4ac}}{2a}$$

$$= \frac{-2 \pm \sqrt{(2)^2 - (4)(1)(2)}}{(2)(1)}$$

$$= \frac{-2 \pm \sqrt{4 - 8}}{2}$$

$$= \frac{-2 \pm \sqrt{-4}}{2}$$

$$= \frac{-2 \pm 2\sqrt{-1}}{2}$$

$$= -1 \pm \sqrt{-1}$$

$$= -1 \pm i$$

$$r_1 = -1 + i \text{ and } r_2 = -1 - i$$

Since the roots are imaginary and of the form $\alpha + i\omega$ and $\alpha - i\omega$ where $\alpha = -1$ and $\omega = 1$, the general form of the solution is given by

$$x(t) = A_1 e^{\alpha t} \cos \omega t + A_2 e^{\alpha t} \sin \omega t$$

$$= A_1 e^{-1t} \cos(1t) + A_2 e^{-1t} \sin(1t)$$

$$= A_1 e^{-t} \cos t + A_2 e^{-t} \sin t$$

Apply the initial conditions $x(0) = 0$ and $x'(0) = 1$ to solve for A_1 and A_2.

First, apply the initial condition $x(0) = 0$.

$$x(t) = A_1 e^0 \cos 0 + A_2 e^0 \sin 0 = 0$$
$$A_1(1)(1) + A_2(1)(0) = 0$$
$$A_1 = 0$$

Substituting, the solution of the differential equation becomes

$$x(t) = A_2 e^{-t} \sin t$$

To apply the second initial condition, take the first derivative.

$$x'(t) = \frac{d}{dt}\left(A_2 e^{-t} \sin t\right)$$

$$= A_2 \frac{d}{dt}\left(e^{-t} \sin t\right)$$

$$= A_2 \left(\sin t \frac{d}{dt}\left(e^{-t}\right) + e^{-t} \frac{d}{dt} \sin t\right)$$

$$= A_2 \left(\sin t\left(-e^{-t}\right) + e^{-t}(\cos t)\right)$$

$$= A_2\left(e^{-t}\right)(-\sin t + \cos t)$$

Apply the initial condition, $x'(0) = 1$.

$$x(0) = A_2\left(e^0\right)(-\sin 0 + \cos 0) = 1$$

$$A_2(1)(0 + 1) = 1$$

$$A_2 = 1$$

The solution is

$$x(t) = A_2 e^{-t} \sin t$$

$$= (1)e^{-t} \sin t$$

$$= \boxed{e^{-t} \sin t}$$

The answer is (C).

(b) To determine the natural frequency, set the damping term to zero. The equation has the form

$$x'' + 2x = 0$$

This equation has a general solution of the form

$$x(t) = x_0 \cos \omega t + \left(\frac{v_0}{\omega}\right) \sin \omega t$$

ω is the natural frequency. Given the equation $x'' + 2x = 0$, the characteristic equation is

$$r^2 + 2 = 0$$

$$r = \sqrt{-2}$$

$$= \pm\sqrt{2}i$$

Since the roots are imaginary and of the form $\alpha + i\omega$ and $\alpha - i\omega$ where $\alpha = 0$ and $\omega = \sqrt{2}$, the general form of the solution is given by

$$x(t) = A_1 e^{\alpha t} \cos \omega t + A_2 e^{\alpha t} \sin \omega t$$

$$= A_1 e^{0t} \cos \sqrt{2}t + A_2 e^{0t} \sin \sqrt{2}t$$

$$= A_1(1) \cos \sqrt{2}t + A_2(1) \sin \sqrt{2}t$$

$$= A_1 \cos \sqrt{2}t + A_2 \sin \sqrt{2}t$$

Apply the initial conditions, $x(0) = 0$ and $x'(0) = 1$ to solve for A_1 and A_2. Applying the initial condition $x(0) = 0$ gives

$$x(0) = A_1 \cos\left((\sqrt{2})(0)\right) + A_2 \sin(\sqrt{2})(0) = 0$$

$$A_1 \cos 0 + A_2 \sin 0 = 0$$

$$A_1(1) + A_2(0) = 0$$

$$A_1 = 0$$

Substituting, the solution of the differential equation becomes

$$x(t) = A_2 \sin \sqrt{2}t$$

To apply the second initial condition, take the first derivative.

$$x'(t) = \frac{d}{dt}(A_2 \sin \sqrt{2}t)$$

$$= A_2\sqrt{2} \cos \sqrt{2}t$$

Apply the second initial condition, $x'(0) = 1$.

$$x'(0) = A_2\sqrt{2} \cos(\sqrt{2})(0) = 1$$

$$A_2\sqrt{2} \cos(0) = 1$$

$$A_2(\sqrt{2})(1) = 1$$

$$A_2\sqrt{2} = 1$$

$$A_2 = \frac{1}{\sqrt{2}} = \frac{\sqrt{2}}{2}$$

Substituting, the undamped solution becomes

$$x(t) = \left(\frac{\sqrt{2}}{2}\right) \sin \sqrt{2}t$$

Therefore, the undamped natural frequency is $\boxed{\omega = \sqrt{2}.}$

The answer is (C).

(c) The amplitude of the oscillation is the maximum displacement.

Take the derivative of the solution, $x(t) = e^{-t} \sin t$.

$$x'(t) = \frac{d}{dt}(e^{-t} \sin t)$$

$$= \sin t \frac{d}{dt}\left(e^{-t}\right) + e^{-t} \frac{d}{dt} \sin t$$

$$= \sin(t)(-e^{-t}) + e^{-t} \cos t$$

$$= e^{-t}(\cos t - \sin t)$$

The maximum displacement occurs at $x'(t) = 0$.

Since $e^{-t} \neq 0$ except as t approaches infinity,

$$\cos t - \sin t = 0$$

$$\tan t = 1$$

$$t = \tan^{-1}(1)$$

$$= 0.785 \text{ rad}$$

At $t = 0.785$ rad, the displacement is maximum. Substitute into the orginal solution to obtain a value for the maximum displacement.

$$x(0.785) = e^{-0.785} \sin(0.785)$$
$$= 0.322$$

The amplitude is $\boxed{0.322.}$

The answer is (A).

(d) (An alternative solution using Laplace transforms follows this solution.) The application of a lateral wind load with the form $\sin t$ revises the differential equation to the form

$$x'' + 2x' + 2x = \sin t$$

Express the solution as the sum of the complementary x_c and particular x_p solutions.

$$x(t) = x_c(t) + x_p(t)$$

From part (a),

$$x_c(t) = A_1 e^{-t} \cos t + A_2 e^{-t} \sin t$$

The general form of the particular solution is given by

$$x_p(t) = x^s(A_3 \cos t + A_4 \sin t)$$

Determine the value of s; check to see if the terms of the particular solution solve the homogeneous equation.

Examine the term $A_3 \cos(t)$.

Take the first derivative.

$$\frac{d}{dx}(A_3 \cos t) = -A_3 \sin t$$

Take the second derivative.

$$\frac{d}{dx}\left(\frac{d}{dx}(A_3 \cos t)\right) = \frac{d}{dx}(-A_3 \sin t)$$
$$= -A_3 \cos t$$

Substitute the terms into the homogeneous equation.

$$x'' + 2x' + 2x = -A_3 \cos t + (2)(-A_3 \sin t)$$
$$+ (2)(-A_3 \cos t)$$
$$= A_3 \cos t - 2A_3 \sin t$$
$$\neq 0$$

Except for the trival solution $A_3 = 0$, the term $A_3 \cos t$ does not solve the homogeneous equation.

Examine the second term $A_4 \sin t$.

Take the first derivative.

$$\frac{d}{dx}(A_4 \sin t) = A_4 \cos t$$

Take the second derivative.

$$\frac{d}{dx}\left(\frac{d}{dx}(A_4 \sin t)\right) = \frac{d}{dx}(A_4 \cos t)$$
$$= -A_4 \sin t$$

Substitute the terms into the homogeneous equation.

$$x'' + 2x' + 2x = -A_4 \sin t + (2)(A_4 \cos t)$$
$$+ (2)(A_4 \sin t)$$
$$= A_4 \sin t + 2A_4 \cos t$$
$$\neq 0$$

Except for the trival solution $A_4 = 0$, the term $A_4 \sin t$ does not solve the homogeneous equation.

Neither of the terms satisfies the homogeneous equation $s = 0$; therefore, the particular solution is of the form

$$x_p(t) = A_3 \cos t + A_4 \sin t$$

Use the method of undetermined coefficients to solve for A_3 and A_4. Take the first derivative.

$$x'_p(t) = \frac{d}{dx}(A_3 \cos t + A_4 \sin t)$$
$$= -A_3 \sin t + A_4 \cos t$$

Take the second derivative.

$$x''_p(t) = \frac{d}{dx}\left(\frac{d}{dx}(A_3 \cos t + A_4 \sin t)\right)$$
$$= \frac{d}{dx}(-A_3 \sin t + A_4 \cos t)$$
$$= -A_3 \cos t - A_4 \sin t$$

Substitute the expressions for the derivatives into the differential equation.

$$x'' + 2x' + 2x = (-A_3 \cos t - A_4 \sin t)$$
$$+ (2)(-A_3 \sin t + A_4 \cos t)$$
$$+ (2)(A_3 \cos t + A_4 \sin t)$$
$$= \sin t$$

Rearranging terms gives

$$(-A_3 + 2A_4 + 2A_3) \cos t$$
$$+ (-A_4 - 2A_3 + 2A_4) \sin t = \sin t$$
$$(A_3 + 2A_4) \cos t + (-2A_3 + A_4) \sin t = \sin t$$

Equating coefficients gives

$$A_3 + 2A_4 = 0$$
$$-2A_3 + A_4 = 1$$

Multiplying the first equation by 2 and adding equations gives

$$
\begin{aligned}
A_3 + 2A_4 &= 0 \\
+(-2A_3 + A_4) &= 1 \\
\hline
5A_4 &= 1 \text{ or } A_4 = \tfrac{1}{5}
\end{aligned}
$$

From the first equation for $A_4 = \,^1/_5$, $A_3 + (2)(^1/_5) = 0$ and $A_3 = -\,^2/_5$.

Substituting for the coefficients, the particular solution becomes

$$x_p(t) = -\tfrac{2}{5}\cos t + \tfrac{1}{5}\sin t$$

Combining the complementary and particular solutions gives

$$
\begin{aligned}
x(t) &= x_c(t) + x_p(t) \\
&= A_1 e^{-t}\cos t + A_2 e^{-t}\sin t - \tfrac{2}{5}\cos t + \tfrac{1}{5}\sin t
\end{aligned}
$$

Apply the initial conditions to solve for the coefficients A_1 and A_2; then apply the first initial condition, $x(0) = 0$.

$$
\begin{aligned}
x(t) &= A_1 e^0 \cos 0 + A_2 e^0 \sin 0 \\
&\quad - \tfrac{2}{5}\cos 0 + \tfrac{1}{5}\sin 0 = 0 \\
A_1(1)(1) + A_2(1)(0) &+ \left(-\tfrac{2}{5}\right)(1) + \left(\tfrac{1}{5}\right)(0) = 0 \\
A_1 - \tfrac{2}{5} &= 0 \\
A_1 &= \tfrac{2}{5}
\end{aligned}
$$

Substituting for A_1, the solution becomes

$$x(t) = \tfrac{2}{5}e^{-t}\cos t + A_2 e^{-t}\sin t - \tfrac{2}{5}\cos t + \tfrac{1}{5}\sin t$$

Take the first derivative.

$$
\begin{aligned}
x'(t) &= \frac{d}{dx}\left(\tfrac{2}{5}e^{-t}\cos t + A_2 e^{-t}\sin t\right) \\
&\quad + \left(\left(-\tfrac{2}{5}\right)\cos t + \tfrac{1}{5}\sin t\right) \\
&= \left(\tfrac{2}{5}\right)\left(-e^{-t}\cos t - e^{-t}\sin t\right) \\
&\quad + A_2\left(-e^{-t}\sin t + e^{-t}\cos t\right) \\
&\quad + \left(-\tfrac{2}{5}\right)(-\sin t) + \tfrac{1}{5}\cos t
\end{aligned}
$$

Apply the second initial condition, $x'(0) = 1$.

$$
\begin{aligned}
x'(0) &= \left(\tfrac{2}{5}\right)\left(-e^0\cos 0 - e^0\sin 0\right) \\
&\quad + A_2\left(-e^0\sin 0 + e^0\cos 0\right) \\
&\quad + \left(-\tfrac{2}{5}\right)(-\sin 0) + \tfrac{1}{5}\cos 0 \\
&= 1
\end{aligned}
$$

$$
\begin{aligned}
\left(\tfrac{2}{5}\right)\left(-(1)(1) - (1)(0)\right) &+ A_2\left(-(1)(0)\right. \\
+ (1)(1)\right) + \left(-\tfrac{2}{5}\right)(0) &+ \left(\tfrac{1}{5}\right)(1) = 1 \\
\left(\tfrac{2}{5}\right)(-1) + A_2(1) + \left(\tfrac{1}{5}\right) &= 1 \\
A_2 &= \tfrac{6}{5}
\end{aligned}
$$

Substituting for A_2, the solution becomes

$$
x(t) = \boxed{\begin{aligned} &\tfrac{2}{5}e^{-t}\cos t + \tfrac{6}{5}e^{-t}\sin t \\ &-\tfrac{2}{5}\cos t + \tfrac{1}{5}\sin t \end{aligned}}
$$

The answer is (D).

(d) *Alternate solution:*

Use the Laplace transform method.

$$
\begin{aligned}
x'' + 2x' + 2x &= \sin t \\
\mathcal{L}(x'') + 2\mathcal{L}(x') + 2\mathcal{L}(x) &= \mathcal{L}(\sin t) \\
s^2\mathcal{L}(x) - 1 + 2s\mathcal{L}(x) + 2\mathcal{L}(x) &= \frac{1}{s^2 + 1} \\
\mathcal{L}(x)(s^2 + 2s + 2) - 1 &= \frac{1}{s^2 + 1}
\end{aligned}
$$

$$
\begin{aligned}
\mathcal{L}(x) &= \frac{1}{s^2 + 2s + 2} + \frac{1}{(s^2 + 1)(s^2 + 2s + 2)} \\
&= \frac{1}{(s+1)^2 + 1} + \frac{1}{(s^2 + 1)(s^2 + 2s + 2)}
\end{aligned}
$$

Use partial fractions to expand the second term.

$$
\frac{1}{(s^2 + 1)(s^2 + 2s + 2)} = \frac{A_1 + B_1 s}{s^2 + 1} + \frac{A_2 + B_2 s}{s^2 + 2s + 2}
$$

Cross multiply.

$$
\begin{aligned}
&= \frac{\begin{aligned} A_1 s^2 + 2A_1 s + 2A_1 + B_1 s^3 + 2B_1 s^2 \\ + 2B_1 s + A_2 s^2 + A_2 + B_2 s^3 + B_2 s \end{aligned}}{(s^2 + 1)(s^2 + 2s + 2)} \\
&= \frac{\begin{aligned} s^3(B_1 + B_2) + s^2(A_1 + A_2 + 2B_1) \\ + s(2A_1 + 2B_1 + B_2) + 2A_1 + A_2 \end{aligned}}{(s^2 + 1)(s^2 + 2s + 2)}
\end{aligned}
$$

Compare numerators to obtain the following four simultaneous equations.

$$
\begin{aligned}
B_1 + B_2 &= 0 \\
A_1 + A_2 + 2B_1 &= 0 \\
2A_1 + 2B_1 + B_2 &= 0 \\
2A_1 + A_2 &= 1
\end{aligned}
$$

Use Cramer's rule to find A_1.

$$
A_1 = \frac{\begin{vmatrix} 0 & 0 & 1 & 1 \\ 0 & 1 & 2 & 0 \\ 0 & 0 & 2 & 1 \\ 1 & 1 & 0 & 0 \end{vmatrix}}{\begin{vmatrix} 0 & 0 & 1 & 1 \\ 1 & 1 & 2 & 0 \\ 2 & 0 & 2 & 1 \\ 2 & 1 & 0 & 0 \end{vmatrix}} = \frac{-1}{-5} = \frac{1}{5}
$$

The rest of the coefficients are found similarly.

$$A_1 = \tfrac{1}{5}$$
$$A_2 = \tfrac{3}{5}$$
$$B_1 = -\tfrac{2}{5}$$
$$B_2 = \tfrac{2}{5}$$

Then,

$$\mathcal{L}(x) = \frac{1}{(s+1)^2+1} + \frac{\tfrac{1}{5}}{s^2+1} + \frac{-\tfrac{2}{5}s}{s^2+1}$$
$$+ \frac{\tfrac{3}{5}}{s^2+2s+2} + \frac{\tfrac{2}{5}s}{s^2+2s+2}$$

Take the inverse transform.

$$x(t) = \mathcal{L}^{-1}\{\mathcal{L}(x)\}$$
$$= e^{-t}\sin t + \tfrac{1}{5}\sin t - \tfrac{2}{5}\cos t + \tfrac{3}{5}e^{-t}\sin t$$
$$+ \tfrac{2}{5}(e^{-t}\cos t - e^{-t}\sin t)$$
$$= \boxed{\tfrac{6}{5}e^{-t}\sin t + \tfrac{2}{5}e^{-t}\cos t + \tfrac{1}{5}\sin t - \tfrac{2}{5}\cos t}$$

The answer is (D).

4. *Customary U.S. Solution*

The differential equation is given as

$$m'(t) = a(t) - \frac{m(t)o(t)}{V(t)}$$

$$a(t) = \text{rate of addition of chemical}$$
$$m(t) = \text{mass of chemical at time } t$$
$$o(t) = \text{volumetric flow out of the lagoon}$$
$$(= 30 \text{ gal/min})$$
$$V(t) = \text{volume in the lagoon at time } t$$

Water flows into the lagoon at a rate of 30 gal/min, and a water-chemical mix flows out of the lagoon at rate of 30 gal/min. Therefore, the volume of the lagoon at time t is equal to the initial volume.

$$V(t) = \left(\frac{\pi}{4}\right)(\text{diameter of lagoon})^2(\text{depth of lagoon})$$
$$= \left(\frac{\pi}{4}\right)(120 \text{ ft})^2(10 \text{ ft})$$
$$= 113{,}097 \text{ ft}^3$$

Use a conversion factor of 7.48 gal/ft^3.

$$o(t) = \frac{30 \ \dfrac{\text{gal}}{\text{min}}}{7.48 \ \dfrac{\text{gal}}{\text{ft}^3}}$$
$$= 4.01 \text{ ft}^3/\text{min}$$

Substituting into the general form of the differential equation gives

$$m'(t) = a(t) - \frac{m(t)o(t)}{V(t)}$$
$$= (0) - m(t)\left(\frac{4.01 \ \dfrac{\text{ft}^3}{\text{min}}}{113{,}097 \text{ ft}^3}\right)$$
$$= -\left(\frac{3.55 \times 10^{-5}}{\text{min}}\right)m(t)$$

$$m'(t) + \left(\frac{3.55 \times 10^{-5}}{\text{min}}\right)m(t) = 0$$

The differential equation of the problem has a characteristic equation.

$$r + \frac{3.55 \times 10^{-5}}{\text{min}} = 0$$
$$r = -3.55 \times 10^{-5}/\text{min}$$

The general form of the solution is given by

$$m(t) = Ae^{rt}$$

Substituting for the root, r, gives

$$m(t) = Ae^{\left(\frac{-3.55\times10^{-5}}{\text{min}}\right)t}$$

Apply the initial condition $m(0) = 90$ lbm at time $t = 0$.

$$m(0) = Ae^{\left(\frac{-3.55\times10^{-5}}{\text{min}}\right)(0)} = 90 \text{ lbm}$$
$$Ae^0 = 90 \text{ lbm}$$
$$A = 90 \text{ lbm}$$

Therefore,

$$m(t) = (90 \text{ lbm})\, e^{\left(\frac{-3.55\times10^{-5}}{\text{min}}\right)t}$$

Solve for t.

$$\frac{m(t)}{90 \text{ lbm}} = e^{\left(\frac{-3.55\times10^{-5}}{\text{min}}\right)t}$$

$$\ln\left(\frac{m(t)}{90 \text{ lbm}}\right) = \ln\left(e^{\left(\frac{-3.55\times10^{-5}}{\text{min}}\right)t}\right)$$
$$= \left(\frac{-3.55 \times 10^{-5}}{\text{min}}\right)t$$

$$t = \frac{\ln\left(\dfrac{m(t)}{90 \text{ lbm}}\right)}{\dfrac{-3.55 \times 10^{-5}}{\text{min}}}$$

The initial mass of the water in the lagoon is given by

$$m_i = V\rho$$

$$= (113{,}097 \text{ ft}^3)\left(62.4 \; \frac{\text{lbm}}{\text{ft}^3}\right)$$

$$= 7.05 \times 10^6 \text{ lbm}$$

The final mass of chemicals is achieved at a concentration of 1 ppb or

$$m_f = \frac{7.06 \times 10^6 \text{ lbm}}{1 \times 10^9}$$

$$= 7.06 \times 10^{-3} \text{ lbm}$$

Find the time required to achieve a mass of 7.06×10^{-3} lbm.

$$t = \left(\frac{\ln\left(\dfrac{m(t)}{90 \text{ lbm}}\right)}{\dfrac{-3.55 \times 10^{-5}}{\text{min}}}\right)\left(\frac{1 \text{ hr}}{60 \text{ min}}\right)\left(\frac{1 \text{ day}}{24 \text{ hr}}\right)$$

$$= \left(\frac{\ln\left(\dfrac{7.06 \times 10^{-3} \text{ lbm}}{90 \text{ lbm}}\right)}{\dfrac{-3.55 \times 10^{-5}}{\text{min}}}\right)\left(\frac{1 \text{ hr}}{60 \text{ min}}\right)\left(\frac{1 \text{ day}}{24 \text{ hr}}\right)$$

$$= \boxed{185 \text{ days}}$$

The answer is (D).

SI Solution

The differential equation is given as

$$m'(t) = a(t) - \frac{m(t)o(t)}{V(t)}$$

$$a(t) = \text{rate of addition of chemical}$$
$$m(t) = \text{mass of chemical at time } t$$
$$o(t) = \text{volumetric flow out of the lagoon}$$
$$(= 115 \text{ L/min})$$
$$V(t) = \text{volume in the lagoon at time } t$$

Water flows into the lagoon at a rate of 115 L/min, and a water-chemical mix flows out of the lagoon at a rate of 115 L/min. Therefore, the volume of the lagoon at time t is equal to the initial volume.

$$V(t) = \left(\frac{\pi}{4}\right)(\text{diameter of lagoon})^2(\text{depth of lagoon})$$

$$= \left(\frac{\pi}{4}\right)(35 \text{ m})^2(3 \text{ m})$$

$$= 2886 \text{ m}^3$$

Using a conversion factor of 1 m³/1000 L gives

$$o(t) = \left(115 \; \frac{\text{L}}{\text{min}}\right)\left(\frac{1 \text{ m}^3}{1000 \text{ L}}\right)$$

$$= 0.115 \text{ m}^3/\text{min}$$

Substitute into the general form of the differential equation.

$$m'(t) = a(t) - \frac{m(t)o(t)}{V(t)}$$

$$= 0 - m(t)\left(\frac{0.115 \; \dfrac{\text{m}^3}{\text{min}}}{2886 \text{ m}^3}\right)$$

$$= -\left(\frac{3.985 \times 10^{-5}}{\text{min}}\right)m(t)$$

$$m'(t) + \left(\frac{3.985 \times 10^{-5}}{\text{min}}\right)m(t) = 0$$

The differential equation of the problem has the following characteristic equation.

$$r + \frac{3.985 \times 10^{-5}}{\text{min}} = 0$$

$$r = -3.985 \times 10^{-5}/\text{min}$$

The general form of the solution is given by

$$m(t) = Ae^{rt}$$

Substituting in for the root, r, gives

$$m(t) = Ae^{\left(\frac{-3.985 \times 10^{-5}}{\text{min}}\right)t}$$

Apply the initial condition $m(0) = 40$ kg at time $t = 0$.

$$m(0) = Ae^{\left(\frac{-3.985 \times 10^{-5}}{\text{min}}\right)(0)} = 40 \text{ kg}$$

$$Ae^0 = 40 \text{ kg}$$

$$A = 40 \text{ kg}$$

Therefore,

$$m(t) = (40 \text{ kg})e^{\left(\frac{-3.985 \times 10^{-5}}{\text{min}}\right)t}$$

Solve for t.

$$\frac{m(t)}{40 \text{ kg}} = e^{\left(\frac{-3.985 \times 10^{-5}}{\text{min}}\right)t}$$

$$\ln\left(\frac{m(t)}{40 \text{ kg}}\right) = \ln\left(e^{\left(\frac{-3.985 \times 10^{-5}}{\text{min}}\right)t}\right)$$

$$= \left(\frac{-3.985 \times 10^{-5}}{\text{min}}\right)t$$

$$t = \ln\left(\frac{\dfrac{m(t)}{40 \text{ kg}}}{\dfrac{-3.985 \times 10^{-5}}{\text{min}}}\right)$$

The initial mass of water in the lagoon is given by

$$m_i = V\rho$$

$$= (2886 \text{ m}^3)\left(1000 \, \frac{\text{kg}}{\text{m}^3}\right)$$

$$= 2.886 \times 10^6 \text{ kg}$$

The final mass of chemicals is achieved at a concentration of 1 ppb or

$$m_f = \frac{2.886 \times 10^6 \text{ kg}}{1 \times 10^9}$$

$$= 2.886 \times 10^{-3} \text{ kg}$$

Find the time required to achieve a mass of 2.886×10^{-3} kg.

$$t = \left(\frac{\ln\dfrac{m(t)}{40 \text{ kg}}}{\dfrac{-3.985 \times 10^{-5}}{\text{min}}}\right)\left(\frac{1 \text{ h}}{60 \text{ min}}\right)\left(\frac{1 \text{ day}}{24 \text{ h}}\right)$$

$$= \left(\frac{\ln\left(\dfrac{2.886 \times 10^{-3} \text{ kg}}{40 \text{ kg}}\right)}{\dfrac{-3.985 \times 10^{-5}}{\text{min}}}\right)\left(\frac{1 \text{ h}}{60 \text{ min}}\right)\left(\frac{1 \text{ day}}{24 \text{ h}}\right)$$

$$= \boxed{166 \text{ days}}$$

The answer is (D).

5. Let

$$m(t) = \text{mass of salt in tank at time } t$$

$$m_0 = 60 \text{ mass units}$$

$$m'(t) = \text{rate at which salt content is changing}$$

2 mass units of salt enter each minute, and 3 volumes leave each minute. The amount of salt leaving each minute is

$$\left(3 \, \frac{\text{vol}}{\text{min}}\right)\left(\text{concentration in } \frac{\text{mass}}{\text{vol}}\right)$$

$$= \left(3 \, \frac{\text{vol}}{\text{min}}\right)\left(\frac{\text{salt content}}{\text{volume}}\right)$$

$$= \left(3 \, \frac{\text{vol}}{\text{min}}\right)\left(\frac{m(t)}{100 - t}\right)$$

$$m'(t) = 2 - (3)\left(\frac{m(t)}{100 - t}\right) \text{ or } m'(t) + \frac{3m(t)}{100 - t}$$

$$= 2 \text{ mass/min}$$

This is a first-order linear differential equation. The integrating factor is

$$m = \exp\left(3\int \frac{dt}{100 - t}\right)$$

$$= \exp\Big((3)\big(-\ln(100 - t)\big)\Big)$$

$$= (100 - t)^{-3}$$

$$m(t) = (100 - t)^3\left(2\int \frac{dt}{(100 - t)^3} + k\right)$$

$$= 100 - t + (k)(100 - t)^3$$

But $m = 60$ mass units at $t = 0$, so $k = -0.00004$.

$$m(t) = 100 - t - (0.00004)(100 - t)^3$$

At $t = 60$ min,

$$m = 100 - 60 \text{ min} - (0.00004)(100 - 60 \text{ min})^3$$

$$= \boxed{37.44 \text{ mass units}}$$

The answer is (B).

11 Probability and Statistical Analysis of Data

PRACTICE PROBLEMS

Probability

1. Four military recruits whose respective shoe sizes are 7, 8, 9, and 10 report to the supply clerk to be issued boots. The supply clerk selects one pair of boots in each of the four required sizes and hands them at random to the recruits.

(a) What is the probability that all recruits will receive boots of an incorrect size?
- (A) 0.25
- (B) 0.38
- (C) 0.45
- (D) 0.61

(b) What is the probability that exactly three recruits will receive boots of the correct size?
- (A) 0
- (B) 0.063
- (C) 0.17
- (D) 0.25

The answer is (A).

Probability Distributions

2. The time spent by a toll taker in collecting the toll from vehicles crossing a bridge is an exponential distribution with a mean of 23 sec. What is the probability that a random vehicle will be processed in 25 sec or more (i.e., will take longer than 25 sec)?
- (A) 0.17
- (B) 0.25
- (C) 0.34
- (D) 0.52

3. The number of cars entering a toll plaza on a bridge during the hour after midnight follows a Poisson distribution with a mean of 20.

(a) What is the probability that 17 cars will pass through the toll plaza during that hour on any given night?
- (A) 0.076
- (B) 0.12
- (C) 0.16
- (D) 0.23

(b) What is the probability that three or fewer cars will pass through the toll plaza at that hour on any given night?
- (A) 0.0000032
- (B) 0.0019
- (C) 0.079
- (D) 0.11

4. A mechanical component exhibits a negative exponential failure distribution with a mean time to failure of 1000 hr. What is the maximum operating time such that the reliability remains above 99%?
- (A) 3.3 hr
- (B) 5.6 hr
- (C) 8.1 hr
- (D) 10 hr

5. (*Time limit: one hour*) A survey field crew measures one leg of a traverse four times. The following results are obtained.

repetition	measurement	direction
1	1249.529	forward
2	1249.494	backward
3	1249.384	forward
4	1249.348	backward

The crew chief is under orders to obtain readings with confidence limits of 90%.

(a) Which readings are acceptable?
- (A) No readings are acceptable.
- (B) Two readings are acceptable.
- (C) Three readings are acceptable.
- (D) All four readings are acceptable.

(b) Which readings are not acceptable?
- (A) No readings are unacceptable.
- (B) One reading is unacceptable.
- (C) Two readings are unacceptable.
- (D) All four readings are unacceptable.

(c) Explain how to determine which readings are not acceptable.

- (A) Readings inside the 90% confidence limits are unacceptable.
- (B) Readings outside the 90% confidence limits are unacceptable.
- (C) Readings outside the upper 90% confidence limit are unacceptable.
- (D) Readings outside the lower 90% confidence limit are unacceptable.

(d) What is the most probable value of the distance?

- (A) 1249.399
- (B) 1249.410
- (C) 1249.439
- (D) 1249.452

(e) What is the error in the most probable value (at 90% confidence)?

- (A) 0.08
- (B) 0.11
- (C) 0.14
- (D) 0.19

(f) If the distance is one side of a square traverse whose sides are all equal, what is the most probable closure error?

- (A) 0.14
- (B) 0.20
- (C) 0.28
- (D) 0.35

(g) What is the probable error of part (f) expressed as a fraction?

- (A) 1:17,600
- (B) 1:14,200
- (C) 1:12,500
- (D) 1:10,900

(h) What is the order of accuracy of the closure?

- (A) first order
- (B) second order
- (C) third order
- (D) fourth order

(i) Define accuracy and distinguish it from precision.

- (A) If an experiment can be repeated with identical results, the results are considered accurate.
- (B) If an experiment has a small bias, the results are considered precise.
- (C) If an experiment is precise, it cannot also be accurate.
- (D) If an experiment is unaffected by experimental error, the results are accurate.

(j) Give an example of systematic error.

- (A) measuring river depth as a motorized ski boat passes by
- (B) using a steel tape that is too short to measure consecutive distances
- (C) locating magnetic north near a large iron ore deposit along an overland route
- (D) determining local wastewater BOD after a toxic spill

Statistical Analysis

6. *(Time limit: one hour)* California law requires a statistical analysis of the average speed driven by motorists on a road prior to the use of radar speed control. The following speeds (all in mi/hr) were observed in a random sample of 40 cars.

44, 48, 26, 25, 20, 43, 40, 42, 29, 39, 23, 26, 24, 47, 45, 28, 29, 41, 38, 36, 27, 44, 42, 43, 29, 37, 34, 31, 33, 30, 42, 43, 28, 41, 29, 36, 35, 30, 32, 31

(a) Tabulate the frequency distribution of the data.

(b) Draw the frequency histogram.

(c) Draw the frequency polygon.

(d) Tabulate the cumulative frequency distribution.

(e) Draw the cumulative frequency graph.

(f) What is the upper quartile speed?

- (A) 30 mph
- (B) 35 mph
- (C) 40 mph
- (D) 45 mph

(g) What is the mean speed?

- (A) 31 mph
- (B) 33 mph
- (C) 35 mph
- (D) 37 mph

(h) What is the standard deviation of the sample data?

- (A) 2.1 mph
- (B) 6.1 mph
- (C) 6.8 mph
- (D) 7.4 mph

(i) What is the sample standard deviation?

- (A) 7.5 mph
- (B) 18 mph
- (C) 35 mph
- (D) 56 mph

(j) What is the sample variance?

- (A) $56 \text{ mi}^2/\text{hr}^2$
- (B) $324 \text{ mi}^2/\text{hr}^2$
- (C) $1225 \text{ mi}^2/\text{hr}^2$
- (D) $3136 \text{ mi}^2/\text{hr}^2$

7. A spot speed study is conducted for a stretch of roadway. During a normal day, the speeds were found to be normally distributed with a mean of 46 and a standard deviation of 3.

(a) What is the 50th percentile speed?
- (A) 39
- (B) 43
- (C) 46
- (D) 49

(b) What is the 85th percentile speed?
- (A) 47.1
- (B) 48.3
- (C) 49.1
- (D) 52.7

(c) What is the upper two standard deviation speed?
- (A) 47.2
- (B) 49.3
- (C) 51.1
- (D) 52.0

(d) The daily average speeds for the same stretch of roadway on consecutive normal days were determined by sampling 25 vehicles each day. What is the upper two-standard deviation average speed?
- (A) 46.6
- (B) 47.2
- (C) 52.0
- (D) 54.7

8. The diameters of bolt holes drilled in structural steel members are normally distributed with a mean of 0.502 in and a standard deviation of 0.005 in. Holes are out of specification if their diameters are less than 0.497 in or more than 0.507 in.

(a) What is the probability that a hole chosen at random will be out of specification?
- (A) 0.16
- (B) 0.22
- (C) 0.32
- (D) 0.68

(b) What is the probability that 2 holes out of a sample of 15 will be out of specification?
- (A) 0.074
- (B) 0.12
- (C) 0.15
- (D) 0.32

9. Two resistances, the meter resistance and a shunt resistor, are connected in parallel in an ammeter. Most of the current passing through the meter goes through the shunt resistor. In order to determine the accuracy of the resistance of shunt resistors being manufactured for a line of ammeters, a manufacturer tests a sample of 100 shunt resistors. The numbers of shunt resistors with the resistance indicated (to the nearest hundredth of an ohm) are as follows.

$0.200 \ \Omega$, 1; $0.210 \ \Omega$, 3; $0.220 \ \Omega$, 5; $0.230 \ \Omega$, 10; $0.240 \ \Omega$, 17; $0.250 \ \Omega$, 40; $0.260 \ \Omega$, 13; $0.270 \ \Omega$, 6; $0.280 \ \Omega$, 3; $0.290 \ \Omega$, 2

(a) What is the mean resistance?
- (A) $0.235 \ \Omega$
- (B) $0.247 \ \Omega$
- (C) $0.251 \ \Omega$
- (D) $0.259 \ \Omega$

(b) What is the sample standard deviation?
- (A) 0.0003
- (B) 0.010
- (C) 0.016
- (D) 0.24

(c) What is the median resistance?
- (A) $0.22 \ \Omega$
- (B) $0.24 \ \Omega$
- (C) $0.25 \ \Omega$
- (D) $0.26 \ \Omega$

(d) What is the sample variance?
- (A) 0.00027
- (B) 0.0083
- (C) 0.0114
- (D) 0.0163

Hypothesis Testing

10. 100 bearings were tested to failure. The average life was 1520 hr, and the standard deviation was 120 hr. The manufacturer claims a 1600 hr life. Evaluate using confidence limits of 95% and 99%.
- (A) The claim is accurate at both 95% and 99% confidence.
- (B) The claim is inaccurate only at 95%.
- (C) The claim is inaccurate only at 99%.
- (D) The claim is inaccurate at both 95% and 99% confidence.

Curve Fitting

11. (a) Find the best equation for a line passing through the points given. (b) Find the correlation coefficient.

x	y
400	370
800	780
1250	1210
1600	1560
2000	1980
2500	2450
4000	3950

12. Find the best equation for a line passing through the points given.

s	t
20	43
18	141
16	385
14	1099

13. The number of vehicles lining up behind a flashing railroad crossing has been observed for five trains of different lengths, as given. What is the mathematical formula that relates the two variables?

no. of cars in train	no. of vehicles
2	14.8
5	18.0
8	20.4
12	23.0
27	29.9

14. The following yield data are obtained from five identical treatment plants.

(a) Develop a mathematical equation to correlate the yield and average temperature.

(b) What is the correlation coefficient?

treatment plant	average temperature (T)	average yield (Y)
1	207.1	92.30
2	210.3	92.58
3	200.4	91.56
4	201.1	91.63
5	203.4	91.83

15. The following data are obtained from a soil compaction test. What is the mathematical formula that relates the two variables?

x	y
−1	0
0	1
1	1.4
2	1.7
3	2
4	2.2
5	2.4
6	2.6
7	2.8
8	3

SOLUTIONS

1. (a) There are 4! = 24 different possible outcomes. By enumeration (i.e., by making a table of all possible combinations), there are 9 completely wrong combinations.

$$p\{\text{all wrong}\} = \frac{9}{24} = \boxed{0.375}$$

	sizes				
correct →	7	8	9	10	all wrong
sizes issued	7	8	9	10	
	7	8	10	9	
	7	9	8	10	
	7	9	10	8	
	7	10	8	9	
	7	10	9	8	
	8	9	10	7	X
	8	9	7	10	
	8	10	9	7	
	8	10	7	9	X
	8	7	9	10	
	8	7	10	9	X
	9	10	7	8	X
	9	10	8	7	X
	9	7	10	8	X
	9	7	8	10	
	9	8	7	10	
	9	8	10	7	
	10	7	8	9	X
	10	7	9	8	
	10	8	7	9	
	10	8	9	7	
	10	9	8	7	X
	10	9	7	8	X

The answer is (B).

(b) If three recruits get the correct size, the fourth recruit will also since there will be only one pair remaining.

$$p\{\text{exactly 3}\} = \boxed{0}$$

The answer is (A).

2. For an exponential distribution function, the mean is given as

$$\mu = \frac{1}{\lambda}$$

For a mean of 23,

$$\mu = 23 = \frac{1}{\lambda}$$
$$\lambda = 0.0435$$

For an exponential distribution function,

$$p\{X < x\} = F(x) = 1 - e^{-\lambda x}$$
$$p\{x > X\} = 1 - p\{X < x\}$$
$$= 1 - F(x)$$
$$= 1 - \left(1 - e^{-\lambda x}\right)$$
$$= e^{-\lambda x}$$

The probability of a random vehicle being processed in 25 sec or more is given by

$$p\{x > 25\} = e^{-(0.0435)(25)}$$
$$= e^{-1.0875}$$

$$= \boxed{0.337}$$

The answer is (C).

3. (a) The distribution is a Poisson distribution with an average of $\lambda = 20$.

The probability for a Poisson distribution is given by

$$p\{x\} = f(x) = \frac{e^{-\lambda}\lambda^x}{x!}$$

The probability of 17 cars is

$$p\{x = 17\} = f(17) = \frac{e^{-20} \times 20^{17}}{17!}$$

$$= \boxed{0.076 \ \ (7.6\%)}$$

The answer is (A).

(b) The probability of three or fewer cars is given by

$$p\{x \le 3\} = p\{x = 0\} + p\{x = 1\} + p\{x = 2\}$$
$$+ p\{x = 3\}$$
$$= f(0) + f(1) + f(2) + f(3)$$
$$= \frac{e^{-20} \times 20^0}{0!} + \frac{e^{-20} \times 20^1}{1!}$$
$$+ \frac{e^{-20} \times 20^2}{2!} + \frac{e^{-20} \times 20^3}{3!}$$
$$= 2 \times 10^{-9} + 4.1 \times 10^{-8}$$
$$+ 4.12 \times 10^{-7} + 2.75 \times 10^{-6}$$
$$= 3.2 \times 10^{-6}$$

$$= \boxed{0.0000032 \ \ (0.00032\%)}$$

The answer is (A).

4.
$$\lambda = \frac{1}{\text{MTTF}}$$
$$= \frac{1}{1000} = 0.001$$

The reliability function is

$$R\{t\} = e^{-\lambda t} = e^{-0.001t}$$

Since the reliability is greater than 99%,

$$e^{-0.001t} > 0.99$$
$$\ln(e^{-0.001t}) > \ln(0.99)$$
$$-0.001t > \ln(0.99)$$
$$t < -1000 \ln(0.99)$$

$$\boxed{t < 10.05}$$

The maximum operating time such that the reliability remains above 99% is 10.05 hr.

The answer is (D).

5. Find the average.

$$\bar{x} = \frac{\sum x_i}{n}$$
$$= \frac{1249.529 + 1249.494 + 1249.384 + 1249.348}{4}$$
$$= 1249.439$$

Since the sample population is small, use the sample standard deviation.

$$s = \sqrt{\frac{\sum(x_i - \bar{x})^2}{n - 1}}$$
$$= \sqrt{\frac{\substack{(1249.529 - 1249.439)^2 + (1249.494 - 1249.439)^2 \\ + (1249.384 - 1249.439)^2 + (1249.348 - 1249.439)^2}}{4 - 1}}$$
$$= 0.08647$$

From the standard deviation table, a two-tail 90% confidence limit falls within $1.645s$ of $\bar{x}$.

$$1249.439 \pm (1.645)(0.08647) = 1249.439 \pm 0.142$$

Therefore, $(1249.297, 1249.581)$ is the 90% confidence range.

(a) By observation, all the readings fall within the 90% confidence range.

The answer is (D).

(b) No readings are unacceptable.

The answer is (A).

(c) Readings outside the 90% confidence limits are unacceptable.

The answer is (B).

(d) The unbiased estimate of the most probable distance is 1249.439.

The answer is (C).

(e) The error for the 90% confidence range is 0.142.

The answer is (C).

(f) If the surveying crew places a marker, measures a distance x, places a second marker, and then measures the same distance x back to the original marker, the ending point should coincide with the original marker. If, due to measurement errors, the ending and starting points do not coincide, the difference is the closure error.

In this example, the survey crew moves around the four sides of a square, so there are two measurements in the x-direction and two measurements in the y-direction. If the errors E_1 and E_2 are known for two measurements, x_1 and x_2, the error associated with the sum or difference $x_1 \pm x_2$ is

$$E\{x_1 \pm x_2\} = \sqrt{E_1^2 + E_2^2}$$

In this case, the error in the x-direction is

$$E_x = \sqrt{(0.1422)^2 + (0.1422)^2}$$
$$= 0.2011$$

The error in the y-direction is calculated the same way and is also 0.2011. E_x and E_y are combined by the Pythagorean theorem to yield

$$E_{\text{closure}} = \sqrt{(0.2011)^2 + (0.2011)^2}$$
$$= \boxed{0.2844}$$

The answer is (C).

(g) In surveying, error may be expressed as a fraction of one or more legs of the traverse. Assume that the total of all four legs is to be used as the basis.

$$\frac{0.2844}{(4)(1249)} = \boxed{\frac{1}{17,567}}$$

The answer is (A).

(h) In surveying, a class 1 third-order error is smaller than 1/10,000. The error of 1/17,567 is smaller than the third-order error; therefore, the error is within the third-order accuracy.

The answer is (C).

(i) An experiment is accurate if it is unchanged by experimental error. Precision is concerned with the repeatability of the experimental results. If an experiment is repeated with identical results, the experiment is said to be precise. However, it is possible to have a highly precise experiment with a large bias.

The answer is (D).

(j) A systematic error is one that is always present and is unchanged from sample to sample. For example, a steel tape that is 0.02 ft short introduces a systematic error.

The answer is (B).

6. (a) and (d) Tabulate the frequency distribution data.

(Note that the lowest speed is 20 mi/hr and the highest speed is 48 mi/hr; therefore, the range is 28 mi/hr. Choose 10 cells with a width of 3 mi/hr.)

midpoint	interval (mi/hr)	frequency	cumulative frequency	cumulative percent
21	20–22	1	1	3
24	23–25	3	4	10
27	26–28	5	9	23
30	29–31	8	17	43
33	32–34	3	20	50
36	35–37	4	24	60
39	38–40	3	27	68
42	41–43	8	35	88
45	44–46	3	38	95
48	47–49	2	40	100

(b)

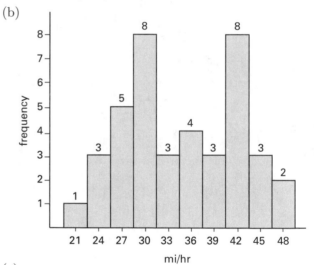

(c)

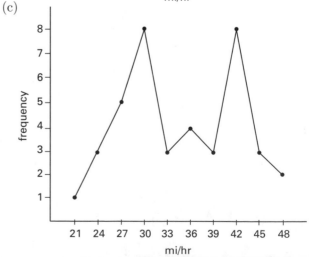

(e)

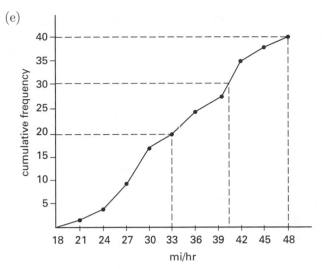

(f) From the cumulative frequency graph in part (e), the upper quartile speed occurs at 30 cars or 75%, which corresponds to approximately 40 mi/hr.

The answer is (C).

(g)
$$\sum x_i = 1390 \text{ mi/hr}$$
$$n = 40$$

The mean is computed as

$$\overline{x} = \frac{\sum x_i}{n}$$

$$= \frac{1390 \ \dfrac{\text{mi}}{\text{hr}}}{40}$$

$$= \boxed{34.75 \text{ mi/hr}}$$

The answer is (C).

(h) The standard deviation of the sample data is given as

$$\sigma = \sqrt{\frac{\sum x^2}{n} - \mu^2}$$

$$\sum x^2 = 50{,}496 \text{ mi}^2/\text{hr}^2$$

Use the sample mean as an unbiased estimator of the population mean, μ.

$$\sigma = \sqrt{\frac{\sum x^2}{n} - \mu^2}$$

$$= \sqrt{\frac{50{,}496 \ \dfrac{\text{mi}^2}{\text{hr}^2}}{40} - \left(34.75 \ \frac{\text{mi}}{\text{hr}}\right)^2}$$

$$= \boxed{7.405 \text{ mi/hr}}$$

The answer is (D).

(i) The sample standard deviation is given by

$$s = \sqrt{\frac{\sum x^2 - \dfrac{\left(\sum x\right)^2}{n}}{n-1}}$$

$$= \sqrt{\frac{50{,}496 \ \dfrac{\text{mi}^2}{\text{hr}^2} - \dfrac{\left(1390 \ \dfrac{\text{mi}}{\text{hr}}\right)^2}{40}}{40 - 1}}$$

$$= \boxed{7.500 \text{ mi/hr}}$$

The answer is (A).

(j) The sample variance is given by the square of the sample standard deviation.

$$s^2 = \left(7.500 \ \frac{\text{mi}}{\text{hr}}\right)^2$$

$$= \boxed{56.25 \text{ mi}^2/\text{hr}^2}$$

The answer is (A).

7. (a) The 50th percentile speed is the median speed, $\boxed{46,}$ which for a symmetrical normal distribution is the mean speed.

The answer is (C).

(b) The 85th percentile speed is the speed that is exceeded by only 15% of the measurements. Since this is a normal distribution, App. 11.A can be used. 15% in the upper tail corresponds to 35% between the mean and the 85th percentile. This occurs at approximately $= 1.04\sigma$. The 85th percentile speed is

$$x_{85\%} = \mu + 1.04\sigma$$

$$= 46 + (1.04)(3) = \boxed{49.12}$$

The answer is (C).

(c) The upper 2σ speed is

$$x_{2\sigma} = \mu + 2\sigma$$

$$= 46 + (2)(3) = \boxed{52}$$

The answer is (D).

(d) According to the central limit theorem, the mean of the average speeds is the same as the distribution mean, and the standard deviation of sample means is

$$\sigma_{\overline{x}} = \frac{\sigma_x}{\sqrt{K}}$$

$$= \frac{3}{\sqrt{25}} = 0.6$$

$$\overline{x}_{2\sigma} = \mu + 2\sigma_{\overline{x}}$$

$$= 46 + (2)(0.6) = \boxed{47.2}$$

The answer is (B).

8. (a) From Eq. 11.43,

$$z_{upper} = \frac{0.507 \text{ in} - 0.502 \text{ in}}{0.005 \text{ in}} = +1$$

From App. 11.A, the area outside $z = +1$ is

$$0.5 - 0.3413 = 0.1587$$

Since these are symmetrical limits, $z_{lower} = -1$.

total fraction defective $= (2)(0.1587) = \boxed{0.3174}$

The answer is (C).

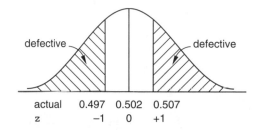

actual	0.497	0.502	0.507
z	−1	0	+1

(b) This is a binomial problem.

$$p = p\{\text{defective}\} = 0.3174$$

$$q = 1 - p = 0.6826$$

From Eq. 11.28,

$$f(2) = \binom{15}{2}(0.3174)^2(0.6826)^{13}$$

$$= \left(\frac{15!}{13!2!}\right)(0.3174)^2(0.6826)^{13} = \boxed{0.0739}$$

The answer is (A).

9. (a)

R	f	fR	fR^2
0.200	1	0.200	0.0400
0.210	3	0.630	0.1323
0.220	5	1.100	0.2420
0.230	10	2.300	0.5290
0.240	17	4.080	0.9792
0.250	40	10.000	2.5000
0.260	13	3.380	0.8788
0.270	6	1.620	0.4374
0.280	3	0.840	0.2352
0.290	2	0.580	0.1682
	100	24.730	6.1421

$$\overline{R} = \frac{\sum fR}{\sum f} = \frac{24.730 \ \Omega}{100} = \boxed{0.2473 \ \Omega}$$

The answer is (B).

(b) The sample standard deviation is given by Eq. 11.61.

$$s = \sqrt{\frac{\sum fR^2 - \frac{(\sum fR)^2}{n}}{n-1}}$$

$$= \sqrt{\frac{6.1421 \ \Omega - \frac{(24.73 \ \Omega)^2}{100}}{99}}$$

$$= \boxed{0.0163 \ \Omega}$$

The answer is (C).

(c) The 50th and 51st values are both 0.250 Ω. The median is 0.250 Ω.

The answer is (C).

(d) $\qquad s^2 = (0.0163 \ \Omega)^2 = \boxed{0.0002656 \ \Omega^2}$

The answer is (A).

10. This is a typical hypothesis test of two sample population means. The two populations are the original population the manufacturer used to determine the 1600 hr average life value and the new population the sample was taken from. The mean ($\overline{x} = 1520$ hr) of the sample and its standard deviation ($s = 120$ hr) are known, but the mean and standard deviation of a population of average lifetimes are unknown.

Assume that the average lifetime population mean and the sample mean are identical.

$$\overline{x} = \mu = 1520 \text{ hr}$$

The standard deviation of the average lifetime population is

$$s = \frac{\sigma_{\overline{x}}}{\sqrt{n}} = \frac{120 \text{ hr}}{\sqrt{100}} = 12 \text{ hr}$$

The manufacturer can be reasonably sure that the claim of a 1600 hr average life is justified if the average test life is near 1600 hr. "Reasonably sure" must be evaluated based on acceptable probability of being incorrect. If the manufacturer is willing to be wrong with a 5% probability, then a 95% confidence level is required.

Since the direction of bias is known, a one-tailed test is required. To determine if the mean has shifted downward, test the hypothesis that 1600 hr is within the 95% limit of a distribution with a mean of 1520 hr and a standard deviation of 12 hr. From a standard normal table, 5% of a standard normal distribution is outside of $z = 1.645$. Therefore, the 95% confidence limit is

$$1520 \text{ hr} + (1.645)(12 \text{ hr}) = 1540 \text{ hr}$$

The manufacturer can be 95% certain that the average lifetime of the bearings is less than 1600 hr, not as much as 1600 hr as claimed.

If the manufacturer is willing to be wrong with a probability of only 1%, then a 99% confidence limit is required. From the normal table, $z = 2.33$ and the 99% confidence limit is

$$1520 \text{ hr} + (2.33)(12 \text{ hr}) = 1548 \text{ hr}$$

The manufacturer can be 99% certain that the average bearing life is less than 1600 hr.

The answer is (D).

11. (a) Plot the data points to determine if the relationship is linear.

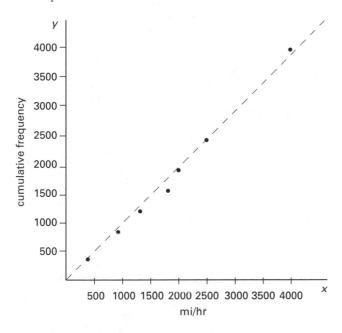

The data appear to be essentially linear. The slope, m, and the y-intercept, b, can be determined using linear regression.

The individual terms are

$$n = 7$$

$$\sum x_i = 400 + 800 + 1250 + 1600 + 2000 + 2500$$
$$+ 4000$$
$$= 12{,}550$$

$$\left(\sum x_i\right)^2 = (12{,}550)^2$$
$$= 1.575 \times 10^8$$

$$\bar{x} = \frac{\sum x_i}{n}$$
$$= \frac{12{,}550}{7}$$
$$= 1792.9$$

$$\sum x_i^2 = (400)^2 + (800)^2 + (1250)^2 + (1600)^2$$
$$+ (2000)^2 + (2500)^2 + (4000)^2$$
$$= 3.117 \times 10^7$$

Similarly,

$$\sum y_i = 370 + 780 + 1210 + 1560 + 1980$$
$$+ 2450 + 3950$$
$$= 12{,}300$$

$$\left(\sum y_i\right)^2 = (12{,}300)^2 = 1.513 \times 10^8$$

$$\bar{y} = \frac{\sum y_i}{n} = \frac{12{,}300}{7} = 1757.1$$

$$\sum y_i^2 = (370)^2 + (780)^2 + (1210)^2 + (1560)^2$$
$$+ (1980)^2 + (2450)^2 + (3950)^2$$
$$= 3.017 \times 10^7$$

Also,

$$\sum x_i y_i = (400)(370) + (800)(780) + (1250)(1210)$$
$$+ (1600)(1560) + (2000)(1980)$$
$$+ (2500)(2450) + (4000)(3950)$$
$$= 3.067 \times 10^7$$

The slope is

$$m = \frac{n\sum x_i y_i - \sum x_i \sum y_i}{n\sum x_i^2 - \left(\sum x_i\right)^2}$$
$$= \frac{(7)(3.067 \times 10^7) - (12{,}550)(12{,}300)}{(7)(3.117 \times 10^7) - (12{,}550)^2}$$
$$= 0.994$$

The y intercept is

$$b = \bar{y} - m\bar{x}$$
$$= 1757.1 - (0.994)(1792.9)$$
$$= -25.0$$

The least squares equation of the line is

$$y = mx + b$$
$$= \boxed{0.994x - 25.0}$$

(b) The correlation coefficient is

$$r = \frac{n\sum (x_i y_i) - \left(\sum x_i\right)\left(\sum y_i\right)}{\sqrt{\left(n\sum x_i^2 - \left(\sum x_i\right)^2\right)\left(n\sum y_i^2 - \left(\sum y_i\right)^2\right)}}$$

$$= \frac{(7)(3.067 \times 10^7) - (12{,}500)(12{,}300)}{\sqrt{\begin{array}{c}\left((7)(3.117 \times 10^7) - (12{,}500)^2\right)\\ \times (7)(3.017 \times 10^7) - (12{,}300)^2)\end{array}}}$$

$$\approx \boxed{1.00}$$

12. Plotting the data shows that the relationship is nonlinear.

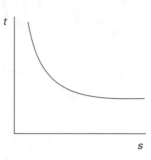

This appears to be an exponential with the form

$$t = ae^{bs}$$

Take the natural log of both sides.

$$\ln t = \ln(ae^{bs})$$
$$= \ln a + \ln(e^{bs})$$
$$= \ln a + bs$$

But, $\ln a$ is just a constant, c.

$$\ln t = c + bs$$

Make the transformation $R = \ln t$.

$$R = c + bs$$

s	R
20	3.76
18	4.95
16	5.95
14	7.00

This is linear.

$$n = 4$$
$$\sum s_i = 20 + 18 + 16 + 14 = 68$$
$$\bar{s} = \frac{\sum s}{n} = \frac{68}{4} = 17$$
$$\sum s_i^2 = (20)^2 + (18)^2 + (16)^2 + (14)^2 = 1176$$
$$\left(\sum s_i\right)^2 = (68)^2 = 4624$$
$$\sum R_i = 3.76 + 4.95 + 5.95 + 7.00 = 21.66$$
$$\bar{R} = \frac{\sum R_i}{n} = \frac{21.66}{4} = 5.415$$
$$\sum R_i^2 = (3.76)^2 + (4.95)^2 + (5.95)^2 + (7.00)^2$$
$$= 123.04$$
$$\left(\sum R_i\right)^2 = (21.66)^2 = 469.16$$
$$\sum s_i R_i = (20)(3.76) + (18)(4.95) + (16)(5.95)$$
$$+ (14)(7.00)$$
$$= 357.5$$

The slope, b, of the transformed line is

$$b = \frac{n \sum s_i R_i - \sum s_i \sum R_i}{n \sum s_i^2 - \left(\sum s_i\right)^2}$$
$$= \frac{(4)(357.5) - (68)(21.66)}{(4)(1176) - (68)^2} = -0.536$$

The intercept is

$$c = \bar{R} - b\bar{s} = 5.415 - (-0.536)(17)$$
$$= 14.527$$

The transformed equation is

$$R = c + bs$$
$$= 14.527 - 0.536s$$

$$\boxed{\ln t = 14.527 - 0.536s}$$

13. The first step is to graph the data.

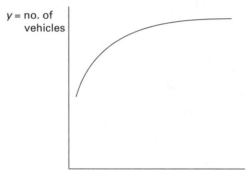

It is assumed that the relationship between the variables has the form $y = a + b \log x$. Therefore, the variable change $z = \log x$ is made, resulting in the following set of data.

z	y
0.301	14.8
0.699	18.0
0.903	20.4
1.079	23.0
1.431	29.9

$$\sum z_i = 4.413$$
$$\sum y_i = 106.1$$
$$\sum z_i^2 = 4.6082$$
$$\sum y_i^2 = 2382.2$$
$$\left(\sum z_i\right)^2 = 19.475$$
$$\left(\sum y_i\right)^2 = 11{,}257.2$$
$$\bar{z} = 0.8826$$
$$\bar{y} = 21.22$$
$$\sum z_i y_i = 103.06$$
$$n = 5$$

The slope is

$$m = \frac{n\sum z_i y_i - \sum z_i \sum y_i}{n\sum z_i^2 - \left(\sum z_i\right)^2}$$

$$= \frac{(5)(103.06) - (4.413)(106.1)}{(5)(4.6082) - 19.475}$$

$$= 13.20$$

The y-intercept is

$$b = \bar{y} - m\bar{z}$$

$$= 21.22 - (13.20)(0.8826)$$

$$= 9.570$$

The resulting equation is

$$y = 9.570 + 13.20z$$

The relationship between x and y is approximately

$$\boxed{y = 9.570 + 13.20 \log x}$$

This is not an optimal correlation, as better correlation coefficients can be obtained if other assumptions about the form of the equation are made. For example, $y = 9.1 + 4\sqrt{x}$ has a better correlation coefficient.

14. (a) Plot the data to verify that they are linear.

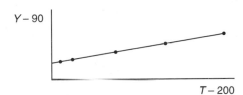

x	y
$T - 200$	$Y - 90$
7.1	2.30
10.3	2.58
0.4	1.56
1.1	1.63
3.4	1.83

step 1:

$$\sum x_i = 22.3 \qquad \sum y_i = 9.9$$
$$\sum x_i^2 = 169.43 \qquad \sum y_i^2 = 20.39$$
$$\left(\sum x_i\right)^2 = 497.29 \qquad \left(\sum y_i\right)^2 = 98.01$$
$$\bar{x} = \frac{22.3}{5} = 4.46 \qquad \bar{y} = 1.98$$
$$\sum x_i y_i = 51.54$$

step 2: From Eq. 11.70, the slope is

$$m = \frac{(5)(51.54) - (22.3)(9.9)}{(5)(169.43) - 497.29} = 0.1055$$

step 3: From Eq. 11.71, the y-intercept is

$$b = 1.98 - (0.1055)(4.46) = 1.509$$

The equation of the line is

$$y = 0.1055x + 1.509$$
$$Y - 90 = (0.1055)(T - 200) + 1.509$$
$$Y = \boxed{0.1055T + 70.409}$$

(b) *step 4:* Use Eq. 11.72 to get the correlation coefficient.

$$r = \frac{(5)(51.54) - (22.3)(9.9)}{\sqrt{\big((5)(169.43) - 497.29\big)\big((5)(20.39) - 98.01\big)}}$$

$$= \boxed{0.995}$$

15. Plot the data to see if they are linear.

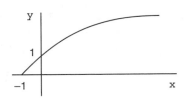

This looks like it could be of the form

$$y = a + b\sqrt{x}$$

However, when x is negative (as in the first point), the function is imaginary. Try shifting the curve to the right, replacing x with $x + 1$.

$$y = a + bz$$
$$z = \sqrt{x + 1}$$

z	y
0	0
1	1
1.414	1.4
1.732	1.7
2	2
2.236	2.2
2.45	2.4
2.65	2.6
2.83	2.8
3	3

Since $y \approx z$, the relationship is

$$\boxed{y = \sqrt{x + 1}}$$

In this problem, the answer was found accidentally. Usually, regression would be necessary.

12 Numbering Systems

PRACTICE PROBLEMS

Calculations in Other Numbering Systems

1. Perform the following binary operations. Check your work by converting to decimal.

   (a) $101 + 011$ (d) $0100 - 1100$ (g) 111×11
   (b) $101 + 110$ (e) $1110 - 1000$ (h) 100×11
   (c) $101 + 100$ (f) $010 - 101$ (i) 1011×1101

2. Perform the following octal operations. Check your work by converting to decimal.

   (a) $466 + 457$ (d) $71 - 27$ (g) 77×66
   (b) $1007 + 6661$ (e) $1143 - 367$ (h) 325×36
   (c) $321 + 465$ (f) $646 - 677$ (i) 3251×161

3. Perform the following hexadecimal operations. Check your work by converting to decimal.

   (a) $BA + C$ (d) $FF - E$ (g) $4A \times 3E$
   (b) $BB + A$ (e) $74 - 4A$ (h) $FE \times EF$
   (c) $BE + 10 + 1A$ (f) $FB - BF$ (i) $17 \times 7A$

Conversions Between Bases

4. Convert the following numbers to decimal numbers.

(a) $(674)_8$

(b) $(101101)_2$

(c) $(734.262)_8$

(d) $(1011.11)_2$

5. Convert the following numbers to octal numbers.

(a) $(75)_{10}$

(b) $(0.375)_{10}$

(c) $(121.875)_{10}$

(d) $(1011100.01110)_2$

6. Convert the following numbers to binary numbers.

(a) $(83)_{10}$

(b) $(100.3)_{10}$

(c) $(0.97)_{10}$

(d) $(321.422)_8$

SOLUTIONS

1. (a)
$$\begin{array}{r} 101 \\ +\ \ 011 \\ \hline \boxed{1000} \end{array}$$

(b)
$$\begin{array}{r} 101 \\ +\ \ 110 \\ \hline \boxed{1011} \end{array}$$

(c)
$$\begin{array}{r} 101 \\ +\ \ 100 \\ \hline \boxed{1001} \end{array}$$

(d)
$$-\left(\begin{array}{r} 1100 \\ -\ \ 0100 \\ \hline 1000 \end{array}\right) = \boxed{-1000}$$

(e)
$$\begin{array}{r} 1110 \\ -\ \ 1000 \\ \hline \boxed{0110} \end{array}$$

(f)
$$-\left(\begin{array}{r} 101 \\ -\ \ 010 \\ \hline 011 \end{array}\right) = \boxed{-011}$$

(g)
$$\begin{array}{r} 111 \\ \times\ \ \ 11 \\ \hline 111 \\ 111\ \ \\ \hline \boxed{10101} \end{array}$$

(h)
$$
\begin{array}{r}
100 \\
\times \quad 11 \\
\hline
100 \\
100 \\
\hline
\boxed{1100}
\end{array}
$$

(g)
$$
\begin{array}{r}
77 \\
\times \quad 66 \\
\hline
572 \\
572 \\
\hline
\boxed{6512}
\end{array}
$$

(h)
$$
\begin{array}{r}
325 \\
\times \quad 36 \\
\hline
2376 \\
1177 \\
\hline
\boxed{14366}
\end{array}
$$

(i)
$$
\begin{array}{r}
1011 \\
\times \quad 1101 \\
\hline
1011 \\
1011 \\
1011 \\
\hline
\boxed{10001111}
\end{array}
$$

(i)
$$
\begin{array}{r}
3251 \\
\times \quad 161 \\
\hline
3251 \\
23766 \\
3251 \\
\hline
\boxed{570231}
\end{array}
$$

2. (a)
$$
\begin{array}{r}
466 \\
+ \quad 457 \\
\hline
\boxed{1145}
\end{array}
$$

(b)
$$
\begin{array}{r}
1007 \\
+ \quad 6661 \\
\hline
\boxed{7670}
\end{array}
$$

3. (a)
$$
\begin{array}{r}
BA \\
+ \quad C \\
\hline
\boxed{C6}
\end{array}
$$

(c)
$$
\begin{array}{r}
321 \\
+ \quad 465 \\
\hline
\boxed{1006}
\end{array}
$$

(b)
$$
\begin{array}{r}
BB \\
+ \quad A \\
\hline
\boxed{C5}
\end{array}
$$

(d)
$$
\begin{array}{rcr}
71 & & 6\ 11 \\
-\ 27 & = & -\ 2\ \ 7 \\
\hline
& & \boxed{4\ \ 2}
\end{array}
$$

(c)
$$
\begin{array}{r}
BE \\
1\,0 \\
+ \quad 1A \\
\hline
\boxed{E8}
\end{array}
$$

(e)
$$
\begin{array}{rcccr}
1143 & & 113\ 13 & & 10\ 13\ 13 \\
-\ 367 & = & -\ 36\ \ 7 & = & -\ 3\ \ 6\ \ 7 \\
\hline
& & & & \boxed{5\ \ 5\ \ 4}
\end{array}
$$

(d)
$$
\begin{array}{r}
FF \\
-\quad E \\
\hline
\boxed{F1}
\end{array}
$$

(f)
$$
-\left(
\begin{array}{r}
677 \\
-\ 646 \\
\hline
31
\end{array}
\right) = \boxed{-31}
$$

(e)
$$
\begin{array}{rcr}
74 & & 6\ 14 \\
-\ 4A & = & -\ 4\ \ A \\
\hline
& & \boxed{2\ \ A}
\end{array}
$$

(f)
$$\begin{array}{r} \text{FB} \\ -\ \text{BF} \end{array} = \begin{array}{r} \text{E 1B} \\ -\ \text{B\ \ F} \end{array}$$

$$\boxed{3\ \ \text{C}}$$

(g)
$$\begin{array}{r} 4\text{A} \\ \times\ \ \ 3\text{E} \\ \hline 40\text{C} \\ \text{DE} \\ \hline \end{array}$$

$$\boxed{11\text{EC}}$$

(h)
$$\begin{array}{r} \text{FE} \\ \times\ \ \ \text{EF} \\ \hline \text{EE2} \\ \text{DE4} \\ \hline \end{array}$$

$$\boxed{\text{ED22}}$$

(i)
$$\begin{array}{r} 17 \\ \times\ \ \ 7\text{A} \\ \hline \text{E6} \\ \text{A1} \\ \hline \end{array}$$

$$\boxed{\text{AF6}}$$

4. (a) $(674)_8 = (6)(8)^2 + (7)(8)^1 + (4)(8)^0$

$$= \boxed{444}$$

(b) $(101101)_2 = (1)(2)^5 + (0)(2)^4 + (1)(2)^3$
$$+ (1)(2)^2 + (0)(2)^1 + (1)(2)^0$$

$$= \boxed{45}$$

(c) $(734.262)_8 = (7)(8)^2 + (3)(8)^1 + (4)(8)^0$
$$+ (2)(8)^{-1} + (6)(8)^{-2} + (2)(8)^{-3}$$

$$= \boxed{476.348}$$

(d) $(1011.11)_2 = (1)(2)^3 + (0)(2)^2 + (1)(2)^1$
$$+ (1)(2)^0 + (1)(2)^{-1} + (1)(2)^{-2}$$

$$= \boxed{11.75}$$

5. (a)
$$75 \div 8 = 9 \quad \text{remainder } 3$$
$$9 \div 8 = 1 \quad \text{remainder } 1$$
$$1 \div 8 = 0 \quad \text{remainder } 1$$
$$(75)_{10} = \boxed{(113)_8}$$

(b)
$$0.375 \times 8 = 3 \quad \text{remainder } 0$$
$$(0.375)_{10} = \boxed{(0.3)_8}$$

(c)
$$121 \div 8 = 15 \quad \text{remainder } 1$$
$$15 \div 8 = 1 \quad \text{remainder } 7$$
$$1 \div 8 = 0 \quad \text{remainder } 1$$
$$0.875 \times 8 = 7 \quad \text{remainder } 0$$
$$(121.875)_{10} = \boxed{(171.7)_8}$$

(d) Since $(2)^3 = 8$, break the bits into groups of three starting at the decimal point and working outward in both directions.

$$001011100.011100 = 001\ 011\ 100.011\ 100$$

Convert each of the groups into its octal equivalent.

$$001\ 011\ 100.011\ 100 = 1\ 3\ 4.3\ 4$$

$$\boxed{(134.34)_8}$$

6. (a)
$$83 \div 2 = 41 \quad \text{remainder } 1$$
$$41 \div 2 = 20 \quad \text{remainder } 1$$
$$20 \div 2 = 10 \quad \text{remainder } 0$$
$$10 \div 2 = 5 \quad \text{remainder } 0$$
$$5 \div 2 = 2 \quad \text{remainder } 1$$
$$2 \div 2 = 1 \quad \text{remainder } 0$$
$$1 \div 2 = 0 \quad \text{remainder } 1$$
$$(83)_{10} = \boxed{(1010011)_2}$$

(b)
$$100 \div 2 = 50 \quad \text{remainder } 0$$
$$50 \div 2 = 25 \quad \text{remainder } 0$$
$$25 \div 2 = 12 \quad \text{remainder } 1$$
$$12 \div 2 = 6 \quad \text{remainder } 0$$
$$6 \div 2 = 3 \quad \text{remainder } 0$$
$$3 \div 2 = 1 \quad \text{remainder } 1$$

$$1 \div 2 = 0 \quad \text{remainder } 1$$
$$0.3 \times 2 = 0 \quad \text{remainder } 0.6$$
$$0.6 \times 2 = 1 \quad \text{remainder } 0.2$$
$$0.2 \times 2 = 0 \quad \text{remainder } 0.4$$
$$0.4 \times 2 = 0 \quad \text{remainder } 0.8$$
$$0.8 \times 2 = 1 \quad \text{remainder } 0.6$$
$$0.6 \times 2 = 1 \quad \text{remainder } 0.2$$
$$\vdots$$

$$(100.3)_{10} = \boxed{(1100100.010011\cdots)_2}$$

(c)
$$0.97 \times 2 = 1 \quad \text{remainder } 0.94$$
$$0.94 \times 2 = 1 \quad \text{remainder } 0.88$$
$$0.88 \times 2 = 1 \quad \text{remainder } 0.76$$
$$0.76 \times 2 = 1 \quad \text{remainder } 0.52$$
$$0.52 \times 2 = 1 \quad \text{remainder } 0.04$$
$$0.04 \times 2 = 0 \quad \text{remainder } 0.08$$
$$\vdots$$

$$(0.97)_{10} = \boxed{(0.111110\cdots)_2}$$

(d) Since $8 = (2)^3$, convert each octal digit into its binary equivalent.

$$321.422: \quad 3 = 011$$
$$2 = 010$$
$$1 = 001$$
$$4 = 100$$
$$2 = 010$$
$$2 = 010$$

$$\boxed{(321.422)_8 = 011010001.100010010}$$

13 Numerical Analysis

PRACTICE PROBLEMS

1. A function is given as $y = 3x^{0.93} + 4.2$. What is the percent error if the value of y at $x = 2.7$ is found by using straight-line interpolation between $x = 2$ and $x = 3$?

(A) 0.06%
(B) 0.18%
(C) 2.5%
(D) 5.4%

2. Given the following data points, find y by straight-line interpolation for $x = 2.75$.

x	y
1	4
2	6
3	2
4	−14

(A) 2.1
(B) 2.4
(C) 2.7
(D) 3.0

3. Using the bisection method, find all of the roots of $f(x) = 0$ to the nearest 0.000005.

$$f(x) = x^3 + 2x^2 + 8x - 2$$

SOLUTIONS

1. The actual value at $x = 2.7$ is given by

$$y(x) = 3x^{0.93} + 4.2$$
$$y(2.7) = (3)(2.7)^{0.93} + 4.2$$
$$= 11.756$$

At $x = 3$,

$$y(3) = (3)(3)^{0.93} + 4.2$$
$$= 12.534$$

At $x = 2$,

$$y(2) = (3)(2)^{0.93} + 4.2$$
$$= 9.916$$

Use straight-line interpolation.

$$\frac{x_2 - x}{x_2 - x_1} = \frac{y_2 - y}{y_2 - y_1}$$
$$\frac{3 - 2.7}{3 - 2} = \frac{12.534 - y}{12.534 - 9.916}$$
$$y = 11.749$$

The relative error is given by

$$\frac{\text{actual value} - \text{predicted value}}{\text{actual value}} = \frac{11.756 - 11.749}{11.756}$$

$$= \boxed{0.0006 \quad (0.06\%)}$$

The answer is (A).

2. Let $x_1 = 2$; therefore, from the table of data points, $y_1 = 6$. Let $x_2 = 3$; therefore, from the table of data points, $y_2 = 2$.

Let $x = 2.75$. By straight-line interpolation,

$$\frac{x_2 - x}{x_2 - x_1} = \frac{y_2 - y}{y_2 - y_1}$$
$$\frac{3 - 2.75}{3 - 2} = \frac{2 - y}{2 - 6}$$
$$\boxed{y = 3}$$

The answer is (D).

3. $f(x) = x^3 + 2x^2 + 8x - 2$

Try to find an interval in which there is a root.

x	$f(x)$
0	-2
1	9

A root exists in the interval $[0,1]$.

Try $x = \left(\frac{1}{2}\right)(0+1) = 0.5$.

$$f(0.5) = (0.5)^3 + (2)(0.5)^2 + (8)(0.5) - 2 = 2.625$$

A root exists in $[0, 0.5]$.

Try $x = 0.25$.

$$f(0.25) = (0.25)^3 + (2)(0.25)^2 + (8)(0.25) - 2 = 0.1406$$

A root exists in $[0, 0.25]$.

Try $x = 0.125$.

$$f(0.125) = (0.125)^3 + (2)(0.125)^2 + (8)(0.125) - 2$$
$$= -0.967$$

A root exists in $[0.125, 0.25]$.

Try $x = \left(\frac{1}{2}\right)(0.125 + 0.25) = 0.1875$.

Continuing,

$$f(0.1875) = -0.42 \quad [0.1875, 0.25]$$
$$f(0.21875) = -0.144 \quad [0.21875, 0.25]$$
$$f(0.234375) = -0.002 \quad \text{[This is close enough.]}$$

One root is $x_1 \approx \boxed{0.234375}$.

Try to find the other two roots. Use long division to factor the polynomial.

$$
\begin{array}{r}
x^2 + 2.234375x + 8.52368 \\
x - 0.234375 \enclose{longdiv}{x^3 + \quad 2x^2 + \quad\quad 8x - 2} \\
\underline{-(x^3 - 0.234375x^2)} \quad\quad\quad\quad\quad \\
2.234375x^2 + \quad\quad 8x \quad\quad \\
\underline{-(2.234375x^2 - 0.52368x)} \quad\quad \\
8.52368x - 2 \\
\underline{-(8.52368x - 1.9977)} \\
\approx 0
\end{array}
$$

Use the quadratic equation to find the roots of $x^2 + 2.234375x + 8.52368$.

$$x_2, x_3 = \frac{-2.234375 \pm \sqrt{(2.234375)^2 - (4)(1)(8.52368)}}{(2)(1)}$$

$$= \boxed{-1.117189 \pm i2.697327} \quad \text{[both imaginary]}$$

14 Fluid Properties

PRACTICE PROBLEMS

(Use $g = 32.2$ ft/sec^2 or 9.81 m/s^2 unless told to do otherwise in the problem.)

Pressure

1. What is the absolute pressure if a gauge reads 8.7 psi (60 kPa) vacuum?

(A) 4 psi (27 kPa)

(B) 6 psi (41 kPa)

(C) 8 psi (55 kPa)

(D) 10 psi (68 kPa)

Viscosity

2. Calculate the kinematic viscosity of air at 80°F (27°C) and 70 psia (480 kPa).

(A) 3.54×10^{-5} ft^2/sec (3.30×10^{-6} m^2/s)

(B) 4.25×10^{-5} ft^2/sec (3.96×10^{-6} m^2/s)

(C) 4.96×10^{-5} ft^2/sec (4.62×10^{-6} m^2/s)

(D) 6.37×10^{-5} ft^2/sec (5.94×10^{-6} m^2/s)

Solutions

3. Volumes of an 8% solution, a 10% solution, and a 20% solution of nitric acid are to be mixed in order to get 100 mL of a 12% solution. The 8% solution will contribute half the volume of nitric acid contributed by the 10% and 20% solutions combined. What volume of 10% acid solution is required?

(A) 20 mL

(B) 30 mL

(C) 50 mL

(D) 80 mL

SOLUTIONS

1. *Customary U.S. Solution*

$$p_{\text{gage}} = -8.7 \text{ lbf/in}^2$$

$$p_{\text{atmospheric}} = 14.7 \text{ lbf/in}^2$$

The relationship between absolute, gage, and atmospheric pressure is given by

$$p_{\text{absolute}} = p_{\text{gage}} + p_{\text{atmospheric}}$$

$$= -8.7 \frac{\text{lbf}}{\text{in}^2} + 14.7 \frac{\text{lbf}}{\text{in}^2}$$

$$= \boxed{6 \text{ lbf/in}^2 \quad (6 \text{ psi})}$$

The answer is (B).

SI Solution

$$p_{\text{gage}} = -60 \text{ kPa}$$

$$p_{\text{atmospheric}} = 101.3 \text{ kPa}$$

The relationship between absolute, gage, and atmospheric pressure is given by

$$p_{\text{absolute}} = p_{\text{gage}} + p_{\text{atmospheric}}$$

$$= -60 \text{ kPa} + 101.3 \text{ kPa}$$

$$= \boxed{41.3 \text{ kPa}}$$

The answer is (B).

2. *Customary U.S. Solution*

For air at 14.7 psia and 80°F, the absolute viscosity independent of pressure is $\mu = 3.85 \times 10^{-7}$ lbf-sec/ft^2.

Determine the density of air at 70 psia and 80°F. (Assume an ideal gas.)

$$\rho = \frac{p}{RT}$$

For air, $R = 53.3$ lbf-ft/lbm-°R.

Substituting gives

$$\rho = \frac{\left(70 \frac{\text{lbf}}{\text{in}^2}\right)\left(144 \frac{\text{in}^2}{\text{ft}^2}\right)}{\left(53.3 \frac{\text{lbf-ft}}{\text{lbm-°R}}\right)(80°\text{F} + 460)}$$

$$= 0.350 \text{ lbm/ft}^3$$

Fluids

The kinematic viscosity, ν, is related to the absolute viscosity by

$$\nu = \frac{\mu g_c}{\rho}$$

$$= \frac{\left(3.85 \times 10^{-7} \; \dfrac{\text{lbf- sec}}{\text{ft}^2}\right)\left(32.2 \; \dfrac{\text{lbm-ft}}{\text{lbf-sec}^2}\right)}{0.350 \; \dfrac{\text{lbm}}{\text{ft}^3}}$$

$$= \boxed{3.54 \times 10^{-5} \; \text{ft}^2/\text{sec}}$$

The answer is (A).

SI Solution

For air at 480 kPa and 27°C, the absolute viscosity independent of pressure is $\mu = 1.84 \times 10^{-5}$ Pa·s.

Determine the density of air at 480 kPa and 27°C. (Assume an ideal gas.)

$$\rho = \frac{p}{RT}$$

For air, $R = 287$ J/kg·K.

Substituting gives

$$\rho = \frac{(480 \text{ kPa})\left(1000 \; \dfrac{\text{Pa}}{\text{kPa}}\right)}{\left(287 \; \dfrac{\text{J}}{\text{kg·K}}\right)(27°\text{C} + 273)}$$

$$= 5.575 \text{ kg/m}^3$$

The kinematic viscosity, ν, is related to the absolute viscosity by

$$\nu = \frac{\mu}{\rho}$$

$$= \frac{1.84 \times 10^{-5} \text{ Pa·s}}{5.575 \; \dfrac{\text{kg}}{\text{m}^3}}$$

$$= \boxed{3.30 \times 10^{-6} \; \text{m}^2/\text{s}}$$

The answer is (A).

3. Let

$$x = \text{volume of 8\% solution}$$
$$y = \text{volume of 10\% solution}$$
$$z = \text{volume of 20\% solution}$$

The three conditions that must be satisfied are

$$x + y + z = 100 \text{ mL}$$
$$0.08x + 0.10y + 0.20z = (0.12)(100 \text{ mL}) = 12 \text{ mL}$$
$$0.08x = \left(\tfrac{1}{2}\right)(0.10y + 0.20z)$$

Note that "nitric acid" is HNO_3, not HNO_3 diluted to an 8% solution.

Simplifying these equations,

$$\begin{array}{rrrrr} x & + & y & + & z & = & 100 \\ 4x & + & 5y & + & 10z & = & 600 \\ 8x & - & 5y & - & 10z & = & 0 \end{array}$$

Adding the second and third equations gives

$$12x = 600$$

$$x = \boxed{50 \text{ mL}}$$

Work with the first two equations to get

$$y + z = 100 - 50 = 50$$
$$5y + 10z = 600 - (4)(50) = 400$$

Multiplying the top equation by -5 and adding to the bottom equation,

$$5z = 150$$
$$z = 30 \text{ mL}$$

From the first equation,

$$y = 20 \text{ mL}$$

The answer is (A).

15 Fluid Statics

PRACTICE PROBLEMS

(Use $g = 32.2$ ft/sec^2 or 9.81 m/s^2 unless told to do otherwise in the problem.)

Buoyancy

1. A 4000 lbm blimp contains 10,000 lbm (4500 kg) of hydrogen (specific gas constant = 766.5 ft-lbf/lbm-°R (4124 J/kg·K)) at 56°F (13°C) and 30.2 in Hg (770 mm Hg). What is its lift if the hydrogen and air are in thermal and pressure equilibrium?

    (A) 7.6×10^3 lbf (3.4×10^4 N)
    (B) 1.2×10^4 lbf (5.3×10^4 N)
    (C) 1.3×10^5 lbf (5.9×10^5 N)
    (D) 1.7×10^5 lbf (7.7×10^5 N)

2. A hollow 6 ft (1.8 m) diameter sphere floats half-submerged in seawater. What mass of concrete is required as an external anchor to just submerge the sphere completely?

    (A) 2700 lbm (1200 kg)
    (B) 4200 lbm (1900 kg)
    (C) 5500 lbm (2500 kg)
    (D) 6300 lbm (2700 kg)

SOLUTIONS

1. *Customary U.S. Solution*

The lift of the hydrogen-filled blimp (F_{lift}) is equal to the difference between the buoyant force (F_b) and the weight of the hydrogen contained in the blimp (W_H).

$$F_{\text{lift}} = F_b - W_H$$

The weight of the hydrogen is calculated from the mass of hydrogen by

$$W_H = \frac{mg}{g_c}$$

$$= \frac{(10{,}000 \text{ lbm})\left(32.2 \; \dfrac{\text{ft}}{\text{sec}^2}\right)}{32.2 \; \dfrac{\text{lbm-ft}}{\text{lbf-sec}^2}}$$

$$= 10{,}000 \text{ lbf}$$

The buoyant force is equal to the weight of the displaced air. The weight of the displaced air is calculated by knowing that the volume of the air displaced is equal to the volume of hydrogen enclosed in the blimp. Compute the volume of the hydrogen contained in the blimp by assuming the hydrogen behaves like an ideal gas.

$$V_H = \frac{mRT}{p}$$

For hydrogen, $R = 766.5$ ft-lbf/lbm-°R.

The temperature of the hydrogen is given as 56°F. Convert to absolute temperature (°R).

$$T = 56°\text{F} + 460 = 516°\text{R}$$

The pressure of the hydrogen is given as 30.2 in Hg. Convert the pressure to units of pounds per square foot.

$$p = (30.2 \text{ in Hg})\left(\frac{1 \; \dfrac{\text{lbf}}{\text{in}^2}}{2.036 \text{ in Hg}}\right)\left(144 \; \dfrac{\text{in}^2}{\text{ft}^2}\right)$$

$$= 2136 \text{ lbf/ft}^2$$

Compute the volume of hydrogen.

$$V_{\rm H} = \frac{mRT}{p}$$

$$= \frac{(10{,}000\ {\rm lbm}) \left(766.5\ \dfrac{\text{ft-lbf}}{\text{lbm-}^\circ{\rm R}}\right) (516^\circ{\rm R})}{2136\ \dfrac{\text{lbf}}{\text{ft}^2}}$$

$$= 1.85 \times 10^6\ {\rm ft}^3$$

Since the volume of the hydrogen contained in the blimp is equal to the air displaced, the air displaced can be computed from the ideal gas equation by assuming the air behaves like an ideal gas.

$$m = \frac{pV_{\rm H}}{RT}$$

Since the air and hydrogen are assumed to be in thermal and pressure equilibrium, the temperature and pressure are equal to the value given for the hydrogen.

For air, $R = 53.35$ ft-lbf/lbm-$^\circ$R.

Substituting gives

$$m_{\rm air} = \frac{pV_{\rm H}}{RT}$$

$$= \frac{\left(2136\ \dfrac{\text{lbf}}{\text{ft}^2}\right) (1.85 \times 10^6\ {\rm ft}^3)}{\left(53.35\ \dfrac{\text{ft-lbf}}{\text{lbm-}^\circ{\rm R}}\right) (516^\circ{\rm R})}$$

$$= 1.435 \times 10^5\ {\rm lbm}$$

Recall that the buoyant force is equal to the weight of the air.

$$F_b = W_{\rm air} = \frac{mg}{g_c}$$

$$= \frac{(1.435 \times 10^5\ {\rm lbm}) \left(32.2\ \dfrac{\text{ft}}{\text{sec}^2}\right)}{32.2\ \dfrac{\text{lbm-ft}}{\text{lbf-sec}^2}}$$

$$= 1.435 \times 10^5\ {\rm lbf}$$

Therefore, the lift can be calculated as

$$F_{\rm lift} = F_b - W_{\rm H} - W_{\rm blimp}$$

$$= 1.435 \times 10^5\ {\rm lbf} - 10{,}000\ {\rm lbf} - 4000\ {\rm lbf}$$

$$= \boxed{1.295 \times 10^5\ {\rm lbf}}$$

The answer is (C).

SI Solution

Assume the mass of the blimp structure is small (negligible) compared with the mass of the hydrogen.

The lift of the hydrogen-filled blimp ($F_{\rm lift}$) is equal to the difference between the buoyant force (F_b) and the weight of the hydrogen contained in the blimp ($W_{\rm H}$).

$$F_{\rm lift} = F_b - W_{\rm H}$$

The weight of the hydrogen is calculated from the mass of hydrogen by

$$W_{\rm H} = mg$$

$$= (4500\ {\rm kg}) \left(9.81\ \dfrac{\text{m}}{\text{s}^2}\right)$$

$$= 44\,145\ {\rm N}$$

The buoyant force is equal to the weight of the displaced air. The weight of the displaced air is calculated by knowing that the volume of the air displaced is equal to the volume of hydrogen enclosed in the blimp. Compute the volume of the hydrogen contained in the blimp by assuming the hydrogen behaves like an ideal gas.

$$V_{\rm H} = \frac{mRT}{p}$$

For hydrogen, $R = 4124$ J/kg·K.

The temperature of the hydrogen is given as 13°C. Convert to absolute temperature (K).

$$T = 13^\circ{\rm C} + 273 = 286{\rm K}$$

The pressure of the hydrogen is given as 770 mm Hg. Convert the pressure to units of pascals.

$$p = \frac{(770\ {\rm mm\ Hg}) \left(133.4\ \dfrac{\text{kPa}}{\text{m}}\right)}{1000\ \dfrac{\text{mm}}{\text{m}}}$$

$$= 102.7\ {\rm kPa}$$

The volume of hydrogen is

$$V_{\rm H} = \frac{mRT}{p}$$

$$= \frac{(4500\ {\rm kg}) \left(4124\ \dfrac{\text{J}}{\text{kg·K}}\right) (286{\rm K})}{(102.7\ {\rm kPa}) \left(1000\ \dfrac{\text{Pa}}{\text{kPa}}\right)}$$

$$= 5.168 \times 10^4\ {\rm m}^3$$

Since the volume of the hydrogen contained in the blimp is equal to the air displaced, the air displaced can be computed from the ideal gas equation assuming the air behaves like an ideal gas.

$$m = \frac{pV_{\rm H}}{RT}$$

Since the air and hydrogen are assumed to be in thermal and pressure equilibrium, the temperature and pressure are equal to the value given for the hydrogen.

For air, $R = 287$ J/kg·K.

Substituting gives

$$m_{air} = \frac{pV_H}{RT}$$

$$= \frac{(102.7 \text{ kPa}) \left(1000 \dfrac{\text{Pa}}{\text{kPa}}\right) (5.168 \times 10^4 \text{ m}^3)}{\left(287 \dfrac{\text{J}}{\text{kg·K}}\right)(286\text{K})}$$

$$= 6.466 \times 10^4 \text{ kg}$$

The buoyant force is equal to the weight of the air, so

$$F_b = W_{air} = mg$$

$$= (6.466 \times 10^4 \text{ kg}) \left(9.81 \dfrac{\text{m}}{\text{s}^2}\right)$$

$$= 6.34 \times 10^5 \text{ N}$$

Therefore, the lift can be calculated as

$$F_{lift} = F_b - W_H$$

$$= 6.34 \times 10^5 \text{ N} - 44\,145 \text{ N}$$

$$= \boxed{5.90 \times 10^5 \text{ N}}$$

The answer is (C).

2. *Customary U.S. Solution*

The weight of the sphere is equal to the weight of the displaced volume of water when the sphere is floating.

The buoyant force is given by

$$F_b = \frac{\rho g V_{displaced}}{g_c}$$

Since the sphere is half submerged,

$$W_{sphere} = \left(\tfrac{1}{2}\right)\left(\frac{\rho g V_{sphere}}{g_c}\right)$$

For seawater, $\rho = 64.0$ lbm/ft^3.

The volume of the sphere is given by

$$V_{sphere} = \left(\frac{\pi}{6}\right) d^3$$

$$= \left(\frac{\pi}{6}\right) (6 \text{ ft})^3$$

$$= 113.1 \text{ ft}^3$$

The weight of the sphere is

$$W_{sphere} = \left(\tfrac{1}{2}\right)\left(\frac{\rho g V_{sphere}}{g_c}\right)$$

$$= \left(\tfrac{1}{2}\right)\left(\frac{\left(64.0 \dfrac{\text{lbm}}{\text{ft}^3}\right)\left(32.2 \dfrac{\text{ft}}{\text{sec}^2}\right) \times (113.1 \text{ ft}^3)}{32.2 \dfrac{\text{lbm-ft}}{\text{lbf-sec}^2}}\right)$$

$$= 3619 \text{ lbf}$$

The buoyant force equation for a fully submerged sphere and anchor can be solved for the concrete volume.

$$W_{sphere} + W_{concrete} = (V_{sphere} + V_{concrete})\gamma_{water}$$

$$W_{sphere} + \rho_{concrete} V_{concrete}\left(\frac{g}{g_c}\right)$$
$$= (V_{sphere} + V_{concrete})\rho_{water}$$
$$\times \left(\frac{g}{g_c}\right)$$

$$3619 \text{ lbf} + \left(150 \dfrac{\text{lbm}}{\text{ft}^3}\right)(V_{concrete})\left(\dfrac{32.2 \dfrac{\text{ft}}{\text{sec}^2}}{32.2 \dfrac{\text{ft-lbm}}{\text{lbf-sec}^2}}\right)$$

$$= (113.1 \text{ ft}^3 + V_{concrete})$$

$$\times \left(64.0 \dfrac{\text{lbm}}{\text{ft}^3}\right)$$

$$\times \left(\dfrac{32.2 \dfrac{\text{ft}}{\text{sec}^2}}{32.2 \dfrac{\text{ft-lbm}}{\text{lbf-sec}^2}}\right)$$

$$V_{concrete} = 42.09 \text{ ft}^3$$

$$m_{concrete} = \rho_{concrete} V_{concrete}$$

$$= \left(150 \dfrac{\text{lbm}}{\text{ft}^3}\right)(42.09 \text{ ft}^3)$$

$$= \boxed{6314 \text{ lbm}}$$

The answer is (D).

SI Solution

The weight of the sphere is equal to the weight of the displaced volume of water when the sphere is floating.

The buoyant force is given by

$$F_b = \rho g V_{displaced}$$

Since the sphere is half submerged,

$$W_{sphere} = \tfrac{1}{2}\rho g V_{sphere}$$

For seawater, $\rho = 1025$ kg/m^3.

The volume of the sphere is given by

$$V_{sphere} = \left(\frac{\pi}{6}\right) d^3$$

$$= \left(\frac{\pi}{6}\right) (1.8 \text{ m})^3$$

$$= 3.054 \text{ m}^3$$

The weight of the sphere required is

$$W_{sphere} = \tfrac{1}{2}\rho g V_{sphere}$$

$$= \left(\tfrac{1}{2}\right)\left(1025 \dfrac{\text{kg}}{\text{m}^3}\right)\left(9.81 \dfrac{\text{m}}{\text{s}^2}\right)(3.054 \text{ m}^3)$$

$$= 15\,354 \text{ N}$$

The buoyant force equation for a fully submerged sphere and anchor can be solved for the concrete volume.

$$W_{\text{sphere}} + W_{\text{concrete}} = (V_{\text{sphere}} + V_{\text{concrete}})\rho_{\text{water}}g$$

$$W_{\text{sphere}} + \rho_{\text{concrete}}gV_{\text{concrete}} = g(V_{\text{sphere}} + V_{\text{concrete}})\rho_{\text{water}}$$

$$15\,354 \text{ N} + \left(2400 \ \frac{\text{kg}}{\text{m}^3}\right)\left(9.81 \ \frac{\text{m}}{\text{s}^2}\right)(V_{\text{concrete}})$$

$$= (3.054 \text{ m}^3 + V_{\text{concrete}})$$

$$\times \left(1025 \ \frac{\text{kg}}{\text{m}^3}\right)\left(9.81 \ \frac{\text{m}}{\text{s}^2}\right)$$

$$V_{\text{concrete}} = 1.138 \text{ m}^3$$

$$m_{\text{concrete}} = \rho_{\text{concrete}}V_{\text{concrete}}$$

$$= \left(2400 \ \frac{\text{kg}}{\text{m}^3}\right)(1.138 \text{ m}^3)$$

$$= \boxed{2731 \text{ kg}}$$

The answer is (D).

16 Fluid Flow Parameters

PRACTICE PROBLEMS

Use the following values unless told to do otherwise in the problem:

$$g = 32.2 \text{ ft/sec}^2 \ (9.81 \text{ m/s}^2)$$

$$\rho_{\text{water}} = 62.4 \text{ lbm/ft}^3 \ (1000 \text{ kg/m}^3)$$

$$p_{\text{atmospheric}} = 14.7 \text{ psia} \ (101.3 \text{ kPa})$$

Hydraulic Radius

1. A 10 in (25 cm) composition pipe is compressed by a tree root until its inside height is only 7.2 in (18 cm). What is its approximate hydraulic radius when flowing half full? State your assumptions.

- (A) 2.2 in (5.5 cm)
- (B) 2.7 in (6.9 cm)
- (C) 3.2 in (8.1 cm)
- (D) 4.5 in (11.4 cm)

2. A pipe with an inside diameter of 18.812 in contains water to a depth of 15.7 in. What is the hydraulic radius? (Work in customary U.S. units only.)

- (A) 4.39 in
- (B) 5.08 in
- (C) 5.72 in
- (D) 6.51 in

SOLUTIONS

1. *Customary U.S. Solution*

The perimeter of the pipe is

$$p = \pi d = \pi (10 \text{ in})$$
$$= 31.42 \text{ in}$$

If the pipe is flowing half-full, the wetted perimeter becomes

$$\text{wetted perimeter} = \tfrac{1}{2} p = \left(\tfrac{1}{2}\right)(31.42 \text{ in})$$
$$= 15.71 \text{ in}$$

Assume the compressed pipe is an elliptical cross section. The ellipse will have a minor axis, b, equal to one-half the height of the compressed pipe or

$$b = \frac{7.2 \text{ in}}{2} = 3.6 \text{ in}$$

When the pipe is compressed, the perimeter of the pipe will remain constant. The perimeter of an ellipse is given by

$$p \approx 2\pi \sqrt{\tfrac{1}{2}(a^2 + b^2)}$$

Solve for the major axis.

$$a = \sqrt{2\left(\frac{p}{2\pi}\right)^2 - b^2}$$

$$= \sqrt{(2)\left(\frac{31.42 \text{ in}}{2\pi}\right)^2 - (3.6 \text{ in})^2}$$

$$= 6.09 \text{ in}$$

The flow area or area of the ellipse is given by

$$\text{flow area} = \tfrac{1}{2}\pi ab$$
$$= \tfrac{1}{2}\pi (6.09 \text{ in})(3.6 \text{ in})$$
$$= 34.4 \text{ in}^2$$

The hydraulic radius is

$$r_h = \frac{\text{area in flow}}{\text{wetted perimeter}} = \frac{34.4 \text{ in}^2}{15.7 \text{ in}}$$
$$= \boxed{2.19 \text{ in}}$$

The answer is (A).

SI Solution

The perimeter of the pipe is

$$p = \pi d = \pi(25 \text{ cm})$$
$$= 78.54 \text{ cm}$$

If the pipe is flowing half-full, the wetted perimeter becomes

$$\text{wetted perimeter} = \tfrac{1}{2}p = \left(\tfrac{1}{2}\right)(78.54 \text{ cm})$$
$$= 39.27 \text{ cm}$$

Assume the compressed pipe is an elliptical cross section. The ellipse will have a minor axis, b, equal to one-half the height of the compressed pipe or

$$b = \frac{18 \text{ cm}}{2} = 9 \text{ cm}$$

When the pipe is compressed, the perimeter of the pipe will remain constant. The perimeter of an ellipse is given by

$$p \approx 2\pi\sqrt{\tfrac{1}{2}(a^2 + b^2)}$$

Solve for the major axis.

$$a = \sqrt{2\left(\frac{p}{2\pi}\right)^2 - b^2}$$
$$= \sqrt{(2)\left(\frac{78.54 \text{ cm}}{2\pi}\right)^2 - (9 \text{ cm})^2}$$
$$= 15.2 \text{ cm}$$

The flow area or area of the ellipse is given by

$$\text{flow area} = \tfrac{1}{2}\pi ab = \tfrac{1}{2}\pi(15.2 \text{ cm})(9 \text{ cm})$$
$$= 214.9 \text{ cm}^2$$

The hydraulic radius is

$$r_h = \frac{\text{area in flow}}{\text{wetted perimeter}}$$
$$= \frac{214.9 \text{ cm}^2}{39.27 \text{ cm}}$$
$$= \boxed{5.47 \text{ cm}}$$

The answer is (A).

2. *method 1:* Use App. 7.A for a circular segment.

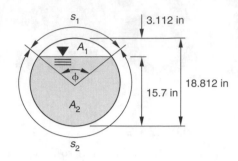

$$r = \frac{d}{2} = \frac{18.812 \text{ in}}{2} = 9.406 \text{ in}$$
$$\phi = (2)\left(\arccos \frac{r - d}{r}\right)$$
$$= (2)\left(\arccos \frac{9.406 \text{ in} - 3.112 \text{ in}}{9.406 \text{ in}}\right) = 1.675 \text{ rad}$$
$$\sin \phi = 0.9946$$
$$A_1 = \tfrac{1}{2}r^2(\phi - \sin \phi)$$
$$= \left(\tfrac{1}{2}\right)(9.406 \text{ in})^2(1.675 - 0.9946) = 30.1 \text{ in}^2$$
$$A_{\text{total}} = A_1 + A_2 = \frac{\pi}{4}D^2 = \left(\frac{\pi}{4}\right)(18.812 \text{ in})^2$$
$$= 277.95 \text{ in}^2$$
$$A_2 = A_{\text{total}} - A_1 = 277.95 \text{ in}^2 - 30.1 \text{ in}^2$$
$$= 247.85 \text{ in}^2$$
$$s_1 = r\phi = (9.406 \text{ in})(1.675) = 15.76 \text{ in}$$
$$s_{\text{total}} = s_1 + s_2 = \pi D = \pi(18.812 \text{ in}) = 59.1 \text{ in}$$
$$s_2 = s_{\text{total}} - s_1 = 59.1 \text{ in} - 15.76 \text{ in} = 43.34 \text{ in}$$
$$r_h = \frac{A_2}{s_2} = \frac{247.85 \text{ in}^2}{43.34 \text{ in}} = \boxed{5.719 \text{ in}}$$

method 2: Use App. 16.A.

$$\frac{d}{D} = \frac{15.7 \text{ in}}{18.812 \text{ in}} = 0.83$$

From App. 16.A, $r_h/D = 0.3041$.

$$r_h = (0.3041)(18.812 \text{ in}) = \boxed{5.72 \text{ in}}$$

The answer is (C).

17 Fluid Dynamics

PRACTICE PROBLEMS

(Use $g = 32.2$ ft/sec^2 (9.81 m/s^2) and 60°F (16°C) water unless told to do otherwise.)

Conservation of Energy

1. 5.0 ft^3/sec (130 L/s) of water flows through a schedule-40 pipe that changes gradually in diameter from 6 in at point A to 18 in at point B. Point B is 15 ft (4.6 m) higher than point A. The respective pressures at points A and B are 10 psia (70 kPa) and 7 psia (48.3 kPa). Unknown mechanical processes may act on the water between A and B. All minor losses are insignificant. What are the direction of flow and velocity at point A?

    (A) 3.2 ft/sec (1 m/s); from A to B

    (B) 25 ft/sec (7.0 m/s); from A to B

    (C) 3.2 ft/sec (1 m/s); from B to A

    (D) 25 ft/sec (7.5 m/s); from B to A

2. Points A and B are separated by 3000 ft of new 6 in schedule-40 steel pipe. 750 gal/min of 60°F water flow from point A to point B. Point B is 60 ft above point A. What must be the pressure at point A if the pressure at B must be 50 psig?

    (A) 87 psig

    (B) 103 psig

    (C) 125 psig

    (D) 167 psig

3. *(Time limit: one hour)* A pipe network connects junctions A, B, C, and D as shown. All pipe sections have a C-value of 150. Water can be added and removed at any of the junctions to achieve the flows listed. Water flows from point A to point D. No flows are backward. The minimum allowable pressure anywhere in the system is 20 psi. All minor losses are insignificant. For simplification, use the nominal pipe sizes.

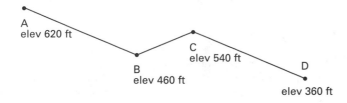

A
elev 620 ft

B
elev 460 ft

C
elev 540 ft

D
elev 360 ft

pipe section	length	diameter (in)	flow
A to B	20,000 ft	6	120 gal/min
B to C	10,000 ft	6	160 gal/min
C to D	30,000 ft	4	120 gal/min

(a) What is the pressure at point A?

    (A) 14 psig

    (B) 23 psig

    (C) 31 psig

    (D) 47 psig

(b) What is the elevation of the hydraulic grade line at point A referenced to point D?

    (A) 330 ft

    (B) 470 ft

    (C) 610 ft

    (D) 750 ft

Friction Loss

4. 1.5 ft^3/sec (40 L/s) of 70°F (20°C) water flows through 1200 ft (355 m) of 6 in (nominal) diameter new schedule-40 steel pipe. What is the friction loss?

    (A) 4 ft (1.2 m)

    (B) 18 ft (5.2 m)

    (C) 36 ft (9.5 m)

    (D) 70 ft (2.1 m)

5. 500 gal/min (30 L/s) of 100°F (40°C) water flows through 300 ft (90 m) of 6 in schedule-40 pipe. The pipe contains two 6 in flanged steel elbows, two full-open gate valves, a full-open 90° angle valve, and a swing check valve. The discharge is located 20 ft (6 m) higher than the entrance. What is the pressure difference between the two ends of the pipe?

    (A) 12 psi (78 kPa)

    (B) 21 psi (140 kPa)

    (C) 45 psi (310 kPa)

    (D) 87 psi (600 kPa)

6. 70°F (20°C) air is flowing at 60 ft/sec (18 m/s) through 300 ft (90 m) of 6 in schedule-40 pipe. The pipe contains two 6 in flanged steel elbows, two full-open gate valves, a full-open 90° angle valve, and a swing check valve. The discharge is located 20 ft (6 m) higher than the entrance. The average air density is 0.075 lbm/ft³. What is the pressure difference between the two ends of the pipe?

(A) 0.26 psi (1.8 kPa)

(B) 0.49 psi (3.2 kPa)

(C) 1.5 psi (10 kPa)

(D) 13 psi (90 kPa)

Reservoirs

7. Three reservoirs (A, B, and C) are interconnected with a common junction (point D) at elevation 25 ft above an arbitrary reference point. The water levels for reservoirs A, B, and C are at elevations of 50, 40, and 22 ft, respectively. The pipe from reservoir A to the junction is 800 ft of 3 in (nominal) steel pipe. The pipe from reservoir B to the junction is 500 ft of 10 in (nominal) steel pipe. The pipe from reservoir C to the junction is 1000 ft of 4 in (nominal) steel pipe. All pipes are schedule 40 with a friction factor of 0.02. All minor losses and velocity heads can be neglected. What are the direction of flow and pressure at point D?

(A) out of reservoir B; 500 psf

(B) out of reservoir B; 930 psf

(C) into reservoir B; 1100 psf

(D) into reservoir B; 1260 psf

Water Hammer

8. (*Time limit: one hour*) A cast-iron pipe has an inside diameter of 24 in (600 mm) and a wall thickness of 0.75 in (20 mm). The pipe's modulus of elasticity is 20×10^6 psi (140 GPa). The pipeline is 500 ft (150 m) long. 70°F (20°C) water is flowing at 6 ft/sec (2 m/s).

(a) If a valve is closed instantaneously, what will be the maximum pressure experienced in the pipe?

(A) 48 psi (330 kPa)

(B) 140 psi (970 kPa)

(C) 320 psi (2.5 MPa)

(D) 470 psi (3.2 MPa)

(b) If the pipe is 500 ft (150 m) long, over what length of time must the valve be closed to create a pressure equivalent to instantaneous closure?

(A) 0.25 sec

(B) 0.68 sec

(C) 1.6 sec

(D) 2.1 sec

Parallel Pipe Systems

9. 8 MGD (millions of gallons per day) (350 L/s) of 70°F (20°C) water flows into the new schedule-40 steel pipe network shown. Minor losses are insignificant.

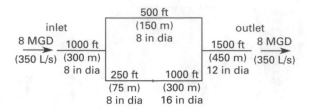

(a) What quantity of water is flowing in the upper branch?

(A) 1.2 ft³/sec (0.034 m³/s)

(B) 2.9 ft³/sec (0.081 m³/s)

(C) 4.1 ft³/sec (0.11 m³/s)

(D) 5.3 ft³/sec (0.15 m³/s)

(b) What is the energy loss per unit mass between the inlet and the outlet?

(A) 120 ft-lbf/lbm (0.37 kJ/kg)

(B) 300 ft-lbf/lbm (0.90 kJ/kg)

(C) 480 ft-lbf/lbm (1.4 kJ/kg)

(D) 570 ft-lbf/lbm (1.7 kJ/kg)

Pipe Networks

10. A single-loop pipe network is shown. The distance between each junction is 1000 ft. All junctions are on the same elevation. All pipes have a C-value of 100. The volumetric flow rates are to be determined to within 2 gal/min.

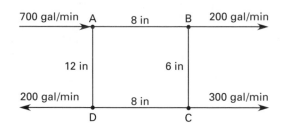

What is the flow rate between junctions B and C?

(A) 36 gal/min from B to C

(B) 58 gal/min from B to C

(C) 84 gal/min from C to B

(D) 112 gal/min from C to B

11. (*Time limit: one hour*) A double-loop pipe network is shown. The distance between each junction is 1000 ft. The water temperature is 60°F. Elevations and some pressure are known for the junctions. All pipes have a C-value of 100.

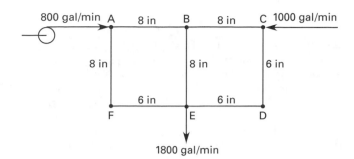

point	pressure	elevation
A		200 ft
B		150 ft
C	40 psig	300 ft
D		150 ft
E		200 ft
F		150 ft

(a) What is the flow rate between junctions B and E?

    (A) 540 gal/min

    (B) 620 gal/min

    (C) 810 gal/min

    (D) 980 gal/min

(b) What is the pressure at point D?

    (A) 51 psig

    (B) 73 psig

    (C) 96 psig

    (D) 120 psig

(c) If the pump receives water at 20 psig (140 kPa), what hydraulic power is required?

    (A) 14 hp

    (B) 28 hp

    (C) 35 hp

    (D) 67 hp

12. *(Time limit: one hour)* The water distribution network shown consists of class A cast-iron pipe installed 15 years ago. 1.5 MGD of 50°F water enters at junction A and leaves at junction B. The minimum acceptable pressure at point D is 40 psig (280 kPa).

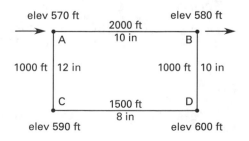

(a) What is the flow rate between junctions A and B?

    (A) 320 gal/min

    (B) 480 gal/min

    (C) 590 gal/min

    (D) 660 gal/min

(b) What is the pressure at point B?

    (A) 32 psig

    (B) 48 psig

    (C) 57 psig

    (D) 88 psig

Tank Discharge

13. The velocity of discharge from a fire hose is 50 ft/sec (15 m/s). The hose is oriented 45° from the horizontal. Disregarding air friction, what is the maximum range of the discharge?

    (A) 45 ft (14 m)

    (B) 78 ft (23 m)

    (C) 91 ft (24 m)

    (D) 110 ft (33 m)

14. A full cylindrical tank 40 ft (12 m) high has a constant diameter of 20 ft. The tank has a 4 in (100 mm) diameter hole in its bottom. The coefficient of discharge for the hole is 0.98. How long will it take for the water level to drop from 40 ft to 20 ft (12 m to 6 m)?

    (A) 950 sec

    (B) 1200 sec

    (C) 1450 sec

    (D) 1700 sec

Venturi Meters

15. A venturi meter with an 8 in diameter throat is installed in a 12 in diameter water line. The venturi is perfectly smooth, so that the discharge coefficient is 1.00. An attached mercury manometer registers a 4 in differential. What is the volumetric flow rate?

    (A) 1.7 ft^3/sec

    (B) 5.2 ft^3/sec

    (C) 6.4 ft^3/sec

    (D) 18 ft^3/sec

16. 60°F (15°C) benzene (specific gravity at 60°F (15°C) of 0.885) flows through an 8 in/3^1/$_2$ in (200 mm/90 mm) venturi meter whose coefficient of discharge is 0.99. A mercury manometer indicates a 4 in difference in the heights of the mercury columns. What is the volumetric flow rate of the benzene?

    (A) 1.2 ft^3/sec (34 L/s)

    (B) 9.1 ft^3/sec (250 L/s)

    (C) 13 ft^3/sec (360 L/s)

    (D) 27 ft^3/sec (760 L/s)

Orifice Meters

17. A sharp-edged orifice meter with a 0.2 ft diameter opening is installed in a 1.0 ft diameter pipe. 70°F water approaches the orifice at 2 ft/sec. What is the indicated pressure drop across the orifice meter?

  (A)   5.9 psi
  (B)   13 psi
  (C)   22 psi
  (D)   50 psi

18. A mercury manometer is used to measure a pressure difference across an orifice meter in a water line. The difference in mercury levels is 7 in (17.8 cm). What is the pressure differential?

  (A)   1.7 psi (12 kPa)
  (B)   3.2 psi (22 kPa)
  (C)   7.9 psi (55 kPa)
  (D)   23 psi (160 kPa)

19. A sharp-edged ISA orifice is used in a 12 in (300 mm) schedule-40 water line. (Figure 17.28 is applicable.) The water temperature is 70°F (20°C), and the flow rate is 10 ft³/sec (250 L/s). The differential pressure change across the orifice (to vena contracta) must not exceed 25 ft (7.5 m). What is most nearly the smallest orifice that can be used?

  (A)   5.5 in (14 cm)
  (B)   7.3 in (19 cm)
  (C)   8.1 in (20 cm)
  (D)   8.9 in (23 cm)

Impulse-Momentum

20. A pipe necks down from 24 in at point A to 12 in at point B. 8 ft³/sec of 60°F water flow from point A to point B. The pressure head at point A is 20 ft. Friction is insignificant over the distance between points A and B. What are the magnitude and direction of the resultant force on the water?

  (A)   2900 lbf; toward A
  (B)   3500 lbf; toward A
  (C)   2900 lbf; toward B
  (D)   3500 lbf; toward B

21. A 2 in (50 mm) diameter horizontal water jet has an absolute velocity (with respect to a stationary point) of 40 ft/sec (12 m/s) as it strikes a curved blade. The blade is moving horizontally away with an absolute velocity of 15 ft/sec (4.5 m/s). Water is deflected 60° from the horizontal. What is the force on the blade?

  (A)   18 lbf (80 N)
  (B)   26 lbf (110 N)
  (C)   35 lbf (160 N)
  (D)   47 lbf (210 N)

Pumps

22. 2000 gal/min (125 L/s) of brine with a specific gravity of 1.2 pass through an 85% efficient pump. The centerlines of the pump's 12 in (300 mm) inlet and 8 in (200 mm) outlet are at the same elevation. The inlet suction gauge indicates 6 in (150 mm) of mercury below atmospheric. The discharge pressure gauge is located 4 ft (1.2 m) above the centerline of the pump's outlet and indicates 20 psig (138 kPa). What is the input power to the pump?

  (A)   12 hp (8.9 kW)
  (B)   36 hp (26 kW)
  (C)   52 hp (39 kW)
  (D)   87 hp (65 kW)

Turbines

23. 100 ft³/sec (2.6 kL/s) of water pass through a horizontal turbine. The water's pressure is reduced from 30 psig (210 kPa) to 5 psi (35 kPa) vacuum. Disregarding friction, velocity, and other factors, what power is generated?

  (A)   110 hp (82 kW)
  (B)   380 hp (280 kW)
  (C)   730 hp (540 kW)
  (D)   920 hp (640 kW)

Drag

24. (*Time limit: one hour*) A car traveling through 70°F (20°C) air has the following characteristics.

frontal area	28 ft² (2.6 m²)
mass	3300 lbm (1500 kg)
drag coefficient	0.42
rolling resistance	1% of weight
engine thermal efficiency	28%
fuel heating value	115,000 Btu/gal (460 MJ/L)

(a) Considering only the drag and rolling resistance, what is the fuel consumption when the car is traveling at 55 mi/hr (90 km/h)?

  (A)   0.026 gal/mi (0.0043 L/km)
  (B)   0.038 gal/mi (0.0065 L/km)
  (C)   0.051 gal/mi (0.0097 L/km)
  (D)   0.13 gal/mi (0.022 L/km)

(b) What is the percentage increase in fuel consumption at 65 mi/hr (105 km/h) compared to at 55 mi/hr (90 km/h)?

  (A)   10%
  (B)   20%
  (C)   30%
  (D)   40%

Similarity

25. A 1/20th airplane model is tested in a wind tunnel at full velocity and temperature. What is the ratio of the wind tunnel pressure to normal ambient pressure?

    (A) 5
    (B) 10
    (C) 20
    (D) 40

26. 68°F (20°C) castor oil (kinematic viscosity at 68°F (20°C) of 0.0111 ft²/sec (0.00103 m²/s)) flows through a pump whose impeller turns at 1000 rpm. A similar pump twice the first pump's size is tested with 68°F (20°C) air. What should be the speed of the second pump to ensure similarity?

    (A) 3.6 rpm
    (B) 88 rpm
    (C) 250 rpm
    (D) 1600 rpm

SOLUTIONS

1. *Customary U.S. Solution*

Assume schedule-40 pipe.

$$D_A = 0.5054 \text{ ft}$$
$$D_B = 1.4063 \text{ ft}$$

Let point A be at zero elevation.

The total energy at point A from Bernoulli's equation is

$$E_{tA} = E_p + E_v + E_z$$
$$= \frac{p_A}{\rho} + \frac{v_A^2}{2g_c} + \frac{z_A g}{g_c}$$

At point A, the diameter is 6 in. The velocity at point A is

$$\dot{V} = v_A A_A$$
$$= v_A \left(\frac{\pi}{4}\right) D_A^2$$
$$v_A = \left(\frac{4}{\pi}\right) \left(\frac{\dot{V}}{D_A^2}\right)$$
$$= \left(\frac{4}{\pi}\right) \left(\frac{5.0 \frac{\text{ft}^3}{\text{sec}}}{(0.5054 \text{ ft})^2}\right)$$
$$= \boxed{24.9 \text{ ft/sec}}$$

$$p_A = \left(10 \frac{\text{lbf}}{\text{in}^2}\right) \left(\frac{144 \text{ in}^2}{1 \text{ ft}^2}\right) = 1440 \text{ lbf/ft}^2$$
$$z_A = 0$$

For water, $\rho \approx 62.4 \text{ lbm/ft}^3$.

$$E_{tA} = \frac{p_A}{\rho} + \frac{v_A^2}{2g_c} + \frac{z_A g}{g_c}$$
$$= \frac{1440 \frac{\text{lbf}}{\text{ft}^2}}{62.4 \frac{\text{lbm}}{\text{ft}^3}} + \frac{\left(24.9 \frac{\text{ft}}{\text{sec}}\right)^2}{(2)\left(32.2 \frac{\text{lbm-ft}}{\text{lbf-sec}^2}\right)} + 0$$
$$= 32.7 \text{ ft-lbf/lbm}$$

Similarly, the total energy at point B is

$$v_B = \left(\frac{4}{\pi}\right) \left(\frac{\dot{V}}{D_B^2}\right)$$
$$= \left(\frac{4}{\pi}\right) \left(\frac{5.0 \frac{\text{ft}^3}{\text{sec}}}{(1.4063 \text{ ft})^2}\right)$$
$$= 3.22 \text{ ft/sec}$$

$$p_B = \left(7 \; \frac{\text{lbf}}{\text{in}^2}\right)\left(\frac{144 \; \text{in}^2}{1 \; \text{ft}^2}\right) = 1008 \; \text{lbf/ft}^2$$

$$z_B = 15 \; \text{ft}$$

$$E_{tB} = \frac{p_B}{\rho} + \frac{v_B^2}{2g_c} + \frac{z_B g}{g_c}$$

$$= \frac{1008 \; \frac{\text{lbf}}{\text{ft}^2}}{62.4 \; \frac{\text{lbm}}{\text{ft}^3}} + \frac{\left(3.22 \; \frac{\text{ft}}{\text{sec}}\right)^2}{(2)\left(32.2 \; \frac{\text{lbm-ft}}{\text{lbf-sec}^2}\right)}$$

$$+ \frac{(15 \; \text{ft})\left(32.2 \; \frac{\text{ft}}{\text{sec}^2}\right)}{32.2 \; \frac{\text{lbm-ft}}{\text{lbf-sec}^2}}$$

$$= 31.3 \; \text{ft-lbf/lbm}$$

Since $E_{tA} > E_{tB}$, $\boxed{\text{the flow is from point A to point B.}}$

The answer is (B).

SI Solution

Let point A be at zero elevation.

The total energy at point A from Bernoulli's equation is

$$E_{tA} = E_p + E_v + E_z$$

$$= \frac{p_A}{\rho} + \frac{v_A^2}{2} + z_A g$$

From App. 16.C, the diameter at point A is 154 mm (0.154 m). The velocity at point A is

$$\dot{V} = v_A A_A$$

$$= v_A \left(\frac{\pi}{4}\right) D_A^2$$

$$v_A = \left(\frac{4}{\pi}\right)\left(\frac{\dot{V}}{D_A^2}\right)$$

$$= \left(\frac{4}{\pi}\right)\left(\frac{\left(130 \; \frac{\text{L}}{\text{s}}\right)\left(\frac{1 \; \text{m}^3}{1000 \; \text{L}}\right)}{(0.154 \; \text{m})^2}\right)$$

$$= \boxed{6.98 \; \text{m/s}}$$

$$p_A = 69 \; \text{kPa} \; (69\,000 \; \text{Pa})$$

$$z_A = 0$$

For water, $\rho = 1000 \; \text{kg/m}^3$.

$$E_{tA} = \frac{p_A}{\rho} + \frac{v_A^2}{2} + z_A g$$

$$= \frac{70\,000 \; \text{Pa}}{1000 \; \frac{\text{kg}}{\text{m}^3}} + \frac{\left(6.98 \; \frac{\text{m}}{\text{s}}\right)^2}{2} + 0$$

$$= 94.36 \; \text{J/kg}$$

Similarly, the diameter at B is 0.429 m. The total energy at point B is

$$v_B = \left(\frac{4}{\pi}\right)\left(\frac{\dot{V}}{D_B^2}\right)$$

$$= \left(\frac{4}{\pi}\right)\left(\frac{\left(130 \; \frac{\text{L}}{\text{s}}\right)\left(\frac{1 \; \text{m}^3}{1000 \; \text{L}}\right)}{(0.429 \; \text{m})^2}\right)$$

$$= 0.90 \; \text{m/s}$$

$$p_B = 48.3 \; \text{kPa} \; (48\,300 \; \text{Pa})$$

$$z_B = 4.6 \; \text{m}$$

$$E_{tB} = \frac{p_B}{\rho} + \frac{v_B^2}{2} + z_B g$$

$$= \frac{48\,300 \; \text{Pa}}{1000 \; \frac{\text{kg}}{\text{m}^3}} + \frac{\left(0.90 \; \frac{\text{m}}{\text{s}}\right)^2}{2}$$

$$+ (4.6 \; \text{m})\left(9.81 \; \frac{\text{m}}{\text{s}^2}\right)$$

$$= 93.8 \; \text{J/kg}$$

Since $E_{tA} > E_{tB}$, $\boxed{\text{the flow is from point A to point B.}}$

The answer is (B).

2.
$$\dot{V} = \left(750 \; \frac{\text{gal}}{\text{min}}\right)\left(0.002228 \; \frac{\text{ft}^3\text{-min}}{\text{sec-gal}}\right)$$

$$= 1.671 \; \text{ft}^3/\text{sec}$$

$D = 0.5054 \; \text{ft}$ and $A = 0.2006 \; \text{ft}^2$.

$$v = \frac{\dot{V}}{A} = \frac{1.671 \; \frac{\text{ft}^3}{\text{sec}}}{0.2006 \; \text{ft}^2} = 8.33 \; \text{ft/sec}$$

For 60°F water,

$$\nu = 1.217 \times 10^{-5} \; \text{ft}^2/\text{sec} \quad [\text{App. 14.A}]$$

$$\text{Re} = \frac{vD}{\nu} = \frac{\left(8.33 \; \frac{\text{ft}}{\text{sec}}\right)(0.5054 \; \text{ft})}{1.217 \times 10^{-5} \; \frac{\text{ft}^2}{\text{sec}}} = 3.46 \times 10^5$$

For steel,

$$\epsilon = 0.0002 \quad [\text{App. 17.A}]$$

$$\frac{\epsilon}{D} = \frac{0.0002 \text{ ft}}{0.5054 \text{ ft}} \approx 0.0004$$

$$f = 0.0175 \quad [\text{App. 17.B}]$$

From Eq. 17.22,

$$h_f = \frac{(0.0175)(3000 \text{ ft})\left(8.33 \frac{\text{ft}}{\text{sec}}\right)^2}{(2)(0.5054 \text{ ft})\left(32.2 \frac{\text{ft}}{\text{sec}^2}\right)} = 111.9 \text{ ft}$$

Use the Bernoulli equation. Since velocity is the same at points A and B, it may be omitted.

$$\frac{\left(144 \frac{\text{in}^2}{\text{ft}^2}\right) p_1}{62.37 \frac{\text{lbf}}{\text{ft}^3}} = \frac{\left(50 \frac{\text{lbf}}{\text{in}^2}\right)\left(144 \frac{\text{in}^2}{\text{ft}^2}\right)}{62.37 \frac{\text{lbf}}{\text{ft}^3}}$$
$$+ 60 \text{ ft} + 111.9 \text{ ft}$$

$$\boxed{p_1 = \; 124.5 \text{ lbf/in}^2 \text{ (psig)}}$$

The answer is (C).

3. From Eq. 17.31,

$$h_{f,A-B} = \frac{(10.44)(20{,}000 \text{ ft})\left(120 \frac{\text{gal}}{\text{min}}\right)^{1.85}}{(150)^{1.85}(6 \text{ in})^{4.8655}}$$
$$= 22.6 \text{ ft}$$

Investigate velocity head.

$$\text{v} = \frac{\dot{V}}{A}$$
$$= \frac{\left(120 \frac{\text{gal}}{\text{min}}\right)\left(0.002228 \frac{\text{ft}^3\text{-min}}{\text{sec-gal}}\right)}{\left(\frac{\pi}{4}\right)\left(\frac{6 \text{ in}}{12 \frac{\text{in}}{\text{ft}}}\right)^2} = 1.36 \text{ ft/sec}$$

$$h_\text{v} = \frac{\text{v}^2}{2g}$$
$$= \frac{\left(1.36 \frac{\text{ft}}{\text{sec}}\right)^2}{(2)\left(32.2 \frac{\text{ft}}{\text{sec}^2}\right)} = 0.029 \text{ ft}$$

Velocity heads are low and can be disregarded.

$$h_{f,B-C} = \frac{(10.44)(10{,}000 \text{ ft})\left(160 \frac{\text{gal}}{\text{min}}\right)^{1.85}}{(150)^{1.85}(6 \text{ in})^{4.8655}}$$
$$= 19.25 \text{ ft}$$

$$h_{f,C-D} = \frac{(10.44)(30{,}000 \text{ ft})\left(120 \frac{\text{gal}}{\text{min}}\right)^{1.85}}{(150)^{1.85}(4 \text{ in})^{4.8655}}$$
$$= 243.9 \text{ ft}$$

Assume a pressure of 20 psig at point A.

$$h_{p,A} = \frac{\left(20 \frac{\text{lbf}}{\text{in}^2}\right)\left(144 \frac{\text{in}^2}{\text{ft}^2}\right)}{62.4 \frac{\text{lbf}}{\text{ft}^3}} = 46.2 \text{ ft}$$

From the Bernoulli equation, ignoring velocity head,

$$h_{p,A} + z_A = h_{p,B} + z_B + h_{f,A-B}$$
$$46.2 \text{ ft} + 620 \text{ ft} = h_{p,B} + 460 \text{ ft} + 22.6 \text{ ft}$$
$$h_{p,B} = 183.6 \text{ ft}$$

$$p_B = \gamma h = \frac{\left(62.4 \frac{\text{lbf}}{\text{ft}^3}\right)(183.6 \text{ ft})}{144 \frac{\text{in}^2}{\text{ft}^2}}$$
$$= 79.6 \text{ lbf/in}^2 \text{ (psig)}$$

For B to C,

$$183.6 \text{ ft} + 460 \text{ ft} = h_{p,C} + 540 \text{ ft} + 19.25 \text{ ft}$$
$$h_{p,C} = 84.35 \text{ ft}$$

$$p_C = \frac{(84.35 \text{ ft})\left(62.4 \frac{\text{lbf}}{\text{ft}^3}\right)}{144 \frac{\text{in}^2}{\text{ft}^2}}$$
$$= 36.6 \text{ lbf/in}^2 \text{ (psig)}$$

For C to D,

$$84.35 \text{ ft} + 540 \text{ ft} = h_{p,D} + 360 \text{ ft} + 243.9 \text{ ft}$$
$$h_{p,D} = 20.45 \text{ ft}$$

$$p_D = \frac{(20.45 \text{ ft})\left(62.4 \frac{\text{lbf}}{\text{ft}^3}\right)}{144 \frac{\text{in}^2}{\text{ft}^2}}$$
$$= 8.9 \text{ lbf/in}^2 \text{ (psig)} \quad [\text{too low}]$$

Since p_D is too low, add $20 - 8.9 = 11.1$ lbf/in^2 (psig) to each point.

(a) $\quad p_A = 20.0 \dfrac{\text{lbf}}{\text{in}^2} + 11.1 \dfrac{\text{lbf}}{\text{in}^2} = \boxed{31.1 \text{ lbf/in}^2 \text{ (psig)}}$

$$p_B = 79.6 \ \frac{\text{lbf}}{\text{in}^2} + 11.1 \ \frac{\text{lbf}}{\text{in}^2} = 90.7 \ \text{lbf/in}^2 \ (\text{psig})$$

$$p_C = 36.6 \ \frac{\text{lbf}}{\text{in}^2} + 11.1 \ \frac{\text{lbf}}{\text{in}^2} = 47.7 \ \text{lbf/in}^2 \ (\text{psig})$$

$$p_D = 8.9 \ \frac{\text{lbf}}{\text{in}^2} + 11.1 \ \frac{\text{lbf}}{\text{in}^2} = 20.0 \ \text{lbf/in}^2 \ (\text{psig})$$

The answer is (C).

(b) The elevation of the hydraulic grade line above point D is the sum of the potential and static heads.

$$\Delta h_{A-D} = z_A - z_D + p_A$$
$$= 620 \ \text{ft} - 360 \ \text{ft}$$
$$+ \frac{\left(31.1 \ \frac{\text{lbf}}{\text{in}^2}\right)\left(144 \ \frac{\text{in}^2}{\text{ft}^2}\right)}{62.4 \ \frac{\text{lbf}}{\text{ft}^3}}$$
$$= \boxed{331.8 \ \text{ft}}$$

If the hydraulic grade line was with respect to point A, only the static head would be included.

The answer is (A).

4. *Customary U.S. Solution*

For 6 in schedule-40 pipe, the internal diameter, D_i, is 0.5054 ft. The internal area is 0.2006 ft^2.

The velocity, v, is calculated from the volumetric flow, $\dot{V}$, and the flow area, A_i, by

$$\text{v} = \frac{\dot{V}}{A_i}$$
$$= \frac{1.5 \ \frac{\text{ft}^3}{\text{sec}}}{0.2006 \ \text{ft}^2}$$
$$= 7.48 \ \text{ft/sec}$$

For water at 70°F, the kinematic viscosity, ν, is 1.059×10^{-5} ft^2/sec [App. 14.A].

Calculate the Reynolds number.

$$\text{Re} = \frac{D_i \text{v}}{\nu}$$
$$= \frac{(0.5054 \ \text{ft})\left(7.48 \ \frac{\text{ft}}{\text{sec}}\right)}{1.059 \times 10^{-5} \ \frac{\text{ft}^2}{\text{sec}}}$$
$$= 3.57 \times 10^5$$

Since Re > 2100, the flow is turbulent. The friction loss coefficient can be determined from the Moody diagram.

For new steel pipe, the specific roughness, ϵ, is 0.0002 ft.

The relative roughness is

$$\frac{\epsilon}{D_i} = \frac{0.0002 \ \text{ft}}{0.5054 \ \text{ft}}$$
$$= 0.0004$$

From the Moody diagram with Re = 3.57×10^5 and $e/D_i = 0.0004$, the friction factor, f, can be determined as 0.0174.

Use Darcy's equation to compute the frictional loss.

$$h_f = \frac{fL\text{v}^2}{2D_i g}$$
$$= \frac{(0.0174)(1200 \ \text{ft})\left(7.48 \ \frac{\text{ft}}{\text{sec}}\right)^2}{(2)(0.5054 \ \text{ft})\left(32.2 \ \frac{\text{ft}}{\text{sec}^2}\right)}$$
$$= \boxed{35.9 \ \text{ft}}$$

The answer is (C).

SI Solution

For 6 in pipe, the internal diameter is 154.1 mm and the internal area is 186.5×10^{-4} m^2.

The velocity, v, is calculated from the volumetric flow, $\dot{V}$, and the flow area, A_i, by

$$\text{v} = \frac{\dot{V}}{A_i}$$
$$= \frac{\left(40 \ \frac{\text{L}}{\text{s}}\right)\left(0.001 \ \frac{\text{m}^3}{\text{L}}\right)}{186.5 \times 10^{-4} \ \text{m}^2}$$
$$= 2.145 \ \text{m/s}$$

For water at 20°C, the kinematic viscosity is

$$\nu = \frac{\mu}{\rho} = \frac{1.0050 \times 10^{-3} \ \text{Pa·s}}{998.23 \ \frac{\text{kg}}{\text{m}^3}}$$
$$= 1.007 \times 10^{-6} \ \text{m}^2/\text{s}$$

Calculate the Reynolds number.

$$\text{Re} = \frac{D_i \text{v}}{\nu}$$
$$= \frac{(154.1 \ \text{mm})\left(0.001 \ \frac{\text{m}}{\text{mm}}\right)\left(2.145 \ \frac{\text{m}}{\text{s}}\right)}{1.007 \times 10^{-6} \ \frac{\text{m}^2}{\text{s}}}$$
$$= 3.282 \times 10^5$$

Since Re > 2100, the flow is turbulent. The friction loss coefficient can be determined from the Moody diagram.

For new steel pipe, the specific roughness, ϵ, is 6.0×10^{-5} m.

The relative roughness is

$$\frac{\epsilon}{D_i} = \frac{6.0 \times 10^{-5} \text{ m}}{0.1541 \text{ m}}$$
$$= 0.0004$$

From the Moody diagram with Re = 3.28×10^5 and $e/D_i = 0.0004$, the friction factor, f, can be determined as 0.0175.

Use Darcy's equation to compute the frictional loss.

$$h_f = \frac{fLv^2}{2D_i g}$$
$$= \frac{(0.0175)(355 \text{ m})\left(2.145 \, \frac{\text{m}}{\text{s}}\right)^2}{(2)(0.1541 \text{ m})\left(9.81 \, \frac{\text{m}}{\text{s}^2}\right)}$$
$$= \boxed{9.45 \text{ m}}$$

The answer is (C).

5. *Customary U.S. Solution*

For 6 in schedule-40 pipe, the internal diameter, D_i, is 0.5054 ft. The internal area is 0.2006 ft^2.

The velocity, v, is calculated from the volumetric flow, $\dot{V}$, and the flow area, A_i.

Convert the volumetric flow rate from gal/min to ft^3/sec.

$$\dot{V} = \left(500 \, \frac{\text{gal}}{\text{min}}\right)\left(\frac{1 \text{ ft}^3}{7.48 \text{ gal}}\right)\left(\frac{1 \text{ min}}{60 \text{ sec}}\right)$$
$$= 1.114 \text{ ft}^3/\text{sec}$$

The velocity is

$$v = \frac{\dot{V}}{A_i}$$
$$= \frac{1.114 \, \frac{\text{ft}^3}{\text{sec}}}{0.2006 \text{ ft}^2}$$
$$= 5.55 \text{ ft/sec}$$

For water at 100°F, the kinematic viscosity, ν, is 0.739×10^{-5} ft^2/sec and the density is 62.0 lbm/ft^3.

Calculate the Reynolds number.

$$\text{Re} = \frac{D_i v}{\nu}$$
$$= \frac{(0.5054 \text{ ft})\left(5.55 \, \frac{\text{ft}}{\text{sec}}\right)}{0.739 \times 10^{-5} \, \frac{\text{ft}^2}{\text{sec}}}$$
$$= 3.80 \times 10^5$$

Since Re > 2100, the flow is turbulent. The friction loss coefficient can be determined from the Moody diagram.

For new steel pipe, the specific roughness, ϵ, is 0.0002 ft.

The relative roughness is

$$\frac{\epsilon}{D_i} = \frac{0.0002 \text{ ft}}{0.5054 \text{ ft}}$$
$$= 0.0004$$

From the Moody diagram with Re = 3.80×10^5 and $e/D_i = 0.0004$, the friction factor, f, can be determined as 0.0173.

The equivalent lengths of the valves and fittings are

standard radius elbow	2×8.9 ft $=$	17.8 ft
gate valve (fully open)	2×3.2 ft $=$	6.4 ft
90° angle valve (fully open)	1×63.0 ft $=$	63.0 ft
swing check valve	1×63.0 ft $=$	63.0 ft
		150.2 ft

The equivalent pipe length is the sum of the straight run of pipe and the equivalent length of pipe for the valves and fittings.

$$L_e = L + L_{\text{fittings}}$$
$$= 300 \text{ ft} + 150.2 \text{ ft}$$
$$= 450.2 \text{ ft}$$

Use Darcy's equation to compute the frictional loss.

$$h_f = \frac{fL_e v^2}{2D_i g}$$
$$= \frac{(0.0173)(450.2 \text{ ft})\left(5.55 \, \frac{\text{ft}}{\text{sec}}\right)^2}{(2)(0.5054 \text{ ft})\left(32.2 \, \frac{\text{ft}}{\text{sec}^2}\right)}$$
$$= 7.37 \text{ ft}$$

The total difference in head is the sum of the head loss through the pipe, valves, and fittings and the change in elevation.

$$\Delta h = h_f + \Delta z$$
$$= 7.37 \text{ ft} + 20 \text{ ft}$$
$$= 27.37 \text{ ft}$$

The pressure difference between the entrance and discharge can be determined from

$$\Delta p = \gamma \Delta h = \rho h \times \left(\frac{g}{g_c} \right)$$

$$= \frac{\left(62.0 \, \dfrac{\text{lbm}}{\text{ft}^3} \right) (27.37 \, \text{ft}) \left(32.2 \, \dfrac{\text{ft}}{\text{sec}^2} \right) \left(\dfrac{1 \, \text{ft}^2}{144 \, \text{in}^2} \right)}{32.2 \, \dfrac{\text{lbm-ft}}{\text{lbf-sec}^2}}$$

$$= \boxed{11.8 \, \text{lbf/in}^2}$$

The answer is (A).

SI Solution

For 6 in pipe, the internal diameter is 154.1 mm (0.1541 m).

The internal area is $186.5 \times 10^{-4} \, \text{m}^2$.

The velocity, v, is calculated from the volumetric flow, $\dot{V}$, and the flow area, A_i.

$$\text{v} = \frac{\dot{V}}{A_i}$$

$$= \frac{\left(30 \, \dfrac{\text{L}}{\text{s}} \right) \left(0.001 \, \dfrac{\text{m}^3}{\text{L}} \right)}{186.5 \times 10^{-4} \, \text{m}^2}$$

$$= 1.61 \, \text{m/s}$$

For water at 40°C, the density is 992.25 kg/m³ and the kinematic viscosity is

$$\nu = \frac{\mu}{\rho} = \frac{0.6560 \times 10^{-3} \, \text{Pa·s}}{992.25 \, \dfrac{\text{kg}}{\text{m}^3}}$$

$$= 6.611 \times 10^{-7} \, \text{m}^2/\text{s}$$

Calculate the Reynolds number.

$$\text{Re} = \frac{D_i \text{v}}{\nu}$$

$$= \frac{(0.1541 \, \text{m}) \left(1.61 \, \dfrac{\text{m}}{\text{s}} \right)}{6.611 \times 10^{-7} \, \dfrac{\text{m}^2}{\text{s}}}$$

$$= 3.75 \times 10^5$$

Since Re > 2100, the flow is turbulent. The friction loss coefficient can be determined from the Moody diagram.

For new steel pipe, the specific roughness is 6.0×10^{-5} m.

The relative roughness is

$$\frac{\epsilon}{D_i} = \frac{6.0 \times 10^{-5} \, \text{m}}{0.1541 \, \text{m}}$$

$$= 0.0004$$

From the Moody diagram with Re $= 3.75 \times 10^5$ and $e/D_i = 0.0004$, the friction factor, f, can be determined as 0.0173.

The equivalent lengths of the valves and fittings are

standard radius elbow	2×2.7 m $=$	5.4 m
gate valve (fully open)	2×1.0 m $=$	2.0 m
90° angle valve (fully open)	1×18.9 m $=$	18.9 m
swing check valve	1×18.9 m $=$	18.9 m
		45.2 m

The equivalent pipe length is the sum of the straight run of pipe and the equivalent length of pipe for the valves and fittings.

$$L_e = L + L_{\text{fittings}}$$

$$= 90 \, \text{m} + 45.2 \, \text{m}$$

$$= 135.2 \, \text{m}$$

Use Darcy's equation to compute the frictional loss.

$$h_f = \frac{f L \text{v}^2}{2 D_i g}$$

$$= \frac{(0.0173)(135.2 \, \text{m}) \left(1.61 \, \dfrac{\text{m}}{\text{s}^2} \right)^2}{(2)(0.1541 \, \text{m}) \left(9.81 \, \dfrac{\text{m}}{\text{s}^2} \right)}$$

$$= 2.01 \, \text{m}$$

The total difference in head is the sum of the head loss through the pipe, valves, and fittings and the change in elevation.

$$\Delta h = h_f + \Delta z$$

$$= 2.01 \, \text{m} + 6 \, \text{m}$$

$$= 8.01 \, \text{m}$$

The pressure difference between the entrance and discharge can be determined from

$$\Delta p = \rho h g$$

$$= \left(992.25 \, \dfrac{\text{kg}}{\text{m}^3} \right) (8.01 \, \text{m}) \left(9.81 \, \dfrac{\text{m}}{\text{s}^2} \right)$$

$$= \boxed{77\,969 \, \text{Pa} \quad (78 \, \text{kPa})}$$

The answer is (A).

6. *Customary U.S. Solution*

For 6 in schedule-40 pipe, the internal diameter, D_i, is 0.5054 ft. The internal area is 0.2006 ft².

For air at 70°F, the kinematic viscosity is 16.15×10^{-5} ft²/sec.

Calculate the Reynolds number.

$$\text{Re} = \frac{D_i \text{v}}{\nu}$$

$$= \frac{(0.5054 \text{ ft})\left(60 \dfrac{\text{ft}}{\text{sec}}\right)}{16.15 \times 10^{-5} \dfrac{\text{ft}^2}{\text{sec}}}$$

$$= 1.88 \times 10^5$$

Since Re > 2100, the flow is turbulent. The friction loss coefficient can be determined from the Moody diagram.

For new steel pipe, the specific roughness, ϵ, is 0.0002 ft.

The relative roughness is

$$\frac{\epsilon}{D_i} = \frac{0.0002 \text{ ft}}{0.5054 \text{ ft}}$$

$$= 0.0004$$

From the Moody diagram with Re $= 1.88 \times 10^5$ and $e/D_i = 0.0004$, the friction factor, f, can be determined as 0.0184.

From App. 17.D, the equivalent length of the valves and fittings is

standard radius elbow	2×8.9 ft $=$	17.8 ft
gate valve (fully open)	2×3.2 ft $=$	6.4 ft
90° angle valve (fully open)	1×63.0 ft $=$	63.0 ft
swing check valve	1×63.0 ft $=$	63.0 ft
		150.2 ft

The equivalent pipe length is the sum of the straight run of pipe and the equivalent length of pipe for the valves and fittings.

$$L_e = L + L_{\text{fittings}}$$

$$= 300 \text{ ft} + 150.2 \text{ ft}$$

$$= 450.2 \text{ ft}$$

Use Darcy's equation to compute the frictional loss.

$$h_f = \frac{f L_e \text{v}^2}{2 D_i g}$$

$$= \frac{(0.0184)(450.2 \text{ ft})\left(60 \dfrac{\text{ft}}{\text{sec}}\right)^2}{(2)(0.5054 \text{ ft})\left(32.2 \dfrac{\text{ft}}{\text{sec}^2}\right)}$$

$$= 916.2 \text{ ft}$$

The difference in head is the sum of the head loss through the pipe, valves, and fittings and the change in elevation.

$$h = h_f + \Delta z$$

$$= 916.2 \text{ ft} + 20 \text{ ft}$$

$$= 936.2 \text{ ft}$$

The pressure difference between the entrance and discharge can be determined from

$$\Delta p = \gamma h = \rho h \times \left(\frac{g}{g_c}\right)$$

$$= \left(0.075 \dfrac{\text{lbm}}{\text{ft}^3}\right)(936.2 \text{ ft})$$

$$\times \left(\frac{32.2 \dfrac{\text{ft}}{\text{sec}^2}}{32.2 \dfrac{\text{lbm-ft}}{\text{lbf-sec}^2}}\right)\left(\frac{1 \text{ ft}^2}{144 \text{ in}^2}\right)$$

$$= \boxed{0.49 \text{ lbf/in}^2 \quad (0.49 \text{ psi})}$$

The answer is (B).

SI Solution

For 6 in pipe, the internal diameter, D_i, is 154.1 mm (0.1541 m) and the internal area is $186.5 \times 10^{-4} \text{ m}^2$ [App. 16.C].

For air at 20°C, the kinematic viscosity, ν, is $1.51 \times 10^{-5} \text{ m}^2/\text{s}$ [App. 14.E].

Calculate the Reynolds number.

$$\text{Re} = \frac{D_i \text{v}}{\nu}$$

$$= \frac{(0.1541 \text{ m})\left(18 \dfrac{\text{m}}{\text{s}}\right)}{1.51 \times 10^{-5} \dfrac{\text{m}^2}{\text{s}}}$$

$$= 1.84 \times 10^5$$

Since Re > 2100, the flow is turbulent. The friction loss coefficient can be determined from the Moody diagram.

For new steel pipe, the specific roughness, ϵ, is 6.0×10^{-5} m.

The relative roughness is

$$\frac{\epsilon}{D_i} = \frac{6.0 \times 10^{-5}}{0.1541 \text{ m}}$$

$$= 0.0004$$

From the Moody diagram with Re $= 1.84 \times 10^5$ and $e/D_i = 0.0004$, the friction factor, f, can be determined as 0.0185.

Compute the equivalent lengths of the valves and fittings. (Convert from App. 17.D.)

standard radius elbow	2×2.7 m $=$	5.4 m
gate valve (fully open)	2×1.0 m $=$	2.0 m
90° angle valve (fully open)	1×18.9 m $=$	18.9 m
swing check valve	1×18.9 m $=$	18.9 m
		45.2 m

Fluids

The equivalent pipe length is the sum of the straight run of pipe and the equivalent length of pipe for the valves and fittings.

$$L_e = L + L_{\text{fittings}}$$
$$= 90 \text{ m} + 45.2 \text{ m}$$
$$= 135.2 \text{ m}$$

Use Darcy's equation to compute the frictional loss.

$$h_f = \frac{fLv^2}{2D_i g}$$
$$= \frac{(0.0185)(135.2 \text{ m})\left(18 \dfrac{\text{m}}{\text{s}}\right)^2}{(2)(0.1541 \text{ m})\left(9.81 \dfrac{\text{m}}{\text{s}^2}\right)}$$
$$= 268.0 \text{ m}$$

The difference in head is the sum of the head loss through the pipe, valves, and fittings and the change in elevation.

$$\Delta h = h_f + \Delta z$$
$$= 268.0 \text{ m} + 6 \text{ m}$$
$$= 274.0 \text{ m}$$

Assume the density of the air, ρ, is approximately 1.20 kg/m^3.

The pressure difference between the entrance and discharge can be determined from

$$\Delta p = \rho h g$$
$$= \left(1.20 \dfrac{\text{kg}}{\text{m}^3}\right)(274.0 \text{ m})\left(9.81 \dfrac{\text{m}}{\text{s}^2}\right)$$
$$= \boxed{3226 \text{ Pa} \quad (3.23 \text{ kPa})}$$

The answer is (B).

7. Assume that flows from reservoirs A and B are toward D and then toward C. Then,

$$\dot{V}_{\text{A–D}} + \dot{V}_{\text{B–D}} = \dot{V}_{\text{D–C}}$$

or

$$A_A v_{\text{A–D}} + A_B v_{\text{B–D}} - A_C v_{\text{D–C}} = 0$$

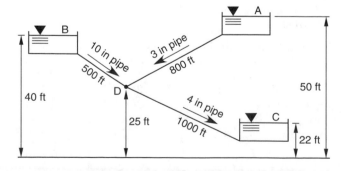

From App. 16.B, for schedule-40 pipe,

$$\begin{array}{ll} A_A = 0.05134 \text{ ft}^2 & D_A = 0.2557 \text{ ft} \\ A_B = 0.5476 \text{ ft}^2 & D_B = 0.8350 \text{ ft} \\ A_C = 0.08841 \text{ ft}^2 & D_C = 0.3355 \text{ ft} \end{array}$$

$$0.05134 v_{\text{A–D}} + 0.5476 v_{\text{B–D}} - 0.08841 v_{\text{D–C}} = 0 \quad \text{[Eq. 1]}$$

Ignoring the velocity heads, the conservation of energy equation between A and D is

$$z_A = \frac{p_D}{\gamma} + z_D + h_{f,\text{A–D}}$$

$$50 \text{ ft} = \frac{p_D}{62.4 \dfrac{\text{lbf}}{\text{ft}^3}} + 25 \text{ ft} + \frac{(0.02)(800 \text{ ft})(v_{\text{A–D}})^2}{(2)(0.2557 \text{ ft})\left(32.2 \dfrac{\text{ft}}{\text{sec}^2}\right)}$$

or

$$v_{\text{A–D}} = \sqrt{25.73 - 0.0165 p_D} \quad \text{[Eq. 2]}$$

Similarly, for B–D,

$$40 \text{ ft} = \frac{p_D}{62.4 \dfrac{\text{lbf}}{\text{ft}^3}} + 25 \text{ ft} + \frac{(0.02)(500 \text{ ft})(v_{\text{B–D}})^2}{(2)(0.8350 \text{ ft})\left(32.2 \dfrac{\text{ft}}{\text{sec}^2}\right)}$$

or

$$v_{\text{B–D}} = \sqrt{80.66 - 0.0862 p_D} \quad \text{[Eq. 3]}$$

For D–C,

$$22 \text{ ft} = \frac{p_D}{62.4 \dfrac{\text{lbf}}{\text{ft}^3}} + 25 \text{ ft} - \frac{(0.02)(1000 \text{ ft})(v_{\text{D–C}})^2}{(2)(0.3355 \text{ ft})\left(32.2 \dfrac{\text{ft}}{\text{sec}^2}\right)}$$

or

$$v_{\text{D–C}} = \sqrt{3.24 + 0.0173 p_D} \quad \text{[Eq. 4]}$$

Equations 1, 2, 3, and 4 must be solved simultaneously. To do this, assume a value for p_D. This value then determines all three velocities in Eqs. 2, 3, and 4. These velocities are substituted into Eq. 1. A trial and error solution yields

$$v_{\text{A–D}} = 3.21 \text{ ft/sec}$$
$$v_{\text{B–D}} = 0.408 \text{ ft/sec}$$
$$v_{\text{D–C}} = 4.40 \text{ ft/sec}$$

$$\boxed{p_D = 933.8 \text{ lbf/ft}^2 \quad (\text{psf})}$$

$$\boxed{\text{Flow is from B to D.}}$$

The answer is (B).

8. *Customary U.S. Solution*

The composite modulus of elasticity of the pipe and water is given by Eq. 17.158.

For water at 70°F, $E_{\text{water}} = 320 \times 10^3 \text{ lbf/in}^2$.

For cast-iron pipe, $E_{pipe} = 20 \times 10^6$ lbf/in^2.

$$E = \frac{E_{water} t_{pipe} E_{pipe}}{t_{pipe} E_{pipe} + d_{pipe} E_{water}}$$

$$= \frac{\left(320 \times 10^3 \; \frac{lbf}{in^2}\right)(0.75 \text{ in})\left(20 \times 10^6 \; \frac{lbf}{in^2}\right)}{(0.75 \text{ in})\left(20 \times 10^6 \; \frac{lbf}{in^2}\right)}$$
$$+ (24 \text{ in})\left(320 \times 10^3 \; \frac{lbf}{in^2}\right)$$

$$= 2.12 \times 10^5 \text{ lbf/in}^2$$

The speed of sound in the pipe is

$$a = \sqrt{\frac{E g_c}{\rho}}$$

$$= \sqrt{\frac{\left(2.12 \times 10^5 \; \frac{lbf}{in^2}\right)\left(\frac{144 \text{ in}^2}{1 \text{ ft}^2}\right)\left(32.2 \; \frac{lbm\text{-}ft}{lbf\text{-}sec^2}\right)}{62.3 \; \frac{lbm}{ft^3}}}$$

$$= 3972 \text{ ft/sec}$$

(a) The maximum pressure is given by Eq. 17.157.

$$\Delta p = \frac{\rho a \Delta v}{g_c}$$

$$= \left(\frac{\left(62.3 \; \frac{lbm}{ft^3}\right)\left(3972 \; \frac{ft}{sec}\right)\left(6 \; \frac{ft}{sec}\right)}{32.2 \; \frac{lbm\text{-}ft}{lbf\text{-}sec^2}}\right)$$
$$\times \left(\frac{1 \text{ ft}^2}{144 \text{ in}^2}\right)$$

$$= \boxed{320.2 \text{ lbf/in}^2 \;\; (320.2 \text{ psi})}$$

The answer is (C).

(b) The length of time the pressure is constant at the valve is given by

$$t = \frac{2L}{a}$$

$$= \frac{(2)(500 \text{ ft})}{3972 \; \frac{ft}{sec}}$$

$$= \boxed{0.252 \text{ sec}}$$

The answer is (A).

SI Solution

The composite modulus of elasticity of the pipe and water is given by Eq. 17.158.

For water at 20°C, $E_{water} = 2.2 \times 10^9$ Pa.

For cast-iron pipe, $E_{pipe} = 1.4 \times 10^{11}$ Pa.

$$E = \frac{E_{water} t_{pipe} E_{pipe}}{t_{pipe} E_{pipe} + d_{pipe} E_{water}}$$

$$= \frac{(2.2 \times 10^9 \text{ Pa})(0.02 \text{ m})(1.4 \times 10^{11} \text{ Pa})}{(0.02 \text{ m})(1.4 \times 10^{11} \text{ Pa}) + (0.6 \text{ m})(2.2 \times 10^9 \text{ Pa})}$$

$$= 1.50 \times 10^9 \text{ Pa}$$

The speed of sound in the pipe is

$$a = \sqrt{\frac{E}{\rho}}$$

$$= \sqrt{\frac{1.50 \times 10^9 \text{ Pa}}{1000 \; \frac{kg}{m^3}}}$$

$$= 1225 \text{ m/s}$$

(a) The maximum pressure is given by

$$\Delta p = \rho a \Delta v$$

$$= \left(1000 \; \frac{kg}{m^3}\right)\left(1225 \; \frac{m}{s}\right)\left(2 \; \frac{m}{s}\right)$$

$$= \boxed{2.45 \times 10^6 \text{ Pa} \;\; (2450 \text{ kPa})}$$

The answer is (C).

(b) The length of time the pressure is constant at the valve is given by

$$t = \frac{2L}{a}$$

$$= \frac{(2)(150 \text{ m})}{1225 \; \frac{m}{s}}$$

$$= \boxed{0.245 \text{ s}}$$

The answer is (A).

9. *Customary U.S. Solution*

First it is necessary to collect data on schedule-40 pipe and water. The fluid viscosity, pipe dimensions, and other parameters can be found in various appendices in Chs. 14 and 16. At 70°F water, $\nu = 1.059 \times 10^{-5}$ ft^2/sec.

From Table 17.2, $\epsilon = 0.0002$ ft.

8 in pipe	$D = 0.6651$ ft $A = 0.3474$ ft^2
12 in pipe	$D = 0.9948$ ft $A = 0.7773$ ft^2
16 in pipe	$D = 1.25$ ft $A = 1.2272$ ft^2

The flow quantity is converted from gallons per minute to cubic feet per second.

$$\dot{V} = \frac{(8 \text{ MGD})\left(10^6 \ \frac{\frac{\text{gal}}{\text{day}}}{\text{MGD}}\right)\left(0.002228 \ \frac{\frac{\text{ft}^3}{\text{sec}}}{\frac{\text{gal}}{\text{min}}}\right)}{\left(24 \ \frac{\text{hr}}{\text{day}}\right)\left(60 \ \frac{\text{min}}{\text{hr}}\right)}$$

$$= 12.378 \text{ ft}^3/\text{sec}$$

For the inlet pipe, the velocity is

$$\mathrm{v} = \frac{\dot{V}}{A} = \frac{12.378 \ \frac{\text{ft}^3}{\text{sec}}}{0.3474 \text{ ft}^2} = 35.63 \text{ ft}/\text{sec}$$

The Reynolds number is

$$\text{Re} = \frac{D\mathrm{v}}{\nu} = \frac{(0.6651 \text{ ft})\left(35.63 \ \frac{\text{ft}}{\text{sec}}\right)}{1.059 \times 10^{-5} \ \frac{\text{ft}^2}{\text{sec}}}$$

$$= 2.24 \times 10^6$$

The relative roughness is

$$\frac{\epsilon}{D} = \frac{0.0002 \text{ ft}}{0.6651 \text{ ft}} = 0.0003$$

From the Moody diagram, $f = 0.015$.

Equation 17.23(b) is used to calculate the frictional energy loss.

$$E_{f,1} = h_f \times \left(\frac{g}{g_c}\right) = \frac{fL\mathrm{v}^2}{2Dg_c}$$

$$= \frac{(0.015)(1000 \text{ ft})\left(35.63 \ \frac{\text{ft}}{\text{sec}}\right)^2}{(2)(0.6651 \text{ ft})\left(32.2 \ \frac{\text{lbm-ft}}{\text{lbf-sec}^2}\right)}$$

$$= 444.6 \text{ ft-lbf/lbm}$$

For the outlet pipe, the velocity is

$$\mathrm{v} = \frac{\dot{V}}{A} = \frac{12.378 \ \frac{\text{ft}^3}{\text{sec}}}{0.7773 \text{ ft}^2} = 15.92 \text{ ft}/\text{sec}$$

The Reynolds number is

$$\text{Re} = \frac{D\mathrm{v}}{\nu} = \frac{(0.9948 \text{ ft})\left(15.92 \ \frac{\text{ft}}{\text{sec}}\right)}{1.059 \times 10^{-5} \ \frac{\text{ft}^2}{\text{sec}}}$$

$$= 1.5 \times 10^6$$

The relative roughness is

$$\frac{\epsilon}{D} = \frac{0.0002 \text{ ft}}{0.9948 \text{ ft}} = 0.0002$$

From the Moody diagram, $f = 0.014$.

Equation 17.23(b) is used to calculate the frictional energy loss.

$$E_{f,2} = h_f \times \left(\frac{g}{g_c}\right) = \frac{fL\mathrm{v}^2}{2Dg_c}$$

$$= \frac{(0.014)(1500 \text{ ft})\left(15.92 \ \frac{\text{ft}}{\text{sec}}\right)^2}{(2)(0.9948 \text{ ft})\left(32.2 \ \frac{\text{lbm-ft}}{\text{lbf-sec}^2}\right)}$$

$$= 83.1 \text{ ft-lbf/lbm}$$

Assume a 50% split through the two branches. In the upper branch, the velocity is

$$\mathrm{v} = \frac{\dot{V}}{A} = \frac{\left(\frac{1}{2}\right)\left(12.378 \ \frac{\text{ft}^3}{\text{sec}}\right)}{0.3474 \text{ ft}^2} = 17.81 \text{ ft}/\text{sec}$$

The Reynolds number is

$$\text{Re} = \frac{D\mathrm{v}}{\nu} = \frac{(0.6651 \text{ ft})\left(17.81 \ \frac{\text{ft}}{\text{sec}}\right)}{1.059 \times 10^{-5} \ \frac{\text{ft}^2}{\text{sec}}}$$

$$= 1.1 \times 10^6$$

The relative roughness is

$$\frac{\epsilon}{D} = \frac{0.0002 \text{ ft}}{0.6651 \text{ ft}} = 0.0003$$

From the Moody diagram, $f = 0.015$.

For the 16 in pipe in the lower branch, the velocity is

$$\mathrm{v} = \frac{\dot{V}}{A} = \frac{\left(\frac{1}{2}\right)\left(12.378 \ \frac{\text{ft}^3}{\text{sec}}\right)}{1.2272 \text{ ft}^2} = 5.04 \text{ ft}/\text{sec}$$

The Reynolds number is

$$
\text{Re} = \frac{D\text{v}}{\nu} = \frac{(1.25 \text{ ft})\left(5.04 \dfrac{\text{ft}}{\text{sec}}\right)}{1.059 \times 10^{-5} \dfrac{\text{ft}^2}{\text{sec}}}
$$

$$
= 5.95 \times 10^5
$$

The relative roughness is

$$
\frac{\epsilon}{D} = \frac{0.0002 \text{ ft}}{1.25 \text{ ft}} = 0.00016
$$

From the Moody diagram, $f = 0.015$.

These values of f for the two branches are fairly insensitive to changes in $\dot{V}$, so they will be used for the rest of the problem in the upper branch.

Eq. 17.23(b) is used to calculate the frictional energy loss in the upper branch.

$$
E_{f,\text{upper}} = h_f \times \left(\frac{g}{g_c}\right) = \frac{fL\text{v}^2}{2Dg_c}
$$

$$
= \frac{(0.015)(500 \text{ ft})\left(17.81 \dfrac{\text{ft}}{\text{sec}}\right)^2}{(2)(0.6651 \text{ ft})\left(32.2 \dfrac{\text{lbm-ft}}{\text{lbf-sec}^2}\right)}
$$

$$
= 55.5 \text{ ft-lbf/lbm}
$$

To calculate a loss for any other flow in the upper branch,

$$
E_{f,\text{upper 2}} = E_{f,\text{upper}}\left(\frac{\dot{V}}{\left(\frac{1}{2}\right)\left(12.378 \dfrac{\text{ft}^3}{\text{sec}}\right)}\right)^2
$$

$$
= \left(55.5 \dfrac{\text{ft-lbf}}{\text{lbm}}\right)\left(\frac{\dot{V}}{6.189 \dfrac{\text{ft}^3}{\text{sec}}}\right)^2
$$

$$
= 1.45\dot{V}^2
$$

Similarly, for the lower branch, in the 8 in section,

$$
E_{f,\text{lower, 8 in}} = \frac{(0.015)(250 \text{ ft})\left(17.81 \dfrac{\text{ft}}{\text{sec}}\right)^2}{(2)(0.6651 \text{ ft})\left(32.2 \dfrac{\text{lbm-ft}}{\text{lbf-sec}^2}\right)}
$$

$$
= 27.8 \text{ ft-lbf/lbm}
$$

For the lower branch, in the 16 in section,

$$
E_{f,\text{lower,16 in}} = \frac{(0.015)(1000 \text{ ft})\left(5.04 \dfrac{\text{ft}}{\text{sec}}\right)^2}{(2)(1.25 \text{ ft})\left(32.2 \dfrac{\text{lbm-ft}}{\text{lbf-sec}^2}\right)}
$$

$$
= 4.7 \text{ ft-lbf/lbm}
$$

The total loss in the lower branch is

$$
E_{f,\text{lower}} = E_{f,\text{lower,8 in}} + E_{f,\text{lower,16 in}}
$$

$$
= 27.8 \dfrac{\text{ft-lbf}}{\text{lbm}} + 4.7 \dfrac{\text{ft-lbf}}{\text{lbm}}
$$

$$
= 32.5 \text{ ft-lbf/lbm}
$$

To calculate a loss for any other flow in the lower branch,

$$
E_{f,\text{lower 2}} = E_{f,\text{lower}}\left(\frac{\dot{V}}{\left(\frac{1}{2}\right)\left(12.378 \dfrac{\text{ft}^3}{\text{sec}}\right)}\right)^2
$$

$$
= \left(32.5 \dfrac{\text{ft-lbf}}{\text{lbm}}\right)\left(\frac{\dot{V}}{6.189 \dfrac{\text{ft}^3}{\text{sec}}}\right)^2
$$

$$
= 0.85\dot{V}^2
$$

Let x be the fraction flowing in the upper branch. Then, because the friction losses are equal,

$$
E_{f,\text{upper 2}} = E_{f,\text{lower 2}}
$$

$$
1.45x^2 = (0.85)(1-x)^2
$$

$$
x = 0.432
$$

(a) $\qquad \dot{V}_{\text{upper}} = (0.432)\left(12.378 \dfrac{\text{ft}^3}{\text{sec}}\right)$

$$
= \boxed{5.347 \text{ ft}^3/\text{sec}}
$$

The answer is (D).

(b) $\qquad \dot{V}_{\text{lower}} = (1 - 0.432)\left(12.378 \dfrac{\text{ft}^3}{\text{sec}}\right)$

$$
= 7.03 \text{ ft}^3/\text{sec}
$$

$$
E_{f,\text{total}} = E_{f,1} + E_{f,\text{lower 2}} + E_{f,2}
$$

$$
E_{f,\text{lower 2}} = 0.85\dot{V}_{\text{lower}}^2
$$

$$
= (0.85)\left(7.03 \dfrac{\text{ft}^3}{\text{sec}}\right)^2
$$

$$
= 42.0 \text{ ft}
$$

$$
E_{f,\text{total}} = 444.6 \dfrac{\text{ft-lbf}}{\text{lbm}} + 42.0 \dfrac{\text{ft-lbf}}{\text{lbm}} + 83.1 \dfrac{\text{ft-lbf}}{\text{lbm}}
$$

$$
= \boxed{569.7 \text{ ft-lbf/lbm}}
$$

The answer is (D).

SI Solution

First it is necessary to collect data on schedule-40 pipe and water. The fluid viscosity, pipe dimensions, and other parameters can be found in various appendices in Chs. 14 and 16. At 20°C water, $\nu = 1.007 \times 10^{-6}$ m^2/s.

From Table 17.2, $\epsilon = 6 \times 10^{-5}$ m.

8 in pipe	$D = 202.7$ mm
	$A = 322.7 \times 10^{-4}$ m^2
12 in pipe	$D = 303.2$ mm
	$A = 721.9 \times 10^{-4}$ m^2
16 in pipe	$D = 381$ mm
	$A = 1104 \times 10^{-4}$ m^2

For the inlet pipe, the velocity is

$$v = \frac{\dot{V}}{A} = \frac{\left(350 \frac{\text{L}}{\text{s}}\right)\left(\frac{1 \text{ m}^3}{1000 \text{ L}}\right)}{322.7 \times 10^{-4} \text{ m}^2} = 10.85 \text{ m/s}$$

The Reynolds number is

$$\text{Re} = \frac{D v}{\nu} = \frac{(0.2027 \text{ m})\left(10.85 \frac{\text{m}}{\text{s}}\right)}{1.007 \times 10^{-6} \frac{\text{m}^2}{\text{s}}}$$

$$= 2.18 \times 10^6$$

The relative roughness is

$$\frac{\epsilon}{D} = \frac{6 \times 10^{-5} \text{ m}}{0.2027 \text{ m}} = 0.0003$$

From the Moody diagram, $f = 0.015$.

Equation 17.23(a) is used to calculate the frictional energy loss.

$$E_{f,1} = h_f g = \frac{f L v^2}{2D}$$

$$= \frac{(0.015)(300 \text{ m})\left(10.85 \frac{\text{m}}{\text{s}}\right)^2}{(2)(0.2027 \text{ m})}$$

$$= 1307 \text{ J/kg}$$

For the outlet pipe, the velocity is

$$v = \frac{\dot{V}}{A} = \frac{\left(350 \frac{\text{L}}{\text{s}}\right)\left(\frac{1 \text{ m}^3}{1000 \text{ L}}\right)}{721.9 \times 10^{-4} \text{ m}^2} = 4.848 \text{ m/s}$$

The Reynolds number is

$$\text{Re} = \frac{D v}{\nu} = \frac{(0.3032 \text{ m})\left(4.848 \frac{\text{m}}{\text{s}}\right)}{1.007 \times 10^{-6} \frac{\text{m}^2}{\text{s}}}$$

$$= 1.46 \times 10^6$$

The relative roughness is

$$\frac{\epsilon}{D} = \frac{6 \times 10^{-5} \text{ m}}{0.3032 \text{ m}} = 0.0002$$

From the Moody diagram, $f = 0.014$.

Equation 17.23(a) is used to calculate the frictional energy loss.

$$E_{f,2} = h_f g = \frac{f L v^2}{2D}$$

$$= \frac{(0.014)(450 \text{ m})\left(4.848 \frac{\text{m}}{\text{s}}\right)^2}{(2)(0.3032 \text{ m})}$$

$$= 244.2 \text{ J/kg}$$

Assume a 50% split through the two branches. In the upper branch, the velocity is

$$v = \frac{\dot{V}}{A} = \frac{\left(\frac{1}{2}\right)\left(350 \frac{\text{L}}{\text{s}}\right)\left(\frac{1 \text{ m}^3}{1000 \text{ L}}\right)}{322.7 \times 10^{-4} \text{ m}^2} = 5.423 \text{ m/s}$$

The Reynolds number is

$$\text{Re} = \frac{D v}{\nu} = \frac{(0.2027 \text{ m})\left(5.423 \frac{\text{m}}{\text{s}}\right)}{1.007 \times 10^{-6} \frac{\text{m}^2}{\text{s}}}$$

$$= 1.1 \times 10^6$$

The relative roughness is

$$\frac{\epsilon}{D} = \frac{6 \times 10^{-5} \text{ m}}{0.2027 \text{ m}} = 0.0003$$

From the Moody diagram, $f = 0.015$.

For the 16 in pipe in the lower branch, the velocity is

$$v = \frac{\dot{V}}{A} = \frac{\left(\frac{1}{2}\right)\left(350 \frac{\text{L}}{\text{s}}\right)\left(\frac{1 \text{ m}^3}{1000 \text{ L}}\right)}{1104 \times 10^{-4} \text{ m}^2} = 1.585 \text{ m/s}$$

The Reynolds number is

$$\text{Re} = \frac{D v}{\nu} = \frac{(0.381 \text{ m})\left(1.585 \frac{\text{m}}{\text{s}}\right)}{1.007 \times 10^{-6} \frac{\text{m}^2}{\text{s}}}$$

$$= 6.00 \times 10^5$$

The relative roughness is

$$\frac{\epsilon}{D} = \frac{6 \times 10^{-5} \text{ m}}{0.381 \text{ m}} = 0.00016$$

From the Moody diagram, $f = 0.015$.

These values of f for the two branches are fairly insensitive to changes in $\dot{V}$, so they will be used for the rest of the problem in the upper branch.

Equation 17.23(a) is used to calculate the frictional energy loss in the upper branch.

$$E_{f,\text{upper}} = h_f g = \frac{fL\text{v}^2}{2D}$$

$$= \frac{(0.015)(150 \text{ m})\left(5.423 \dfrac{\text{m}}{\text{s}}\right)^2}{(2)(0.2027 \text{ m})}$$

$$= 163.2 \text{ J/kg}$$

To calculate a loss for any other flow in the upper branch,

$$E_{f,\text{upper 2}} = E_{f,\text{upper}}\left(\frac{\dot{V}}{\left(\frac{1}{2}\right)\left(0.350 \dfrac{\text{m}^3}{\text{s}}\right)}\right)^2$$

$$= \left(163.2 \dfrac{\text{J}}{\text{kg}}\right)\left(\frac{\dot{V}}{0.175 \dfrac{\text{m}^3}{\text{s}}}\right)^2$$

$$= 5329\dot{V}^2$$

Similarly, for the lower branch, in the 8 in section,

$$E_{f,\text{lower, 8 in}} = \frac{(0.015)(75 \text{ m})\left(5.423 \dfrac{\text{m}}{\text{s}}\right)^2}{(2)(0.2027 \text{ m})}$$

$$= 81.61 \text{ J/kg}$$

For the lower branch, in the 16 in section,

$$E_{f,\text{lower,16 in}} = \frac{(0.015)(300 \text{ m})\left(1.585 \dfrac{\text{m}}{\text{s}}\right)^2}{(2)(0.381 \text{ m})}$$

$$= 14.84 \text{ J/kg}$$

The total loss in the lower branch is

$$E_{f,\text{lower}} = E_{f,\text{lower,8 in}} + E_{f,\text{lower,16 in}}$$

$$= 81.61 \dfrac{\text{J}}{\text{kg}} + 14.84 \dfrac{\text{J}}{\text{kg}}$$

$$= 96.45 \text{ J/kg}$$

To calculate a loss for any other flow in the lower branch,

$$E_{f,\text{lower 2}} = E_{f,\text{lower}}\left(\frac{\dot{V}}{\left(\frac{1}{2}\right)\left(0.350 \dfrac{\text{m}^3}{\text{s}}\right)}\right)^2$$

$$= \left(96.45 \dfrac{\text{J}}{\text{kg}}\right)\left(\frac{\dot{V}}{0.175 \dfrac{\text{m}^3}{\text{s}}}\right)^2$$

$$= 3149\dot{V}^2$$

Let x be the fraction flowing in the upper branch. Then, because the friction losses are equal,

$$E_{f,\text{upper 2}} = E_{f,\text{lower 2}}$$

$$5329x^2 = (3149)(1-x)^2$$

$$x = 0.435$$

(a) $$\dot{V}_{\text{upper}} = (0.435)\left(0.350 \dfrac{\text{m}^3}{\text{s}}\right)$$

$$= \boxed{0.152 \text{ m}^3/\text{s}}$$

The answer is (D).

(b) $$\dot{V}_{\text{lower}} = (1 - 0.435)\left(0.350 \dfrac{\text{m}^3}{\text{s}}\right)$$

$$= 0.198 \text{ m}^3/\text{s}$$

$$E_{f,\text{total}} = E_{f,1} + E_{f,\text{lower 2}} + E_{f,2}$$

$$E_{f,\text{lower 2}} = 3149\dot{V}_{\text{lower}}^2$$

$$= (3149)\left(0.198 \dfrac{\text{m}^3}{\text{s}}\right)^2$$

$$= 123.5 \text{ J/kg}$$

$$E_{f,\text{total}} = 1307 \dfrac{\text{J}}{\text{kg}} + 123.5 \dfrac{\text{J}}{\text{kg}} + 244.2 \dfrac{\text{J}}{\text{kg}}$$

$$= \boxed{1675 \text{ J/kg } (1.7 \text{ kJ/kg})}$$

The answer is (D).

10. Use the Hardy-Cross method, *steps 1, 2, and 3:*

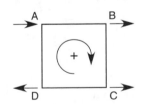

step 4: There is only one loop: ABCD.

step 5: pipe AB: $\quad K' = \dfrac{(10.44)(1000 \text{ ft})}{(100)^{1.85}(8 \text{ in})^{4.87}}$

$$= 8.33 \times 10^{-5}$$

pipe BC: $\quad K' = \dfrac{(10.44)(1000 \text{ ft})}{(100)^{1.85}(6 \text{ in})^{4.87}}$

$$= 3.38 \times 10^{-4}$$

pipe CD: $\quad K' = 8.33 \times 10^{-5}$ [same as AB]

pipe DA: $\quad K' = \dfrac{(10.44)(1000 \text{ ft})}{(100)^{1.85}(12 \text{ in})^{4.87}}$

$$= 1.16 \times 10^{-5}$$

Fluids

step 6:   Assume the flows are as shown in the figure.

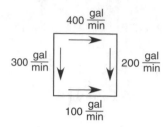

step 7:

$$\delta = \left(\frac{-1}{1.85}\right)$$

$$\times \left(\frac{\begin{array}{c}(8.33 \times 10^{-5})(400)^{1.85} + (3.38 \times 10^{-4})(200)^{1.85} \\ - (8.33 \times 10^{-5})(100)^{1.85} \\ -(1.16 \times 10^{-5})(300)^{1.85}\end{array}}{\begin{array}{c}(8.33 \times 10^{-5})(400)^{0.85} + (3.38 \times 10^{-4})(200)^{0.85} \\ + (8.33 \times 10^{-5})(100)^{0.85} \\ +(1.16 \times 10^{-5})(300)^{0.85}\end{array}} \right)$$

$$= \left(\frac{-1}{1.85}\right)\left(\frac{10.67}{4.98 \times 10^{-2}}\right) = -116 \text{ gal/min}$$

step 8:   The adjusted flows are shown.

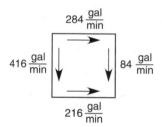

step 7: $\delta = -24$ gal/min

step 8:   The adjusted flows are shown.

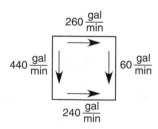

step 7: $\delta = -2$ gal/min   [small enough]

step 8:   The final adjusted flows are shown.

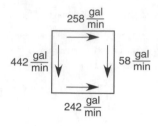

The answer is (B).

11. (a) This is a Hardy Cross problem. The pressure at point C does not change the solution procedure.

step 1:   The Hazen-Williams roughness coefficient is given.

step 2:   Choose clockwise as positive.

step 3:   Nodes are already numbered.

step 4:   Choose the loops as shown.

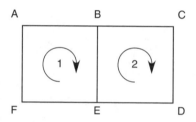

step 5: $\dot{V}$ is in gal/min.
d is in inches.
L is in feet.

Each pipe has the same length.

$$K'_{8 \text{ in}} = \frac{(10.44)(1000 \text{ ft})}{(100)^{1.85}(8 \text{ in})^{4.87}} = 8.33 \times 10^{-5}$$

$$K'_{6 \text{ in}} = \frac{(10.44)(1000 \text{ ft})}{(100)^{1.85}(6 \text{ in})^{4.87}} = 3.38 \times 10^{-4}$$

$$K'_{CE} = 2K'_{6 \text{ in}} = 6.76 \times 10^{-4}$$

$$K'_{EA} = K'_{6 \text{ in}} + K'_{8 \text{ in}} = 4.21 \times 10^{-4}$$

step 6:   Assume the flows shown.

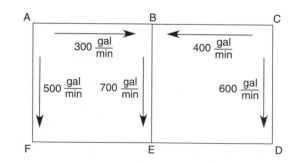

$$AB = 300 \text{ gal/min}$$
$$BE = 700 \text{ gal/min}$$
$$AE = 500 \text{ gal/min}$$
$$CE = 600 \text{ gal/min}$$
$$CB = 400 \text{ gal/min}$$

step 7: If the elevations are included as part of the head loss,

$$\Sigma h = \Sigma K' \dot{V}_a^n + \delta \Sigma n K' \dot{V}_a^{n-1} + z_2 - z_1 = 0$$

However, since the loop closes on itself, $z_2 = z_1$, and the elevations can be omitted.

- First iteration:

Loop 1:

$$\delta_1 = \frac{\begin{array}{c} -\left((8.33 \times 10^{-5})(300)^{1.85} \right. \\ +(8.33 \times 10^{-5})(700)^{1.85} \\ \left. -(4.21 \times 10^{-4})(500)^{1.85}\right) \end{array}}{(1.85)\left((8.33 \times 10^{-5})(300)^{0.85} \right. \\ +(8.33 \times 10^{-5})(700)^{0.85} \\ \left. +(4.21 \times 10^{-4})(500)^{0.85}\right)}$$

$$= \frac{-(-22.97)}{0.213} = +108 \text{ gal/min}$$

Loop 2:

$$\delta_2 = \frac{\begin{array}{c} -\left((6.76 \times 10^{-4})(600)^{1.85} \right. \\ -(8.33 \times 10^{-5})(700)^{1.85} \\ \left. -(8.33 \times 10^{-5})(400)^{1.85}\right) \end{array}}{(1.85)\left((6.76 \times 10^{-4})(600)^{0.85} \right. \\ +(8.33 \times 10^{-5})(700)^{0.85} \\ \left. +8.33 \times 10^{-5})(400)^{0.85}\right)}$$

$$= \frac{-(72.5)}{0.353} = -205 \text{ gal/min}$$

- Second iteration:

AB:	300 +	108		= 408 gal/min
BE:	700 +	108 − (−205)		= 1013 gal/min
AE:	500 −	108		= 392 gal/min
CE:	600 + (−205)			= 395 gal/min
CB:	400 − (−205)			= 605 gal/min

Loop 1:

$$\delta_1 = \frac{-(9.48)}{0.205} = -46 \text{ gal/min}$$

Loop 2:

$$\delta_2 = \frac{-(1.08)}{0.292} = -3.7 \text{ gal/min} \quad [\text{round to } -4]$$

- Third iteration:

AB:	408 + (−46)		= 362 gal/min
BE:	1013 + (−46) −(−4)	= 971 gal/min
AE:	392 − (−46)		= 438 gal/min
CE:	395 + (−4)		= 391 gal/min
CB:	605 − (−4)		= 609 gal/min

Loop 1:

$$\delta_1 = \frac{-(0.066)}{0.213} = -0.31 \text{ gal/min} \quad [\text{round to } 0]$$

Loop 2:

$$\delta_2 = \frac{-2.4}{0.29} = -8.3 \text{ gal/min} \quad [\text{round to } -8]$$

Use the following flows.

AB:	362 + 0	=	362 gal/min
BE:	971 + 0 − (−8) =		979 gal/min
AE:	438 − 0	=	438 gal/min
CE:	391 + (−8)	=	383 gal/min
CB:	609 − (−8)	=	617 gal/min

The answer is (D).

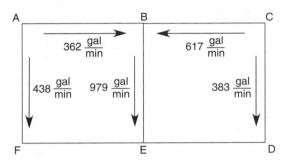

(b) The friction loss in each section is

$$h_{f,AB} = (8.33 \times 10^{-5})(362)^{1.85} = 4.5 \text{ ft}$$
$$h_{f,BE} = (8.33 \times 10^{-5})(979)^{1.85} = 28.4 \text{ ft}$$
$$h_{f,AF} = (8.33 \times 10^{-5})(438)^{1.85} = 6.4 \text{ ft}$$
$$h_{f,FE} = (3.38 \times 10^{-4})(438)^{1.85} = 26.0 \text{ ft}$$
$$h_{f,CD} = h_{f,DE} = (3.38 \times 10^{-4})(383)^{1.85} = 20.3 \text{ ft}$$
$$h_{f,CB} = (8.33 \times 10^{-5})(617)^{1.85} = 12.1 \text{ ft}$$

$$\gamma = 62.37 \text{ lbf/ft}^3 \quad [\text{at } 60°F]$$

The pressure at C is 40 psig.

$$h_C = \frac{\left(40 \dfrac{\text{lbf}}{\text{in}^2}\right)\left(144 \dfrac{\text{in}^2}{\text{ft}^2}\right)}{62.37 \dfrac{\text{lbf}}{\text{ft}^3}} = 92.4 \text{ ft}$$

Next, use the energy continuity equation between adjacent points.

$$h = h_C + z_C - z - h_f$$

$$h_D = 92.4 \text{ ft} + 300 \text{ ft} - 150 \text{ ft} - 20.3 \text{ ft} = 222.1 \text{ ft}$$

$$h_E = 222.1 \text{ ft} + 150 \text{ ft} - 200 \text{ ft} - 20.3 \text{ ft} = 151.8 \text{ ft}$$

$$h_F = 151.8 \text{ ft} + 200 \text{ ft} - 150 \text{ ft} + 26 \text{ ft} = 227.8 \text{ ft}$$

$$h_B = 92.4 \text{ ft} + 300 \text{ ft} - 150 \text{ ft} - 12.1 \text{ ft} = 230.3 \text{ ft}$$

$$h_A = 230.3 \text{ ft} + 150 \text{ ft} - 200 \text{ ft} + 4.5 \text{ ft} = 184.8 \text{ ft}$$

Using $p = \gamma h$,

$$p_A = \frac{\left(62.37 \, \frac{\text{lbf}}{\text{ft}^3}\right)(184.8 \text{ ft})}{144 \, \frac{\text{in}^2}{\text{ft}^2}} = 80.0 \text{ lbf/in}^2 \text{ (psig)}$$

$$p_B = \frac{\left(62.37 \, \frac{\text{lbf}}{\text{ft}^3}\right)(230.3 \text{ ft})}{144 \, \frac{\text{in}^2}{\text{ft}^2}} = 99.7 \text{ psig}$$

$$p_C = 40 \text{ psig} \quad \text{[given]}$$

$$p_D = \frac{\left(62.37 \, \frac{\text{lbf}}{\text{ft}^3}\right)(222.1 \text{ ft})}{144 \, \frac{\text{in}^2}{\text{ft}^2}} = \boxed{96.2 \text{ psig}}$$

$$p_E = \frac{\left(62.37 \, \frac{\text{lbf}}{\text{ft}^3}\right)(151.8 \text{ ft})}{144 \, \frac{\text{in}^2}{\text{ft}^2}} = 65.7 \text{ psig}$$

$$p_F = \frac{\left(62.37 \, \frac{\text{lbf}}{\text{ft}^3}\right)(227.8 \text{ ft})}{144 \, \frac{\text{in}^2}{\text{ft}^2}} = 98.7 \text{ psig}$$

The answer is (C).

(c) The pressure increase across the pump is

$$\left(80 \, \frac{\text{lbf}}{\text{in}^2} - 20 \, \frac{\text{lbf}}{\text{in}^2}\right)\left(144 \, \frac{\text{in}^2}{\text{ft}^2}\right) = 8640 \text{ lbf/ft}^2$$

The hydraulic horsepower is

$$P = \frac{\left(8640 \, \frac{\text{lbf}}{\text{ft}^2}\right)\left(800 \, \frac{\text{gal}}{\text{min}}\right)}{2.468 \times 10^5 \, \frac{\text{lbf-gal}}{\text{min-ft}^2\text{-hp}}} = \boxed{28 \text{ hp}}$$

The answer is (B).

12. This could be solved as a parallel pipe problem or as a pipe network problem.

(a) • Pipe network solution:

step 1: Use the Darcy equation since Hazen-Williams coefficients are not given (or assume C-values based on the age of the pipe).

step 2: Clockwise is positive.

step 3: All nodes are lettered.

step 4: There is only one loop.

step 5: $\epsilon \approx 0.0008$ ft. Assume full turbulence. For class A cast-iron pipe, $D_i = 10.1$ in.

$$\frac{\epsilon}{D} = \frac{0.0008 \text{ ft}}{\dfrac{10.1 \text{ in}}{12 \, \frac{\text{in}}{\text{ft}}}} \approx 0.001$$

$f \approx 0.020$ for full turbulence.

$$K'_{AB} = (1.251 \times 10^{-7})\left(\frac{(0.02)(2000 \text{ ft})}{\left(\dfrac{10.1 \text{ in}}{12 \, \frac{\text{in}}{\text{ft}}}\right)^5}\right) = 11.8 \times 10^{-6}$$

$$K'_{BD} = (1.251 \times 10^{-7})\left(\frac{(0.02)(1000 \text{ ft})}{\left(\dfrac{10.1 \text{ in}}{12 \, \frac{\text{in}}{\text{ft}}}\right)^5}\right) = 5.92 \times 10^{-6}$$

$$K'_{DC} = (1.251 \times 10^{-7})\left(\frac{(0.02)(1500 \text{ ft})}{\left(\dfrac{8.13 \text{ in}}{12 \, \frac{\text{in}}{\text{ft}}}\right)^5}\right) = 26.3 \times 10^{-6}$$

$$K'_{CA} = (1.251 \times 10^{-7})\left(\frac{(0.02)(1000 \text{ ft})}{\left(\dfrac{12.12 \text{ in}}{12 \, \frac{\text{in}}{\text{ft}}}\right)^5}\right) = 2.38 \times 10^{-6}$$

step 6: Assume $\dot{V}_{AB} = 1$ MGD.

$$\dot{V}_{ACDB} = 0.5 \text{ MGD}$$

Convert $\dot{V}$ to gal/min.

$$\dot{V}_{AB} = \frac{(1 \text{ MGD})(1 \times 10^6)}{\left(24 \dfrac{\text{hr}}{\text{day}}\right)\left(60 \dfrac{\text{min}}{\text{hr}}\right)} = 694 \text{ gal/min}$$

$$\dot{V}_{ACDB} = \frac{(0.5 \text{ MGD})(1 \times 10^6)}{\left(24 \dfrac{\text{hr}}{\text{day}}\right)\left(60 \dfrac{\text{min}}{\text{hr}}\right)} = 347 \text{ gal/min}$$

step 7: There is only one loop.

$$694 \frac{\text{gal}}{\text{min}} = (2)\left(347 \frac{\text{gal}}{\text{min}}\right)$$

$$\left(694 \frac{\text{gal}}{\text{min}}\right)^2 = (4)\left(347 \frac{\text{gal}}{\text{min}}\right)^2$$

$$\delta = \frac{(-1)(1 \times 10^{-6})(347)^2\big((11.8)(4) - 5.92 - 26.3 - 2.38\big)}{(2)(1 \times 10^{-6})(347)\big((11.8)(2) + 5.92 + 26.3 + 2.38\big)}$$

$$= -37.6 \text{ gal/min} \quad [\text{use } -38 \text{ gal/min}]$$

step 8: $\dot{V}_{AB} = 694 \dfrac{\text{gal}}{\text{min}} + \left(-38 \dfrac{\text{gal}}{\text{min}}\right)$

$$= 656 \text{ gal/min}$$

$$\dot{V}_{ACDB} = 347 \frac{\text{gal}}{\text{min}} - \left(-38 \frac{\text{gal}}{\text{min}}\right)$$

$$= 385 \text{ gal/min}$$

Repeat step 7.

$$656 \frac{\text{gal}}{\text{min}} = (1.70)\left(385 \frac{\text{gal}}{\text{min}}\right)$$

$$\left(656 \frac{\text{gal}}{\text{min}}\right)^2 = (2.90)\left(385 \frac{\text{gal}}{\text{min}}\right)^2$$

$$\delta = \frac{\begin{aligned}&(-1)(1 \times 10^{-6})(385)^2 \\ &\times \big((11.8)(2.90) - 5.92 - 26.3 - 2.38\big)\end{aligned}}{\begin{aligned}&(2)(1 \times 10^{-6})(385) \\ &\times \big((12.5)(1.70) + 5.92 + 26.3 + 2.38\big)\end{aligned}}$$

$$\approx 0$$

$$\boxed{\dot{V}_{AB} = 656 \text{ gal/min}}$$

$$\dot{V}_{ACDB} = 385 \text{ gal/min}$$

Check the Reynolds number in leg AB to verify that $f = 0.02$ was a good choice.

$$A_{10 \text{ in pipe}} = \left(\frac{\pi}{4}\right)\left(\frac{10.1 \text{ in}}{12 \dfrac{\text{in}}{\text{ft}}}\right)^2 = 0.5564 \text{ ft}^2 \ [\text{cast-iron pipe}]$$

The flow rate is

$$\left(656 \frac{\text{gal}}{\text{min}}\right)\left(0.002228 \frac{\text{ft}^3\text{-min}}{\text{sec-gal}}\right) = 1.46 \text{ ft}^3/\text{sec}$$

$$v = \frac{\dot{V}}{A} = \frac{1.46 \dfrac{\text{ft}^3}{\text{sec}}}{0.5564 \text{ ft}^2} = 2.62 \text{ ft/sec} \quad [\text{reasonable}]$$

For 50°F water,

$$\nu = 1.410 \times 10^{-5} \text{ ft}^2/\text{sec}$$

$$\text{Re} = \frac{\left(\dfrac{10.1 \text{ in}}{12 \dfrac{\text{in}}{\text{ft}}}\right)\left(2.62 \dfrac{\text{ft}}{\text{sec}}\right)}{1.410 \times 10^{-5} \dfrac{\text{ft}^2}{\text{sec}}} = 1.56 \times 10^5 \ [\text{turbulent}]$$

The answer is (D).

• Alternative closed-form solution:

Use the Darcy equation. Assume $\epsilon = 0.0008$ ft. The relative roughness is

$$\frac{\epsilon}{D} = \frac{0.0008 \text{ ft}}{\dfrac{10.1 \text{ in}}{12 \dfrac{\text{in}}{\text{ft}}}} = 0.00095 \quad [\text{use } 0.001]$$

Assume $v_{\max} = 5$ ft/sec and 50°F temperature. The Reynolds number is

$$\text{Re} = \frac{vD}{\nu} = \frac{\left(5 \dfrac{\text{ft}}{\text{sec}}\right)\left(\dfrac{10.1 \text{ in}}{12 \dfrac{\text{in}}{\text{ft}}}\right)}{1.41 \times 10^{-5} \dfrac{\text{ft}^2}{\text{sec}}} = 2.98 \times 10^5$$

From the Moody diagram, $f \approx 0.0205$.

$$h_{f,AB} = \frac{fLv^2}{2Dg}$$

$$= \frac{(0.0205)(2000 \text{ ft})\dot{V}_{AB}^2}{(2)\left(\dfrac{10.1 \text{ in}}{12 \dfrac{\text{in}}{\text{ft}}}\right)(0.556 \text{ ft}^2)^2\left(32.2 \dfrac{\text{ft}}{\text{sec}^2}\right)}$$

$$= 2.446\dot{V}_{AB}^2$$

$$h_{f,\text{ACDB}} = \frac{(0.0205)(1000)\dot{V}^2_{\text{ACDB}}}{(2)\left(\dfrac{12.12}{12}\right)(0.801)^2(32.2)}$$

$$+ \frac{(0.0205)(1500)\dot{V}^2_{\text{ACDB}}}{(2)\left(\dfrac{8.13}{12}\right)(0.360)^2(32.2)}$$

$$+ \frac{(0.0205)(1000)\dot{V}^2_{\text{ACDB}}}{(2)\left(\dfrac{10.1}{12}\right)(0.556 \text{ ft}^2)^2(32.2)}$$

$$= 0.4912\dot{V}^2_{\text{ACDB}} + 5.438\dot{V}^2_{\text{ACDB}} + 1.223\dot{V}^2_{\text{ACDB}}$$

$$= 7.152\dot{V}^2_{\text{ACDB}}$$

$$h_{f,\text{AB}} = h_{f,\text{ACDB}}$$

$$2.446\dot{V}^2_{\text{AB}} = 7.152\dot{V}^2_{\text{ACDB}}$$

$$\dot{V}_{\text{AB}} = \sqrt{\frac{7.152}{2.446}}\,\dot{V}_{\text{ACDB}} = 1.71\dot{V}_{\text{ACDB}} \qquad \text{[Eq. 1]}$$

The total flow rate is

$$\frac{1.5 \text{ MGD}}{\left(24\,\dfrac{\text{hr}}{\text{day}}\right)\left(60\,\dfrac{\text{min}}{\text{hr}}\right)} = 1041.7 \text{ gal/min}$$

$$\dot{V}_{\text{AB}} + \dot{V}_{\text{ACDB}} = 1041.7 \text{ gal/min} \quad \text{[Eq. 2]}$$

Solving Eqs. 1 and 2 simultaneously,

$$\dot{V}_{\text{AB}} = \boxed{657.3 \text{ gal/min}}$$

$$\dot{V}_{\text{ACDB}} = 384.4 \text{ gal/min}$$

The answer is (D).

This answer is insensitive to the $v_{\text{max}} = 5$ ft/sec assumption. A second iteration using actual velocities from these flow rates does not change the answer.

• The same technique can be used with the Hazen-Williams equation and an assumed value of C. If $C = 100$ is used, then

$$\dot{V}_{\text{AB}} = \boxed{725.4 \text{ gal/min}}$$

$$\dot{V}_{\text{ACDB}} = 316.3 \text{ gal/min}$$

(b) $f \approx 0.021$.

$$h_{f,\text{AB}} = \frac{(0.021)(2000 \text{ ft})\left(2.62\,\dfrac{\text{ft}}{\text{sec}}\right)^2}{(2)\left(\dfrac{10.1 \text{ in}}{12\,\dfrac{\text{in}}{\text{ft}}}\right)\left(32.2\,\dfrac{\text{ft}}{\text{sec}^2}\right)} = 5.32 \text{ ft}$$

For leg BD, use $f = 0.021$ (assumed).

$$v = \frac{\dot{V}}{A} = \frac{\left(385\,\dfrac{\text{gal}}{\text{min}}\right)\left(0.002228\,\dfrac{\text{ft}^3\text{-min}}{\text{sec-gal}}\right)}{0.5564 \text{ ft}^2}$$

$$= 1.54 \text{ ft/sec}$$

$$h_{f,\text{DB}} = \frac{(0.021)(1000 \text{ ft})\left(1.54\,\dfrac{\text{ft}}{\text{sec}}\right)^2}{(2)\left(\dfrac{10 \text{ in}}{12\,\dfrac{\text{in}}{\text{ft}}}\right)\left(32.2\,\dfrac{\text{ft}}{\text{sec}^2}\right)} = 0.9 \text{ ft}$$

At 50°F, $\gamma = 62.4$ lbf/ft^3. From the Bernoulli equation (omitting the velocity term),

$$\frac{p_{\text{B}}}{\gamma} + z_{\text{B}} + h_{f,\text{DB}} = \frac{p_{\text{D}}}{\gamma} + z_{\text{D}}$$

$$p_{\text{B}} = \left(\frac{62.4\,\dfrac{\text{lbf}}{\text{ft}^3}}{144\,\dfrac{\text{in}^2}{\text{ft}^2}}\right)$$

$$\times \left(\frac{\left(40\,\dfrac{\text{lbf}}{\text{in}^2}\right)\left(144\,\dfrac{\text{in}^2}{\text{ft}^2}\right)}{62.4\,\dfrac{\text{lbf}}{\text{ft}^3}} + 600 \text{ ft} - 0.9 \text{ ft} - 580 \text{ ft}\right)$$

$$= \boxed{48.3 \text{ lbf/in}^2 \text{ (psi)}}$$

The answer is (B).

$$\frac{p_{\text{A}}}{\gamma} + z_{\text{A}} = \frac{p_{\text{B}}}{\gamma} + z_{\text{B}} + h_{f,\text{AB}}$$

$$p_{\text{A}} = \left(\frac{62.4\,\dfrac{\text{lbf}}{\text{ft}^3}}{144\,\dfrac{\text{in}^2}{\text{ft}^2}}\right)$$

$$\times \left(\frac{\left(48.3\,\dfrac{\text{lbf}}{\text{in}^2}\right)\left(144\,\dfrac{\text{in}^2}{\text{ft}^2}\right)}{62.4\,\dfrac{\text{lbf}}{\text{ft}^3}}\right.$$

$$\left. + 580 \text{ ft} + 5.32 \text{ ft} - 540 \text{ ft}\right)$$

$$= 67.9 \text{ lbf/in}^2 \text{ (psi)}$$

13. *Customary U.S. Solution*

Use projectile equations.

The maximum range of the discharge is given by

$$R = v_o^2 \left(\frac{\sin 2\phi}{g} \right)$$

$$= \left(50 \frac{\text{ft}}{\text{sec}} \right)^2 \left(\frac{\sin ((2)(45°))}{32.2 \frac{\text{ft}}{\text{sec}^2}} \right)$$

$$= \boxed{77.64 \text{ ft}}$$

The answer is (B).

SI Solution

Use projectile equations.

The maximum range of the discharge is given by

$$R = v_o^2 \left(\frac{\sin 2\phi}{g} \right)$$

$$= \frac{\left(15 \frac{\text{m}}{\text{s}} \right)^2 \sin ((2)(45°))}{9.81 \frac{\text{m}}{\text{s}^2}}$$

$$= \boxed{22.94 \text{ m}}$$

The answer is (B).

14. $A_o = \left(\frac{\pi}{4} \right) \left(\frac{4 \text{ in}}{12 \frac{\text{in}}{\text{ft}}} \right)^2 = 0.08727 \text{ ft}^2$

$A_t = \left(\frac{\pi}{4} \right) (20 \text{ ft})^2 = 314.16 \text{ ft}^2$

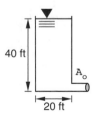

The time to drop from 40 ft to 20 ft is

$$t = \frac{(2)(314.16 \text{ ft}^2) \left(\sqrt{40 \text{ ft}} - \sqrt{20 \text{ ft}} \right)}{(0.98)(0.08727 \text{ ft}^2) \sqrt{(2) \left(32.2 \frac{\text{ft}}{\text{sec}^2} \right)}}$$

$$= \boxed{1696 \text{ sec}}$$

The answer is (D).

15. $C_d = 1.00$ [given]

$$F_{\text{va}} = \frac{1}{\sqrt{1 - \left(\frac{D_2}{D_1} \right)^4}} = \frac{1}{\sqrt{1 - \left(\frac{8 \text{ in}}{12 \text{ in}} \right)^4}}$$

$$= 1.116$$

$$A_2 = \left(\frac{\pi}{4} \right) \left(\frac{8 \text{ in}}{12 \frac{\text{in}}{\text{ft}}} \right)^2 = 0.3491 \text{ ft}^2$$

$$p_1 - p_2 = \left(\left(0.491 \frac{\text{lbf}}{\text{in}^3} \right) (4 \text{ in}) \right.$$

$$\left. - \left(0.0361 \frac{\text{lbf}}{\text{in}^3} \right) (4 \text{ in}) \right) \left(144 \frac{\text{in}^2}{\text{ft}^2} \right)$$

$$= 262.0 \text{ lbf/ft}^2$$

$$Q = F_{\text{va}} C_d A_2 \sqrt{\frac{2g(p_1 - p_2)}{\gamma}}$$

$$= (1.116)(1)(0.3491 \text{ ft}^2)$$

$$\times \sqrt{\frac{(2) \left(32.2 \frac{\text{ft}}{\text{sec}^2} \right) \left(262 \frac{\text{lbf}}{\text{ft}^2} \right)}{62.4 \frac{\text{lbf}}{\text{ft}^3}}}$$

$$= \boxed{6.406 \text{ ft}^3/\text{sec (cfs)}}$$

The answer is (C).

16. *Customary U.S. Solution*

The volumetric flow rate of benzene through the venturi meter is given by

$$\dot{V} = C_f A_2 \sqrt{\frac{2g(\rho_m - \rho)h}{\rho}}$$

The density of mercury, ρ_m, at 60°F is approximately 848 lbm/ft³.

The density of the benzene at 60°F is

$$\rho = (\text{SG}) \rho_{\text{water}}$$

$$= (0.885) \left(62.4 \frac{\text{lbm}}{\text{ft}^3} \right)$$

$$= 55.22 \text{ lbm/ft}^3$$

The throat area is

$$A_2 = \left(\frac{\pi}{4} \right) D_2^2$$

$$= \left(\frac{\pi}{4} \right) (3.5 \text{ in})^2 \left(\frac{1 \text{ ft}^2}{144 \text{ in}^2} \right) = 0.0668 \text{ ft}^2$$

The flow coefficient is defined as

$$C_f = \frac{C_d}{\sqrt{1 - \beta^4}}$$

β is the ratio of the throat to inlet diameters.

$$\beta = \frac{3.5 \text{ in}}{8 \text{ in}}$$
$$= 0.4375$$

$$C_f = \frac{C_d}{\sqrt{1 - \beta^4}}$$
$$= \frac{0.99}{\sqrt{1 - (0.4375)^4}}$$
$$= 1.00865$$

Find the volumetric flow of benzene.

$$\dot{V} = C_f A_2 \sqrt{\frac{2g(\rho_m - \rho)h}{\rho}}$$

$$= (1.00865)(0.0668 \text{ ft}^2)$$

$$\times \sqrt{\frac{(2)\left(32.2 \dfrac{\text{ft}}{\text{sec}^2}\right)\left(848 \dfrac{\text{lbm}}{\text{ft}^3} - 55.22 \dfrac{\text{lbm}}{\text{ft}^3}\right)(4 \text{ in})\left(\dfrac{1 \text{ ft}}{12 \text{ in}}\right)}{55.22 \dfrac{\text{lbm}}{\text{ft}^3}}}$$

$$= \boxed{1.183 \text{ ft}^3/\text{sec}}$$

The answer is (A).

SI Solution

The volumetric flow rate of benzene through the venturi meter is given by

$$\dot{V} = C_f A_2 \sqrt{\frac{2g(\rho_m - \rho)h}{\rho}}$$

ρ_m is the density of mercury at 15°C; ρ_m is approximately $13\,600 \text{ kg/m}^3$.

The density of the benzene at 15°C is

$$\rho = (\text{SG})\rho_{\text{water}}$$
$$= (0.885)\left(1000 \dfrac{\text{kg}}{\text{m}^3}\right)$$
$$= 885 \text{ kg/m}^3$$

The throat area is

$$A_2 = \left(\frac{\pi}{4}\right) D_2^2$$
$$= \left(\frac{\pi}{4}\right)(0.09 \text{ m})^2$$
$$= 0.0064 \text{ m}^2$$

The flow coefficient is defined as

$$C_f = \frac{C_d}{\sqrt{1 - \beta^4}}$$

β is the ratio of the throat to inlet diameters.

$$\beta = \frac{9 \text{ cm}}{20 \text{ cm}}$$
$$= 0.45$$

$$C_f = \frac{C_d}{\sqrt{1 - \beta^4}}$$
$$= \frac{0.99}{\sqrt{1 - (0.45)^4}}$$
$$= 1.01094$$

Find the volumetric flow of benzene.

$$\dot{V} = C_f A_2 \sqrt{\frac{2g(\rho_m - \rho)h}{\rho}}$$

$$= (1.01094)(0.0064 \text{ m}^2)$$

$$\times \sqrt{\frac{(2)\left(9.81 \dfrac{\text{m}}{\text{s}^2}\right) \times \left(13\,600 \dfrac{\text{kg}}{\text{m}^3} - 885 \dfrac{\text{kg}}{\text{m}^3}\right)(0.1 \text{ m})}{885 \dfrac{\text{kg}}{\text{m}^3}}}$$

$$= \boxed{0.0344 \text{ m}^3/\text{s} \quad (34.4 \text{ L/s})}$$

The answer is (A).

17.
$$\Delta p = \left(\frac{\gamma}{2g}\right)\left(\frac{\dot{V}}{C_f A_o}\right)^2$$

For 70°F water,

$$\nu = 1.059 \times 10^{-5} \text{ ft}^2/\text{sec}$$
$$\gamma = 62.3 \text{ lbf/ft}^3$$

$$\text{Re}_{\text{pipe}} = \frac{D\text{v}}{\nu} = \frac{(1 \text{ ft})\left(2 \dfrac{\text{ft}}{\text{sec}}\right)}{1.059 \times 10^{-5} \dfrac{\text{ft}^2}{\text{sec}}} = 1.89 \times 10^5$$

$$\text{Re}_{\text{bore}} = \frac{\text{Re}_{\text{pipe}}}{\beta} = \frac{\text{Re}_{\text{pipe}} D_{\text{pipe}}}{D_{\text{bore}}}$$
$$= (1.89 \times 10^5)\left(\frac{1.0 \text{ ft}}{0.2 \text{ ft}}\right)$$
$$= 9.45 \times 10^5$$

$$A_o = \left(\frac{\pi}{4}\right)(0.2 \text{ ft})^2 = 0.0314 \text{ ft}^2$$

$$A_p = \left(\frac{\pi}{4}\right)(1\text{ ft})^2 = 0.7854\text{ ft}^2$$

$$\frac{A_o}{A_p} = \frac{0.0314\text{ ft}^2}{0.7854\text{ ft}^2} = 0.040$$

$$C_f \approx 0.58$$

$$\dot{V} = Av = (0.7854\text{ ft}^2)\left(2\ \frac{\text{ft}}{\text{sec}}\right) = 1.571\text{ ft}^3/\text{sec}$$

$$\Delta p = \left(\frac{62.3\ \frac{\text{lbf}}{\text{ft}^3}}{(2)\left(32.2\ \frac{\text{ft}}{\text{sec}^2}\right)}\right)\left(\frac{1.571\ \frac{\text{ft}^3}{\text{sec}}}{(0.58)(0.0314\text{ ft}^2)}\right)^2$$

$$= 7198\text{ lbf/ft}^2\text{ (psf)}$$

$$\Delta p = \frac{7198\ \frac{\text{lbf}}{\text{ft}^2}}{144\ \frac{\text{in}^2}{\text{ft}^2}} = \boxed{50.0\text{ lbf/in}^2\text{ (psi)}}$$

The answer is (D).

18. *Customary U.S. Solution*

The manometer tube is filled with water above the mercury column. The pressure differential across the orifice meter is given by

$$\Delta p = p_1 - p_2 = (\rho_{\text{mercury}} - \rho_{\text{water}})h \times \left(\frac{g}{g_c}\right)$$

The densities of mercury and water are

$$\rho_{\text{mercury}} = 848\text{ lbm/ft}^3$$

$$\rho_{\text{water}} = 62.4\text{ lbm/ft}^3$$

Substituting gives

$$\Delta p = (\rho_{\text{mercury}} - \rho_{\text{water}})h \times \left(\frac{g}{g_c}\right)$$

$$= \frac{\left(848\ \frac{\text{lbm}}{\text{ft}^3} - 62.4\ \frac{\text{lbm}}{\text{ft}^3}\right)(7\text{ in})}{32.2\ \frac{\text{lbm-ft}}{\text{lbf-sec}^2}}$$

$$\times \left(\frac{1\text{ ft}}{12\text{ in}}\right)\left(32.2\ \frac{\text{ft}}{\text{sec}^2}\right)$$

$$= \boxed{458.3\text{ lbf/ft}^2\quad(3.18\text{ psi})}$$

The answer is (B).

SI Solution

The manometer tube is filled with water above the mercury column. The pressure differential across the orifice meter is given by

$$\Delta p = p_1 - p_2 = (\rho_{\text{mercury}} - \rho_{\text{water}})hg$$

The densities of mercury and water are

$$\rho_{\text{mercury}} = 13\,600\text{ kg/m}^3$$

$$\rho_{\text{water}} = 1000\text{ kg/m}^3$$

Substituting gives

$$\Delta p = (\rho_{\text{mercury}} - \rho_{\text{water}})gh$$

$$= \left(13\,600\ \frac{\text{kg}}{\text{m}^3} - 1000\ \frac{\text{kg}}{\text{m}^3}\right)(0.178\text{ m})\left(9.81\ \frac{\text{m}}{\text{s}^2}\right)$$

$$= \boxed{22\,002\text{ Pa}\quad(22.0\text{ kPa})}$$

The answer is (B).

19. *Customary U.S. Solution*

For 12 in pipe (assuming schedule-40),

$$D_i = 0.99483\text{ ft}\quad[\text{App. 16.B}]$$

$$A_i = 0.7773\text{ ft}^2$$

The velocity is

$$v = \frac{\dot{V}}{A}$$

$$= \frac{10\ \frac{\text{ft}^3}{\text{sec}}}{0.7773\text{ ft}^2}$$

$$= 12.87\text{ ft/sec}$$

For water at 70°F, $\nu = 1.059 \times 10^{-5}\text{ ft}^2/\text{sec}$.

The orifice diameter is not yet known. The Reynolds number in the pipe is

$$\text{Re} = \frac{vD_i}{\nu}$$

$$= \frac{\left(12.87\ \frac{\text{ft}}{\text{sec}}\right)(0.99483\text{ ft})}{1.059 \times 10^{-5}\ \frac{\text{ft}^2}{\text{sec}}}$$

$$= 1.21 \times 10^6$$

This is fully turbulent. The Reynolds number in the orifice will be even higher.

The volumetric flow rate through a sharp-edged orifice is given by

$$\dot{V} = C_f A_o\sqrt{\frac{2g(\rho_m - \rho)h}{\rho}}$$

In terms of pressure,

$$\dot{V} = C_f A_o\sqrt{\frac{2g_c(p_1 - p_2)}{\rho}}$$

Rearranging gives

$$C_f A_o = \frac{\dot{V}}{\sqrt{\frac{2g_c(p_1 - p_2)}{\rho}}}$$

p_1 is the upstream pressure, and p_2 is the downstream pressure.

The maximum head loss must not exceed 25 ft; therefore,

$$\frac{\left(\frac{g_c}{g}\right) \times (p_1 - p_2)}{\rho} = 25 \text{ ft}$$

$$\frac{g_c(p_1 - p_2)}{\rho} = (25 \text{ ft})g$$

Substituting gives

$$C_f A_0 = \frac{10 \frac{\text{ft}^3}{\text{sec}}}{\sqrt{(2)\left(32.2 \frac{\text{ft}}{\text{sec}^2}\right)(25 \text{ ft})}}$$

$$= 0.249 \text{ ft}^2$$

Both C_f and A_o depend on the orifice diameter.

For a 7 in diameter orifice,

$$A_o = \frac{\pi D_o^2}{4} = \frac{\pi\left((7 \text{ in})\left(\frac{1 \text{ ft}}{12 \text{ in}}\right)\right)^2}{4}$$

$$= 0.267 \text{ ft}^2$$

$$\frac{A_o}{A_1} = \frac{0.267 \text{ ft}^2}{0.7768 \text{ ft}^2} = 0.344$$

From a chart of flow coefficients (Fig. 17.28), for $A_o/A_1 = 0.344$ and full turbulence,

$$C_f = 0.645$$

$$C_f A_o = (0.645)(0.267 \text{ ft}^2) = 0.172 \text{ ft}^2 < 0.249 \text{ ft}^2$$

Therefore, a 7 in diameter orifice is too small.

Try a 9 in diameter orifice.

$$A_o = \frac{\pi D_o^2}{4} = \frac{\pi\left((9 \text{ in})\left(\frac{1 \text{ ft}}{12 \text{ in}}\right)\right)^2}{4}$$

$$= 0.442 \text{ ft}^2$$

$$\frac{A_o}{A_1} = \frac{0.442 \text{ ft}^2}{0.7768 \text{ ft}^2} = 0.569$$

From Fig. 17.28, for $A_o/A_1 = 0.568$ and full turbulence,

$$C_f = 0.73$$

$$C_f A_o = (0.73)(0.442 \text{ ft}^2) = 0.323 \text{ ft}^2 > 0.249 \text{ ft}^2$$

Therefore, a 9 in orifice is too large.

Interpolating gives

$$D_o = 7 \text{ in} + \frac{(9 \text{ in} - 7 \text{ in})(0.249 \text{ ft}^2 - 0.172 \text{ ft}^2)}{0.323 \text{ ft}^2 - 0.172 \text{ ft}^2}$$

$$= 8.0 \text{ in}$$

Further iterations yield

$$D_o = \boxed{8.1 \text{ in}}$$

$$C_f A_o = 0.243 \text{ ft}^2$$

The answer is (C).

SI Solution

For 300 mm pipe (assume the nominal diameter is the inner diameter), $D_i = 0.30$ m.

The velocity is

$$\text{v} = \frac{\dot{V}}{A} = \frac{\dot{V}}{\frac{\pi D_i^2}{4}}$$

$$= \frac{\left(250 \frac{\text{L}}{\text{s}}\right)\left(\frac{1 \text{ m}^3}{1000 \text{ L}}\right)}{\frac{\pi(0.3 \text{ m})^2}{4}}$$

$$= 3.54 \text{ m/s}$$

From App. 14.B, for water at 20°C,

$$\nu = 1.007 \times 10^{-6} \text{ m}^2/\text{s}$$

The orifice diameter is not yet known. The Reynolds number in the pipe is

$$\text{Re} = \frac{\text{v}D_i}{\nu}$$

$$= \frac{\left(3.54 \frac{\text{m}}{\text{s}}\right)(0.3 \text{ m})}{1.007 \times 10^{-6} \frac{\text{m}^2}{\text{s}}}$$

$$= 1.05 \times 10^6$$

This is fully turbulent. The Reynolds number in the orifice will be even higher.

The volumetric flow rate through a sharp-edged orifice is given by

$$\dot{V} = C_f A_o \sqrt{\frac{2g(\rho_m - \rho)h}{\rho}}$$

In terms of pressure,

$$\dot{V} = C_f A_o \sqrt{\frac{2(p_1 - p_2)}{\rho}}$$

Rearranging gives

$$C_f A_o = \frac{\dot{V}}{\sqrt{\dfrac{2(p_1 - p_2)}{\rho}}}$$

p_1 is the upstream pressure, and p_2 is the downstream pressure.

The maximum head loss must not exceed 7.5 m; therefore,

$$\frac{p_1 - p_2}{g\rho} = 7.5 \text{ m}$$

$$\frac{p_1 - p_2}{\rho} = (7.5 \text{ m})g$$

Substituting gives

$$C_f A_o = \frac{0.25 \ \dfrac{\text{m}^3}{\text{s}}}{\sqrt{(2)\left(9.81 \ \dfrac{\text{m}}{\text{s}^2}\right)(7.5 \text{ m})}}$$

$$= 0.021 \text{ m}^2$$

Both C_f and A_o depend on the orifice diameter.

For an 18 cm diameter orifice,

$$A_o = \frac{\pi D_o^2}{4} = \frac{\pi (0.18 \text{ m})^2}{4} = 0.0254 \text{ m}^2$$

$$\frac{A_o}{A_1} = \frac{0.0254 \text{ m}^2}{0.0707 \text{ m}^2} = 0.359$$

From a chart of flow coefficients (Fig. 17.28), for $A_o/A_1 = 0.359$ and full turbulence,

$$C_f = 0.65$$

$$C_f A_o = (0.65)(0.0254 \text{ m}^2) = 0.0165 \text{ m}^2 < 0.021 \text{ m}^2$$

Therefore, an 18 cm diameter orifice is too small.

Try a 23 cm diameter orifice.

$$A_o = \frac{\pi D_o^2}{4} = \frac{\pi (0.23 \text{ m})^2}{4} = 0.0415 \text{ m}^2$$

$$\frac{A_o}{A_1} = \frac{0.0415 \text{ m}^2}{0.0707 \text{ m}^2} = 0.587$$

From Fig. 17.28, for $A_o/A_1 = 0.587$ and full turbulence,

$$C_f = 0.73$$

$$C_f A_o = (0.73)(0.0415 \text{ m}^2) = 0.0303 \text{ m}^2 > 0.021 \text{ m}^2$$

Therefore, a 23 cm orifice is too large.

Interpolating gives

$$D_o = 18 \text{ cm}$$

$$+ (23 \text{ cm} - 18 \text{ cm}) \left(\frac{0.021 \text{ m}^2 - 0.0165 \text{ m}^2}{0.0303 \text{ m}^2 - 0.0165 \text{ m}^2} \right)$$

$$= 19.6 \text{ cm}$$

Further iteration yields

$$D_o = \boxed{20.0 \text{ cm}}$$

$$C_f = 0.675$$

$$C_f A_o = 0.021 \text{ m}^2$$

The answer is (C).

20. $A_\text{A} = \left(\dfrac{\pi}{4}\right)\left(\dfrac{24 \text{ in}}{12 \ \frac{\text{in}}{\text{ft}}}\right)^2 = 3.142 \text{ ft}^2$

$A_\text{B} = \left(\dfrac{\pi}{4}\right)\left(\dfrac{12 \text{ in}}{12 \ \frac{\text{in}}{\text{ft}}}\right)^2 = 0.7854 \text{ ft}^2$

$\text{v}_\text{A} = \dfrac{\dot{V}}{A} = \dfrac{8 \ \frac{\text{ft}^3}{\text{sec}}}{3.142 \text{ ft}^2} = 2.546 \text{ ft/sec}$

$p_\text{A} = \gamma h = \left(62.4 \ \dfrac{\text{lbf}}{\text{ft}^3}\right)(20 \text{ ft}) = 1248 \text{ lbf/ft}^2$

Using the Bernoulli equation to solve for p_B,

$\text{v}_\text{B} = \dfrac{\dot{V}}{A} = \dfrac{8 \ \frac{\text{ft}^3}{\text{sec}}}{0.7854 \text{ ft}^2} = 10.19 \text{ ft/sec}$

$p_\text{B} = 1248 \ \dfrac{\text{lbf}}{\text{ft}^2} - \left(\dfrac{\left(10.19 \ \frac{\text{ft}}{\text{sec}}\right)^2 - \left(2.546 \ \frac{\text{ft}}{\text{sec}}\right)^2}{(2)\left(32.2 \ \frac{\text{ft}}{\text{sec}^2}\right)} \right)$

$\times \left(62.4 \ \dfrac{\text{lbf}}{\text{ft}^3}\right)$

$= 1153.67 \text{ lbf/ft}^2$

With $\phi = 0$,

$F_x = \left(1153.67 \ \dfrac{\text{lbf}}{\text{ft}^2}\right)(0.7854 \text{ ft}^2) - \left(1248 \ \dfrac{\text{lbf}}{\text{ft}^2}\right)(3.142 \text{ ft}^2)$

$+ \left(\dfrac{\left(8 \ \frac{\text{ft}^3}{\text{sec}}\right)\left(62.4 \ \frac{\text{lbf}}{\text{ft}^3}\right)}{32.2 \ \frac{\text{ft}}{\text{sec}^2}} \right)\left(10.19 \ \dfrac{\text{ft}}{\text{sec}} - 2.546 \ \dfrac{\text{ft}}{\text{sec}}\right)$

$= \boxed{-2897 \text{ lbf on the fluid (toward A)}}$

$F_y = 0$

The answer is (A).

21. *Customary U.S. Solution*

The mass flow rate of the water is

$$\dot{m} = \rho\dot{V} = \rho v A = \frac{\rho v \pi D^2}{4}$$

$$= \left(62.4 \ \frac{\text{lbm}}{\text{ft}^3}\right)\left(40 \ \frac{\text{ft}}{\text{sec}}\right)\left(\frac{\pi\left((2 \ \text{in})\left(\frac{1 \ \text{ft}}{12 \ \text{in}}\right)\right)^2}{4}\right)$$

$$= 54.45 \ \text{lbm/sec}$$

The effective mass flow rate of the water is

$$\dot{m}_{\text{eff}} = \left(\frac{v - v_b}{v}\right)\dot{m}$$

$$= \left(\frac{40 \ \dfrac{\text{ft}}{\text{sec}} - 15 \ \dfrac{\text{ft}}{\text{sec}}}{40 \ \dfrac{\text{ft}}{\text{sec}}}\right)\left(54.45 \ \frac{\text{lbm}}{\text{sec}}\right)$$

$$= 34.0 \ \text{lbm/sec}$$

The force in the (horizontal) x-direction is given by

$$F_x = \frac{\dot{m}_{\text{eff}}(v - v_b)(\cos\theta - 1)}{g_c}$$

$$= \frac{\left(34.0 \ \dfrac{\text{lbm}}{\text{sec}}\right)\left(40 \ \dfrac{\text{ft}}{\text{sec}} - 15 \ \dfrac{\text{ft}}{\text{sec}}\right)(\cos 60° - 1)}{32.2 \ \dfrac{\text{lbm-ft}}{\text{lbf-sec}^2}}$$

$$= -13.2 \ \text{lbf} \quad \text{[the force is acting to the left]}$$

The force in the (vertical) y-direction is given by

$$F_y = \frac{\dot{m}_{\text{eff}}(v - v_b)\sin\theta}{g_c}$$

$$= \frac{\left(34.0 \ \dfrac{\text{lbm}}{\text{sec}}\right)\left(40 \ \dfrac{\text{ft}}{\text{sec}} - 15 \ \dfrac{\text{ft}}{\text{sec}}\right)(\sin 60°)}{32.2 \ \dfrac{\text{lbm-ft}}{\text{lbf-sec}^2}}$$

$$= 22.9 \ \text{lbf} \quad \text{[the force is acting upward]}$$

The net resultant force is

$$F = \sqrt{F_x^2 + F_y^2}$$

$$= \sqrt{(-13.2 \ \text{lbf})^2 + (22.9 \ \text{lbf})^2}$$

$$= \boxed{26.4 \ \text{lbf}}$$

The answer is (B).

SI Solution

The mass flow rate of the water is

$$\dot{m} = \rho\dot{V} = \rho v A = \frac{\rho v \pi D^2}{4}$$

$$= \left(1000 \ \frac{\text{kg}}{\text{m}^3}\right)\left(12 \ \frac{\text{m}}{\text{s}}\right)\left(\frac{\pi(0.05 \ \text{m})^2}{4}\right)$$

$$= 23.56 \ \text{kg/s}$$

The effective mass flow rate of the water is

$$\dot{m}_{\text{eff}} = \left(\frac{v - v_b}{v}\right)\dot{m}$$

$$= \left(\frac{12 \ \dfrac{\text{m}}{\text{s}} - 4.5 \ \dfrac{\text{m}}{\text{s}}}{12 \ \dfrac{\text{m}}{\text{s}}}\right)\left(23.56 \ \frac{\text{kg}}{\text{s}}\right)$$

$$= 14.73 \ \text{kg/s}$$

The force in the (horizontal) x-direction is given by

$$F_x = \dot{m}_{\text{eff}}(v - v_b)(\cos\theta - 1)$$

$$= \left(14.73 \ \frac{\text{kg}}{\text{s}}\right)\left(12 \ \frac{\text{m}}{\text{s}} - 4.5 \ \frac{\text{m}}{\text{s}}\right)(\cos 60° - 1)$$

$$= -55.2 \ \text{N} \quad \text{[the force is acting to the left]}$$

The force in the (vertical) y-direction is given by

$$F_y = \dot{m}_{\text{eff}}(v - v_b)\sin\theta$$

$$= \left(14.73 \ \frac{\text{kg}}{\text{s}}\right)\left(12 \ \frac{\text{m}}{\text{s}} - 4.5 \ \frac{\text{m}}{\text{s}}\right)(\sin 60°)$$

$$= 95.7 \ \text{N} \quad \text{[the force is acting upward]}$$

The net resultant force is

$$F = \sqrt{F_x^2 + F_y^2}$$

$$= \sqrt{(-55.2 \ \text{N})^2 + (95.7 \ \text{N})^2}$$

$$= \boxed{110.5 \ \text{N}}$$

The answer is (B).

22. *Customary U.S. Solution*

The power that must be added to the pump is given by

$$P = \frac{\Delta h\dot{m} \times \left(\dfrac{g}{g_c}\right)}{\eta}$$

Assume schedule-40 pipe.

$$D_i = 0.9948 \text{ ft} \quad [\text{App. 16.B}]$$

$$A_i = 0.7773 \text{ ft}^2$$

$$\text{v} = \frac{\dot{V}}{A_i}$$

$$= \frac{\left(2000 \, \dfrac{\text{gal}}{\text{min}}\right)\left(\dfrac{0.002228 \, \dfrac{\text{ft}^3}{\text{sec}}}{1 \, \dfrac{\text{gal}}{\text{min}}}\right)}{0.7773 \text{ ft}^2}$$

$$= 5.73 \text{ ft/sec}$$

(Note that the pressures are in terms of gage pressure, and the density of mercury is 0.491 lbm/in^3.)

$$p_i = \left(14.7 \, \frac{\text{lbf}}{\text{in}^2} - \frac{(6 \text{ in})\left(0.491 \, \dfrac{\text{lbm}}{\text{in}^3}\right)\left(32.2 \, \dfrac{\text{ft}}{\text{sec}^2}\right)}{32.2 \, \dfrac{\text{lbm-ft}}{\text{lbf-sec}^2}}\right)$$

$$\times \left(\frac{144 \text{ in}^2}{1 \text{ ft}^2}\right)$$

$$= 1692.6 \text{ lbf/ft}^2$$

$$E_{ti} = \frac{p_i}{\rho} + \frac{\text{v}_i^2}{2g_c} + \frac{z_i g}{g_c}$$

Since the pump inlet and outlet are at the same elevation, use $z = 0$ and $\rho = (\text{SG})\rho_{\text{water}}$.

$$E_{ti} = \frac{p_i}{(\text{SG})\rho_{\text{water}}} + \frac{\text{v}_i^2}{2g_c} + 0$$

$$= \frac{1692.6 \, \dfrac{\text{lbf}}{\text{ft}^2}}{(1.2)\left(62.4 \, \dfrac{\text{lbm}}{\text{ft}^3}\right)} + \frac{\left(5.73 \, \dfrac{\text{ft}}{\text{sec}}\right)^2}{(2)\left(32.2 \, \dfrac{\text{lbm-ft}}{\text{lbf-sec}^2}\right)}$$

$$= 23.11 \text{ ft-lbf/lbm}$$

Calculate the total head at the inlet.

$$h_{ti} = E_{ti} \times \left(\frac{g_c}{g}\right)$$

$$= \frac{\left(23.11 \, \dfrac{\text{ft-lbf}}{\text{lbm}}\right)\left(32.2 \, \dfrac{\text{lbm-ft}}{\text{lbf-sec}^2}\right)}{32.2 \, \dfrac{\text{ft}}{\text{sec}^2}}$$

$$= 23.11 \text{ ft}$$

At the outlet side of the pump,

$$D_o = 0.6651 \text{ ft}$$

$$A_o = 0.3474 \text{ ft}^2$$

$$Q = \text{v}_o A_o$$

$$\text{v}_o = \frac{Q}{A_o}$$

$$= \frac{\left(2000 \, \dfrac{\text{gal}}{\text{min}}\right)\left(\dfrac{0.002228 \, \dfrac{\text{ft}^3}{\text{sec}}}{1 \, \dfrac{\text{gal}}{\text{min}}}\right)}{0.3474 \text{ ft}^2}$$

$$= 12.83 \text{ ft/sec}$$

(Note that the pressures are in terms of gage pressure and the gauge is located 4 ft above the pump outlet, which adds 4 ft of pressure head at the pump outlet.)

$$p_o = \left(14.7 \, \frac{\text{lbf}}{\text{in}^2} + 20 \, \frac{\text{lbf}}{\text{in}^2}\right)\left(\frac{144 \text{ in}^2}{1 \text{ ft}^2}\right)$$

$$+ 4 \text{ ft} \left(\frac{(1.2)\left(62.4 \, \dfrac{\text{lbm}}{\text{ft}^3}\right)\left(32.2 \, \dfrac{\text{ft}}{\text{sec}^2}\right)}{32.2 \, \dfrac{\text{lbm-ft}}{\text{lbf-sec}^2}}\right)$$

$$= 5296 \text{ lbf/ft}^2$$

$$E_{to} = \frac{p_o}{\rho} + \frac{\text{v}_o^2}{2g_c} + \frac{z_o g}{g_c}$$

Since the pump inlet and outlet are at the same elevation, use $z = 0$ and $\rho = (\text{SG})\rho_{\text{water}}$.

$$E_{to} = \frac{p_o}{(\text{SG})\rho_{\text{water}}} + \frac{\text{v}_o^2}{2g_c} + 0$$

$$= \frac{5296 \, \dfrac{\text{lbf}}{\text{ft}^2}}{(1.2)\left(62.4 \, \dfrac{\text{lbm}}{\text{ft}^3}\right)} + \frac{\left(12.83 \, \dfrac{\text{ft}}{\text{sec}}\right)^2}{(2)\left(32.2 \, \dfrac{\text{lbm-ft}}{\text{lbf-sec}^2}\right)}$$

$$= 73.28 \text{ ft-lbf/lbm}$$

Calculate the total head at the outlet.

$$h_{to} = E_{to} \times \left(\frac{g_c}{g}\right)$$

$$= \left(73.28 \, \frac{\text{ft-lbf}}{\text{lbm}}\right)\left(\frac{32.2 \, \dfrac{\text{lbm-ft}}{\text{lbf-sec}^2}}{32.2 \, \dfrac{\text{ft}}{\text{sec}^2}}\right)$$

$$= 73.28 \text{ ft}$$

Compute the total head required across the pump.

$$\Delta h = h_{to} - h_{ti}$$
$$= 73.28 \text{ ft} - 23.11 \text{ ft}$$
$$= 50.17 \text{ ft}$$

The mass flow rate is

$$\dot{m} = \rho \dot{V}$$

In terms of the specific gravity, the mass flow rate is

$$\dot{m} = (SG)\rho_{\text{water}}\dot{V}$$

$$= (1.2)\left(62.4 \ \frac{\text{lbm}}{\text{ft}^3}\right)\left(2000 \ \frac{\text{gal}}{\text{min}}\right)\left(\frac{0.002228 \ \frac{\text{ft}^3}{\text{sec}}}{1 \ \frac{\text{gal}}{\text{min}}}\right)$$

$$= 333.7 \text{ lbm/sec}$$

The power that must be added to the pump is

$$P = \frac{\Delta h \dot{m} \times \left(\frac{g}{g_c}\right)}{\eta}$$

$$= \frac{(50.17 \text{ ft})\left(333.7 \ \frac{\text{lbm}}{\text{sec}}\right)\left(\frac{32.2 \ \frac{\text{ft}}{\text{sec}^2}}{32.2 \ \frac{\text{lbm-ft}}{\text{lbf-sec}^2}}\right)}{(0.85)\left(550 \ \frac{\text{ft-lbf}}{\text{hp-sec}}\right)}$$

$$= \boxed{35.8 \text{ hp}}$$

(Note that it is not necessary to use absolute pressures as has been done in this solution.)

The answer is (B).

SI Solution

The power that must be added to the pump is given by

$$P = \frac{\Delta h \dot{m} g}{\eta}$$

Assume the pipe nominal diameter is equal to the internal diameter.

$$D_i = 0.30 \text{ m}$$

$$A_i = \frac{\pi D_i^2}{4} = x\frac{\pi(0.30 \text{ m})^2}{4} = 0.0707 \text{ m}^2$$

$$v = \frac{\dot{V}}{A_i}$$

$$= \frac{0.125 \ \frac{\text{m}^3}{\text{s}}}{0.0707 \text{ m}^2}$$

$$= 1.77 \text{ m/s}$$

(Note that the pressures are in terms of gage pressure, and the density of mercury is 13 600 kg/m³.)

$$p_i = 1.013 \times 10^5 \text{ Pa}$$
$$- (0.15 \text{ m})\left(13\,600 \ \frac{\text{kg}}{\text{m}^3}\right)\left(9.81 \ \frac{\text{m}}{\text{s}^2}\right)$$
$$= 8.13 \times 10^4 \text{ Pa}$$

$$E_{ti} = \frac{p}{\rho} + \frac{v_i^2}{2} + z_i g$$

Since the pump inlet and outlet are at the same elevation, use $z = 0$ and $\rho = (SG)\rho_{\text{water}}$.

$$E_{ti} = \frac{p}{(SG)\rho_{\text{water}}} + \frac{v_i^2}{2} + 0$$

$$= \frac{8.13 \times 10^4 \text{ Pa}}{(1.2)\left(1000 \ \frac{\text{kg}}{\text{m}^3}\right)} + \frac{\left(1.77 \ \frac{\text{m}}{\text{s}}\right)^2}{2}$$

$$= 69.3 \text{ J/kg}$$

The total head at the inlet is

$$h_{ti} = \frac{E_{ti}}{g}$$

$$= \frac{69.3 \ \frac{\text{J}}{\text{kg}}}{9.81 \ \frac{\text{m}}{\text{s}^2}}$$

$$= 7.06 \text{ m}$$

Assume the pipe nominal diameter is equal to the internal diameter. On the outlet side of the pump,

$$D_i = 0.20 \text{ m}$$

$$A_o = \frac{\pi D_o^2}{4} = \frac{\pi(0.20 \text{ m})^2}{4} = 0.0314 \text{ m}^2$$

$$v_o = \frac{\dot{V}}{A_o}$$

$$= \frac{0.125 \ \frac{\text{m}^3}{\text{s}}}{0.0314 \text{ m}^2}$$

$$= 3.98 \text{ m/s}$$

(Note that the pressures are in terms of gage pressure and the gauge is located 1.2 m above the pump outlet, which adds 1.2 m of pressure head at the pump outlet.)

$$p_o = 1.013 \times 10^5 \text{ Pa} + 138 \times 10^3 \text{ Pa}$$
$$+ (1.2 \text{ m})\left((1.2)\left(1000 \ \frac{\text{kg}}{\text{m}^3}\right)\left(9.81 \ \frac{\text{m}}{\text{s}^2}\right)\right)$$
$$= 2.53 \times 10^5 \text{ Pa}$$

$$E_{to} = \frac{p_o}{\rho} + \frac{v_o^2}{2} + z_o g$$

Since the pump inlet and outlet are at the same elevation, use $z = 0$ and $\rho = (SG)\rho_{water}$.

$$E_{to} = \frac{p_o}{(SG)\rho_{water}} + \frac{v_o^2}{2} + 0$$

$$= \frac{2.53 \times 10^5 \text{ Pa}}{(1.2)\left(1000 \frac{\text{kg}}{\text{m}^3}\right)} + \frac{\left(3.98 \frac{\text{m}}{\text{s}}\right)^2}{2}$$

$$= 218.8 \text{ J/kg}$$

The total head at the outlet is

$$h_{to} = \frac{E_{to}}{g}$$

$$= \frac{218.8 \frac{\text{J}}{\text{kg}}}{22.30 \frac{\text{m}}{\text{s}^2}}$$

$$= 11.76 \text{ m}$$

The total head required across the pump is

$$\Delta h = h_{to} - h_{ti}$$

$$= 22.30 \text{ m} - 7.04 \text{ m}$$

$$= 15.26 \text{ m}$$

The mass flow rate is

$$\dot{m} = \rho \dot{V}$$

In terms of the specific gravity, the mass flow rate is

$$\dot{m} = (SG)\rho_{water}Q$$

$$= (1.2)\left(1000 \frac{\text{kg}}{\text{m}^3}\right)\left(0.125 \frac{\text{m}^3}{\text{s}}\right)$$

$$= 150 \text{ kg/s}$$

The power that must be added to the pump is

$$P = \frac{\Delta h \dot{m} g}{\eta}$$

$$= \frac{(15.26 \text{ m})\left(150 \frac{\text{kg}}{\text{s}}\right)\left(9.81 \frac{\text{m}}{\text{s}^2}\right)}{0.85}$$

$$= \boxed{26\,417 \text{ W} \quad (26.4 \text{ kW})}$$

(Note that it is not necessary to use absolute pressures as has been done in this solution.)

The answer is (B).

23. *Customary U.S. Solution*

The power developed by the horizontal turbine is given by

$$P = \dot{m}h_{loss} \times \left(\frac{g}{g_c}\right)$$

The mass flow rate is

$$\dot{m} = \dot{V}\rho$$

$$= \left(100 \frac{\text{ft}^3}{\text{sec}}\right)\left(62.4 \frac{\text{lbm}}{\text{ft}^3}\right)$$

$$= 6240 \text{ lbm/sec}$$

The head loss across the horizontal turbine is given by

$$h_{loss} = \left(\frac{\Delta p}{\rho}\right) \times \left(\frac{g_c}{g}\right)$$

$$= \left(\frac{\left(30 \frac{\text{lbf}}{\text{in}^2} - \left(-5 \frac{\text{lbf}}{\text{in}^2}\right)\right)\left(\frac{144 \text{ in}^2}{1 \text{ ft}^2}\right)}{62.4 \frac{\text{lbm}}{\text{ft}^3}}\right)$$

$$\times \left(\frac{32.2 \frac{\text{lbm-ft}}{\text{lbf-sec}^2}}{32.2 \frac{\text{ft}}{\text{sec}^2}}\right)$$

$$= 80.77 \text{ ft}$$

The power developed is

$$P = \dot{m}h_{loss} \times \left(\frac{g}{g_c}\right)$$

$$= \frac{\left(6240 \frac{\text{lbm}}{\text{sec}}\right)(80.77 \text{ ft})\left(32.2 \frac{\text{ft}}{\text{sec}^2}\right)}{\left(32.2 \frac{\text{lbm-ft}}{\text{lbf-sec}^2}\right)\left(550 \frac{\text{ft-lbf}}{\text{hp-sec}}\right)}$$

$$= \boxed{916 \text{ hp}}$$

The answer is (D).

SI Solution

The power developed by the horizontal turbine is given by

$$P = \dot{m}h_{loss}g$$

The mass flow rate is

$$\dot{m} = \dot{V}\rho$$

$$= \left(2.6 \frac{\text{m}^3}{\text{s}}\right)\left(1000 \frac{\text{kg}}{\text{m}^3}\right)$$

$$= 2600 \text{ kg/s}$$

The head loss across the horizontal turbine is given by

$$h_{loss} = \frac{\Delta p}{\rho g}$$

$$= \frac{(210 \text{ kPa} - (-35 \text{ kPa}))\left(1000 \ \dfrac{\text{Pa}}{\text{kPa}}\right)}{\left(1000 \ \dfrac{\text{kg}}{\text{m}^3}\right)\left(9.81 \ \dfrac{\text{m}}{\text{s}^2}\right)}$$

$$= 25.0 \text{ m}$$

The power developed is

$$P = \dot{m}h_{loss}g$$

$$= \left(2600 \ \frac{\text{kg}}{\text{s}}\right)(25.0 \text{ m})\left(9.81 \ \frac{\text{m}}{\text{s}^2}\right)$$

$$= \boxed{637\,650 \text{ W} \quad (638 \text{ kW})}$$

The answer is (D).

24. *Customary U.S. Solution*

(a) The drag on the car is given by

$$F_D = \frac{C_D A \rho \text{v}^2}{2g_c}$$

For air at 70°F,

$$\rho = \frac{p}{RT} = \frac{\left(14.7 \ \dfrac{\text{lbf}}{\text{in}^2}\right)\left(144 \ \dfrac{\text{in}^2}{\text{ft}^2}\right)}{\left(53.35 \ \dfrac{\text{ft-lbf}}{\text{lbm-°R}}\right)(70°F + 460)}$$

$$= 0.0749 \text{ lbm/ft}^3$$

$$\text{v} = \left(55 \ \frac{\text{mi}}{\text{hr}}\right)\left(5280 \ \frac{\text{ft}}{\text{mi}}\right)\left(\frac{1 \text{ hr}}{3600 \text{ sec}}\right)$$

$$= 80.67 \text{ ft/sec}$$

Substituting gives

$$F_D = \frac{C_D A \rho \text{v}^2}{2g_c}$$

$$= \frac{(0.42)\left(28 \text{ ft}^2\right)\left(0.0749 \ \dfrac{\text{lbm}}{\text{ft}^3}\right)\left(80.67 \ \dfrac{\text{ft}}{\text{sec}}\right)^2}{(2)\left(32.2 \ \dfrac{\text{lbm-ft}}{\text{lbf-sec}^2}\right)}$$

$$= 89.0 \text{ lbf}$$

The total resisting force is

$$F = F_D + \text{rolling resistance}$$

$$= 89.0 \text{ lbf} + (0.01)(3300 \text{ lbm}) \times \left(\frac{g}{g_c}\right)$$

$$= 89.0 \text{ lbf} + \frac{(0.01)(3300 \text{ lbm})\left(32.2 \ \dfrac{\text{ft}}{\text{sec}^2}\right)}{32.2 \ \dfrac{\text{lbm-ft}}{\text{lbf-sec}^2}}$$

$$= 122.0 \text{ lbf}$$

The power consumed is

$$P = F\text{v}$$

$$= \frac{(122.0 \text{ lbf})\left(80.67 \ \dfrac{\text{ft}}{\text{sec}}\right)}{778 \ \dfrac{\text{ft-lbf}}{\text{Btu}}}$$

$$= 12.65 \text{ Btu/sec}$$

The energy available from the fuel is

$$E = (\text{engine thermal efficiency})(\text{fuel heating value})$$

$$= (0.28)\left(115{,}000 \ \frac{\text{Btu}}{\text{gal}}\right)$$

$$= 32{,}200 \text{ Btu/gal}$$

The fuel consumption at 55 mi/hr is

$$\frac{P}{E} = \frac{12.65 \ \dfrac{\text{Btu}}{\text{sec}}}{32{,}200 \ \dfrac{\text{Btu}}{\text{gal}}}$$

$$= 3.93 \times 10^{-4} \text{ gal/sec}$$

The fuel consumption is

$$\frac{3.93 \times 10^{-4} \ \dfrac{\text{gal}}{\text{sec}}}{\left(55 \ \dfrac{\text{mi}}{\text{hr}}\right)\left(\dfrac{1 \text{ hr}}{3600 \text{ sec}}\right)} = \boxed{0.0257 \text{ gal/mi}}$$

The answer is (A).

(b) Similarly, the fuel consumption at 65 mi/hr is

$$\text{v} = \left(65 \ \frac{\text{mi}}{\text{hr}}\right)\left(5280 \ \frac{\text{ft}}{\text{mi}}\right)\left(\frac{1 \text{ hr}}{3600 \text{ sec}}\right)$$

$$= 95.33 \text{ ft/sec}$$

$$F_D = \frac{C_D A \rho \text{v}^2}{2g_c}$$

$$= \frac{(0.42)(28 \text{ ft}^2)\left(0.0749 \ \dfrac{\text{lbm}}{\text{ft}^3}\right)\left(95.33 \ \dfrac{\text{ft}}{\text{sec}}\right)^2}{(2)\left(32.2 \ \dfrac{\text{lbm-ft}}{\text{lbf-sec}^2}\right)}$$

$$= 124.3 \text{ lbf}$$

The total resisting force is

$$F = F_D + \text{rolling resistance}$$

$$= 124.3 \text{ lbf} + (0.01)(3300 \text{ lbm}) \times \left(\frac{g}{g_c}\right)$$

$$= 124.3 \text{ lbf} + \frac{(0.01)(3300 \text{ lbm})\left(32.2 \ \dfrac{\text{ft}}{\text{sec}^2}\right)}{32.2 \ \dfrac{\text{lbm-ft}}{\text{lbf-sec}^2}}$$

$$= 157.3 \text{ lbf}$$

The power consumed is

$$P = F\text{v}$$

$$= \frac{(157.3 \text{ lbf}) \left(95.33 \frac{\text{ft}}{\text{sec}} \right)}{778 \frac{\text{ft-lbf}}{\text{Btu}}}$$

$$= 19.27 \text{ Btu/sec}$$

The fuel consumption at 65 mi/hr is

$$\frac{P}{E} = \frac{19.27 \frac{\text{Btu}}{\text{sec}}}{32{,}200 \frac{\text{Btu}}{\text{gal}}}$$

$$= 5.98 \times 10^{-4} \text{ gal/sec}$$

The fuel consumption is

$$\frac{5.98 \times 10^{-4} \frac{\text{gal}}{\text{sec}}}{\left(65 \frac{\text{mi}}{\text{hr}} \right) \left(\frac{1 \text{ hr}}{3600 \text{ sec}} \right)} = 0.0331 \text{ gal/mi}$$

The relative difference between the fuel consumption at 55 mi/hr and 65 mi/hr is

$$\frac{0.0331 \frac{\text{gal}}{\text{mi}} - 0.0257 \frac{\text{gal}}{\text{mi}}}{0.0257 \frac{\text{gal}}{\text{mi}}} = \boxed{0.288 \quad (28.8\%)}$$

The answer is (C).

SI Solution

(a) The drag on the car is given by

$$F_D = \frac{C_D A \rho \text{v}^2}{2}$$

For air at 20°C,

$$\rho = \frac{p}{RT} = \frac{1.013 \times 10^5 \text{ Pa}}{\left(287 \frac{\text{J}}{\text{kg·K}} \right) (20°\text{C} + 273)}$$

$$= 1.205 \text{ kg/m}^3$$

$$\text{v} = \left(90 \frac{\text{km}}{\text{h}} \right) \left(1000 \frac{\text{m}}{\text{km}} \right) \left(\frac{1 \text{ h}}{3600 \text{ s}} \right)$$

$$= 25.0 \text{ m/s}$$

Substituting gives

$$F_D = \frac{C_D A \rho \text{v}^2}{2}$$

$$= \left(\tfrac{1}{2} \right) (0.42)(2.6 \text{ m}^2) \left(1.205 \frac{\text{kg}}{\text{m}^3} \right) \left(25.0 \frac{\text{m}}{\text{s}} \right)^2$$

$$= 411.2 \text{ N}$$

The total resisting force is

$$F = F_D + \text{rolling resistance}$$

$$= 411.2 \text{ N} + (0.01)(1500 \text{ kg})g$$

$$= 411.2 \text{ N} + (0.01)(1500 \text{ kg}) \left(9.81 \frac{\text{m}}{\text{s}^2} \right)$$

$$= 558.4 \text{ N}$$

The power consumed is

$$P = F\text{v}$$

$$= (558.4 \text{ N}) \left(25 \frac{\text{m}}{\text{s}} \right)$$

$$= 13\,960 \text{ W}$$

The energy available from the fuel is

$$E = (\text{engine thermal efficiency})(\text{fuel heating value})$$

$$= (0.28) \left(4.6 \times 10^8 \frac{\text{J}}{\text{L}} \right)$$

$$= 1.288 \times 10^8 \text{ J/L}$$

The fuel consumption at 90 km/h is

$$\frac{P}{E} = \frac{13\,960 \text{ W}}{1.288 \times 10^8 \frac{\text{J}}{\text{L}}}$$

$$= 1.08 \times 10^{-4} \text{ L/s}$$

The fuel consumption is

$$\frac{1.08 \times 10^{-4} \frac{\text{L}}{\text{s}}}{\left(90 \frac{\text{km}}{\text{h}} \right) \left(\frac{1 \text{ h}}{3600 \text{ s}} \right)} = \boxed{0.00434 \text{ L/km}}$$

The answer is (A).

(b) Similarly, the fuel consumption at 105 km/h is

$$\text{v} = \left(105 \frac{\text{km}}{\text{h}} \right) \left(1000 \frac{\text{m}}{\text{km}} \right) \left(\frac{1 \text{ h}}{3600 \text{ s}} \right)$$

$$= 29.2 \text{ m/s}$$

$$D = \frac{C_D A \rho \text{v}^2}{2}$$

$$= \left(\tfrac{1}{2} \right) (0.42)(2.6 \text{ m}^2) \left(1.205 \frac{\text{kg}}{\text{m}^3} \right) \left(29.2 \frac{\text{m}}{\text{s}} \right)^2$$

$$= 561.0 \text{ N}$$

The total resisting force is

$$F = F_D + \text{rolling resistance}$$

$$= 561.0 \text{ N} + (0.01)(1500 \text{ kg})g$$

$$= 561.0 \text{ N} + (0.01)(1500 \text{ kg}) \left(9.81 \frac{\text{m}}{\text{s}^2} \right)$$

$$= 708.2 \text{ N}$$

The power consumed is

$$P = Fv$$
$$= (708.2 \text{ N}) \left(29.2 \ \frac{\text{m}}{\text{s}}\right)$$
$$= 20\,679 \text{ W}$$

The fuel consumption at 105 km/h is

$$\frac{P}{E} = \frac{20\,679 \text{ W}}{1.288 \times 10^8 \ \dfrac{\text{J}}{\text{L}}}$$
$$= 1.61 \times 10^{-4} \text{ L/s}$$

The fuel consumption is

$$\frac{1.61 \times 10^{-4} \ \dfrac{\text{L}}{\text{s}}}{\left(105 \ \dfrac{\text{km}}{\text{h}}\right) \left(\dfrac{1 \text{ h}}{3600 \text{ s}}\right)} = 0.00552 \text{ L/km}$$

The relative difference between the fuel consumption at 90 km/h and 105 km/h is

$$\frac{0.00552 \ \dfrac{\text{L}}{\text{km}} - 0.00434 \ \dfrac{\text{L}}{\text{km}}}{0.00434 \ \dfrac{\text{L}}{\text{km}}} = \boxed{0.272 \ \ (27.2\%)}$$

The answer is (C).

25. To ensure similarity between the model and the true conditions of the full-scale airplane, the Reynolds numbers must be equal.

Substitute the definition for the Reynolds number where L is a characteristic length.

$$\left(\frac{vL}{\nu}\right)_{\text{model}} = \left(\frac{vL}{\nu}\right)_{\text{true}}$$

Use the absolute viscosity.

$$\mu = \frac{\rho\nu}{g_c}$$
$$\nu = \frac{\mu g_c}{\rho}$$

Substituting gives

$$\left(\frac{vL\rho}{\mu g_c}\right)_{\text{model}} = \left(\frac{vL\rho}{\mu g_c}\right)_{\text{true}}$$

Since g_c is a constant,

$$\left(\frac{vL\rho}{\mu}\right)_{\text{model}} = \left(\frac{vL\rho}{\mu}\right)_{\text{true}}$$

Assume the air behaves as an ideal gas.

$$\rho = \frac{p}{RT}$$

Substituting gives

$$\left(\frac{vLp}{\mu RT}\right)_{\text{model}} = \left(\frac{vLp}{\mu RT}\right)_{\text{true}}$$

Since the tunnel operates at true velocity and temperature,

$$v_{\text{model}} = v_{\text{true}}$$
$$T_{\text{model}} = T_{\text{true}}$$

Since both tunnels operate with air, R is a constant. The expression reduces to

$$\left(\frac{Lp}{\mu}\right)_{\text{model}} = \left(\frac{Lp}{\mu}\right)_{\text{true}}$$

Recall that the absolute viscosity is independent of pressure, so $\mu_{\text{model}} = \mu_{\text{true}}$.

Therefore,

$$(Lp)_{\text{model}} = (Lp)_{\text{true}}$$

Since the scale of the model is 1/20,

$$L_{\text{model}} = \frac{L_{\text{true}}}{20}$$

Substituting gives

$$(Lp)_{\text{model}} = (Lp)_{\text{true}}$$
$$\left(\frac{L_{\text{true}}}{20}\right) p_{\text{model}} = L_{\text{true}} p_{\text{true}}$$
$$\boxed{p_{\text{model}} = 20 p_{\text{true}}}$$

The answer is (C).

26. To ensure similarity between the two pumps, the Reynolds number, Re, of both pumps must be equal. (Let pump 1 represent the pump for the castor oil and pump 2 represent the pump for air.)

$$\text{Re}_1 = \text{Re}_2$$

Substitute the definition for the Reynolds number.

$$\frac{v_1 D_1}{\nu_1} = \frac{v_2 D_2}{\nu_2}$$
$$\frac{v_1}{v_2} = \left(\frac{\nu_1}{\nu_2}\right)\left(\frac{D_2}{D_1}\right)$$

v is the tangential velocity, and D is the impeller diameter.

$$\text{v} \propto nD$$

$$\frac{\text{v}_1}{\text{v}_2} = \left(\frac{n_1}{n_2}\right)\left(\frac{D_1}{D_2}\right)$$

Therefore,

$$\left(\frac{\nu_1}{\nu_2}\right)\left(\frac{D_2}{D_1}\right) = \left(\frac{n_1}{n_2}\right)\left(\frac{D_1}{D_2}\right)$$

$$n_2 = n_1\left(\frac{\nu_2}{\nu_1}\right)\left(\frac{D_1}{D_2}\right)^2$$

Since the second pump has an impeller twice the size of the first pump, $D_2 = 2D_1$.

Substituting gives

$$n_2 = n_1\left(\frac{\nu_2}{\nu_1}\right)\left(\frac{D_1}{D_2}\right)^2$$

$$= n_1\left(\frac{\nu_2}{\nu_1}\right)\left(\frac{D_1}{2D_1}\right)^2$$

$$= \tfrac{1}{4}n_1\left(\frac{\nu_2}{\nu_1}\right)$$

Customary U.S. Solution

For air at 68°F, $\nu_2 = 16.0 \times 10^{-5}$ ft²/sec.

For castor oil at 68°F, $\nu_1 = 1110 \times 10^{-5}$ ft²/sec (given).

Since $n_1 = 1000$ rpm,

$$n_2 = \tfrac{1}{4}n_1\left(\frac{\nu_2}{\nu_1}\right)$$

$$= \left(\tfrac{1}{4}\right)(1000 \text{ rpm})\left(\frac{16.0 \times 10^{-5}\,\frac{\text{ft}^2}{\text{sec}}}{1110 \times 10^{-5}\,\frac{\text{ft}^2}{\text{sec}}}\right)$$

$$= \boxed{3.6 \text{ rpm}}$$

SI Solution

For air at 20°C, $\nu_2 = 1.48 \times 10^{-5}$ m²/s.

For castor oil at 20°C, $\nu_1 = 103 \times 10^{-5}$ m²/s (given).

Since $n_1 = 1000$ rpm,

$$n_2 = \tfrac{1}{4}n_1\left(\frac{\nu_2}{\nu_1}\right)$$

$$= \left(\tfrac{1}{4}\right)(1000 \text{ rpm})\left(\frac{1.48 \times 10^{-5}\,\frac{\text{m}^2}{\text{s}}}{103 \times 10^{-5}\,\frac{\text{m}^2}{\text{s}}}\right)$$

$$= \boxed{3.6 \text{ rpm}}$$

The answer is (A).

Fluids

18 Hydraulic Machines

PRACTICE PROBLEMS

Pumping Power

1. 2000 gal/min of 60°F thickened sludge with a specific gravity of 1.2 flow through a pump with an inlet diameter of 12 in and an outlet of 8 in. The centerlines of the inlet and outlet are at the same elevation. The inlet pressure is 8 in of mercury (vacuum). A discharge pressure gauge located 4 ft above the pump discharge centerline reads 20 psig. The pump efficiency is 85%. All pipes are schedule-40. What is the input power of the pump?

(A) 26 hp
(B) 31 hp
(C) 37 hp
(D) 53 hp

2. 1.25 ft³/sec (35 L/s) of 70°F (21°C) water are pumped from the bottom of a tank through 700 ft (230 m) of 4 in (10.2 cm) schedule-40 steel pipe. The line includes a 50 ft (15 m) rise in elevation, two right-angle elbows, a wide-open gate valve, and a swing check valve. All fittings and valves are regular screwed. The inlet pressure (at the bottom of the supply tank) is 50 psig (345 kPa), and a working pressure of 20 psig (140 kPa) is needed at the end of the discharge pipe. What is the hydraulic power for this pumping application?

(A) 16 hp (13 kW)
(B) 23 hp (17 kW)
(C) 49 hp (37 kW)
(D) 66 hp (50 kW)

3. 80 gal/min (5 L/s) of 80°F (27°C) water are lifted 12 ft (3.6 m) vertically by a pump through a total length of 50 ft (15 m) of a 2 in (5.1 cm) diameter smooth rubber hose. The discharge end of the hose is submerged in 8 ft (2.4 m) of water as shown. What head is added by the pump?

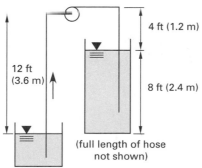

(A) 10 ft (3.0 m)
(B) 13 ft (3.9 m)
(C) 22 ft (6.7 m)
(D) 31 ft (9.3 m)

4. A 20 hp motor drives a centrifugal pump. The pump discharges 60°F (16°C) water at 12 ft/sec (4 m/s) into a 6 in (15.2 cm) steel schedule-40 line. The inlet is 8 in (20.3 cm) schedule-40 steel pipe. The pump suction is 5 psi (35 kPa) below standard atmospheric pressure. The friction head loss in the system is 10 ft (3.3 m). The pump efficiency is 70%. The suction and discharge lines are at the same elevation. What is the maximum height above the pump inlet that water is available at standard atmospheric pressure?

(A) 28 ft (6.9 m)
(B) 37 ft (11 m)
(C) 49 ft (15 m)
(D) 81 ft (25 m)

5. (*Time limit: one hour*) A pump station is used to fill a tank on a hill above from a lake below. The flow rate is 10,000 gal/hr (10.5 L/s) of 60°F (16°C) water. The atmospheric pressure is 14.7 psia (101 kPa). The pump is 12 ft (4 m) above the lake, and the tank surface level is 350 ft (115 m) above the pump. The suction and discharge lines are 4 in (10.2 cm) diameter schedule-40 steel pipe. The equivalent length of the inlet line between the lake and the pump is 300 ft (100 m). The total equivalent length between the lake and the tank is 7000 ft (2300 m), including all fittings, bends, screens, and valves. The cost of electricity is $0.04 per kW-hr. The overall efficiency of the pump and motor set is 70%.

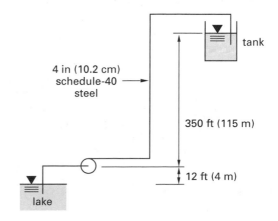

(a) What does it cost to operate the pump for 1 hr?
(A) $0.1
(B) $1
(C) $3
(D) $6

(b) What motor power is required?
(A) 10 hp (7.5 kW)
(B) 30 hp (25 kW)
(C) 50 hp (40 kW)
(D) 75 hp (60 kW)

(c) What is the NPSHA for this application?
(A) 4 ft (1.2 m)
(B) 8 ft (2.4 m)
(C) 12 ft (3.6 m)
(D) 16 ft (4.5 m)

6. (*Time limit: one hour*) A town with a stable, constant population of 10,000 produces sewage at the average rate of 100 gallons per capita day (gpcd), with peak flows of 250 gpcd. The pipe to the pumping station is 5000 ft in length and has a *C*-value of 130. The elevation drop along the length is 48 ft. Minor losses in infiltration are insignificant. The pump's maximum suction lift is 10 ft.

(a) If all diameters are available, what pipe diameter is required?
(A) 8 in
(B) 12 in
(C) 14 in
(D) 18 in

(b) If constant-speed pumps are used, what is the minimum number of pumps (disregarding spares and backups) that should be used?
(A) 2
(B) 3
(C) 4
(D) 5

(c) If variable-speed pumps are used, what is the minimum number of pumps (disregarding spares and backups) that should be used?
(A) 1
(B) 2
(C) 3
(D) 4

(d) If three constant-speed pumps are used, with a fourth as backup, and the pump-motor set efficiency is 60%, what motor power is required?
(A) 2 hp
(B) 3 hp
(C) 5 hp
(D) 8 hp

(e) If two variable-speed pumps are used, and the pump-motor set efficiency is 80%, what motor power is required?
(A) 3 hp
(B) 8 hp
(C) 12 hp
(D) 18 hp

(f) Which of the following are NOT all ways of controlling variable-speed pumps?
  I. detecting sump levels
 II. detecting pressure changes in the sump
III. detecting incoming flow rates
IV. using fixed run times
 V. detecting outgoing flow rates
VI. operating manually
(A) I, II, and III
(B) I, IV, and V
(C) I, III, and V
(D) I, V, and VI

(g) With intermittent fan operation, approximately how many air changes should the wet well receive per hour?
(A) 6
(B) 12
(C) 20
(D) 30

(h) With continuous fan operation, approximately how many air changes should the dry well receive per hour?
(A) 6
(B) 12
(C) 20
(D) 30

7. (*Time limit: one hour*) A pump transfers 3.5 MGD of filtered water from the clear well of a 10 ft × 20 ft (plan) rapid sand filter to an open channel at a higher elevation. The pump efficiency is 85%, and the motor driving pump has an efficiency of 90%. Minor losses are insignificant. Refer to the following illustration for additional information.

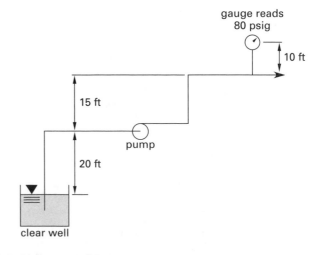

(a) What is the static suction lift?

(A) 15 ft

(B) 20 ft

(C) 35 ft

(D) 40 ft

(b) What is the static discharge head?

(A) 15 ft

(B) 20 ft

(C) 35 ft

(D) 40 ft

(c) Based on the information given, what is the approximate total dynamic head?

(A) 45 ft

(B) 185 ft

(C) 210 ft

(D) 230 ft

(d) What motor power is required?

(A) 50 hp

(B) 100 hp

(C) 150 hp

(D) 200 hp

8. (*Time limit: one hour*) A three-zone heating system uses hot water passing through the network shown. The heater increases the water temperature by 20°F (10°C). All pipes are copper, type L. The pump efficiency is 45%. The pipe equivalent lengths, pipe sizes, and flow rates through each part of the network are known. (Flow-balancing valves are not shown.)

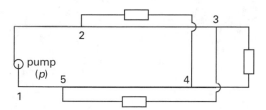

circuit	equivalent length, ft (m)	diameter, in (cm)	flow rate, gpm (L/s)
5–1–p–2	40 (13)	2¹/₂ (6.35)	60 (3.9)
2–4	70 (23)	1¹/₂ (3.81)	20 (1.3)
2–3	55 (18)	2 (5.08)	40 (2.6)
3–4	65 (22)	1¹/₂ (3.81)	20 (1.3)
3–5	60 (20)	1¹/₂ (3.81)	20 (1.3)
4–5	50 (17)	2 (5.08)	40 (2.6)

(a) What is the power delivered to the pump motor?

(A) 0.2 hp (0.2 kW)

(B) 0.8 hp (0.8 kW)

(C) 2.4 hp (2.4 kW)

(D) 4.6 hp (4.6 kW)

(b) What size motor should be chosen to drive the pump?

(A) 0.25 hp (0.25 kW)

(B) 1.0 hp (1.0 kW)

(A) 2.5 hp (2.5 kW)

(A) 5.0 hp (5.0 kW)

(c) How much heat is added to the water each hour?

(A) 200,000 Btu/hr (50 kW)

(B) 400,000 Btu/hr (110 kW)

(C) 600,000 Btu/hr (160 kW)

(D) 800,000 Btu/hr (220 kW)

Pumping Other Fluids

9. (*Time limit: one hour*) Gasoline with a specific gravity of 0.7 and viscosity of 6×10^{-6} ft²/sec (5.6 × 10^{-7} m²/s) is transferred from a tanker to a storage tank. The interior of the storage tank is maintained at atmospheric pressure by a vapor-recovery system. The free surface in the storage tank is 60 ft (20 m) above the tanker's free surface. The pipe consists of 500 ft (170 m) of 3 in (7.62 cm) schedule-40 steel pipe with six flanged elbows and two wide-open gate valves. The pump and motor both have individual efficiencies of 88%. Electricity costs $0.045 per kW-hr. The pump's performance data (based on cold, clear water) are known.

flow rate gpm (L/s)	head ft (m)
0 (0)	127 (42)
100 (6.3)	124 (41)
200 (12)	117 (39)
300 (18)	108 (36)
400 (24)	96 (32)
500 (30)	80 (27)
600 (36)	55 (18)

(a) What is the transfer rate?

(A) 150 gal/min (9.2 L/s)

(B) 180 gal/min (11 L/s)

(C) 200 gal/min (12 L/s)

(D) 220 gal/min (14 L/s)

(b) What is the total cost of operating the pump for 1 hr?

(A) $0.20

(B) $0.80

(C) $1.30

(D) $2.70

10. 37 gal/min (65 L/s) of 80°F (27°C) SAE 40 oil are increased in pressure from 1 atm to 40 psig (275 kPa). What hydraulic power is required?

(A) 0.45 hp (9 kW)

(B) 0.9 hp (18 kW)

(C) 1.8 hp (36 kW)

(D) 3.6 hp (72 kW)

Specific Speed

11. A double-suction water pump moving 300 gal/sec (1.1 kL/s) turns at 900 rpm. The pump adds 20 ft (7 m) of head to the water. What is the specific speed?

- (A)   3000 rpm (52 rpm)
- (B)   6000 rpm (100 rpm)
- (C)   9000 rpm (160 rpm)
- (D)   12,000 rpm (210 rpm)

12. A two-stage centrifugal pump draws water from an inlet 10 ft below its eye. Each stage of the pump adds 150 ft of head. What is the approximate maximum suggested speed for this application?

- (A)   900 rpm
- (B)   1200 rpm
- (C)   1700 rpm
- (D)   2000 rpm

Cavitation

13. 100 gal/min (6.3 L/s) of pressurized hot water at 281°F and 80 psia (138°C and 550 kPa) is drawn through 30 ft (10 m) of 1.5 in (3.81 cm) schedule-40 steel pipe into a 2 psig (14 kPa) tank. The inlet and outlet are both 20 ft (6 m) below the surface of the water when the tank is full. The inlet line contains a square-mouth inlet, two wide-open gate valves, and two long-radius elbows. All components are regular screwed. The pump's NPSHR is 10 ft (3 m) for this application. The kinematic viscosity of 281°F (138°C) water is 0.239×10^{-5} ft^2/sec (0.222×10^{-6} m^2/s) and the vapor pressure is 50.02 psia (3.431 bar). Will the pump cavitate?

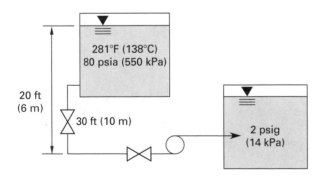

- (A)   yes; NPSHA = 4 ft (1.2 m)
- (B)   yes; NPSHA = 9 ft (2.7 m)
- (C)   no; NPSHA = 24 ft (7.2 m)
- (D)   no; NPSHA = 68 ft (21 m)

14. The velocity of the tip of a marine propeller is 4.2 times the boat velocity. The propeller is located 8 ft (3 m) below the surface. The temperature of the seawater is 68°F (20°C). The density is approximately 64.0 lbm/ft^3 (1024 kg/m^3), and the salt content is 2.5% by weight. What is the practical maximum boat velocity, as limited strictly by cavitation?

- (A)   9.1 ft/sec (2.7 m/s)
- (B)   12 ft/sec (3.8 m/s)
- (C)   15 ft/sec (4.5 m/s)
- (D)   22 ft/sec (6.6 m/s)

Pump and System Curves

15. The inlet of a centrifugal water pump is 7 ft (2.3 m) above the free surface from which it draws. The suction point is a submerged pipe. The supply line consists of 12 ft (4 m) of 2 in (5.08 cm) schedule-40 steel pipe and contains one long-radius elbow and one check valve. The discharge line is 2 in (5.08 cm) schedule-40 steel pipe and includes two long-radius elbows and an 80 ft (27 m) run. The discharge is 20 ft (6.3 m) above the free surface and is open to the atmosphere. All components are regular screwed. The water temperature is 70°F (21°C). The following pump curve data are applicable.

flow rate gpm (L/s)	head ft (m)
0 (0)	110 (37)
10 (0.6)	108 (36)
20 (1.2)	105 (35)
30 (1.8)	102 (34)
40 (2.4)	98 (33)
50 (3.2)	93 (31)
60 (3.6)	87 (29)
70 (4.4)	79 (26)
80 (4.8)	66 (22)
90 (5.7)	50 (17)

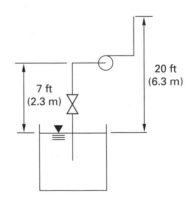

(a) What is the flow rate?

- (A)   44 gal/min (2.9 L/s)
- (B)   69 gal/min (4.5 L/s)
- (C)   82 gal/min (5.5 L/s)
- (D)   95 gal/min (6.2 L/s)

(b) What can be said about the use of this pump in this installation?

(A) A different pump should be used.

(B) The pump is operating near its most efficient point.

(C) Pressure fluctuations could result from surging.

(D) Overloading will not be a problem.

Affinity Laws

16. A pump was intended to run at 1750 rpm when driven by a 0.5 hp (0.37 kW) motor. What is the required power rating of a motor that will turn the pump at 2000 rpm?

(A) 0.25 hp (0.19 kW)

(B) 0.45 hp (0.34 kW)

(C) 0.65 hp (0.49 kW)

(D) 0.75 hp (0.55 kW)

17. (*Time limit: one hour*) A centrifugal pump running at 1400 rpm has the curve shown. The pump will be installed in an existing pipeline with known head requirements given by the formula $H = 30 + 2Q^2$. H is the system head in feet of water. Q is the flow rate in ft^3/sec.

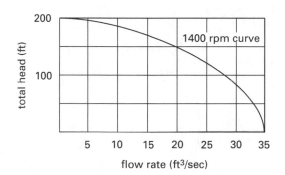

(a) What is the flow rate if the pump is turned at 1400 rpm?

(A) 2000 gal/min

(B) 3500 gal/min

(C) 4000 gal/min

(D) 4500 gal/min

(b) What power is required to drive the pump?

(A) 100 hp

(B) 210 hp

(C) 230 hp

(D) 260 hp

(c) What is the flow rate if the pump is turned at 1200 rpm?

(A) 2000 gal/min

(B) 3500 gal/min

(C) 4000 gal/min

(D) 4500 gal/min

Turbines

18. A horizontal turbine reduces 100 ft^3/sec of water from 30 psia to 5 psia. Friction is negligible. What power is developed?

(A) 350 hp

(B) 500 hp

(C) 650 hp

(D) 800 hp

19. 1000 ft^3/sec of 60°F water flow from a high reservoir through a hydroelectric turbine installation, exiting 625 ft lower. The head loss due to friction is 58 ft. The turbine efficiency is 89%. What power is developed in the turbines?

(A) 40 kW

(B) 18 MW

(C) 43 MW

(D) 71 MW

20. Water at 500 psig and 60°F (3.5 MPa and 16°C) drives a 250 hp (185 kW) turbine at 1750 rpm against a back pressure of 30 psig (210 kPa). The water discharges through a 4 in (100 mm) diameter nozzle at 35 ft/sec (10.5 m/s). The water is deflected 80° by a single blade moving directly away at 10 ft/sec (3 m/s).

(a) What type of turbine is described in this application?

(A) impulse turbine

(B) low-head turbine

(C) reaction turbine

(D) radial-flow turbine

(b) What is the total force acting on the blade?

(A) 100 lbf (450 N)

(B) 140 lbf (570 N)

(C) 160 lbf (720 N)

(D) 280 lbf (1300 N)

21. (*Time limit: one hour*) A Francis-design hydraulic reaction turbine with 22 in (560 mm) diameter blades runs at 610 rpm. The turbine develops 250 hp (185 kW) when 25 ft^3/sec (700 L/s) of water flow through it. The pressure head at the turbine entrance is 92.5 ft (30.8 m). The elevation of the turbine above the tailwater level is 5.26 ft (1.75 m). The inlet and outlet velocity are both 12 ft/sec (3.6 m/s).

(a) What is the effective head?

(A) 90 ft (30 m)

(B) 95 ft (31 m)

(C) 100 ft (33 m)

(D) 105 ft (35 m)

(b) What is the overall turbine efficiency?
- (A) 81%
- (B) 88%
- (C) 93%
- (D) 96%

(c) What will be the turbine speed if the effective head is 225 ft (75 m)?
- (A) 600 rpm
- (B) 920 rpm
- (C) 1100 rpm
- (D) 1400 rpm

(d) What horsepower is developed if the effective head is 225 ft (75 m)?
- (A) 560 hp (420 kW)
- (B) 630 hp (470 kW)
- (C) 750 hp (560 kW)
- (D) 840 hp (630 kW)

(e) What is the flow rate if the effective head is 225 ft (75 m)?
- (A) 25 ft^3/sec (700 L/s)
- (B) 38 ft^3/sec (1100 L/s)
- (C) 56 ft^3/sec (1600 L/s)
- (D) 64 ft^3/sec (1800 L/s)

SOLUTIONS

1. $\left(2000 \dfrac{\text{gal}}{\text{min}} \right) \left(0.002228 \dfrac{\frac{\text{ft}^3}{\text{sec}}}{\frac{\text{gal}}{\text{min}}} \right) = 4.456 \text{ ft}^3/\text{sec}$

Assume schedule-40 steel pipe. From App. 16.B,

$$12'' : \quad D_1 = 0.9948 \text{ ft} \quad A_1 = 0.7773 \text{ ft}^2$$
$$8'' : \quad D_2 = 0.6651 \text{ ft} \quad A_2 = 0.3473 \text{ ft}^2$$

$$p_1 = \left(14.7 \frac{\text{lbf}}{\text{in}^2} - (8 \text{ in}) \left(0.491 \frac{\text{lbf}}{\text{in}^3} \right) \right) \left(144 \frac{\text{in}^2}{\text{ft}^2} \right)$$
$$= 1551.2 \text{ lbf/ft}^2$$

$$p_2 = \left(14.7 \frac{\text{lbf}}{\text{in}^2} + 20 \frac{\text{lbf}}{\text{in}^2} \right) \left(144 \frac{\text{in}^2}{\text{ft}^2} \right)$$
$$\quad + (4 \text{ ft})(1.2) \left(62.4 \frac{\text{lbf}}{\text{ft}^3} \right)$$
$$= 5296.3 \text{ lbf/ft}^2$$

$$v_1 = \frac{4.456 \frac{\text{ft}^3}{\text{sec}}}{0.7773 \text{ ft}^2} = 5.73 \text{ ft/sec}$$

$$v_2 = \frac{4.456 \frac{\text{ft}^3}{\text{sec}}}{0.3473 \text{ ft}^2} = 12.83 \text{ ft/sec}$$

From Eq. 18.9, the total heads (in feet of sludge) at points 1 and 2 are

$$h_{t,1} = \frac{1551.2 \frac{\text{lbf}}{\text{ft}^2}}{\left(62.4 \frac{\text{lbf}}{\text{ft}^3} \right) (1.2)} + \frac{\left(5.73 \frac{\text{ft}}{\text{sec}} \right)^2}{(2) \left(32.2 \frac{\text{ft}}{\text{sec}^2} \right)} = 21.23 \text{ ft}$$

$$h_{t,2} = \frac{5296.3 \frac{\text{lbf}}{\text{ft}^2}}{\left(62.4 \frac{\text{lbf}}{\text{ft}^3} \right) (1.2)} + \frac{\left(12.83 \frac{\text{ft}}{\text{sec}} \right)^2}{(2) \left(32.2 \frac{\text{ft}}{\text{sec}^2} \right)} = 73.29 \text{ ft}$$

The pump must add 73.29 ft − 21.23 ft = 52.06 ft of head (sludge head).

The power required is given in Table 18.4.

$$P_{\text{ideal}} = \frac{(52.06 \text{ ft})(1.2) \left(4.456 \frac{\text{ft}^3}{\text{sec}} \right) \left(62.4 \frac{\text{lbf}}{\text{ft}^3} \right)}{550 \frac{\text{ft-lbf}}{\text{hp-sec}}}$$

$$= 31.58 \text{ hp}$$

The input horsepower is

$$P_{\text{in}} = \frac{P_{\text{ideal}}}{\eta} = \frac{31.58 \text{ hp}}{0.85} = \boxed{37.15 \text{ hp}}$$

The answer is (C).

2. *Customary U.S. Solution*

From App. 16.B, data for 4 in schedule-40 steel pipe are

$$D_i = 0.3355 \text{ ft}$$
$$A_i = 0.08841 \text{ ft}^2$$

The velocity in the pipe is

$$v = \frac{\dot{V}}{A} = \frac{1.25 \dfrac{\text{ft}^3}{\text{sec}}}{0.08841 \text{ ft}^2} = 14.139 \text{ ft/sec}$$

From Table 17.3, typical equivalent lengths for schedule-40, screwed steel fittings for 4 in pipes are

90° elbow: 13 ft

gate valve: 2.5 ft

check valve: 38.0 ft

The total equivalent length is

$$(2)(13 \text{ ft}) + (1)(2.5 \text{ ft}) + (1)(38 \text{ ft}) = 66.5 \text{ ft}$$

At 70°F, from App. 14.A, the density of water is 62.3 lbm/ft³ and the kinematic viscosity of water, ν, is 1.059× 10^{-5} ft²/sec. The Reynolds number is

$$\text{Re} = \frac{Dv}{\nu} = \frac{(0.3355 \text{ ft})\left(14.139 \dfrac{\text{ft}}{\text{sec}}\right)}{1.059 \times 10^{-5} \dfrac{\text{ft}^2}{\text{sec}}}$$
$$= 4.479 \times 10^5$$

From App. 17.A, for steel, $\epsilon = 0.0002$ ft.
So,

$$\frac{\epsilon}{D} = \frac{0.0002 \text{ ft}}{0.3355 \text{ ft}} \approx 0.0006$$

From App. 17.B, the friction factor is $f = 0.01835$.
The friction head is given by Eq. 18.6.

$$h_f = \frac{fLv^2}{2Dg}$$

$$= \frac{(0.01835)(700 \text{ ft} + 66.5 \text{ ft})\left(14.139 \dfrac{\text{ft}}{\text{sec}}\right)^2}{(2)(0.3355 \text{ ft})\left(32.2 \dfrac{\text{ft}}{\text{sec}^2}\right)}$$

$$= 130.1 \text{ ft}$$

The total dynamic head is given by Eq. 18.9. Point s is taken as the bottom of the supply tank.

$$h = \frac{(p_d - p_s)g_c}{\rho g} + \frac{v_d^2 - v_s^2}{2g} + z_d - z_s$$

$$v_s \approx 0$$

$$z_d - z_s = 50 \text{ ft} \quad \text{[given as rise in elevation]}$$

The discharge and suction pressures are

$$p_d = 20 \text{ psig}$$
$$p_s = 50 \text{ psig}$$

$$h = \frac{\begin{pmatrix} \left(20 \dfrac{\text{lbf}}{\text{in}^2} - 50 \dfrac{\text{lbf}}{\text{in}^2}\right) \\ \times \left(144 \dfrac{\text{in}^2}{\text{ft}^2}\right)\left(32.2 \dfrac{\text{ft-lbm}}{\text{lbf-sec}^2}\right) \end{pmatrix}}{\left(62.3 \dfrac{\text{lbm}}{\text{ft}^3}\right)\left(32.2 \dfrac{\text{ft}}{\text{sec}^2}\right)}$$

$$+ \frac{\left(14.139 \dfrac{\text{ft}}{\text{sec}}\right)^2}{(2)\left(32.2 \dfrac{\text{ft}}{\text{sec}^2}\right)} + 50 \text{ ft}$$

$$= -16.2 \text{ ft}$$

The head added is

$$h_A = h + h_f = -16.2 \text{ ft} + 130.1 \text{ ft}$$
$$= 113.9 \text{ ft}$$

The mass flow rate is

$$\dot{m} = \rho \dot{V}$$

$$= \left(62.3 \frac{\text{lbm}}{\text{ft}^3}\right)\left(1.25 \frac{\text{ft}^3}{\text{sec}}\right)$$

$$= 77.875 \text{ lbm/sec}$$

From Table 18.4, the hydraulic horsepower is

$$\text{WHP} = \left(\frac{h_A \dot{m}}{550}\right) \times \left(\frac{g}{g_c}\right)$$

$$= \left(\frac{(113.9 \text{ ft})\left(77.875 \dfrac{\text{lbm}}{\text{sec}}\right)}{550 \dfrac{\text{ft-lbf}}{\text{hp-sec}}}\right)$$

$$\times \left(\frac{32.2 \dfrac{\text{ft}}{\text{sec}^2}}{32.2 \dfrac{\text{ft-lbm}}{\text{lbf-sec}^2}}\right)$$

$$= \boxed{16.13 \text{ hp}}$$

The answer is (A).

SI Solution

From App. 16.C, data for 4 in schedule-40 steel pipe are

$$D_i = 102.3 \text{ mm}$$
$$A_i = 82.19 \times 10^{-4} \text{ m}^2$$

The velocity in the pipe is

$$v = \frac{\dot{V}}{A} = \frac{\left(35 \frac{L}{s}\right)\left(\frac{1\ m^3}{1000\ L}\right)}{82.19 \times 10^{-4}\ m^2} = 4.26\ m/s$$

From Table 17.3, typical equivalent lengths for schedule-40, screwed steel fittings for 4 in pipes are

> 90° elbow: 13 ft
> gate valve: 2.5 ft
> check valve: 38.0 ft

The total equivalent length is

$$(2)(13\ ft) + (1)(2.5\ ft) + (1)(38\ ft) = 66.5\ ft$$
$$(66.5\ ft)\left(0.3048 \frac{m}{ft}\right) = 20.27\ m$$

At 21°C, from App. 14.B, the water properties are

$$\rho = 998\ kg/m^3$$
$$\mu = 0.9827 \times 10^{-3}\ Pa\cdot s$$
$$\nu = \frac{\mu}{\rho} = \frac{0.9827 \times 10^{-3}\ Pa\cdot s}{998 \frac{kg}{m^3}}$$
$$= 9.85 \times 10^{-7}\ m^2/s$$

The Reynolds number is

$$Re = \frac{Dv}{\nu} = \frac{(102.3\ mm)\left(\frac{1\ m}{1000\ mm}\right)\left(4.26 \frac{m}{s}\right)}{9.85 \times 10^{-7} \frac{m^2}{s}}$$
$$= 4.424 \times 10^5$$

From Table 17.2, for steel, $\epsilon = 6 \times 10^{-5}$ m.
So,

$$\frac{\epsilon}{D} = \frac{6.0 \times 10^{-5}\ m}{(102.3\ mm)\left(\frac{1\ m}{1000\ mm}\right)} = 0.0006$$

From App. 17.B, the friction factor is $f = 0.01836$.
From Eq. 18.6, the friction head is

$$h_f = \frac{fLv^2}{2Dg}$$
$$= \frac{(0.01836)(230\ m + 20.27\ m)\left(4.26 \frac{m}{s}\right)^2}{(2)(102.3\ mm)\left(\frac{1\ m}{1000\ mm}\right)\left(9.81 \frac{m}{s^2}\right)}$$
$$= 41.5\ m$$

The total dynamic head is given by Eq. 18.9. Point s is taken as the bottom of the supply tank.

$$h = \frac{p_d - p_s}{\rho g} + \frac{v_d^2 - v_s^2}{2g} + z_d - z_s$$
$$v_s \approx 0$$
$$z_d - z_s = 15\ m \quad [\text{given as rise in elevation}]$$

The difference between discharge and suction pressure is

$$p_d - p_s = 140\ kPa - 345\ kPa = -205\ kPa$$

$$h = \frac{(-205\ kPa)\left(1000 \frac{Pa}{kPa}\right)}{\left(998 \frac{kg}{m^3}\right)\left(9.81 \frac{m}{s^2}\right)}$$
$$+ \frac{\left(4.26 \frac{m}{s}\right)^2}{(2)\left(9.81 \frac{m}{s^2}\right)} + 15\ m$$
$$= -5.0\ m$$

The head added by the pump is

$$h_A = h + h_f = -5.0\ m + 41.5\ m$$
$$= 36.5\ m$$

The mass flow rate is

$$\dot{m} = \rho \dot{V}$$
$$= \left(998 \frac{kg}{m^3}\right)\left(35 \frac{L}{s}\right)\left(\frac{1\ m^3}{1000\ L}\right)$$
$$= 34.93\ kg/s$$

From Table 18.6, the hydraulic power is

$$WkW = \frac{(9.81)h_A\dot{m}}{1000}$$
$$= \frac{\left(9.81 \frac{m}{s^2}\right)(36.5\ m)\left(34.93 \frac{kg}{s}\right)}{1000 \frac{W}{kW}}$$
$$= \boxed{12.51\ kW}$$

The answer is (A).

3. *Customary U.S. Solution*

The area of the rubber hose is

$$A = \left(\frac{\pi}{4}\right)D^2 = \left(\frac{\pi}{4}\right)\left(\frac{2\ in}{12 \frac{in}{ft}}\right)^2 = 0.0218\ ft^2$$

Fluids

The velocity of water in the hose is

$$v = \frac{\dot{V}}{A} = \frac{\left(80 \ \frac{\text{gal}}{\text{min}}\right)\left(0.002228 \ \frac{\frac{\text{ft}^3}{\text{sec}}}{\frac{\text{gal}}{\text{min}}}\right)}{0.0218 \ \text{ft}^2}$$

$$= 8\,176 \ \text{ft/sec}$$

At 80°F from App. 14.A, the kinematic viscosity of water is $\nu = 0.93 \times 10^{-5} \ \text{ft}^2/\text{sec}$.

The Reynolds number is

$$\text{Re} = \frac{vD}{\nu} = \frac{\left(8.176 \ \frac{\text{ft}}{\text{sec}}\right)(2 \ \text{in})\left(\frac{1 \ \text{ft}}{12 \ \text{in}}\right)}{0.93 \times 10^{-5} \ \frac{\text{ft}^2}{\text{sec}}}$$

$$= 1.47 \times 10^5$$

Assume that the rubber hose is smooth. From App. 17.B, the friction factor is $f = 0.0166$.

From Eq. 18.6, the friction head is

$$h_f = \frac{fLv^2}{2Dg}$$

$$= \frac{(0.0166)(50 \ \text{ft})\left(8.176 \ \frac{\text{ft}}{\text{sec}}\right)^2}{(2)(2 \ \text{in})\left(\frac{1 \ \text{ft}}{12 \ \text{in}}\right)\left(32.2 \ \frac{\text{ft}}{\text{sec}^2}\right)}$$

$$= 5.17 \ \text{ft}$$

Neglecting entrance and exit losses, the head added by the pump is

$$h_A = h_f + h_z$$

$$= 5.17 \ \text{ft} + 12 \ \text{ft} - 4 \ \text{ft}$$

$$= \boxed{13.17 \ \text{ft}}$$

The answer is (B).

SI Solution

The area of the rubber hose is

$$A = \left(\frac{\pi}{4}\right)D^2 = \left(\frac{\pi}{4}\right)(5.1 \ \text{cm})^2 \left(\frac{1 \ \text{m}}{100 \ \text{cm}}\right)^2$$

$$= 0.00204 \ \text{m}^2$$

The velocity of water in the hose is

$$v = \frac{\dot{V}}{A} = \frac{\left(5 \ \frac{\text{L}}{\text{s}}\right)\left(\frac{1 \ \text{m}^3}{1000 \ \text{L}}\right)}{0.00204 \ \text{m}^2} = 2.45 \ \text{m/s}$$

At 27°C from App. 14.B, the water data are

$$\rho = 996.5 \ \text{kg/m}^3$$
$$\mu = 0.8565 \times 10^{-3} \ \text{Pa·s}$$
$$\nu = \frac{\mu}{\rho} = \frac{0.8565 \times 10^{-3} \ \text{Pa·s}}{996.5 \ \frac{\text{kg}}{\text{m}^3}}$$
$$= 8.60 \times 10^{-7} \ \text{m}^2/\text{s}$$

The Reynolds number is

$$\text{Re} = \frac{vD}{\nu} = \frac{\left(2.45 \ \frac{\text{m}}{\text{s}}\right)(5.1 \ \text{cm})\left(\frac{1 \ \text{m}}{100 \ \text{cm}}\right)}{8.60 \times 10^{-7} \ \frac{\text{m}^2}{\text{s}}}$$

$$= 1.45 \times 10^5$$

The rubber hose is smooth. From App. 17.B, the friction factor is $f \approx 0.0166$.

From Eq. 18.6, the friction head is

$$h_f = \frac{fLv^2}{2Dg}$$

$$= \frac{(0.0166)(15 \ \text{m})\left(2.45 \ \frac{\text{m}}{\text{s}}\right)^2}{(2)(5.1 \ \text{cm})\left(\frac{1 \ \text{m}}{100 \ \text{cm}}\right)\left(9.81 \ \frac{\text{m}}{\text{s}^2}\right)}$$

$$= 1.49 \ \text{m}$$

Neglecting entrance and exit losses, the head added by the pump is

$$h_A = h_f + h_z$$

$$= 1.49 \ \text{m} + 3.6 \ \text{m} - 1.2 \ \text{m}$$

$$= \boxed{3.89 \ \text{m}}$$

The answer is (B).

4. *Customary U.S. Solution*

From App. 16.B, the diameters (inside) for 8 in and 6 in schedule-40 steel pipe are

$$D_1 = 7.981 \ \text{in}$$
$$D_2 = 6.065 \ \text{in}$$

At 60°F from App. 14.A, the density of water is 62.37 lbm/ft³.

The mass flow rate through 6 in pipe is

$$\dot{m} = A_2 v_2 \rho$$

$$= \left(\frac{\pi}{4}\right)(6.065 \ \text{in})^2 \left(\frac{1 \ \text{ft}^2}{144 \ \text{in}^2}\right)\left(12 \ \frac{\text{ft}}{\text{sec}}\right)\left(62.37 \ \frac{\text{lbm}}{\text{ft}^3}\right)$$

$$= 150.2 \ \text{lbm/sec}$$

The inlet (suction) pressure is

$$14.7 \text{ psia} - 5 \text{ psig} = 9.7 \text{ psia}$$

$$= \left(9.7 \frac{\text{lbf}}{\text{in}^2}\right)\left(144 \frac{\text{in}^2}{\text{ft}^2}\right)$$

$$= 1397 \text{ lbf/ft}^2$$

Assume a motor efficiency of 100%. From Table 18.4, the head added by the pump is

$$h_A = \left(\frac{(550)(\text{BHP})\eta}{\dot{m}}\right) \times \left(\frac{g_c}{g}\right)$$

$$= \left(\frac{\left(550 \frac{\text{ft-lbf}}{\text{hp-sec}}\right)(20 \text{ hp})(0.70)}{150.2 \frac{\text{lbm}}{\text{sec}}}\right)\left(\frac{32.2 \frac{\text{ft-lbm}}{\text{lbf-sec}^2}}{32.2 \frac{\text{ft}}{\text{sec}^2}}\right)$$

$$= 51.26 \text{ ft}$$

At 1:

$$p_1 = 1397 \text{ lbf/ft}^2$$

$$z_1 = 0$$

$$v_1 = \frac{v_2 A_2}{A_1}$$

$$= v_2 \left(\frac{D_2}{D_1}\right)^2$$

$$= \left(12 \frac{\text{ft}}{\text{sec}}\right)\left(\frac{6.065 \text{ in}}{7.981 \text{ in}}\right)^2 = 6.93 \text{ ft/sec}$$

At 2:

$$p_2 = \left(14.7 \frac{\text{lbf}}{\text{in}^2}\right)\left(144 \frac{\text{in}^2}{\text{ft}^2}\right) = 2117 \text{ lbf/ft}^2$$

$$v_2 = 12 \text{ ft/sec} \quad [\text{given}]$$

From Eq. 18.9(b), the head added by the pump is

$$h_A = \frac{(p_2 - p_1)g_c}{\rho g} + \frac{v_2^2 - v_1^2}{2g} + z_2 - z_1 + h_f + z_3$$

$$51.26 \text{ ft} = \left(\frac{2117 \frac{\text{lbf}}{\text{ft}^2} - 1397 \frac{\text{lbf}}{\text{ft}^2}}{62.37 \frac{\text{lbm}}{\text{ft}^3}}\right) \times \left(\frac{32.2 \frac{\text{ft-lbm}}{\text{lbf-sec}^2}}{32.2 \frac{\text{ft}}{\text{sec}^2}}\right)$$

$$+ \frac{\left(12 \frac{\text{ft}}{\text{sec}}\right)^2 - \left(6.93 \frac{\text{ft}}{\text{sec}}\right)^2}{(2)\left(32.2 \frac{\text{ft}}{\text{sec}^2}\right)}$$

$$+ 0 - 0 + 10 \text{ ft} + z_3$$

$$\boxed{z_3 = 28.2 \text{ ft}}$$

The answer is (A).

SI Solution

From App. 16.C, the inside diameters for 8 in and 6 in steel schedule-40 pipe are

$$D_1 = 202.7 \text{ mm}$$

$$D_2 = 154.1 \text{ mm}$$

At 16°C from App. 14.B, the density of water is 998.83 kg/m³.

The mass flow rate through the 6 in pipe is

$$\dot{m} = A_2 v_2 \rho$$

$$= \left(\frac{\pi}{4}\right)(154.1 \text{ mm})^2\left(\frac{1 \text{ m}}{1000 \text{ mm}}\right)^2\left(4 \frac{\text{m}}{\text{s}}\right)$$

$$\times \left(998.83 \frac{\text{kg}}{\text{m}^3}\right)$$

$$= 74.5 \text{ kg/s}$$

The inlet (suction) pressure is

$$101.3 \text{ kPa} - 35 \text{ kPa} = 66.3 \text{ kPa}$$

Assume a motor efficiency of 100%. From Table 18.5, the head added by the pump is

$$h_A = \frac{(1000)(\text{BkW})\eta}{(9.81)\dot{m}}$$

$$= \frac{\left(1000 \frac{\text{W}}{\text{kW}}\right)(20 \text{ hp})\left(\frac{0.7457 \text{ kW}}{\text{hp}}\right)(0.70)}{\left(9.81 \frac{\text{m}}{\text{s}^2}\right)\left(74.5 \frac{\text{kg}}{\text{s}}\right)}$$

$$= 14.28 \text{ m}$$

At 1:

$$p_1 = 66.3 \text{ kPa}$$

$$z_1 = 0$$

$$v_1 = v_2\left(\frac{A_2}{A_1}\right) = v_2\left(\frac{D_2}{D_1}\right)^2$$

$$= \left(4 \frac{\text{m}}{\text{s}}\right)\left(\frac{154.1 \text{ mm}}{202.7 \text{ mm}}\right)^2 = 2.31 \text{ m/s}$$

At 2:

$$p_2 = 101.3 \text{ kPa}$$

$$v_2 = 4 \text{ m/s} \quad [\text{given}]$$

From Eq. 18.9(a), the head added by the pump is

$$h_A = \frac{p_2 - p_1}{\rho g} + \frac{v_2^2 - v_1^2}{2g} + z_2 - z_1 + h_f + z_3$$

$$14.28 \text{ m} = \frac{(101.3 \text{ kPa} - 66.3 \text{ kPa})\left(1000 \frac{\text{Pa}}{\text{kPa}}\right)}{\left(998.83 \frac{\text{kg}}{\text{m}^3}\right)\left(9.81 \frac{\text{m}}{\text{s}^2}\right)}$$

$$+ \frac{\left(4 \frac{\text{m}}{\text{s}}\right)^2 - \left(2.31 \frac{\text{m}}{\text{s}}\right)^2}{(2)\left(9.81 \frac{\text{m}}{\text{s}^2}\right)}$$

$$+ 0 - 0 + 3.3 \text{ m} + z_3$$

$$z_3 = \boxed{6.86 \text{ m}}$$

The answer is (A).

5. *Customary U.S. Solution*

The flow rate is

$$\dot{V} = \left(10{,}000 \frac{\text{gal}}{\text{hr}}\right)\left(0.1337 \frac{\text{ft}^3}{\text{gal}}\right) = 1337 \text{ ft}^3/\text{hr}$$

From App. 16.B, data for 4 in schedule-40 steel pipe are

$$D_i = 0.3355 \text{ ft}$$
$$A_i = 0.08841 \text{ ft}^2$$

The velocity in the pipe is

$$v = \frac{\dot{V}}{A} = \frac{\left(1337 \frac{\text{ft}^3}{\text{hr}}\right)\left(\frac{1 \text{ hr}}{3600 \text{ sec}}\right)}{0.08841 \text{ ft}^2} = 4.20 \text{ ft/sec}$$

At 60°F from App. 14.A, the kinematic viscosity of water is

$$\nu = 1.217 \times 10^{-5} \text{ ft}^2/\text{sec}$$
$$\rho = 62.37 \text{ lbm/ft}^3$$

The Reynolds number is

$$\text{Re} = \frac{Dv}{\nu} = \frac{(0.3355 \text{ ft})\left(4.20 \frac{\text{ft}}{\text{sec}}\right)}{1.217 \times 10^{-5} \frac{\text{ft}^2}{\text{sec}}}$$

$$= 1.16 \times 10^5$$

From App. 17.A, for welded and seamless steel, $\epsilon = 0.0002$ ft.

$$\frac{\epsilon}{D} = \frac{0.0002 \text{ ft}}{0.3355 \text{ ft}} \approx 0.0006$$

From App. 17.B, the friction factor, f, is 0.0205. The friction head is

$$h_f = \frac{fLv^2}{2Dg}$$

$$= \frac{(0.0205)(7000 \text{ ft})\left(4.2 \frac{\text{ft}}{\text{sec}}\right)^2}{(2)(0.3355 \text{ ft})\left(32.2 \frac{\text{ft}}{\text{sec}^2}\right)}$$

$$= 117.2 \text{ ft}$$

The head added by the pump is

$$h_A = h_f + h_z$$
$$= 117.2 \text{ ft} + 12 \text{ ft} + 350 \text{ ft}$$
$$= 479.5 \text{ ft}$$

From Table 18.4, the hydraulic horsepower is

$$\text{WHP} = \frac{h_A Q(\text{SG})}{3956}$$

$$= \frac{(479.2 \text{ ft})\left(10{,}000 \frac{\text{gal}}{\text{hr}}\right)\left(\frac{1 \text{ hr}}{60 \text{ min}}\right)(1)}{3956 \frac{\text{ft-gal}}{\text{hp-min}}}$$

$$= 20.2 \text{ hp}$$

From Eq. 18.16, the overall efficiency of the pump is

$$\eta = \frac{\text{WHP}}{\text{EHP}}$$

$$\text{EHP} = \frac{20.2 \text{ hp}}{0.7}$$
$$= 28.9 \text{ hp}$$

(a) At \$0.04/kW-hr, power costs for 1 hr are

$$(28.9 \text{ hp})\left(\frac{0.7457 \text{ kW}}{\text{hp}}\right)(1 \text{ hr})\left(\frac{\$0.04}{\text{kW-hr}}\right)$$

$$= \boxed{\$0.86 \text{ per hour}}$$

The answer is (B).

(b) The motor horsepower, EHP, is 28.9 hp. Select a

$$\boxed{30 \text{ hp motor.}}$$

The answer is (B).

Fluids

(c) From Eq. 18.5(b),

$$h_{atm} = \left(\frac{p_{atm}}{\rho}\right) \times \left(\frac{g_c}{g}\right)$$

$$= \left(\frac{\left(14.7 \frac{lbf}{in^2}\right)\left(144 \frac{in^2}{ft^2}\right)}{62.37 \frac{lbm}{ft^3}}\right) \times \left(\frac{32.2 \frac{ft}{sec^2}}{32.2 \frac{ft\text{-}lbm}{lbf\text{-}sec^2}}\right)$$

$$= 33.94 \text{ ft}$$

Since friction loss is proportional to length, the friction losses due to 300 ft is

$$h_{f(s)} = \left(\frac{300 \text{ ft}}{7000 \text{ ft}}\right) h_f$$

$$= \left(\frac{300 \text{ ft}}{7000 \text{ ft}}\right)(117.2 \text{ ft})$$

$$= 5.0 \text{ ft}$$

The vapor pressure at 60°F is 0.2564 psia.

From Eq. 18.5(b),

$$h_{vp} = \left(\frac{p_{vp}}{\rho}\right) \times \left(\frac{g_c}{g}\right)$$

$$= \left(\frac{\left(0.2564 \frac{lbf}{in^2}\right)\left(144 \frac{in^2}{ft^2}\right)}{62.37 \frac{lbm}{ft^3}}\right)$$

$$\times \left(\frac{32.2 \frac{ft\text{-}lbm}{lbf\text{-}sec^2}}{32.2 \frac{ft}{sec^2}}\right)$$

$$= 0.59 \text{ ft}$$

The NPSHA from Eq. 18.30(a) is

$$NPSHA = h_{atm} + h_{z(s)} - h_{f(s)} - h_{vp}$$

$$= 33.94 \text{ ft} - 12 \text{ ft} - 5.0 \text{ ft} - 0.59 \text{ ft}$$

$$= \boxed{16.35 \text{ ft}}$$

The answer is (D).

SI Solution

From App. 16.C, data for 4 in schedule-40 steel pipe are

$$D_i = 102.3 \text{ mm}$$

$$A_i = 82.19 \times 10^{-4} \text{ m}^2$$

The velocity in the pipe is

$$v = \frac{\dot{V}}{A} = \frac{\left(10.5 \frac{L}{s}\right)\left(\frac{1 \text{ m}^3}{1000 \text{ L}}\right)}{82.19 \times 10^{-4} \text{ m}^2} = 1.28 \text{ m/s}$$

From App. 14.B, at 16°C the water data are

$$\rho = 998.83 \text{ kg/m}^3$$

$$\mu = 1.1261 \times 10^{-3} \text{ Pa·s}$$

The Reynolds number is

$$Re = \frac{\rho v D}{\mu}$$

$$= \frac{\left(998.83 \frac{kg}{m^3}\right)\left(1.28 \frac{m}{s}\right)}{1.1261 \times 10^{-3} \text{ Pa·s}}$$

$$\times (102.3 \text{ mm})\left(\frac{1 \text{ m}}{1000 \text{ mm}}\right)$$

$$= 1.16 \times 10^5$$

From Table 17.2, for welded and seamless steel, $\epsilon = 6.0 \times 10^{-5}$ m.

$$\frac{\epsilon}{D} = \frac{6.0 \times 10^{-5} \text{ m}}{(102.3 \text{ mm})\left(\frac{1 \text{ m}}{1000 \text{ mm}}\right)} = 0.0006$$

From App. 17.B, the friction factor is $f = 0.0205$.

From Eq. 18.6, the friction head is

$$h_f = \frac{fLv^2}{2Dg}$$

$$= \frac{(0.0205)(2300 \text{ m})\left(1.28 \frac{m}{s}\right)^2}{(2)(102.3 \text{ mm})\left(\frac{1 \text{ m}}{1000 \text{ mm}}\right)\left(9.81 \frac{m}{s^2}\right)}$$

$$= 38.5 \text{ m}$$

The head added by the pump is

$$h_A = h_f + h_z$$

$$= 38.5 \text{ m} + 4 \text{ m} + 115 \text{ m}$$

$$= 157.5 \text{ m}$$

From Table 18.5, the hydraulic power is

$$WkW = \frac{(9.81)h_A Q(SG)}{1000}$$

$$= \frac{\left(9.81 \frac{m}{s^2}\right)(157.5 \text{ m})\left(10.5 \frac{L}{s}\right)(1)}{1000 \frac{W \cdot L}{kW \cdot kg}}$$

$$= 16.22 \text{ kW}$$

From Eq. 18.16,

$$EHP = \frac{WHP}{\eta_{overall}}$$

$$= \frac{16.22 \text{ kW}}{0.7} = 23.2 \text{ kW}$$

(a) At $0.04/kW·h, power costs for 1 h are

$$(23.2 \text{ kW})(1 \text{ h}) \left(\frac{\$0.04}{\text{kW·h}} \right) = \boxed{\$0.93 \text{ per hour}}$$

The answer is (B).

(b) The required motor power is 23.2 kW. Select the next higher standard motor size.

The answer is (B).

(c) From Eq. 18.5(a),

$$h_{\text{atm}} = \frac{p}{\rho g}$$
$$= \frac{(101 \text{ kPa})(1000 \text{ Pa})}{\left(998.83 \dfrac{\text{kg}}{\text{m}^3} \right) \left(9.81 \dfrac{\text{m}}{\text{s}^2} \right)}$$
$$= 10.31 \text{ m}$$

Since friction loss is proportional to length, the friction loss due to 100 m is

$$h_{f(s)} = \left(\frac{100 \text{ m}}{2300 \text{ m}} \right) h_f$$
$$= \left(\frac{100 \text{ m}}{2300 \text{ m}} \right) (38.5 \text{ m})$$
$$= 1.67 \text{ m}$$

The vapor pressure at 16°C is 0.01819 bar.

From Eq. 18.5(a),

$$h_{\text{vp}} = \frac{p_{\text{vp}}}{g\rho}$$
$$= \frac{(0.01819 \text{ bar}) \left(1 \times 10^5 \dfrac{\text{Pa}}{\text{bar}} \right)}{\left(9.81 \dfrac{\text{m}}{\text{s}^2} \right) \left(998.83 \dfrac{\text{kg}}{\text{m}^3} \right)}$$
$$= 0.19 \text{ m}$$

The NPSHA from Eq. 18.30(a) is

$$\text{NPSHA} = h_{\text{atm}} + h_{z(s)} - h_{f(s)} - h_{\text{vp}}$$
$$= 10.31 \text{ m} - 4 \text{ m} - 1.67 \text{ m} - 0.19 \text{ m}$$
$$= \boxed{4.45 \text{ m}}$$

The answer is (D).

6. (a) $h_f = \Delta z = 48$ ft since $\Delta p = 0$ and $\Delta v = 0$ for gravity flow (typical for sewers).

$$Q = \frac{\left(250 \dfrac{\text{gal}}{\text{person·day}} \right) (10{,}000 \text{ people})}{\left(24 \dfrac{\text{hr}}{\text{day}} \right) \left(60 \dfrac{\text{min}}{\text{hr}} \right)}$$
$$= 1736 \text{ gal/min}$$

Given $C = 130$, from Eq. 17.31,

$$d_{\text{in}}^{4.8655} = \frac{(10.44)(5000 \text{ ft}) \left(1736 \dfrac{\text{gal}}{\text{min}} \right)^{1.85}}{(130)^{1.85}(48 \text{ ft})} = 131{,}462$$

$$d = \boxed{11.27 \text{ in} \quad [\text{round to 12 in minimum}]}$$

The answer is (B).

(b) Without having a specific pump curve, the number of pumps can only be specified based on general rules.

- No station will have less than two identical pumps.

- Full capacity must be maintained with one pump out of service.

> Two pumps are required, plus one backup.

The answer is (A).

(c) With a variable speed pump, it will be possible to adjust to the wide variations in flow (100 to 250 gpcd). It may be possible to operate with one pump and one backup.

The answer is (A).

(d) With three constant speed pumps,

$$Q = \frac{1736 \dfrac{\text{gal}}{\text{min}}}{3} = 579 \text{ gal/min at maximum capacity}$$

From Table 18.4, assuming specific gravity ≈ 1.00 and using $\eta_{\text{pump}} = 0.60$,

$$\text{rated motor power} = \frac{(10 \text{ ft}) \left(579 \dfrac{\text{gal}}{\text{min}} \right)}{\left(3956 \dfrac{\text{gal-ft}}{\text{min-hp}} \right) (0.60)}$$
$$= \boxed{2.44 \text{ hp} \quad [\text{use 3 hp}]}$$

The answer is (B).

(e) With two variable-speed pumps,

$$Q = \frac{1736 \frac{\text{gal}}{\text{min}}}{2} = 868 \text{ gal/min}$$

$$\text{rated motor power} = \frac{(10 \text{ ft})\left(868 \frac{\text{gal}}{\text{min}}\right)}{\left(3956 \frac{\text{gal-ft}}{\text{min-hp}}\right)(0.80)}$$

$$= \boxed{2.74 \text{ hp} \quad [\text{use } 3.0 \text{ hp}]}$$

The answer is (A).

(f) Assume the question refers to demand control rather than to motor speed control. Methods of control are

- variable run times
- pressure change in sump
- flow rate incoming
- sump level
- flow rate outgoing
- manual on/off

The answer is (B).

(g) In the wet well, the pump (and perhaps motor) is submerged. Forced ventilation air will prevent a concentration of explosive methane.

$$\boxed{\begin{array}{l} 12 \text{ air changes per hour if continuous;} \\ 30 \text{ per hour if intermittent.} \end{array}}$$

The answer is (D).

(h) In the dry well,

$$\boxed{\begin{array}{l} 6 \text{ air changes per hour if continuous;} \\ 30 \text{ per hour if intermittent.} \end{array}}$$

The answer is (A).

7. (a) $h_{p(s)} = \boxed{20 \text{ ft}}$

The answer is (B).

(b) $h_{p(d)} = \boxed{15 \text{ ft}}$

The answer is (A).

(c) There is no pipe size specified, so h_v cannot be calculated. Even so, v is typically in the 5 to 10 ft/sec range, and $h_v \approx 0$.

Since pipe lengths are not given, assume $h_f \approx 0$. Both ends of the pipe system are exposed to atmospheric pressure, so atmospheric pressure is omitted.

$$20 \text{ ft} + 15 \text{ ft} + \frac{\left(80 \frac{\text{lbf}}{\text{in}^2}\right)\left(144 \frac{\text{in}^2}{\text{ft}^2}\right)}{62.4 \frac{\text{lbf}}{\text{ft}^3}} + 10 \text{ ft}$$

$$= \boxed{229.6 \text{ ft of water}}$$

The answer is (D).

(d) The flow rate is

$$\frac{(3.5 \text{ MGD})\left(62.4 \frac{\text{lbf}}{\text{ft}^3}\right)}{0.64632 \frac{\text{MGD}}{\frac{\text{ft}^3}{\text{sec}}}} = 337.9 \text{ lbf/sec}$$

From Table 18.4, the rated motor output power does not depend on the motor efficiency.

$$P = \frac{(229.6 \text{ ft})\left(337.9 \frac{\text{lbf}}{\text{sec}}\right)}{\left(550 \frac{\text{ft-lbf}}{\text{hp-sec}}\right)(0.85)} = 166.0 \text{ hp}$$

$$\boxed{\text{Use a 200 hp motor.}}$$

The answer is (D).

8. *Customary U.S. Solution*

(a) The equivalent lengths are assumed to be in feet of section's diameters. Since the diameters are different for many sections, the equivalent lengths cannot be added together. It is necessary to convert the lengths to pressure drop using the Darcy equation (Eq. 17.22). (A similar process can be carried out more expediently if a table of standard losses in copper tubing is used.)

From App. 14.A, for 70°F water

$$\rho = 62.3 \text{ lbm/ft}^3$$

$$\nu = 1.059 \times 10^{-5} \text{ ft}^2/\text{sec}$$

From App. 17.A, the pipe roughness is $\epsilon = 0.000005$ ft.

From App. 16.D, the data for $2\frac{1}{2}$ in type L pipe are

$$D_i = 2.465 \text{ in}$$

$$A_i = 4.77 \text{ in}^2$$

The flow rate is converted from gal/min to ft³/sec.

$$\dot{V} = \left(60 \frac{\text{gal}}{\text{min}}\right)\left(0.002228 \frac{\frac{\text{ft}^3}{\text{sec}}}{\frac{\text{gal}}{\text{min}}}\right)$$

$$= 0.13368 \text{ ft}^3/\text{sec}$$

The velocity in the pipe is

$$v = \frac{\dot{V}}{A} = \frac{0.13368 \frac{ft^3}{sec}}{(4.77 \ in^2)\left(\frac{1 \ ft^2}{144 \ in^2}\right)} = 4.04 \ ft/sec$$

The Reynolds number is

$$Re = \frac{Dv}{\nu} = \frac{(2.465 \ in)\left(\frac{1 \ ft}{12 \ in}\right)\left(4.04 \ \frac{ft}{sec}\right)}{1.059 \times 10^{-5} \ \frac{ft^2}{sec}}$$

$$= 7.84 \times 10^4$$

$$\frac{\epsilon}{D} = \frac{0.000005 \ ft}{(2.465 \ in)\left(\frac{1 \ ft}{12 \ in}\right)} \approx 0.000025$$

From the friction factor table (App. 17.B), $f \approx 0.019$.

From Eq. 17.22, the friction head loss is

$$h_f = \frac{fLv^2}{2Dg}$$

$$= \frac{(0.019)(40 \ ft)\left(4.04 \ \frac{ft}{sec}\right)^2}{(2)(2.465 \ in)\left(\frac{1 \ ft}{12 \ in}\right)\left(32.2 \ \frac{ft}{sec^2}\right)}$$

$$= 0.94 \ ft$$

For all other sections, using the similar procedure, the following table can be constructed.

circuit	equivalent length (ft)	total loss (ft)
5-1-P-2	40	0.94
2-4	70	2.44
2-3	55	1.76
3-4	65	2.27
3-5	60	2.10
4-5	50	1.60

These head losses are for cold water and are greater than hot water losses. The pumping power will be less for hot water. However, the system may be tested with cold water. Therefore, horsepower must be adequate for this system.

Redraw the pipe network.

The pump must be able to handle the longest run.

run	total loss
2-4-5	2.44 ft + 1.60 ft = 4.04 ft
2-3-4-5	1.76 ft + 2.27 ft + 1.60 ft = 5.63 ft
2-3-5	1.76 ft + 2.10 ft = 3.86 ft

The longest run is 2-3-4-5 with a loss of 5.63 ft. The total circuit loss is 0.94 ft + 5.63 ft = 6.57 ft. Note that the loss between points 2 and 5 must be the same regardless of the path. Therefore, flow restriction valves must exist in the other two runs in order to match the longest run's loss.

From Table 18.4, the hydraulic horsepower is

$$WHP = \frac{h_A Q(SG)}{3956}$$

$$= \frac{(6.57 \ ft)\left(60 \ \frac{gal}{min}\right)(1)}{3956 \ \frac{ft\text{-}gal}{hp\text{-}min}}$$

$$= 0.10 \ hp$$

From Eq. 18.16, the motor horsepower is

$$P = \frac{WHP}{\eta} = \frac{0.10 \ hp}{0.45} = \boxed{0.22 \ hp}$$

The answer is (A).

(b) Use a $\boxed{1/4 \ hp \ \text{or larger motor.}}$

The answer is (A).

(c) For water, the mass flow rate is

$$\dot{m} = \left(60 \ \frac{gal}{min}\right)\left(0.1337 \ \frac{ft^3}{gal}\right)\left(62.30 \ \frac{lbm}{ft^3}\right)$$

$$= 500 \ lbm/min$$

The heat added is

$$q = \dot{m}c_p\Delta T$$

$$= \left(500 \ \frac{lbm}{min}\right)\left(60 \ \frac{min}{hr}\right)\left(1 \ \frac{Btu}{lbm\text{-}°F}\right)(20°F)$$

$$= \boxed{600{,}000 \ Btu/hr}$$

The answer is (C).

Fluids

SI Solution

(a) Convert the lengths to pressure drop using the Darcy equation (Eq. 17.22).

From App. 14.B, the water data at 20°C are

$$\rho = 998.23 \text{ kg/m}^3$$
$$\mu = 1.0050 \times 10^{-3} \text{ Pa·s}$$

From Table 17.2, the pipe roughness is $\epsilon = 1.5 \times 10^{-6}$ m.

From App. 16.D, the data for $2^1/_2$ in type-L pipe are

$$D_i = (2.465 \text{ in}) \left(2.540 \frac{\text{cm}}{\text{in}}\right) \left(\frac{1 \text{ m}}{100 \text{ cm}}\right)$$
$$= 0.0626 \text{ m}$$

$$A_i = (4.77 \text{ in}^2) \left(\frac{1 \text{ ft}^2}{144 \text{ in}^2}\right) \left(0.0929 \frac{\text{m}^2}{\text{ft}^2}\right)$$
$$= 3.08 \times 10^{-3} \text{ m}^2$$

The velocity in the pipe is

$$\text{v} = \frac{\dot{V}}{A} = \frac{\left(3.9 \frac{\text{L}}{\text{s}}\right) \left(\frac{1 \text{ m}^3}{1000 \text{ L}}\right)}{3.08 \times 10^{-3} \text{ m}^2} = 1.27 \text{ m/s}$$

The Reynolds number is

$$\text{Re} = \frac{\rho \text{v} D}{\mu}$$
$$= \frac{\left(998.23 \frac{\text{kg}}{\text{m}^3}\right) \left(1.27 \frac{\text{m}}{\text{s}}\right) (0.0626 \text{ m})}{1.0050 \times 10^{-3} \text{ Pa·s}}$$
$$= 7.9 \times 10^4$$

$$\frac{\epsilon}{D} = \frac{1.5 \times 10^{-6} \text{ m}}{0.0626 \text{ m}} \approx 0.000025$$

From the friction factor table (App. 17.B), $f \approx 0.019$.

From Eq. 17.22, the friction head loss is

$$h_f = \frac{fL\text{v}^2}{2Dg} = \frac{(0.019)(13 \text{ m}) \left(1.27 \frac{\text{m}}{\text{s}}\right)^2}{(2)(0.0626 \text{ m}) \left(9.81 \frac{\text{m}}{\text{s}^2}\right)}$$
$$= 0.324 \text{ m}$$

For all other sections, using the similar procedure, the following table can be constructed.

circuit	equivalent length (m)	total loss (m)
5-1-P-2	13	0.324
2-4	23	0.850
2-3	18	0.612
3-4	22	0.813
3-5	20	0.739
4-5	17	0.578

Redraw the pipe network.

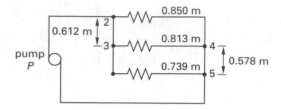

The pump must be able to handle the longest run.

run	total loss
2-4-5	0.850 m + 0.578 m = 1.43 m
2-3-4-5	0.612 m + 0.813 m + 0.578 m = 2.00 m
2-3-5	0.612 m + 0.739 m = 1.35 m

The longest run is 2-3-4-5 with a loss of 2.00 m. The total circuit loss is 0.324 m + 2.00 m = 2.324 m. Note that the loss between points 2 and 5 must be the same regardless of the path. Therefore, flow restriction valves must exist in the other two runs in order to match the longest run's loss.

From Table 18.5, hydraulic power is

$$\text{WkW} = \frac{9.81 h_A Q (\text{SG})}{1000}$$
$$= \frac{\left(9.81 \frac{\text{m}}{\text{s}^2}\right) (2.324 \text{ m}) \left(3.9 \frac{\text{L}}{\text{s}}\right) (1)}{1000 \frac{\text{W}}{\text{kW}}}$$
$$= 0.0889 \text{ kW}$$

From Eq. 18.16, the motor power is

$$P = \frac{\text{WkW}}{\eta} = \frac{0.0889 \text{ kW}}{0.45}$$
$$= \boxed{0.198 \text{ kW}}$$

The answer is (A).

(b) Use a $\boxed{0.2 \text{ kW or larger motor.}}$

The answer is (A).

(c) For water, the mass flow rate is

$$\dot{m} = \left(3.9 \frac{\text{L}}{\text{s}}\right) \left(\frac{1 \text{ m}^3}{1000 \text{ L}}\right) \left(998.23 \frac{\text{kg}}{\text{m}^3}\right)$$
$$= 3.89 \text{ kg/s}$$

The heat added is

$$q = \dot{m} c_p \Delta T$$
$$= \left(3.89 \frac{\text{kg}}{\text{s}}\right) \left(4.187 \frac{\text{kJ}}{\text{kg·C}}\right) (10°\text{C})$$
$$= \boxed{162.9 \text{ kW}}$$

The answer is (C).

9. *Customary U.S. Solution*

From App. 16.B, the pipe data for 3 in schedule-40 steel pipe are

$$D_i = 0.2557 \text{ ft}$$
$$A_i = 0.05134 \text{ ft}^2$$

From App. 17.D, the equivalent length for various fittings is

flanged elbow, $L_e = 4.4$ ft

wide-open gate valve, $L_e = 2.8$ ft

The total equivalent length of pipe and fittings is

$$L_e = 500 \text{ ft} + (6)(4.4 \text{ ft}) + (2)(2.8 \text{ ft})$$
$$= 532 \text{ ft}$$

As a first estimate, assume the flow rate is 100 gal/min.

The velocity in the pipe is

$$v = \frac{\dot{V}}{A} = \frac{\left(100 \; \frac{\text{gal}}{\text{min}}\right)\left(0.002228 \; \frac{\frac{\text{ft}^3}{\text{sec}}}{\frac{\text{gal}}{\text{min}}}\right)}{0.05134 \text{ ft}^2}$$
$$= 4.34 \text{ ft/sec}$$

The Reynolds number is

$$\text{Re} = \frac{vD}{\nu} = \frac{\left(4.34 \; \frac{\text{ft}}{\text{sec}}\right)(0.2557 \text{ ft})}{6 \times 10^{-6} \; \frac{\text{ft}^2}{\text{sec}}}$$
$$= 1.85 \times 10^5$$

From App. 17.A, $\epsilon = 0.0002$ ft.

So,

$$\frac{\epsilon}{D} = \frac{0.0002 \text{ ft}}{0.2557 \text{ ft}} \approx 0.0008$$

From the friction factor table, $f \approx 0.0204$.

For higher flow rates, f approaches 0.0186. Since the chosen flow rate was almost the lowest, $f = 0.0186$ should be used.

From Eq. 18.6, the friction head loss is

$$h_f = \frac{fLv^2}{2Dg}$$
$$= \frac{(0.0186)(532 \text{ ft})\left(4.34 \; \frac{\text{ft}}{\text{sec}}\right)^2}{(2)(0.2557 \text{ ft})\left(32.2 \; \frac{\text{ft}}{\text{sec}^2}\right)}$$
$$= 11.3 \text{ ft of gasoline}$$

This neglects the small velocity head. The other system points can be found using

$$\frac{h_{f_1}}{h_{f_2}} = \left(\frac{Q_1}{Q_2}\right)^2$$

$$h_{f_2} = h_{f_1}\left(\frac{Q_2}{100 \; \frac{\text{gal}}{\text{min}}}\right)^2$$

$$= (11.3 \text{ ft})\left(\frac{Q_2}{100 \; \frac{\text{gal}}{\text{min}}}\right)^2$$

$$= 0.00113Q_2^2$$

Q (gal/min)	h_f (ft)	$h_f + 60$ (ft)
100	11.3	71.3
200	45.2	105.2
300	101.7	161.7
400	180.8	240.8
500	282.5	342.5
600	406.8	466.8

(a) Plot the system and pump curves.

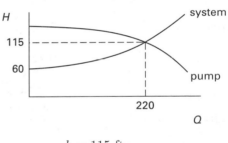

$$h = 115 \text{ ft}$$

$$\boxed{Q = 227 \text{ gal/min}}$$

The answer is (D).

(b) From Table 18.4, the hydraulic horsepower is

$$\text{WHP} = \frac{h_A Q(\text{SG})}{3956} = \frac{(115 \text{ ft})\left(220 \; \frac{\text{gal}}{\text{min}}\right)(0.7)}{3956 \; \frac{\text{ft-gal}}{\text{hp-min}}}$$

$$= 4.48 \text{ hp}$$

$$\frac{(4.48 \text{ hp})\left(0.7457 \; \frac{\text{kW}}{\text{hp}}\right)}{\times (1 \text{ hr})\left(0.045 \; \frac{\$}{\text{kW-hr}}\right)}{(0.88)(0.88)} = \boxed{\$0.19}$$

The answer is (A).

SI Solution

From App. 16.C, the pipe data for 3 in schedule-40 pipe are

$$D_i = 77.92 \text{ mm}$$
$$A_i = 47.69 \times 10^{-4} \text{ m}^2$$

From App. 17.D, the equivalent length for various fittings is

flanged elbow, $L_e = 4.4$ ft

wide-open gate valve, $L_e = 2.8$ ft

The total equivalent length of pipe and fittings is

$$L_e = 170 \text{ m} + (6)(4.4 \text{ ft})\left(0.3048 \frac{\text{m}}{\text{ft}}\right)$$
$$+ (2)(2.8 \text{ ft})\left(0.3048 \frac{\text{m}}{\text{ft}}\right)$$
$$= 180 \text{ m}$$

As a first estimate, assume flow rate is 6.3 L/s. The velocity in the pipe is

$$v = \frac{\dot{V}}{A} = \frac{\left(6.3 \frac{\text{L}}{\text{s}}\right)\left(\frac{1 \text{ m}^3}{1000 \text{ L}}\right)}{47.69 \times 10^{-4} \text{ m}^2} = 1.32 \text{ m/s}$$

The Reynolds number is

$$\text{Re} = \frac{vD}{\nu}$$

$$= \frac{\left(1.32 \frac{\text{m}}{\text{s}}\right)(77.92 \text{ mm})\left(\frac{1 \text{ m}}{1000 \text{ mm}}\right)}{5.6 \times 10^{-7} \frac{\text{m}^2}{\text{s}}}$$

$$= 1.75 \times 10^5$$

From Table 17.2, $\epsilon = 6.0 \times 10^{-5}$ m.

$$\frac{\epsilon}{D} = \frac{6.0 \times 10^{-5} \text{ m}}{(77.92 \text{ mm})\left(\frac{1 \text{ m}}{1000 \text{ mm}}\right)} \approx 0.0008$$

From the friction factor table (App. 17.B), $f = 0.0205$.

For higher flow rates, f approaches 0.0186. Since the chosen flow rate was almost the lowest, $f = 0.0186$ should be used.

From Eq. 17.22, the friction head loss is

$$h_f = \frac{fLv^2}{2Dg} = \frac{(0.0186)(180 \text{ m})\left(1.32 \frac{\text{m}}{\text{s}}\right)^2}{(2)(77.92 \text{ mm})\left(\frac{1 \text{ m}}{1000 \text{ mm}}\right)\left(9.81 \frac{\text{m}}{\text{s}^2}\right)}$$

$$= 3.82 \text{ m of gasoline}$$

This neglects the small velocity head. The other system points can be found using

$$\frac{h_{f_1}}{h_{f_2}} = \left(\frac{Q_1}{Q_2}\right)^2$$

$$h_{f_2} = h_{f_1}\left(\frac{Q_2}{Q_1}\right)^2 = (3.82 \text{ m})\left(\frac{Q_2}{6.3 \frac{\text{L}}{\text{s}}}\right)^2$$

$$= 0.0962 Q_2^2$$

Q (L/s)	h_f (m)	$h_f + 20$ (m)
6.3	3.82	23.82
12	13.85	33.85
18	31.2	51.2
24	55.4	75.4
30	86.6	106.6
36	124.7	144.7

(a) Plot the system and pump curves.

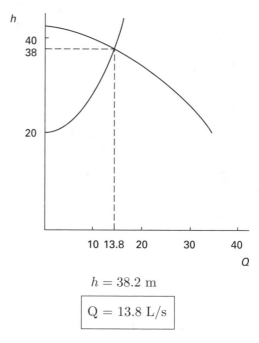

$$h = 38.2 \text{ m}$$

$$\boxed{Q = 13.8 \text{ L/s}}$$

The answer is (D).

(b) From Table 18.5, the hydraulic power is

$$\text{WkW} = \frac{9.81 h_A Q(\text{SG})}{1000}$$

$$= \frac{\left(9.81 \frac{\text{m}}{\text{s}^2}\right)(38.2 \text{ m})\left(13.8 \frac{\text{L}}{\text{s}}\right)(0.7)}{1000 \frac{\text{W}}{\text{kW}}}$$

$$= 3.62 \text{ kW}$$

The cost per hour is

$$= \left(\frac{3.62 \text{ kW}}{(0.88)(0.88)} \right) (1 \text{ h}) \left(0.045 \text{ } \frac{\$}{\text{kW·h}} \right)$$

$$= \boxed{\$0.21}$$

The answer is (A).

10. *Customary U.S. Solution*

From Table 18.4, the hydraulic horsepower is

$$\text{WHP} = \frac{\Delta p Q}{1714}$$

$$\Delta p = p_d - p_s$$

The absolute pressures are

$$p_d = 40 \text{ psig} + 14.7 = 54.7 \text{ psia}$$

$$p_s = 1 \text{ atm} = 14.7 \text{ psia}$$

$$\Delta p = 54.7 \text{ psia} - 14.7 \text{ psia} = 40 \text{ psia}$$

$$\text{WHP} = \frac{\left(40 \text{ } \frac{\text{lbf}}{\text{in}^2} \right) \left(37 \text{ } \frac{\text{gal}}{\text{min}} \right)}{1714 \text{ } \frac{\text{lbf-gal}}{\text{in}^2\text{-min-hp}}} = \boxed{0.863 \text{ hp}}$$

The answer is (B).

SI Solution

From Table 18.5, the hydraulic kilowatts are

$$\text{WkW} = \frac{\Delta p Q}{1000}$$

$$\Delta p = p_d - p_s$$

The absolute pressures are

$$p_d = 275 \text{ kPa} + 101.3 \text{ kPa} = 376.3 \text{ kPa}$$

$$p_s = 1 \text{ atm} = 101.3 \text{ kPa}$$

$$\Delta p = 376.3 \text{ kPa} - 101.3 \text{ kPa} = 275 \text{ kPa}$$

$$\text{WkW} = \frac{(275 \text{ kPa}) \left(65 \text{ } \frac{\text{L}}{\text{s}} \right)}{1000 \text{ } \frac{\text{W}}{\text{kW}}} = \boxed{17.88 \text{ kW}}$$

The answer is (B).

11. *Customary U.S. Solution*

From Eq. 18.28(b), the specific speed is

$$n_s = \frac{n\sqrt{Q}}{h_A^{0.75}}$$

For a double-suction pump, Q in the preceding equation is half of the full flow rate.

$$n_s = \frac{(900 \text{ rpm})\sqrt{\left(300 \text{ } \frac{\text{gal}}{\text{sec}} \right) \left(60 \text{ } \frac{\text{sec}}{\text{min}} \right) \left(\frac{1}{2} \right)}}{(20 \text{ ft})^{0.75}}$$

$$= \boxed{9028 \text{ rpm}}$$

The answer is (C).

SI Solution

From Eq. 18.28(a), the specific speed is

$$n_s = \frac{n\sqrt{\dot{V}}}{h_A^{0.75}}$$

For a double-suction pump, $\dot{V}$ in the preceding equation is half of the full flow rate.

$$n_s = \frac{(900 \text{ rpm})\sqrt{\left(1.1 \text{ } \frac{\text{kL}}{\text{s}} \right) \left(\frac{1 \text{ m}^3}{1 \text{ kL}} \right) \left(\frac{1}{2} \right)}}{(7 \text{ m})^{0.75}}$$

$$= \boxed{155.1 \text{ rpm}}$$

The answer is (C).

12. *Customary U.S. Solution*

This problem is solved graphically using charts from *Standards of the Hydraulic Institute*.

Since each stage adds 150 ft of head and the suction lift is 10 ft, for a single-suction pump, $n_s \approx 2050$ rpm.

The answer is (D).

13.

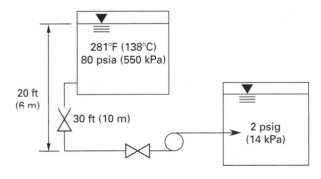

Customary U.S. Solution

From App. 16.B, data for 1.5 in schedule-40 steel pipe are

$$D_i = 0.1342 \text{ ft}$$

$$A_i = 0.01414 \text{ ft}^2$$

The velocity in the pipe is

$$v = \frac{\dot{V}}{A} = \frac{\left(100 \; \frac{\text{gal}}{\text{min}}\right)\left(0.002228 \; \frac{\text{ft}^3\text{-min}}{\text{sec-gal}}\right)}{0.01414 \; \text{ft}^2}$$

$$= 15.76 \; \text{ft/sec}$$

From App. 17.D, for screwed steel fittings, the approximate equivalent lengths for fittings are

inlet (square mouth): $L_e = 3.1$ ft

long radius 90° ell: $L_e = 3.4$ ft

wide-open gate valves: $L_e = 1.2$ ft

The total equivalent length is

$$30 \; \text{ft} + 3.1 \; \text{ft} + (2)(3.4 \; \text{ft}) + (2)(1.2 \; \text{ft}) = 42.3 \; \text{ft}$$

From App. 17.A, for steel, $\epsilon = 0.0002$ ft, so

$$\frac{\epsilon}{D} = \frac{0.0002 \; \text{ft}}{0.1342 \; \text{ft}} = 0.0015$$

At 281°F, from App. 35.A, $\nu = 0.239 \times 10^{-5} \; \text{ft}^2/\text{sec}$. The Reynolds number is

$$Re = \frac{Dv}{\nu} = \frac{(0.1342 \; \text{ft})\left(15.76 \; \frac{\text{ft}}{\text{sec}}\right)}{0.239 \times 10^{-5} \; \frac{\text{ft}^2}{\text{sec}}}$$

$$= 8.85 \times 10^5$$

From App. 17.B, the friction factor is $f = 0.022$.

From Eq. 18.6, the friction head is

$$h_f = \frac{fLv^2}{2Dg}$$

$$= \frac{(0.022)(42.3 \; \text{ft})\left(15.76 \; \frac{\text{ft}}{\text{sec}}\right)^2}{(2)(0.1342 \; \text{ft})\left(32.2 \; \frac{\text{ft}}{\text{sec}^2}\right)}$$

$$= 26.74 \; \text{ft}$$

At 281°F,

$$p_{\text{vapor}} = 50.06 \; \text{psia} \quad \text{[from steam tables]}$$

$$\rho = \frac{1}{v_f} = \frac{1}{0.01727 \; \frac{\text{ft}^3}{\text{lbm}}} = 57.9 \; \text{lbm/ft}^3$$

From Eq. 18.5(b),

$$h_{\text{vp}} = \left(\frac{p_{\text{vapor}}}{\rho}\right) \times \left(\frac{g_c}{g}\right)$$

$$= \left(\frac{\left(50.06 \; \frac{\text{lbf}}{\text{in}^2}\right)\left(144 \; \frac{\text{in}^2}{\text{ft}^2}\right)}{57.9 \; \frac{\text{lbm}}{\text{ft}^3}}\right) \times \left(\frac{32.2 \; \frac{\text{ft-lbm}}{\text{lbf-sec}^2}}{32.2 \; \frac{\text{ft}}{\text{sec}^2}}\right)$$

$$= 124.5 \; \text{ft}$$

From Eq. 18.5(b), the pressure head is

$$h_p = \left(\frac{p}{\rho}\right) \times \left(\frac{g_c}{g}\right)$$

$$= \left(\frac{\left(80 \; \frac{\text{lbf}}{\text{in}^2}\right)\left(144 \; \frac{\text{in}^2}{\text{ft}^2}\right)}{57.9 \; \frac{\text{lbm}}{\text{ft}^3}}\right) \times \left(\frac{32.2 \; \frac{\text{ft-lbm}}{\text{lbf-sec}^2}}{32.2 \; \frac{\text{ft}}{\text{sec}^2}}\right)$$

$$= 199.0 \; \text{ft}$$

From Eq. 18.30(a), the NPSHA is

$$NPSHA = h_p + h_{z(s)} - h_{f(s)} - h_{\text{vp}}$$

$$= 199.0 \; \text{ft} + 20 \; \text{ft} - 26.74 \; \text{ft} - 124.5 \; \text{ft}$$

$$= 67.8 \; \text{ft}$$

Since NPSHR = 10 ft, the pump will not cavitate.

(Note that a pump may not actually be needed in this configuration.)

The answer is (D).

SI Solution

From App. 16.C, data for 1.5 in schedule-40 steel pipe are

$$D_i = 40.89 \; \text{mm}$$

$$A_i = 13.13 \times 10^{-4} \; \text{m}^2$$

The velocity in the pipe is

$$v = \frac{\dot{V}}{A} = \frac{\left(6.3 \; \frac{\text{L}}{\text{s}}\right)\left(\frac{1 \; \text{m}^3}{1000 \; \text{L}}\right)}{13.13 \times 10^{-4} \; \text{m}^2}$$

$$= 4.80 \; \text{m/s}$$

From App. 17.D, for screwed steel fittings, the approximate equivalent lengths for fittings are

inlet (square mouth): $L_e = 3.1$ ft

long radius 90° ell: $L_e = 3.4$ ft

wide-open gate valves: $L_e = 1.2$ ft

The total equivalent length is

$$30 \; \text{ft} + 3.1 \; \text{ft} + (2)(3.4 \; \text{ft}) + (2)(1.2 \; \text{ft}) = 42.3 \; \text{ft}$$

$$(42.3 \; \text{ft})\left(0.3048 \; \frac{\text{m}}{\text{ft}}\right) = 12.89 \; \text{m}$$

From Table 17.2, for steel, $\epsilon = 6.0 \times 10^{-5} \; \text{m}$.

$$\frac{\epsilon}{D} = \frac{6.0 \times 10^{-5} \; \text{m}}{(40.89 \; \text{mm})\left(\frac{1 \; \text{m}}{1000 \; \text{mm}}\right)} \approx 0.0015$$

At 138°C, from App. 35.B, $\nu = 0.222 \times 10^{-6}$ m^2/s. The Reynolds number is

$$\text{Re} = \frac{D\text{v}}{\nu}$$

$$= \frac{(40.89 \text{ mm})\left(\dfrac{1 \text{ m}}{1000 \text{ mm}}\right)\left(4.80 \dfrac{\text{m}}{\text{s}}\right)}{0.222 \times 10^{-6} \dfrac{\text{m}^2}{\text{s}}}$$

$$= 8.84 \times 10^5$$

From App. 17.B, the friction factor is $f = 0.022$.

From Eq. 18.6, the friction head is

$$h_f = \frac{fL\text{v}^2}{2Dg}$$

$$= \frac{(0.022)(12.89 \text{ m})\left(4.8 \dfrac{\text{m}}{\text{s}}\right)^2}{(2)(40.89 \text{ mm})\left(\dfrac{1 \text{ m}}{1000 \text{ mm}}\right)\left(9.81 \dfrac{\text{m}}{\text{s}^2}\right)}$$

$$= 8.14 \text{ m}$$

From Eq. 18.5(a),

$$h_{\text{vp}} = \frac{p_{\text{vapor}}}{\rho g}$$

At 138°C,

$$p_{\text{vapor}} = 3.422 \text{ bar} \quad \text{[from steam tables]}$$

$$\rho = \frac{1}{v_f}$$

$$= \frac{1}{\left(1.0777 \dfrac{\text{cm}^3}{\text{g}}\right)\left(1000 \dfrac{\text{g}}{\text{kg}}\right)\left(\dfrac{1 \text{ m}^3}{(100 \text{ cm})^3}\right)}$$

$$= 927.9 \text{ kg/m}^3$$

$$h_{\text{vp}} = \frac{(3.422 \text{ bar})\left(10^5 \dfrac{\text{Pa}}{\text{bar}}\right)}{\left(927.9 \dfrac{\text{kg}}{\text{m}^3}\right)\left(9.81 \dfrac{\text{m}}{\text{s}^2}\right)}$$

$$= 37.59 \text{ m}$$

From Eq. 18.5(a), the pressure head is

$$h_p = \frac{p}{\rho g}$$

$$= \frac{(550 \text{ kPa})\left(1000 \dfrac{\text{Pa}}{\text{kPa}}\right)}{\left(927.9 \dfrac{\text{kg}}{\text{m}^3}\right)\left(9.81 \dfrac{\text{m}}{\text{s}^2}\right)}$$

$$= 60.42 \text{ m}$$

From Eq. 18.30(a), the NPSHA is

$$\text{NPSHA} = h_p + h_{z(s)} - h_{f(s)} - h_{\text{vp}}$$

$$= 60.42 \text{ m} + 6 \text{ m} - 8.14 \text{ m} - 37.59 \text{ m}$$

$$= 20.7 \text{ m}$$

Since NPSHR is 3 m, $\boxed{\text{the pump will not cavitate.}}$

(Note that a pump may not actually be needed in this configuration.)

The answer is (D).

14. The solvent is the water (fresh), and the solution is the seawater. Since seawater contains approximately $2\frac{1}{2}$% salt by weight, 100 lbm of seawater will yield 2.5 lbm salt and 97.5 lbm water. The molecular weight of salt is $23.0 + 35.5 = 58.5$ lbm/lbmol. The number of moles of salt in 100 lbm of seawater is

$$n_{\text{salt}} = \frac{2.5 \text{ lbm}}{58.5 \dfrac{\text{lbm}}{\text{lbmol}}} = 0.043 \text{ lbmol}$$

Similarly, the molecular weight of water is 18.016 lbm/lbmol. The number of moles of water is

$$n_{\text{water}} = \frac{97.5 \text{ lbm}}{18.016 \dfrac{\text{lbm}}{\text{lbmol}}} = 5.412 \text{ lbmol}$$

The mole fraction of water is

$$\frac{5.412 \text{ lbmol}}{5.412 \text{ lbmol} + 0.043 \text{ lbmol}} = 0.992$$

Customary U.S. Solution

Cavitation will occur when

$$h_{\text{atm}} - h_{\text{v}} < h_{\text{vp}}$$

The density of seawater is 64.0 lbm/ft^3.

From Eq. 18.5(b), the atmospheric head is

$$h_{\text{atm}} = \left(\frac{p}{\rho}\right) \times \left(\frac{g_c}{g}\right)$$

$$= \left(\frac{\left(14.7 \dfrac{\text{lbf}}{\text{in}^2}\right)\left(144 \dfrac{\text{in}^2}{\text{ft}^2}\right)}{64.0 \dfrac{\text{lbm}}{\text{ft}^3}}\right) \times \left(\frac{32.2 \dfrac{\text{ft-lbm}}{\text{lbf-sec}^2}}{32.2 \dfrac{\text{ft}}{\text{sec}^2}}\right)$$

$$= 33.075 \text{ ft} \quad \text{[ft of seawater]}$$

$$h_{\text{depth}} = 8 \text{ ft} \quad \text{[given]}$$

From Eq. 18.7, the velocity head is

$$h_{\text{v}} = \frac{\text{v}_{\text{propeller}}^2}{2g} = \frac{(4.2\text{v}_{\text{boat}})^2}{(2)\left(32.2 \dfrac{\text{ft}}{\text{sec}^2}\right)}$$

$$= 0.2739\text{v}_{\text{boat}}^2$$

The vapor pressure of 68°F freshwater is $p_{vp} = 0.3393$ psia.

From App. 14.A, the density of water at 68°F is 62.32 lbm/ft³. Raoult's law predicts the actual vapor pressure of the solution.

$$p_{vapor,solution} = (p_{vapor,solvent})\begin{pmatrix} \text{mole fraction} \\ \text{of the solvent} \end{pmatrix}$$

$$p_{vapor,seawater} = (0.992)(0.3393 \text{ psia})$$
$$= 0.3366 \text{ psia}$$

From Eq. 18.5(b), the vapor pressure head is

$$h_{vapor,seawater} = \left(\frac{p}{\rho}\right) \times \left(\frac{g_c}{g}\right)$$

$$= \left(\frac{\left(0.3366 \ \frac{lbf}{in^2}\right)\left(144 \ \frac{in^2}{ft^2}\right)}{64.0 \ \frac{lbm}{ft^3}}\right)$$

$$\times \left(\frac{32.2 \ \frac{ft\text{-}lbm}{lbf\text{-}sec^2}}{32.2 \ \frac{ft}{sec^2}}\right)$$

$$= 0.7574 \text{ ft}$$

Then,

$$8 \text{ ft} + 33.075 \text{ ft} - 0.2739v_{boat}^2 = 0.7574 \text{ ft}$$

$$\boxed{v_{boat} = 12.13 \text{ ft/sec}}$$

The answer is (B).

SI Solution

Cavitation will occur when

$$h_{atm} - h_v < h_{vp}$$

The density of seawater is 1024 kg/m³.

From Eq. 18.5(a), the atmospheric head is

$$h_{atm} = \frac{p}{\rho g}$$

$$= \frac{(101.3 \text{ kPa})\left(1000 \ \frac{Pa}{kPa}\right)}{\left(1024 \ \frac{kg}{m^3}\right)\left(9.81 \ \frac{m}{s^2}\right)}$$

$$= 10.08 \text{ m}$$

$$h_{depth} = 3 \text{ m} \quad \text{[given]}$$

From Eq. 18.7, the velocity head is

$$h_v = \frac{v_{propeller}^2}{2g} = \frac{(4.2v_{boat})^2}{(2)\left(9.81 \ \frac{m}{s^2}\right)}$$

$$= 0.899v_{boat}^2$$

The vapor pressure of 20°C freshwater is

$$p_{vp} = (0.02339 \text{ bar})\left(100 \ \frac{kPa}{bar}\right)$$

$$= 2.339 \text{ kPa}$$

From App. 14.B, the density of water at 20°C is 998.23 kg/m³. Raoult's law predicts the actual vapor pressure of the solution.

$$p_{vapor,solution} = (p_{vapor,solvent})\begin{pmatrix} \text{mole fraction} \\ \text{of the solvent} \end{pmatrix}$$

The solvent is the freshwater and the solution is the seawater.

The mole fraction of water is 0.992.

$$p_{vapor,seawater} = (0.992)(2.339 \text{ kPa}) = 2.320 \text{ kPa}$$

From Eq. 18.5(a), the vapor pressure head is

$$h_{vapor,seawater} = \frac{(2.320 \text{ kPa})\left(1000 \ \frac{Pa}{kPa}\right)}{\left(9.81 \ \frac{m}{s^2}\right)\left(1024 \ \frac{kg}{m^3}\right)}$$

$$= 0.231 \text{ m}$$

Then,

$$3 \text{ m} + 10.08 \text{ m} - 0.899v_{boat}^2 = 0.231 \text{ m}$$

$$\boxed{v_{boat} = 3.78 \text{ m/s}}$$

The answer is (B).

15.

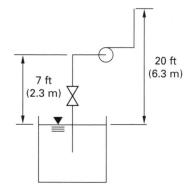

Customary U.S. Solution

From App. 17.D, the approximate equivalent lengths of various screwed steel fittings are

inlet: $L_e = 8.5$ ft   [essentially a reentrant inlet]

check valve: $L_e = 19$ ft

long radius elbows: $L_e = 3.6$ ft

Fluids

The total equivalent length of the 2 in line is

$$L_e = 12 \text{ ft} + 8.5 \text{ ft} + 19.0 \text{ ft} + (3)(3.6 \text{ ft}) + 80 \text{ ft}$$
$$= 130.3 \text{ ft}$$

From App. 16.B, for schedule-40 2 in pipe, the pipe data are

$$D_i = 0.1723 \text{ ft}$$
$$A_i = 0.0233 \text{ ft}^2$$

Since the flow rate is unknown, it must be assumed in order to find velocity. Assume 90 gal/min.

$$\dot{V} = \left(90 \ \frac{\text{gal}}{\text{min}}\right)\left(0.002228 \ \frac{\frac{\text{ft}^3}{\text{sec}}}{\frac{\text{gal}}{\text{min}}}\right) = 0.2005 \ \text{ft}^3/\text{sec}$$

The velocity is

$$\text{v} = \frac{\dot{V}}{A_i} = \frac{0.2005 \ \frac{\text{ft}^3}{\text{sec}}}{0.0233 \ \text{ft}^2} = 8.605 \ \text{ft}/\text{sec}$$

From App. 14.A, the kinematic viscosity of water at 70°F is $\nu = 1.059 \times 10^{-5} \ \text{ft}^2/\text{sec}$.

The Reynolds number is

$$\text{Re} = \frac{D\text{v}}{\nu} = \frac{(0.1723 \text{ ft})\left(8.605 \ \frac{\text{ft}}{\text{sec}}\right)}{1.059 \times 10^{-5} \ \frac{\text{ft}^2}{\text{sec}}}$$
$$= 1.4 \times 10^5$$

From App. 17.A, the specific roughness of steel pipe is

$$\epsilon = 0.0002 \text{ ft}$$

$$\frac{\epsilon}{D} = \frac{0.0002 \text{ ft}}{0.1723 \text{ ft}} = 0.0012$$

From App. 17.B, $f = 0.022$. At 90 gal/min, the friction loss in the line from Eq. 18.6 is

$$h_f = \frac{fL\text{v}^2}{2Dg}$$

$$= \frac{(0.022)(130.3 \text{ ft})\left(8.605 \ \frac{\text{ft}}{\text{sec}}\right)^2}{(2)(0.1723 \text{ ft})\left(32.2 \ \frac{\text{ft}}{\text{sec}^2}\right)} = 19.1 \text{ ft}$$

From Eq. 18.7, the velocity head at 90 gal/min is

$$h_\text{v} = \frac{\text{v}^2}{2g} = \frac{\left(8.605 \ \frac{\text{ft}}{\text{sec}}\right)^2}{(2)\left(32.2 \ \frac{\text{ft}}{\text{sec}^2}\right)} = 1.1 \text{ ft}$$

In general, the friction head and velocity head are

$$h_f = (19.1 \text{ ft})\left(\frac{Q_2}{90 \ \frac{\text{gal}}{\text{min}}}\right)^2$$

$$h_\text{v} = (1.1 \text{ ft})\left(\frac{Q_2}{90 \ \frac{\text{gal}}{\text{min}}}\right)^2$$

(Note that the 7 ft dimension is included in the 20 ft dimension.)

The total system head is

$$h = h_z + h_\text{v} + h_f$$

$$= 20 \text{ ft} + (1.1 + 19.1 \text{ ft})\left(\frac{Q_2}{90 \ \frac{\text{gal}}{\text{min}}}\right)^2$$

From this equation, the following table for system head can be generated.

Q_2 (gal/min)	system head, h (ft)
0	20.0
10	20.2
20	21.0
30	22.2
40	24.0
50	26.2
60	29.0
70	32.2
80	36.0
90	40.2
100	44.9
110	50.2

(a) The intersection point of the system curve and the pump curve defines the operating flow rate. The flow rate is $\boxed{95 \text{ gal/min.}}$

The answer is (D).

(b) The intersection point is not in an efficient range for the pump because it is so far down on the system curve that the pumping efficiency will be low.

$$\boxed{\text{A different pump should be used.}}$$

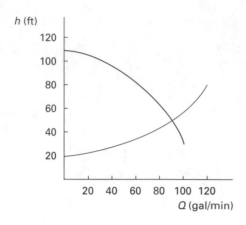

The answer is (A).

SI Solution

Use the approximate quivalent lengths of various screwed steel fittings from the customary U.S. solution. The total equivalent length of 5.08 cm schedule-40 pipe is

$$L_e = 4 \text{ m} + (8.5 \text{ ft} + 19.0 \text{ ft} + (3)(3.6 \text{ ft})) \left(0.3048 \, \frac{\text{m}}{\text{ft}}\right)$$
$$+ \, 27 \text{ m}$$
$$= 42.67 \text{ m}$$

From App. 16.C, for 2 in schedule-40 pipe, the pipe data are

$$D_i = 52.50 \text{ mm}$$
$$A_i = 21.65 \times 10^{-4} \text{ m}^2$$

Since the flow rate is unknown, it must be assumed in order to find velocity. Assume 6 L/s.

$$\dot{V} = \left(6 \, \frac{\text{L}}{\text{s}}\right) \left(\frac{1 \text{ m}^3}{1000 \text{ L}}\right) = 6 \times 10^{-3} \text{ m}^3/\text{s}$$

The velocity is

$$v = \frac{\dot{V}}{A_i} = \frac{6 \times 10^{-3} \, \frac{\text{m}^3}{\text{s}}}{21.65 \times 10^{-4} \text{ m}^2} = 2.77 \text{ m/s}$$

From App. 14.B, the absolute viscosity of water at 21°C is $\mu = 0.9827 \times 10^{-3}$ Pa·s.

The density of water is $\rho = 998$ kg/m³.

The Reynolds number is

$$\text{Re} = \frac{\rho v D_i}{\mu} = \frac{\left(998 \, \frac{\text{kg}}{\text{m}^3}\right) \left(2.77 \, \frac{\text{m}}{\text{s}}\right)}{0.9827 \times 10^{-3} \text{ Pa·s}}$$
$$\times \, (52.50 \text{ mm}) \left(\frac{1 \text{ m}}{1000 \text{ mm}}\right)$$

$$= 1.5 \times 10^5$$

From Table 17.2, the specific roughness of steel pipe is

$$\epsilon = 6.0 \times 10^{-5} \text{ m}$$

$$\frac{\epsilon}{D} = \frac{6.0 \times 10^{-5} \text{ m}}{(52.50 \text{ mm}) \left(\frac{1 \text{ m}}{1000 \text{ mm}}\right)} \approx 0.0012$$

From App. 17.B, $f = 0.022$. At 6 L/s, the friction loss in the line from Eq. 18.6 is

$$h_f = \frac{fLv^2}{2Dg}$$
$$= \frac{(0.022)(42.67 \text{ m}) \left(2.77 \, \frac{\text{m}}{\text{s}}\right)^2}{(2)(52.50 \text{ mm}) \left(\frac{1 \text{ m}}{1000 \text{ mm}}\right) \left(9.81 \, \frac{\text{m}}{\text{s}^2}\right)}$$
$$= 6.99 \text{ m}$$

At 6 L/s, the velocity head from Eq. 18.7 is

$$h_v = \frac{v^2}{2g} = \frac{\left(2.77 \, \frac{\text{m}}{\text{s}}\right)^2}{(2) \left(9.81 \, \frac{\text{m}}{\text{s}^2}\right)} = 0.39 \text{ m}$$

In general, the friction head and velocity head are

$$h_f = (6.99 \text{ m}) \left(\frac{Q_2}{6 \, \frac{\text{L}}{\text{s}}}\right)^2$$

$$h_v = (0.39 \text{ m}) \left(\frac{Q_2}{6 \, \frac{\text{L}}{\text{s}}}\right)^2$$

(Note that the 2.3 m dimension is included in the 6.3 m dimension.)

The total system head is

$$h = h_z + h_v + h_f$$
$$= 6.3 \text{ m} + (0.39 \text{ m} + 6.99 \text{ m}) \left(\frac{Q_2}{6 \, \frac{\text{L}}{\text{s}}}\right)^2$$

From this equation the following table for the system head can be generated.

Q_2 (L/s)	h (m)
0	6.3
0.6	6.37
1.2	6.60
1.8	6.96
2.4	7.48
3.2	8.40
3.6	8.96
4.4	10.27
4.8	11.02
5.7	12.96
6.5	14.96
7.0	16.35
7.5	17.83

The intersection point of the system curve and the pump curve will define the operating flow rate.

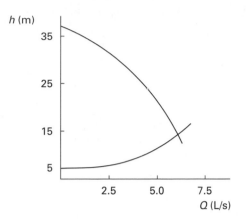

(a) The flow rate is $\boxed{6.2 \text{ L/s.}}$

The answer is (D).

(b) The intersection point is not in an efficient range of the pump because it is so far down on the system curve that the pumping efficiency will be low.

$$\boxed{\text{A different pump should be used.}}$$

The answer is (A).

16. From Eq. 18.52,

$$P_2 = P_1 \left(\frac{\rho_2 n_2^3 D_2^5}{\rho_1 n_1^3 D_1^5}\right)$$

$$= P_1 \left(\frac{n_2}{n_1}\right)^3 \quad [\rho_2 = \rho_1 \text{ and } D_2 = D_1]$$

Customary U.S. Solution

$$P_2 = (0.5 \text{ hp}) \left(\frac{2000 \text{ rpm}}{1750 \text{ rpm}}\right)^3 = \boxed{0.746 \text{ hp}}$$

The answer is (D).

SI Solution

$$P_2 = P_1 \left(\frac{n_2}{n_1}\right)^3$$

$$= (0.37 \text{ kW}) \left(\frac{2000 \text{ rpm}}{1750 \text{ rpm}}\right)^3$$

$$= \boxed{0.55 \text{ kW}}$$

The answer is (D).

17. (a) Random values of Q are chosen, and the corresponding values of H are determined by the formula $H = 30 + 2Q^2$.

Q (ft^3/sec)	H (ft)
0	30
2.5	42.5
5	80
7.5	142.5
10	230
15	480
20	830
25	1280
30	1830

The intersection of the system curve and the 1400 rpm pump curve defines the operating point at that rpm.

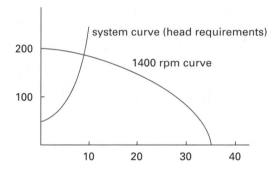

From the intersection of the graphs, at 1400 rpm the flow rate is approximately 9 ft^3/sec and the corresponding head is $30 + (2)(9)^2 \approx 192$ ft.

$$Q = \left(9 \, \frac{\text{ft}^3}{\text{sec}}\right) \left(448.8 \, \frac{\frac{\text{gal}}{\text{min}}}{\frac{\text{ft}^3}{\text{sec}}}\right) = \boxed{4039 \text{ gal/min}}$$

The answer is (C).

(b) From Table 18.5, the hydraulic horsepower is

$$\text{WHP} = \frac{h_A \dot{V} (\text{SG})}{8.814}$$

$$= \frac{(192 \text{ ft}) \left(9 \, \frac{\text{ft}^3}{\text{sec}}\right) (1)}{8.814 \, \frac{\text{ft}^4}{\text{hp sec}}}$$

$$= 196 \text{ hp}$$

From Eq. 18.28(b), the specific speed is

$$n_s = \frac{n\sqrt{Q}}{h_A^{0.75}}$$

$$= \frac{(1400 \text{ rpm})\sqrt{4039 \, \frac{\text{gal}}{\text{min}}}}{(192 \text{ ft})^{0.75}}$$

$$= 1725$$

From Fig. 18.7 with curve E, $\eta \approx 86\%$.

The minimum pump motor power should be

$$\frac{196 \text{ hp}}{0.86} = \boxed{228 \text{ hp}}$$

The answer is (C).

(c) From Eq. 18.41,

$$Q_2 = Q_1 \left(\frac{n_2}{n_1}\right) = \left(4039 \frac{\text{gal}}{\text{min}}\right) \left(\frac{1200 \text{ rpm}}{1400 \text{ rpm}}\right)$$

$$= \boxed{3462 \text{ gal/min}}$$

The answer is (B).

18. Since turbines are essentially pumps running backward, use Table 18.4.

$$\Delta p = \left(30 \frac{\text{lbf}}{\text{in}^2} - 5 \frac{\text{lbf}}{\text{in}^2}\right) \left(144 \frac{\text{in}^2}{\text{ft}^2}\right) = 3600 \text{ lbf/ft}^2$$

$$P = \frac{\left(3600 \frac{\text{lbf}}{\text{ft}^2}\right) \left(100 \frac{\text{ft}^3}{\text{sec}}\right)}{550 \frac{\text{ft-lbf}}{\text{hp-sec}}} = \boxed{654.5 \text{ hp}}$$

The answer is (C).

19. The flow rate is

$$\gamma \dot{V} = \left(62.4 \frac{\text{lbf}}{\text{ft}^3}\right) \left(1000 \frac{\text{ft}^3}{\text{sec}}\right) = 6.24 \times 10^4 \text{ lbf/sec}$$

The head available for work is

$$\Delta h = 625 \text{ ft} - 58 \text{ ft} = 567 \text{ ft}$$

The power is

$$P = (0.89) \left(6.24 \times 10^4 \frac{\text{lbf}}{\text{sec}}\right) (567 \text{ ft})$$

$$= 3.149 \times 10^7 \text{ ft-lbf/sec}$$

Convert from hp to kW.

$$\frac{\left(3.149 \times 10^7 \frac{\text{ft-lbf}}{\text{sec}}\right) \left(0.7457 \frac{\text{kW}}{\text{hp}}\right)}{550 \frac{\text{ft-lbf}}{\text{hp-sec}}}$$

$$= \boxed{4.27 \times 10^4 \text{ kW (43 MW)}}$$

The answer is (C).

20. *Customary U.S. Solution*

(a) The description of the turbine (i.e., of a water jet hitting a blade) is consistent with an impulse turbine.

The answer is (A).

(b) The flow rate, $\dot{V}$, is

$$\dot{V} = A\text{v} = \left(\frac{\pi}{4}\right) \left(\frac{4 \text{ in}}{12 \frac{\text{in}}{\text{ft}}}\right)^2 \left(35 \frac{\text{ft}}{\text{sec}}\right)$$

$$= 3.054 \text{ ft}^3/\text{sec}$$

The flow rate, considering the blade's moving away at 10 ft/sec, is

$$\dot{V}' = \frac{\left(35 \frac{\text{ft}}{\text{sec}} - 10 \frac{\text{ft}}{\text{sec}}\right) \left(3.054 \frac{\text{ft}^3}{\text{sec}}\right)}{35 \frac{\text{ft}}{\text{sec}}}$$

$$= 2.181 \text{ ft}^3/\text{sec}$$

The forces in the x-direction and the y-direction are

$$F_x = \left(\frac{\dot{V}'\rho}{g_c}\right) (\text{v}_j - \text{v}_b)(\cos\theta - 1)$$

$$= \left(\frac{\left(2.181 \frac{\text{ft}^3}{\text{sec}}\right) \left(62.37 \frac{\text{lbm}}{\text{ft}^3}\right)}{32.2 \frac{\text{lbm-ft}}{\text{lbf-sec}^2}}\right)$$

$$\times \left(35 \frac{\text{ft}}{\text{sec}} - 10 \frac{\text{ft}}{\text{sec}}\right) (\cos 80° - 1)$$

$$= -87.27 \text{ lbf}$$

$$F_y = \left(\frac{\dot{V}'\rho}{g_c}\right) (\text{v}_j - \text{v}_b) \sin\theta$$

$$= \left(\frac{\left(2.181 \frac{\text{ft}^3}{\text{sec}}\right) \left(62.37 \frac{\text{lbm}}{\text{ft}^3}\right)}{32.2 \frac{\text{lbm-ft}}{\text{lbf-sec}^2}}\right)$$

$$\times \left(35 \frac{\text{ft}}{\text{sec}} - 10 \frac{\text{ft}}{\text{sec}}\right) \sin 80°$$

$$= 104.0 \text{ lbf}$$

$$R = \sqrt{F_x^2 + F_y^2}$$

$$= \sqrt{(-87.27 \text{ lbf})^2 + (104.0 \text{ lbf})^2} = \boxed{135.8 \text{ lbf}}$$

The answer is (B).

SI Solution

(a) The description of the turbine (i.e., of a water jet hitting a blade) is consistent with an impulse turbine.

The answer is (A).

(b) The flow rate, $\dot{V}$, is

$$\dot{V} = A\mathrm{v} = \left(\frac{\pi}{4}\right)\left((100 \text{ mm})\left(\frac{1 \text{ m}}{1000 \text{ mm}}\right)\right)^2\left(10.5 \frac{\text{m}}{\text{s}}\right)$$
$$= 0.08247 \text{ m}^3/\text{s}$$

The flow rate, considering the blade's moving away at 3 m/s, is

$$\dot{V}' = \frac{\left(10.5 \frac{\text{m}}{\text{s}} - 3 \frac{\text{m}}{\text{s}}\right)\left(0.08247 \frac{\text{m}^3}{\text{s}}\right)}{10.5 \frac{\text{m}}{\text{s}}}$$
$$= 0.05891 \text{ m}^3/\text{s}$$

The forces in the x-direction and the y-direction are

$$F_x = \dot{V}'\rho(\mathrm{v}_j - \mathrm{v}_b)(\cos\theta - 1)$$
$$= \left(0.05891 \frac{\text{m}^3}{\text{s}}\right)\left(998.83 \frac{\text{kg}}{\text{m}^3}\right)$$
$$\times \left(10.5 \frac{\text{m}}{\text{s}} - 3 \frac{\text{m}}{\text{s}}\right)(\cos 80° - 1)$$
$$= -364.7 \text{ N}$$

$$F_y = \dot{V}'\rho(\mathrm{v}_j - \mathrm{v}_b)\sin\theta$$
$$= \left(0.05891 \frac{\text{m}^3}{\text{s}}\right)\left(998.83 \frac{\text{kg}}{\text{m}^3}\right)$$
$$\times \left(10.5 \frac{\text{m}}{\text{s}} - 3 \frac{\text{m}}{\text{s}}\right)\sin 80°$$
$$= 434.6 \text{ N}$$

$$R = \sqrt{F_x^2 + F_y^2}$$
$$= \sqrt{(-364.7 \text{ N})^2 + (434.6 \text{ N})^2} = \boxed{567.3 \text{ N}}$$

The answer is (B).

21. *Customary U.S. Solution*

(a) The total effective head is due to the pressure head, velocity head, and tailwater head.

$$h_{\text{eff}} = h_p + h_\mathrm{v} - h_{z(\text{tailwater})}$$
$$= 92.5 \text{ ft} + \frac{\left(12 \frac{\text{ft}}{\text{sec}}\right)^2}{(2)\left(32.2 \frac{\text{ft}}{\text{sec}^2}\right)} - (-5.26 \text{ ft})$$
$$= \boxed{100 \text{ ft}}$$

The answer is (C).

(b) From Table 18.4, the theoretical hydraulic horsepower is

$$P_{\text{th}} = \frac{h_A\dot{V}(\text{SG})}{8.814} = \frac{(100 \text{ ft})\left(25 \frac{\text{ft}^3}{\text{sec}}\right)(1)}{8.814 \frac{\text{ft}^4}{\text{hp-sec}}}$$
$$= 283.6 \text{ hp}$$

The overall turbine efficiency is

$$\eta = \frac{P_{\text{brake}}}{P_{\text{th}}} = \frac{250 \text{ hp}}{283.6 \text{ hp}} = \boxed{0.882 \quad (88.2\%)}$$

The answer is (B).

(c) From Eq. 18.42 (the affinity laws),

$$n_2 = n_1\sqrt{\frac{h_2}{h_1}} = (610 \text{ rpm})\sqrt{\frac{225 \text{ ft}}{100 \text{ ft}}}$$
$$= \boxed{915 \text{ rpm}}$$

The answer is (B).

(d) Combine Eqs. 18.42 and 18.43.

$$P_2 = P_1\left(\frac{h_2}{h_1}\right)^{1.5} = (250 \text{ hp})\left(\frac{225 \text{ ft}}{100 \text{ ft}}\right)^{1.5}$$
$$= \boxed{843.8 \text{ hp}}$$

The answer is (D).

(e) Combine Eqs. 18.41 and 18.42.

$$Q_2 = Q_1\sqrt{\frac{h_2}{h_1}} = \left(25 \frac{\text{ft}^3}{\text{sec}}\right)\sqrt{\frac{225 \text{ ft}}{100 \text{ ft}}}$$
$$= \boxed{37.5 \text{ ft}^3/\text{sec}}$$

The answer is (B).

SI Solution

(a) The total effective head is due to the pressure head, velocity head, and tailwater head.

$$h_{\text{eff}} = h_p + h_\mathrm{v} - h_{z(\text{tailwater})}$$
$$= 30.8 \text{ m} + \frac{\left(3.6 \frac{\text{m}}{\text{s}}\right)^2}{(2)\left(9.81 \frac{\text{m}}{\text{s}^2}\right)} - (-1.75 \text{ m})$$
$$= \boxed{33.21 \text{ m}}$$

The answer is (C).

(b) From Table 18.5, the theoretical hydraulic kilowatt is

$$P_{\text{th}} = \frac{9.81 h_A Q (\text{SG})}{1000}$$

$$= \frac{\left(9.81 \ \dfrac{\text{m}}{\text{s}^2}\right)(33.21 \ \text{m})\left(700 \ \dfrac{\text{L}}{\text{s}}\right)(1)}{1000 \ \dfrac{\text{W}}{\text{kW}}}$$

$$= 228.1 \ \text{kW}$$

The overall turbine efficiency is

$$\eta = \frac{P_{\text{brake}}}{P_{\text{th}}} = \frac{185 \ \text{kW}}{228.1 \ \text{kW}}$$

$$= \boxed{0.811 \quad (81.1\%)}$$

The answer is (A).

(c) From Eq. 18.42 (the affinity laws),

$$h_2 = n_1 \sqrt{\frac{h_2}{h_1}} = (610 \ \text{rpm})\sqrt{\frac{75 \ \text{m}}{33.21 \ \text{m}}}$$

$$= \boxed{917 \ \text{rpm}}$$

The answer is (B).

(d) Combine Eqs. 18.42 and 18.43.

$$P_2 = P_1 \left(\frac{h_2}{h_1}\right)^{1.5} = (185 \ \text{kW})\left(\frac{75 \ \text{m}}{33.21 \ \text{m}}\right)^{1.5}$$

$$= \boxed{627.3 \ \text{kW}}$$

The answer is (D).

(e) Combine Eqs. 18.41 and 18.42.

$$Q_2 = Q_1 \sqrt{\frac{h_2}{h_1}} = \left(700 \ \frac{\text{L}}{\text{s}}\right)\sqrt{\frac{75 \ \text{m}}{33.21 \ \text{m}}}$$

$$= \boxed{1051.9 \ \text{L/s}}$$

The answer is (B).

19 Fluid Power

PRACTICE PROBLEMS

1. A valve is controlled by a pump and actuator connected by a tube with an internal diameter of 3.5 mm. The nominal working pressure in the tube is 28 MPa. The density of the hydraulic fluid is 860 kg/m³. The effective bulk modulus is 1.5×10^9 Pa. When closing, the speed of a valve actuator is 10 m/s. The actuator piston diameter is 40 mm. The line distance between the pump and the actuator is 11 m. What is the characteristic impedance per unit of actuator piston area?

    (A) 4×10^8 kg/s·m⁴

    (B) 9×10^8 kg/s·m⁴

    (C) 17×10^8 kg/s·m⁴

    (D) 31×10^8 kg/s·m⁴

2. Given the situation described in Prob. 1, what will be the maximum pressure in the transmission line if the valve is suddenly closed?

    (A) 11 MPa

    (B) 32 MPa

    (C) 40 MPa

    (D) 52 MPa

3. Given the situation described in Prob. 1, and assuming a lossless line, what is the fundamental frequency of the hydraulic line?

    (A) 380 rad/s

    (B) 620 rad/s

    (C) 1700 rad/s

    (D) 2500 rad/s

SOLUTIONS

1. The internal area of the piston is

$$A_p = \left(\frac{\pi}{4}\right) d_p^2 = \left(\frac{\pi}{4}\right)\left(\frac{40 \text{ mm}}{1000 \frac{\text{mm}}{\text{m}}}\right)^2$$

$$= 1.25664 \times 10^{-3} \text{ m}^2$$

The characteristic impedance per unit area of piston is

$$z_o = \frac{1}{A_p}\sqrt{\rho B}$$

$$= \left(\frac{1}{1.25664 \times 10^{-3} \text{ m}^2}\right)$$

$$\times \sqrt{\left(860 \frac{\text{kg}}{\text{m}^3}\right)(1.5 \times 10^9 \text{ Pa})}$$

$$= \boxed{9.03824 \times 10^8 \text{ kg/s·m}^4}$$

The answer is (B).

2. The flow rate of control fluid in the valve is

$$\Delta q = \Delta \text{v}_p A_p = \left(10 \frac{\text{m}}{\text{s}}\right)(0.00125664 \text{ m}^2)$$

$$= 0.0125664 \text{ m}^3/\text{s}$$

Use Jukowski's equation.

$$\Delta p = z_o \Delta q$$

$$= \left(9.03824 \times 10^8 \frac{\text{kg}}{\text{s}}\cdot\text{m}^4\right)\left(\frac{0.0125664 \text{ m}^3}{\text{s}}\right)$$

$$= 1.13578 \times 10^7 \text{ Pa}$$

The maximum pressure includes the nominal pressure.

$$p_{\text{max}} = p_o + \Delta p = 28 \times 10^6 \text{ Pa} + 1.13578 \times 10^7 \text{ Pa}$$

$$= \boxed{3.94 \times 10^7 \text{ Pa}}$$

The answer is (C).

3. The effective speed of sound in the line is

$$c = \sqrt{\frac{B}{\rho}} = \sqrt{\frac{1.5 \times 10^9 \text{ Pa}}{860 \frac{\text{kg}}{\text{m}^3}}} = 1320.7 \text{ m/s}$$

The time (i.e., the period) required for a pressure wave to travel from the pump to the actuator and return is

$$T = \frac{2L}{c} = \frac{(2)(11 \text{ m})}{1320.7 \frac{\text{m}}{\text{s}}} = 1.6658 \times 10^{-2} \text{ s}$$

The fundamental frequency is

$$\omega = 2\pi f = \frac{2\pi}{T} = \frac{2\pi}{1.6658 \times 10^{-2} \text{ s}}$$

$$= \boxed{377.2 \text{ rad/s}}$$

The answer is (A).

20 Fans and Ductwork

PRACTICE PROBLEMS

(Note: Round all duct dimensions to the next larger whole inch or multiples of 25 mm.)

1. A round 18 in (457 mm) duct is to be replaced by a rectangular duct with an aspect ratio of 4:1. What should be the dimensions of the rectangular duct?

- (A) 8 in × 30 in (200 mm × 760 mm)
- (B) 9 in × 35 in (220 mm × 880 mm)
- (C) 12 in × 18 in (300 mm × 460 mm)
- (D) 18 in × 72 in (460 mm × 1800 mm)

2. Draw the system curve for a fan that moves 10,000 SFCM (4700 L/s) against a pressure of 4 in wg (1 kPa).

3. A $\frac{1}{8}$ scale model fan is tested at 300 rpm. At standard air conditions and at that speed, the model fan moves 40 cfm (19 L/s) against 0.5 in wg (125 Pa). What is the air power if a full-sized fan operates at 5000 ft (1500 m) altitude at half the model's speed?

- (A) 11 hp (8.6 kW)
- (B) 14 hp (11 kW)
- (C) 19 hp (15 kW)
- (D) 24 hp (19 kW)

4. A duct system consists of 750 ft (230 m) of galvanized 20 in (508 mm) duct with four round elbows (radius-to-diameter ratio of 1.5). A 2 in (51 mm) diameter pipe passes perpendicularly through the duct at two locations. The flow rate is 6000 SCFM (4200 L/s). What is the friction loss?

- (A) 1.7 in wg (1.1 kPa)
- (B) 2.9 in wg (2.0 kPa)
- (C) 3.7 in wg (2.4 kPa)
- (D) 4.8 in wg (3.2 kPa)

5. 1500 SCFM (700 L/s) of air flows in an 18 in (457 mm) round duct. After a branch reduction of 300 SCFM (140 L/s), the fitting reduces to 14 in (356 mm) in the through direction. What is the static regain in the through direction?

- (A) −0.037 in wg (−8.7 Pa)
- (B) −0.073 in wg (−17 Pa)
- (C) −0.11 in wg (−26 Pa)
- (D) −0.19 in wg (−45 Pa)

6. (*Time limit: one hour*) A fan in a small theater delivers 1500 cfm (700 L/s) through the system shown. The duct is rectangular with an aspect ratio of 1.5:1. All elbows have a radius-to-diameter ratio of 1.5. The terminal pressure at each outlet must be 0.25 in wg (63 Pa) or higher. Takeoff fitting losses are to be disregarded.

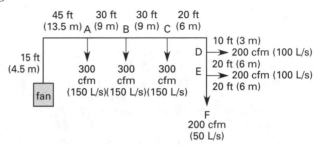

(a) Use the equal-friction method to size the system.

(b) What is the static pressure at the fan?
- (A) 0.47 in wg (110 Pa)
- (B) 0.69 in wg (160 Pa)
- (C) 0.82 in wg (210 Pa)
- (D) 1.1 in wg (260 Pa)

(c) What is the total pressure at the fan?
- (A) 1.0 in wg (250 Pa)
- (B) 1.2 in wg (280 Pa)
- (C) 1.6 in wg (380 Pa)
- (D) 2.0 in wg (500 Pa)

7. (*Time limit: one hour*) A fan in a theater delivers 300 cfm (140 L/s) of air through round ducts to each of the twelve outlets shown. The minimum terminal pressure at the outlets is 0.15 in wg (38 Pa). All elbows have a radius-to-diameter ratio of 1.25. Branch takeoff fitting losses are to be disregarded.

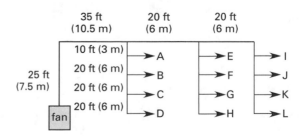

(a) Use the equal-friction method to size the system.

(b) What is the static pressure at the fan?
   (A) 0.49 in wg (130 Pa)
   (B) 0.69 in wg (160 Pa)
   (C) 0.82 in wg (210 Pa)
   (D) 1.1 in wg (260 Pa)

(c) What is the total pressure at the fan?
   (A) 0.47 in wg (110 Pa)
   (B) 0.65 in wg (170 Pa)
   (C) 0.82 in wg (210 Pa)
   (D) 1.1 in wg (260 Pa)

8. (*Time limit: one hour*) Size the longest run in Prob. 7 using the static regain method with a regain coefficient of 0.75. (Customary U.S. solution only required.)

9. (*Time limit: one hour*) 3000 cfm (1400 L/s) of air with a density of 0.075 lbm/ft³ (1.2 kg/m³) enters a 12 in (305 mm) diameter duct at section A. The four-piece 90° elbows have a radius-to-diameter ratio of 1.5 and an equivalent length of 14 diameters. The Darcy friction factor is 0.02 everywhere in the system. The static regain coefficient for this system is 0.65.

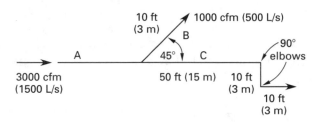

(a) Use the static regain method to calculate the diameters at sections B and C.

(b) Specify if dampers should be placed in sections A, B, and/or C.

(c) What is the regain in sections B and C?

SOLUTIONS

1. *Customary U.S. Solution*

With aspect ratio $R = 4$, the short side is given by Eq. 20.31.

$$
\begin{aligned}
\text{short side} &= \frac{D(1+R)^{\frac{1}{4}}}{1.3R^{\frac{5}{8}}} \\
&= \frac{(18 \text{ in})(1+4)^{\frac{1}{4}}}{(1.3)(4)^{\frac{5}{8}}} = 8.7 \text{ in} \approx \boxed{9.0 \text{ in}}
\end{aligned}
$$

From Eq. 20.32, the long side is

$$
\begin{aligned}
\text{long side} &= R(\text{short side}) \\
&= (4)(8.7 \text{ in}) = 34.8 \text{ in} \approx \boxed{35 \text{ in}}
\end{aligned}
$$

The answer is (B).

SI Solution

With aspect ratio $R = 4$, the short side is given by Eq. 20.31.

$$
\begin{aligned}
\text{short side} &= \frac{D(1+R)^{\frac{1}{4}}}{1.3R^{\frac{5}{8}}} \\
&= \frac{(457 \text{ mm})(1+4)^{\frac{1}{4}}}{(1.3)(4)^{\frac{5}{8}}} = \boxed{221 \text{ mm}}
\end{aligned}
$$

From Eq. 20.32, the long side is

$$
\begin{aligned}
\text{long side} &= R(\text{short side}) \\
&= (4)(221 \text{ mm}) = \boxed{884 \text{ mm}}
\end{aligned}
$$

The answer is (B).

2. *Customary U.S. Solution*

Disregarding terminal pressure, the remainder of the points on the system curve can be found from Eq. 20.19.

$$
\frac{p_2}{p_1} = \left(\frac{Q_2}{Q_1}\right)^2
$$

$$
\begin{aligned}
Q_2 &= Q_1 \sqrt{\frac{p_2}{p_1}} \\
&= (10{,}000 \text{ SCFM})\sqrt{\frac{p_2}{4 \text{ in wg}}} \\
&= 5000\sqrt{p_2}
\end{aligned}
$$

The following table can be generated from the preceding equation.

p_2 (in wg)	Q (SCFM)
1	5000
2	7071
3	8660
4	10,000
5	11,180
6	12,247

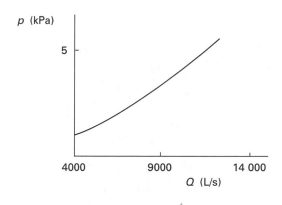

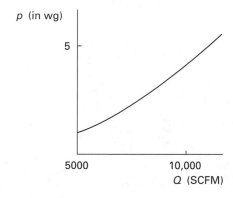

SI Solution

Disregarding terminal pressure, the remainder of the points on the system curve can be found from Eq. 20.19.

$$\frac{p_2}{p_1} = \left(\frac{Q_2}{Q_1}\right)^2$$

$$Q_2 = Q_1\sqrt{\frac{p_2}{p_1}}$$

$$= \left(4700 \ \frac{\text{L}}{\text{s}}\right)\sqrt{\frac{p_2}{1 \ \text{kPa}}}$$

$$= 4700\sqrt{p_2}$$

The following table can be generated from the preceding equation.

p_2 (kPa)	Q (L/s)
1	4700
2	6647
3	8141
4	9400
5	10 510
6	11 513

3. *Customary U.S. Solution*

The air horsepower is given by Eq. 20.11(b).

$$(\text{AHP})_A = \frac{Q_{\text{cfm}}(\text{TP}_{\text{in wg}})}{6356}$$

$$= \frac{(40 \ \text{cfm})(0.5 \ \text{in wg})}{6356 \ \dfrac{\text{in-ft}^3}{\text{hp-min}}}$$

$$= 3.147 \times 10^{-3} \ \text{hp}$$

In order to predict the performance of a dynamically similar fan, use Eq. 20.26.

$$\frac{(\text{AHP})_1}{(\text{AHP})_2} = \left(\frac{D_1}{D_2}\right)^5 \left(\frac{n_1}{n_2}\right)^3 \left(\frac{\gamma_1}{\gamma_2}\right)$$

$$(\text{AHP})_2 = (\text{AHP})_1 \left(\frac{D_2}{D_1}\right)^5 \left(\frac{n_2}{n_1}\right)^3 \left(\frac{\gamma_2}{\gamma_1}\right)$$

$$\frac{D_2}{D_1} = \frac{1}{\frac{1}{8}} = 8$$

$$\frac{n_2}{n_1} = \frac{1}{2}$$

At standard air conditions, $\rho_1 = 0.075 \ \text{lbm/ft}^3$. Use standard atmospheric data from App. 26.E for 5000 ft altitude.

$$\rho_{5000} = \rho_2 = \frac{p}{RT} = \frac{\left(12.225 \ \dfrac{\text{lbf}}{\text{in}^2}\right)\left(144 \ \dfrac{\text{in}^2}{\text{ft}^2}\right)}{\left(53.3 \ \dfrac{\text{ft-lbf}}{\text{lbm-}^\circ\text{R}}\right)(500.9^\circ\text{R})}$$

$$= 0.06594 \ \text{lbm/ft}^3$$

$$\frac{\gamma_2}{\gamma_1} = \frac{\rho_2 g}{\rho_1 g} = \frac{\rho_2}{\rho_1} = \frac{0.06594 \ \dfrac{\text{lbm}}{\text{ft}^3}}{0.075 \ \dfrac{\text{lbm}}{\text{ft}^3}}$$

$$= 0.8792$$

$$(\text{AHP})_2 = (3.147 \times^{-3} \ \text{hp})(8)^5 \left(\tfrac{1}{2}\right)^3 (0.8792)$$

$$= \boxed{11.33 \ \text{hp}}$$

The answer is (A).

SI Solution

The air power is given by Eq. 20.11(a).

$$(\text{AkW})_1 = \frac{Q_{\text{L/s}}(\text{TP}_{\text{Pa}})}{10^6}$$

$$= \frac{\left(19 \, \dfrac{\text{L}}{\text{s}}\right)(125 \text{ Pa})}{10^6 \, \dfrac{\text{L}\cdot\text{W}}{\text{m}^3\cdot\text{kW}}}$$

$$= 2.375 \times 10^{-3} \text{ kW}$$

In order to predict the performance of a dynamically similar fan, use Eq. 20.26.

$$\frac{(\text{AkW})_1}{(\text{AkW})_2} = \left(\frac{D_1}{D_2}\right)^5 \left(\frac{n_1}{n_2}\right)^3 \left(\frac{\gamma_1}{\gamma_2}\right)$$

$$(\text{AkW})_2 = (\text{AkW})_1 \left(\frac{D_2}{D_1}\right)^5 \left(\frac{n_2}{n_1}\right)^3 \left(\frac{\gamma_2}{\gamma_1}\right)$$

$$\frac{D_2}{D_1} = \frac{1}{\frac{1}{8}} = 8$$

$$\frac{n_2}{n_1} = \frac{1}{2}$$

At standard air conditions, $\rho = 1.2$ kg/m^3. Use standard atmospheric data from App. 26.E for 1500 m altitude.

$$\rho_{1525} = \rho_2 = \frac{p}{RT} = \frac{(0.8456 \text{ bar})\left(10^5 \, \dfrac{\text{Pa}}{\text{bar}}\right)}{\left(287 \, \dfrac{\text{J}}{\text{kg·K}}\right)(278.4\text{K})}$$

$$= 1.058 \text{ kg/m}^3$$

$$\frac{\gamma_2}{\gamma_1} = \frac{\rho_2 g}{\rho_1 g} = \frac{\rho_2}{\rho_1} = \frac{1.058 \, \dfrac{\text{kg}}{\text{m}^3}}{1.2 \, \dfrac{\text{kg}}{\text{m}^3}} = 0.88$$

$$(\text{AkW})_2 = (2.375 \times 10^{-3} \text{ kW})(8)^5 \left(\tfrac{1}{2}\right)^3 (0.88)$$

$$= \boxed{8.56 \text{ kW}}$$

The answer is (A).

4. *Customary U.S. Solution*

From the standard friction loss in the standard duct chart (Fig. 20.4), the friction loss is 0.42 in of water per 100 ft of duct and v = 2700 ft/min. The friction loss due to the 750 ft of duct is

$$\text{FP}_{f,1} = (0.42 \text{ in wg})\left(\frac{750 \text{ ft}}{100 \text{ ft}}\right)$$

$$= 3.15 \text{ in wg}$$

The equivalent length of each round elbow (radius-to-diameter ratio of 1.5) can be found from Table 20.4.

$$L_e = 12D$$

For four round elbows,

$$L_e = (4)(12D)$$

$$= \frac{(4)(12)(20 \text{ in})}{12 \, \dfrac{\text{in}}{\text{ft}}} = 80 \text{ ft}$$

The friction loss due to the 80 ft of equivalent length of the duct is

$$\text{FP}_{f,2} = (0.42 \text{ in wg})\left(\frac{80 \text{ ft}}{100 \text{ ft}}\right) = 0.336 \text{ in wg}$$

For a cross duct,

$$\frac{D_1}{D_2} = \frac{2 \text{ in}}{20 \text{ in}} = 0.1$$

From Table 20.3, the loss coefficient is $K_{\text{up}} = 0.2$. For two cross pipes, the friction loss is given by Eq. 20.34.

$$\text{FP}_{f,3} = (2)K_{\text{up}}\left(\frac{\text{v}}{4005}\right)^2$$

$$= (2)(0.2)\left(\frac{2700 \, \dfrac{\text{ft}}{\text{min}}}{4005 \, \dfrac{\text{ft}}{\text{min}}}\right)^2 = 0.182 \text{ in wg}$$

The total loss is

$$\text{FP}_{f,1} + \text{FP}_{f,2} + \text{FP}_{f,3} = 3.15 \text{ in wg} + 0.336 \text{ in wg}$$
$$+ 0.182 \text{ in wg}$$

$$= \boxed{3.668 \text{ in wg}}$$

The answer is (C).

SI Solution

From the standard friction loss in the standard duct chart (Fig. 20.5), the friction loss is 9 Pa/m and v = 21.5 m/s. The friction loss due to the 230 m of the duct is

$$\text{FP}_{f,1} = \left(9 \, \frac{\text{Pa}}{\text{m}}\right)(230 \text{ m}) = 2070 \text{ Pa}$$

The equivalent length of each round elbow (radius-to-diameter ratio of 1.5) can be found in Table 20.4.

$$L_e = 12D$$

For four round elbows,

$$L_e = (4)(12D)$$

$$= \frac{(4)(12D)(508 \text{ mm})}{1000 \text{ mm}} = 24.38 \text{ m}$$

The friction loss due to the 24.38 m of equivalent length of duct is

$$\text{FP}_{f,2} = \left(9 \frac{\text{Pa}}{\text{m}}\right)(24.38 \text{ m}) = 219.4 \text{ Pa}$$

For a cross duct,

$$\frac{D_1}{D_2} = \frac{51 \text{ mm}}{508 \text{ mm}} = 0.1$$

From Table 20.3, the loss coefficient K_{up} is 0.2. For two cross pipes, the friction loss is given by Eq. 20.34.

$$\text{FP}_{f,3} = (2)K_{\text{up}}(\text{v}_{\text{m/s}})^2$$

$$= (2)(0.2)(0.6)\left(21.5 \frac{\text{m}}{\text{s}}\right)^2 = 110.9 \text{ Pa}$$

The total loss is

$$\text{FP}_{f,1} + \text{FP}_{f,2} + \text{FP}_{f,3} = 2070 \text{ Pa} + 219.4 \text{ Pa} + 110.9 \text{ Pa}$$

$$= \boxed{2400.3 \text{ Pa}}$$

The answer is (C).

5. *Customary U.S. Solution*

For the 18 in duct,

$$A = \frac{\pi}{4}D^2 = \left(\frac{\pi}{4}\right)\left(\frac{18 \text{ in}}{12 \frac{\text{in}}{\text{ft}}}\right)^2$$

$$= 1.767 \text{ ft}^2$$

$$\text{v}_1 = \frac{Q}{A} = \frac{1500 \frac{\text{ft}^3}{\text{min}}}{1.767 \text{ ft}^2}$$

$$= 848.9 \text{ ft/min}$$

For the 14 in duct,

$$A = \frac{\pi}{4}D^2 = \left(\frac{\pi}{4}\right)\left(\frac{14 \text{ in}}{12 \frac{\text{in}}{\text{ft}}}\right)^2 = 1.069 \text{ ft}^2$$

$$\text{v}_2 = \frac{Q}{A} = \frac{1500 \text{ cfm} - 300 \text{ cfm}}{1.069 \text{ ft}^2}$$

$$= 1122.5 \text{ ft/min}$$

Since $\text{v}_2 > \text{v}_1$, there is no regain. The regain will be a static pressure loss. Use $R = 1.1$ in that case. From Eq. 20.39, the static pressure loss is

$$\text{SR}_{\text{actual}} = R\left(\frac{\text{v}_{\text{up}}^2 - \text{v}_{\text{down}}^2}{(4005)^2}\right)$$

$$= (1.1)\left(\frac{\left(848.9 \frac{\text{ft}}{\text{min}}\right)^2 - \left(1122.5 \frac{\text{ft}}{\text{min}}\right)^2}{(4005)^2}\right)$$

$$= \boxed{-0.037 \text{ in wg loss}}$$

The answer is (A).

SI Solution

For the 457 mm duct,

$$A = \frac{\pi}{4}D^2 = \left(\frac{\pi}{4}\right)\left(\frac{457 \text{ mm}}{1000 \frac{\text{mm}}{\text{m}}}\right)^2 = 0.164 \text{ m}^2$$

$$\text{v}_1 = \frac{Q}{A} = \frac{\left(700 \frac{\text{L}}{\text{s}}\right)\left(\frac{1 \text{ m}^3}{1000 \text{ L}}\right)}{0.164 \text{ m}^2}$$

$$= 4.27 \text{ m/s}$$

For 356 mm duct,

$$A = \frac{\pi}{4}D^2 = \left(\frac{\pi}{4}\right)\left(\frac{356 \text{ mm}}{1000 \frac{\text{mm}}{\text{m}}}\right)^2 = 0.1 \text{ m}^2$$

$$\text{v}_2 = \frac{Q}{A} = \left(\frac{700 \frac{\text{L}}{\text{s}} - 140 \frac{\text{L}}{\text{s}}}{0.1 \text{ m}^2}\right)\left(\frac{1 \text{ m}^3}{1000 \text{ L}}\right)$$

$$= 5.60 \text{ m/s}$$

Since $\text{v}_2 > \text{v}_1$, there is no regain. The regain will be a static pressure loss. Use $R = 1.1$ in that case. From Eq. 20.39, the static pressure loss is

$$\text{SR} = R\left(0.6 \frac{\text{Pa·s}}{\text{m}}\right)(\text{v}_{\text{up}}^2 - \text{v}_{\text{down}}^2)$$

$$= (1.1)\left(0.6 \frac{\text{Pa·s}^2}{\text{m}^2}\right)\left(\left(4.27 \frac{\text{m}}{\text{s}}\right)^2 - \left(5.60 \frac{\text{m}}{\text{s}}\right)^2\right)$$

$$= \boxed{-8.66 \text{ Pa loss}}$$

The answer is (A).

6. *Customary U.S. Solution*

(a) *step 1:* From Table 20.7, choose the main duct velocity as 1600 fpm.

step 2: The total air flow from the fan is 1500 cfm. From Fig. 20.4, the main duct diameter is 13 in. The friction loss is 0.27 in wg per 100 ft.

step 3: After the first takeoff at A, the flow rate in section A–B is

$$1500 \text{ cfm} - 300 \text{ cfm} = 1200 \text{ cfm}$$

From Fig. 20.4, for 1200 cfm and 0.27 in wg per 100 ft, the diameter is 11.8 in (say 12 in) and the velocity is 1500 fpm. Similarly, the diameter and the velocity for other sections are obtained and listed in the following table.

section	Q (cfm)	D (in)	v (fpm)
fan–A	1500	13	1600
A–B	1200	12	1500
B–C	900	11	1400
C–D	600	9.5	1300
D–E	400	8	1150
E–F	200	6.3	975

These diameters are for round duct. The equivalent rectangular duct sides with an aspect ratio of 1.5 are found as follows.

$$a = \frac{D(1.5+1)^{0.25}}{(1.3)(1.5)^{0.625}} = 0.75D$$

$$b = Ra = (1.5)(0.75D) = 1.125D$$

Convert the diameters to sides a and b.

$$a_{\text{fan–A}} = (0.75)(13 \text{ in}) = 9.75 \text{ in}$$

$$b_{\text{fan–A}} = (1.125)(13 \text{ in}) = 14.63 \text{ in}$$

Use a 10×15 in duct.

The following table is prepared similarly, rounding up as appropriate.

section	D (in)	a (in)	b (in)
fan–A	13	10	15
A–B	12	9	14
B–C	11	9	13
C–D	9.5	8	11
D–E	8	6	9
E–F	6.3	5	8

(b) *step 4:* By inspection, the longest run is fan–F. From Table 20.4, the equivalent length of each bend is 12D. For two elbows,

$$L_{e,\text{bend}} = 12D = (12)\left(\frac{13 \text{ in} + 9.5 \text{ in}}{12 \frac{\text{in}}{\text{ft}}}\right)$$

$$= 22.5 \text{ ft}$$

The equivalent length of the entire run is

15 ft	(fan to first bend)
45 ft	(first bend to point A)
30 ft	(section A–B)
30 ft	(section B–C)
20 ft	(point A to second bend)
22.5 ft	(equivalent length of two bends)
10 ft	(second bend to point D)
20 ft	(section D–E)
20 ft	(section E–F)

total: 212.5 ft

The straight-through friction loss in the longest run is

$$\left(\frac{212.5 \text{ ft}}{100 \text{ ft}}\right)\left(0.27 \frac{\text{in wg}}{100 \text{ ft}}\right) = 0.57 \text{ in wg}$$

The fan must be able to supply a static pressure of

$$\text{SP}_{\text{fan}} = 0.57 \text{ in wg} + 0.25 \text{ in wg}$$

$$= \boxed{0.82 \text{ in wg}}$$

The answer is (C).

(c) The total pressure supplied by the fan is

$$\text{TP}_{\text{fan}} = 0.82 \text{ in wg} + \left(\frac{1600 \text{ fpm}}{4005}\right)^2$$

$$= \boxed{0.98 \text{ in wg}}$$

The answer is (A).

SI Solution

(a) *step 1:* From Table 20.7, choose the main duct velocity as 1600 cfm. From the table footnote (*a*), the SI velocity is

$$v_{\text{main}} = (1600 \text{ fpm})(0.00508) = 8.1 \text{ m/s}$$

step 2: The total air flow from the fan is 700 L/s. From Fig. 20.5, the main duct diameter is approximately 340 mm. The friction loss is 2.3 Pa/m.

step 3: After the first takeoff at A, the flow rate in section A–B is

$$700 \; \frac{L}{s} - 150 \; \frac{L}{s} = 550 \; L/s$$

From Fig. 20.5 for 550 L/s and 2.3 Pa/m, the diameter is 300 mm and the velocity is 7.7 m/s. Similarly, the diameter and the velocity for other sections are obtained and listed in the following table.

section	Q (L/s)	D (mm)	v (m/s)
fan–A	700	340	8.1
A–B	550	300	7.7
B–C	400	275	7
C–D	250	225	6.3
D–E	150	180	5.5
E–F	50	125	4.4

These diameters are for round duct. The equivalent rectangular duct sides with an aspect ratio of 1.5 are found as follows.

$$a = \frac{D(1.5+1)^{0.25}}{(1.3)(1.5)^{0.625}} = 0.75D$$

$$b = Ra = (1.5)(0.75D) = 1.125D$$

Convert the diameters to sides a and b.

$$a_{\text{fan–A}} = (0.75)(340 \text{ mm}) = 255 \text{ mm}$$

$$b_{\text{fan–A}} = (1.125)(340 \text{ mm}) = 383 \text{ mm}$$

Use a 275 × 400 mm duct.

The following table is prepared similarly, rounding up as appropriate.

section	D (mm)	a (mm)	b (mm)
fan–A	340	275	400
A–B	300	225	350
B–C	275	225	325
C–D	225	175	275
D–E	180	150	225
E–F	125	100	150

(b) *step 4:* By inspection, the longest run is fan–F. From Table 20.4, the equivalent length of each bend is 12D. For two elbows,

$$L_{e,\text{bend}} = 12D$$

$$= (12) \left(\frac{340 \text{ mm} + 225 \text{ mm}}{1000 \; \frac{\text{mm}}{\text{m}}} \right)$$

$$= 6.8 \text{ m}$$

The equivalent length of the entire run is

4.5 m	(fan to first bend)
13.5 m	(first bend to point A)
9 m	(section A–B)
9 m	(section B–C)
6 m	(point C to second bend)
6.8 m	(equivalent length of two bends)
3 m	(second bend to point D)
6 m	(section D–E)
6 m	(section E–F)

total: 63.8 m

The straight-through friction loss in the longest run is

$$(63.8 \text{ m}) \left(2.3 \; \frac{\text{Pa}}{\text{m}} \right) = 147 \text{ Pa}$$

The fan must be able to supply a static pressure of

$$\text{SP}_{\text{fan}} = 147 \text{ Pa} + 63 \text{ Pa} = \boxed{210 \text{ Pa}}$$

The answer is (C).

(c) The total pressure supplied by the fan is

$$\text{TP}_{\text{fan}} = 210 \text{ Pa} + \left(0.6 \; \frac{\text{Pa·s}^2}{\text{m}^2} \right) \left(8.1 \; \frac{\text{m}}{\text{s}} \right)^2$$

$$= \boxed{249 \text{ Pa}}$$

The answer is (A).

7. *Customary U.S. Solution*

(a) *step 1:* From Table 20.7, choose the main duct velocity as 1600 fpm.

step 2: The total air flow from the fan is

$$(12)(300 \text{ cfm}) = 3600 \text{ cfm}$$

From Fig. 20.4, the main duct diameter is 20 in. The friction loss is 0.16 in wg per 100 ft.

step 3: After the first takeoff, the flow rate between first and second is

$$3600 \text{ cfm} - (4)(300 \text{ cfm}) = 2400 \text{ cfm}$$

From Fig. 20.4, for 2400 cfm and 0.16 in wg per 100 ft, the diameter is 17.2 in and the velocity is 1440 fpm. Similarly, the diameter and the velocity for other sections are obtained and listed in the following table.

section	Q (cfm)	D (in)	v (fpm)
fan–first	3600	20	1600
first–second	2400	17.2	1440
second–third	1200	13.2	1210
first–A	1200	13.2	1210
A–B	900	12	1150
B–C	600	10.2	1030
C–D	300	7.9	860
second–E	1200	13.2	1210
E–F	900	12	1150
F–G	600	10.2	1030
G–H	300	7.9	860
I–J	900	12	1150
J–K	600	10.2	1030
K–L	300	7.9	860

(b) *step 4:* By inspection, the longest run is fan–L. From Table 20.4, the equivalent length of each bend is $14.5D$ (interpolated). For two elbows,

$$L_{e,\text{bend}} = 14.5D = (14.5)\left(\dfrac{20\text{ in} + 13.2\text{ in}}{12\ \frac{\text{in}}{\text{ft}}}\right)$$

$$= 40.1\text{ ft}$$

The equivalent length of the entire run is

25 ft	(fan to first bend)
35 ft	(first bend to first section)
20 ft	(first section to second section)
20 ft	(second section to third section)
10 ft	(third section to point I)
40.1 ft	(equivalent length of two bends)
20 ft	(point I to point J)
20 ft	(point J to point K)
20 ft	(point K to point L)

total: 210.1 ft

The straight-through friction loss in the longest run is

$$\left(\dfrac{210.1\text{ ft}}{100\text{ ft}}\right)\left(0.16\ \dfrac{\text{in wg}}{100\text{ ft}}\right) = 0.34\text{ in wg}$$

The fan must be able to supply a static pressure of

$$\text{SP}_{\text{fan}} = 0.34\text{ in wg} + 0.15\text{ in wg}$$

$$= \boxed{0.49\text{ in wg}}$$

The answer is (A).

(c) The total pressure supplied by the fan is

$$\text{TP}_{\text{fan}} = 0.49\text{ in wg} + \left(\dfrac{1600\text{ fpm}}{4005}\right)^2$$

$$= \boxed{0.65\text{ in wg}}$$

The answer is (B).

SI Solution

(a) *step 1:* From Table 20.7, choose the main duct velocity as 1600 fpm. From the table footnote (a), the SI velocity is

$$v_{\text{main}} = (1600\text{ fpm})(0.00508) = 8.1\text{ m/s}$$

step 2: The total air flow rate from the fan is

$$(12)\left(140\ \dfrac{\text{L}}{\text{s}}\right) = 1680\text{ L/s}$$

From Fig. 20.5, the main duct diameter is 500 mm. The friction loss is 1.5 Pa/m.

step 3: After the first takeoff, the flow rate between first and second is

$$1680\ \dfrac{\text{L}}{\text{s}} - (4)\left(140\ \dfrac{\text{L}}{\text{s}}\right) = 1120\text{ L/s}$$

From Fig. 20.5, for 1120 L/s and 1.5 Pa/m, the diameter is 440 mm and the velocity is 7.5 m/s. Similarly, the diameter and the velocity for other sections are obtained and listed in the following table.

section	Q (L/s)	D (mm)	v (m/s)
fan–first	1680	500	8.1
first–second	1120	440	7.5
second–third	560	335	6.4
first–A	560	335	6.4
A–B	420	305	5.9
B–C	280	260	5.4
C–D	140	195	4.5
second–E	560	335	6.4
E–F	420	305	5.9
F–G	280	260	5.4
G–H	140	195	4.5
I–J	420	305	5.9
J–K	280	260	5.4
K–L	140	195	4.5

(b) *step 4:* By inspection, the longest run is fan–L. From Table 20.4, the equivalent length of each bend is $14.5D$ (interpolated). For two elbows,

$$L_{e,\text{bend}} = 14.5D$$

$$= (14.5)\left(\frac{500 \text{ mm} + 335 \text{ mm}}{1000 \frac{\text{mm}}{\text{m}}}\right)$$

$$= 12.1 \text{ m}$$

The equivalent length of the entire run is

7.5 m	(fan to first bend)
10.5 m	(first bend to first section)
6 m	(first section to second section)
6 m	(second section to third section)
3 m	(third section to point I)
12.1 m	(equivalent length of two bends)
6 m	(point I to point J)
6 m	(point J to point K)
6 m	(point K to point L)

total: 63.1 m

The straight-through friction loss in the longest run is

$$(63.1 \text{ m})\left(1.5 \frac{\text{Pa}}{\text{m}}\right) = 95 \text{ Pa}$$

The fan must be able to supply a static pressure of

$$\text{SP}_{\text{fan}} = 95 \text{ Pa} + 38 \text{ Pa} = \boxed{133 \text{ Pa}}$$

The answer is (A).

(c) The total pressure supplied by the fan is

$$\text{TP}_{\text{fan}} = 133 \text{ Pa} + (0.6)\left(8.1 \frac{\text{m}}{\text{s}}\right)^2$$

$$= \boxed{172 \text{ Pa}}$$

The answer is (B).

8. *step 1:* From Table 20.7, choose the main duct velocity as 1600 fpm.

step 2: The area and diameter of the main duct are

$$A = \frac{Q}{\text{v}} = \frac{3600 \text{ cfm}}{1600 \text{ fpm}} = 2.25 \text{ ft}^2$$

$$D = \sqrt{\frac{4A}{\pi}} = \sqrt{\frac{(4)(2.25 \text{ ft}^2)}{\pi}}\left(12 \frac{\text{in}}{\text{ft}}\right)$$

$$= 20.3 \text{ in}$$

step 3: From Table 20.4, the equivalent length of each bend is $14.5D$ (interpolated). For the first elbow,

$$L_{e,\text{bend}} = 14.5D = (14.5)\left(\frac{20.3 \text{ in}}{12 \frac{\text{in}}{\text{ft}}}\right) = 24.5 \text{ ft}$$

The equivalent length of the main duct from the fan to the first takeoff and bend is

$$L = 25 \text{ ft} + 24.5 \text{ ft} + 35 \text{ ft} = 84.5 \text{ ft}$$

step 4: From Fig. 20.4, the friction loss in the main run up to the branch takeoff is approximately 0.16 in wg per 100 ft. The actual friction loss is

$$\text{FP}_{\text{main}} = (0.16 \text{ in wg})\left(\frac{84.5 \text{ ft}}{100 \text{ ft}}\right)$$

$$= 0.135 \text{ in wg}$$

step 5: After the first takeoff,

$$Q = 3600 \text{ cfm} - 1200 \text{ cfm} = 2400 \text{ cfm}$$

$$L = 20 \text{ ft}$$

$$\frac{L}{Q^{0.61}} = \frac{20 \text{ ft}}{(2400 \text{ cfm})^{0.61}} = 0.173$$

From Fig. 20.9, the velocity, v_2, is 1390 fpm.

step 6: Solve for the duct size from $A = Q/\text{v}$.

$$D_2 = \sqrt{\frac{4A}{\pi}} = \sqrt{\frac{4Q}{\pi \text{v}_2}}$$

$$= \sqrt{\frac{(4)(2400 \text{ cfm})}{\pi(1390 \text{ fpm})}}\left(12 \frac{\text{in}}{\text{ft}}\right)$$

$$= 18 \text{ in}$$

step 7: After the second takeoff,

$$Q = 2400 \text{ cfm} - 1200 \text{ cfm} = 1200 \text{ cfm}$$

$$L = 20 \text{ ft} + 14.5D + 10 \text{ ft}$$

$$= 30 \text{ ft} + 14.5D$$

Since L contains diameter, D, as an unknown, this will require an iterative procedure. Using velocity $\text{v}_1 = 1390$ fpm before the takeoff and using an iterative procedure, $D = 15$ in and $\text{v} = 980$ fpm.

step 8: Proceeding similarly, the following table is developed for the remaining sizes.

section	L (ft)	Q (cfm)	$\dfrac{L}{Q^{0.61}}$	v (fpm)
I–J	20	900	0.32	830
J–K	20	600	0.40	660
K–L	20	300	0.62	500

step 9: Solve for the duct size from $A = Q/\mathrm{v}$.

$$D = \sqrt{\frac{4A}{\pi}} = \sqrt{\frac{4Q}{\pi \mathrm{v}}}$$

$$D_{\text{I–J}} = \sqrt{\frac{(4)(900 \text{ cfm})}{\pi(830 \text{ fpm})}}\left(12 \; \frac{\text{in}}{\text{ft}}\right)$$

$$= 14 \text{ in}$$

$$D_{\text{J–K}} = \sqrt{\frac{(4)(600 \text{ cfm})}{\pi(660 \text{ fpm})}}\left(12 \; \frac{\text{in}}{\text{ft}}\right)$$

$$= 13 \text{ in}$$

$$D_{\text{K–L}} = \sqrt{\frac{(4)(300 \text{ cfm})}{\pi(500 \text{ fpm})}}\left(12 \; \frac{\text{in}}{\text{ft}}\right)$$

$$= 11 \text{ in}$$

step 10: The total pressure supplied by the fan is

$$0.135 \text{ in wg} + 0.15 \text{ in wg} = \boxed{0.285 \text{ in wg}}$$

9. *Customary U.S. Solution*

((a) and (b)) No dampers are needed in duct A. The area of section A is

$$A = \frac{\pi}{4}D^2 = \left(\frac{\pi}{4}\right)\left(\frac{12 \text{ in}}{12 \; \frac{\text{in}}{\text{ft}}}\right)^2 = 0.7854 \text{ ft}^2$$

The velocity in section A is

$$\mathrm{v_A} = \frac{Q}{A} = \frac{3000 \text{ cfm}}{(0.7854 \text{ ft}^2)\left(60 \; \frac{\text{sec}}{\text{min}}\right)}$$

$$= \frac{3820 \text{ fpm}}{60 \; \frac{\text{sec}}{\text{min}}} = 63.66 \text{ ft/sec}$$

For a four-piece elbow with a radius-to-diameter ratio of 1.5, $L_e \approx 14D$.

Assume $D = 10$ in. The total equivalent length of run C is

$$L_e = 50 \text{ ft} + 10 \text{ ft} + 10 \text{ ft} + (2)\left((14)\left(\frac{10 \text{ in}}{12 \; \frac{\text{in}}{\text{ft}}}\right)\right)$$

$$= 93.3 \text{ ft}$$

For any diameter, D, in inches of section C, the velocity will be

$$\mathrm{v_C} = \frac{Q}{A} = \frac{2000 \text{ cfm}}{\left(\frac{\pi}{4}\right)\left(\frac{D}{12 \; \frac{\text{in}}{\text{ft}}}\right)^2\left(60 \; \frac{\text{sec}}{\text{min}}\right)}$$

$$= 6111.5/D^2$$

From Eq. 17.28, the friction loss in section C will be

$$h_{f,\text{C}} = \frac{fL\mathrm{v}^2}{2Dg}$$

$$= \frac{(0.02)(93.3 \text{ ft})\left(\frac{6111.5}{D^2}\right)^2}{(2)\left(\frac{D}{12 \; \frac{\text{in}}{\text{ft}}}\right)\left(32.2 \; \frac{\text{ft}}{\text{sec}^2}\right)}$$

$$= \frac{1.3 \times 10^7}{D^5} \quad [\text{ft of air}]$$

The regain between A and C will be

$$h_{\text{regain}} = R\left(\frac{\mathrm{v}_A^2 - \mathrm{v}_G^2}{2g}\right)$$

$$= 0.65\left(\frac{\left(63.66 \; \frac{\text{ft}}{\text{sec}}\right)^2 - \left(\frac{6111.5}{D^2}\right)^2}{(2)\left(32.2 \; \frac{\text{ft}}{\text{sec}^2}\right)}\right)$$

$$= 40.9 - \frac{3.77 \times 10^5}{D^4} \quad [\text{ft of air}]$$

The principle of static regain is that

$$h_{f,\text{C}} = h_{\text{regain}}$$

$$\frac{1.3 \times 10^7}{D^5} = 40.9 - \frac{3.77 \times 10^5}{D^4}$$

By trial and error, $D \approx 13.5$ in. Since the assumed value of D is different from the calculated value, this process should be repeated.

$$L_e = 50 \text{ ft} + 10 \text{ ft} + 10 \text{ ft} + (2)\left((14)\left(\frac{13.5 \text{ in}}{12 \; \frac{\text{in}}{\text{ft}}}\right)\right)$$

$$= 101.5 \text{ ft}$$

From Eq. 17.28, the friction loss in section C will be

$$h_{f,\text{C}} = \frac{fLv^2}{2Dg}$$

$$= \frac{(0.02)(101.5\text{ ft})\left(\dfrac{6111.5}{D^2}\right)^2}{(2)\left(\dfrac{D}{12\,\frac{\text{in}}{\text{ft}}}\right)\left(32.2\,\dfrac{\text{ft}}{\text{sec}^2}\right)}$$

$$= \frac{1.41 \times 10^7}{D^5} \quad [\text{ft of air}]$$

The principle of static regain is that

$$h_{f,\text{C}} = h_{\text{regain}}$$

$$\frac{1.41 \times 10^7}{D^5} = 40.9 - \frac{3.77 \times 10^5}{D^4}$$

By trial and error,

$$D_\text{C} = \boxed{13.63\text{ in}\quad[\text{say }14\text{ in}]}$$

This results in a friction loss of

$$h_{f,\text{C}} = \frac{1.41 \times 10^7}{(14\text{ in})^5} = 26.2\text{ ft of air}$$

Because the regain cancels friction loss, the pressure loss from A to C is zero. No dampers are needed in duct C. A damper is needed in section B, however.

For any diameter, D, in inches in section B, the velocity will be

$$v_\text{B} = \frac{Q}{A} = \frac{1000\text{ cfm}}{\left(\dfrac{\pi}{4}\right)\left(\dfrac{D}{12\,\frac{\text{in}}{\text{ft}}}\right)^2\left(60\,\dfrac{\text{sec}}{\text{min}}\right)}$$

$$= \frac{3055.8}{D^2} \quad [\text{ft/sec}]$$

The friction loss in section B for Eq. 17.28 will be

$$h_{f,\text{B}} = \frac{fLv^2}{2Dg}$$

$$= \frac{(0.02)(10\text{ ft})\left(\dfrac{3055.8}{D^2}\right)^2}{(2)\left(\dfrac{D}{12\,\frac{\text{in}}{\text{ft}}}\right)\left(32.2\,\dfrac{\text{ft}}{\text{sec}^2}\right)}$$

$$= \frac{3.48 \times 10^5}{D^5} \quad [\text{ft of air}]$$

From Eq. 20.42, the friction loss in the branch takeoff between sections A and B will be

$$\Delta \text{TP}_{\text{A–B}} = K_{\text{br}}(\text{VP}_{\text{up}})$$

At this point, assume $v_\text{B}/v_\text{A} = 1.0$. Then from Table 20.6, for a 45° angle of takeoff, $K_{\text{br}} = 0.5$.

$$\Delta \text{TP}_{\text{A–B}} = (0.5)\left(\frac{3820\text{ fpm}}{4005}\right)^2 = 0.455\text{ in wg}$$

From Eq. 20.2,

$$p_{\text{psig}} = (0.455\text{ in wg})\left(0.0361\,\frac{\text{lbf}}{\text{in}^3}\right) = 0.01643\text{ psig}$$

From Eq. 20.1,

$$h_f = \frac{p_{\text{psig}}}{\gamma} = \frac{\left(0.01643\,\dfrac{\text{lbf}}{\text{in}^2}\right)\left(12\,\dfrac{\text{in}}{\text{ft}}\right)^2}{0.075\,\dfrac{\text{lbf}}{\text{ft}^3}}$$

$$= 31.5\text{ ft of air}$$

The regain between A and B will be

$$h_{\text{regain}} = R\left(\frac{v_\text{A}^2 - v_\text{B}^2}{2g}\right)$$

$$= (0.65)\left(\frac{\left(63.66\,\dfrac{\text{ft}}{\text{sec}}\right)^2 - \left(\dfrac{3055.8}{D^2}\right)^2}{(2)\left(32.2\,\dfrac{\text{ft}}{\text{sec}^2}\right)}\right)$$

$$= 40.9\text{ ft} - \frac{9.42 \times 10^4}{D^4} \quad [\text{ft of air}]$$

Set the regain equal to the loss.

$$40.9\text{ ft} - \frac{9.42 \times 10^4}{D^4} = \frac{3.48 \times 10^5}{D^5} + 31.5\text{ ft}$$

By trial and error,

$$D_\text{B} = \boxed{10.77\text{ in}\quad[\text{say }11\text{ in}]}$$

(c)
$$v_\text{B} = \frac{3055.8}{D^2}$$

$$= \frac{3055.8}{(11\text{ in})^2} = 25.25\text{ ft/sec}$$

$$\frac{v_\text{B}}{v_\text{A}} = \frac{25.25\,\dfrac{\text{ft}}{\text{sec}}}{63.6\,\dfrac{\text{ft}}{\text{sec}}} \approx 0.4$$

From Table 20.6, for a 45° angle of takeoff, $K_{\mathrm{br}} = 0.5$. Since the value of K_{br} remains the same, there is no need to repeat the preceding procedure. For section B, the friction canceled by the regain is

$$h_{f,\mathrm{B}} = 31.5 \text{ ft} + \frac{3.48 \times 10^5}{D^5}$$

$$= 31.5 \text{ ft} + \frac{3.48 \times 10^5}{(11 \text{ in})^5}$$

$$= \boxed{33.7 \text{ ft of air}}$$

SI Solution

((a) and (b)) No dampers are needed in duct A. The area of section A is

$$A = \frac{\pi}{4} D^2 = \left(\frac{\pi}{4}\right)\left(\frac{305 \text{ mm}}{1000 \frac{\text{mm}}{\text{m}}}\right)^2 = 0.0731 \text{ m}^2$$

The velocity in section A is

$$v_{\mathrm{A}} = \frac{Q}{A} = \frac{\left(1400 \frac{\text{L}}{\text{s}}\right)\left(\frac{1 \text{ m}^3}{1000 \text{ L}}\right)}{0.0731 \text{ m}^2} = 19.15 \text{ m/s}$$

For a four-piece elbow with a radius-to-diameter ratio of 1.5, $L_e \approx 14D$.

Assume $D = 250$ mm. The total equivalent length of run C is

$$L_e = 15 \text{ m} + 3 \text{ m} + 3 \text{ m} + (2)\left((14)\left(\frac{250 \text{ mm}}{1000 \frac{\text{mm}}{\text{m}}}\right)\right)$$

$$= 28 \text{ m}$$

For any diameter, D, in mm of section C, the velocity will be

$$v_{\mathrm{C}} = \frac{Q}{A} = \frac{\left(900 \frac{\text{L}}{\text{s}}\right)\left(\frac{1 \text{ m}^3}{1000 \text{ L}}\right)}{\left(\frac{\pi}{4}\right)\left(\frac{D}{1000 \frac{\text{mm}}{\text{m}}}\right)^2}$$

$$= \frac{1.146 \times 10^6}{D^2} \quad [\text{m/s}]$$

From Eq. 17.28, the friction loss in section C will be

$$h_{f,\mathrm{C}} = \frac{fLv^2}{2Dg}$$

$$= \frac{(0.02)(28 \text{ m})\left(\frac{1.146 \times 10^6}{D^2}\right)^2}{(2)\left(\frac{D}{1000 \frac{\text{mm}}{\text{m}}}\right)\left(9.81 \frac{\text{m}}{\text{s}^2}\right)}$$

$$= \frac{3.75 \times 10^{13}}{D^5} \quad [\text{m of air}]$$

The regain between A and C will be

$$h_{\mathrm{regain}} = R\left(\frac{v_{\mathrm{A}}^2 - v_{\mathrm{C}}^2}{2g}\right)$$

$$= (0.65)\left(\frac{\left(19.15 \frac{\text{m}}{\text{s}}\right)^2 - \left(\frac{1.146 \times 10^6}{D^2}\right)^2}{(2)\left(9.81 \frac{\text{m}}{\text{s}^2}\right)}\right)$$

$$= 12.15 \text{ m} - \frac{4.35 \times 10^{10}}{D^4} \quad [\text{m of air}]$$

The principle of static regain is that

$$h_{f,\mathrm{C}} = h_{\mathrm{regain}}$$

$$\frac{3.75 \times 10^{13}}{D^5} = 12.15 \text{ m} - \frac{4.35 \times 10^{10}}{D^4}$$

By trial and error, $D \approx 335$ mm. Since the assumed value of D is different from the calculated value, this process should be repeated.

$$L_e = 15 \text{ m} + 3 \text{ m} + 3 \text{ m} + (2)\left((14)\left(\frac{335 \text{ mm}}{1000 \frac{\text{mm}}{\text{m}}}\right)\right)$$

$$= 30.4 \text{ m}$$

From Eq. 17.28, the friction loss in section C will be

$$h_{f,\mathrm{C}} = \frac{fLv^2}{2Dg}$$

$$= \frac{(0.02)(30.4 \text{ m})\left(\frac{1.146 \times 10^6}{D^2}\right)^2}{(2)\left(\frac{D}{1000 \frac{\text{mm}}{\text{m}}}\right)\left(9.81 \frac{\text{m}}{\text{s}^2}\right)}$$

$$= \frac{4.07 \times 10^{13}}{D^5} \quad [\text{m of air}]$$

The principle of static regain is that

$$h_{f,\mathrm{C}} = h_{\mathrm{regain}}$$

$$\frac{4.07 \times 10^{13}}{D^5} = 12.15 - \frac{4.35 \times 10^{10}}{D^4}$$

By trial and error, $D = 340$ mm. This results in a friction loss of

$$h_{f,\mathrm{C}} = \frac{4.07 \times 10^{13}}{(340 \text{ mm})^5} = 8.96 \text{ m of air}$$

Because the regain cancels this friction loss, the pressure loss from A to C is zero. No dampers are needed in duct C. A damper is needed at B, however.

For any diameter, D, in mm in section B, the velocity will be

$$v_B = \frac{Q}{A} = \frac{\left(500 \dfrac{L}{s}\right)\left(\dfrac{1 \text{ m}^3}{1000 \text{ L}}\right)}{\left(\dfrac{\pi}{4}\right)\left(\dfrac{D}{1000 \dfrac{\text{mm}}{\text{m}}}\right)^2}$$

$$= \frac{6.366 \times 10^5}{D^2} \quad [\text{m/s}]$$

From Eq. 17.28, the friction loss in section B will be

$$h_{f,B} = \frac{fLv^2}{2Dg}$$

$$= \frac{(0.02)(3 \text{ m})\left(\dfrac{6.366 \times 10^5}{D^2}\right)^2}{(2)\left(\dfrac{D}{1000 \dfrac{\text{mm}}{\text{m}}}\right)\left(9.81 \dfrac{\text{m}}{\text{s}^2}\right)}$$

$$= \frac{1.239 \times 10^{12}}{D^5} \quad [\text{m of air}]$$

From Eq. 20.42, the friction loss in the branch takeoff between sections A and B will be

$$\Delta TP_{A-B} = K_{br}(VP_{up})$$

At this point, assume $v_B/v_A = 1.0$. Then from Table 20.6, for a 45° angle of takeoff, $K_{br} = 0.5$.

$$\Delta TP_{A-B} = (0.5)(0.6)\left(19.15 \dfrac{\text{m}}{\text{s}}\right)^2$$

$$= 110 \text{ Pa}$$

From Eq. 20.1,

$$h_f = \frac{p}{\rho g} = \frac{110 \text{ Pa}}{\left(1.2 \dfrac{\text{kg}}{\text{m}^3}\right)\left(9.81 \dfrac{\text{m}}{\text{s}^2}\right)}$$

$$= 9.34 \text{ m of air}$$

The regain between A and B will be

$$h_{\text{regain}} = R\left(\frac{v_A^2 - v_B^2}{2g}\right)$$

$$= (0.65)\left(\frac{\left(19.15 \dfrac{\text{m}}{\text{s}}\right)^2 - \left(\dfrac{6.366 \times 10^5}{D^2}\right)^2}{(2)\left(9.81 \dfrac{\text{m}}{\text{s}^2}\right)}\right)$$

$$= \left(12.15 \text{ m} - \frac{1.343 \times 10^{10}}{D^4}\right) \quad [\text{m of air}]$$

Set the regain equal to the loss.

$$12.15 \text{ m} - \frac{1.343 \times 10^{10}}{D^4} = \frac{1.239 \times 10^{12}}{D^5} + 9.34 \text{ m}$$

By trial and error,

$$D_B = 280 \text{ mm}$$
$$v_B = 8.12 \text{ m/s}$$

(c)
$$\frac{v_B}{v_A} = \frac{8.12 \dfrac{\text{m}}{\text{s}}}{19.15 \dfrac{\text{m}}{\text{s}}} \approx 0.4$$

From Table 20.6, for a 45° angle of takeoff, $K_{br} = 0.5$. Since the value of K_{br} remains the same, there is no need to repeat the preceding procedure. For section B, the friction canceled by the regain is

$$h_{f,B} = 9.34 \text{ m} + \frac{1.239 \times 10^{12}}{D^5}$$

$$= 9.34 \text{ m} + \frac{1.239 \times 10^{12}}{(280 \text{ mm})^5}$$

$$= \boxed{10.06 \text{ m of air}}$$

Fluids

21 Inorganic Chemistry

PRACTICE PROBLEMS

Empirical Formula Development

1. The gravimetric analysis of a compound is 40% carbon, 6.7% hydrogen, and 53.3% oxygen. What is the simplest formula for the compound?

(A) HCO

(B) HCO_2

(C) CH_2O

(D) CHO_2

Precipitation Softening

2. A town's water supply has the following ionic concentrations.

Al^{+++}	0.5 mg/L
Ca^{++}	80.2 mg/L
Cl^-	85.9 mg/L
CO_2	19 mg/L
CO_3^{--}	0
Fe^{++}	1.0 mg/L
Fl^-	0
HCO_3^-	185 mg/L
Mg^{++}	24.3 mg/L
Na^+	46.0 mg/L
NO_3^-	0
SO_4^{--}	125 mg/L

(a) What is the total hardness?

(A) 160 mg/L as $CaCO_3$

(B) 200 mg/L as $CaCO_3$

(C) 260 mg/L as $CaCO_3$

(D) 300 mg/L as $CaCO_3$

(b) How much slaked lime is required to combine with the carbonate hardness?

(A) 45 mg/L as substance

(B) 90 mg/L as substance

(C) 130 mg/L as substance

(D) 150 mg/L as substance

(c) How much soda ash is required to react with the carbonate hardness?

(A) none

(B) 15 mg/L as substance

(C) 35 mg/L as substance

(D) 60 mg/L as substance

3. A city's water supply contains the following ionic concentrations.

$Ca(HCO_3)_2$	137 mg/L as $CaCO_3$
$MgSO_4$	72 mg/L as $CaCO_3$
CO_2	0

(a) How much slaked lime is required to soften 1,000,000 gal of this water to a hardness of 100 mg/L if 30 mg/L of excess lime is used?

(A) 930 lbm

(B) 1200 lbm

(C) 1300 lbm

(D) 1700 lbm

(b) How much soda ash is required to soften 1,000,000 gal of this water to a hardness of 100 mg/L if 30 mg/L of excess lime is used?

(A) none

(B) 300 lbm

(C) 500 lbm

(D) 700 lbm

SOLUTIONS

1. Calculate the mole ratios of the atoms by assuming there are 100 g of sample.

For 100 g of sample,

substance	mass	$\dfrac{m}{AW}$ = no. moles	mole ratio
C	40 g	$\dfrac{40}{12} = 3.33$	1
H	6.7 g	$\dfrac{6.7}{1} = 6.7$	2
O	53.3 g	$\dfrac{53.3}{16} = 3.33$	1

The empirical formula is $\boxed{CH_2O.}$

The answer is (C).

2. (a)

	mg/L as substance	factor from App. 21.C		
Ca^{++}:	80.2	$\times$	2.5	= 200.5 mg/L
Mg^{++}:	24.3	$\times$	4.1	= 99.63 mg/L
Fe^{++}:	1	$\times$	1.79	= 1.79 mg/L
Al^{+++}:	0.5	$\times$	5.56	= 2.78 mg/L
			hardness =	$\boxed{304.7 \text{ mg/L}}$

The answer is (D).

(b) To remove the carbonate hardness,

$$CO_2: \quad 19 \, \frac{mg}{L} \times 2.27 = 43.13 \text{ mg/L as } CaCO_3$$

Add lime to remove the carbonate hardness. It does not matter whether the HCO_3^- comes from Mg^{++}, Ca^{++}, or Fe^{++}; adding lime will remove it.

There may be Mg^{++}, Ca^{++}, or Fe^{++} ions left over in the form of noncarbonate hardness, but the problem asked for carbonate hardness. Converting from mg/L of substance to mg/L as $CaCO_3$,

$$HCO_3^-: \quad 185 \, \frac{mg}{L} \times 0.82 = 151.7 \text{ mg/L}$$

The total equivalents to be neutralized are

$$43.13 \, \frac{mg}{L} + 151.7 \, \frac{mg}{L} = 194.83 \text{ mg/L}$$

Convert $Ca(OH)_2$ using App. 21.C.

$$\frac{mg}{L} \text{ of } Ca(OH)_2 = \frac{194.83 \, \dfrac{mg}{L}}{1.35}$$

$$= \boxed{144.3 \text{ mg/L as substance}}$$

The answer is (D).

(c) | No soda ash is required since it is used to remove noncarbonate hardness.

The answer is (A).

3. (a) $Ca(HCO_3)_2$ and $MgSO_4$ both contribute to hardness. Since 100 mg/L of hardness is the goal, leave all $MgSO_4$ in the water. Take out 137 mg/L + 72 mg/L − 100 mg/L = 109 mg/L of $Ca(HCO_3)_2$. From App. 21.C (including the excess even though the reaction is not complete),

$$\text{pure } Ca(OH)_2 = 30 \, \frac{mg}{L} + \frac{109 \, \dfrac{mg}{L}}{1.35}$$

$$= 110.74 \text{ mg/L}$$

$$\left(110.74 \, \frac{mg}{L}\right)\left(8.345 \, \frac{\text{lbm-L}}{\text{mg-MG}}\right) = \boxed{924 \text{ lbm/MG}}$$

The answer is (A).

(b) No soda ash is required.

The answer is (A).

22 Fuels and Combustion

PRACTICE PROBLEMS

1. Methane (MW = 16.043) with a higher heating value of 24,000 Btu/lbm (55.8 MJ/kg) is burned in a furnace with a 50% efficiency based on the higher heating value. How much water can be heated from 60 to 200°F (15 to 95°C) when 7 ft^3 (200 L) at 60°F and 14.73 psia are burned?

(A) 25 lbm (11 kg)
(B) 35 lbm (16 kg)
(C) 50 lbm (23 kg)
(D) 95 lbm (43 kg)

2. 15 lbm/hr (6.8 kg/h) of propane (C_3H_8, MW = 44.097) are burned stoichiometrically in air. What volume of dry carbon dioxide (CO_2) is formed after cooling to 70°F (21°C) and 14.7 psia (101 kPa)?

(A) 180 ft^3/hr (5.0 m^3/h)
(B) 270 ft^3/hr (7.6 m^3/h)
(C) 390 ft^3/hr (11 m^3/h)
(D) 450 ft^3/hr (13 m^3/h)

3. In a particular installation, 30% by volume excess air at 15 psia (103 kPa) and 100°F (40°C) is needed for the combustion of methane. How much nitrogen (MW = 28.016) passes through the furnace if methane at the same conditions is burned at the rate of 4000 ft^3/hr (31 L/s)?

(A) 270 lbm/hr (0.033 kg/s)
(B) 930 lbm/hr (0.11 kg/s)
(C) 1800 lbm/hr (0.22 kg/s)
(D) 2700 lbm/hr (0.34 kg/s)

4. How much air is required to completely burn one unit mass of a fuel that is 84% carbon, 15.3% hydrogen, 0.3% sulfur, and 0.4% nitrogen by weight?

(A) 9 lbm air/lbm fuel (9 kg air/kg fuel)
(B) 12 lbm air/lbm fuel (12 kg air/kg fuel)
(C) 15 lbm air/lbm fuel (15 kg air/kg fuel)
(D) 18 lbm air/lbm fuel (18 kg air/kg fuel)

5. Propane (C_3H_8) is burned with 20% excess air. What is the gravimetric percentage of carbon dioxide in the flue gas?

(A) 8%
(B) 12%
(C) 15%
(D) 22%

6. The ultimate analysis of a coal is 80% carbon, 4% hydrogen, 2% oxygen, and the rest ash. The flue gases are 60°F and 14.7 psia (15.6°C and 101.3 kPa) when sampled, and are 80% nitrogen, 12% carbon dioxide, 7% oxygen, and 1% carbon monoxide by volume. How much air is required to burn 1 lbm (1 kg) of coal under these conditions?

(A) 11 lbm (11 kg)
(B) 15 lbm (15 kg)
(C) 19 lbm (19 kg)
(D) 23 lbm (23 kg)

7. What is the approximate heating value of an oil with a specific gravity of 40° API?

(A) 20,000 Btu/lbm (46 MJ/kg)
(B) 25,000 Btu/lbm (58 MJ/kg)
(C) 30,000 Btu/lbm (69 MJ/kg)
(D) 35,000 Btu/lbm (81 MJ/kg)

8. The ultimate analysis of a coal is 75% carbon, 5% hydrogen, 3% oxygen, 2% nitrogen, and the rest ash. Atmospheric air is 60°F (16°C) and at standard pressure.

(a) What is the theoretical temperature of the combustion products?

(A) 3500°F (1900°C)
(B) 4000°F (2200°C)
(C) 4500°F (2300°C)
(D) 5000°F (2500°C)

(b) Estimate the actual temperature of the combustion products at the boiler outlet. Neglect dissociation. Assume 40% excess air is used and 75% of the heat is transferred to the boiler.

(A) 650°F (340°C)
(B) 880°F (470°C)
(C) 970°F (520°C)
(D) 1300°F (700°C)

9. A fuel oil has the following ultimate analysis: 85.43% carbon, 11.31% hydrogen, 2.7% oxygen, 0.34% sulfur, and 0.22% nitrogen. The oil is burned with 60% excess air. Evaluate flue gas volumes at 600°F (320°C).

(a) What volume of wet flue gases will be produced?
- (A) 450 ft³ (28 m³)
- (B) 500 ft³ (32 m³)
- (C) 550 ft³ (35 m³)
- (D) 600 ft³ (38 m³)

(b) What volume of dry flue gases will be produced?
- (A) 480 ft³ (30 m³)
- (B) 560 ft³ (35 m³)
- (C) 630 ft³ (40 m³)
- (D) 850 ft³ (79 m³)

(c) What will be the volumetric fraction of carbon dioxide?
- (A) 6%
- (B) 8%
- (C) 10%
- (D) 13%

10. The ultimate analysis of a coal is 51.45% carbon, 16.69% ash, 15.71% moisture, 7.28% oxygen, 4.02% hydrogen, 3.92% sulfur, and 0.93% nitrogen. 15,395 lbm (6923 kg) of the coal are burned, and 2816 lbm (1267 kg) of ash containing 20.9% carbon (by weight) are recovered. 13.3 lbm (kg) of dry gases are produced per pound (kilogram) of fuel burned. How much air was used per unit mass of fuel?
- (A) 13 lbm air/lbm fuel (13 kg air/kg fuel)
- (B) 14 lbm air/lbm fuel (14 kg air/kg fuel)
- (C) 15 lbm air/lbm fuel (15 kg air/kg fuel)
- (D) 17 lbm air/lbm fuel (17 kg air/kg fuel)

11. A coal is 65% carbon by weight. During combustion, 3% of the coal is lost in the ash pit. Combustion uses 9.87 lbm (kg) of air per pound (kg) of fuel. The flue gas analysis is 81.5% nitrogen, 9.5% carbon dioxide, and 9% oxygen. What is the percentage of excess air by mass?
- (A) 10%
- (B) 30%
- (C) 70%
- (D) 140%

12. A coal has an ultimate analysis of 67.34% carbon, 4.91% oxygen, 4.43% hydrogen, 4.28% sulfur, 1.08% nitrogen, and the rest ash. 3% of the carbon is lost during combustion. The flue gases are 81.9% nitrogen, 15.5% carbon dioxide, 1.6% carbon monoxide, and 1% oxygen by volume. What is the heat loss due to the formation of carbon monoxide?

- (A) 600 Btu/lbm (1.4 MJ/kg)
- (B) 800 Btu/lbm (1.9 MJ/kg)
- (C) 1000 Btu/lbm (2.3 MJ/kg)
- (D) 1200 Btu/lbm (2.8 MJ/kg)

13. (*Time limit: one hour*) A natural gas is 93% methane, 3.73% nitrogen, 1.82% hydrogen, 0.45% carbon monoxide, 0.35% oxygen, 0.25% ethylene, 0.22% carbon dioxide, and 0.18% hydrogen sulfide by volume. The gas is burned with 40% excess air. Atmospheric air is 60°F and at standard atmospheric pressure.

(a) What is the gas density?
- (A) 0.017 lbm/ft³
- (B) 0.043 lbm/ft³
- (C) 0.069 lbm/ft³
- (D) 0.110 lbm/ft³

(b) What are the theoretical air requirements?
- (A) 5 ft³ air/ft³ fuel
- (B) 7 ft³ air/ft³ fuel
- (C) 9 ft³ air/ft³ fuel
- (D) 13 ft³ air/ft³ fuel

(c) What is the percentage of CO_2 in the flue gas (wet basis)?
- (A) 6.9%
- (B) 7.7%
- (C) 8.1%
- (D) 11%

(d) What is the percentage of CO_2 in the flue gas (dry basis)?
- (A) 6.9%
- (B) 7.7%
- (C) 8.1%
- (D) 11%

14. Coal enters a steam generator at 73°F (23°C). The coal has an ultimate analysis of 78.42% carbon, 8.25% oxygen, 5.68% ash, 5.56% hydrogen, 1.09% nitrogen, and 1.0% sulfur. The coal's heating value is 14,000 Btu/lbm (32.6 MJ/kg). 7.03% of the coal's weight is lost in the ash pit. The ash contains 31.5% carbon. Air at 67°F (19°C) wet bulb and 73°F (23°C) dry bulb is supplied. The flue gases consist of 80.08% nitrogen, 14.0% carbon dioxide, 5.5% oxygen, and 0.42% carbon monoxide. The flue gases are at a temperature of 575°F (300°C). Saturated water at 212°F (100°C) and 1.0 atm enters the steam generator. 11.12 lbm (kg) of water are evaporated in the boiler per pound (kilogram) of dry coal consumed. What is the complete heat balance, enumerating all combustion losses in the furnace?

15. Coal has a gravimetric analysis of 83% carbon, 5% hydrogen, 5% oxygen, and 7% noncombustible matter. 10% of the fired coal mass, including all of the noncombustible matter, is recovered in the ash pit. 26 lbm (26 kg) of air at 70°F (21°C) and 1 atmosphere are used per lbm (kg) of coal burned. When loaded into the furnace, the coal temperature is 60°F (15.6°C). The stack gas from coal combustion has a temperature of 550°F (290°C). What percentage of the combustion heat is carried away by the stack gases?

   (A) 18%
   (B) 24%
   (C) 37%
   (D) 49%

16. (*Time limit: one hour*) A utility boiler burns coal with an ultimate analysis of 76.56% carbon, 7.7% oxygen, 6.1% silicon, 5.5% hydrogen, 2.44% sulfur, and 1.7% nitrogen. 410 lbm/hr of refuse are removed with a composition of 30% carbon and 0% sulfur. All the sulfur and the remaining carbon is burned. The power plant has the following characteristics.

- coal feed rate: 15,300 lbm/hr

- electric power rating: 17 MW

- generator efficiency: 95%

- steam generator efficiency: 86%

- cooling water rate: 225 ft^3/sec

(a) What is the emission rate of solid particulates in lbm/hr?

   (A) 23 lbm/hr
   (B) 150 lbm/hr
   (C) 810 lbm/hr
   (D) 1700 lbm/hr

(b) How much sulfur dioxide is produced per hour?

   (A) 220 lbm/hr
   (B) 340 lbm/hr
   (C) 750 lbm/hr
   (D) 1100 lbm/hr

(c) What is the temperature rise of the cooling water?

   (A) 2.4°F
   (B) 6.5°F
   (C) 9.8°F
   (D) 13°F

(d) What efficiency must the flue gas particulate collectors have in order to meet a limit of 0.1 lbm of particulates per million Btu per hour (0.155 kg/MW)?

   (A) 93.1%
   (B) 97.4%
   (C) 98.8%
   (D) 99.1%

17. (*Time limit: one hour*) 250 SCFM (118 L/s) of propane are mixed with an oxidizer consisting of 60% oxygen and 40% nitrogen by volume in a proportion allowing 40% excess oxygen by weight. The maximum velocity for the two reactants when combined is 400 ft/min (2 m/s) at 14.7 psia (101 kPa) and 80°F (27°C). Maximum velocity for the products is 800 ft/min (4 m/s) at 8 psia (55 kPa) and 460°F (240°C).

(a) Size the inlet pipe.

   (A) 8 ft^2 (0.8 m^2)
   (B) 11 ft^2 (1.1 m^2)
   (C) 14 ft^2 (1.3 m^2)
   (D) 17 ft^2 (1.6 m^2)

(b) Size the stack.

   (A) 8 ft^2 (0.8 m^2)
   (B) 11 ft^2 (1.1 m^2)
   (C) 14 ft^2 (1.3 m^2)
   (D) 17 ft^2 (1.6 m^2)

(c) What is the actual flow of oxygen?

   (A) 150 lbm/min (1.1 kg/s)
   (B) 180 lbm/min (1.3 kg/s)
   (C) 220 lbm/min (1.6 kg/s)
   (D) 270 lbm/min (2.0 kg/s)

(d) What is the volume of flue gases?

   (A) 4000 ft^3/min (1.9 m^3/s)
   (B) 7000 ft^3/min (3.3 m^3/s)
   (C) 9000 ft^3/min (4.3 m^3/s)
   (D) 11,000 ft^3/min (5.3 m^3/s)

(e) What is the dew point of the flue gases?

   (A) 40°F (4°C)
   (B) 100°F (38°C)
   (C) 130°F (55°C)
   (D) 180°F (80°C)

18. (*Time limit: one hour*) An industrial process uses hot gas at 3600°R (1980°C) and 14.7 psia (101 kPa). It is proposed that propane be burned stoichiometrically in a mixture of nitrogen and oxygen. After passing through the process, gas will be exhausted through a duct, being cooled slowly to 100°F (38°C) and 14.7 psia (101 kPa) before discharge. The following data are available.

- The enthalpies of formation (at the standard reference temperature) are

   $C_3H_8(g)$ $\Delta H_f = +28,800$ Btu/lbmol
                          (+67.0 GJ/kmol)

   $CO_2(g)$ $\Delta H_f = -169,300$ Btu/lbmol
                          (−393.8 GJ/kmol)

   $H_2O(g)$ $\Delta H_f = -104,040$ Btu/lbmol
                          (−242 GJ/kmol)

- The enthalpy increases from the standard reference temperature to 3600°R (1980°C) are

  CO_2 : 39,791 Btu/lbmol (92.6 GJ/kmol)
  H_2O : 31,658 Btu/lbmol (73.6 GJ/kmol)
  N_2 : 24,471 Btu/lbmol (56.9 GJ/kmol)

(a) What are the proportions by weight of the oxygen and nitrogen?

(b) What is the amount of liquid water, if any, present in the stack gas?

(c) How much water is removed from the stack gas?

19. (*Time limit: one hour*) An internal combustion engine is normally fuel-injected with gasoline (C_8H_{18}; lower heating value of 23,200 Btu/lbm; 54 MJ/kg). It is desired to switch to alcohol (C_2H_5OH; lower heating value of 11,930 Btu/lbm; 27.7 MJ/kg). Minimal changes to the engine are to be made. The indicated and mechanical efficiencies are unchanged. Stoichiometric air/fuel ratios are being used for initial calculations.

(a) If the power output is to be unchanged, what must be the percentage change in specific fuel consumption?
- (A) 45%
- (B) 65%
- (C) 85%
- (D) 95%

(b) If the fuel-injection velocity is to be unchanged, what is the percentage change in jet size?
- (A) 35%
- (B) 55%
- (C) 80%
- (D) 95%

(c) If no changes are made to the engine, what will be the percentage change in power output?
- (A) −25%
- (B) −45%
- (C) −60%
- (D) −75%

SOLUTIONS

1. *Customary U.S. Solution*

$$T = 60°F + 460 = 520°R$$

$$p = 14.73 \text{ psia}$$

$$R = \frac{R^*}{MW} = \frac{1545.33 \dfrac{\text{ft-lbf}}{\text{lbmol-°R}}}{16.043 \dfrac{\text{lbm}}{\text{lbmol}}}$$

$$= 96.32 \text{ ft-lbf/lbm-°R}$$

$$m = \frac{pV}{RT}$$

$$= \frac{\left(14.73 \dfrac{\text{lbf}}{\text{in}^2}\right)\left(144 \dfrac{\text{in}^2}{\text{ft}^2}\right)(7 \text{ ft}^3)}{\left(96.32 \dfrac{\text{ft-lbf}}{\text{lbm-°R}}\right)(520°R)}$$

$$= 0.296 \text{ lbm}$$

The energy available from methane is

$$Q = \eta m(\text{HHV})$$

$$= (0.5)(0.296 \text{ lbm})\left(24{,}000 \dfrac{\text{Btu}}{\text{lbm}}\right)$$

$$= 3552 \text{ Btu}$$

This energy is used by water to heat from 60°F to 200°F.

$$Q = m_{\text{water}} c_p (T_2 - T_1)$$

$$m_{\text{water}} = \frac{3552 \text{ Btu}}{\left(1 \dfrac{\text{Btu}}{\text{lbm-°F}}\right)(200°F - 60°F)} = \boxed{25.37 \text{ lbm}}$$

The answer is (A).

SI Solution

$$T = (60°F + 460)\left(\frac{1 \text{ K}}{1.8°R}\right) = 288.89\text{K}$$

$$p = (14.73 \text{ psia})\left(\frac{101.325 \text{ kPa}}{14.696 \text{ psia}}\right) = 101.56 \text{ kPa}$$

$$R = \frac{R^*}{MW} = \frac{8314.3 \dfrac{\text{J}}{\text{kmol·K}}}{16.043 \dfrac{\text{kg}}{\text{kmol}}}$$

$$= 518.25 \text{ J/kg·K}$$

$$m = \frac{pV}{RT}$$

$$= \frac{(101.56 \text{ kPa})\left(1000 \dfrac{\text{Pa}}{\text{kPa}}\right)(200 \text{ L})\left(\dfrac{1 \text{ m}^3}{1000 \text{ L}}\right)}{\left(518.25 \dfrac{\text{J}}{\text{kg·K}}\right)(288.89\text{K})}$$

$$= 0.136 \text{ kg}$$

The energy available from methane is

$$Q = \eta m(\text{HHV})$$

$$= (0.5)(0.136 \text{ kg}) \left(55.8 \; \frac{\text{MJ}}{\text{kg}}\right) \left(1000 \; \frac{\text{kJ}}{\text{MJ}}\right)$$

$$= 3794 \text{ kJ}$$

This energy is used by water to heat it from 15°C to 95°C.

$$Q = m_{\text{water}} c_p (T_2 - T_1)$$

$$m_{\text{water}} = \frac{3794 \text{ kJ}}{\left(4.1868 \; \frac{\text{kJ}}{\text{kg·C}}\right)(95°\text{C} - 15°\text{C})} = \boxed{11.33 \text{ kg}}$$

The answer is (A).

2. *Customary U.S. Solution*

From Table 22.7,

$$\begin{array}{cccccc} & \text{C}_3\text{H}_8 & + & 5\text{O}_2 & \longrightarrow & 3\text{CO}_2 & + & 4\text{H}_2\text{O} \\ \text{MW} & 44.097 & & (5)(32) & & (3)(44.011) \\ & 44.097 & & 160 & & 132.033 \end{array}$$

The amount of carbon dioxide produced is 132.033 lbm/ 44.097 lbm propane. For 15 lbm/hr of propane, the amount of carbon dioxide produced is

$$\left(\frac{132.033 \text{ lbm}}{44.097 \text{ lbm}}\right)\left(15 \; \frac{\text{lbm}}{\text{hr}}\right) = 44.91 \text{ lbm/hr}$$

$$R = \frac{R^*}{\text{MW}} = \frac{1545.33 \; \frac{\text{ft-lbf}}{\text{lbmol-°R}}}{44.011 \; \frac{\text{lbm}}{\text{lbmol}}}$$

$$= 35.11 \text{ ft-lbf/lbm-°R}$$

$$T = 70°\text{F} + 460 = 530°\text{R}$$

$$\dot{V} = \frac{\dot{m}RT}{p}$$

$$= \frac{\left(44.91 \; \frac{\text{lbm}}{\text{hr}}\right)\left(35.11 \; \frac{\text{ft-lbf}}{\text{lbm-°R}}\right)(530°\text{R})}{\left(14.7 \; \frac{\text{lbf}}{\text{in}^2}\right)\left(144 \; \frac{\text{in}^2}{\text{ft}^2}\right)}$$

$$= \boxed{394.7 \text{ ft}^3/\text{hr}}$$

The answer is (C).

SI Solution

From Table 22.7,

$$\begin{array}{cccccc} & \text{C}_3\text{H}_8 & + & 5\text{O}_2 & \longrightarrow & 3\text{CO}_2 & + & 4\text{H}_2\text{O} \\ \text{MW} & 44.097 & & (5)(32) & & (3)(44.011) \\ & 44.097 & & 160 & & 132.033 \end{array}$$

The amount of carbon dioxide produced is 132.033 kg/ 44.097 kg propane. For 6.8 kg/h of propane, the amount of carbon dioxide produced is

$$\left(\frac{132.033 \text{ kg}}{44.097 \text{ kg}}\right)\left(6.8 \; \frac{\text{kg}}{\text{h}}\right) = 20.36 \text{ kg/h}$$

$$R = \frac{R^*}{\text{MW}} = \frac{8314.3 \; \frac{\text{J}}{\text{kmol·K}}}{44.011 \; \frac{\text{kg}}{\text{kmol}}}$$

$$= 188.91 \text{ J/kg·K}$$

$$T = 21°\text{C} + 273 = 294\text{K}$$

$$V = \frac{mRT}{p}$$

$$= \frac{\left(20.36 \; \frac{\text{kg}}{\text{h}}\right)\left(188.91 \; \frac{\text{J}}{\text{kg·K}}\right)(294\text{K})}{(101 \text{ kPa})\left(1000 \; \frac{\text{Pa}}{\text{kPa}}\right)}$$

$$= \boxed{11.20 \text{ m}^3/\text{h}}$$

The answer is (C).

3. Use the balanced chemical reaction equation from Table 22.7.

$$\text{CH}_4 + 2\text{O}_2 \longrightarrow \text{CO}_2 + 2\text{H}_2\text{O}$$

With 30% excess air and considering there are 3.773 volumes of nitrogen for every volume of oxygen (Table 22.6), the reaction equation is

$$\text{CH}_4 + (1.3)(2)\text{O}_2 + (1.3)(2)(3.773)\text{N}_2$$
$$\longrightarrow \text{CO}_2 + 2\text{H}_2\text{O} + (1.3)(2)(3.773)\text{N}_2 + 0.6\text{O}_2$$

$$\text{CH}_4 + 2.6\text{O}_2 + 9.81\text{N}_2$$
$$\longrightarrow \text{CO}_2 + 2\text{H}_2\text{O} + 9.81\text{N}_2 + 0.6\text{O}_2$$

Customary U.S. Solution

The volume of nitrogen that accompanies 4000 ft³/hr of entering methane is

$$V_{\text{N}_2} = \left(\frac{9.81 \text{ ft}^3 \text{ N}_2}{1 \text{ ft}^3 \text{ CH}_4}\right)\left(4000 \; \frac{\text{ft}^3}{\text{hr}} \text{ CH}_4\right)$$

$$= 39{,}240 \text{ ft}^3 \text{ N}_2/\text{hr}$$

This is the "partial volume" of nitrogen in the input stream.

$$R = \frac{R^*}{\text{MW}} = \frac{1545.33 \; \frac{\text{ft-lbf}}{\text{lbmol-°R}}}{28.016 \; \frac{\text{lbm}}{\text{lbmol}}}$$

$$= 55.16 \text{ ft-lbf/lbm-°R}$$

The absolute temperature is

$$T = 100°F + 460 = 560°R$$

$$m_{N_2} = \frac{p_{N_2} V_{N_2}}{RT}$$

$$= \frac{\left(15 \dfrac{lbf}{in^2}\right)\left(144 \dfrac{in^2}{ft^2}\right)\left(39{,}240 \dfrac{ft^3}{hr}\right)}{\left(55.16 \dfrac{ft\text{-}lbf}{lbm\text{-}°R}\right)(560°R)}$$

$$= \boxed{2744 \ lbm/hr}$$

The answer is (D).

SI Solution

The volume of nitrogen that accompanies 31 L/s of entering methane is

$$\left(\frac{9.81 \ m^3 \ N_2}{1 \ m^3 \ CH_4}\right)\left(31 \ \frac{L}{s}\right)\left(\frac{1 \ m^3}{1000 \ L}\right) = 0.3041 \ m^3/s$$

This is the "partial volume" of nitrogen in the input stream.

$$R = \frac{R^*}{MW} = \frac{8314.3 \ \dfrac{J}{kmol \cdot K}}{28.016 \ \dfrac{kg}{kmol}}$$

$$= 296.8 \ J/kg \cdot K$$

The absolute temperature is

$$T = 40°C + 273 = 313K$$

$$m_{N_2} = \frac{p_{N_2} V_{N_2}}{RT}$$

$$= \frac{(103 \ kPa)\left(1000 \ \dfrac{Pa}{kPa}\right)\left(0.3041 \ \dfrac{m^3}{s}\right)}{\left(296.8 \ \dfrac{J}{kg \cdot K}\right)(313K)}$$

$$= \boxed{0.337 \ kg/s}$$

The answer is (D).

4. From Table 22.7, combustion reactions are

$$\begin{array}{ccccc} & C & + & O_2 & \longrightarrow & CO_2 \\ MW & 12 & & 32 \end{array}$$

The mass of oxygen required per unit mass of carbon is

$$\frac{32}{12} = 2.67$$

$$\begin{array}{ccccc} & 2H_2 & + & O_2 & \longrightarrow & 2H_2O \\ MW & (2)(2) & & 32 \end{array}$$

The mass of oxygen required per unit mass of hydrogen is

$$\frac{32}{(2)(2)} = 8.0$$

$$\begin{array}{ccccc} & S & + & O_2 & \longrightarrow & SO_2 \\ MW & 32.1 & & 32 \end{array}$$

The mass of oxygen required per unit mass of sulfur is

$$\frac{32}{32.1} = 1.0$$

Nitrogen does not burn.

The mass of oxygen required per unit mass of fuel is

$$(0.84)(2.67) + (0.153)(8) + (0.003)(1)$$
$$= 3.47 \ \text{unit of mass of } O_2/\text{unit mass fuel}$$

From Table 22.6, air is 0.2315 O_2/unit mass, so the air required is

$$\frac{3.47}{0.2315} = \boxed{15.0 \ lbm \ air/lbm \ fuel}$$

(Customary U.S. Solution)

$$= \boxed{15.0 \ kg \ air/kg \ fuel}$$

(SI Solution)

(Equation 22.6 can also be used to solve this problem.)

The answer is (C).

5. The balanced chemical reaction equation is

$$C_3H_8 + 5O_2 \longrightarrow 3CO_2 + 4H_2O$$

With 20% excess air, the oxygen volume is $(1.2)(5) = 6$.

$$C_3H_8 + 6O_2 \longrightarrow 3CO_2 + 4H_2O + O_2$$

From Table 22.6, there are 3.773 volumes of nitrogen for every volume of oxygen.

$$(6)(3.773) = 22.6$$
$$C_3H_8 + 6O_2 + 22.6 \, N_2 \longrightarrow 3CO_2 + 4H_2O + O_2 + 22.6 \, N_2$$

The percentage of carbon dioxide by weight in flue gas is

$$G_{CO_2} = \frac{(3)(44.011)}{(3)(44.011) + (4)(18.016)}$$
$$+ 32 + (22.6)(28.016)$$

$$= \boxed{0.152 \ (15.2\%)}$$

The answer is (C).

6. The actual air/fuel ratio can be estimated from the flue gas analysis and the fraction of carbon in fuel.

From Eq. 22.9,

$$\frac{m_{\text{air}}}{m_{\text{fuel}}} = \frac{3.04 B_{\text{N}_2} G_{\text{C}}}{B_{\text{CO}_2} + B_{\text{CO}}}$$

$$= \frac{(3.04)(80\%)(0.80)}{12\% + 1\%}$$

$$= 14.97$$

The air required to burn 1 lbm of coal is $\boxed{14.97 \text{ lbm.}}$

(Customary U.S. Solution)

The air required to burn 1 kg of coal is $\boxed{14.97 \text{ kg.}}$

(SI Solution)

The answer is (B).

Alternate Solution

The use of Eq. 22.9 obscures the process of finding the air/fuel ratio. (The SI solution is similar but is not presented here.)

step 1: Find the mass of oxygen in the stack gases.

$$R_{\text{CO}_2} - 35.11 \text{ ft-lbf/lbm-}^\circ\text{R}$$

$$R_{\text{CO}} = 55.17 \text{ ft-lbf/lbm-}^\circ\text{R}$$

$$R_{\text{O}_2} = 48.29 \text{ ft-lbf/lbm-}^\circ\text{R}$$

The partial densities are

$$\rho_{\text{CO}_2} = \frac{p}{RT} = \frac{(0.12)\left(14.7 \, \frac{\text{lbf}}{\text{in}^2}\right)\left(144 \, \frac{\text{in}^2}{\text{ft}^2}\right)}{\left(35.11 \, \frac{\text{ft-lbf}}{\text{lbm-}^\circ\text{R}}\right)(60^\circ\text{F} + 460)}$$

$$= 1.391 \times 10^{-2} \text{ lbm/ft}^3$$

$$\rho_{\text{CO}} = \frac{(0.01)\left(14.7 \, \frac{\text{lbf}}{\text{in}^2}\right)\left(144 \, \frac{\text{in}^2}{\text{ft}^2}\right)}{\left(55.11 \, \frac{\text{ft-lbf}}{\text{lbm-}^\circ\text{R}}\right)(60^\circ\text{F} + 460)}$$

$$= 7.387 \times 10^{-4} \text{ lbm/ft}^3$$

$$\rho_{\text{O}_2} = \frac{(0.07)\left(14.7 \, \frac{\text{lbf}}{\text{in}^2}\right)\left(144 \, \frac{\text{in}^2}{\text{ft}^2}\right)}{\left(48.29 \, \frac{\text{ft-lbf}}{\text{lbm-}^\circ\text{R}}\right)(60^\circ\text{F} + 460)}$$

$$= 5.901 \times 10^{-3} \text{ lbm/ft}^3$$

The fraction of oxygen in the three components is

$$\text{CO}_2: \quad \frac{32.0}{44} = 0.7273$$

$$\text{CO}: \quad \frac{16}{28} = 0.5714$$

$$\text{O}_2: \quad 1.00$$

In 100 ft^3 of stack gases, the total oxygen mass will be

$$(100 \text{ ft}^3)\left((0.7273)\left(1.391 \times 10^{-2} \, \frac{\text{lbm}}{\text{ft}^3}\right)\right.$$

$$+ (0.5714)\left(7.387 \times 10^{-4} \, \frac{\text{lbm}}{\text{ft}^3}\right)$$

$$\left.+ (1.00)\left(5.901 \times 10^{-3} \, \frac{\text{lbm}}{\text{ft}^3}\right)\right) = 1.644 \text{ lbm}$$

step 2: Since air is 23.15% oxygen by weight, the mass of air per 100 ft^3 of stack gases is

$$\frac{1.644 \text{ lbm}}{0.2315} = 7.102 \text{ lbm}$$

step 3: Find the mass of carbon in the stack gases by a similar process.

$$\text{CO}_2: \quad \frac{12}{44} = 0.2727$$

$$\text{CO}: \quad \frac{12}{28} = 0.4286$$

$$(100 \text{ ft}^3)\left((0.2727)\left(1.391 \times 10^{-2} \, \frac{\text{lbm}}{\text{ft}^3}\right)\right.$$

$$\left.+ (0.4286)\left(7.387 \times 10^{-4} \, \frac{\text{lbm}}{\text{ft}^3}\right)\right)$$

$$= 0.4110 \text{ lbm}$$

step 4: The coal is 80% carbon, so the air per lbm of coal for combustion of the carbon is

$$\left(\frac{0.80 \, \frac{\text{lbm carbon}}{\text{lbm coal}}}{0.4110 \, \frac{\text{lbm carbon}}{100 \text{ ft}^3}}\right)(7.102 \text{ lbm})$$

$$= 13.824 \text{ lbm air/lbm coal}$$

step 5: This does not include air to burn hydrogen, since Orsat is a dry analysis.

The theoretical air for the hydrogen is given by Eq. 22.6.

$$R_{a/f,\text{H}} = \left(34.5 \ \frac{\text{lbm}}{\text{lbm}}\right)\left(G_{\text{H}} - \frac{G_{\text{O}}}{8}\right)$$

$$= \left(34.05 \ \frac{\text{lbm}}{\text{lbm}}\right)\left(0.04 - \frac{0.02}{8}\right)$$

$$= 1.277 \ \text{lbm air/lbm fuel}$$

Ignoring any excess air for the hydrogen, the total air per pound of coal is

$$13.824 \ \frac{\text{lbm air}}{\text{lbm coal}} + 1.277 \ \frac{\text{lbm air}}{\text{lbm coal}}$$

$$= \boxed{15.10 \ \text{lbm air/lbm coal}}$$

The answer is (B).

7. From Eq. 14.11,

$$\text{SG} = \frac{141.5}{^{\circ}\text{API} + 131.5}$$

$$= \frac{141.5}{40 + 131.5} = 0.825$$

Customary U.S. Solution

From Eq. 22.18(b),

$$\text{HHV} = 22{,}320 - (3780)(\text{SG})^2$$

$$= 22{,}320 - (3780)(0.825)^2$$

$$= \boxed{19{,}747 \ \text{Btu/lbm}}$$

The answer is (A).

SI Solution

From Eq. 22.18(a),

$$\text{HHV} = 51.92 - (8.792)(\text{SG})^2$$

$$= 51.92 - (8.792)(0.825)^2$$

$$= \boxed{45.94 \ \text{MJ/kg}}$$

The answer is (A).

8. *Customary U.S. Solution*

(a) *step 1:* From Eq. 22.16(b), substituting the lower heating value of hydrogen from App. 22.A, the lower heating value of coal is

$$\text{LHV} = 14{,}093 G_{\text{C}} + (51{,}623)\left(G_{\text{H}} - \frac{G_{\text{O}}}{8}\right)$$

$$+ 3983 G_{\text{S}}$$

$$= (14{,}093)(0.75)$$

$$+ (51{,}623)\left(0.05 - \frac{0.03}{8}\right) + (3983)(0)$$

$$= 12{,}957 \ \text{Btu/lbm}$$

step 2: The gravimetric analysis of 1 lbm of coal is

carbon: 0.75 lbm

free hydrogen: $\left(G_{\text{H,total}} - \dfrac{G_{\text{O}}}{8}\right) = 0.05 - \dfrac{0.03}{8}$

$$= 0.0463$$

The ratio of the molecular weight of water (18) to that of a hydrogen molecule (2) is 9.

water: $(9)(0.05 - 0.0463) = 0.0333$

nitrogen: 0.02

step 3: From Table 22.8, the theoretical stack gases per lbm coal for 0.75 lbm of carbon are

$$\text{CO}_2 = (0.75)(3.667 \ \text{lbm}) = 2.750 \ \text{lbm}$$

$$\text{N}_2 = (0.75)(8.883 \ \text{lbm}) = 6.662 \ \text{lbm}$$

All products are calculated similarly (as the following table summarizes).

	CO_2	N_2	H_2O
from C:	2.750 lbm	6.662 lbm	
from H$_2$:		1.217 lbm	0.414 lbm
from H$_2$O:			0.0333 lbm
from O$_2$:	shows up in	CO$_2$ and H$_2$	
from N$_2$:		0.02 lbm	
total:	2.750 lbm	7.899 lbm	0.4473 lbm

step 4: Assume the combustion gases leave at 1000°F.

$$T_{\text{ave}} = \left(\tfrac{1}{2}\right)(60^{\circ}\text{F} + 1000^{\circ}\text{F}) = 530^{\circ}\text{F}$$

$$T_{\text{ave}} = 530^{\circ}\text{F} + 460 = 990^{\circ}\text{R}$$

The specific heat values are given in Table 22.1.

$$c_{p\text{CO}_2} = 0.251 \ \text{Btu/lbm-}^{\circ}\text{F}$$

$$c_{p\text{N}_2} = 0.255 \ \text{Btu/lbm-}^{\circ}\text{F}$$

$$c_{p\text{H}_2\text{O}} = 0.475 \ \text{Btu/lbm-}^{\circ}\text{F}$$

The energy required to raise the combustion products for 1 lbm of coal 1°F is

$$m_{\text{CO}_2} c_{p\text{CO}_2} + m_{\text{N}_2} c_{p\text{N}_2} + m_{\text{H}_2\text{O}} c_{p\text{H}_2\text{O}}$$

$$= (2.750 \ \text{lbm})\left(0.251 \ \frac{\text{Btu}}{\text{lbm-}^{\circ}\text{F}}\right)$$

$$+ (7.899 \ \text{lbm})\left(0.255 \ \frac{\text{Btu}}{\text{lbm-}^{\circ}\text{F}}\right)$$

$$+ (0.4473 \ \text{lbm})\left(0.475 \ \frac{\text{Btu}}{\text{lbm-}^{\circ}\text{F}}\right)$$

$$= 2.92 \ \text{Btu/}^{\circ}\text{F}$$

step 5: Assuming all combustion heat goes into the stack gases, the temperature is given by Eq. 22.19.

$$T_{max} = T_i + \frac{\text{lower heat of combustion}}{\text{energy required}}$$

$$= 60°F + \frac{12{,}957 \ \dfrac{Btu}{lbm}}{2.92 \ \dfrac{Btu}{lbm\text{-}°F}} = \boxed{4497°F}$$

[unreasonable]

The answer is (C).

(b) *step 6:* In reality, assuming 40% excess air and 75% of heat absorbed by the boiler, the excess air (based on 76.85% N_2 by weight) is

$$(0.40) \left(\frac{7.899 \ lbm}{0.7685} \right) = 4.111 \ lbm$$

From Table 22.1, $c_{p_{air}} = 0.249$ Btu/lbm-°R. Therefore,

$$T_{max} = 60°F + \frac{\left(12{,}957 \ \dfrac{Btu}{lbm} \right)(1 - 0.75)}{2.92 \ \dfrac{Btu}{°F}}$$

$$+ (4.111 \ lbm) \left(0.249 \ \dfrac{Btu}{lbm\text{-}°R} \right)$$

$$= \boxed{881°F}$$

The answer is (B).

SI Solution

(a) *step 1:* From Eq. 22.16(a), and substituting the lower heating value of hydrogen from App. 22.A, the lower heating value of coal is

$$LHV = 32.78 G_C + \left(51{,}623 \ \frac{Btu}{lbm} \right) \left(2.326 \ \frac{kJ\text{-}lbm}{kg\text{-}Btu} \right)$$

$$\times \left(\frac{1 \ MJ}{1000 \ kJ} \right) \left(G_H - \frac{G_O}{8} \right)$$

$$+ 9.264 \ G_S$$

$$= (32.78)(0.75) + (120.1) \left(0.05 - \frac{0.03}{8} \right)$$

$$+ (9.264)(0)$$

$$= 30.14 \ MJ/kg$$

Steps 2 and 3 are the same as for the U.S. solution except that all masses are in kg.

step 4: Assume the combustion gases leave at 550°C.

$$T_{ave} = \left(\tfrac{1}{2} \right) (16°C + 550°C) + 273$$

$$= 283°C + 273 = 556K$$

Specific heat values are given in Table 22.1. Using the footnote for SI units,

$$c_{p_{CO_2}} = 1.051 \ kJ/kg\text{·}K$$

$$c_{p_{N_2}} = 1.068 \ kJ/kg\text{·}K$$

$$c_{p_{H_2O}} = 1.989 \ kJ/kg\text{·}K$$

The energy required to raise the combustion products for 1 kg of coal 1°C is

$$m_{CO_2} c_{p_{CO_2}} + m_{N_2} c_{p_{N_2}} + m_{H_2O} c_{p_{H_2O}}$$

$$= (2.750 \ kg) \left(1.051 \ \frac{kJ}{kg\text{·}K} \right)$$

$$+ (7.899 \ kg) \left(1.068 \ \frac{kJ}{kg\text{·}K} \right)$$

$$+ (0.4473 \ kg) \left(1.989 \ \frac{kJ}{kg\text{·}K} \right)$$

$$= 12.22 \ kJ/K$$

step 5: Assuming all combustion heat goes into the stack gases, the temperature is given by Eq. 22.19.

$$T_{max} = T_i + \frac{\text{lower heat of combustion}}{\text{energy required}}$$

$$= 16°C + \frac{\left(30.14 \ \dfrac{MJ}{kg} \right) \left(1000 \ \dfrac{kJ}{MJ} \right)}{12.22 \ \dfrac{kJ}{K}}$$

$$= \boxed{2482°C} \quad \text{[unreasonable]}$$

The answer is (D).

(b) *step 6:* In reality, assuming 40% excess air and 75% of heat absorbed by the boiler, the excess air (based on 76.85% N_2 by weight) is

$$(0.40) \left(\frac{7.899 \ kg}{0.7685} \right) = 4.111 \ kg$$

From Table 22.1, using the table footnote, $c_{p_{air}} = 10\,043$ kJ/kg·K.

Therefore,

$$T_{max} = 16°C + \frac{\left(30.14 \ \dfrac{MJ}{kg} \right)}{\times \left(1000 \ \dfrac{kJ}{MJ} \right)(1 - 0.75)}$$
$$T_{max} = 16°C + \frac{}{12.22 \ \dfrac{kJ}{K} + (4.111 \ kg) \left(1.043 \ \dfrac{kJ}{kg\text{·}K} \right)}$$

$$= \boxed{472.5°C}$$

The answer is (B).

Thermodynamics

9. Assume the oxygen is in the form of moisture in the fuel.

The available hydrogen is

$$G_{H,free} = G_H - \frac{G_O}{8}$$

$$= 0.1131 - \frac{0.027}{8}$$

$$= 0.1097$$

Customary U.S. Solution

step 1:   From Table 22.8, find the stoichiometric oxygen required per lbm of fuel oil.

$$C \longrightarrow CO_2: \ O_2 \text{ required} = (0.8543)(2.667 \text{ lbm})$$

$$= 2.2784 \text{ lbm}$$

$$H_2 \longrightarrow H_2O: \ O_2 \text{ required} = (0.1097)(7.936 \text{ lbm})$$

$$= 0.8706 \text{ lbm}$$

$$S \longrightarrow SO_2: \ O_2 \text{ required} = (0.0034)(0.998 \text{ lbm})$$

$$= 0.0034 \text{ lbm}$$

The total amount of oxygen required per lbm of fuel oil is

$$2.2784 \text{ lbm} + 0.8706 \text{ lbm}$$

$$+ \ 0.0034 \text{ lbm} = 3.1524 \text{ lbm}$$

step 2:   With 60% excess air, the excess oxygen is

$$(0.6)(3.1524 \text{ lbm}) = 1.8914 \text{ lbm } O_2/\text{lbm fuel}$$

From Eq. 24.47, this oxygen occupies a volume of

$$V = \frac{mRT}{p}$$

At standard conditions,

$$p = 14.7 \text{ psia}$$

$$T = 60°F + 460 = 520°R$$

From Table 24.7, for oxygen, $R = 48.29$ ft-lbf/lbm-°R.

$$V = \frac{(1.8914 \text{ lbm}) \left(48.29 \ \dfrac{\text{ft-lbf}}{\text{lbm-°R}} \right) (520°R)}{\left(14.7 \ \dfrac{\text{lbf}}{\text{in}^2} \right) \left(144 \ \dfrac{\text{in}^2}{\text{ft}^2} \right)}$$

$$= 22.44 \text{ ft}^3$$

step 3:   The theoretical nitrogen based on Table 22.6 is

$$\left(\frac{3.1524 \text{ lbm}}{0.2315} \right) (0.7685) = 10.465 \text{ lbm } N_2/\text{lbm fuel}$$

The actual nitrogen with 60% excess air and nitrogen in the fuel is

$$(10.465 \text{ lbm})(1.6) + 0.0022 \text{ lbm}$$

$$= 16.746 \text{ lbm } N_2/\text{lbm fuel}$$

From Eq. 24.47, this nitrogen occupies a volume of

$$V = \frac{mRT}{p}$$

From Table 24.7, R for nitrogen $= 55.16$ ft-lbf/lbm-°R.

$$V = \frac{(16.746 \text{ lbm}) \left(55.16 \ \dfrac{\text{ft-lbf}}{\text{lbm-°R}} \right) (520°R)}{\left(14.7 \ \dfrac{\text{lbf}}{\text{in}^2} \right) \left(144 \ \dfrac{\text{in}^2}{\text{ft}^2} \right)}$$

$$= 226.91 \text{ ft}^3$$

step 4:   From Table 22.8, the 60°F combustion product volumes per lbm of fuel will be

CO_2:	$(0.8543)(31.63 \text{ ft}^3)$	$=$	27.02 ft^3
H_2O:	$(0.1131)(188.25 \text{ ft}^3)$	$=$	21.29 ft^3
SO_2:	$(0.0034)(11.84 \text{ ft}^3)$	$=$	0.040 ft^3
N_2:	from step 3	$=$	226.91 ft^3
O_2:	from step 2	$=$	22.44 ft^3
		total $=$	297.7 ft^3

(a) At 60°F, the wet volume will be 297.7 ft^3.

At 600°F, the wet volume is

$$V_{wet,600°F} = (297.7 \text{ ft}^3) \left(\frac{600°F + 460}{60°F + 460} \right)$$

$$= \boxed{606.9 \text{ ft}^3}$$

The answer is (D).

(b) At 60°F, the dry volume will be

$$297.7 \text{ ft}^3 - 21.29 \text{ ft}^3 = 276.4 \text{ ft}^3$$

At 600°F, the dry volume is

$$V_{dry,600°F} = (276.4 \text{ ft}^3) \left(\frac{600°F + 460}{60°F + 460} \right)$$

$$= \boxed{563.4 \text{ ft}^3}$$

The answer is (B).

(c) The volumetric fraction of dry carbon dioxide is

$$\frac{27.02 \text{ ft}^3}{276.41 \text{ ft}^3} = \boxed{0.098 \ \ (9.8\%)}$$

The answer is (C).

SI Solution

step 1: From Table 22.8, find the stoichiometric oxygen required per lbm of fuel oil.

$$C \longrightarrow CO_2 : \quad O_2 \text{ required} = (0.8543)(2.667 \text{ kg})$$
$$= 2.2784 \text{ kg}$$
$$H_2 \longrightarrow H_2O : \quad O_2 \text{ required} = (0.1097)(7.936 \text{ kg})$$
$$= 0.8706 \text{ kg}$$
$$S \longrightarrow SO_2 : \quad O_2 \text{ required} = (0.0034)(0.998 \text{ kg})$$
$$= 0.0034 \text{ kg}$$

The total amount of oxygen required per kg of fuel oil is

$$2.2784 \text{ kg} + 0.8706 \text{ kg}$$
$$+ 0.0034 \text{ kg} = 3.1524 \text{ kg}$$

step 2: With 60% excess air, the excess oxygen is

$$(0.6)(3.1524 \text{ kg}) = 1.8914 \text{ kg O}_2/\text{kg fuel}$$

From Eq. 24.47, this oxygen occupies a volume of

$$V = \frac{mRT}{p}$$

At standard conditions,

$$p = 101.3 \text{ kPa}$$
$$T = 16°C + 273 = 289K$$

From Table 24.7, for oxygen, $R = 259.82$ J/kg·K.

$$V = \frac{(1.8914 \text{ kg})\left(259.82 \dfrac{\text{J}}{\text{kg·K}}\right)(289K)}{(101.3 \text{ kPa})(1000 \text{ kPa})}$$
$$= 1.402 \text{ m}^3$$

step 3: The theoretical nitrogen based on Table 22.6 is

$$\left(\frac{3.1524 \text{ kg}}{0.2315}\right)(0.7685) = 10.465 \text{ kg N}_2/\text{kg fuel}$$

The actual nitrogen with 60% excess air and nitrogen in the fuel is

$$(10.465 \text{ kg})(1.6) + 0.0022 \text{ kg}$$
$$= 16.746 \text{ kg N}_2/\text{kg fuel}$$

From Eq. 24.47, this nitrogen occupies a volume of

$$V = \frac{mRT}{p}$$

From Table 24.7, R for nitrogen $= 296.77$ J/kg·K.

$$V = \frac{(16.746 \text{ kg})\left(296.77 \dfrac{\text{J}}{\text{kg·K}}\right)(289K)}{(101.3 \text{ kPa})\left(1000 \dfrac{\text{Pa}}{\text{kPa}}\right)}$$
$$= 14.178 \text{ m}^3$$

step 4: From Table 22.8, the 16°C combustion product volumes per kg of fuel will be

CO_2:	$(0.8543)(31.63 \text{ ft}^3)(0.06243)$	=	1.687 m^3
H_2O:	$(0.1131)(188.25 \text{ ft}^3)(0.06243)$	=	1.329 m^3
SO_2:	$(0.0034)(11.84 \text{ ft}^3)(0.06243)$	=	0.003 m^3
N_2:	from step 3	=	14.178 m^3
O_2:	from step 2	=	1.402 m^3
		total =	18.599 m^3

(a) At 16°C, the wet volume will be 18.599 m³.

At 320°C, the wet volume is

$$V_{\text{wet},320°C} = (18.599 \text{ m}^3)\left(\frac{320°C + 273}{16°C + 273}\right)$$
$$= \boxed{38.16 \text{ m}^3}$$

The answer is (D).

(b) At 16°C, the dry volume will be

$$18.599 \text{ m}^3 - 1.329 \text{ m}^3 = 17.27 \text{ m}^3$$

At 320°C, the dry volume is

$$V_{\text{dry},320°C} = (17.27 \text{ m}^3)\left(\frac{320°C + 273}{16°C + 273}\right)$$
$$= \boxed{35.44 \text{ m}^3}$$

The answer is (B).

(c) The volumetric fraction of dry carbon dioxide is

$$\frac{1.687 \text{ m}^3}{17.27 \text{ m}^3} = \boxed{0.098 \quad (9.8\%)}$$

The answer is (C).

Thermodynamics

10. *Customary U.S. Solution*

step 1: Based on the 15,395 lbm of coal burned producing 2816 lbm of ash containing 20.9% carbon by weight, the usable percentage of carbon per lbm of fuel is

$$0.5145 - \frac{(2816 \text{ lbm})(0.209)}{15,395 \text{ lbm}} = 0.4763$$

step 2: Since moisture is reported separately, assume all of the oxygen and hydrogen are free. (This is not ordinarily the case.) From Table 22.8, find the stoichiometric oxygen required per lbm of fuel.

$$C \longrightarrow CO_2: \ O_2 \text{ required} = (0.4763)(2.667 \text{ lbm})$$
$$= 1.2703 \text{ lbm}$$
$$H_2 \longrightarrow H_2O: \ O_2 \text{ required} = (0.0402)(7.936 \text{ lbm})$$
$$= 0.3190 \text{ lbm}$$
$$S \longrightarrow SO_2: \ O_2 \text{ required} = (0.0392)(0.998 \text{ lbm})$$
$$= 0.0391 \text{ lbm}$$

The total amount of O_2 required per lbm of fuel is

$$1.2703 \text{ lbm} + 0.3190 \text{ lbm}$$
$$+ 0.0391 \text{ lbm} - 0.0728 \text{ lbm} = 1.5556 \text{ lbm}$$

step 3: The theoretical air based on Table 22.6 is

$$\frac{1.5556 \text{ lbm}}{0.2315} = 6.720 \text{ lbm air/lbm fuel}$$

step 4: Ignoring fly ash, the theoretical dry products are given from Table 22.8.

$$
\begin{aligned}
CO_2: \ &(0.4763)(3.667 \text{ lbm}) = 1.7466 \text{ lbm} \\
SO_2: \ &(0.0392)(1.998 \text{ lbm}) = 0.0783 \text{ lbm} \\
N_2: \ &0.0093 \text{ lbm} \\
&+ 6.720 \text{ lbm} \\
&\times 0.7685 = \underline{5.1736 \text{ lbm}} \\
&\text{total} = 6.999 \text{ lbm}
\end{aligned}
$$

step 5: The excess air is

$$13.3 \text{ lbm} - 6.999 \text{ lbm} = 6.301 \text{ lbm}$$

step 6: The total air supplied per pound of fuel is

$$6.301 \text{ lbm} + 6.720 \text{ lbm} = \boxed{13.02 \text{ lbm}}$$

The answer is (A).

SI Solution

step 1: Based on the 6923 kg of coal burned producing 1267 kg of ash containing 20.9% carbon by weight, the usable percentage of carbon per kg of fuel is

$$0.5145 - \frac{(1267 \text{ kg})(0.209)}{6923 \text{ kg}} = 0.4763$$

step 2: From Table 22.8, find the stoichiometric oxygen required per kg of fuel.

$$C \longrightarrow CO_2: \ O_2 \text{ required} = (0.4763)(2.667 \text{ kg})$$
$$= 1.2703 \text{ kg}$$
$$H_2 \longrightarrow H_2O: \ O_2 \text{ required} = (0.0402)(7.936 \text{ kg})$$
$$= 0.3190 \text{ kg}$$
$$S \longrightarrow SO_2: \ O_2 \text{ required} = (0.0392)(0.998 \text{ kg})$$
$$= 0.0391 \text{ kg}$$

The total amount of O_2 required per kg of fuel is

$$1.2703 \text{ kg} + 0.3190 \text{ kg}$$
$$+ 0.0391 \text{ kg} - 0.0728 \text{ kg} = 1.5556 \text{ kg}$$

step 3: The theoretical air based on Table 22.6 is

$$\frac{1.5556 \text{ kg}}{0.2315} = 6.720 \text{ kg air/kg fuel}$$

step 4: Ignoring fly ash, the theoretical dry products are given from Table 22.8.

$$
\begin{aligned}
CO_2: \ &(0.4763)(3.667 \text{ kg}) = 1.7466 \text{ kg} \\
SO_2: \ &(0.0392)(1.998 \text{ kg}) = 0.0783 \text{ kg} \\
N_2: \ &0.0093 \text{ kg} \\
&+ (6.720 \text{ kg}) \\
&\times (0.7685 \text{ kg}) = \underline{5.1736 \text{ kg}} \\
&\text{total} = 6.999 \text{ kg}
\end{aligned}
$$

step 5: The excess air is

$$13.3 \text{ kg} - 6.999 \text{ kg} = 6.301 \text{ kg}$$

step 6: The total air supplied is

$$6.301 \text{ kg} + 6.720 \text{ kg} = \boxed{13.02 \text{ kg}}$$

The answer is (A).

11. (The customary U.S. and SI solutions are essentially identical.)

From Eq. 22.9, the actual air-fuel ratio can be estimated as

$$R_{a/f,\text{actual}} = \frac{3.04 B_{N_2} G_C}{B_{CO_2} + B_{CO}}$$

A fraction of carbon is reduced due to the percentage of coal lost in the ash pit.

$$G_C = (1 - 0.03)(0.65) = 0.6305$$

$$R_{a/f,\text{actual}} = \frac{(3.04)(0.815)(0.6305)}{0.095 + 0}$$

$$= 16.44 \text{ lbm air/lbm fuel (kg air/kg fuel)}$$

Combustion uses 9.87 lbm of air/lbm of fuel.

$$= (9.87 \text{ lbm})(1 - 0.03)$$

$$= 9.57 \text{ lbm of air/lbm of fuel}$$

$$\% \text{ of excess air} = \left(\frac{16.44 \text{ lbm} - 9.57 \text{ lbm}}{9.57 \text{ lbm}}\right)(100\%)$$

$$= \boxed{71.8\%}$$

The answer is (C).

12. From Eq. 22.23, the heat loss due to the formation of carbon monoxide is

$$q = \frac{(\text{HHV}_C - \text{HHV}_{CO}) G_C B_{CO}}{B_{CO_2} + B_{CO}}$$

From App. 22.A, the difference in the two heating values is

$$\text{HHV}_C - \text{HHV}_{CO} = 14{,}093 \frac{\text{Btu}}{\text{lbm}} - 4347 \frac{\text{Btu}}{\text{lbm}}$$

$$= 9746 \text{ Btu/lbm} \quad (22.67 \text{ MJ/kg})$$

$$G_C = (1 - 0.03)(0.6734) = 0.6532$$

$$q = \frac{\left(9746 \frac{\text{Btu}}{\text{lbm}}\right)(0.6532)(1.6\%)}{15.5\% + 1.6\%}$$

$$= \boxed{596 \text{ Btu/lbm}}$$

The answer is (A).

SI Solution

$$q = \frac{\left(22.67 \frac{\text{MJ}}{\text{kg}}\right)(0.6532)(1.6\%)}{15.5\% + 1.6\%}$$

$$= \boxed{1.39 \text{ MJ/kg}}$$

The answer is (A).

13. (a) For methane, $B = 0.93$.

From ideal gas laws (R for methane $= 96.32$ ft-lbf/lbm-°R),

$$\rho = \frac{p}{RT}$$

$$= \frac{\left(14.7 \frac{\text{lbf}}{\text{in}^2}\right)\left(144 \frac{\text{in}^2}{\text{ft}^2}\right)}{\left(96.32 \frac{\text{ft-lbf}}{\text{lbm-°R}}\right)(60°\text{F} + 460)}$$

$$= 0.0422 \text{ lbm/ft}^3$$

From Table 22.9, $K = 9.55$ ft^3 air/ft^3 fuel.

From Table 22.8,

$$\text{products: } 1 \text{ ft}^3 \text{ CO}_2, \ 2 \text{ ft}^3 \text{ H}_2\text{O}$$

From App. 22.A, HHV = 1013 Btu/lbm.

Reaction Products for Prob. 13

gas	B	ρ $\left(\frac{\text{lbm}}{\text{ft}^3}\right)$	ft^3 air	HHV $\left(\frac{\text{Btu}}{\text{lbm}}\right)$	CO$_2$	H$_2$O	other
CH$_4$	0.93	0.0422	9.556	1013	1	2	–
N$_2$	0.0373	0.0738	–	–	–	–	1 N$_2$
CO	0.0045	0.0738	2.389	322	1	–	–
H$_2$	0.0182	0.0053	2.389	325	–	1	–
C$_2$H$_4$	0.0025	0.0739	14.33	1614	2	2	–
H$_2$S	0.0018	0.0900	7.167	647	–	1	1 SO$_2$
O$_2$	0.0035	0.0843	–	–	–	–	–
CO$_2$	0.0022	0.1160	–	–	1	–	–

Similar results for all the other fuel components are tabulated in the table on the previous page.

The composite density is

$$\rho = \sum B_i \rho_i$$

$$= (0.93)\left(0.0422\,\frac{\text{lbm}}{\text{ft}^3}\right) + (0.0373)\left(0.0738\,\frac{\text{lbm}}{\text{ft}^3}\right)$$

$$+ (0.0045)\left(0.0738\,\frac{\text{lbm}}{\text{ft}^3}\right) + (0.0182)\left(0.0053\,\frac{\text{lbm}}{\text{ft}^3}\right)$$

$$+ (0.0025)\left(0.0739\,\frac{\text{lbm}}{\text{ft}^3}\right) + (0.0018)\left(0.0900\,\frac{\text{lbm}}{\text{ft}^3}\right)$$

$$+ (0.0035)\left(0.0843\,\frac{\text{lbm}}{\text{ft}^3}\right) + (0.0022)\left(0.1160\,\frac{\text{lbm}}{\text{ft}^3}\right)$$

$$= \boxed{0.0433\,\text{lbm/ft}^3}$$

The answer is (B).

(b) The air is 20.9% oxygen by volume. The theoretical air requirements are

$$\sum B_i V_{\text{air},i} - \frac{\text{O}_2 \text{ in fuel}}{0.209}$$

$$= (0.93)(9.556\,\text{ft}^3) + (0.0373)(0)$$

$$+ (0.0045)(2.389\,\text{ft}^3) + (0.0182)(2.389\,\text{ft}^3)$$

$$+ (0.0025)(14.33\,\text{ft}^3) + (0.0018)(7.167\,\text{ft}^3)$$

$$+ (0.0035)(0) + (0.0022)(0) - \frac{0.0035}{0.209}$$

$$= 8.990\,\text{ft}^3 - 0.01675\,\text{ft}^3$$

$$= \boxed{8.9733\,\text{ft}^3\ \text{air/ft}^3\ \text{fuel}}$$

The answer is (C).

((c) and (d)) The theoretical oxygen will be

$$(8.9733\,\text{ft}^3)(0.209) = 1.875\,\text{ft}^3/\text{ft}^3$$

The excess oxygen will be

$$(0.4)(1.875\,\text{ft}^3) = 0.75\,\text{ft}^3/\text{ft}^3$$

Similarly, with 40% excess air, the total nitrogen in the stack gases is

$$(1.4)(0.791)(8.9733\,\text{ft}^3) + 0.0373 = 9.974\,\text{ft}^3/\text{ft}^3\ \text{fuel}$$

The stack gases per ft^3 of fuel are

excess O$_2$:	$= 0.7500\,\text{ft}^3$
excess N$_2$:	$= 9.974\,\text{ft}^3$
excess SO$_2$:	$= 0.0018\,\text{ft}^3$
excess CO$_2$: $(0.93)(1) + (0.0045)(1)$ $+ (0.0025)(2) + (0.0022)(1)$	$= 0.942\,\text{ft}^3$
excess H$_2$O: $(0.93)(2) + (0.0182)(1)$ $+ (0.0025)(2) + (0.0018)(1)$	$= 1.885\,\text{ft}^3$
total	$= 13.57\,\text{ft}^3$

The total wet volume is 13.57 ft^3/ft^3 fuel.

The total dry volume is 11.68 ft^3/ft^3 fuel.

The volumetric analyses are

	O$_2$	N$_2$	SO$_2$	CO$_2$	H$_2$O
wet:	$\dfrac{0.7500\,\text{ft}^3}{13.57\,\text{ft}^3}$	$\dfrac{9.974\,\text{ft}^3}{13.57\,\text{ft}^3}$	$\dfrac{0.0018\,\text{ft}^3}{13.57\,\text{ft}^3}$	$\dfrac{0.942\,\text{ft}^3}{13.57\,\text{ft}^3}$	$\dfrac{1.885}{13.57}$
	$= 0.0553$	0.735	$-$	0.069	0.139
dry:	$\dfrac{0.7500\,\text{ft}^3}{11.68\,\text{ft}^3}$	$\dfrac{9.974\,\text{ft}^3}{11.68\,\text{ft}^3}$	$\dfrac{0.0018\,\text{ft}^3}{11.68\,\text{ft}^3}$	$\dfrac{0.942\,\text{ft}^3}{11.68\,\text{ft}^3}$	$-$
	$= 0.0642$	0.854	$-$	0.081	$-$

$$B_{\text{CO}_2,\text{wet}} = \boxed{6.9\%}$$

The answer is (A) for Prob. 13(c).

$$B_{\text{CO}_2,\text{dry}} = \boxed{8.1\%}$$

The answer is (C) for Prob. 13(d).

14. *Customary U.S. Solution*

step 1: From App. 24.B, the heat of vaporization at 212°F is $h_{fg} = 970.1$ Btu/lbm.

The heat absorbed in the boiler is

$$m_{\text{H}_2\text{O}} h_{fg} = (11.12\,\text{lbm H}_2\text{O})\left(970.1\,\frac{\text{Btu}}{\text{lbm}}\right)$$

$$= 10{,}787.5\,\text{Btu/lbm fuel}$$

step 2: The losses for heating stack gases can be found as follows.

The burned carbon per lbm of fuel is

$$0.7842 - (0.315)(0.0703) = 0.7621\,\text{lbm/lbm fuel}$$

The mass ratio of dry flue gases to solid fuel is given by Eq. 22.12.

$$\frac{\substack{\text{mass of} \\ \text{flue gas}}}{\substack{\text{mass of} \\ \text{solid fuel}}} = \frac{\begin{pmatrix} 11 B_{\text{CO}_2} + 8 B_{\text{O}_2} \\ + (7)(B_{\text{CO}} + B_{\text{N}_2}) \end{pmatrix}\left(G_\text{C} + \dfrac{G_\text{S}}{1.833}\right)}{(3)(B_{\text{CO}_2} + B_{\text{CO}})}$$

$$= \frac{\begin{pmatrix} (11)(14.0) + (8)(5.5) \\ + (7)(0.42 + 80.08) \end{pmatrix}\left(0.7621 + \dfrac{0.01}{1.833}\right)}{(3)(14.0 + 0.42)}$$

$$= 13.51\,\text{lbm stack gases/lbm fuel}$$

Properties of nitrogen frequently are assumed for dry flue gas. From Table 22.1, for nitrogen at an average temperature of $(575°F + 73°F)/2 = 324°F$ $(784°R)$, $c_p \approx 0.252$ Btu/lbm-°F.

The losses for heating stack gases are given by Eq. 22.20.

$$q_1 = m_{\text{flue gas}} c_p (T_{\text{flue gas}} - T_{\text{incoming air}})$$

$$= (13.51 \text{ lbm}) \left(0.252 \frac{\text{Btu}}{\text{lbm-°F}}\right)(575°F - 67°F)$$

$$= 1729.5 \text{ Btu/lbm fuel}$$

The heat loss in the vapor formed during the combustion of hydrogen is given by Eq. 22.21.

$$q_2 = 8.94 G_H (h_g - h_f)$$

Assume that the partial pressure of the water vapor is below 1 psia (the lowest App. 24.C goes).

h_g at 575°F and 1 psia can be found from the superheat tables, App. 24.C.

$$h_g \approx 1324.3 \text{ Btu/lbm}$$

h_f at 73°F can be found from App. 24.A.

$$h_f = 41.07 \text{ Btu/lbm}$$

$$G_{H,\text{available}} = G_{H,\text{total}} - \frac{G_O}{8}$$

$$= 0.0556 - \frac{0.0825}{8}$$

$$= 0.0453$$

$$q_2 = (8.94)(0.0453)\left(1324.3 \frac{\text{Btu}}{\text{lbm}} - 41.07 \frac{\text{Btu}}{\text{lbm}}\right)$$

$$= 519.7 \text{ Btu/lbm fuel}$$

Heat is also lost when it is absorbed by the moisture originally in the combustion air.

$$q_3 = \omega m_{\text{combustion air}}(h_g - h_g')$$

Assume that the partial pressure of the water vapor is below 1 psia (the lowest App. 24.C goes).

From App. 24.B, at 73°F and 0.4 psia, $h_g' \approx 1093$ Btu/lbm. From the psychrometric chart,

$$\omega = 90 \text{ grains/lbm air}$$

$$= 0.0129 \text{ lbm water/lbm air}$$

Considering the sulfur content, find the air/fuel ratio.

$$\frac{\text{lbm air}}{\text{lbm fuel}} = \frac{3.04 B_{N_2}\left(G_C + \dfrac{G_S}{1.833}\right)}{B_{CO_2} + B_{CO}}$$

$$= \frac{(3.04)(80.08\%)\left(0.7621 + \dfrac{0.01}{1.833}\right)}{14\% + 0.42\%}$$

$$= 12.96 \text{ lbm air/lbm fuel}$$

$$q_3 = \left(0.0129 \frac{\text{lbm water}}{\text{lbm air}}\right)\left(12.96 \frac{\text{lbm air}}{\text{lbm fuel}}\right)$$

$$\times \left(1324.3 \frac{\text{Btu}}{\text{lbm}} - 1093 \frac{\text{Btu}}{\text{lbm}}\right)$$

$$= 38.7 \text{ Btu/lbm fuel}$$

The energy lost in incomplete combustion is given by Eq. 22.23.

$$q_4 = \frac{(\text{HHV}_C - \text{HHV}_{CO})\, G_C B_{CO}}{B_{CO_2} + B_{CO}}$$

$$= \frac{\left(9746 \dfrac{\text{Btu}}{\text{lbm}}\right)(0.7621)(0.42\%)}{14\% + 0.42\%}$$

$$= 216.3 \text{ Btu/lbm fuel}$$

The energy lost in unburned carbon is given by Eq. 22.24.

$$q_5 = \left(14{,}093 \frac{\text{Btu}}{\text{lbm}}\right) m_{\text{ash}} G_{C,\text{ash}}$$

$$= \left(14{,}093 \frac{\text{Btu}}{\text{lbm}}\right)(0.0703)(0.315)$$

$$= 312.1 \text{ Btu/lbm fuel}$$

The energy lost in radiation and unaccounted for is

$$14{,}000 \frac{\text{Btu}}{\text{lbm}} - 10{,}787.5 \frac{\text{Btu}}{\text{lbm}} - 1729.5 \frac{\text{Btu}}{\text{lbm}}$$

$$- 519.7 \frac{\text{Btu}}{\text{lbm}} - 38.7 \frac{\text{Btu}}{\text{lbm}} - 216.3 \frac{\text{Btu}}{\text{lbm}}$$

$$- 312.1 \frac{\text{Btu}}{\text{lbm}}$$

$$= \boxed{396.2 \text{ Btu/lbm fuel}}$$

SI Solution

step 1: From App. 24.N, the heat of vaporization at 100°C is $h_{fg} = 2256.4$ kJ/kg.

The heat absorbed in the boiler is

$$m_{H_2O}h_{fg} = (11.12 \text{ kg})\left(2256.4 \ \frac{kJ}{kg}\right)$$

$$= 25\,091 \text{ kJ/kg fuel}$$

step 2: From step 2 of the U.S. solution,

$$\frac{\text{mass of flue gas}}{\text{mass of solid fuel}} = 13.51 \text{ kg stack gases/kg fuel}$$

Properties of nitrogen frequently are assumed for dry flue gas. For an average nitrogen temperature of $(^1/_2)(300°C + 23°C) + 273 = 434.5$ K (782°R), $c_p \approx 0.252$ Btu/lbm-°R.

c_p from Table 22.1 can be found (using the table footnote) as

$$c_p = \left(0.252 \ \frac{Btu}{lbm\text{-}°R}\right)\left(4.187 \ \frac{\frac{kJ}{kg\cdot K}}{\frac{Btu}{lbm\text{-}°R}}\right)$$

$$= 1.055 \text{ kJ/kg·K}$$

The losses for heating stack gases are given by Eq. 22.20.

$$q_1 = m_{\text{flue gas}}c_p(T_{\text{flue gas}} - T_{\text{incoming air}})$$

$$= (13.51 \text{ kg})\left(1.055 \ \frac{kJ}{kg\cdot K}\right)(300°C - 23°C)$$

$$= 3948.1 \text{ kJ/kg fuel}$$

The heat loss in the vapor formed during the combustion of hydrogen is given by Eq. 22.21.

$$q_2 = 8.94 G_H(h_g - h_f)$$

Assume that the partial pressure of the water vapor is below 10 kPa (the lowest App. 24.P goes).

h_g at 300°C and 10 kPa can be found from the superheat tables, App. 24.P.

$$h_g = 3076.7 \text{ kJ/kg}$$

h_f at 23°C can be found from App. 24.N.

$$h_f = 96.47 \text{ kJ/kg}$$

$$G_{H,\text{available}} = G_{H,\text{total}} - \frac{G_O}{8}$$

$$= 0.0556 - \frac{0.0825}{8}$$

$$= 0.0453$$

$$q_2 = (8.94)(0.0453)\left(3076.7 \ \frac{kJ}{kg} - 96.47 \ \frac{kJ}{kg}\right)$$

$$= 1206.9 \text{ kJ/kg fuel}$$

Heat is also lost when it is absorbed by the moisture originally in the combustion air.

$$q_3 = \omega m_{\text{combustion air}}(h_g - h'_g)$$

Assume that the partial pressure of the water vapor is low. At 23°C, from App. 24.N, $h'_g \approx 2542.9$ kJ/kg.

From the psychrometric chart for 19°C wet bulb and 23°C dry bulb, $\omega = 12.2$ g/kg dry air.

The air-fuel ratio from the customary U.S. solution is

$$\frac{\text{kg air}}{\text{kg fuel}} = 12.96 \text{ kg air/kg fuel}$$

$$q_3 = \left(12.2 \ \frac{g}{kg}\right)\left(\frac{1 \text{ kg}}{1000 \text{ g}}\right)\left(12.96 \ \frac{kg}{kg}\right)$$

$$\times \left(3076.7 \ \frac{kJ}{kg} - 2542.9 \ \frac{kJ}{kg}\right)$$

$$= 84.80 \text{ kJ/kg}$$

Energy lost in incomplete combustion is given by Eq. 22.23.

$$q_4 = \frac{(HHV_C - HHV_{CO})G_C B_{CO}}{B_{CO_2} + B_{CO}}$$

$$= \left(\frac{\left(22.67 \ \frac{MJ}{kg}\right)(0.7621)(0.42)}{14 + 0.42}\right)\left(1000 \ \frac{kJ}{MJ}\right)$$

$$= \left(0.5032 \ \frac{MJ}{kg}\right)\left(1000 \ \frac{kJ}{MJ}\right)$$

$$= 503.2 \text{ kJ/kg fuel}$$

The energy lost in unburned carbon is given by Eq. 22.24.

$$q_5 = \left(32.8 \ \frac{MJ}{kg}\right)m_{\text{ash}}G_{C,\text{ash}}$$

$$= \left(32.8 \ \frac{MJ}{kg}\right)\left(1000 \ \frac{kJ}{MJ}\right)(0.0703)(0.315)$$

$$= 726.3 \text{ kJ/kg fuel}$$

Energy lost in radiation and unaccounted for is

$$\left(32.6 \ \frac{MJ}{kg}\right)\left(1000 \ \frac{kJ}{MJ}\right) - 25\,091 \ \frac{kJ}{kg}$$

$$- 3948.1 \ \frac{kJ}{kg} - 1206.9 \ \frac{kJ}{kg} - 84.40 \ \frac{kJ}{kg}$$

$$- 503.2 \ \frac{kJ}{kg} - 726.3 \ \frac{kJ}{kg} = \boxed{1040.1 \text{ kJ/kg}}$$

15. *Customary U.S. Solution*

step 1: The incoming reactants on a per-pound basis are

> 0.07 lbm ash
> 0.05 lbm hydrogen
> 0.05 lbm oxygen
> 0.83 lbm carbon

This is an ultimate analysis. Assume that only the hydrogen that is not locked up with oxygen in the form of water is combustible. From Eq. 22.15, the available hydrogen is

$$G_{H,available} = G_{H,total} - \frac{G_O}{8}$$

$$= 0.05 \text{ lbm} - \frac{0.05 \text{ lbm}}{8}$$

$$= 0.04375 \text{ lbm}$$

The mass of water produced is the hydrogen mass plus eight times as much oxygen. The locked hydrogen is

$$0.05 \text{ lbm} - 0.04375 \text{ lbm} = 0.00625 \text{ lbm}$$

$$\frac{\text{lbm of}}{\text{moisture}} = G_H + G_O$$

$$= G_H + 8G_H$$

$$= 0.00625 \text{ lbm} + (8)(0.00625 \text{ lbm})$$

$$= 0.05625 \text{ lbm}$$

The air is 23.15% oxygen by weight (Table 22.6), so other reactants for 26 lbm of air are

$$(0.2315)(26 \text{ lbm}) = 6.019 \text{ lbm O}_2$$

$$(0.7685)(26 \text{ lbm}) = 19.981 \text{ lbm N}_2$$

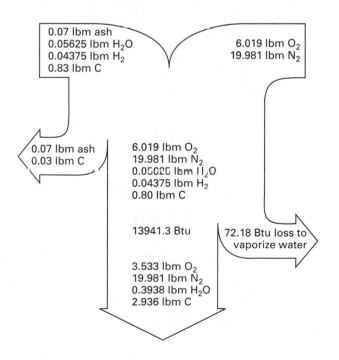

step 2: Ash pit material losses are 10% or 0.1 lbm, which includes all of the ash.

> 0.07 lbm ash (noncombustible matter)
> 0.03 lbm unburned carbon

step 3: Determine what remains.

> 6.019 lbm oxygen
> 19.981 lbm nitrogen
> 0.05625 lbm water
> 0.04375 lbm hydrogen
> 0.80 lbm carbon

step 4: Determine the energy loss in vaporizing the moisture.

$$q = (\text{moisture})(h_g - h_f)$$

From App. 24.C, h_g at 550°F and 14.7 psia is 1311.4 Btu/lbm.

From App. 24.A, h_f at 60°F is 28.08 Btu/lbm.

$$q = (0.05625 \text{ lbm})\left(1311.4 \frac{\text{Btu}}{\text{lbm}} - 28.08 \frac{\text{Btu}}{\text{lbm}}\right)$$

$$= 72.19 \text{ Btu}$$

step 5: Calculate the heating value of the remaining fuel components using App. 22.A.

$$HV_C = (0.80 \text{ lbm})\left(14{,}093 \frac{\text{Btu}}{\text{lbm}}\right)$$

$$= 11274.4 \text{ Btu}$$

$$HV_H = (0.04375 \text{ lbm})\left(60{,}958 \frac{\text{Btu}}{\text{lbm}}\right)$$

$$= 2666.9 \text{ Btu}$$

The heating value after the coal moisture is evaporated is

$$11{,}274.4 \text{ Btu} + 2666.9 \text{ Btu} - 72.19 \text{ Btu}$$

$$= 13{,}869 \text{ Btu}$$

step 6: Using Table 22.8, determine the combustion products.

$$\frac{\text{oxygen required}}{\text{by carbon}} = (0.80)(2.667 \text{ lbm})$$

$$= 2.134 \text{ lbm}$$

$$\frac{\text{oxygen required}}{\text{by hydrogen}} = (0.04375)(7.936 \text{ lbm})$$

$$= 0.3472 \text{ lbm}$$

$$\frac{\text{carbon dioxide}}{\text{produced by carbon}} = (0.8)(3.667 \text{ lbm})$$

$$= 2.934 \text{ lbm}$$

$$\frac{\text{water produced}}{\text{by hydrogen}} = (0.04375)(8.936 \text{ lbm})$$

$$= 0.3910 \text{ lbm}$$

The remaining oxygen is

$$6.019 \text{ lbm} - 2.134 \text{ lbm} - 0.3472 \text{ lbm} = 3.538 \text{ lbm}$$

step 7: The gaseous products must be heated from 70°F to 550°F. The average temperature is

$$\left(\tfrac{1}{2}\right)(70°\text{F} + 550°\text{F}) = 310°\text{F} \ (770°\text{R})$$

From Table 22.1, the specific heat of gaseous products is

gas	c_p $\left(\dfrac{\text{Btu}}{\text{lbm-°R}}\right)$
oxygen	0.228
nitrogen	0.252
water	0.460
carbon dioxide	0.225

$$
\begin{aligned}
Q_{\text{heating}} = \bigg(& (3.538 \text{ lbm})\left(0.228 \ \frac{\text{Btu}}{\text{lbm-°R}}\right) \\
& + (19.981 \text{ lbm})\left(0.252 \ \frac{\text{Btu}}{\text{lbm-°R}}\right) \\
& + (0.3910 \text{ lbm})\left(0.460 \ \frac{\text{Btu}}{\text{lbm-°R}}\right) \\
& + (2.934 \text{ lbm})\left(0.225 \ \frac{\text{Btu}}{\text{lbm-°R}}\right) \bigg) \\
& \times (550°\text{F} - 70°\text{F}) \\
= \ & 3207.3 \text{ Btu}
\end{aligned}
$$

step 8: The percent loss is

$$\frac{3207.3 \text{ Btu} + 72.19 \text{ Btu}}{13{,}869 \text{ Btu}} = \boxed{0.236 \ (23.6\%)}$$

The answer is (B).

SI Solution

Steps 1 through 3 are the same as for the customary U.S. solution except that everything is based on kg.

step 4: Determine the energy loss in the vaporizing moisture.

$$q = (\text{moisture})(h_g - h_f)$$

From App. 24.P, h_g at 290°C and 101.3 kPa is 3054.5 kJ/kg.

h_f at 15.6°C from App. 24.N is 65.49 kJ/kg.

$$q = (0.05625 \text{ kg})\left(3054.5 \ \frac{\text{kJ}}{\text{kg}} - 65.49 \ \frac{\text{kJ}}{\text{kg}}\right)$$

$$= 168.1 \text{ kJ}$$

step 5: Calculate the heating value of the remaining fuel components using App. 22.A and the table footnote.

$$\text{HV}_\text{C} = (0.80 \text{ kg})\left(14\,093 \ \frac{\text{Btu}}{\text{lbm}}\right)\left(2.326 \ \frac{\frac{\text{kJ}}{\text{kg}}}{\frac{\text{Btu}}{\text{lbm}}}\right)$$

$$= 26\,224 \text{ kJ}$$

$$\text{HV}_\text{H} = (0.04375 \text{ kg})\left(60\,958 \ \frac{\text{Btu}}{\text{lbm}}\right)\left(2.326 \ \frac{\frac{\text{kJ}}{\text{kg}}}{\frac{\text{Btu}}{\text{lbm}}}\right)$$

$$= 6203 \text{ kJ}$$

The heating value after the coal moisture is evaporated is

$$26\,224 \text{ kJ} + 6203 \text{ kJ} - 168.1 \text{ kJ} = 32\,259 \text{ kJ}$$

step 6: This step is the same as for the customary U.S. solution except that all quantities are in kg.

step 7: The gaseous products must be heated from 21°C to 290°C. The average temperature is

$$\left(\tfrac{1}{2}\right)(21°\text{C} + 290°\text{C}) = 156°\text{C} \ (429°\text{K})$$

$$(429\text{K})\left(1.8 \ \frac{°\text{R}}{\text{K}}\right) = 771°\text{R}$$

From Table 22.1, the specific heat of gaseous products is calculated using the table footnote.

gas	c_p $\left(\dfrac{\text{kJ}}{\text{kg·K}}\right)$
oxygen	0.955
nitrogen	1.055
water	1.926
carbon dioxide	0.942

$$
\begin{aligned}
Q_{\text{heating}} = \bigg(& (3.538 \text{ kg})\left(0.955 \ \frac{\text{kJ}}{\text{kg·K}}\right) \\
& + (19.981 \text{ lbm})\left(1.055 \ \frac{\text{kJ}}{\text{kg·K}}\right) \\
& + (0.3910 \text{ kg})\left(1.926 \ \frac{\text{kJ}}{\text{kg·K}}\right) \\
& + (2.934 \text{ kg})\left(0.942 \ \frac{\text{kJ}}{\text{kg·K}}\right) \bigg) \\
& \times (290°\text{C} - 21°\text{C}) \\
= \ & \boxed{7525.4 \text{ kJ}}
\end{aligned}
$$

step 8: The percentage loss is

$$\frac{7525.4 \text{ kJ} + 168.1 \text{ kJ}}{32\,259 \text{ kJ}} = \boxed{0.238 \quad (23.8\%)}$$

The answer is (B).

16. (a) Silicon in ash is SiO_2 with a molecular weight of

$$28.09 \frac{\text{lbm}}{\text{lbmol}} + (2)\left(16 \frac{\text{lbm}}{\text{lbmol}}\right) = 60.09 \text{ lbm/lbmol}$$

The oxygen used with 6.1% by mass silicon is

$$\left(\frac{(2)(16 \text{ lbm})}{28.09 \text{ lbm}}\right)\left(0.061 \frac{\text{lbm}}{\text{lbm coal}}\right)$$
$$= 0.0695 \text{ lbm/lbm coal}$$

Silicon ash produced per lbm of coal is

$$0.061 \frac{\text{lbm}}{\text{lbm coal}} + 0.0695 \frac{\text{lbm}}{\text{lbm coal}}$$
$$= 0.1305 \text{ lbm/lbm coal}$$

Silicon ash produced per hour is

$$\left(0.1305 \frac{\text{lbm}}{\text{lbm coal}}\right)\left(15{,}300 \frac{\text{lbm coal}}{\text{hr}}\right)$$
$$= 1996.7 \text{ lbm/hr}$$

The silicon in 410 lbm/hr refuse is

$$\left(410 \frac{\text{lbm}}{\text{hr}}\right)(1 - 0.3) = 287 \text{ lbm/hr}$$

The emission rate is

$$1996.7 \text{ lbm/hr} - 287 \text{ lbm/hr} = \boxed{1709.7 \text{ lbm/hr}}$$

The answer is (D).

(b) From Table 22.7, the stoichiometric reaction for sulfur is

$$\begin{array}{ccccc} & S & + & O_2 & \longrightarrow & SO_2 \\ \text{MW} & 32 & & 32 & & 64 \end{array}$$

Sulfur dioxide produced for 15,300 lbm/hr of coal feed is

$$\left(15{,}300 \frac{\text{lbm}}{\text{hr}}\right)(0.0244 \text{ lbm S})\left(\frac{64 \text{ lbm SO}_2}{32 \text{ lbm S}}\right)$$
$$= \boxed{746.6 \text{ lbm/hr}}$$

The answer is (C).

(c) From Eq. 22.16(b), the heating value of the fuel is

$$\text{HHV} = 14{,}093 G_C + (60{,}958)\left(G_H - \frac{G_O}{8}\right) + 3983 G_S$$
$$= \left(14{,}093 \frac{\text{Btu}}{\text{lbm}}\right)(0.7656)$$
$$+ \left(60{,}958 \frac{\text{Btu}}{\text{lbm}}\right)\left(0.055 - \frac{0.077}{8}\right)$$
$$+ \left(3983 \frac{\text{Btu}}{\text{lbm}}\right)(0.0244)$$
$$= 13{,}653 \text{ Btu/lbm}$$

The gross available combustion power is

$$\dot{m}_f(\text{HV}) = \left(15{,}300 \frac{\text{lbm}}{\text{hr}}\right)\left(13{,}653 \frac{\text{Btu}}{\text{lbm}}\right)$$
$$= 2.089 \times 10^8 \text{ Btu/hr}$$

The carbon in 410 lbm/hr refuse is

$$\left(410 \frac{\text{lbm}}{\text{hr}}\right)(0.3) = 123 \text{ lbm/hr}$$

Power lost in unburned carbon in refuse is $\dot{m}_C(\text{HV})$.

From App. 22.A, the gross heat of combustion for carbon is 14,093 Btu/lbm.

$$\left(123 \frac{\text{lbm}}{\text{hr}}\right)\left(14{,}093 \frac{\text{Btu}}{\text{lbm}}\right) = 1.733 \times 10^6 \text{ Btu/hr}$$

The remaining combustion power is

$$2.089 \times 10^8 \frac{\text{Btu}}{\text{hr}} - 1.733 \times 10^6 \frac{\text{Btu}}{\text{hr}}$$
$$= 2.072 \times 10^8 \text{ Btu/hr}$$

Losses in the steam generator and electrical generator will further reduce this to

$$(0.86)\left(2.072 \times 10^8 \frac{\text{Btu}}{\text{hr}}\right) = 1.782 \times 10^8 \text{ Btu/hr}$$

With an electrical output of 17 MW, thermal energy removed by cooling water is

$$Q = 1.782 \times 10^8 \frac{\text{Btu}}{\text{hr}} - (17 \text{ MW})\left(1000 \frac{\text{kW}}{\text{MW}}\right)$$
$$\times \left(3413 \frac{\frac{\text{Btu}}{\text{hr}}}{\text{kW}}\right)$$
$$= 1.202 \times 10^8 \text{ Btu/hr}$$

The temperature rise of the cooling water is

$$\Delta T = \frac{Q}{\dot{m}c_p}$$

At 60°F, the specific heat of water is $c_p = 1$ Btu/lbm-°F.

$$\Delta T = \frac{1.202 \times 10^8 \; \dfrac{\text{Btu}}{\text{hr}}}{\left(225 \; \dfrac{\text{ft}^3}{\text{sec}}\right)\left(62.4 \; \dfrac{\text{lbm}}{\text{ft}^3}\right)\left(3600 \; \dfrac{\text{sec}}{\text{hr}}\right)\left(1 \; \dfrac{\text{Btu}}{\text{lbm-}°\text{F}}\right)}$$

$$= \boxed{2.38°\text{F}}$$

The answer is (A).

The electrical generation is not cooled by the cooling water. Therefore, it is not correct to include the generation efficiency in the calculation of losses.

(d) Limiting 0.1 lbm of particulates per million Btu per hour, the allowable emission rate is

$$\left(0.1 \; \frac{\text{lbm}}{\text{MBtu}}\right)\left(15{,}300 \; \frac{\text{lbm}}{\text{hr}}\right)$$
$$\times \left(13{,}653 \; \frac{\text{Btu}}{\text{lbm}}\right)\left(\frac{1 \text{ MBtu}}{10^6 \text{ Btu}}\right) = 20.89 \text{ lbm/hr}$$

The efficiency of the flue gas particulate collectors is

$$\eta = \frac{\text{actual emission rate} - \text{allowable emission rate}}{\text{actual emission rate}}$$

$$= \frac{1709.7 \; \dfrac{\text{lbm}}{\text{hr}} - 20.89 \; \dfrac{\text{lbm}}{\text{hr}}}{1709.7 \; \dfrac{\text{lbm}}{\text{hr}}}$$

$$= \boxed{0.988 \; (98.8\%)}$$

The answer is (C).

17. *Customary U.S. Solution*

(c) The stoichiometric reaction for propane is given in Table 22.7.

$$\begin{array}{cccccc} & \text{C}_3\text{H}_8 + & 5\text{O}_2 & \longrightarrow & 3\text{CO}_2 & + & 4\text{H}_2\text{O} \\ \text{MW} & 44.097 & (5)(32.000) & & (3)(44.011) & & (4)(18.016) \end{array}$$

With 40% excess O_2 by weight,

$$\begin{array}{cc} \text{C}_3\text{H}_8 & (1.4)(5)\text{O}_2 \\ \text{MW} \quad 44.097 \; + & (7)(32.000) \\ 44.097 & 224 \end{array}$$

$$\begin{array}{ccc} 3\text{CO}_2 & 4\text{H}_2\text{O} & 2\text{O}_2 \\ \longrightarrow (3)(44.011) + & (4)(18.016) + & (2)(32.000) \\ 132.033 & 72.064 & 64 \end{array}$$

The excess oxygen is

$$(2)(32) = 64 \text{ lbm/lbmol C}_3\text{H}_8$$

The mass ratio of nitrogen to oxygen is

$$\begin{aligned} \frac{G_N}{G_O} &= \left(\frac{B_N}{R_N}\right)\left(\frac{R_O}{B_O}\right) \\ &= \left(\frac{0.40}{55.16 \; \dfrac{\text{ft-lbf}}{\text{lbm-}°\text{R}}}\right)\left(\frac{48.29 \; \dfrac{\text{ft-lbf}}{\text{lbm-}°\text{R}}}{0.60}\right) \\ &= 0.584 \end{aligned}$$

(Values of R_N and R_O are taken from Table 24.7.)

The nitrogen accompanying the oxygen is

$$(7)(32)(0.584) = 130.8 \text{ lbm}$$

The mass balance per mole of propane is

$$\text{C}_3\text{H}_8 + \text{O}_2 + \text{N}_2 \longrightarrow \text{CO}_2 + \text{H}_2\text{O} + \text{O}_2 + \text{N}_2$$

mass
per $44.097 + 224 + 130.8 \longrightarrow 132.033 + 72.064 + 64 + 130.8$
mole

At standard conditions (60°F, 1 atm), the propane density is given by Eq. 24.50.

$$\rho = \frac{p}{RT}$$

The absolute temperature, T, is

$$60°\text{F} + 460 = 520°\text{R}$$

R for propane from Table 24.7 is 35.04 ft-lbf/lbm-°R.

$$\rho = \frac{\left(14.7 \; \dfrac{\text{lbf}}{\text{in}^2}\right)\left(144 \; \dfrac{\text{in}^2}{\text{ft}^2}\right)}{\left(35.04 \; \dfrac{\text{ft-lbf}}{\text{lbm-}°\text{R}}\right)(520°\text{R})}$$

$$= 0.1162 \text{ lbm/ft}^3$$

The mass flow rate of propane based on 250 SCFM of propane is

$$\left(250 \; \frac{\text{ft}^3}{\text{min}}\right)\left(0.1162 \; \frac{\text{lbm}}{\text{ft}^3}\right) = 29.05 \text{ lbm/min}$$

Scaling the other mass balance factors down by

$$\frac{29.05 \; \dfrac{\text{lbm}}{\text{min}}}{44.097 \; \dfrac{\text{lbm}}{\text{lbmol}}} = 0.6588 \text{ lbmol/min}$$

$$
\begin{array}{cccccc}
 & C_3H_8 + & O_2 & + & N_2 \\
\dfrac{lbm}{min} & 29.05 + & 147.57 + & 86.17
\end{array}
$$

$$
\longrightarrow \quad CO_2 + H_2O + O_2 + N_2 \\
86.98 + \ 47.48 \ + 42.16 + 86.17
$$

The oxygen flow rate is $\boxed{\begin{array}{c} 147.57 \ lbm/min \\ (150 \ lbm/min) \end{array}}$ [part (c)].

The answer is (A).

(a) Using R from Table 24.7, the specific volumes of the reactants are given by Eq. 24.50.

$$
\begin{aligned}
v_{C_3H_8} &= \frac{RT}{p} \\
&= \frac{\left(35.04 \ \dfrac{\text{ft-lbf}}{\text{lbm-}^\circ R}\right)(80^\circ F + 460)}{\left(14.7 \ \dfrac{lbf}{in^2}\right)\left(144 \ \dfrac{in^2}{ft^2}\right)} \\
&= 8.939 \ ft^3/lbm
\end{aligned}
$$

$$
\begin{aligned}
v_{O_2} &= \frac{\left(48.29 \ \dfrac{\text{ft-lbf}}{\text{lbm-}^\circ R}\right)(80^\circ F + 460)}{\left(14.7 \ \dfrac{lbf}{in^2}\right)\left(144 \ \dfrac{in^2}{ft^2}\right)} \\
&= 12.319 \ ft^3/lbm
\end{aligned}
$$

$$
\begin{aligned}
v_{N_2} &= \frac{\left(55.16 \ \dfrac{\text{ft-lbf}}{\text{lbm-}^\circ R}\right)(80^\circ F + 460)^\circ R}{\left(14.7 \ \dfrac{lbf}{in^2}\right)\left(144 \ \dfrac{in^2}{ft^2}\right)} \\
&= 14.071 \ ft^3/lbm
\end{aligned}
$$

The total incoming volume is

$$
\begin{aligned}
\dot{V} &= \left(29.05 \ \frac{lbm}{min}\right)\left(8.939 \ \frac{ft^3}{lbm}\right) \\
&\quad + \left(147.57 \ \frac{lbm}{min}\right)\left(12.319 \ \frac{ft^3}{lbm}\right) \\
&\quad + \left(86.17 \ \frac{lbm}{min}\right)\left(14.071 \ \frac{ft^3}{lbm}\right) \\
&= 3290 \ ft^3/min
\end{aligned}
$$

Since the velocity must be kept below 400 ft/min, the area of inlet pipe is

$$
A_{in} = \frac{\dot{V}}{v} = \frac{3290 \ \dfrac{ft^3}{min}}{400 \ \dfrac{ft}{min}} = \boxed{8.23 \ ft^2}
$$

The answer is (A).

(d) Similarly, the specific volumes of the products are

$$
\begin{aligned}
v_{CO_2} &= \frac{\left(35.11 \ \dfrac{\text{ft-lbf}}{\text{lbm-}^\circ R}\right)(460^\circ F + 460)}{\left(8 \ \dfrac{lbf}{in^2}\right)\left(144 \ \dfrac{in^2}{ft^2}\right)} \\
&= 28.04 \ ft^3/lbm
\end{aligned}
$$

$$
\begin{aligned}
v_{H_2O} &= \frac{\left(85.78 \ \dfrac{\text{ft-lbf}}{\text{lbm-}^\circ R}\right)(460^\circ F + 460)}{\left(8 \ \dfrac{lbf}{in^2}\right)\left(144 \ \dfrac{in^2}{ft^2}\right)} \\
&= 68.50 \ ft^3/lbm
\end{aligned}
$$

$$
\begin{aligned}
v_{O_2} &= \frac{\left(48.29 \ \dfrac{\text{ft-lbf}}{\text{lbm-}^\circ R}\right)(460^\circ F + 460)}{\left(8 \ \dfrac{lbf}{in^2}\right)\left(144 \ \dfrac{in^2}{ft^2}\right)} \\
&= 38.56 \ ft^3/lbm
\end{aligned}
$$

$$
\begin{aligned}
v_{N_2} &= \frac{\left(55.16 \ \dfrac{\text{ft-lbf}}{\text{lbm-}^\circ R}\right)(460^\circ F + 460)}{\left(8 \ \dfrac{lbf}{in^2}\right)\left(144 \ \dfrac{in^2}{ft^2}\right)} \\
&= 44.05 \ ft^3/lbm
\end{aligned}
$$

The total exhaust volume is

$$
\begin{aligned}
\dot{V} &= \left(86.98 \ \frac{lbm}{min}\right)\left(28.04 \ \frac{ft^3}{lbm}\right) \\
&\quad + \left(47.48 \ \frac{lbm}{min}\right)\left(68.50 \ \frac{ft^3}{lbm}\right) \\
&\quad + \left(42.16 \ \frac{lbm}{min}\right)\left(38.56 \ \frac{ft^3}{lbm}\right) \\
&\quad + \left(86.98 \ \frac{lbm}{min}\right)\left(44.05 \ \frac{ft^3}{min}\right) \\
&= \boxed{11{,}148 \ ft^3/min} \quad \text{[part (d)]}
\end{aligned}
$$

The answer is (D).

(b) Since the velocity of the products must be kept below 800 ft/min, the area of the stack is

$$
\begin{aligned}
A_{stack} &= \frac{Q}{v} = \frac{11{,}148 \ \dfrac{ft^3}{min}}{800 \ \dfrac{ft}{min}} \\
&= \boxed{13.94 \ ft^2} \quad \text{[part (b)]}
\end{aligned}
$$

The answer is (C).

Thermodynamics

(e) For ideal gases, the partial pressure is volumetrically weighted. The water vapor partial pressure in the stack is

$$(8 \text{ psia}) \left(\frac{\left(47.48 \dfrac{\text{lbm}}{\text{min}}\right) \left(68.50 \dfrac{\text{ft}^3}{\text{lbm}}\right)}{11{,}148 \dfrac{\text{ft}^3}{\text{lbm}}} \right) = 2.33 \text{ psia}$$

The saturation temperature corresponding to 2.33 psia is $T_{\text{dp}} = \boxed{131°\text{F.}}$

The answer is (C).

SI Solution

(c) Following the procedure for the customary U.S. solution, the mass balance per mole of propane is

$$\begin{array}{c} \text{C}_3\text{H}_8 + \text{O}_2 + \text{N}_2 \longrightarrow \text{CO}_2 + \text{H}_2\text{O} + \text{O}_2 + \text{N}_2 \\ \frac{\text{kg}}{\text{mole}} \quad 44.097 + 224 + 130.8 \longrightarrow 132.033 + 72.064 + 64 + 130.8 \end{array}$$

At standard conditions (16°C, 101.3 kPa), the propane density is given by Eq. 24.50.

$$\rho = \frac{p}{RT}$$

The absolute temperature is

$$T = 16°\text{C} + 273 = 289\text{K}$$

From Table 24.7, R for propane is 188.55 J/kg·K.

$$\rho = \frac{(101.3 \text{ kPa}) \left(1000 \dfrac{\text{Pa}}{\text{kPa}}\right)}{\left(188.55 \dfrac{\text{J}}{\text{kg·K}}\right)(289\text{K})}$$

$$= 1.86 \text{ kg/m}^3$$

The mass flow rate of propane based on 118 L/s of propane is

$$\left(118 \dfrac{\text{L}}{\text{s}}\right) \left(\dfrac{1 \text{ m}^3}{1000 \text{ L}}\right) \left(1.86 \dfrac{\text{kg}}{\text{m}^3}\right) = 0.2195 \text{ kg/s}$$

Scale the other mass balance factors down.

$$\frac{0.2195 \dfrac{\text{kg}}{\text{s}}}{44.097 \dfrac{\text{kg}}{\text{kmol}}} = 0.004978 \text{ kmol/s}$$

$$\begin{array}{c} \text{C}_3\text{H}_8 + \text{O}_2 + \text{N}_2 \longrightarrow \text{CO}_2 + \text{H}_2\text{O} \\ \frac{\text{kg}}{\text{s}} \quad 0.2195 \quad 1.115 \quad 0.6511 \qquad 0.6572 \quad 0.3587 \end{array}$$

$$\begin{array}{c} + \text{O}_2 + \text{N}_2 \\ 0.3186 \quad 0.6511 \end{array}$$

The oxygen flow rate is $\boxed{1.115 \text{ kg/s.}}$

The answer is (A).

(a) Using R from Table 24.7, the specific volumes of the reactants are given by Eq. 24.50.

$$v_{\text{C}_3\text{H}_8} = \frac{RT}{p} = \frac{\left(188.55 \dfrac{\text{J}}{\text{kg·K}}\right)(27°\text{C} + 273)}{(101 \text{ kPa}) \left(1000 \dfrac{\text{Pa}}{\text{kPa}}\right)}$$

$$= 0.5600 \text{ m}^3/\text{kg}$$

$$v_{\text{O}_2} = \frac{\left(259.82 \dfrac{\text{J}}{\text{kg·K}}\right)(27°\text{C} + 273)}{(101 \text{ kPa}) \left(1000 \dfrac{\text{Pa}}{\text{kPa}}\right)}$$

$$= 0.7717 \text{ m}^3/\text{kg}$$

$$v_{\text{N}_2} = \frac{\left(296.77 \dfrac{\text{J}}{\text{kg·K}}\right)(27°\text{C} + 273)}{(101 \text{ kPa}) \left(1000 \dfrac{\text{Pa}}{\text{kPa}}\right)}$$

$$= 0.8815 \text{ m}^3/\text{kg}$$

The total incoming volume, $\dot{V}$, is

$$\begin{aligned} \dot{V} &= \left(0.2195 \dfrac{\text{kg}}{\text{s}}\right) \left(0.5600 \dfrac{\text{m}^3}{\text{kg}}\right) \\ &\quad + \left(1.115 \dfrac{\text{kg}}{\text{s}}\right) \left(0.7717 \dfrac{\text{m}^3}{\text{kg}}\right) \\ &\quad + \left(0.6511 \dfrac{\text{kg}}{\text{s}}\right) \left(0.8815 \dfrac{\text{m}^3}{\text{kg}}\right) \\ &= 1.557 \text{ m}^3/\text{s} \end{aligned}$$

Since the velocity for the reactants must be kept below 2 m/s, the area of inlet pipe is

$$A_{\text{in}} = \frac{\dot{V}}{\text{v}} = \frac{1.557 \dfrac{\text{m}^3}{\text{s}}}{2 \dfrac{\text{m}}{\text{s}}} = \boxed{0.779 \text{ m}^2}$$

The answer is (A).

(d) Similarly, the specific volumes of the products are

$$v_{CO_2} = \frac{\left(188.92 \ \frac{J}{kg\cdot}\right)(240°C + 273)}{(55 \text{ kPa})\left(1000 \ \frac{Pa}{kPa}\right)}$$

$$= 1.762 \ \text{m}^3/\text{kg}$$

$$v_{H_2O} = \frac{\left(461.5 \ \frac{J}{kg\cdot}\right)(240°C + 273)}{(55 \text{ kPa})\left(1000 \ \frac{Pa}{kPa}\right)}$$

$$= 4.305 \ \text{m}^3/\text{kg}$$

$$v_{O_2} = \frac{\left(259.82 \ \frac{J}{kg\cdot K}\right)(240°C + 273)}{(55 \text{ kPa})\left(1000 \ \frac{Pa}{kPa}\right)}$$

$$= 2.423 \ \text{m}^3/\text{kg}$$

$$v_{N_2} = \frac{\left(296.77 \ \frac{J}{kg\cdot K}\right)(240°C + 273)}{(55 \text{ kPa})\left(1000 \ \frac{Pa}{kPa}\right)}$$

$$= 2.768 \ \text{m}^3/\text{kg}$$

The total exhaust volume is

$$\dot{V} = \left(0.6572 \ \frac{kg}{s}\right)\left(1.762 \ \frac{m^3}{kg}\right)$$

$$+ \left(0.3587 \ \frac{kg}{s}\right)\left(4.305 \ \frac{m^3}{kg}\right)$$

$$+ \left(0.3186 \ \frac{kg}{s}\right)\left(2.423 \ \frac{m^3}{kg}\right)$$

$$+ \left(0.6511 \ \frac{kg}{s}\right)\left(2.768 \ \frac{m^3}{kg}\right)$$

$$= \boxed{5.276 \ \text{m}^3/\text{s}}$$

The answer is (D).

(b) Since the velocity of products must be kept below 4 m/s, the area of stack is

$$A_{\text{stack}} = \frac{\dot{V}}{v} = \frac{5.276 \ \frac{m^3}{s}}{4 \ \frac{m}{s}}$$

$$= \boxed{1.319 \ \text{m}^2}$$

The answer is (C).

(e) For ideal gases, the partial pressure is volumetrically weighted. The water vapor partial pressure in the stack is

$$(55 \text{ kPa})\left(\frac{\left(0.3584 \ \frac{kg}{s}\right)\left(4.305 \ \frac{m^3}{kg}\right)}{5.273 \ \frac{m^3}{s}}\right)$$

$$= 16.093 \text{ kPa}$$

The saturation temperature corresponding to 16.093 kPa is found from App. 24.O to be $T_{\text{dp}} = \boxed{54.5°C.}$

The answer is (C).

18. *Customary U.S. Solution*

(a) Since atmospheric air is not used, the nitrogen and oxygen can be varied independently. Furthermore, since enthalpy increase information is not given for oxygen, a 0% excess oxygen can be assumed.

From Table 22.7,

$$C_3H_8 + 5O_2 \longrightarrow 3CO_2 + 4H_2O$$
$$\text{moles} \quad (1) \qquad (5) \qquad\quad (3) \qquad (4)$$

Subtract the reactant enthalpies from the product enthalpies to calculate the heat of reaction. The enthalpy of formation of oxygen is zero, since it is an element in its natural state.

$$n_{CO_2}(\Delta H_f)_{CO_2} + n_{H_2O}(\Delta H_f)_{H_2O}$$
$$- n_{C_3H_8}(\Delta H_f)_{C_3H_8} - n_{O_2}(\Delta H_f)_{O_2}$$

$$= (3 \text{ lbmol})\left(-169,300 \ \frac{\text{Btu}}{\text{lbmol}}\right)$$

$$+ (4 \text{ lbmol})\left(-104,040 \ \frac{\text{Btu}}{\text{lbmol}}\right)$$

$$- (1 \text{ lbmol})\left(28,800 \ \frac{\text{Btu}}{\text{lbmol}}\right)$$

$$- (5 \text{ lbmol})(0)$$

$$= -952,860 \text{ Btu/lbmol of fuel}$$

The negative sign indicates an exothermal reaction.

Let x be the number of moles of nitrogen per mole of propane. Use the nitrogen to cool the combustion. The above heat of reaction will increase the enthalpy of products from the standard reference temperature to 3600°R. Therefore,

$$952,860 \ \frac{\text{Btu}}{\text{lbmol}} = (3 \text{ lbmol})\left(39,791 \ \frac{\text{Btu}}{\text{lbmol}}\right)$$

$$+ (4 \text{ lbmol})\left(31,658 \ \frac{\text{Btu}}{\text{lbmol}}\right)$$

$$+ x\left(24,471 \ \frac{\text{Btu}}{\text{lbmol}}\right)$$

$$x = 28.89 \text{ lbmol/lbmol fuel}$$

The mass of nitrogen per lbmole of propane is

$$M_{N_2} = \left(28.89 \ \frac{\text{lbmol}}{\text{lbmol fuel}}\right)\left(28.016 \ \frac{\text{lbm}}{\text{lbmol}}\right)$$

$$= \boxed{809.4 \text{ lbm/lbmol propane}}$$

The mass of oxygen per lbmole of propane is

$$M_{O_2} = (5 \text{ lbmol})\left(32 \ \frac{\text{lbm}}{\text{lbmol}}\right)$$

$$= \boxed{160 \text{ lbm/lbmol fuel}}$$

(b) The partial pressure is volumetrically weighted. This is the same as molar weighting.

product	lbmol	volumetric fraction
CO_2	3	$3/35.89 = 0.0836$
H_2O	4	$4/35.89 = 0.1115$
N_2	28.89	$28.89/35.89 = 0.8049$
O_2	0	$0/35.89 = 0$
	35.89 lbmol	1.000

The partial pressure of water vapor is

$$p_{H_2O} = \left(\frac{n_{H_2O}}{n}\right)p = (0.1115)(14.7 \text{ psia})$$

$$= 1.64 \text{ psia}$$

From App. 24.B, this corresponds to approximately 118°F. Since the stack temperature is 100°F, some of the water will condense. From App. 24.A, the maximum vapor pressure of water is 0.9505 psia. Let n be the number of moles of water in the stack gas.

$$n_{H_2O} = \left(\frac{p_{H_2O}}{p}\right)n = \left(\frac{0.9505 \text{ psia}}{14.7 \text{ psia}}\right)(35.89 \text{ lbmol})$$

$$= \boxed{2.321 \text{ lbmol/lbmol } C_3H_8}$$

(c) The water removed is

$$4 - n_{H_2O} = 4 \text{ lbmol} - 2.321 \text{ lbmol}$$

$$= 1.679 \text{ lbmol of } H_2O/\text{lbmol of } C_3H_8$$

$$m = (1.679 \text{ lbmol})\left(18.016 \ \frac{\text{lbm}}{\text{lbmol}}\right)$$

$$= \boxed{30.26 \text{ lbm } H_2O/\text{lbmol } C_3H_8}$$

SI Solution

(a) From the customary U.S. solution, the heat of reaction is

$$n_{CO_2}(\Delta H_f)_{CO_2} + n_{H_2O}(\Delta H_f)_{H_2O}$$
$$- n_{C_3H_8}(\Delta H_f)_{C_3H_8} - n_{O_2}(\Delta H_f)_{O_2}$$

$$= (3 \text{ kmol})\left(-393.8 \ \frac{\text{GJ}}{\text{kmol}}\right)$$

$$+ (4 \text{ kmol})\left(-242 \ \frac{\text{GJ}}{\text{kmol}}\right)$$

$$- (1 \text{ kmol})\left(67.0 \ \frac{\text{GJ}}{\text{kmol}}\right) - (5 \text{ kmol})(0)$$

$$= -2216.4 \text{ GJ/kmol fuel}$$

The negative sign indicates an exothermal reaction.

Let x be the number of moles of nitrogen per mole of propane. Use the nitrogen to cool the combustion. The above heat of reaction will increase the enthalpy of products from the standard reference temperature to 1980°C. Therefore,

$$2216.4 \ \frac{\text{GJ}}{\text{mol}} = (3 \text{ kmol})\left(92.6 \ \frac{\text{GJ}}{\text{kmol}}\right)$$

$$+ (4 \text{ kmol})\left(73.6 \ \frac{\text{GJ}}{\text{kmol}}\right)$$

$$+ x\left(56.9 \ \frac{\text{GJ}}{\text{kmol}}\right)$$

$$x = 28.90 \text{ kmol/kmol fuel}$$

The mass of nitrogen per mole of propane is

$$M_{N_2} = (28.90 \text{ kmol})\left(28.016 \ \frac{\text{kg}}{\text{kmol}}\right)$$

$$= \boxed{809.7 \text{ kg/kmol fuel}}$$

The mass of oxygen per kmol of propane is

$$M_{O_2} = (5 \text{ kmol})\left(32 \ \frac{\text{kg}}{\text{kmol}}\right)$$

$$= \boxed{160 \text{ kg/kmol propane}}$$

(b) The partial pressure is volumetrically weighted. This is the same as molar weighting.

product	kmol	volumetric fraction
CO_2	3	$3/35.90 = 0.0836$
H_2O	4	$4/35.90 = 0.1114$
N_2	28.90	$28.90/35.90 = 0.8050$
O_2	0	$0/35.90 = 0$
	35.90 kmol	1.000

The partial pressure of water vapor is

$$p_{H_2O} = \left(\frac{n_{H_2O}}{n}\right) p = (0.1114)(101 \text{ kPa})$$
$$= 11.25 \text{ kPa}$$

From App. 24.O, this corresponds to approximately 47.6°C. Since the stack temperature is 38°C, some of the water will condense. From App. 24.N, the maximum vapor pressure of water at the stack temperature is 6.633 kPa. Let n be the number of moles of water in the stack gas.

$$n_{H_2O} = \left(\frac{p_{H_2O}}{p}\right) n = \left(\frac{6.633 \text{ kPa}}{101 \text{ kPa}}\right)(35.90 \text{ kmol})$$

$$\boxed{= 2.358 \text{ kmol H}_2\text{O/kmol C}_3\text{H}_8}$$

(c) The liquid water removed is

$$4 - n_{H_2O} = 4 \text{ kmol} - 2.358 \text{ kmol}$$
$$= 1.642 \text{ kmol of H}_2\text{O/kmol propane}$$
$$m = (1.642 \text{ kmol})\left(18.016 \frac{\text{kg}}{\text{kmol}}\right)$$

$$\boxed{= 29.59 \text{ kg H}_2\text{O/kmol C}_3\text{H}_8}$$

19. *Customary U.S. Solution*

(a) If the power output is to be unchanged,

$$\text{BHP}_1 = \text{BHP}_2$$
$$(\dot{m}_{F,1})(\text{LHV}_1) = (\dot{m}_{F,2})(\text{LHV}_2)$$

From Eq. 29.8,

$$\dot{m}_F = (\text{BSFC})(\text{BHP})$$
$$(\text{BSFC}_1)(\text{LHV}_1) = (\text{BSFC}_2)(\text{LHV}_2)$$

$$\frac{\text{BSFC}_2}{\text{BSFC}_1} = \frac{\text{LHV}_1}{\text{LHV}_2} = \frac{23{,}200 \dfrac{\text{Btu}}{\text{lbm}}}{11{,}930 \dfrac{\text{Btu}}{\text{lbm}}}$$
$$= 1.945$$
$$\frac{\text{BSFC}_2 - \text{BSFC}_1}{\text{BSFC}_1} = \frac{\text{BSFC}_2}{\text{BSFC}_1} - 1 = 1.945 - 1$$

$$\boxed{= 0.945 \quad [94.5\% \text{ increase}]}$$

The answer is (D).

(b) If the fuel injection velocity is to be unchanged ($v_2 = v_1$),

$$\dot{m} = \rho A v$$
$$A = \frac{\dot{m}}{\rho v}$$
$$\dot{m}_2 = 1.945 \dot{m}_1 \quad [\text{part (a)}]$$

$$\frac{A_2 - A_1}{A_1} = \frac{\dfrac{\dot{m}_2}{\rho_2 v_2} - \dfrac{\dot{m}_1}{\rho_1 v_1}}{\dfrac{\dot{m}_1}{\rho_1 v_1}} = \frac{\dfrac{\dot{m}_2}{\rho_2} - \dfrac{\dot{m}_1}{\rho_1}}{\dfrac{\dot{m}_1}{\rho_1}}$$

$$= \frac{\dfrac{1.945 m_1}{\rho_2} - \dfrac{\dot{m}_1}{\rho_1}}{\dfrac{\dot{m}_1}{\rho_1}} = \frac{\dfrac{1.945}{\rho_2} - \dfrac{1}{\rho_1}}{\dfrac{1}{\rho_1}}$$

$$\frac{A_2 - A_1}{A_1} = (1.945)\left(\frac{\rho_1}{\rho_2}\right) - 1$$

From Table 22.3, for gasoline, $SG_1 = 0.74$. For ethanol, $SG_2 = 0.794$.

$$\frac{\rho_1}{\rho_2} = \frac{SG_1}{SG_2} = \frac{0.74}{0.794}$$
$$\frac{A_2 - A_1}{A_1} = (1.945)\left(\frac{0.74}{0.794}\right) - 1$$

$$\boxed{= 0.810 \quad [81\% \text{ increase}]}$$

The answer is (C).

(c) If no changes are made to the engine, power output is proportional to the weight flow and heating value.

$$\frac{P_2 - P_1}{P_1} = \frac{\dot{m}_{F2}(\text{LHV}_2) - \dot{m}_{F1}(\text{LHV}_1)}{\dot{m}_{F1}(\text{LHV}_1)}$$
$$\dot{m}_F = \rho A v$$
$$\frac{P_2 - P_1}{P_1} = \frac{(\rho_2 A_2 v_2)(\text{LHV}_2) - (\rho_1 A_1 v_1)(\text{LHV}_1)}{(\rho_1 A_1 v_1)(\text{LHV}_1)}$$

Since no changes are made to the engine, $A_2 = A_1$ and $v_2 = v_1$.

$$\frac{P_2 - P_1}{P_1} = \frac{\rho_2(\text{LHV}_2) - \rho_1(\text{LHV}_1)}{\rho_1(\text{LHV}_1)}$$

$$= \frac{\text{LHV}_2 - \left(\dfrac{\rho_1}{\rho_2}\right)(\text{LHV}_1)}{\left(\dfrac{\rho_1}{\rho_2}\right)(\text{LHV}_1)}$$

From part (b),

$$\frac{\rho_1}{\rho_2} = \frac{0.74}{0.794}$$

$$\frac{P_2 - P_1}{P_1} = \frac{11,930 \; \frac{\text{Btu}}{\text{lbm}} - \left(\frac{0.74}{0.794}\right)\left(23,200 \; \frac{\text{Btu}}{\text{lbm}}\right)}{\left(\frac{0.74}{0.794}\right)\left(23,200 \; \frac{\text{Btu}}{\text{lbm}}\right)}$$

$$= \boxed{-0.45 \quad [45\% \text{ decrease}]}$$

The answer is (B).

SI Solution

(a) From part (a) of the customary U.S. solution,

$$\frac{\text{BSFC}_2}{\text{BSFC}_1} = \frac{\text{LHV}_1}{\text{LHV}_2} = \frac{54 \; \frac{\text{MJ}}{\text{kg}}}{27.7 \; \frac{\text{MJ}}{\text{kg}}} = 1.949$$

$$\frac{\text{BSFC}_2 - \text{BSFC}_1}{\text{BSFC}_1} = \frac{\text{BSFC}_2}{\text{BSFC}_1} - 1 = 1.949 - 1$$

$$= \boxed{0.949 \quad [94.9\% \text{ increase}]}$$

The answer is (D).

(b) From part (b) of the customary U.S. solution,

$$\frac{A_2 - A_1}{A_1} = \frac{\dfrac{\dot{m}_2}{\rho_2} - \dfrac{\dot{m}_1}{\rho_1}}{\dfrac{\dot{m}_1}{\rho_1}}$$

$$\dot{m}_2 = 1.949 \dot{m}_1 \quad [\text{part (a)}]$$

$$\frac{A_2 - A_1}{A_1} = \frac{\dfrac{1.949 \dot{m}_1}{\rho_2} - \dfrac{\dot{m}_1}{\rho_1}}{\dfrac{\dot{m}_1}{\rho_1}}$$

$$= (1.949)\left(\frac{\rho_1}{\rho_2}\right) - 1$$

$$= (1.949)\left(\frac{0.74}{0.794}\right) - 1$$

$$= \boxed{0.816 \quad [81.6\% \text{ increase}]}$$

The answer is (C).

(c) From part (c) of the customary U.S. solution,

$$\frac{P_2 - P_1}{P_1} = \frac{\text{LHV}_2 - \left(\dfrac{\rho_1}{\rho_2}\right)(\text{LHV}_1)}{\left(\dfrac{\rho_1}{\rho_2}\right)(\text{LHV}_1)}$$

$$= \frac{27.7 \; \frac{\text{MJ}}{\text{kg}} - \left(\frac{0.74}{0.794}\right)\left(54 \; \frac{\text{MJ}}{\text{kg}}\right)}{\left(\frac{0.74}{0.795}\right)\left(54 \; \frac{\text{MJ}}{\text{kg}}\right)}$$

$$= \boxed{-0.45 \quad [45\% \text{ decrease}]}$$

The answer is (B).

23 Energy, Work, and Power

PRACTICE PROBLEMS

Energy

1. A solid cast-iron sphere ($\rho = 0.256$ lbm/in^3 (7090 kg/m^3)) of 10 in (25 cm) diameter travels without friction at 30 ft/sec (9 m/s) horizontally. What is its kinetic energy?

(A) 900 ft-lbf (12 kJ)

(B) 1200 ft-lbf (1.6 kJ)

(C) 1600 ft-lbf (2.0 kJ)

(D) 1900 ft-lbf (2.4 kJ)

Work

2. What work is done when a balloon carries a 12 lbm (5.2 kg) load to 40,000 ft (12 000 m) height?

(A) 2.4×10^5 ft-lbf (300 kJ)

(B) 4.8×10^5 ft-lbf (610 kJ)

(C) 7.7×10^5 ft-lbf (980 kJ)

(D) 9.9×10^5 ft-lbf (1.3 MJ)

3. Find the compression of a spring if a 100 lbm (50 kg) weight is dropped from 8 ft (2 m) onto a spring with a constant of 33.33 lbf/in (5.837×10^3 N/m).

(A) 27 in (0.67 m)

(B) 34 in (0.85 m)

(C) 39 in (0.90 m)

(D) 45 in (1.1 m)

4. A punch press flywheel operates at 300 rpm with a moment of inertia of 15 slug-ft^2 (20 kg·m^2). Find the speed in rpm to which the wheel will be reduced after a sudden punching requiring 4500 ft-lbf (6100 J) of work.

(A) 160 rpm

(B) 190 rpm

(C) 220 rpm

(D) 310 rpm

5. A force of 550 lbf (2500 N) making a 40° angle (upward) from the horizontal pushes a box 20 ft (6 m) across the floor. What work is done?

(A) 2200 ft-lbf (3.0 kJ)

(B) 3700 ft-lbf (5.2 kJ)

(C) 4200 ft-lbf (6.0 kJ)

(D) 8400 ft-lbf (12 kJ)

6. A 1000 ft long (300 m long) cable has a mass of 2 lbm per foot (3 kg/m) and is suspended from a winding drum down into a vertical shaft. What work must be done to rewind the cable?

(A) 0.50×10^6 ft-lbf (0.6 MJ)

(B) 0.75×10^6 ft-lbf (0.9 MJ)

(C) 1×10^6 ft-lbf (1.3 MJ)

(D) 2×10^6 ft-lbf (2.6 MJ)

Power

7. What volume in ft^3 (m^3) of water can be pumped to a 130 ft (40 m) height in one hour by a 7 hp (5 kW) pump? Assume 85% efficiency.

(A) 1500 ft^3 (40 m^3)

(B) 1800 ft^3 (49 m^3)

(C) 2000 ft^3 (54 m^3)

(D) 2400 ft^3 (65 m^3)

8. What power in horsepower (kW) is required to lift a 3300 lbm (1500 kg) mass 250 feet (80 m) in 14 seconds?

(A) 40 hp (30 kW)

(B) 70 hp (53 kW)

(C) 90 hp (68 kW)

(D) 110 hp (84 kW)

SOLUTIONS

1. *Customary U.S. Solution*

Since there is no friction, there is no rotation. The sphere slides.

$$E_{\text{kinetic}} = \frac{1}{2}\left(\frac{m}{g_c}\right)v^2 = \frac{1}{2}\left(V\left(\frac{\rho}{g_c}\right)\right)v^2$$

$$= \left(\frac{1}{2}\right)\left(\frac{4}{3}\pi r^3\right)\left(\frac{\rho}{g_c}\right)v^2 = \left(\frac{2}{3}\pi\right)\left(\frac{10}{2}\text{ in}\right)^3\left(\frac{\rho}{g_c}\right)v^2$$

$$= \left(\frac{2}{3}\pi\right)\left(\left(\frac{10}{2}\text{ in}\right)\left(\frac{1\text{ ft}}{12\text{ in}}\right)\right)^3$$

$$\times\left(0.256\,\frac{\text{lbm}}{\text{in}^3}\right)\left(1728\,\frac{\text{in}^3}{\text{ft}^3}\right)$$

$$\times\left(30\,\frac{\text{ft}}{\text{sec}}\right)^2\left(\frac{1}{32.2}\,\frac{\text{lbf-sec}^2}{\text{lbm-ft}}\right)$$

$$= \boxed{1873\text{ ft-lbf}}$$

The answer is (D).

SI Solution

Since there is no friction, there is no rotation. The sphere slides.

$$E_{\text{kinetic}} = \frac{1}{2}mv^2 = \frac{1}{2}(\rho V)v^2$$

$$= \left(\frac{1}{2}\right)\rho\left(\frac{4}{3}\pi r^3\right)v^2 = \left(\frac{2}{3}\pi\right)\rho\left(\frac{0.25\text{ m}}{2}\right)^3 v^2$$

$$= \left(\frac{2}{3}\pi\right)\left(\frac{0.25\text{ m}}{2}\right)^3\left(7.09\times10^3\,\frac{\text{kg}}{\text{m}^3}\right)\left(9\,\frac{\text{m}}{\text{s}}\right)^2$$

$$= 2.35\times10^3\text{ J} = \boxed{2.35\text{ kJ}}$$

The answer is (D).

2. *Customary U.S. Solution*

$$W = \Delta E_{\text{potential}} = m\frac{g}{g_c}\Delta h$$

$$= (12\text{ lbm})\left(\frac{32.2\,\frac{\text{ft}}{\text{sec}^2}}{32.2\,\frac{\text{lbm-ft}}{\text{sec}^2\text{-lbf}}}\right)(40{,}000\text{ ft})$$

$$= \boxed{4.8\times10^5\text{ ft-lbf}}$$

The answer is (B).

SI Solution

$$W = \Delta E_{\text{potential}} = mg\Delta h$$

$$= (5.2\text{ kg})\left(9.81\,\frac{\text{m}}{\text{s}^2}\right)(12\,000\text{ m})\left(\frac{\text{kJ}}{1000\text{ J}}\right)$$

$$= \boxed{612.1\text{ kJ}}$$

The answer is (B).

3. *Customary U.S. Solution*

$$\Delta E_{\text{potential}} = \Delta E_{\text{spring}}$$

$$W(\Delta h + \Delta x) = \frac{1}{2}k(\Delta x)^2$$

Rearranging,

$$\frac{1}{2}k(\Delta x)^2 - W\Delta x - W\Delta h = 0$$

$$\left(\frac{1}{2}\right)\left(33.33\,\frac{\text{lbf}}{\text{in}}\right)(\Delta x)^2 - (100\text{ lbf})\Delta x$$

$$- (100\text{ lbf})(8\text{ ft})\left(12\,\frac{\text{in}}{\text{ft}}\right) = 0$$

$$16.665\Delta x^2 - 100\Delta x = 9600$$

$$\Delta x^2 - 6\Delta x = 576$$

$$(\Delta x - 3)^2 = 576 + 9$$

$$\Delta x - 3 = \sqrt{585} = \pm24.2$$

$$\Delta x = \boxed{27.2\text{ in}}$$

The answer is (A).

SI Solution

$$\Delta E_{\text{potential}} = \Delta E_{\text{spring}}$$

$$mg(\Delta h + \Delta x) = \frac{1}{2}k(\Delta x)^2$$

Rearranging,

$$\frac{1}{2}k(\Delta x)^2 - mg\Delta x - mg\Delta h = 0$$

$$\left(\frac{1}{2}\right)\left(5.837\times10^3\,\frac{\text{N}}{\text{m}}\right)(\Delta x)^2 - (50\text{ kg})\left(9.81\,\frac{\text{m}}{\text{s}^2}\right)\Delta x$$

$$- (50\text{ kg})\left(9.81\,\frac{\text{m}}{\text{s}^2}\right)(2\text{ m}) = 0$$

$$2918.5\Delta x^2 - 490.5\Delta x - 981.0 = 0$$

$$\Delta x^2 - 0.1681\Delta x = 0.3361$$

$$(\Delta x - 0.08403)^2 = 0.3361 + (0.08403)^2$$

$$= 0.3432$$

$$\Delta x - 0.08403 = \sqrt{0.3432} = \pm0.5858$$

$$\Delta x = \boxed{0.6699\text{ m}}$$

The answer is (A).

4. *Customary U.S. Solution*

$$W_{\text{done by wheel}} = \Delta E_{\text{rotational}}$$

$$= \frac{1}{2}I\omega_{\text{initial}}^2 - \frac{1}{2}I\omega_{\text{final}}^2$$

$$\omega_{\text{final}} = \sqrt{(\omega_{\text{initial}})^2 - \frac{2W}{I}} = 2\pi f$$

$$f_{\text{final}} = \left(\frac{1}{2\pi}\right)\left(\frac{60 \text{ rpm}}{\frac{\text{rev}}{\text{sec}}}\right)$$

$$\times \sqrt{\left(\left(2\pi \frac{\text{rad}}{\text{rev}}\right)(300 \text{ rpm})\left(\frac{\frac{\text{rev}}{\text{sec}}}{60 \text{ rpm}}\right)\right)^2 - \frac{(2)(45 \times 10^2 \text{ ft-lbf})}{15 \text{ slug-ft}^2}}$$

$$= \boxed{187.8 \text{ rpm}}$$

The answer is (B).

SI Solution

$$W_{\text{done by wheel}} = \Delta E_{\text{rotational}}$$
$$= \tfrac{1}{2} I \omega_{\text{initial}}^2 - \tfrac{1}{2} I \omega_{\text{final}}^2$$

$$\omega_{\text{final}} = \sqrt{(\omega_{\text{initial}})^2 - \frac{2W}{I}} = 2\pi f$$

$$f_{\text{final}} = \left(\frac{1}{2\pi}\right)\left(\frac{60 \text{ rpm}}{\text{rps}}\right)$$

$$\times \sqrt{\left((2\pi)\left(\frac{300 \text{ rpm}}{\frac{\text{s}}{\text{min}}}\right)\right)^2 - \frac{(2)(6.1 \times 10^3 \text{ J})}{20 \text{ kg·m}^2}}$$

$$= \boxed{185.4 \text{ rpm}}$$

The answer is (B).

5. *Customary U.S. Solution*

$$W_{\text{done on box}} = F_x \Delta x = (F)(\cos\theta)\Delta x$$
$$= (550 \text{ lbf})(\cos 40°)(20 \text{ ft})$$
$$= \boxed{8430 \text{ ft-lbf}}$$

The answer is (D).

SI Solution

$$W_{\text{done on box}} = F_x \Delta x = (F)(\cos\theta)\Delta x$$
$$= (2500 \text{ N})(\cos 40°)(6 \text{ m})\left(\frac{\text{kJ}}{1000 \text{ J}}\right)$$
$$= \boxed{11.5 \text{ kJ}}$$

The answer is (D).

6. *Customary U.S. Solution*

$$W_{\text{to retrieve cable}} = \int_0^l F\, dh$$
$$= \int_0^l ((l-h)w)\, dh$$
$$= \tfrac{1}{2}wl^2 = \left(\tfrac{1}{2}\right)\left(2 \frac{\text{lbf}}{\text{ft}}\right)(1000 \text{ ft})^2$$
$$= \boxed{10^6 \text{ ft-lbf}}$$

The answer is (C).

SI Solution

$$W_{\text{to retrieve cable}} = \int_0^l F\, dh$$
$$= \int_0^l ((l-h)m_l g)\, dh$$
$$= \tfrac{1}{2}m_l g l^2$$
$$= \left(\tfrac{1}{2}\right)\left(3 \frac{\text{kg}}{\text{m}}\right)\left(9.81 \frac{\text{m}}{\text{s}^2}\right)(300 \text{ m})^2$$
$$= 1.32 \times 10^6 \text{ J} = \boxed{1.32 \text{ MJ}}$$

The answer is (C).

7. *Customary U.S. Solution*

$$P_{\text{actual}}\Delta t = W_{\text{done by pump}}$$
$$\eta P_{\text{ideal}}\Delta t = \Delta E_{\text{potential}}$$
$$= m\frac{g}{g_c}\Delta h$$
$$= (\rho V)\frac{g}{g_c}\Delta h$$

$$V = \frac{\eta P_{\text{ideal}}\Delta t}{\rho\frac{g}{g_c}\Delta h}$$

$$= \frac{(0.85)(7 \text{ hp})\left(550 \frac{\text{ft-lbf}}{\text{hp-sec}}\right)(3600 \text{ sec})}{\left(62.4 \frac{\text{lbm}}{\text{ft}^3}\right)\left(\frac{32.2 \frac{\text{ft}}{\text{sec}^2}}{32.2 \frac{\text{lbm-ft}}{\text{sec}^2\text{-lbf}}}\right)(130 \text{ ft})}$$

$$= \boxed{1450 \text{ ft}^3}$$

The answer is (A).

SI Solution

$$P_{\text{actual}}\Delta t = W_{\text{done by pump}}$$
$$\eta P_{\text{ideal}}\Delta t = \Delta E_{\text{potential}}$$
$$= mg\Delta h$$
$$= (\rho V)g\Delta h$$

$$V = \frac{\eta P_{\text{ideal}} \Delta t}{\rho g \Delta h}$$

$$= \frac{(0.85)(5 \times 10^3 \text{ W})(3600 \text{ s})}{\left(1000 \dfrac{\text{kg}}{\text{m}^3}\right)\left(9.81 \dfrac{\text{m}}{\text{s}^2}\right)(40 \text{ m})}$$

$$= \boxed{39.0 \text{ m}^3}$$

The answer is (A).

8. *Customary U.S. Solution*

$$P\Delta t = W = m\frac{g}{g_c}\Delta h$$

$$P = \frac{mg\Delta h}{g_c \Delta t}$$

$$= \frac{(3300 \text{ lbm})\left(32.2 \dfrac{\text{ft}}{\text{sec}^2}\right)(250 \text{ ft})}{\left(32.2 \dfrac{\text{lbm-ft}}{\text{sec}^2\text{-lbf}}\right)(14 \text{ sec})\left(550 \dfrac{\text{ft-lbf}}{\text{hp-sec}}\right)}$$

$$= \boxed{107 \text{ hp}}$$

The answer is (D).

SI Solution

$$P\Delta t = W = mg\Delta h$$

$$P = \frac{mg\Delta h}{\Delta t}$$

$$= \left(\frac{(1500 \text{ kg})\left(9.81 \dfrac{\text{m}}{\text{s}^2}\right)(80 \text{ m})}{14 \text{ s}}\right)\left(\frac{\text{kW}}{1000 \text{ W}}\right)$$

$$= \boxed{84.1 \text{ kW}}$$

The answer is (D).

24 Thermodynamic Properties of Substances

PRACTICE PROBLEMS

1. What is the molar enthalpy of 250°F (120°C) steam with a quality of 92%?

    (A) 16,000 Btu/lbmol (37 MJ/kmol)

    (B) 18,000 Btu/lbmol (41 MJ/kmol)

    (C) 20,000 Btu/lbmol (46 MJ/kmol)

    (D) 22,000 Btu/lbmol (51 MJ/kmol)

2. What is the ratio of specific heats for air at 600°F (300°C)?

    (A) 1.33

    (B) 1.38

    (C) 1.41

    (D) 1.67

3. What is the density of helium at 600°F (300°C) and one standard atmosphere?

    (A) 0.0052 lbm/ft^3 (0.085 kg/m^3)

    (B) 0.0061 lbm/ft^3 (0.098 kg/m^3)

    (C) 0.0076 lbm/ft^3 (0.12 kg/m^3)

    (D) 0.0095 lbm/ft^3 (0.15 kg/m^3)

SOLUTIONS

1. *Customary U.S. Solution*

From App. 24.A, for 250°F steam, the enthalpy of saturated liquid, h_f, is 218.6 Btu/lbm. The heat of vaporization, h_{fg}, is 945.4 Btu/lbm. The enthalpy is given by Eq. 24.40.

$$h = h_f + x h_{fg}$$
$$= 218.6 \ \frac{\text{Btu}}{\text{lbm}} + (0.92) \left(945.4 \ \frac{\text{Btu}}{\text{lbm}} \right)$$
$$= 1088.4 \ \text{Btu/lbm}$$

The molecular weight of water is 18 lbm/lbmol. The molar enthalpy is given by Eq. 24.14.

$$H = (\text{MW}) h$$
$$= \left(18 \ \frac{\text{lbm}}{\text{lbmol}} \right) \left(1088.4 \ \frac{\text{Btu}}{\text{lbm}} \right)$$
$$= \boxed{19{,}591 \ \text{Btu/lbmol}}$$

The answer is (C).

SI Solution

From App. 24.N, for 120°C steam, the enthalpy of saturated liquid, h_f, is 503.81 kJ/kg. The heat of vaporization, h_{fg}, is 2202.1 kJ/kg. The enthalpy is given by Eq. 24.40.

$$h = h_f + x h_{fg}$$
$$= 503.81 \ \frac{\text{kJ}}{\text{kg}} + (0.92) \left(2202.1 \ \frac{\text{kJ}}{\text{kg}} \right)$$
$$= 2529.7 \ \text{kJ/kg}$$

The molecular weight of water is 18 kg/kmol. Molar enthalpy is given by Eq. 24.14.

$$H = (\text{MW}) h$$
$$= \left(18 \ \frac{\text{kg}}{\text{kmol}} \right) \left(2529.7 \ \frac{\text{kJ}}{\text{kg}} \right)$$
$$= \boxed{45\,535 \ \text{kJ/kmol}}$$

The answer is (C).

2. *Customary U.S. Solution*

The absolute temperature is

$$600°F + 460 = 1060°R$$

From Table 22.1, the specific heat at constant pressure for air at 1060°R is $c_p = 0.250$ Btu/lbm-°R.

From Eq. 24.48, the specific gas constant is

$$R = \frac{R^*}{MW} = \frac{1545.33 \; \frac{\text{ft-lbf}}{\text{lbmol-°R}}}{28.967 \; \frac{\text{lbm}}{\text{lbmol}}}$$

$$= 53.35 \text{ ft-lbf/lbm-°R}$$

From Eq. 24.95(b),

$$c_v = c_p - \frac{R}{J}$$

$$= 0.250 \; \frac{\text{Btu}}{\text{lbm-°R}} - \frac{53.35 \; \frac{\text{ft-lbf}}{\text{lbm-°R}}}{778 \; \frac{\text{ft-lbf}}{\text{Btu}}}$$

$$= 0.1814 \text{ Btu/lbm-°R}$$

The ratio of specific heats is given by Eq. 24.28.

$$k = \frac{c_p}{c_v} = \frac{0.250 \; \frac{\text{Btu}}{\text{lbm-°R}}}{0.1814 \; \frac{\text{Btu}}{\text{lbm-°R}}}$$

$$= \boxed{1.378}$$

The answer is (B).

SI Solution

From Table 22.1, the specific heat at constant pressure for air is 0.250 Btu/lbm-°R. From the table footnote, the SI specific heat at constant pressure for air is

$$c_p = \left(0.250 \; \frac{\text{Btu}}{\text{lbm·°R}}\right) \left(4.187 \; \frac{\frac{\text{kJ}}{\text{kg·K}}}{\frac{\text{Btu}}{\text{lbm-°R}}}\right)$$

$$= 1.047 \text{ kJ/kg·K}$$

From Eq. 24.48, the specific gas constant is

$$R = \frac{R^*}{MW} = \frac{8314.3 \; \frac{\text{J}}{\text{kmol·K}}}{28.967 \; \frac{\text{kg}}{\text{kmol}}}$$

$$= 287.0 \text{ J/kg·K}$$

From Eq. 24.95(a),

$$c_v = c_p - R$$

$$= \left(1.047 \; \frac{\text{kJ}}{\text{kg·K}}\right) \left(1000 \; \frac{\text{J}}{\text{kJ}}\right) - 287.0 \; \frac{\text{J}}{\text{kg·K}}$$

$$= 760 \text{ J/kg·K}$$

The ratio of specific heats is given by Eq. 24.28.

$$k = \frac{c_p}{c_v} = \frac{\left(1.047 \; \frac{\text{kJ}}{\text{kg·K}}\right) \left(1000 \; \frac{\text{J}}{\text{kJ}}\right)}{760 \; \frac{\text{J}}{\text{kg·K}}}$$

$$= \boxed{1.377}$$

The answer is (B).

3. *Customary U.S. Solution*

From Eq. 24.48, the specific gas constant is

$$R = \frac{R^*}{MW} = \frac{1545.33 \; \frac{\text{ft-lbf}}{\text{lbmol-°R}}}{4 \; \frac{\text{lbm}}{\text{lbmol}}}$$

$$= 386.3 \text{ ft-lbf/lbm-°R}$$

The absolute temperature is

$$600°F + 460 = 1060°R$$

From Eq. 24.50, the density of helium is

$$\rho = \frac{p}{RT}$$

$$= \frac{\left(14.7 \; \frac{\text{lbf}}{\text{in}^2}\right) \left(144 \; \frac{\text{in}^2}{\text{ft}^2}\right)}{\left(386.3 \; \frac{\text{ft-lbf}}{\text{lbm-°R}}\right) (1060°R)}$$

$$= \boxed{0.00517 \text{ lbm/ft}^3}$$

The answer is (A).

SI Solution

From Eq. 24.48, the specific gas constant is

$$R = \frac{R^*}{MW} = \frac{8314.3 \; \frac{\text{J}}{\text{kmol·K}}}{4 \; \frac{\text{kg}}{\text{kmol}}}$$

$$= 2079 \text{ J/kg·K}$$

The absolute temperature is

$$300°C + 273 = 573K$$

From Eq. 24.50, the density of helium is

$$\rho = \frac{p}{RT}$$

$$= \frac{1.013 \times 10^5 \text{ Pa}}{2079 \ \dfrac{\text{J}}{\text{kg·K}} (573 \text{K})}$$

$$= \boxed{0.0850 \text{ kg/m}^3}$$

The answer is (A).

25 Changes in Thermodynamic Properties

PRACTICE PROBLEMS

1. Cast iron is heated from 80°F to 780°F (27°C to 416°C). What heat is required per unit mass?

  (A) 70 Btu/lbm (160 kJ/kg)

  (B) 120 Btu/lbm (280 kJ/kg)

  (C) 170 Btu/lbm (390 kJ/kg)

  (D) 320 Btu/lbm (740 kJ/kg)

2. The ventilation rate in a building is 3×10^5 ft³/hr (2.4 m³/s). The air is heated from 35°F to 75°F (2°C to 24°C) by water whose temperature decreases from 180°F to 150°F (82°C to 66°C). What is the water flow rate in gal/min (L/s)?

  (A) 9 gal/min (0.58 L/s)

  (B) 13 gal/min (0.83 L/s)

  (C) 15 gal/min (0.96 L/s)

  (D) 22 gal/min (1.4 L/s)

3. 8.0 ft³ (0.25 m³) of 180°F, 14.7 psia (82°C, 101.3 kPa) air are cooled to 100°F (38°C) in a constant-pressure process. What work is done?

  (A) −900 ft-lbf (−1.3 kJ)

  (B) −1100 ft-lbf (−1.5 kJ)

  (C) −1500 ft-lbf (−2.3 kJ)

  (D) −2100 ft-lbf (−3.1 kJ)

4. What is the availability of an isentropic process using steam with an initial quality of 95% and operating between 300 psia and 50 psia (2 MPa and 0.35 MPa)?

  (A) 100 Btu/lbm (230 kJ/kg)

  (B) 130 Btu/lbm (300 kJ/kg)

  (C) 210 Btu/lbm (480 kJ/kg)

  (D) 340 Btu/lbm (780 kJ/kg)

5. (*Time limit: one hour*) A closed air heater receives 540°F, 100 psia (280°C, 700 kPa) air and heats it to 1540°F (840°C). The outside temperature is 100°F (40°C). The pressure of the air drops 20 psi (150 kPa) as it passes through the heater. What is the percentage loss in available energy due to the pressure drop?

  (A) 5%

  (B) 12%

  (C) 18%

  (D) 34%

6. (*Time limit: one hour*) Xenon gas at 20 psia and 70°F (150 kPa and 21°C) is compressed to 3800 psia and 70°F (25 MPa and 21°C) by a compressor/heat exchanger combination. The compressed gas is stored in a 100 ft³ (3 m³) rigid tank initially charged with xenon gas at 20 psia (150 kPa).

(a) What is the mass of the xenon gas initially in the tank?

  (A) 35 lbm (18 kg)

  (B) 42 lbm (22 kg)

  (C) 46 lbm (24 kg)

  (D) 51 lbm (27 kg)

(b) What is the average mass flow rate of xenon gas into the tank if the compressor fills the tank in exactly one hour?

  (A) 6300 lbm/hr (0.88 kg/s)

  (B) 9700 lbm/hr (1.3 kg/s)

  (C) 12,000 lbm/hr (1.6 kg/s)

  (D) 14,000 lbm/hr (1.9 kg/s)

(c) If filling takes exactly one hour and electricity costs $0.045 per kW-hr, what is the cost of filling the tank?

  (A) $8

  (B) $14

  (C) $27

  (D) $35

7. (*Time limit: one hour*) The mass of an insulated 20 ft³ (0.6 m³) steel tank is 40 lbm (20 kg). The steel has a specific heat of 0.11 Btu/lbm-°R (0.46 kJ/kg·K). The tank is placed in a room where the surrounding air is 70°F and 14.7 psia (21°C and 101.3 kPa). After the tank is evacuated to 1 psia and 70°F (7 kPa and 21°C), a valve is suddenly opened, allowing the tank to fill with room air. The air enters the tank in a well-mixed, turbulent condition. Find the air temperature inside the tank after filling and after the gas and tank have reached thermal equilibrium, but before any heat loss from the tank to the room occurs.

  (A) 73°F (20°C)

  (B) 80°F (30°C)

  (C) 103°F (40°C)

  (D) 190°F (90°C)

SOLUTIONS

1. *Customary U.S. Solution*

From Table 24.2, the approximate value of specific heat for cast iron is $c_p = 0.10$ Btu/lbm-°F.

The heat required per unit mass is

$$q = c_p(T_2 - T_1)$$
$$= \left(0.10 \; \frac{\text{Btu}}{\text{lbm-°F}}\right)(780°\text{F} - 80°\text{F})$$
$$= \boxed{70 \; \text{Btu/lbm}}$$

The answer is (A).

SI Solution

From Table 24.2, the approximate value of specific heat of cast iron is $c_p = 0.42$ kJ/kg·K.

The heat required per unit mass is

$$q = c_p(T_2 - T_1)$$
$$= \left(0.42 \; \frac{\text{kJ}}{\text{kg·K}}\right)(416°\text{C} - 27°\text{C})$$
$$= \boxed{163.4 \; \text{kJ/kg}}$$

The answer is (A).

2. *Customary U.S. Solution*

First calculate the mass flow rate of air to be heated by using the ideal gas law. (Usually the air mass would be evaluated at the entering conditions. This problem is ambiguous. The building conditions are used because a building ventilation rate was specified.)

$$\dot{m}_{\text{air}} = \frac{p\dot{V}}{RT}$$

The absolute temperature is

$$T = 75°\text{F} + 460 = 535°\text{R}$$

$$\dot{m}_{\text{air}} = \frac{\left(14.7 \; \frac{\text{lbf}}{\text{in}^2}\right)\left(144 \; \frac{\text{in}^2}{\text{ft}^2}\right)\left(3 \times 10^5 \; \frac{\text{ft}^3}{\text{hr}}\right)}{\left(53.3 \; \frac{\text{ft-lbf}}{\text{lbm-°R}}\right)(535°\text{R})}$$

$$= 2.227 \times 10^4 \; \text{lbm/hr}$$

The heat lost by the water is equal to the heat gained by the air.

$$\dot{m}_w c_{p,w}(T_{1,w} - T_{2,w}) = \dot{m}_{\text{air}} c_{p,\text{air}}(T_{2,\text{air}} - T_{1,\text{air}})$$

$$\dot{m}_w \left(1 \; \frac{\text{Btu}}{\text{lbm-°F}}\right)(180°\text{F} - 150°\text{F})$$
$$= \left(2.227 \times 10^4 \; \frac{\text{lbm}}{\text{hr}}\right)\left(0.241 \; \frac{\text{Btu}}{\text{lbm-°F}}\right)(75°\text{F} - 35°\text{F})$$

$$\dot{m}_w = 7156.1 \; \text{lbm/hr}$$

From App. 35.A, the density of water at 165°F is approximately 61 lbm/ft³. The water volume flow rate is

$$\dot{V}_w = \frac{\dot{m}}{\rho} = \frac{\left(7156.1 \; \frac{\text{lbm}}{\text{hr}}\right)\left(\frac{1 \; \text{hr}}{60 \; \text{min}}\right)\left(7.481 \; \frac{\text{gal}}{\text{ft}^3}\right)}{\left(61 \; \frac{\text{lbm}}{\text{ft}^3}\right)}$$

$$= \boxed{14.63 \; \text{gal/min}}$$

The answer is (C).

SI Solution

First calculate the mass flow rate of air to be heated by using the ideal gas law. (Usually the air mass would be evaluated at the entering conditions. This problem is ambiguous. The building conditions are used because a building ventilation rate was specified.)

$$\dot{m}_{\text{air}} = \frac{p\dot{V}}{RT}$$

The absolute temperature is

$$T = 24°\text{C} + 273 = 297\text{K}$$

$$\dot{m}_{\text{air}} = \frac{(1.013 \times 10^5 \; \text{Pa})\left(2.4 \; \frac{\text{m}^3}{\text{s}}\right)}{\left(287 \; \frac{\text{J}}{\text{kg·K}}\right)(297\text{K})}$$

$$= 2.85 \; \text{kg/s}$$

The heat lost by the water is equal to the heat gained by the air.

$$\dot{m}_w c_{p,w}(T_{1,w} - T_{2,w}) = \dot{m}_{\text{air}} c_{p,\text{air}}(T_{2,\text{air}} - T_{1,\text{air}})$$

$$\dot{m}_w \left(4.190 \; \frac{\text{kJ}}{\text{kg·K}}\right)(82°\text{C} - 66°\text{C})$$
$$= \left(2.85 \; \frac{\text{kg}}{\text{s}}\right)\left(1.005 \; \frac{\text{kJ}}{\text{kg·K}}\right)(24°\text{C} - 2°\text{C})$$

$$\dot{m}_w = 0.940 \; \text{kg/s}$$

From App. 35.B, the density of water at 74°C is approximately 976 kg/m³. The water volume flow rate is

$$\dot{V}_w = \left(\frac{\dot{m}}{\rho}\right) = \left(0.940 \ \frac{\text{kg}}{\text{s}}\right)\left(976 \ \frac{\text{kg}}{\text{m}^3}\right)\left(1000 \ \frac{\text{L}}{\text{m}^3}\right)$$

$$= \boxed{0.963 \ \text{L/s}}$$

The answer is (C).

3. *Customary U.S. Solution*

The mass of air is

$$m = \frac{p_1 V_1}{RT_1}$$

$$= \frac{\left(14.7 \ \frac{\text{lbf}}{\text{in}^2}\right)\left(144 \ \frac{\text{in}^2}{\text{ft}^2}\right)(8.0 \ \text{ft}^3)}{\left(53.3 \ \frac{\text{ft-lbf}}{\text{lbm-°R}}\right)(180°\text{F} + 460)}$$

$$= 0.4964 \ \text{lbm}$$

For a constant pressure process from Eq. 25.51, on a per unit mass basis,

$$W = R(T_2 - T_1)$$

Since $\Delta T_{°\text{R}} = \Delta T_{°\text{F}}$, the total work for m in lbm is

$$W = mR(T_2 - T_1)$$

$$= (0.4964 \ \text{lbm})\left(53.3 \ \frac{\text{ft-lbf}}{\text{lbm-°R}}\right)(100°\text{F} - 180°\text{F})$$

$$= \boxed{-2116.6 \ \text{ft-lbf}}$$

This is negative because work is done by the system.

The answer is (D).

SI Solution

The mass of air is

$$m = \frac{p_1 V_1}{RT_1}$$

$$= \frac{(101.3 \ \text{kPa})\left(1000 \ \frac{\text{Pa}}{\text{kPa}}\right)(0.25 \ \text{m}^3)}{\left(287 \ \frac{\text{J}}{\text{kg·K}}\right)(82°\text{C} + 273)}$$

$$= 0.2486 \ \text{kg}$$

For a constant pressure process from Eq. 25.51, on a per unit mass basis,

$$W = R(T_2 - T_1)$$

The total work for m in kg is

$$W = mR(T_2 - T_1)$$

$$= (0.2486 \ \text{kg})\left(287 \ \frac{\text{J}}{\text{kg·K}}\right)(38°\text{C} - 82°\text{C})$$

$$= \boxed{-3139.3 \ \text{J}}$$

The answer is (D).

4. *Customary U.S. Solution*

From App. 24.B, for 300 psia, the enthalpy of saturated liquid, h_f, is 394.0 Btu/lbm. The heat of vaporization, h_{fg}, is 809.4 Btu/lbm. The enthalpy is given by Eq. 24.40.

$$h_1 = h_f + x h_{fg}$$

$$= 394.0 \ \frac{\text{Btu}}{\text{lbm}} + (0.95)\left(809.4 \ \frac{\text{Btu}}{\text{lbm}}\right)$$

$$= 1162.9 \ \text{Btu/lbm}$$

From the Mollier diagram, for an isentropic process from 300 psia to 50 psia, $h_2 = 1031$ Btu/lbm.

The availability is calculated from Eq. 25.164 using an isentropic process $(s_1 = s_2)$ for unit mass.

$$\text{availability} = h_1 - h_2$$

$$= 1162.9 \ \frac{\text{Btu}}{\text{lbm}} - 1031 \ \frac{\text{Btu}}{\text{lbm}}$$

$$= \boxed{131.9 \ \text{Btu/lbm}}$$

The answer is (B).

SI Solution

From App. 24.0, for 2 MPa, the enthalpy of saturated liquid, h_f, is 908.50 kJ/kg. The heat of vaporization, h_{fg}, is 1889.8 kJ/kg. The enthalpy is given by Eq. 24.40.

$$h_1 = h_f + x h_{fg}$$

$$= 908.50 \ \frac{\text{kJ}}{\text{kg}} + (0.95)\left(1889.8 \ \frac{\text{kJ}}{\text{kg}}\right)$$

$$= 2703.8 \ \text{kJ/kg}$$

From the Mollier diagram, for an isentropic process from 2 MPa to 0.35 MPa, $h_2 = 2405$ kJ/kg.

The availability is calculated from Eq. 25.164 using an isentropic process $(s_1 = s_2)$ for unit mass.

$$\text{availability} = h_1 - h_2$$

$$= 2703.8 \ \frac{\text{kJ}}{\text{kg}} - 2405 \ \frac{\text{kJ}}{\text{kg}}$$

$$= \boxed{298.8 \ \text{kJ/kg}}$$

The answer is (B).

Thermodynamics

5. *Customary U.S. Solution*

The absolute temperature at the inlet of the air heater is

$$T_1 = 540°F + 460 = 1000°R$$

The absolute temperature at the outlet of the air heater is

$$T_2 = 1540°F + 460 = 2000°R$$

Since pressures are low and temperatures are high, use an air table.

From App. 24.F at 1000°R,

$$h_1 = 240.98 \text{ Btu/lbm}$$

$$\phi_1 = 0.75042 \text{ Btu/lbm-°R}$$

From App. 24.F at 2000°R,

$$h_2 = 504.71 \text{ Btu/lbm}$$

$$\phi_2 = 0.93205 \text{ Btu/lbm-°R}$$

The availability per unit mass is calculated from Eq. 25.164 using $T_L = 100°F + 460 = 560°R$.

$$W_{\max} = h_1 - h_2 + T_L(s_2 - s_1)$$

For no pressure drop,

$$s_2 - s_1 = \phi_2 - \phi_1$$

$$W_{\max} = h_1 - h_2 + T_L(\phi_2 - \phi_1)$$

$$= 240.98 \frac{\text{Btu}}{\text{lbm}} - 504.71 \frac{\text{Btu}}{\text{lbm}}$$

$$+ (560°R)\left(0.93205 \frac{\text{Btu}}{\text{lbm-°R}}\right.$$

$$\left. -0.75042 \frac{\text{Btu}}{\text{lbm-°R}}\right)$$

$$= -162.02 \text{ Btu/lbm}$$

With a pressure drop from 100 psia to 80 psia, from Eq. 24.39,

$$s_2 - s_1 = \phi_2 - \phi_1 - \left(\frac{R}{J}\right) \ln\left(\frac{p_2}{p_1}\right)$$

$$W_{\max,p \text{ loss}} = h_1 - h_2 + T_L\left(\phi_2 - \phi_1\right.$$

$$\left. -\left(\frac{R}{J}\right) \ln\left(\frac{p_2}{p_1}\right)\right)$$

$$= 240.98 \frac{\text{Btu}}{\text{lbm}} - 504.71 \frac{\text{Btu}}{\text{lbm}} + (560°R)$$

$$\times \left(0.93205 \frac{\text{Btu}}{\text{lbm-°R}} - 0.75042 \frac{\text{Btu}}{\text{lbm-°R}}\right.$$

$$\left. -\left(\frac{53.3 \frac{\text{ft-lbf}}{\text{lbm-°R}}}{778 \frac{\text{ft-lbf}}{\text{Btu}}}\right) \ln\left(\frac{80 \text{ psia}}{100 \text{ psia}}\right)\right)$$

$$= -153.46 \text{ Btu/lbm}$$

The percentage loss in available energy is

$$\frac{W_{\max} - W_{\max,p \text{ loss}}}{W_{\max}} \times 100\%$$

$$= \frac{-162.02 \frac{\text{Btu}}{\text{lbm}} - \left(-153.46 \frac{\text{Btu}}{\text{lbm}}\right)}{-162.02 \frac{\text{Btu}}{\text{lbm}}} \times 100\%$$

$$= \boxed{5.28\%}$$

The answer is (A).

SI Solution

The absolute temerature at the inlet of the air heater is

$$T_1 = 280°C + 273 = 553K$$

The absolute temperature at the outlet of the air heater is

$$T_2 = 840°C + 273 = 1113K$$

Since pressures are low and temperatures are high, use an air table.

From App. 24.S at 553K,

$$h_1 = 557.9 \text{ kJ/kg}$$

$$\phi_1 = 2.32372 \text{ kJ/kg·K}$$

From App. 24.S at 1113K,

$$h_2 = 1176.2 \text{ kJ/kg}$$

$$\phi_2 = 3.09092 \text{ kJ/kg·K}$$

The availability per unit mass is calculated from Eq. 25.164 using $T_L = 40°C + 273 = 313K$.

$$W_{\max} = h_1 - h_2 + T_L(s_2 - s_1)$$

For no pressure drop,

$$s_2 - s_1 = \phi_2 - \phi_1$$

$$W_{\max} = h_1 - h_2 + T_L(\phi_2 - \phi_1)$$

$$= 557.9 \frac{\text{kJ}}{\text{kg}} - 1176.2 \frac{\text{kJ}}{\text{kg}}$$

$$+ (313K)\left(3.09092 \frac{\text{kJ}}{\text{kg·K}} - 2.32372 \frac{\text{kJ}}{\text{kg·K}}\right)$$

$$= -378.17 \text{ kJ/kg}$$

With a pressure drop from 700 kPa to 550 kPa, from Eq. 24.39,

$$s_2 - s_1 = \phi_2 - \phi_1 - R \ln \left(\frac{p_2}{p_1} \right)$$

$$W_{\text{max},p \text{ loss}} = h_1 - h_2$$
$$+ T_L \left(\phi_2 - \phi_1 - R \ln \left(\frac{p_2}{p_1} \right) \right)$$
$$= 557.9 \; \frac{\text{kJ}}{\text{kg}} - 1176.2 \; \frac{\text{kJ}}{\text{kg}} + 313\text{K}$$
$$\times \left(3.09092 \; \frac{\text{kJ}}{\text{kg·K}} - 2.32372 \; \frac{\text{kJ}}{\text{kg·K}} \right.$$
$$\left. - \left(\frac{287 \; \frac{\text{J}}{\text{kg·K}}}{1000 \; \frac{\text{J}}{\text{kJ}}} \right) \ln \left(\frac{550 \; \text{kPa}}{700 \; \text{kPa}} \right) \right)$$
$$= -356.50 \; \text{kJ/kg}$$

The percentage loss in available energy is

$$\frac{W_{\text{max}} - W_{\text{max},p \text{ loss}}}{W_{\text{max}}} \times 100\%$$

$$= \frac{-378.17 \; \frac{\text{kJ}}{\text{kg}} - \left(-356.50 \; \frac{\text{kJ}}{\text{kg}} \right)}{-378.17 \; \frac{\text{kJ}}{\text{kg}}} \times 100\%$$

$$= \boxed{5.73\%}$$

The answer is (A).

6. *Customary U.S. Solution*

(a) Assume the tank is originally at 70°F. The absolute temperature is

$$T = 70°\text{F} + 460 = 530°\text{R}$$

From Table 24.7, $R = 11.77$ ft-lbf/lbm-°R. From Eq. 24.47,

$$m = \frac{pV}{RT}$$
$$= \frac{\left(20 \; \frac{\text{lbf}}{\text{in}^2} \right) \left(144 \; \frac{\text{in}^2}{\text{ft}^2} \right) (100 \; \text{ft}^3)}{\left(11.77 \; \frac{\text{ft-lbf}}{\text{lbm-°R}} \right) (530°\text{R})}$$
$$= \boxed{46.17 \; \text{lbm}}$$

The answer is (C).

(b) From Table 24.4, the critical temperature and pressure of xenon are 521.9°R and 58.2 atm, respectively. The reduced variables are

$$T_r = \frac{T}{T_c} = \frac{530°\text{R}}{521.9°\text{R}} = 1.02$$

$$p_r = \frac{p}{p_c} = \frac{3800 \; \text{psia}}{(58.2 \; \text{atm}) \left(14.7 \; \frac{\text{psia}}{\text{atm}} \right)} = 4.44$$

From App. 24.Z, z is read as 0.61. Using Eq. 24.93,

$$m = \frac{pV}{zRT} = \frac{\left(3800 \; \frac{\text{lbf}}{\text{in}^2} \right) \left(144 \; \frac{\text{in}^2}{\text{ft}^2} \right) (100 \; \text{ft}^3)}{(0.61) \left(11.77 \; \frac{\text{ft-lbf}}{\text{lbm-°R}} \right) (530°\text{R})}$$
$$= 14{,}380 \; \text{lbm}$$

The average mass flow rate of xenon is

$$\dot{m} = \frac{14{,}380 \; \text{lbm} - 46.17 \; \text{lbm}}{1 \; \text{hr}}$$
$$= \boxed{14{,}334 \; \text{lbm/hr}}$$

The answer is (D).

(c) For isothermal compression, the work per unit mass is calculated from Eq. 25.79.

$$W = mRT \ln \left(\frac{p_1}{p_2} \right)$$

$$= \frac{\begin{array}{c} (14{,}334 \; \text{lbm}) \left(11.77 \; \frac{\text{ft-lbf}}{\text{lbm-°R}} \right) \\ \times (530°\text{R}) \ln \left(\frac{20 \; \text{psia}}{3800 \; \text{psia}} \right) \end{array}}{\left(778 \; \frac{\text{ft-lbf}}{\text{Btu}} \right) \left(3413 \; \frac{\text{Btu}}{\text{kW·hr}} \right)}$$
$$= -176.7 \; \text{kW-hr} \quad \text{(for 1 hr)}$$

The cost of electricity is

$$\left(\frac{\$0.045}{\text{kW-hr}} \right) (176.7 \; \text{kW-hr}) = \boxed{\$7.95}$$

The answer is (A).

SI Solution

(a) Assume the tank is originally at 21°C. The absolute temperature is

$$T = 21°\text{C} + 273 = 294\text{K}$$

The answer is (C).

From Table 24.7, $R = 63.32$ J/kg·K. From Eq. 24.47,

$$m = \frac{pV}{RT}$$

$$= \frac{(150 \text{ kPa}) \left(1000 \frac{\text{Pa}}{\text{kPa}}\right) (3 \text{ m}^3)}{\left(63.32 \frac{\text{J}}{\text{kg·K}}\right)(294\text{K})}$$

$$= \boxed{24.17 \text{ kg}}$$

(b) From Table 24.4, the critical temperature and pressure of xenon are 289.9K and 58.2 atm, respectively. The reduced variables are

$$T_r = \frac{T}{T_c} = \frac{294\text{K}}{289.9\text{K}} = 1.01$$

$$p_r = \frac{p}{p_c} = \frac{25 \text{ MPa}}{(58.2 \text{ atm})\left(0.1013 \frac{\text{MPa}}{\text{atm}}\right)} = 4.24$$

From App. 24.Z, z is read as 0.59. Using Eq. 24.93,

$$m = \frac{pV}{zRT}$$

$$= \frac{(25 \text{ MPa})\left(10^6 \frac{\text{Pa}}{\text{MPa}}\right)(3 \text{ m}^3)}{(0.59)\left(63.32 \frac{\text{J}}{\text{kg·K}}\right)(294\text{K})}$$

$$= 6.828 \text{ kg}$$

The average mass flow rate of xenon is

$$\dot{m} = \frac{6828 \text{ kg} - 24.17 \text{ kg}}{(1 \text{ h})\left(3600 \frac{\text{s}}{\text{h}}\right)}$$

$$= \boxed{1.89 \text{ kg/s}}$$

The answer is (D).

(c) For isothermal compression, the work per unit mass is calculated from Eq. 25.79.

$$W = mRT \ln\left(\frac{p_1}{p_2}\right)$$

$$= (6828 \text{ kg} - 24.17 \text{ kg})\left(63.32 \frac{\text{J}}{\text{kg·K}}\right)(294\text{K})$$

$$\times \ln\left(\frac{150 \text{ kPa}}{(25 \text{ MPa})\left(1000 \frac{\text{kPa}}{\text{MPa}}\right)}\right)$$

$$\times \left(\frac{1 \text{ kJ}}{1000 \text{ J}}\right)\left(\frac{1 \text{ h}}{3600 \text{ s}}\right)$$

$$= -180 \text{ kW·h}$$

The cost of electricity is

$$\left(\frac{\$0.045}{\text{kW·h}}\right)(180 \text{ kW·h}) = \boxed{\$8.10}$$

The answer is (A).

7. Choose the control volume to include the air outside the tank that is pushed into the tank (subscript "e" for "entering"), as well as the tank volume.

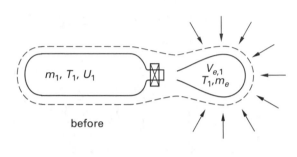

before

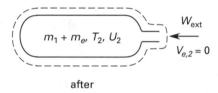

after

Customary U.S. Solution

The absolute temperature of the air in the tank when evacuated is

$$T_1 = 70°\text{F} + 460 = 530°\text{R}$$

From Table 24.7, $R = 53.3$ ft-lbf/lbm-°R. From Eq. 24.47,

$$m = \frac{p_1 V_1}{R T_1}$$

$$= \frac{\left(1 \frac{\text{lbf}}{\text{in}^2}\right)\left(144 \frac{\text{in}^2}{\text{ft}^2}\right)(20 \text{ ft}^3)}{\left(53.3 \frac{\text{ft-lbf}}{\text{lbm-°R}}\right)(530°\text{R})}$$

$$= 0.102 \text{ lbm}$$

Assume $T_2 = 300°\text{F}$. The absolute temperature is

$$T_2 = 300°\text{F} + 460 = 760°\text{R}$$

From Eq. 24.47,

$$m_2 = m_1 + m_e$$

$$= \frac{p_2 V_2}{R T_2}$$

$$= \frac{\left(14.7 \ \frac{\text{lbf}}{\text{in}^2}\right)\left(144 \ \frac{\text{in}^2}{\text{ft}^2}\right)(20 \ \text{ft}^3)}{\left(53.3 \ \frac{\text{ft-lbf}}{\text{lbm-}^\circ\text{R}}\right)(760^\circ\text{R})}$$

$$m_1 + m_e \approx 1.045 \ \text{lbm}$$

$$m_e \approx 1.045 \ \text{lbm} - m_1$$

$$= 1.045 \ \text{lbm} - 0.102 \ \text{lbm}$$

$$= 0.943 \ \text{lbm}$$

From Eq. 24.47, the initial volume of the external air is

$$V_{e,1} = \frac{mRT}{p}$$

$$= \frac{(0.943 \ \text{lbm})\left(53.3 \ \frac{\text{ft-lbf}}{\text{lbm-}^\circ\text{R}}\right)(530^\circ\text{R})}{\left(14.7 \ \frac{\text{lbf}}{\text{in}^2}\right)\left(144 \ \frac{\text{in}^2}{\text{ft}^2}\right)}$$

$$= 12.58 \ \text{ft}^3$$

From Eq. 25.23, for a closed system,

$$Q = \Delta U + W$$

For an adiabatic system, $Q = 0$.

$$W_{\text{ext}} = \Delta U$$

For a constant pressure, closed system, from Eq. 25.50, the total work is

$$W_{\text{ext}} = p(\text{V}_{e,2} - \text{V}_{e,1})$$

$$= \frac{\left(14.7 \ \frac{\text{lbf}}{\text{in}^2}\right)\left(144 \ \frac{\text{in}^2}{\text{ft}^2}\right)(0 - 12.58 \ \text{ft}^3)}{778 \ \frac{\text{ft-lbf}}{\text{Btu}}}$$

$$= -34.23 \ \text{Btu} \quad \begin{bmatrix} \text{surroundings do work} \\ \text{on the system} \end{bmatrix}$$

This energy is used to raise the temperature of the air and tank. Consider air as an ideal gas. The process inside the tank is not constant-pressure, as was the compression of the external air mass. However, the process is adiabatic, so $\Delta U = -W$. From Eq. 25.37, $\Delta U = c_v \Delta T$.

$$W_{\text{ext}} = \left((m_1 + m_e)c_v\right)(T_2 - T_1)$$

$$34.23 \ \text{Btu} = \left((1.045 \ \text{lbm})\left(0.171 \ \frac{\text{Btu}}{\text{lbm-}^\circ\text{F}}\right)\right.$$

$$\left. \times (T_2 - 70^\circ\text{F})\right)$$

$$T_2 = 261.6^\circ\text{F}$$

(A second iteration may be required.)

Now, allow the 261.6°F air and the 70°F tank to reach thermal equilibrium. The process will be constant-pressure.

$$q_{\text{air}} = q_{\text{tank}}$$

$$c_{p,\text{air}} m_{\text{air}}(T_{\text{air}} - T_{\text{equilibrium}})$$

$$= c_{p,\text{tank}} m_{\text{tank}}(T_{\text{equilibrium}} - T_1)$$

$$\left(0.24 \ \frac{\text{Btu}}{\text{lbm-}^\circ\text{F}}\right)(1.045 \ \text{lbm})(261.6^\circ\text{F} - T_{\text{equilibrium}})$$

$$= \left(0.11 \ \frac{\text{Btu}}{\text{lbm-}^\circ\text{F}}\right)(40 \ \text{lbm})(T_{\text{equilibrium}} - 70^\circ\text{F})$$

$$T_{\text{equilibrium}} = \boxed{80.3^\circ\text{F}}$$

The answer is (B).

SI Solution

The absolute temperature of the air in the tank when evacuated is

$$T_1 = 21^\circ\text{C} + 273 = 294\text{K}$$

From Table 24.7, $R = 287 \ \text{J/kg·K}$. From Eq. 24.47,

$$m = \frac{p_1 V_1}{R T_1}$$

$$= \frac{(7 \ \text{kPa})\left(1000 \ \frac{\text{Pa}}{\text{kPa}}\right)(0.6 \ \text{m}^3)}{\left(287 \ \frac{\text{J}}{\text{kg·K}}\right)(294\text{K})}$$

$$= 0.0498 \ \text{kg}$$

Assume $T_2 = 127^\circ\text{C}$. The absolute temperature is

$$T_2 = 127^\circ\text{C} + 273 = 400\text{K}$$

From Eq. 24.47,

$$m_2 = m_1 + m_e = \frac{p_2 V_2}{R T_2}$$

$$= \frac{(101.3 \ \text{kPa})\left(1000 \ \frac{\text{Pa}}{\text{kPa}}\right)(0.6 \ \text{m}^3)}{\left(287 \ \frac{\text{J}}{\text{kg·K}}\right)(400\text{K})}$$

$$= 0.5294 \ \text{kg}$$

$$m_e \approx 0.5294 \ \text{kg} - m_1$$

$$= 0.5294 \ \text{kg} - 0.0498 \ \text{kg}$$

$$= 0.4796 \ \text{kg}$$

From Eq. 24.47, the initial volume of the external air is

$$V_{e,1} = \frac{mRT}{p}$$

$$= \frac{(0.4796 \ \text{kg})\left(287 \ \frac{\text{J}}{\text{kg·K}}\right)(294\text{K})}{(101.3 \ \text{kPa})\left(1000 \ \frac{\text{Pa}}{\text{kPa}}\right)}$$

$$= 0.3995 \ \text{m}^3$$

From Eq. 25.23, for a closed system,

$$Q = \Delta U + W$$

For an adiabatic system, $Q = 0$.

$$W_{\text{ext}} = \Delta U$$

For a constant pressure, closed system, from Eq. 25.50, the total work is

$$W_{\text{ext}} = p(V_{e,2} - V_{e,1})$$

$$= (101.3 \text{ kPa}) \left(1000 \, \frac{\text{Pa}}{\text{kPa}} \right) (0 - 0.3995 \text{ m}^3)$$

$$= -40\,469 \text{ J} \quad \begin{bmatrix} \text{surroundings do work} \\ \text{on the system} \end{bmatrix}$$

This energy is used to raise the temperature of the air and tank. Consider air as an ideal gas. The process inside the tank is not constant-pressure, as was the compression of the external air mass. However, the process is adiabatic, so $\Delta U = -W$. From Eq. 25.37, $\Delta U = c_{\text{v}} \Delta T$.

$$W_{\text{ext}} = \left((m_1 + m_e)c_{\text{v}} \right)(T_2 - T_1)$$

$$40\,469 \text{ J} = \left((0.5294 \text{ kg}) \left(718 \, \frac{\text{J}}{\text{kg·K}} \right) (T_2 - 21°\text{C}) \right)$$

$$T_2 = 127.5°\text{C}$$

This is close enough to the assumed value of T_2 that a second iteration is not necessary.

Now, allow the 127.5°C air and the 21°C tank to reach thermal equilibrium. The process will be constant-pressure.

$$q_{\text{air}} = q_{\text{tank}}$$

$$c_{p,\text{air}} m_{\text{air}}(T_{\text{air}} - T_{\text{equilibrium}})$$

$$= c_{p,\text{tank}} m_{\text{tank}}(T_{\text{equilibrium}} - T_1)$$

$$\left(1005 \, \frac{\text{J}}{\text{kg·K}} \right) (0.5294 \text{ kg})(127.5°\text{C} - T_{\text{equilibrium}})$$

$$= \left(460 \, \frac{\text{J}}{\text{kg·K}} \right) (20 \text{ kg})(T_{\text{equilibrium}} - 21°\text{C})$$

$$T_{\text{equilibrium}} = \boxed{26.8°\text{C}}$$

The answer is (B).

26 Compressible Fluid Dynamics

PRACTICE PROBLEMS

1. 150°F, 10 psia (65°C, 70 kPa) air flows at 750 ft/sec (225 m/s) through a converging section into a chamber whose back pressure is 5.5 psia (40 kPa).

(a) What is the entrance Mach number?
- (A) 0.5
- (B) 0.6
- (C) 0.7
- (D) 0.8

(b) What is the throat static pressure?
- (A) 6.9 psia (48 kPa)
- (B) 10 psia (69 kPa)
- (C) 14 psia (98 kPa)
- (D) 19 psia (130 kPa)

(c) What is the throat static temperature?
- (A) 450°R (250K)
- (B) 500°R (280K)
- (C) 550°R (300K)
- (D) 600°R (330K)

2. A round-nosed bullet travels at 2000 ft/sec (600 m/s) through 32°F, 14.7 psia (0°C, 101.3 kPa) air.

(a) Is the bullet speed subsonic or supersonic?
- (A) subsonic (M = 0.9)
- (B) supersonic (M = 1.1)
- (C) supersonic (M = 1.4)
- (D) supersonic (M = 1.8)

(b) What is the stagnation value of pressure?
- (A) 50 psia (350 kPa)
- (B) 70 psia (480 kPa)
- (C) 90 psia (620 kPa)
- (D) 110 psia (750 kPa)

(c) What is the stagnation value of temperature?
- (A) 770°R (430K)
- (B) 800°R (440K)
- (C) 830°R (450K)
- (D) 860°R (480K)

(d) What is the stagnation value of enthalpy at the bullet face?
- (A) 200 Btu/lbm (450 kJ/kg)
- (B) 250 Btu/lbm (560 kJ/kg)
- (C) 300 Btu/lbm (680 kJ/kg)
- (D) 350 Btu/lbm (790 kJ/kg)

3. 4.5 lbm/sec (2 kg/s) of air with total properties of 160 psia and 240°F (1.1 MPa and 120°C) expands through a converging-diverging nozzle to 20 psia (140 kPa).

(a) What is the throat Mach number?
- (A) 0.7
- (B) 0.8
- (C) 0.9
- (D) 1.0

(b) What is the throat area?
- (A) 0.005 ft² (0.0005 m²)
- (B) 0.01 ft² (0.0009 m²)
- (C) 0.02 ft² (0.002 m²)
- (D) 0.04 ft² (0.004 m²)

(c) What is the exit Mach number?
- (A) 1.4
- (B) 1.7
- (C) 2.0
- (D) 2.6

(d) What is the exit area?
- (A) 0.017 ft² (0.0015 m²)
- (B) 0.022 ft² (0.0019 m²)
- (C) 0.038 ft² (0.0033 m²)
- (D) 0.062 ft² (0.0054 m²)

4. A wedge-shaped leading edge with a semivertex angle of 20° travels at 2700 ft/sec (800 m/s) through 60°F, 14.7 psia (16°C, 101.3 kPa) air. What are the semivertex shock angles?
- (A) 26° (weak); 50° (strong)
- (B) 33° (weak); 60° (strong)
- (C) 39° (weak); 70° (strong)
- (D) 46° (weak); 80° (strong)

5. A large tank contains 100 psia, 80°F (0.7 MPa, 27°C) air. The tank feeds a converging-diverging nozzle with a throat area of 1 in² (6.45 × 10⁻⁴ m²). At a particular point in the nozzle, the Mach number is 2.

(a) What is the temperature at that point?
- (A) 250°R (140K)
- (B) 300°R (170K)
- (C) 350°R (190K)
- (D) 400°R (220K)

(b) What is the area at that point?
- (A) 1.2 in² (0.0008 m²)
- (B) 1.7 in² (0.0011 m²)
- (C) 2.3 in² (0.0015 m²)
- (D) 2.8 in² (0.0018 m²)

(c) What is the mass flow rate at that point?
- (A) 2.3 lbm/sec (1.1 kg/s)
- (B) 2.7 lbm/sec (1.2 kg/s)
- (C) 3.1 lbm/sec (1.4 kg/s)
- (D) 3.6 lbm/sec (1.6 kg/s)

6. The Mach number before a shock wave is 2. The temperature before the shock wave is 500°R (280K). What is the air velocity behind the shock wave?
- (A) 730 ft/sec (220 m/s)
- (B) 820 ft/sec (250 m/s)
- (C) 900 ft/sec (270 m/s)
- (D) 940 ft/sec (280 m/s)

7. Air with a total pressure of 100 psia (0.7 MPa) and a total temperature of 70°F (21°C) flows through a converging-diverging nozzle. At a particular point, the ratio of flow area to the critical throat area is 1.555.

(a) What is the Mach number at that point?
- (A) 1.1
- (B) 1.3
- (C) 1.9
- (D) 2.4

(b) What is the temperature at that point?
- (A) 290R (160K)
- (B) 310R (170K)
- (C) 340R (190K)
- (D) 360R (200K)

(c) What is the pressure at that point?
- (A) 15 psia (0.10 MPa)
- (B) 20 psia (0.14 MPa)
- (C) 25 psia (0.18 MPa)
- (D) 30 psia (0.21 MPa)

8. 20 lbm/sec (9 kg/s) of air with a static pressure of 10 psia (70 kPa) and a total temperature of 40°F (4°C) flow through a 1 ft² (0.09 m²) round duct. What is the smallest area to which the duct can be reduced?
- (A) 0.26 ft² (0.023 m²)
- (B) 0.54 ft² (0.049 m²)
- (C) 1.0 ft² (0.091 m²)
- (D) 1.3 ft² (0.17 m²)

9. The static pressure in a supersonic wind tunnel is 1.38 psia (9.51 kPa). A pitot tube records a total pressure of 20 psia (140 kPa) behind the shock wave that forms at its entrance. What is the Mach number in the tunnel?
- (A) 1.9
- (B) 2.2
- (C) 2.7
- (D) 3.3

10. A boiler produces 100 psia, 800°F (0.7 MPa, 425°C) steam. An isentropic nozzle expands the steam to 60 psia (400 kPa). What is the steam velocity?
- (A) 1100 ft/sec (330 m/s)
- (B) 1500 ft/sec (450 m/s)
- (C) 1800 ft/sec (580 m/s)
- (D) 2400 ft/sec (720 m/s)

11. Air flows through a converging section. At a particular point where the flow area is 0.1 ft² (0.0009 m²), the static pressure is 50 psia (350 kPa), the static temperature is 1000°R (560K), and the velocity is 600 ft/sec (180 m/s).

(a) What is the Mach number?
- (A) 0.39
- (B) 0.46
- (C) 0.66
- (D) 0.78

(b) What is the total temperature?
- (A) 700°R (380K)
- (B) 800°R (470K)
- (C) 900°R (500K)
- (D) 1000°R (580K)

(c) What is the total pressure?
- (A) 53 psia (370 kPa)
- (B) 56 psia (390 kPa)
- (C) 65 psia (450 kPa)
- (D) 82 psia (570 kPa)

(d) What is the critical exit area?
- (A) 0.062 ft² (0.00054 m²)
- (B) 0.080 ft² (0.00072 m²)
- (C) 0.093 ft² (0.00085 m²)
- (D) 0.110 ft² (0.0010 m²)

(e) What is the critical pressure?
  (A) 23 psia (160 kPa)
  (B) 29 psia (200 kPa)
  (C) 38 psia (260 kPa)
  (D) 63 psia (440 kPa)

(f) What is the critical temperature?
  (A) 790°R (440K)
  (B) 860°R (480K)
  (C) 910°R (510K)
  (D) 980°R (540K)

12. 3 lbm/sec (1.4 kg/s) of steam ($k = 1.31$) enters an 85% efficient transonic nozzle at 300 ft/sec (90 m/s). The steam expands from 200 psia and 600°F (1.5 MPa and 300°C) to 80 psia (500 kPa). What is the throat area?
  (A) 0.0067 ft^2 (0.00060 m^2)
  (B) 0.0081 ft^2 (0.00072 m^2)
  (C) 0.010 ft^2 (0.00090 m^2)
  (D) 0.013 ft^2 (0.0012 m^2)

13. (*Time limit: one hour*) 35,200 lbm/hr (4.4 kg/s) of steam flows ($k = 1.29$) through 95 ft (30 m) of 3 in (7.6 cm) inside diameter pipe. The pipe is insulated to prevent heat loss. The Fanning friction factor is 0.012. The average steam pressure along the pipe length is 200 psia (1.5 MPa). At the exit, the steam is at 115 psia and 540°F (0.7 MPa and 300°C). 30 ft (9 m) of pipe are added to the end of the pipe. Will the steam flow be maintained?

14. (*Time limit: one hour*) 3600 lbm/hr (0.45 kg/s) of air flow through an adiabatic converging-diverging nozzle. The air enters with total properties of 160 psia and 660°R (1.1 MPa and 370K). The exit pressure is 14.7 psia (101.3 kPa). The ratio of specific heats for the air is 1.4 throughout the nozzle. The nozzle has a coefficient of discharge of 0.90. The total diverging angle is 6°. The length of the converging section is 5% of the diverging section.

(a) Draw and dimension the nozzle.

(b) What is the throat area?

SOLUTIONS

1. *Customary U.S. Solution*

(a)

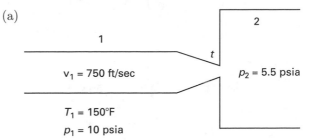

$$T_1 = 150°F + 460$$
$$= 610°R$$

The entrance Mach number is given by Eq. 26.4(b) as

$$M_1 = \frac{v_1}{a_1} = \frac{v_1}{\sqrt{\frac{kg_c R^* T_1}{(MW)}}}$$

$$= \frac{750 \; \frac{ft}{sec}}{\sqrt{\frac{(1.4)\left(32.2 \; \frac{lbm\text{-}ft}{lbf\text{-}sec^2}\right) \times \left(1545 \; \frac{ft\text{-}lbf}{lbmol\text{-}°R}\right)(610°R)}{29.0 \; \frac{lbm}{lbmol}}}}$$

$$= \boxed{0.62}$$

The answer is (B).

(b) From the M = 0.62 line in App. 26.A, the following factors can be read.

$$\frac{T_1}{T_0} = 0.9286$$

$$\frac{p_1}{p_0} = 0.7716$$

$$T_0 = \frac{T_1}{0.9286} = \frac{610°R}{0.9286} = 656.9°R$$

$$p_0 = \frac{p_1}{0.7716} = \frac{10 \; psia}{0.7716} = 12.96 \; psia$$

$$\frac{p_2}{p_0} = \frac{5.5 \; psia}{12.96 \; psia} = 0.424$$

The critical pressure ratio for air is 0.5283. Since $p_2/p_0 < 0.5283$, the flow is choked and the Mach number at the throat is $M_t = 1$. From App. 26.A, for M = 1,

$$\frac{p}{p_0} = 0.5283$$

$$\frac{T}{T_0} = 0.8333$$

$$p = (0.5283)p_0 = (0.5283)(12.96 \text{ psia})$$

$$= \boxed{6.85 \text{ psia}}$$

The answer is (A).

(c) $T = (0.8333)T_0 = (0.8333)(656.9°\text{R})$

$$= \boxed{547.4°\text{R}}$$

The answer is (C).

SI Solution

(a)

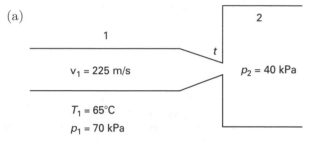

$$T_1 = 65°\text{C} + 273 = 338\text{K}$$

The entrance Mach number is given by Eq. 26.4(a) as

$$M_1 = \frac{v_1}{\sqrt{\dfrac{kR^*T}{(\text{MW})}}}$$

$$= \frac{225 \ \dfrac{\text{m}}{\text{s}}}{\sqrt{\dfrac{(1.4)\left(8314 \ \dfrac{\text{J}}{\text{kmol·k}}\right)(338\text{K})}{29.0 \ \dfrac{\text{kg}}{\text{kmol}}}}}$$

$$= \boxed{0.61}$$

The answer is (B).

(b) From the M = 0.61 line in App. 26.A, the following factors can be read.

$$\frac{T_1}{T_0} = 0.9307$$

$$\frac{p_1}{p_0} = 0.7778$$

$$T_0 = \frac{T_1}{0.9307} = \frac{338.2\text{K}}{0.9307}$$

$$= 363.4\text{K}$$

$$p_0 = \frac{p_1}{0.7778} = \frac{70 \text{ kPa}}{0.7778}$$

$$= 90.0 \text{ kPa}$$

$$\frac{p_2}{p_0} = \frac{40 \text{ kPa}}{90.0 \text{ kPa}} = 0.444$$

The critical pressure ratio for air is 0.5283. Since $p_2/p_0 < 0.5283$, the flow is choked and the Mach number at the throat is $M_t = 1$. From App. 26.A, for M = 1,

$$\frac{p}{p_0} = 0.5283$$

$$\frac{T}{T_0} = 0.8333$$

$$p = (0.5283)p_0$$

$$= (0.5283)(90.0 \text{ kPa})$$

$$= \boxed{47.5 \text{ kPa}}$$

The answer is (A).

(c) $T = (0.8333)T_0$

$$= (0.8333)(363.4\text{K})$$

$$= \boxed{302.8\text{K}}$$

The answer is (C).

2. *Customary U.S. Solution*

The shock wave is normal to the direction of flight, so normal shock tables can be used.

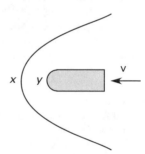

$$T_x = 32°\text{F} + 460$$

$$= 492°\text{R}$$

$$p_x = 14.7 \text{ psia}$$

(a) Using Eq. 26.4(b), the Mach number at y is

$$M_y = \frac{v}{\sqrt{\dfrac{kg_cR^*T}{(MW)}}}$$

$$= \frac{2000 \ \dfrac{\text{ft}}{\text{sec}}}{\sqrt{\dfrac{(1.4)\left(32.2 \ \dfrac{\text{lbm-ft}}{\text{lbf-sec}^2}\right)\left(1545 \ \dfrac{\text{ft-lbf}}{\text{lbmol-°R}}\right)(492°R)}{29.0 \ \dfrac{\text{lbm}}{\text{lbmol}}}}}$$

$$= 1.84 \ \boxed{[\text{supersonic}]}$$

The answer is (D).

(b) Assume that the bullet is stationary and air is moving at M = 1.84.

The ratio of static pressure before the shock to the total pressure after the shock is read from the normal shock table (App. 26.B) for $M_x = 1.84$ (interpolation).

$$\frac{p_x}{p_{0,y}} = 0.2060$$

Therefore, the stagnation (i.e., total) pressure is

$$p_{0,y} = \frac{p_x}{0.2060} = \frac{14.7 \text{ psia}}{0.2060}$$

$$= \boxed{71.36 \text{ psia}}$$

The answer is (B).

(c) The ratio of static temperature before the shock to the total temperature after the shock is read from the normal shock table (App. 26.B) for $M_x = 1.84$ (interpolation).

$$\frac{T_x}{T_0} = 0.5963$$

Therefore, the stagnation (i.e., total) temperature is

$$T_0 = \frac{T_x}{0.5963} = \frac{492°R}{0.5963}$$

$$= \boxed{825.1°R}$$

Since the shock wave is adiabatic, T_0 remains constant.

The answer is (C).

(d) The enthalpy at the bullet face from the air tables is $h \approx \boxed{197.9 \text{ Btu/lbm.}}$

The answer is (A).

SI Solution

The shock wave is normal to the direction of flight, so normal shock tables can be used.

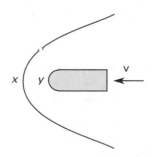

$$T_x = 0°C + 273 = 273K$$

$$p_x = 101.3 \text{ kPa}$$

(a) Using Eq. 26.4(a), the Mach number at y is

$$M_y = \frac{v}{\sqrt{\dfrac{kR^*T}{(MW)}}}$$

$$= \frac{600 \ \dfrac{m}{s}}{\sqrt{\dfrac{(1.4)\left(8314 \ \dfrac{J}{kmol \cdot K}\right)(273K)}{29.0 \ \dfrac{kg}{kmol}}}}$$

$$= 1.81 \ \boxed{[\text{supersonic}]}$$

The answer is (D).

(b) Assume that the bullet is stationary and air is moving at M = 1.81.

The ratio of static pressure before the shock to the total pressure after the shock is read from the normal shock table (App. 26.B) for $M_x = 1.81$ (interpolation).

$$\frac{p_x}{p_{0,y}} = 0.2122$$

Therefore, the stagnation (i.e., total) pressure is

$$p_{0,y} = \frac{p_x}{0.2122} = \frac{101.3 \text{ kPa}}{0.2122} = \boxed{477.4 \text{ kPa}}$$

The answer is (B).

(c) The ratio of static temperature before the shock to the total temperature after the shock is read from the normal shock table (App. 26.B) for $M_x = 1.81$ (interpolation).

$$\frac{T_x}{T_0} - 0.6042$$

Therefore, the stagnation temperature is

$$T_0 = \frac{T_x}{0.6042} = \frac{273K}{0.6042} = \boxed{451.8K}$$

Since the shock wave is adiabatic, T_0 remains constant.

The answer is (C).

(d) Enthalpy at the bullet face from air tables is $h \approx \boxed{454.0 \text{ kJ/kg.}}$

The answer is (A).

3. *Customary U.S. Solution*

(a) The absolute total temperature is

$$T_0 = 240°F + 460 = 700°R$$

The total density of the air is

$$\rho_0 = \frac{p_0}{RT_0}$$

$$= \frac{\left(160 \, \frac{\text{lbf}}{\text{in}^2}\right)\left(144 \, \frac{\text{in}^2}{\text{ft}^2}\right)}{\left(53.3 \, \frac{\text{ft-lbf}}{\text{lbm-}^\circ\text{R}}\right)(700^\circ\text{R})}$$

$$= 0.6175 \, \text{lbm/ft}^3$$

$$\frac{p_{\text{back}}}{p_0} = \frac{20 \, \text{psia}}{160 \, \text{psia}}$$

$$= 0.125$$

Since $p_{\text{back}}/p_0 < 0.5283$, the nozzle is supersonic and the throat flow is sonic. So, $\boxed{\text{M} = 1}$ at throat.

The answer is (D).

(b) Read the property ratios for Mach 1 from the isentropic flow table: $[T/T_0] = 0.8333$ and $[\rho/\rho_0] = 0.6339$. The sonic properties at the throat are

$$T^* = \left[\frac{T}{T_0}\right] T_0 = (0.8333)(700^\circ\text{R})$$

$$= 583.3^\circ\text{R}$$

$$\rho^* = \left[\frac{\rho}{\rho_0}\right] \rho_0 = (0.6339)\left(0.6175 \, \frac{\text{lbm}}{\text{ft}^3}\right)$$

$$= 0.3914 \, \text{lbm/ft}^3$$

The sonic velocity at the throat is

$$a^* = \sqrt{kg_cRT^*}$$

$$= \sqrt{\begin{array}{c}(1.4)\left(32.2 \, \frac{\text{ft-lbm}}{\text{lbf-sec}^2}\right) \\ \times \left(53.3 \, \frac{\text{ft-lbf}}{\text{lbm-}^\circ\text{R}}\right)(583.3^\circ\text{R})\end{array}}$$

$$= 1183.9 \, \text{ft/sec}$$

The throat area is

$$A^* = \frac{\dot{m}}{\rho^* a^*}$$

$$= \frac{4.5 \, \frac{\text{lbm}}{\text{sec}}}{\left(0.3914 \, \frac{\text{lbm}}{\text{ft}^3}\right)\left(1183.9 \, \frac{\text{ft}}{\text{sec}}\right)}$$

$$= \boxed{0.00971 \, \text{ft}^2}$$

The answer is (B).

(c) At the exit,

$$\frac{p_e}{p_0} = \frac{20 \, \text{psia}}{160 \, \text{psia}} = 0.125$$

Searching the $[p/p_0]$ column of the isentropic flow tables for this value gives $\boxed{\text{M} \approx 2.01.}$

The answer is (C).

(d) The corresponding ratio is $[A/A^*] = 1.7024$.

$$A_e = \left[\frac{A_e}{A^*}\right] A^*$$

$$= (1.7024)(0.00971 \, \text{ft}^2)$$

$$= \boxed{0.01653 \, \text{ft}^2}$$

The answer is (A).

SI Solution

(a) The absolute total temperature is

$$T_0 = 120^\circ\text{C} + 273 = 393\text{K}$$

The total density of the air is

$$\rho_0 = \frac{p_0}{RT_0}$$

$$= \frac{(1.1 \, \text{MPa})\left(10^6 \, \frac{\text{Pa}}{\text{MPa}}\right)}{\left(287 \, \frac{\text{J}}{\text{kg·K}}\right)(393\text{K})}$$

$$= 9.753 \, \text{kg/m}^3$$

$$\frac{p_{\text{back}}}{p_0} = \frac{140 \, \text{kPa}}{(1.1 \, \text{MPa})\left(10^3 \, \frac{\text{kPa}}{\text{MPa}}\right)} = 0.127$$

Since $p_{\text{back}}/p_0 < 0.5283$, the nozzle is supersonic and the throat flow is sonic. So, $\boxed{\text{M} = 1}$ at throat.

The answer is (D).

(b) Read the property ratios for Mach 1 from the isentropic flow table: $[T/T_0] = 0.8333$ and $[\rho/\rho_0] = 0.6339$. The sonic properties at the throat are

$$T^* = \left[\frac{T}{T_0}\right] T_0 = (0.8333)(393\text{K}) = 327.5\text{K}$$

$$\rho^* = \left[\frac{\rho}{\rho_0}\right] \rho_0 = (0.6339)\left(9.753 \, \frac{\text{kg}}{\text{m}^3}\right)$$

$$= 6.182 \, \text{kg/m}^3$$

The sonic velocity at the throat is

$$a^* = \sqrt{kRT^*}$$

$$= \sqrt{(1.4)\left(287 \ \frac{\text{J}}{\text{kg·K}}\right)(327.5\text{K})}$$

$$= 362.8 \ \text{m/sec}$$

The throat area is

$$A^* = \frac{\dot{m}}{\rho^* a^*}$$

$$= \frac{2 \ \dfrac{\text{kg}}{\text{s}}}{\left(6.182 \ \dfrac{\text{kg}}{\text{m}^3}\right)\left(362.8 \ \dfrac{\text{m}}{\text{s}}\right)}$$

$$= \boxed{8.92 \times 10^{-4} \ \text{m}^2}$$

The answer is (B).

(c) At the exit,

$$\frac{p_e}{p_0} = \frac{140 \ \text{kPa}}{(1.1 \ \text{MPa})\left(10^3 \ \dfrac{\text{kPa}}{\text{MPa}}\right)} = 0.127$$

Searching the $[p/p_0]$ column of the isentropic flow tables for this value gives $\boxed{\text{M} \approx 2.00.}$

The answer is (C).

(d) The corresponding ratio is $[A/A^*] = 1.6875$.

$$A_e = \left[\frac{A_e}{A^*}\right] A^*$$

$$= (1.6875)(8.92 \times 10^{-4} \ \text{m}^2)$$

$$= \boxed{1.505 \times 10^{-3} \ \text{m}^2}$$

The answer is (A).

4. *Customary U.S. Solution*

(This solution is for a wedge with a 20° semivertex angle.)

The absolute total temperature is

$$T_0 = 60°\text{F} + 460 = 520°\text{R}$$

The sonic velocity is

$$a = \sqrt{kg_c RT_0}$$

$$= \sqrt{\begin{array}{c}(1.4)\left(32.2 \ \dfrac{\text{ft-lbm}}{\text{lbf-sec}^2}\right) \\ \times \left(53.3 \ \dfrac{\text{ft-lbf}}{\text{lbm-°R}}\right)(520°\text{R})\end{array}}$$

$$= 1117.8 \ \text{ft/sec}$$

The Mach number is

$$\text{M} = \frac{\text{v}}{a} = \frac{2700 \ \dfrac{\text{ft}}{\text{sec}}}{1117.8 \ \dfrac{\text{ft}}{\text{sec}}} = 2.42$$

From Fig. 26.3, $\boxed{\theta \approx 46° \ \text{(weak) and } 80° \ \text{(strong).}}$

The answer is (D).

SI Solution

(This solution is for a wedge with a 20° semivertex angle.)

The absolute total temperature is

$$T_0 = 16°\text{C} + 273 = 289\text{K}$$

The sonic velocity is

$$a = \sqrt{kRT_0}$$

$$= \sqrt{(1.4)\left(287 \ \frac{\text{J}}{\text{kg·K}}\right)(289\text{K})}$$

$$= 340.8 \ \text{m/s}$$

The Mach number is

$$\text{M} = \frac{\text{v}}{a} = \frac{800 \ \dfrac{\text{m}}{\text{s}}}{340.8 \ \dfrac{\text{m}}{\text{s}}} = 2.35$$

From Fig. 26.3, $\boxed{\theta \approx 46° \ \text{(weak) and } 80° \ \text{(strong).}}$

The answer is (D).

5. *Customary U.S. Solution*

(a) The absolute total temperature is

$$T_0 = 80°\text{F} + 460 = 540°\text{R}$$

The total pressure is $p_0 = 100$ psia.

From the isentropic flow tables at $\text{M} = 2$, $[p/p_0] = 0.1278$, $[T/T_0] = 0.5556$, and $[A/A^*] = 1.6875$.

The properties at Mach 2 are

$$T = \left[\frac{T}{T_0}\right] T_0 = (0.5556)(540°\text{R}) = \boxed{300°\text{R}}$$

$$p = \left[\frac{p}{p_0}\right] p_0 = (0.1278)(100 \ \text{psia}) = 12.78 \ \text{psia}$$

The answer is (B).

(b) $A = \left[\dfrac{A}{A^*}\right] A^* = (1.6875)(1 \ \text{in}^2) = \boxed{1.6875 \ \text{in}^2}$

The answer is (B).

(c) $a = \sqrt{kg_c RT}$

$$= \sqrt{\begin{array}{c}(1.4)\left(32.2\ \dfrac{\text{ft-lbm}}{\text{lbf-sec}^2}\right)\\[2mm] \times \left(53.3\ \dfrac{\text{ft-lbf}}{\text{lbm-}°\text{R}}\right)(300°\text{R})\end{array}}$$

$= 849\ \text{ft/sec}$

$\text{v} = \text{M}a = (2)\left(849\ \dfrac{\text{ft}}{\text{sec}}\right)$

$= 1698\ \text{ft/sec}$

$\rho = \dfrac{p}{RT}$

$= \dfrac{\left(12.78\ \dfrac{\text{lbf}}{\text{in}^2}\right)\left(144\ \dfrac{\text{in}^2}{\text{ft}^2}\right)}{\left(53.3\ \dfrac{\text{ft-lbf}}{\text{lbm-}°\text{R}}\right)(300°\text{R})}$

$= 0.1151\ \text{lbm/ft}^3$

$\dot{m} = \rho A \text{v}$

$= \left(0.1151\ \dfrac{\text{lbm}}{\text{ft}^3}\right)(1.6875\ \text{in}^2)\left(\dfrac{1\ \text{ft}^2}{144\ \text{in}^2}\right)$

$\times \left(1698\ \dfrac{\text{ft}}{\text{sec}}\right)$

$= \boxed{2.29\ \text{lbm/sec}}$

The answer is (A).

SI Solution

(a) The absolute total temperature is

$$T_0 = 27°\text{C} + 273 = 300\text{K}$$

The total pressure is

$$p_0 = (0.7\ \text{MPa})\left(10^6\ \dfrac{\text{Pa}}{\text{MPa}}\right) = 7 \times 10^5\ \text{Pa}$$

From the isentropic flow tables at M = 2, $[p/p_0] = 0.1278$, $[T/T_0] = 0.5556$, and $[A/A^*] = 1.6875$.

The properties at Mach 2 are

$$T = \left[\dfrac{T}{T_0}\right]T_0 = (0.5556)(300\text{K}) = \boxed{166.7\text{K}}$$

$$p = \left[\dfrac{p}{p_0}\right]p_0 = (0.1278)(7 \times 10^5\ \text{Pa}) = 89\,460\ \text{Pa}$$

The answer is (B).

(b) $A = \left[\dfrac{A}{A^*}\right]A^* = (1.6875)(6.45 \times 10^{-4}\ \text{m}^2)$

$= \boxed{0.00109\ \text{m}^2}$

The answer is (B).

(c) $a = \sqrt{kRT}$

$= \sqrt{(1.4)\left(287\ \dfrac{\text{J}}{\text{kg·K}}\right)(166.7\text{K})}$

$= 258.8\ \text{m/s}$

$\text{v} = \text{M}a = (2)\left(258.8\ \dfrac{\text{m}}{\text{s}}\right)$

$= 517.6\ \text{m/s}$

$\rho = \dfrac{p}{RT}$

$= \dfrac{89\,460\ \text{Pa}}{\left(287\ \dfrac{\text{J}}{\text{kg·K}}\right)(166.7\text{K})}$

$= 1.870\ \text{kg/m}^3$

$\dot{m} = \rho A \text{v}$

$= \left(1.870\ \dfrac{\text{kg}}{\text{m}^3}\right)(0.00109\ \text{m}^2)\left(517.6\ \dfrac{\text{m}}{\text{s}}\right)$

$= \boxed{1.055\ \text{kg/s}}$

The answer is (A).

6. *Customary U.S. Solution*

From the normal shock table at $\text{M}_x = 2$,

$$\text{M}_y = 0.5774$$

$$\dfrac{T_y}{T_x} = 1.687$$

The properties behind the shock wave are

$$T_y = \left[\dfrac{T_y}{T_x}\right]T_x$$

$$= (1.687)(500°\text{R})$$

$$= 843.5°\text{R}$$

$a_y = \sqrt{kg_c RT_y}$

$$= \sqrt{\begin{array}{c}(1.4)\left(32.2\ \dfrac{\text{ft-lbm}}{\text{lbf-sec}^2}\right)\\[2mm] \times \left(53.3\ \dfrac{\text{ft-lbf}}{\text{lbm-}°\text{R}}\right)(843.5°\text{R})\end{array}}$$

$= 1423.6\ \text{ft/sec}$

$\text{v}_y = \text{M}_y a_y$

$= (0.5774)\left(1423.6\ \dfrac{\text{ft}}{\text{sec}}\right)$

$= \boxed{822.0\ \text{ft/sec}}$

The answer is (B).

SI Solution

From the normal shock table at $M_x = 2$,

$$M_y = 0.5774$$

$$\frac{T_y}{T_x} = 1.687$$

The properties behind the shock wave are

$$T_y = \left(\frac{T_y}{T_x}\right) T_x = (1.687)(280\text{K}) = 472.4\text{K}$$

$$a_y = \sqrt{kRT_y}$$

$$= \sqrt{(1.4)\left(287 \frac{\text{J}}{\text{kg·K}}\right)(472.4\text{K})}$$

$$= 435.7 \text{ m/s}$$

$$v_y = M_y a_y = (0.5774)\left(435.7 \frac{\text{m}}{\text{s}}\right)$$

$$= \boxed{251.6 \text{ m/s}}$$

The answer is (B).

7. *Customary U.S. Solution*

(a) The absolute total temperature is

$$T_0 = 70°\text{F} + 460 = 530°\text{R}$$

The total pressure is $p_0 = 100$ psia.

The ratio of flow area to the critical throat area is

$$\frac{A}{A^*} = 1.555$$

From the isentropic flow tables at $A/A^* = 1.555$, $[p/p_0] = 0.1492$, $[T/T_0] = 0.5807$, and $\boxed{M = 1.90.}$

The answer is (C).

(b) The properties at $A/A^* = 1.555$ are

$$T = \left[\frac{T}{T_0}\right] T_0 = (0.5807)(530°\text{R}) = \boxed{307.8°\text{R}}$$

The answer is (B).

(c) $p = \left[\dfrac{p}{p_0}\right] p_0 = (0.1492)(100 \text{ psia}) = \boxed{14.92 \text{ psia}}$

The answer is (A).

SI Solution

(a) The absolute total temperature is

$$T_0 = 21°\text{C} + 273 = 294\text{K}$$

The total pressure is $p_0 = 0.7$ MPa.

The ratio of flow area to the critical throat area is

$$\frac{A}{A^*} = 1.555$$

From the isentropic flow tables at $A/A^* = 1.555$, $[p/p_0] = 0.1492$, $[T/T_0] = 0.5807$, and $\boxed{M = 1.90.}$

The answer is (C).

(b) The properties at $A/A^* = 1.555$ are

$$T = \left[\frac{T}{T_0}\right] T_0 = (0.5807)(294\text{K}) = \boxed{170.7\text{K}}$$

The answer is (B).

(c) $p = \left[\dfrac{p}{p_0}\right] p_0 = (0.1492)(0.7 \text{ MPa}) = \boxed{0.104 \text{ MPa}}$

The answer is (A).

8. *Customary U.S. Solution*

The absolute total temperature is

$$T_0 = 40°\text{F} + 460 = 500°\text{R}$$

Since v is unknown, assume that static temperature is 500°R. An iterative process may be required for this problem.

The density of air is

$$\rho = \frac{p}{RT}$$

$$= \frac{\left(10 \dfrac{\text{lbf}}{\text{in}^2}\right)\left(144 \dfrac{\text{in}^2}{\text{ft}^2}\right)}{\left(53.3 \dfrac{\text{ft-lbf}}{\text{lbm-°R}}\right)(500°\text{R})}$$

$$= 0.054 \text{ lbm/ft}^3$$

The velocity of air is

$$v = \frac{\dot{m}}{A\rho}$$

$$= \frac{20 \dfrac{\text{lbm}}{\text{sec}}}{(1 \text{ ft}^2)\left(0.054 \dfrac{\text{lbm}}{\text{ft}^3}\right)}$$

$$= 370.4 \text{ ft/sec}$$

The sonic velocity is

$$a = \sqrt{kg_c RT}$$

$$= \sqrt{\begin{array}{c}(1.4)\left(32.2 \dfrac{\text{ft-lbm}}{\text{lbf-sec}^2}\right) \\ \times \left(53.3 \dfrac{\text{ft-lbf}}{\text{lbm-°R}}\right)(500°\text{R})\end{array}}$$

$$= 1096.1 \text{ ft/sec}$$

Thermodynamics

The Mach number is

$$M = \frac{v}{a}$$

$$= \frac{370.4 \ \frac{ft}{sec}}{1096.1 \ \frac{ft}{sec}}$$

$$= 0.338$$

From the isentropic flow tables at M = 0.338, $[T/T_0] = 0.9777$. A closer approximation to static temperature is

$$T = \left[\frac{T}{T_0}\right] T_0 = (0.9777)(500°R)$$

$$= 488.9°R$$

Recalculate all the properties at $T = 488.9°R$.

$$\rho = \frac{p}{RT}$$

$$= \frac{\left(10 \ \frac{lbf}{in^2}\right)\left(144 \ \frac{in^2}{ft^2}\right)}{\left(53.3 \ \frac{ft\text{-}lbf}{lbm\text{-}°R}\right)(488.9°R)}$$

$$= 0.0553 \ lbm/ft^3$$

$$v = \frac{\dot{m}}{A\rho} = \frac{20 \ \frac{lbm}{sec}}{(1 \ ft^2)\left(0.0553 \ \frac{lbm}{ft^3}\right)}$$

$$= 361.7 \ ft/sec$$

$$a = \sqrt{kg_c RT}$$

$$= \sqrt{\begin{array}{c}(1.4)\left(32.2 \ \frac{ft\text{-}lbm}{lbf\text{-}sec^2}\right) \\ \times \left(53.3 \ \frac{ft\text{-}lbf}{lbm\text{-}°R}\right)(488.9°R)\end{array}}$$

$$= 1083.8 \ ft/sec$$

$$M = \frac{v}{a} = \frac{361.7 \ \frac{ft}{sec}}{1083.8 \ \frac{ft}{sec}} = 0.334$$

From the isentropic flow tables at M = 0.334, $[A/A^*] = 1.8516$.

$$A_{smallest} = \left[\frac{A^*}{A}\right] A$$

$$= \left(\frac{1}{1.8516}\right)(1 \ ft^2)$$

$$= \boxed{0.540 \ ft^2}$$

The answer is (B).

SI Solution

The absolute total temperature is

$$T_0 = 4°C + 273 = 277K$$

Since v is unknown, assume that static temperature is 277K. An iterative process may be required for this problem.

The density of air is

$$\rho = \frac{p}{RT}$$

$$= \frac{(70 \ kPa)\left(1000 \ \frac{Pa}{kPa}\right)}{\left(287 \ \frac{J}{kg\cdot K}\right)(277K)}$$

$$= 0.8805 \ kg/m^3$$

The velocity of air is

$$v = \frac{\dot{m}}{A\rho} = \frac{9 \ \frac{kg}{s}}{(0.09 \ m^2)\left(0.8805 \ \frac{kg}{m^3}\right)}$$

$$= 113.6 \ m/s$$

The sonic velocity is

$$a = \sqrt{kRT}$$

$$= \sqrt{(1.4)\left(287 \ \frac{J}{kg\cdot K}\right)(277K)}$$

$$= 333.6 \ m/s$$

The Mach number is

$$M = \frac{v}{a} = \frac{113.6 \ \frac{m}{s}}{333.6 \ \frac{m}{s}} = 0.34$$

From the isentropic flow tables at M = 0.34, $[T/T_0] = 0.9774$. A closer approximation to the static temperature is

$$T = \left[\frac{T}{T_0}\right] T_0 = (0.9774)(277K)$$

$$= 270.7K$$

Recalculate all the properties at $T = 270.7$K.

$$\rho = \frac{p}{RT}$$

$$= \frac{(70 \text{ kPa}) \left(1000 \frac{\text{Pa}}{\text{kPa}}\right)}{\left(287 \frac{\text{J}}{\text{kg·K}}\right)(270.7\text{K})}$$

$$= 0.901 \text{ kg/m}^3$$

$$v = \frac{\dot{m}}{A\rho} = \frac{9 \frac{\text{kg}}{\text{s}}}{(0.09 \text{ m}^2)\left(0.901 \frac{\text{kg}}{\text{m}^3}\right)}$$

$$= 111.0 \text{ m/s}$$

$$a = \sqrt{kRT}$$

$$= \sqrt{(1.4)\left(287 \frac{\text{J}}{\text{kg·K}}\right)(270.7\text{K})}$$

$$= 329.8 \text{ m/s}$$

$$M = \frac{v}{a} = \frac{111.0 \frac{\text{m}}{\text{s}}}{329.8 \frac{\text{m}}{\text{s}}} = 0.337$$

From the isentropic flow tables at M = 0.337, $[A/A^*] = 1.8372$.

$$A_{\text{smallest}} = \left[\frac{A^*}{A}\right] A$$

$$= \left(\frac{1}{1.8372}\right)(0.09 \text{ m}^2) = \boxed{0.0490 \text{ m}^2}$$

The answer is (B).

9. *Customary U.S. Solution*

The ratio of static pressure before the shock to total pressure after the shock is

$$\frac{p_x}{p_{0,y}} = \frac{1.38 \text{ psia}}{20 \text{ psia}} = 0.069$$

From the normal shock table (App. 26.B) for $p_x/p_{0,y} = 0.069$, the Mach number in the tunnel is read directly as $\boxed{M_x = 3.3.}$

The answer is (D).

SI Solution

The ratio of static pressure before the shock to total pressure after the shock is

$$\frac{p_x}{p_{0,y}} = \frac{9.51 \text{ kPa}}{140 \text{ kPa}} = 0.068$$

From the normal shock table (App. 26.B) for $p_x/p_{0,y} = 0.068$, the Mach number in the tunnel is interpolated as $\boxed{M_x = 3.32.}$

The answer is (D).

10. *Customary U.S. Solution*

The initial enthalpy is found from the superheat tables at 100 psia and 800°F.

$$h_1 = 1429.8 \text{ Btu/lbm}$$

Using the Mollier diagram and assuming isentropic expansion, the final enthalpy at 60 psia is $h_2 = 1362$ Btu/lbm.

From Eq. 26.2, the steam exit velocity is

$$v_2 = \sqrt{2g_c J(h_1 - h_2)}$$

$$= \sqrt{\begin{array}{c}(2)\left(32.2 \frac{\text{lbm-ft}}{\text{lbf-sec}^2}\right)\left(778 \frac{\text{ft-lbf}}{\text{Btu}}\right) \\ \times \left(1429.8 \frac{\text{Btu}}{\text{lbm}} - 1362 \frac{\text{Btu}}{\text{lbm}}\right)\end{array}}$$

$$= \boxed{1843.1 \text{ ft/sec}}$$

The answer is (C).

SI Solution

The initial enthalpy is found from the superheat tables at 0.7 MPa and 425°C.

$$h_1 = 3322.2 \text{ kJ/kg}$$

Using the Mollier diagram and assuming isentropic expansion, the final enthalpy at 400 kPa is $h_2 = 3154$ kJ/kg.

The steam exit velocity is

$$v_2 = \sqrt{(2)(h_1 - h_2)}$$

$$= \sqrt{(2)\left(3322.2 \frac{\text{kJ}}{\text{kg}} - 3154 \frac{\text{kJ}}{\text{kg}}\right)\left(1000 \frac{\text{J}}{\text{kJ}}\right)}$$

$$= \boxed{580.0 \text{ m/s}}$$

The answer is (C).

11. *Customary U.S. Solution*

(a) At the particular point, the sonic velocity is

$$a = \sqrt{kg_c RT}$$

$$= \sqrt{\begin{array}{c}(1.4)\left(32.2 \frac{\text{ft-lbm}}{\text{lbf-sec}^2}\right) \\ \times \left(53.3 \frac{\text{ft-lbf}}{\text{lbm-°R}}\right)(1000°\text{R})\end{array}}$$

$$= 1550.1 \text{ ft/sec}$$

Thermodynamics

The Mach number is

$$M = \frac{v}{a} = \frac{600 \ \frac{ft}{sec}}{1550.1 \ \frac{ft}{sec}} = \boxed{0.387} \quad \text{[say 0.39]}$$

The answer is (A).

(b) From the isentropic flow tables at M = 0.39, $[p/p_0] = 0.9004$, $[T/T_0] = 0.9705$, and $[A/A^*] = 1.6243$.

The total properties at Mach 0.39 are

$$T_0 = \left[\frac{T_0}{T}\right] T = \left(\frac{1}{0.9705}\right)(1000°R)$$

$$= \boxed{1030.4°R}$$

The answer is (D).

(c) $\qquad p_0 = \left[\frac{p_0}{p}\right] p = \left(\frac{1}{0.9004}\right)(50 \text{ psia})$

$$= \boxed{55.5 \text{ psia}}$$

The answer is (B).

(d) $\qquad A^* = \left[\frac{A^*}{A}\right] A = \left(\frac{1}{1.6243}\right)(0.1 \text{ ft}^2)$

$$= \boxed{0.0616 \text{ ft}^2}$$

The answer is (A).

From the isentropic flow tables at M = 1 (critical conditions), $[p^*/p_0] = 0.5283$ and $[T^*/T_0] = 0.8333$.

Critical properties at Mach 1 are

(e) $\qquad p^* = \left[\frac{p^*}{p_0}\right] p_0 = (0.5283)(55.5 \text{ psia})$

$$= \boxed{29.32 \text{ psia}}$$

The answer is (B).

(f) $\qquad T^* = \left[\frac{T^*}{T_0}\right] T_0 = (0.8333)(1030.4°R)$

$$= \boxed{858.6°R}$$

The answer is (B).

SI Solution

(a) At the particular point, the sonic velocity is

$$a = \sqrt{kRT}$$

$$= \sqrt{(1.4)\left(287 \ \frac{J}{kg \cdot K}\right)(560K)}$$

$$= 474.4 \text{ m/s}$$

The Mach number is

$$M = \frac{v}{a} = \frac{180 \ \frac{m}{s}}{474.4 \ \frac{m}{s}} = \boxed{0.38}$$

The answer is (A).

(b) From the isentropic flow tables at M = 0.38, $[p/p_0] = 0.9052$, $[T/T_0] = 0.9719$, and $[A/A^*] = 1.6587$.

The total properties are

$$T_0 = \left[\frac{T_0}{T}\right] T = \left(\frac{1}{0.9719}\right)(560K) = \boxed{576.2K}$$

The answer is (D).

(c) $\qquad p_0 = \left[\frac{p_0}{p}\right] p = \left(\frac{1}{0.9052}\right)(350 \text{ kPa})$

$$= \boxed{386.7 \text{ kPa}}$$

The answer is (B).

(d) $\qquad A^* = \left[\frac{A^*}{A}\right] A = \left(\frac{1}{1.6587}\right)(0.0009 \text{ m}^2)$

$$= \boxed{5.43 \times 10^{-4} \text{ m}^2}$$

The answer is (A).

From the isentropic flow tables at M = 1 (critical conditions), $[p^*/p_0] = 0.5283$ and $[T^*/T_0] = 0.8333$.

The critical properties are

(e) $\qquad p^* = \left[\frac{p^*}{p_0}\right] p_0 = (0.5283)(386.7 \text{ kPa})$

$$= \boxed{204.3 \text{ kPa}}$$

The answer is (B).

(f) $\qquad T^* = \left[\frac{T^*}{T_0}\right] T_0 = (0.8333)(576.2K)$

$$= \boxed{480.1K}$$

The answer is (B).

12. *Customary U.S. Solution*

The initial enthalpy is found from the superheat tables at 200 psia and 600°F.

$$h_1 = 1322.3 \text{ Btu/lbm}$$

If the expansion through the nozzle had been isentropic, the exit enthalpy would have been approximately 1228 Btu/lbm. The nozzle efficiency is defined by Eq. 26.18.

$$\eta_{\text{nozzle}} = \frac{h_1 - h_2'}{h_1 - h_2}$$

$$h_2' = h_1 - \eta_{\text{nozzle}}(h_1 - h_2)$$

$$= 1322.3 \frac{\text{Btu}}{\text{lbm}}$$

$$- (0.85)\left(1322.3 \frac{\text{Btu}}{\text{lbm}} - 1228 \frac{\text{Btu}}{\text{lbm}}\right)$$

$$= 1242.1 \text{ Btu/lbm}$$

Knowing $h = 1242.1$ Btu/lbm and $p = 80$ psia establishes (from superheat tables) that

$$T_2' = 422.6°\text{F}$$

$$v_2' = 6.837 \text{ ft}^3/\text{lbm}$$

(By double interpolation from App. 24.C. Actual value is closer to 6.386 ft^3/lbm)

$$\rho_2' = \frac{1}{v_2'} = \frac{1}{6.837 \frac{\text{ft}^3}{\text{lbm}}} = 0.1463 \text{ lbm/ft}^3$$

The nozzle exit velocity can be calculated as

$$v_2' = \sqrt{2g_c J(h_1 - h_2') + v_1^2}$$

$$= \sqrt{\begin{array}{c}(2)\left(32.2 \frac{\text{lbm-ft}}{\text{lbf-sec}^2}\right)\left(778 \frac{\text{ft-lbf}}{\text{Btu}}\right) \\ \times \left(1322.3 \frac{\text{Btu}}{\text{lbm}} - 1242.1 \frac{\text{Btu}}{\text{lbm}}\right) + \left(300 \frac{\text{ft}}{\text{sec}}\right)^2\end{array}}$$

$$= 2026.9 \text{ ft/sec}$$

The exit area of the nozzle is

$$A_e = \frac{\dot{m}}{\rho_2' v_2'} = \frac{3 \frac{\text{lbm}}{\text{sec}}}{\left(0.1463 \frac{\text{lbm}}{\text{ft}^3}\right)\left(2026.9 \frac{\text{ft}}{\text{sec}}\right)}$$

$$= 0.01012 \text{ ft}^2$$

The absolute temperature at the nozzle exit is

$$T_1 = 422.6°\text{F} + 460 = 882.6°\text{R}$$

$$k_{\text{steam}} = 1.31$$

$$R_{\text{steam}} = 85.8 \text{ ft-lbf/lbm-}°\text{R}$$

The sonic velocity is

$$a = \sqrt{kg_c RT}$$

$$= \sqrt{\begin{array}{c}(1.31)\left(32.2 \frac{\text{ft-lbm}}{\text{lbf-sec}^2}\right) \\ \times \left(85.8 \frac{\text{ft-lbf}}{\text{lbm-}°\text{R}}\right)(882.6°\text{R})\end{array}}$$

$$= 1787.3 \text{ ft/sec}$$

The Mach number is

$$\text{M} = \frac{v_2'}{a} = \frac{2026.9 \frac{\text{ft}}{\text{sec}}}{1787.3 \frac{\text{ft}}{\text{sec}}} = 1.13$$

Calculate the throat area. Since tables for $k = 1.31$ are not available,

$$\frac{A}{A^*} = \left(\frac{1}{\text{M}}\right)\left[\frac{\left(\frac{1}{2}\right)(k-1)\text{M}^2 + 1}{\left(\frac{1}{2}\right)(k-1) + 1}\right]^{\frac{k+1}{(2)(k-1)}}$$

$$\frac{A_e}{A^*} = \left(\frac{1}{1.13}\right)\left[\frac{\left(\frac{1}{2}\right)(1.31-1)(1.13)^2 + 1}{\left(\frac{1}{2}\right)(1.31-1) + 1}\right]^{\frac{1.31+1}{(2)(1.31-1)}}$$

$$= 1.0138$$

$$A^* = \left[\frac{A^*}{A_e}\right]A_e = \left(\frac{1}{1.0138}\right)(0.01012 \text{ ft}^2)$$

$$= \boxed{0.00998 \text{ ft}^2}$$

The answer is (C).

SI Solution

The initial enthalpy is found from the superheat tables at 1.5 MPa and 300°C.

$$h_1 = 3038.2 \text{ kJ/kg}$$

If the expansion through the nozzle had been isentropic, the exit enthalpy would have been approximately 2790 kJ/kg. The nozzle efficiency is defined by Eq. 26.18.

$$\eta_{\text{nozzle}} = \frac{h_1 - h_2'}{h_1 - h_2}$$

$$h_2' = h_1 - \eta_{\text{nozzle}}(h_1 - h_2)$$

$$= 3038.2 \frac{\text{kJ}}{\text{kg}}$$

$$- (0.85)\left(3038.2 \frac{\text{kJ}}{\text{kg}} - 2790 \frac{\text{kJ}}{\text{kg}}\right)$$

$$= 2827 \text{ kJ/kg}$$

Knowing $h = 2827$ kJ/kg and $p = 500$ kPa establishes (from superheat tables) that

$$T_2' = 187.1°\text{C}$$

$$v_2' = 0.4420 \text{ m}^3/\text{kg}$$

$$\rho_2' = \frac{1}{v_2'} = \frac{1}{0.4420 \frac{\text{m}^3}{\text{kg}}} = 2.262 \text{ kg/m}^3$$

The nozzle exit velocity can be calculated as

$$v_2' = \sqrt{(2)(h_1 - h_2') + v_1^2}$$

$$= \sqrt{\begin{array}{c} (2)\left(3038.2 \dfrac{kJ}{kg} - 2827 \dfrac{kJ}{kg}\right) \\ \times \left(1000 \dfrac{J}{kJ}\right) + \left(90 \dfrac{m}{s}\right)^2 \end{array}}$$

$$= 656.1 \text{ m/s}$$

The exit area of the nozzle is

$$A_e = \frac{\dot{m}}{\rho_2' v_2'} = \frac{1.4 \dfrac{kg}{s}}{\left(2.262 \dfrac{kg}{m^3}\right)\left(656.1 \dfrac{m}{s}\right)}$$

$$= 9.433 \times 10^{-4} \text{ m}^2$$

The absolute temperature at the nozzle exit is

$$T_1 = 187.1°C + 273 = 460.1 \text{K}$$

$$k_{\text{steam}} = 1.31$$

$$R_{\text{steam}} = 461.50 \text{ J/kg·K}$$

The sonic velocity is

$$a = \sqrt{kRT}$$

$$= \sqrt{(1.31)\left(461.50 \dfrac{J}{kg·K}\right)(460.1 \text{K})}$$

$$= 527.4 \text{ m/s}$$

The Mach number is

$$M = \frac{v_2'}{a} = \frac{656.1 \dfrac{m}{s}}{527.4 \dfrac{m}{s}} = 1.24$$

Calculate the throat area. Since tables for $k = 1.31$ are not available,

$$\frac{A}{A^*} = \left(\frac{1}{M}\right)\left[\frac{\left(\frac{1}{2}\right)(k-1)M^2 + 1}{\left(\frac{1}{2}\right)(k-1) + 1}\right]^{\frac{k+1}{(2)(k-1)}}$$

$$\frac{A_e}{A^*} = \left(\frac{1}{1.24}\right)\left[\frac{\left(\frac{1}{2}\right)(1.31-1)(1.24)^2 + 1}{\left(\frac{1}{2}\right)(1.31-1) + 1}\right]^{\frac{1.31+1}{(2)(1.31-1)}}$$

$$= 1.0454$$

$$A^* = \left[\frac{A^*}{A_e}\right]A_e = \left(\frac{1}{1.0454}\right)(9.433 \times 10^{-4} \text{ m}^2)$$

$$= \boxed{9.023 \times 10^{-4} \text{ m}^2}$$

The answer is (C).

13. *Customary U.S. Solution*

This is a Fanno flow problem. Check for choked flow.

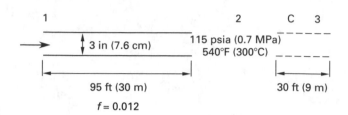

At point 2, from superheat tables at 115 psia and 540°F, $v_2 = 5.244 \text{ ft}^3/\text{lbm}$.

The velocity at point 2 can be calculated as

$$v_2 = \frac{\dot{m}}{A\rho_2} = \frac{\dot{m}v_2}{A}$$

$$= \frac{\left(35,200 \dfrac{lbm}{hr}\right)\left(\dfrac{1 \text{ hr}}{3600 \text{ sec}}\right)\left(5.244 \dfrac{ft^3}{lbm}\right)}{\left(\dfrac{\pi}{4}\right)(3 \text{ in})^2 \left(\dfrac{1 \text{ ft}}{12 \text{ in}}\right)^2}$$

$$= 1045 \text{ ft/sec}$$

The absolute temperature at point 2 is

$$T_2 = 540°F + 460 = 1000°R$$

For 1000°R steam, $k = 1.29$ and $R = 85.8$ ft-lbf/lbm-°R.

The sonic velocity at point 2 is

$$a_2 = \sqrt{kg_c RT}$$

$$= \sqrt{\begin{array}{c} (1.29)\left(32.2 \dfrac{ft\text{-}lbm}{lbf\text{-}sec^2}\right) \\ \times \left(85.8 \dfrac{ft\text{-}lbf}{lbm\text{-}°R}\right)(1000°R) \end{array}}$$

$$= 1888 \text{ ft/sec}$$

The Mach number at point 2 is

$$M_2 = \frac{v_2}{a_2} = \frac{1045 \dfrac{ft}{sec}}{1888 \dfrac{ft}{sec}} = 0.553$$

The distance from point 2 to where the flow becomes choked is calculated from Eq. 26.34.

$$\frac{4fL_{max}}{D} = \frac{1 - M^2}{kM^2}$$
$$+ \left(\frac{1+k}{2k}\right) \ln\left(\frac{(1+k)M^2}{(2)\left(1 + \left(\frac{1}{2}\right)(k-1)M^2\right)}\right)$$

$$L_{max} = \left(\frac{(3 \text{ in})\left(\frac{1 \text{ ft}}{12 \text{ in}}\right)}{(4)(0.012)}\right)$$
$$\times \left(\frac{1 - (0.553)^2}{(1.29)(0.553)^2} + \frac{1 + 1.29}{(2)(1.29)}\right.$$
$$\left. \times \ln\left(\frac{(1 + 1.29)(0.553)^2}{(2)\left(1 + \left(\frac{1}{2}\right)(1.29 - 1)(0.553)^2\right)}\right)\right)$$
$$= 4.11 \text{ ft}$$

> The flow will be choked in less than 30 ft; steam flow is not maintained.

SI Solution

At point 2, from the superheat tables at 0.7 MPa and 300°C, $v_2 = 0.3714 \text{ m}^3/\text{kg}$.

The velocity at point 2 can be calculated as

$$v_2 = \frac{\dot{m}}{\rho_2 A} = \frac{\dot{m}v_2}{A}$$
$$= \frac{\left(4.4 \frac{\text{kg}}{\text{s}}\right)\left(0.3714 \frac{\text{m}^3}{\text{kg}}\right)}{\left(\frac{\pi}{4}\right)(7.6 \text{ cm})^2 \left(\frac{1 \text{ m}}{100 \text{ cm}}\right)^2}$$
$$= 360.2 \text{ m/s}$$

The absolute temperature at point 2 is

$$T_2 = 300°C + 273 = 573K$$

For 573K steam, $k = 1.29$ and $R = 461.5 \text{ J/kg·K}$. The sonic velocity at point 2 is

$$a_2 = \sqrt{kRT}$$
$$= \sqrt{(1.29)\left(461.5 \frac{\text{J}}{\text{kg·K}}\right)(573K)}$$
$$= 584.1 \text{ m/s}$$

The Mach number at point 2 is

$$M_2 = \frac{v_2}{a_2} = \frac{360.2 \frac{\text{m}}{\text{s}}}{584.1 \frac{\text{m}}{\text{s}}} = 0.617$$

The distance from point 2 to where the flow becomes choked is calculated from Eq. 26.34.

$$\frac{4fL_{max}}{D} = \frac{1 - M^2}{kM^2}$$
$$+ \left(\frac{1+k}{2k}\right) \ln\left(\frac{(1+k)M^2}{(2)\left(1 + \left(\frac{1}{2}\right)(k-1)M^2\right)}\right)$$

$$L_{max} = \left(\frac{(7.6 \text{ cm})\left(\frac{1 \text{ m}}{100 \text{ cm}}\right)}{(4)(0.012)}\right)$$
$$\times \left(\frac{1 - (0.617)^2}{(1.29)(0.617)^2} + \frac{1 + 1.29}{(2)(1.29)}\right.$$
$$\left. \times \ln\left(\frac{(1 + 1.29)(0.617)^2}{(2)\left(1 + \left(\frac{1}{2}\right)(1.29 - 1)(0.617)^2\right)}\right)\right)$$
$$= 0.75 \text{ m}$$

> The flow will be choked in less than 9 m; steam flow is not maintained.

14. *Customary U.S. Solution*

Since $p_e/p_0 < 0.5283$, the flow must be supersonic.

The throat flow is sonic. Read the property ratios for Mach 1 from the isentropic flow table: $[T/T_0] = 0.8333$ and $[p/p_0] = 0.5283$.

The sonic properties at the throat are

$$T^* = \left[\frac{T}{T_0}\right] T_0 = (0.8333)(660°R) = 550°R$$

$$p^* = \left[\frac{p}{p_0}\right] p_0 = (0.5283)(160 \text{ psia}) = 84.53 \text{ psia}$$

$$a^* = \sqrt{kg_c RT^*}$$
$$= \sqrt{\begin{array}{c}(1.4)\left(32.2 \dfrac{\text{ft-lbm}}{\text{lbf-sec}^2}\right)\\ \times \left(53.3 \dfrac{\text{ft-lbf}}{\text{lbm-°R}}\right)(550°R)\end{array}}$$
$$= 1150 \text{ ft/sec}$$

$$\rho^* = \frac{p^*}{RT^*} = \frac{\left(84.53 \frac{\text{lbf}}{\text{in}^2}\right)\left(144 \frac{\text{in}^2}{\text{ft}^2}\right)}{\left(53.3 \frac{\text{ft-lbf}}{\text{lbm-°R}}\right)(550°R)}$$
$$= 0.415 \text{ lbm/ft}^3$$

The overall nozzle efficiency is $C_d = 0.90$. From $\dot{m} = C_d A v \rho$, the throat area is

$$A^* = \frac{\dot{m}}{C_d v \rho} = \frac{\left(3600 \dfrac{\text{lbm}}{\text{hr}}\right)\left(\dfrac{1\ \text{hr}}{3600\ \text{sec}}\right)}{(0.90)\left(1150\ \dfrac{\text{ft}}{\text{sec}}\right)\left(0.415\ \dfrac{\text{lbm}}{\text{ft}^3}\right)}$$

$$= \boxed{0.002328\ \text{ft}^2}$$

$$D^* = \sqrt{\frac{4A^*}{\pi}}$$

$$= \sqrt{\frac{(4)(0.002328\ \text{ft}^2)}{\pi}}$$

$$= 0.0544\ \text{ft}$$

At the exit,

$$p_3 = 14.7\ \text{psia}$$

$$\frac{p_e}{p_0} = \frac{14.7\ \text{psia}}{160\ \text{psia}} = 0.0919$$

From the isentropic flow tables at $p/p_0 = 0.0919$, $M = 2.22$ and $[A/A^*] = 2.041$.

$$A_e = \left[\frac{A}{A^*}\right] A^* = (2.041)(0.002328\ \text{ft}^2) = 0.004751\ \text{ft}^2$$

$$D_e = \sqrt{\frac{4A}{\pi}} = \sqrt{\frac{(4)(0.004751\ \text{ft}^2)}{\pi}}$$

$$= 0.0778\ \text{ft}$$

The longitudinal distance from the throat to the exit is

$$x = \frac{0.0778\ \text{ft} - 0.0544\ \text{ft}}{(2)(\tan 3°)} = 0.223\ \text{ft}$$

The entrance velocity is not known, so the entrance area cannot be found. However, the longitudinal distance from the entrance to the throat is

$$(0.05)(0.223\ \text{ft}) = 0.0112\ \text{ft}$$

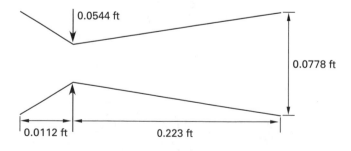

SI Solution

Since $p_e/p_0 < 0.5283$, the flow must be supersonic. The throat flow is sonic. Read the property ratio for Mach 1 from the isentropic flow table: $[T/T_0] = 0.8333$ and $[p/p_0] = 0.5283$.

The sonic properties at the throat are

$$T^* = \left[\frac{T}{T_0}\right] T_0 = (0.8333)(370\text{K}) = 308.3\text{K}$$

$$p^* = \left[\frac{p}{p_0}\right] p_0 = (0.5283)(1.1\ \text{MPa}) = 0.581\ \text{MPa}$$

$$a^* = \sqrt{kRT^*}$$

$$= \sqrt{(1.4)\left(287\ \frac{\text{J}}{\text{kg·K}}\right)(308.3\text{K})}$$

$$= 352\ \text{m/s}$$

$$\rho^* = \frac{p^*}{RT^*} = \frac{(0.581\ \text{MPa})\left(10^6\ \dfrac{\text{Pa}}{\text{MPa}}\right)}{\left(287\ \dfrac{\text{J}}{\text{kg·K}}\right)(308.3\text{K})}$$

$$= 6.57\ \text{kg/m}^3$$

The overall nozzle efficiency is $C_d = 0.90$. From $\dot{m} = C_d A v \rho$,

$$A^* = \frac{\dot{m}}{C_d v \rho} = \frac{0.45\ \dfrac{\text{kg}}{\text{s}}}{(0.90)\left(352\ \dfrac{\text{m}}{\text{s}}\right)\left(6.57\ \dfrac{\text{kg}}{\text{m}^3}\right)}$$

$$= \boxed{0.000216\ \text{m}^2}$$

$$D^* = \sqrt{\frac{4A^*}{\pi}} = \sqrt{\frac{(4)(0.000216\ \text{m}^2)}{\pi}}$$

$$= 0.0166\ \text{m}$$

At the exit,

$$p_e = 101.3\ \text{kPa}$$

$$\frac{p_e}{p_0} = \frac{(101.3\ \text{kPa})\left(10^3\ \dfrac{\text{Pa}}{\text{kPa}}\right)}{(1.1\ \text{MPa})\left(10^6\ \dfrac{\text{Pa}}{\text{MPa}}\right)} = 0.0921$$

From the isentropic flow tables at $p/p_0 = 0.0921$, $M = 2.21$ and $[A/A^*] = 2.024$.

$$A_e = \left[\frac{A}{A^*}\right] A^* = (2.024)(0.000216\ \text{m}^2)$$

$$= 0.000437\ \text{m}^2$$

$$D_e = \sqrt{\frac{4A}{\pi}} = \sqrt{\frac{(4)(0.000437\ \text{m}^2)}{\pi}}$$

$$= 0.0236\ \text{m}$$

The longitudinal distance from the throat to the exit is

$$x = \frac{0.0236 \text{ m} - 0.0166 \text{ m}}{(2)(\tan 3°)} = 0.0668 \text{ m}$$

The entrance velocity is not known, so the entrance area cannot be found. However, the longitudinal distance from the entrance to the throat is

$$(0.05)(0.0668 \text{ m}) = 0.00334 \text{ m}$$

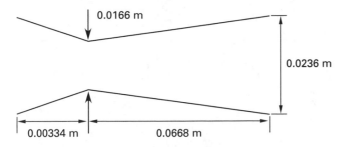

27 Vapor Power Equipment

PRACTICE PROBLEMS

1. What is the isentropic efficiency of a process that expands dry steam from 100 psia to 3 psia (700 kPa to 20 kPa) and 90% quality?
(A) 40%
(B) 50%
(C) 60%
(D) 70%

2. A 5000 kW steam turbine uses 200 psia (1.5 MPa) steam with 100°F (50°C) of superheat. The condenser is at 1 in Hg (3.4 kPa) absolute.

(a) What is the water rate at full load?
(A) 4.7 lbm/kW-hr (0.0006 kg/kW·s)
(B) 8.8 lbm/kW-hr (0.0011 kg/kW·s)
(C) 25 lbm/kW-hr (0.0031 kg/kW·s)
(D) 37 lbm/kW-hr (0.0046 kg/kW·s)

(b) If the actual load is only 2500 kW and the steam undergoes a pressure-reducing throttling process to reduce the availability, what is the loss in available energy per unit mass of steam?
(A) 80 Btu/lbm (190 kJ/kg)
(B) 130 Btu/lbm (310 kJ/kg)
(C) 190 Btu/lbm (460 kJ/kg)
(D) 370 Btu/lbm (890 kJ/kg)

3. A 10,000 kW steam turbine operates on 400 psia (3.0 MPa), 750°F (420°C) dry steam, expanding to 2 in Hg (6.8 kPa) absolute. What is the maximum adiabatic heat drop available for power production?
(A) 450 Btu/lbm (1100 kJ/kg)
(B) 600 Btu/lbm (1400 kJ/kg)
(C) 750 Btu/lbm (1800 kJ/kg)
(D) 900 Btu/lbm (2200 kJ/kg)

4. A 750 kW steam turbine has a water rate of 20 lbm/kW-hr (2.5×10^{-3} kg/kW·s). Steam with 50°F (30°C) of superheat is expanded from 165 psia (1.0 MPa absolute) to 26 in Hg (90 kPa) absolute. 65°F (18°C) cooling water is available. The terminal temperature difference is zero. Find the quantity of cooling water required.
(A) 45,000 lbm/hr (5.9 kg/s)
(B) 60,000 lbm/hr (7.8 kg/s)
(C) 80,000 lbm/hr (10 kg/s)
(D) 95,000 lbm/hr (12 kg/s)

5. 332,000 lbm/hr (41.8 kg/s) of 81°F (27°C) water enters a two-pass, counterflow heat exchanger. The heat exchanger is constructed of 1850 ft² (172 m²) of ⅝ in (15.9 mm) copper tubing. Saturated steam is bled from a turbine at 4.45 psia (30.6 kPa) and condenses to saturated liquid. The heated water leaves at 150°F (65.6°C) with an enthalpy of 1100 Btu/lbm (2.56 MJ/kg).

(a) What is the overall heat transfer coefficient?
(A) 1900 Btu/hr-ft²-°F (11 kW/m²·°C)
(B) 4400 Btu/hr-ft²-°F (25 kW/m²·°C)
(C) 6500 Btu/hr-ft²-°F (37 kW/m²·°C)
(D) 7700 Btu/hr-ft²-°F (44 kW/m²·°C)

(b) What is the steam extraction rate?
(A) 1.7×10^8 Btu/hr (48 MW)
(B) 3.9×10^8 Btu/hr (110 MW)
(C) 9.9×10^8 Btu/hr (270 MW)
(D) 2.5×10^9 Btu/hr (700 MW)

6. A two-pass surface condenser constructed of 1 in (25.4 mm) BWG tubing receives 82,000 lbm/hr (10.3 kg/s) of steam from a turbine. Steam enters the condenser with an enthalpy of 980 Btu/lbm (2.280 MJ/kg). The condenser operates at a pressure of 1 in Hg (3.4 kPa) absolute. Water is circulated at 8 fps (2.4 m/s) through an equivalent length of 120 ft (36 m) of extra strong 30 in (76.2 cm) steel pipe. An additional head loss of 6 in wg (1.5 kPa) is incurred in the intake screens.

(a) What is the head added by the circulating water pump?
(A) 2.1 ft of water (6.1 kPa)
(B) 3.2 ft of water (9.3 kPa)
(C) 6.6 ft of water (19 kPa)
(D) 9.0 ft of water (26 kPa)

(b) If the water temperature increases 10°F (5.6°C) across the condenser, what is the circulation rate of cooling water?
(A) 9000 gal/min (34 kL/min)
(B) 11,000 gal/min (42 kL/min)
(C) 13,000 gal/min (49 kL/min)
(D) 15,000 gal/min (57 kL/min)

7. 100 lbm/hr (0.013 kg/s) of 60°F (16°C) water is turned into 14.7 psia (101.3 kPa) saturated steam in an electric boiler. Radiation losses are 35% of the supplied energy. What is the cost if electricity is $0.04 per kW-hr?

    (A) $2/hr

    (B) $3/hr

    (C) $4/hr

    (D) $5/hr

8. A gas burner produces 250 lbm/hr (0.032 kg/s) of 98% dry steam at 40 psia (300 kPa) from 60°F (16°C) feedwater. The fuel gas enters at 80°F (26°C) and 4 in Hg (13.6 kPa) and has a heating value of 550 Btu/ft³ (20.5 MJ/m³) at standard industrial conditions. The barometric pressure is 30.2 in Hg (102.4 kPa). 13.5 ft³/min (6.4 L/s) of fuel gas is consumed. What is the efficiency of the boiler?

    (A) 37%

    (B) 43%

    (C) 57%

    (D) 66%

9. A boiler evaporates 8.23 lbm (8.23 kg) of 120°F (50°C) water per pound (per kilogram) of coal fired, producing 100 psia (700 kPa) saturated steam. The coal is 2% moisture by weight as fired, and dry coal is 5% ash. 1% of the coal is removed from the ash pit. (The ash pit loss has the same composition as unfired, dry coal.) Coal is initially at 60°F (16°C), and combustion occurs at 14.7 psia (101.3 kPa). The combustion products leave at 600°F (315°C). The heating value of the dry coal is 12,800 Btu/lbm (29.80 MJ/kg). What is the efficiency of the boiler?

    (A) 53%

    (B) 68%

    (C) 73%

    (D) 82%

10. 500 psia (3.5 MPa) steam is superheated to 1000°F (500°C) before expanding through a 75% efficient turbine to 5 psia (30 kPa). No subcooling occurs. The pump work is negligible compared to the 200 MW generated.

(a) What quantity of steam is required?

    (A) 8.6×10^5 lbm/hr (120 kg/s)

    (B) 1.4×10^6 lbm/hr (190 kg/s)

    (C) 2.0×10^6 lbm/hr (270 kg/s)

    (D) 4.2×10^6 lbm/hr (310 kg/s)

(b) What heat is removed by the condenser?

    (A) 3.2×10^8 Btu/hr (99 MW)

    (B) 8.7×10^8 Btu/hr (270 MW)

    (C) 2.1×10^9 Btu/hr (650 MW)

    (D) 8.7×10^9 Btu/hr (2.7 GW)

11. 191,000 lbm/hr (24 kg/s) of 635°F (335°C) combustion gases flow through a 20 ft (6 m) wide boiler stack whose front and back plates are 5 ft-10 in (1.78 m) apart. An integral crossflow economizer is being designed to heat water from 212°F to 285°F (100°C to 140°C) by dropping the stack gas temperature to 470°F (240°C). Layers of 24 tubes with dimensions 0.957 in (24.3 mm) I.D., 1.315 in (33.4 mm) O.D., and 20 ft (6 m) length will be placed on a 2.315 in (58.8 mm) pitch in horizontal banks. The overall coefficient of heat transfer for the tubes is 10 Btu/hr-ft²-°F (57 W/m²·°C). How many 24-tube layers are required?

    (A) 5

    (B) 9

    (C) 12

    (D) 17

12. (*Time limit: one hour*) Water is used in an adiabatic steam desuperheater. 1000 lbm/hr (0.13 kg/s) of 200 psia (1.5 MPa), 600°F (300°C) steam enters with negligible velocity. 50 lbm/hr (0.0063 kg/s) of 82°F (28°C) water enters with negligible velocity. 100 psia (700 kPa) steam leaves the desuperheater at 2000 ft/sec (600 m/s).

(a) What is the temperature of the leaving steam?

    (A) 330°F (165°C)

    (B) 360°F (180°C)

    (C) 400°F (200°C)

    (D) 470°F (240°C)

(b) What is the quality of the leaving steam?

    (A) 85%

    (B) 88%

    (C) 93%

    (D) 99%

13. (*Time limit: one hour*) Waste steam at 400°F (200°C) and 100 psia (700 kPa) was originally used only for heating cold water. Cold water entered at 70°F (21°C) and 60 psia (400 kPa). 2000 lbm/hr (0.25 kg/s) of hot water at 180°F (80°C) and 20 psia (150 kPa) were produced. Now, the same quantity of steam is to be expanded through a low-pressure turbine. The low-pressure turbine has an isentropic efficiency of 60% and a mechanical efficiency of 96%. The steam will then flow through a mixing heater (see diagram). A pressure drop of 5 psi (30 kPa) occurs through the heater. The heater output must remain at 180°F (80°C) and 20 psia (350 kPa), but the output at point F may decrease.

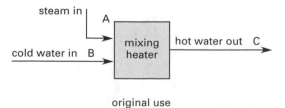

original use

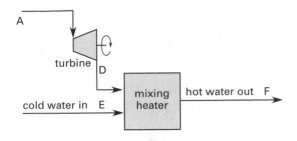

proposed use of waste steam

(a) What power is developed in the turbine?
   (A)   4.7 hp (3.4 kW)
   (B)   8.2 hp (6.3 kW)
   (C)   14 hp (11 kW)
   (D)   27 hp (21 kW)

(b) What is the flow rate at point F?
   (A)   870 lbm/hr (0.10 kg/s)
   (B)   1200 lbm/hr (0.14 kg/s)
   (C)   1900 lbm/hr (0.24 kg/s)
   (D)   3400 lbm/hr (0.41 kg/s)

SOLUTIONS

1. *Customary U.S. Solution*

From App. 24.B, for 100 psia, the enthalpy of dry steam is $h_i = 1187.5$ Btu/lbm.

From the Mollier diagram for an isentropic process from 100 psia to 3 psia, $h_2 = 950$ Btu/lbm.

From App. 24.B, for 3 psia,

$$h_f = 109.4 \text{ Btu/lbm}$$
$$h_{fg} = 1012.8 \text{ Btu/lbm}$$
$$h_2' = h_f + x h_{fg}$$
$$= 109.4 \frac{\text{Btu}}{\text{lbm}} + (0.9) \left(1012.8 \frac{\text{Btu}}{\text{lbm}} \right)$$
$$= 1020.9 \text{ Btu/lbm}$$

From Eq. 27.17, the isentropic efficiency is

$$\eta_s = \frac{h_1 - h_2'}{h_1 - h_2} = \frac{1187.5 \dfrac{\text{Btu}}{\text{lbm}} - 1020.9 \dfrac{\text{Btu}}{\text{lbm}}}{1187.5 \dfrac{\text{Btu}}{\text{lbm}} - 950 \dfrac{\text{Btu}}{\text{lbm}}}$$

$$= \boxed{0.701 \ \ (70.1\%)}$$

The answer is (D).

SI Solution

From App. 24.O, for 700 kPa, the enthalpy of dry steam is $h_1 = 2762.8$ kJ/kg.

From the Mollier diagram for an isentropic process from 700 kPa to 20 kPa, $h_2 = 2245$ kJ/kg.

From App. 24.O, for 20 kPa,

$$h_f = 251.42 \text{ kJ/kg}$$
$$h_{fg} = 2357.5 \text{ kJ/kg}$$
$$h_2' = h_f + x h_{fg}$$
$$= 251.42 \frac{\text{kJ}}{\text{kg}} + (0.9) \left(2357.5 \frac{\text{kJ}}{\text{kg}} \right)$$
$$= 2373.2 \text{ kJ/kg}$$

From Eq. 22.17, the isentropic efficiency is

$$\eta_s = \frac{h_1 - h_2'}{h_1 - h_2}$$

$$= \frac{2762.8 \dfrac{\text{kJ}}{\text{kg}} - 2373.2 \dfrac{\text{kJ}}{\text{kg}}}{2762.8 \dfrac{\text{kJ}}{\text{kg}} - 2245 \dfrac{\text{kJ}}{\text{kg}}}$$

$$= \boxed{0.752 \ \ (75.2\%)}$$

The answer is (D).

Power Cycles

2. *Customary U.S. Solution*

(a) From App. 24.B, T_{sat} for 200 psia is 381.80°F.

The steam temperature is

$$381.80°F + 100°F = 481.80°F$$

From App. 24.C, $h_1 = 1258.4$ Btu/lbm.

1 in Hg is approximately 0.5 psia. From the Mollier diagram, assuming isentropic expansion, dropping straight down to the 0.5 psia line, $h_2 \approx 870$ Btu/lbm.

For isentropic expansion, $\eta_{turbine} = 1$, and the steam mass flow rate through the turbine is given by Eq. 27.22.

$$\dot{m} = \frac{P_{turbine}}{h_1 - h_2}$$

$$= \frac{(5000 \text{ kW})\left(3413 \dfrac{\text{Btu}}{\text{hr-kW}}\right)}{1258.4 \dfrac{\text{Btu}}{\text{lbm}} - 870 \dfrac{\text{Btu}}{\text{lbm}}}$$

$$= 4.394 \times 10^4 \text{ lbm/hr}$$

The water rate is

$$\text{WR} = \frac{\dot{m}}{P_{turbine}} = \frac{4.394 \times 10^4 \dfrac{\text{lbm}}{\text{hr}}}{5000 \text{ kW}}$$

$$= \boxed{8.787 \text{ lbm/kW-hr}}$$

The answer is (B).

(b) The loss in available energy per unit mass is

$$\text{loss} = \left(\tfrac{1}{2}\right)(h_1 - h_2)$$

$$= \left(\tfrac{1}{2}\right)\left(1258.4 \dfrac{\text{Btu}}{\text{lbm}} - 870 \dfrac{\text{Btu}}{\text{lbm}}\right)$$

$$= \boxed{194.2 \text{ Btu/lbm}}$$

The answer is (C).

SI Solution

(a) From App. 24.0, T_{sat} for 1.5 MPa is 198.3°C.

The steam temperature is

$$198.3°C + 50°C = 248.3°C$$

From App. 24.P, $h_1 = 2920$ kJ/kg and $s_1 = 6.7023$ kJ/kg·K.

From App. 24.N, for 3.4 kPa, the entropy of saturated liquid, s_f, the entropy of saturated vapor, s_g, the enthalpy of saturated liquid, h_f, and the enthalpy of vaporization, h_{fg}, are

$$s_f = 0.3837 \text{ kJ/kg·K}$$
$$s_g = 8.5316 \text{ kJ/kg·K}$$
$$h_f = 109.84 \text{ kJ/kg}$$
$$h_{fg} = 2438.9 \text{ kJ/kg}$$

For isentropic expansion, $s_1 = s_2 = 6.7023$ kJ/kg·K.

Since $s_2 < s_g$, the expanded steam is in the liquid-vapor region. The quality of the mixture is given by Eq. 24.41.

$$x = \frac{s - s_f}{s_{fg}} = \frac{s - s_f}{s_g - s_f}$$

$$= \frac{6.7023 \dfrac{\text{kJ}}{\text{kg·K}} - 0.3837 \dfrac{\text{kJ}}{\text{kg·K}}}{8.5316 \dfrac{\text{kJ}}{\text{kg·K}} - 0.3837 \dfrac{\text{kJ}}{\text{kg·K}}}$$

$$= 0.7755$$

The final enthalpy is given by Eq. 24.40.

$$h = h_f + x h_{fg}$$

$$= 109.84 \dfrac{\text{kJ}}{\text{kg}} + (0.7755)\left(2438.9 \dfrac{\text{kJ}}{\text{kg}}\right)$$

$$= 2001.2 \text{ kJ/kg}$$

For isentropic expansion, $\eta_{turbine} = 1$, and the steam mass flow rate through a turbine is given by Eq. 27.22.

$$\dot{m} = \frac{P_{turbine}}{h_1 - h_2}$$

$$= \frac{5000 \text{ kW}}{2920 \dfrac{\text{kJ}}{\text{kg}} - 2001.2 \dfrac{\text{kJ}}{\text{kg}}}$$

$$= 5.442 \text{ kg/s}$$

The water rate is

$$\text{WR} = \frac{\dot{m}}{P_{turbine}} = \frac{5.442 \dfrac{\text{kg}}{\text{s}}}{5000 \text{ kW}}$$

$$= \boxed{1.09 \times 10^{-3} \text{ kg/kW·s}}$$

The answer is (B).

(b) Loss in availability per unit mass is

$$\text{loss} = \left(\tfrac{1}{2}\right)(h_1 - h_2)$$

$$= \left(\tfrac{1}{2}\right)\left(2920 \dfrac{\text{kJ}}{\text{kg}} - 2001.2 \dfrac{\text{kJ}}{\text{kg}}\right)$$

$$= \boxed{459.4 \text{ kJ/kg}}$$

The answer is (C).

3. *Customary U.S. Solution*

From App. 24.C, the enthalpy, h_1, of dry steam at 400 psia and 750°F is $h_1 = 1389.9$ Btu/lbm.

From the Mollier diagram, assuming isentropic expansion to 2 in Hg (about 1 psia), $h_2 = 935$ Btu/lbm.

The maximum adiabatic heat drop is

$$h_1 - h_2 = 1389.9 \ \frac{\text{Btu}}{\text{lbm}} - 935 \ \frac{\text{Btu}}{\text{lbm}}$$

$$= \boxed{454.9 \ \text{Btu/lbm}}$$

The answer is (A).

SI Solution

From App. 24.P, the enthalpy, h_1, of dry steam at 3.0 MPa and 420°C is $h_1 = 3277.1$ kJ/kg.

From the Mollier diagram, assuming isentropic expansion, $h_2 \approx 2210$ kJ/kg.

The maximum adiabatic heat drop is

$$h_1 - h_2 = 3277.1 \ \frac{\text{kJ}}{\text{kg}} - 2210 \ \frac{\text{kJ}}{\text{kg}}$$

$$= \boxed{1067.1 \ \text{kJ/kg}}$$

The answer is (A).

4. *Customary U.S. Solution*

The water rate is

$$\text{WR} = \frac{\dot{m}}{P_{\text{turbine}}}$$

The steam flow rate is

$$\dot{m}_{\text{steam}} = (\text{WR})(P_{\text{turbine}})$$

$$= \left(20 \ \frac{\text{lbm}}{\text{kW-hr}}\right)(750 \ \text{kW})$$

$$= 15{,}000 \ \text{lbm/hr}$$

From App. 24.B, for 165 psia, $T_{\text{sat}} = 366.0°\text{F}$.

The steam temperature is

$$366.0°\text{F} + 50°\text{F} = 416.0°\text{F}$$

From App. 24.C, $h_1 = 1226.0$ Btu/lbm.

From Eq. 27.22,

$$P_{\text{turbine}} = \dot{m}(h_1 - h_2)$$

$$h_1 - h_2 = \frac{P_{\text{turbine}}}{\dot{m}} = \frac{(750 \ \text{kW})\left(3413 \ \dfrac{\text{Btu}}{\text{kW-hr}}\right)}{15{,}000 \ \dfrac{\text{lbm}}{\text{hr}}}$$

$$= 170.6 \ \text{Btu/lbm}$$

$$h_2 = h_1 - 170.6 \ \frac{\text{Btu}}{\text{lbm}}$$

$$= 1226.0 \ \frac{\text{Btu}}{\text{lbm}} - 170.6 \ \frac{\text{Btu}}{\text{lbm}}$$

$$= 1055.4 \ \text{Btu/lbm}$$

$$p_2 = (26 \ \text{in Hg})\left(0.491 \ \frac{\text{lbf}}{\text{in}^3}\right)$$

$$= 12.77 \ \text{psia}$$

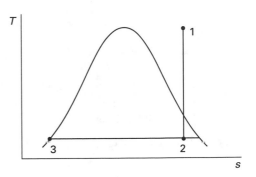

From App. 24.B, for $p_2 = 12.77$ psia,

$$T_{\text{sat}} = T_2 = T_3 \approx 204°\text{F}$$

$$h_{f,3} \approx 172.3 \ \text{Btu/lbm}$$

From App. 35.A, the specific heat of water at 65°F is $c_{p\text{water}} = 0.999$ Btu.

Assuming the water and steam leave in thermal equilibrium, the heat lost by steam is equal to the heat gained by water.

$$\dot{m}_{\text{water}} c_{p\text{water}}(T_{\text{water,out}} - T_{\text{water,in}}) = \dot{m}_{\text{steam}}(h_2 - h_{f,3})$$

Since the terminal temperature difference is zero, $T_{\text{water,out}} = T_3$.

$$\dot{m}_{\text{water}}\left(0.999 \ \frac{\text{Btu}}{\text{lbm-°F}}\right)(204°\text{F} - 65°\text{F})$$

$$= \left(15{,}000 \ \frac{\text{lbm}}{\text{hr}}\right)\left(1055.4 \ \frac{\text{Btu}}{\text{lbm}} - 172.3 \ \frac{\text{Btu}}{\text{lbm}}\right)$$

$$\dot{m}_{\text{water}} = \boxed{9.54 \times 10^4 \ \text{lbm/hr}}$$

The answer is (D).

SI Solution

The steam flow rate is

$$\dot{m}_{\text{steam}} = (\text{WR})(P_{\text{turbine}})$$

$$= \left(2.5 \times 10^{-3} \ \frac{\text{kg}}{\text{kW·s}}\right)(750 \ \text{kW})$$

$$= 1.875 \ \text{kg/s}$$

From App. 24.O for 1 MPa, $T_{\text{sat}} = 179.9°\text{C}$.

The steam temperature is

$$179.9°\text{C} + 30°\text{C} = 209.9°\text{C}$$

From App. 24.P, $h_1 = 2851.0$ kJ/kg.

From Eq. 27.22,

$$P_{\text{turbine}} = \dot{m}(h_1 - h_2)$$

$$h_1 - h_2 = \frac{P_{\text{turbine}}}{\dot{m}} = \frac{750 \text{ kW}}{1.875 \frac{\text{kg}}{\text{s}}}$$

$$= 400 \text{ kJ/kg}$$

$$h_2 = h_1 - 400 \frac{\text{kJ}}{\text{kg}}$$

$$= 2851.0 \frac{\text{kJ}}{\text{kg}} - 400 \frac{\text{kJ}}{\text{kg}}$$

$$= 2451.0 \text{ kJ/kg}$$

From App. 24.O for $p_2 = 90$ kPa,

$$T_{\text{sat}} = T_2 = T_3 = 96.69°C$$

$$h_{f,3} = 405.20 \text{ kJ/kg}$$

From App. 35.B, the specific heat of water at 18°C is $c_p = 4.186$ kJ/kg·K.

Assuming water and steam leave in thermal equilibrium, the heat lost by steam is equal to the heat gained by water.

$$\dot{m}_{\text{water}} c_{p,\text{water}}(T_{\text{water,out}} - T_{\text{water,in}}) = \dot{m}_{\text{steam}}(h_2 - h_{f,3})$$

Since the terminal difference is zero, $T_{\text{water,out}} = T_3$.

$$\dot{m}_{\text{water}}\left(4.186 \frac{\text{kJ}}{\text{kg·K}}\right)(96.71°C - 18°C)$$

$$= \left(1.875 \frac{\text{kg}}{\text{s}}\right)\left(2451.0 \frac{\text{kJ}}{\text{kg}} - 405.20 \frac{\text{kJ}}{\text{kg}}\right)$$

$$\dot{m}_{\text{water}} = \boxed{11.64 \text{ kg/s}}$$

The answer is (D).

5. *Customary U.S. Solution*

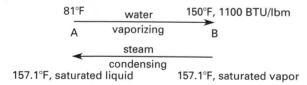

(a) From App. 24.A, at 81°F, $h_{\text{water},1} \approx 49.06$ Btu/lbm.

From App. 24.B, for 4.45 psia, $T_{\text{sat}} = 157.1°F$.

The heat transferred to the water is

$$Q = \dot{m}_{\text{water}}(h_{\text{water},2} - h_{\text{water},1})$$

$$= \left(332{,}000 \frac{\text{lbm}}{\text{hr}}\right)\left(1100 \frac{\text{Btu}}{\text{lbm}} - 49.06 \frac{\text{Btu}}{\text{lbm}}\right)$$

$$= 3.489 \times 10^8 \text{ Btu/hr}$$

The two end temperature differences are

$$\Delta T_A = 157.1°F - 81°F = 76.1°F$$

$$\Delta T_B = 157.1°F - 150°F = 7.1°F$$

The logarithmic mean temperature difference from Eq. 36.67 is

$$\Delta T_{lm} = \frac{\Delta T_A - \Delta T_B}{\ln\left(\dfrac{\Delta T_A}{\Delta T_B}\right)}$$

$$= \frac{76.1°F - 7.1°F}{\ln\left(\dfrac{76.1°F}{7.1°F}\right)} = 29.09°F$$

Since $T_{\text{steam},A} = T_{\text{steam},B}$, the correction factor (F_c) for ΔT_{lm} is 1.

The overall heat transfer coefficient is calculated from Eq. 36.68.

$$Q = UAF_c\Delta T_{lm}$$

$$U = \frac{Q}{AF_c\Delta T_{lm}}$$

$$= \frac{3.489 \times 10^8 \dfrac{\text{Btu}}{\text{hr}}}{(1850 \text{ ft}^2)(1)(29.09°F)}$$

$$= \boxed{6483 \text{ Btu/hr-ft}^2\text{-°F}}$$

The answer is (C).

(b) From App. 24.B, the enthalpy of saturated steam at 4.45 psia is $h_1 = 1128.6$ Btu/lbm.

From App. 24.B, the enthalpy of saturated water at 4.45 psia is $h_2 = 125.1$ Btu/lbm.

The enthalpy change for steam during condensation is

$$\Delta h = h_1 - h_2$$

$$= 1128.6 \frac{\text{Btu}}{\text{lbm}} - 125.1 \frac{\text{Btu}}{\text{lbm}}$$

$$= 1003.5 \text{ Btu/lbm}$$

The heat transferred (Q) from the steam is equal to the heat gained by water.

$$Q = \dot{m}_{\text{steam}}\Delta h$$

$$\dot{m}_{\text{steam}} = \frac{Q}{\Delta h} = \frac{3.489 \times 10^8 \dfrac{\text{Btu}}{\text{hr}}}{1003.5 \dfrac{\text{Btu}}{\text{lbm}}}$$

$$= 3.477 \times 10^5 \text{ lbm/hr}$$

The energy extraction rate is

$$\dot{m}_{\text{steam}} h_1 = \left(3.477 \times 10^5 \ \frac{\text{lbm}}{\text{hr}} \right) \left(1128.6 \ \frac{\text{Btu}}{\text{lbm}} \right)$$

$$= \boxed{3.92 \times 10^8 \ \text{Btu/hr}}$$

The answer is (B).

SI Solution

$$\begin{array}{ccc}
27°\text{C} & \xrightarrow{\quad \text{water} \quad} & 65.6°\text{C, } 2.56 \ \text{MJ/kg} \\
\text{A} & \text{vaporizing} & \text{B} \\
\\
69.51°\text{C} & \xleftarrow[\text{condensing}]{\quad \text{steam} \quad} & 69.51°\text{C}
\end{array}$$

(a) From App. 24.N, at 27°C, $h_{\text{water},1} = 113.19$ kJ/kg.

From App. 24.O, for 30.6 kPa (0.306 bar), $T_{\text{sat}} = 69.51°$C.

The heat transferred to the water is

$$Q = \dot{m}_{\text{water}} (h_{\text{water},2} - h_{\text{water},1})$$

$$= \left(41.8 \ \frac{\text{kg}}{\text{s}} \right) \left(\left(2.56 \ \frac{\text{MJ}}{\text{kg}} \right) \left(1000 \ \frac{\text{kJ}}{\text{MJ}} \right) \right.$$

$$\left. - 113.19 \ \frac{\text{kJ}}{\text{kg}} \right)$$

$$= 102\,277 \ \text{kJ/s}$$

The two end temperature differences are

$$\Delta T_A = 69.51°\text{C} - 27°\text{C} = 42.5°\text{C}$$
$$\Delta T_B = 69.51°\text{C} - 65.6°\text{C} = 3.9°\text{C}$$

The logarithmic mean temperature difference is calculated from Eq. 36.69.

$$\Delta T_{lm} = \frac{\Delta T_A - \Delta T_B}{\ln \left(\dfrac{\Delta T_A}{\Delta T_B} \right)} = \frac{42.5°\text{C} - 3.9°\text{C}}{\ln \left(\dfrac{42.5°\text{C}}{3.9°\text{C}} \right)}$$

$$= 16.16°\text{C}$$

Since $T_{\text{steam},A} = T_{\text{steam},B}$, the correction factor (F_c) for ΔT_{lm} is 1.

The overall heat transfer coefficient is calculated from Eq. 36.70.

$$Q = UAF_c \Delta T_{lm}$$

$$U = \frac{Q}{AF_c \Delta T_{lm}} = \frac{\left(102\,277 \ \dfrac{\text{kJ}}{\text{s}} \right) \left(1000 \ \dfrac{\text{W}}{\text{kW}} \right)}{(172 \ \text{m}^2)(1)(16.16°\text{C})}$$

$$= \boxed{36\,797 \ \text{W/m}^2 \cdot °\text{C}}$$

The answer is (C).

(b) From App. 24.O, the enthalpy of saturated steam at 30.6 kPa is $h_1 = 2625.2$ kJ/kg.

From App. 24.O, the enthalpy of saturated water at 30.6 kPa is $h_2 = 290.97$ kJ/kg.

The enthalpy change for steam during condensation is

$$\Delta h = h_1 - h_2$$

$$= 2625.2 \ \frac{\text{kJ}}{\text{kg}} - 290.97 \ \frac{\text{kJ}}{\text{kg}}$$

$$= 2334.2 \ \text{kJ/kg}$$

The heat transfer (Q) from the steam is equal to the heat gained by water.

$$Q = \dot{m}_{\text{steam}} \Delta h$$

$$\dot{m}_{\text{steam}} = \frac{Q}{\Delta h} = \frac{102\,277 \ \dfrac{\text{kJ}}{\text{s}}}{2334.2 \ \dfrac{\text{kJ}}{\text{kg}}} = 43.82 \ \text{kg/s}$$

The energy extraction rate is

$$\dot{m}_{\text{steam}} h_1 = \left(43.82 \ \frac{\text{kg}}{\text{s}} \right) \left(2625.2 \ \frac{\text{kJ}}{\text{kg}} \right)$$

$$= \boxed{115\,036 \ \text{kW} \quad (115.0 \ \text{MW})}$$

The answer is (B).

6.

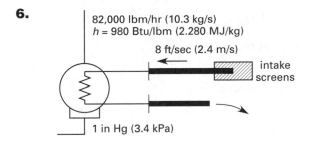

Customary U.S. Solution

(a) From App. 16.B, the dimensions of extra strong 30 in steel pipe are

$$D_i = 2.4167 \ \text{ft}$$
$$A_i = 4.5869 \ \text{ft}^2$$

From Table 17.2, the specific roughness for steel is $\epsilon = 0.0002$ ft.

$$\frac{\epsilon}{D} = \frac{0.0002 \ \text{ft}}{2.4167 \ \text{ft}} = 0.000083$$

Assume fully turbulent flow. (The Reynolds number can also be calculated.)

From the Moody friction factor chart for fully turbulent flow, $f \approx 0.012$.

The head loss from Eq. 17.22 is

$$h_f = \frac{fLv^2}{2Dg}$$

$$= \frac{(0.012)(120 \text{ ft})\left(8 \dfrac{\text{ft}}{\text{sec}}\right)^2}{(2)(2.4187 \text{ ft})\left(32.2 \dfrac{\text{ft}}{\text{sec}^2}\right)} = 0.59 \text{ ft}$$

The screen loss is

$$6 \text{ in wg} = (6 \text{ in})\left(\frac{1 \text{ ft}}{12 \text{ in}}\right)$$

$$= 0.5 \text{ ft}$$

$$\text{velocity head} = \frac{v^2}{2g}$$

$$= \frac{\left(8 \dfrac{\text{ft}}{\text{sec}}\right)^2}{(2)\left(32.2 \dfrac{\text{ft}}{\text{sec}^2}\right)} = 0.99 \text{ ft}$$

There is inadequate information to evaluate pressure and elevation heads. The total head added by a coolant pump (not shown) not including losses inside the condenser is

$$0.59 \text{ ft} + 0.5 \text{ ft} + 0.99 \text{ ft} = \boxed{2.08 \text{ ft}}$$

The answer is (A).

(b) The condenser pressure is

$$(1 \text{ in Hg})\left(0.491 \frac{\text{lbf}}{\text{in}^3}\right) = 0.5 \text{ lbf/in}^2$$

From App. 24.B, the enthalpy of the saturated liquid is $h_f = 47.08$ Btu/lbm.

The specific heat of water is $c_{p,\text{water}} = 1$ Btu/lbm-°F.

The heat lost by the steam is equal to the heat gained by the water.

$$\dot{m}_{\text{water}}c_{p,\text{water}}(\Delta T) = \dot{m}_{\text{steam}}(h_2 - h_f)$$

$$\dot{m}_{\text{water}}\left(1 \frac{\text{Btu}}{\text{lbm-°F}}\right)(10°\text{F})$$

$$= \left(82{,}000 \frac{\text{lbm}}{\text{hr}}\right)\left(980 \frac{\text{Btu}}{\text{lbm}} - 47.08 \frac{\text{Btu}}{\text{lbm}}\right)$$

$$\dot{m}_{\text{water}} = 7.6499 \times 10^6 \text{ lbm/hr}$$

The density of water, ρ, is 62.4 lbm/ft^3.

$$Q = \frac{\dot{m}_{\text{water}}}{\rho}$$

$$= \frac{\left(7.6499 \times 10^6 \dfrac{\text{lbm}}{\text{hr}}\right)\left(\dfrac{1 \text{ hr}}{60 \text{ min}}\right)\left(7.48 \dfrac{\text{gal}}{\text{ft}^3}\right)}{62.4 \dfrac{\text{lbm}}{\text{ft}^3}}$$

$$= \boxed{1.528 \times 10^4 \text{ gal/min}}$$

The answer is (D).

(The flow rate can also be determined from the velocity and pipe area. However, this does not use the 10° data or perform an energy balance.)

SI Solution

(a) From App. 16.B, the dimensions of extra strong 30 in steel pipe, using the footnote from the table, are

$$D_i = (29.00 \text{ in})\left(25.4 \frac{\text{mm}}{\text{in}}\right)\left(\frac{1 \text{ m}}{1000 \text{ mm}}\right)$$

$$= 0.7366 \text{ m}$$

$$A_i = (660.52 \text{ in}^2)\left(645 \frac{\text{mm}^2}{\text{in}^2}\right)\left(\frac{1 \text{ m}}{1000 \text{ mm}}\right)^2$$

$$= 0.4260 \text{ m}^2$$

From Table 17.2, the specific roughness for steel is $\epsilon = 6.0 \times 10^{-5}$ m.

$$\frac{\epsilon}{D} = \frac{6.0 \times 10^{-5} \text{ m}}{0.7366 \text{ m}} = 0.0000815$$

Assume fully turbulent flow. (The Reynolds number can also be calculated.)

From the Moody friction factor chart for fully turbulent flow, $f \approx 0.012$.

The head loss from Eq. 17.22 is

$$h_f = \frac{fLv^2}{2Dg}$$

$$= \frac{(0.012)(36 \text{ m})\left(2.4 \dfrac{\text{m}}{\text{s}}\right)^2}{(2)(0.7366 \text{ m})\left(9.81 \dfrac{\text{m}}{\text{s}^2}\right)} = 0.1722 \text{ m}$$

$$p_f = \rho g h$$

$$= \left(1000 \frac{\text{kg}}{\text{m}^3}\right)\left(9.81 \frac{\text{m}}{\text{s}^2}\right)(0.1722 \text{ m})\left(\frac{1 \text{ kPa}}{1000 \text{ Pa}}\right)$$

$$= 1.69 \text{ kPa}$$

The screen loss is 1.5 kPa.

The velocity head is

$$\frac{v^2}{2g} = \frac{\left(2.4\ \frac{m}{s}\right)^2}{(2)\left(9.81\ \frac{m}{s^2}\right)} = 0.2936\ m$$

$$p_v = \rho g h$$
$$= \left(1000\ \frac{kg}{m^3}\right)\left(9.81\ \frac{m}{s^2}\right)(0.2936\ m)\left(\frac{1\ kPa}{1000\ Pa}\right)$$
$$= 2.88\ kPa$$

There is inadequate information to evaluate pressure and elevation heads. The total head added by a coolant pump (not shown) not including losses inside the condenser is

$$1.69\ kPa + 1.5\ kPa + 2.88\ kPa = \boxed{6.07\ kPa}$$

The answer is (A).

(b) The condenser pressure is 3.4 kPa.

From App. 24.N, the enthalpy of saturated liquid is $h_f = 109.8\ kJ/kg$.

The specific heat of water is $c_{p,\text{water}} = 4.187\ kJ/kg{\cdot}K$.

The heat lost by the steam is equal to the heat gained by the water.

$$\dot{m}_{\text{water}} c_{p,\text{water}}(\Delta T) = \dot{m}_{\text{steam}}(h_2 - h_f)$$

$$\dot{m}_{\text{water}}\left(4.187\ \frac{kJ}{kg{\cdot}K}\right)(5.6°C)$$
$$= \left(10.3\ \frac{kg}{s}\right)\left(\left(2.280\ \frac{MJ}{kg}\right)\left(1000\ \frac{kJ}{MJ}\right) - 109.8\ \frac{kJ}{kg}\right)$$

$$\dot{m}_{\text{water}} = 953.3\ kg/s$$

The density of water, ρ, is $1000\ kg/m^3$.

$$Q = \frac{\dot{m}_{\text{water}}}{\rho}$$
$$= \left(\frac{953.3\ \frac{kg}{s}}{1000\ \frac{kg}{m^3}}\right)\left(1000\ \frac{L}{m^3}\right)\left(60\ \frac{s}{min}\right)$$
$$= \boxed{57\,200\ L/min}$$

The answer is (D).

7. *Customary U.S. Solution*

From App. 24.A, the enthalpy of saturated liquid at 60°F is $h_1 = 28.08\ Btu/lbm$.

From App. 24.B, the enthalpy of saturated steam at 14.7 psia is $h_2 = 1150.3\ Btu/lbm$.

The heat transfer rate to the water is

$$Q = \dot{m}(h_2 - h_1)$$
$$= \left(100\ \frac{lbm}{hr}\right)\left(1150.3\ \frac{Btu}{lbm} - 28.08\ \frac{Btu}{lbm}\right)$$
$$= 1.122 \times 10^5\ Btu/hr$$
$$= \left(1.122 \times 10^5\ \frac{Btu}{hr}\right)\left(0.2931\ \frac{W{\cdot}hr}{Btu}\right)\left(\frac{1\ kW}{1000\ W}\right)$$
$$= 32.89\ kW$$

$$\text{cost} = \frac{(32.89\ kW)\left(\frac{\$0.04}{kW{\cdot}hr}\right)}{1 - 0.35}$$
$$= \boxed{\$2.02/hr}$$

The answer is (A).

SI Solution

From App. 24.N, for saturated liquid at 16°C, $h_1 = 67.17\ kJ/kg$.

From App. 24.N, the enthalpy of saturated steam at 101.3 kPa is $h_2 = 2675.4\ kJ/kg$.

The heat transfer rate to the water is

$$Q = \dot{m}(h_2 - h_1)$$
$$= \left(0.013\ \frac{kg}{s}\right)\left(2675.4\ \frac{kJ}{kg} - 67.17\ \frac{kJ}{kg}\right)$$
$$= 33.907\ kW$$

$$\text{cost} = \frac{(33.907\ kW)\left(\frac{\$0.04}{kW{\cdot}h}\right)}{1 - 0.35}$$
$$= \boxed{\$2.09/h}$$

The answer is (A).

8. *Customary U.S. Solution*

From App. 24.A, the enthalpy of saturated liquid at 60°F is $h_1 = 28.08\ Btu/lbm$.

From App. 24.B, for 40 psia steam, the enthalpy of saturated liquid, h_f, is 236.1 Btu/lbm. The heat of vaporization, h_{fg}, is 933.7 Btu/lbm. The enthalpy is given by Eq. 24.40.

$$h_2 = h_f + x h_{fg}$$
$$= 236.1\ \frac{Btu}{lbm} + (0.98)\left(933.7\ \frac{Btu}{lbm}\right)$$
$$= 1151.13\ Btu/lbm$$

Power Cycles

The heat transfer rate is

$$Q = \dot{m}(h_2 - h_1)$$

$$= \left(250 \ \frac{\text{lbm}}{\text{hr}}\right)\left(1151.13 \ \frac{\text{Btu}}{\text{lbm}} - 28.08 \ \frac{\text{Btu}}{\text{lbm}}\right)$$

$$\times \left(\frac{1 \ \text{hr}}{60 \ \text{min}}\right)$$

$$= 4679.4 \ \text{Btu/min}$$

Find the volume of gas used at standard conditions for a heating gas (60°F).

$$\dot{V}_{\text{std}} = \dot{V}\left(\frac{T_0}{T}\right)\left(\frac{p}{p_0}\right)$$

$$= \left(13.5 \ \frac{\text{ft}^3}{\text{min}}\right)\left(\frac{460 + 60°\text{F}}{460 + 80°\text{F}}\right)$$

$$\times \left(\frac{(4 \ \text{in Hg} + 30.2 \ \text{in Hg})\left(0.491 \ \frac{\text{lbm}}{\text{in}^3}\right)}{14.7 \ \frac{\text{lbf}}{\text{in}^2}}\right)$$

$$= 14.85 \ \text{SCFM}$$

The efficiency of the boiler is

$$\eta = \frac{Q}{\text{heat input}}$$

$$= \frac{4679.4 \ \frac{\text{Btu}}{\text{min}}}{\left(14.85 \ \frac{\text{ft}^3}{\text{min}}\right)\left(550 \ \frac{\text{Btu}}{\text{ft}^3}\right)}$$

$$= \boxed{0.573 \ (57.3\%)}$$

The answer is (C).

SI Solution

From App. 24.N, the enthalpy of saturated liquid at 16°C is $h_1 = 67.17$ kJ/kg.

From App. 24.O, for 300 kPa steam, the enthalpy of saturated liquid, h_f, is 561.43 kJ/kg. The enthalpy of vaporization, h_{fg}, is 2163.5 kJ/kg. The enthalpy is given by Eq. 24.40.

$$h_2 = h_f + x h_{fg}$$

$$= 561.43 \ \frac{\text{kJ}}{\text{kg}} + (0.98)\left(2163.5 \ \frac{\text{kJ}}{\text{kg}}\right)$$

$$= 2681.7 \ \text{kJ/kg}$$

The heat transfer rate is

$$Q = \dot{m}(h_2 - h_1)$$

$$= \left(0.032 \ \frac{\text{kg}}{\text{s}}\right)\left(2681.7 \ \frac{\text{kJ}}{\text{kg}} - 67.17 \ \frac{\text{kJ}}{\text{kg}}\right)$$

$$= 83.66 \ \text{kJ/s}$$

Find the volume of the gas used at standard conditions for a heating gas (16°C).

$$\dot{V}_{\text{std}} = \dot{V}\left(\frac{T_0}{T}\right)\left(\frac{p}{p_0}\right)$$

$$= \left(6.4 \ \frac{\text{L}}{\text{s}}\right)\left(\frac{1 \ \text{m}^3}{1000 \ \text{L}}\right)\left(\frac{16°\text{C} + 273}{26°\text{C} + 273}\right)$$

$$\times \left(\frac{13.6 \ \text{kPa} + 102.4 \ \text{kPa}}{101.3 \ \text{kPa}}\right)$$

$$= 0.00708 \ \text{m}^3/\text{s}$$

The efficiency of the boiler is

$$\eta = \frac{Q}{\text{heat input}}$$

$$= \frac{83.66 \ \frac{\text{kJ}}{\text{s}}}{\left(0.00708 \ \frac{\text{m}^3}{\text{s}}\right)\left(20.5 \ \frac{\text{MJ}}{\text{m}^3}\right)\left(1000 \ \frac{\text{kJ}}{\text{MJ}}\right)}$$

$$= \boxed{0.576 \ (57.6\%)}$$

The answer is (C).

9. *Customary U.S. Solution*

step 1: Determine the actual gravimetric analysis of the coal as fired. 1 lbm of coal contains 0.02 lbm moisture, leaving 0.98 lbm dry coal. 1% is lost to the ash pit. The remainder is 0.98 lbm − 0.01 lbm = 0.97 lbm.

step 2: Calculate the heating value of the remaining coal.

$$(0.97 \ \text{lbm})\left(12{,}800 \ \frac{\text{Btu}}{\text{lbm}}\right) = 12{,}416 \ \text{Btu}$$

step 3: Find the energy (Q) required to produce steam. From App. 24.A, the enthalpy of water at 120°F is $h_1 = 88.00$ Btu/lbm.

From App. 24.B, the enthalpy of saturated steam at 100 psia is $h_2 = 1187.5$ Btu/lbm.

$$Q = m(h_2 - h_1)$$

$$= (8.23 \ \text{lbm})\left(1187.5 \ \frac{\text{Btu}}{\text{lbm}} - 88.00 \ \frac{\text{Btu}}{\text{lbm}}\right)$$

$$= 9048.9 \ \text{Btu}$$

step 4: The combustion efficiency is

$$\eta = \frac{Q}{HV} = \frac{9048.9 \ \text{Btu}}{12{,}416 \ \text{Btu}} = \boxed{0.729 \ (73\%)}$$

The answer is (C).

SI Solution

Since boiler data are based on 1 unit mass of coal fired, step 1 will be the same as for the customary U.S. solution. Repeat the rest of the steps as follows.

step 2: Calculate the heating value of the remaining coal.

$$(0.97 \text{ kg}) \left(29.80 \, \frac{\text{MJ}}{\text{kg}}\right) \left(1000 \, \frac{\text{kJ}}{\text{MJ}}\right)$$
$$= 28\,906 \text{ kJ}$$

step 3: Find the energy (Q) required to produce steam. From App. 24.N, the enthalpy of water at 50°C is $h_1 = 209.34$ kJ/kg.

From App. 24.O, the enthalpy of saturated steam at 700 kPa is $h_2 = 2762.8$ kJ/kg.

$$Q = m(h_2 - h_1)$$
$$= (8.23 \text{ kg}) \left(2762.8 \, \frac{\text{kJ}}{\text{kg}} - 209.34 \, \frac{\text{kJ}}{\text{kg}}\right)$$
$$= 21\,015 \text{ kJ}$$

step 4: The combustion efficiency is

$$\eta = \frac{Q}{\text{HV}} = \frac{21\,015 \text{ kJ}}{28\,906 \text{ kJ}} = \boxed{0.727 \quad (72.7\%)}$$

The answer is (D).

10. *Customary U.S. Solution*

(a) Refer to Fig. 28.1.

At point D (leaving the superheater and entering the turbine), the enthalpy, h_D, and entropy, s_D, can be obtained from App. 24.C.

$$h_D = 1521.0 \text{ Btu/lbm}$$
$$s_D = 1.7376 \text{ Btu/lbm-°F}$$

For isentropic expansion through the turbine, $s_E = s_D$. From App. 24.B, the enthalpy of saturated liquid, h_f, the enthalpy of evaporation, h_{fg}, the entropy of saturated liquid, s_f, and the entropy of evaporation, s_{fg}, are

$$h_f = 130.2 \text{ Btu/lbm}$$
$$h_{fg} = 1000.5 \text{ Btu/lbm}$$
$$s_f = 0.2349 \text{ Btu/lbm-°F}$$
$$s_{fg} = 1.6089 \text{ Btu/lbm-°F}$$

The quality of the mixture for an isentropic process is given by Eq. 24.41 (at point E).

$$x_E = \frac{s_E - s_f}{s_{fg}}$$
$$= \frac{1.7376 \, \frac{\text{Btu}}{\text{lbm-°F}} - 0.2349 \, \frac{\text{Btu}}{\text{lbm-°F}}}{1.6089 \, \frac{\text{Btu}}{\text{lbm-°F}}}$$
$$= 0.9338$$

The isentropic enthalpy is given by Eq. 24.40 (at point E).

$$h_E = h_f + x_E h_{fg}$$
$$= 130.2 \, \frac{\text{Btu}}{\text{lbm}} + (0.9338) \left(1000.5 \, \frac{\text{Btu}}{\text{lbm}}\right)$$
$$= 1064.5 \text{ Btu/lbm}$$

From Eq. 27.17, the actual enthalpy of steam at point E is

$$h'_E = h_D - \eta_s(h_D - h_E)$$
$$= 1521 \, \frac{\text{Btu}}{\text{lbm}} - (0.75) \left(1521 \, \frac{\text{Btu}}{\text{lbm}} - 1064.5 \, \frac{\text{Btu}}{\text{lbm}}\right)$$
$$= 1178.6 \text{ Btu/lbm}$$

Since the pump work is negligible, the mass flow rate of steam is

$$\dot{m} = \frac{P}{W_{\text{turbine}}} = \frac{P}{h_D - h'_E}$$
$$= \frac{(200 \text{ MW}) \left(10^6 \, \frac{\text{W}}{\text{MW}}\right) \left(3.4121 \, \frac{\text{Btu}}{\text{hr}}\right)}{1521 \, \frac{\text{Btu}}{\text{lbm}} - 1178.6 \, \frac{\text{Btu}}{\text{lbm}}}$$
$$= \boxed{1.993 \times 10^6 \text{ lbm/hr}}$$

The answer is (C).

(b) The heat removed by the condenser is given by Eq. 27.30.

$$\dot{Q} = \dot{m}(h'_E - h_F)$$

From App. 24.B, at 5 psia, $h_F = 130.2$ Btu/lbm.

$$\dot{Q} = \left(1.993 \times 10^6 \, \frac{\text{lbm}}{\text{hr}}\right) \left(1178.6 \, \frac{\text{Btu}}{\text{lbm}} - 130.2 \, \frac{\text{Btu}}{\text{lbm}}\right)$$
$$= \boxed{2.090 \times 10^9 \text{ Btu/hr}}$$

The answer is (C).

Power Cycles

SI Solution

(a) Refer to Fig. 28.1.

At point D (leaving the superheater and entering the turbine), the enthalpy, h_D, and entropy, s_D, can be obtained from the Mollier diagram.

$$h_D \approx 3430 \text{ kJ/kg}$$

$$s_D \approx 7.25 \text{ kJ/kg·K}$$

For isentropic expansion through the turbine, $s_E = s_D$. From App. 24.O, the enthalpy of saturated liquid, h_f, the enthalpy of evaporation, h_{fg}, the entropy of saturated liquid, s_f, and the entropy of saturated vapor, s_g, are

$$h_f = 289.27 \text{ kJ/kg}$$

$$h_{fg} = 2335.3 \text{ kJ/kg}$$

$$s_f = 0.9441 \text{ kJ/kg·K}$$

$$s_g = 7.7675 \text{ kJ/kg·K}$$

The quality of the mixture for an isentropic process is given by Eq. 24.41 (at point ϵ) as

$$x_E = \frac{s_E - s_f}{s_{fg}} = \frac{s_E - s_f}{s_g - s_f}$$

$$= \frac{7.25 \dfrac{\text{kJ}}{\text{kg·K}} - 0.9441 \dfrac{\text{kJ}}{\text{kg·K}}}{7.7675 \dfrac{\text{kJ}}{\text{kg·K}} - 0.9441 \dfrac{\text{kJ}}{\text{kg·K}}}$$

$$= 0.9242$$

The isentropic enthalpy is given by Eq. 24.40 (at point E) as

$$h_E = h_f + x_E h_{fg}$$

$$= 289.27 \frac{\text{kJ}}{\text{kg}} + (0.9242)\left(2335.3 \frac{\text{kJ}}{\text{kg}}\right)$$

$$= 2447.6 \text{ kJ/kg}$$

From Eq. 27.17, the actual enthalpy of steam at point E is

$$h'_E = h_D - \eta_s(h_D - h_E)$$

$$= 3430 \frac{\text{kJ}}{\text{kg}} - (0.75)\left(3430 \frac{\text{kJ}}{\text{kg}} - 2447.6 \frac{\text{kJ}}{\text{kg}}\right)$$

$$= 2693.2 \text{ kJ/kg}$$

Since the pump work is negligible, the mass flow rate of steam is

$$\dot{m} = \frac{P}{W_{\text{turbine}}} = \frac{P}{h_D - h'_E}$$

$$= \frac{(200 \text{ MW})\left(10^3 \dfrac{\text{kW}}{\text{MW}}\right)}{3430 \dfrac{\text{kJ}}{\text{kg}} - 2693.2 \dfrac{\text{kJ}}{\text{kg}}}$$

$$= \boxed{271.4 \text{ kg/s}}$$

The answer is (C).

(b) The heat removed by the condenser is given by Eq. 27.30.

$$\dot{Q} = \dot{m}(h'_E - h_F)$$

From App. 24.O at 30 kPa, $h_F = 289.27 \text{ kJ/kg}$.

$$\dot{Q} = \left(271.4 \frac{\text{kg}}{\text{s}}\right)\left(2693.2 \frac{\text{kJ}}{\text{kg}} - 289.27 \frac{\text{kJ}}{\text{kg}}\right)$$

$$= \boxed{6.524 \times 10^5 \text{ kW}}$$

The answer is (C).

11. *Customary U.S. Solution*

The following illustration shows one of N layers. Each layer consists of 24 tubes, only 3 of which are shown.

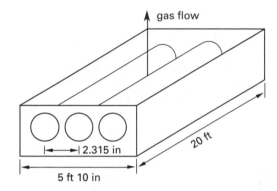

Assume stack gases consist primarily of nitrogen. The average gas temperature is

$$T_{\text{ave}} = \left(\tfrac{1}{2}\right)(635°\text{F} + 470°\text{F}) + 460 = 1012.5°\text{R}$$

The specific heat of nitrogen is calculated from Eq. 24.94 using constants given in Table 24.9.

$$c_p = A + BT + CT^2 + \frac{D}{\sqrt{T}}$$

$$= 0.2510 + (-1.63 \times 10^{-5})(1012.5°\text{R})$$
$$\quad + (20.4 \times 10^{-9})(1012.5°\text{R})^2 + 0$$

$$= 0.2554 \text{ Btu/lbm-°R}$$

For the purpose of calculating the logarithmic temperature difference, assume counterflow operation. The two end temperature differences are

$$\Delta T_A = 635°\text{F} - 285°\text{F} = 350°\text{F}$$

$$\Delta T_B = 470°\text{F} - 212°\text{F} = 258°\text{F}$$

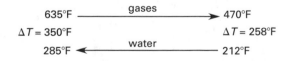

The logarithmic temperature difference (Eq. 36.67) is

$$\Delta T_{lm} = \frac{\Delta T_A - \Delta T_B}{\ln\left(\dfrac{\Delta T_A}{\Delta T_B}\right)} = \frac{350°F - 258°F}{\ln\left(\dfrac{350°F}{258°F}\right)}$$

$$= 301.7°F$$

Determine F_c for crossflow.

$$R = \frac{635°F - 470°F}{285°F - 212°F} = 2.26$$

$$x = \frac{285°F - 212°F}{635°F - 212°F} = 0.17$$

F_c for all heat exchanger configurations is close to 1.0 for this set of parameters.

The heat transfer from the temperature gain of water is equal to the heat transfer based on the logarithmic mean temperature difference. From Eq. 36.68,

$$Q = \dot{m}c_p\Delta T = U_oA_oF_c\Delta T_{lm}$$

$$\left(191{,}000\ \frac{lbm}{hr}\right)\left(0.2554\ \frac{Btu}{lbm\text{-}°R}\right)(635°F - 470°F)$$

$$= \left(10\ \frac{Btu}{hr\text{-}ft^2\text{-}°F}\right)A_o(1.0)(301.7°F)$$

$$A_o = 2667.9\ ft^2$$

The tube area per bank is

$$A_{bank} = N\pi D_o L$$

$$= 24\pi(1.315\ in)\left(\frac{1\ ft}{12\ in}\right)(20\ ft)$$

$$= 165.2\ ft^2$$

$$\text{no. of layers} = \frac{A_o}{A_{bank}}$$

$$= \frac{2667.9\ ft^2}{165.2\ ft^2} = \boxed{16.1 \quad [\text{say } 17]}$$

The answer is (D).

SI Solution

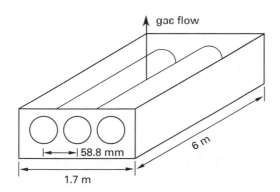

Assume stack gas consists primarily of nitrogen. The average gas temperature is

$$T_{ave} = \left(\tfrac{1}{2}\right)(335°C + 240°C) + 273 = 560.5\text{K} \quad (1009°R)$$

Use the value of the specific heat of nitrogen that was calculated for the customary U.S. solution since the two temperatures are almost the same.

$$c_p = \left(0.2554\ \frac{Btu}{lbm\text{-}°R}\right)\left(4186.8\ \frac{\dfrac{J}{kg\cdot°C}}{\dfrac{Btu}{lbm\text{-}°R}}\right)$$

$$= 1069\ J/kg\cdot°C$$

For the purpose of calculating the logarithmic temperature difference, assume counterflow operation. The two end temperature differences are

$$\Delta T_A = 335°C - 140°C = 195°C$$

$$\Delta T_B = 240°C - 100°C = 140°C$$

	gases	
335°C	⟶	240°C
ΔT = 195°C		ΔT = 140°C
140°C	⟵	100°C
	water	

The logarithmic mean temperature difference (Eq. 36.67) is

$$\Delta T_{lm} = \frac{\Delta T_A - \Delta T_B}{\ln\left(\dfrac{\Delta T_A}{\Delta T_B}\right)} = \frac{195°C - 140°C}{\ln\left(\dfrac{195°C}{140°C}\right)}$$

$$= 166.0°C$$

$$R = \frac{335°C - 240°C}{140°C - 100°C} = 2.38$$

$$x = \frac{140°C - 100°C}{335°C - 100°C} = 0.17$$

F_c for all heat exchanger configurations is close to 1.0 for this set of conditions.

The heat transfer from the temperature gain of water is equal to the heat transfer based on the logarithmic mean temperature difference. From Eq. 36.68,

$$Q = \dot{m}c_p\Delta T = U_oA_oF_c\Delta T_{lm}$$

$$\times\left(24\ \frac{kg}{s}\right)\left(1069\ \frac{J}{kg\cdot°C}\right)(335°C - 240°C)$$

$$= \left(57\ \frac{W}{m^2\cdot°C}\right)A_o(1.0)(166.0°C)$$

$$A_o = 257.6\ m^2$$

The tube area per bank is

$$A_{bank} = N\pi D_o L$$

$$= 24\pi(33.4\ mm)\left(\frac{1\ m}{1000\ mm}\right)(6\ m)$$

$$= 15.1\ m^2$$

$$\text{no. of layers} = \frac{A_o}{A_{\text{bank}}}$$

$$= \frac{257.6 \text{ m}^2}{15.1 \text{ m}^2} = \boxed{17.1} \quad \text{[say 18]}$$

The answer is (D).

12.

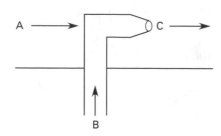

Customary U.S. Solution

(a) The enthalpy, h_A, and entropy, s_A, for steam at $600°F$ and 200 psia can be obtained from App. 24.C.

$$h_A = 1322.3 \text{ Btu/lbm}$$

$$s_A = 1.6771 \text{ Btu/lbm-°R}$$

For water at $82°F$, enthalpy can be obtained for saturated water from App. 24.A and corrected for compression to 200 psia using App. 24.D. (Correction for compression is optional.)

$$h_B = 50.06 \frac{\text{Btu}}{\text{lbm}} + 0.55 \frac{\text{Btu}}{\text{lbm}}$$

$$= 50.61 \text{ Btu/lbm}$$

From App. 24.A, $s_B = 0.09697$ Btu/lbm-°F. From an energy balance equation (with $Q = 0$, $v_1 = 0$, $\Delta z = 0$, and $W = 0$), Eq. 27.34 can be written as

$$0 = h_C(m_A + m_B) - (h_A m_A + h_B m_B)$$
$$+ \left(\frac{v_C^2}{2g_c J}\right)(m_A + m_B)$$

$$= h_C \left(1000 \frac{\text{lbm}}{\text{hr}} + 50 \frac{\text{lbm}}{\text{hr}}\right) - \left(\left(1322.3 \frac{\text{Btu}}{\text{lbm}}\right)\right.$$

$$\times \left(1000 \frac{\text{lbm}}{\text{hr}}\right) + \left(50.61 \frac{\text{Btu}}{\text{lbm}}\right)\left(50 \frac{\text{lbm}}{\text{hr}}\right)\right)$$

$$+ \frac{\left(2000 \frac{\text{ft}}{\text{sec}}\right)^2 \left(1000 \frac{\text{lbm}}{\text{hr}} + 50 \frac{\text{lbm}}{\text{hr}}\right)}{(2)\left(32.2 \frac{\text{lbm-ft}}{\text{lbf-sec}^2}\right)\left(778 \frac{\text{ft-lbf}}{\text{Btu}}\right)}$$

$$h_C = 1181.9 \text{ Btu/lbm}$$

$$p_C = 100 \text{ psia}$$

Since the enthalpy of saturated steam at 100 psia, h_g, from App. 24.B, is greater than h_C, steam leaving the desuperheater is not 100% saturated. From App. 24.B, the temperature of the saturated steam mix is $\boxed{327.81°F.}$

The answer is (A).

(b) The quality of the steam leaving can be determined using Eq. 24.40.

$$h_C = h_f + x h_{fg}$$

From App. 24.B, for 100 psia, $h_f = 298.5$ Btu/lbm and $h_{fg} = 889.0$ Btu/lbm.

$$x = \frac{1181.9 \frac{\text{Btu}}{\text{lbm}} - 298.5 \frac{\text{Btu}}{\text{lbm}}}{889.0 \frac{\text{Btu}}{\text{lbm}}}$$

$$= \boxed{0.994}$$

The answer is (D).

SI Solution

(a) The enthalpy, h_A, and entropy, s_A, for steam at $300°C$ and 1.5 MPa can be obtained from App. 24.P.

$$h_A = 3038.2 \text{ kJ/kg}$$

$$s_A = 6.9198 \text{ kJ/kg·K}$$

App. 24.Q does not go down to 1.5 MPa, so disregard the effect of compression. For water at $28°C$ from App. 24.N, $h_B = 117.37$ kJ/kg.

Similarly, from App. 24.N, $s_B = 0.4091$ kJ/kg·K.

From an energy balance equation (with $Q = 0$, $v_1 = 0$, $\Delta z = 0$, and $W = 0$), Eq. 27.34 can be written as

$$0 = h_C(m_A + m_B) - (m_A h_A + m_B h_B)$$
$$+ \left(\frac{v_C^2}{2}\right)(m_A + m_B)$$

$$= h_C \left(0.13 \frac{\text{kg}}{\text{s}} + 0.0063 \frac{\text{kg}}{\text{s}}\right) - \left(\left(0.13 \frac{\text{kg}}{\text{s}}\right)\right.$$

$$\times \left(3038.2 \frac{\text{kJ}}{\text{kg}}\right) + \left(0.0063 \frac{\text{kg}}{\text{s}}\right)\left(117.37 \frac{\text{kJ}}{\text{kg}}\right)\right)$$

$$+ \left(\frac{\left(600 \frac{\text{m}}{\text{s}}\right)^2}{2}\right)\left(0.13 \frac{\text{kg}}{\text{s}} + 0.0063 \frac{\text{kg}}{\text{s}}\right)$$

$$\times \left(\frac{1 \text{ kJ}}{1000 \text{ J}}\right)$$

$$h_C = 2723.2 \text{ kJ/kg}$$

$$p_C = 700 \text{ kPa}$$

Since the enthalpy of saturated steam at 700 kPa, from App. 24.O, is greater than h_C, steam leaving the desuperheater is not 100% saturated. From App. 24.O, the temperature of the steam mix is $\boxed{164.9°C.}$

The answer is (A).

(h) The quality of steam leaving can be determined using Eq. 24.40.

$$h_C = h_f + xh_{fg}$$

From App. 24.O, for 700 kPa, $h_f = 697.00$ kJ/kg and $h_{fg} = 2065.8$ kJ/kg.

$$x = \frac{2723.2 \ \dfrac{\text{kJ}}{\text{kg}} - 697.00 \ \dfrac{\text{kJ}}{\text{kg}}}{2065.8 \ \dfrac{\text{kJ}}{\text{kg}}} = \boxed{0.9808}$$

The answer is (D).

13. *Customary U.S. Solution*

(a) Work with the original system to find the steam flow.

From App. 24.C, at 100 psia and 400°F, $h_A = 1227.7$ Btu/lbm.

From App. 24.A, at 70°F, $h_B = 38.08$ Btu/lbm.

From App. 24.A, at 180°F, $h_C = 148.04$ Btu/lbm.

Let x = fraction of steam in mixture.

From the energy balance equation,

$$m_A h_A + m_B h_B = m_C h_C$$

$$x \left(1227.7 \ \frac{\text{Btu}}{\text{lbm}} \right)$$
$$+ (1-x) \left(38.08 \ \frac{\text{Btu}}{\text{lbm}} \right) = (1) \left(148.04 \ \frac{\text{Btu}}{\text{lbm}} \right)$$
$$x = 0.0924$$

The steam flow is

$$\dot{m} = x \left(2000 \ \frac{\text{lbm}}{\text{hr}} \right) = (0.0924) \left(2000 \ \frac{\text{lbm}}{\text{hr}} \right)$$
$$= 184.8 \ \text{lbm/hr}$$

Since the pressure drop across the heater is 5 psi,

$$p_D = p_F + 5 \ \text{psi}$$
$$= 20 \ \text{psia} + 5 \ \text{psi} = 25 \ \text{psia}$$

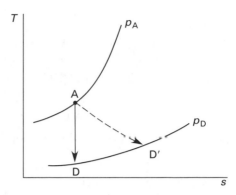

With isentropic expansion through the turbine, using the Mollier diagram, $h_D \approx 1115$ Btu/lbm (liquid-vapor mixture).

From Eq. 27.17,

$$h'_D = h_A - \eta_s(h_A - h_D)$$
$$= 1227.7 \ \frac{\text{Btu}}{\text{lbm}} - (0.60) \left(1227.7 \ \frac{\text{Btu}}{\text{lbm}} - 1115 \ \frac{\text{Btu}}{\text{lbm}} \right)$$
$$= 1160 \ \text{Btu/lbm}$$

The power output is

$$P = \eta \dot{m}(h_A - h'_D)$$

$$= \frac{\begin{array}{c} (0.96) \left(184.8 \ \dfrac{\text{lbm}}{\text{hr}} \right) \\[2mm] \times \left(1227.7 \ \dfrac{\text{Btu}}{\text{lbm}} - 1160 \ \dfrac{\text{Btu}}{\text{lbm}} \right) \left(778 \ \dfrac{\text{ft-lbf}}{\text{Btu}} \right) \end{array}}{\left(3600 \ \dfrac{\text{sec}}{\text{hr}} \right) \left(550 \ \dfrac{\text{ft-lbf}}{\text{hp-sec}} \right)}$$

$$= \boxed{4.72 \ \text{hp}}$$

The answer is (A).

(b) Let x = fraction of steam entering the heater.

From the energy balance equation,

$$m_A h'_D + m_B h_B = m_C h_C$$

$$x \left(1160 \ \frac{\text{Btu}}{\text{lbm}} \right)$$
$$+ (1-x) \left(38.08 \ \frac{\text{Btu}}{\text{lbm}} \right) = (1) \left(148.04 \ \frac{\text{Btu}}{\text{lbm}} \right)$$
$$x = 0.098$$

$$\dot{m}_F = \frac{184.8 \ \dfrac{\text{lbm}}{\text{hr}}}{0.098}$$
$$= \boxed{1886 \ \text{lbm/hr}}$$

The answer is (C).

SI Solution

(a) Work with the original system to find the steam flow.

From App. 24.P, at 700 kPa and 200°C, $h_A = 2845.3$ kJ/kg.

From App. 24.N, at 21°C, $h_B = 88.10$ kJ/kg.

From App. 24.N, at 80°C, $h_C = 335.01$ kJ/kg.

Let x = fraction of steam in mixture.

From the energy balance equation,

$$m_A h_A + m_B h_B = m_C h_C$$

$$x \left(2845.3 \ \frac{\text{kJ}}{\text{kg}} \right)$$

$$+ (1-x) \left(88.10 \ \frac{\text{kJ}}{\text{kg}} \right) = (1) \left(335.01 \ \frac{\text{kJ}}{\text{kg}} \right)$$

$$x = 0.0896$$

The steam flow is

$$\dot{m} = x \left(0.25 \ \frac{\text{kg}}{\text{s}} \right) = (0.0896) \left(0.25 \ \frac{\text{kg}}{\text{s}} \right)$$

$$= 0.0224 \ \text{kg/s}$$

Since the pressure drop across the heater is 30 kPa,

$$p_D = p_F + 30 \ \text{kPa}$$

$$= 150 \ \text{kPa} + 30 \ \text{kPa}$$

$$= 180 \ \text{kPa}$$

For isentropic expansion through the turbine from the Mollier diagram, $h_D \approx 2583$ kJ/kg (liquid-vapor mixture).

From Eq. 27.17,

$$h'_D = h_A - \eta_s (h_A - h_D)$$

$$= 2845.3 \ \frac{\text{kJ}}{\text{kg}} - (0.60) \left(2845.3 \ \frac{\text{kJ}}{\text{kg}} - 2583 \ \frac{\text{kJ}}{\text{kg}} \right)$$

$$= 2687.9 \ \text{kJ/kg}$$

The power output is

$$P = \eta \dot{m}(h_A - h'_D)$$

$$= (0.96) \left(0.0224 \ \frac{\text{kg}}{\text{s}} \right) \left(2845.3 \ \frac{\text{kJ}}{\text{kg}} - 2687.9 \ \frac{\text{kJ}}{\text{kg}} \right)$$

$$= \boxed{3.38 \ \text{kW}}$$

The answer is (A).

(b) Let x = fraction of steam entering the heater.

From the energy balance equation,

$$m_A h'_D + m_B h_B = m_C h_C$$

$$x \left(2687.9 \ \frac{\text{kJ}}{\text{kg}} \right)$$

$$+ (1-x) \left(88.10 \ \frac{\text{kJ}}{\text{kg}} \right) = (1) \left(335.01 \ \frac{\text{kJ}}{\text{kg}} \right)$$

$$x = 0.0950$$

$$\dot{m}_F = \frac{0.0224 \ \dfrac{\text{kg}}{\text{s}}}{0.0950}$$

$$= \boxed{0.236 \ \text{kg/s}}$$

The answer is (C).

28 Vapor Power Cycles

PRACTICE PROBLEMS

1. A steam cycle operates between 650°F and 100°F (340°C and 38°C). What is the maximum possible thermal efficiency?

(A) 42%

(B) 49%

(C) 54%

(D) 58%

2. A steam Carnot cycle operates between 650°F and 100°F (340°C and 38°C). The turbine and compressor (pump) isentropic efficiencies are 90% and 80%, respectively. What is the thermal efficiency?

(A) 31%

(B) 36%

(C) 41%

(D) 47%

3. A steam turbine cycle produces 600 MWe (megawatts of electrical power). The condenser load is 3.07×10^9 Btu/hr (900 MW). What is the thermal efficiency?

(A) 32%

(B) 37%

(C) 40%

(D) 44%

4. A steam Rankine cycle operates with 100 psia (700 kPa) saturated steam that is reduced to 1 atm through expansion in a turbine with an isentropic efficiency of 80%. There is significant subcooling, and water is at 80°F (27°C) and 1 atm when it enters the boilerfeed pump. The pump's isentropic efficiency is 60%. What is the cycle's thermal efficiency?

(A) 10%

(B) 14%

(C) 21%

(D) 26%

5. A turbine and the boilerfeed pumps in a reheat cycle have isentropic efficiencies of 88% and 96%, respectively. The cycle starts with water at 60°F (16°C) at the entrance to the boilerfeed pump and produces 600°F (300°C), 600 psia (4 MPa) steam. The steam is reheated when its pressure drops during the first expansion to 20 psia (150 kPa). What is the thermal efficiency of the cycle?

(A) 28%

(B) 34%

(C) 39%

(D) 42%

6. The cycle in Prob. 5 is modified to include a bleed of 4.5 psia (37.73 kPa) steam (within the second expansion) for feedwater heating in a closed feedwater heater. The condenser pressure is unchanged. Steam leaves the feedwater heater as saturated liquid. The terminal temperature difference in the heater is 6°F (3°C). Condensate and drip pump work are both negligible. What is the thermal efficiency of the cycle?

(A) 23%

(B) 28%

(C) 35%

(D) 44%

7. (*Time limit: one hour*) A reheat steam cycle operates as shown. The pump work between points F and G is 0.15 Btu/lbm (0.3 kJ/kg).

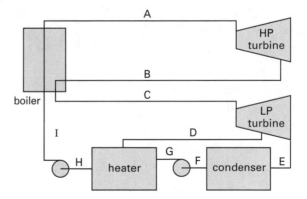

At A: 900 psia (6.2 MPa)
800°F (420°C)

At B: 200 psia (1.5 MPa)
1270 Btu/lbm (2960 kJ/kg)

At C: 190 psia (1.4 MPa)
800°F (420°C)

At D: 50 psia (350 kPa)
1280 Btu/lbm (2980 kJ/kg)

At E: 2 in Hg absolute
1075 Btu/lbm (2500 kJ/kg)

At F: 69.73 Btu/lbm (162.5 kJ/kg)

At H: 250.2 Btu/lbm (583.0 kJ/kg)
0.0173 ft^3/lbm

At I: 253.1 Btu/lbm (589.7 kJ/kg)

(a) What is the isentropic efficiency of the high-pressure turbine?
- (A) 54%
- (B) 61%
- (C) 69%
- (D) 75%

(b) What is the thermal efficiency of the cycle?
- (A) 22%
- (B) 29%
- (C) 34%
- (D) 41%

8. (*Time limit: one hour*) A precision air turbine is used to drive a small dentist's drill. 140°F (60°C) air enters the turbine at the rate of 15 lbm/hr (1.9 g/s). The output of the turbine is 0.25 hp (0.19 kW). The turbine exhausts to 15 psia (103.5 kPa). The flow is steady and the process is adiabatic. The isentropic efficiency of the expansion process is 60%.

(a) What is the exhaust temperature?
- (A) 370°R (210K)
- (B) 420°R (230K)
- (C) 450°R (240K)
- (D) 610°R (340K)

(b) What is the inlet air pressure?
- (A) 160 psia (1.2 MPa)
- (B) 210 psia (1.5 MPa)
- (C) 240 psia (1.7 MPa)
- (D) 290 psia (2.0 MPa)

(c) What is the change in entropy through the turbine?
- (A) 0.078 Btu/lbm-°R (0.36 kJ/kg·K)
- (B) 0.10 Btu/lbm-°R (0.46 kJ/kg·K)
- (C) 0.18 Btu/lbm-°R (0.82 kJ/kg·K)
- (D) 0.32 Btu/lbm-°R (1.5 kJ/kg·K)

SOLUTIONS

1. *Customary U.S. Solution*

The maximum possible thermal efficiency is given by the Carnot cycle. From Eq. 28.8,

$$\eta_{th} = \frac{T_{high} - T_{low}}{T_{high}}$$

The absolute temperatures are

$$T_{high} = 650°F + 460 = 1110°R$$
$$T_{low} = 100°F + 460 = 560°R$$
$$\eta_{th} = \frac{1110°R - 560°R}{1110°R} = \boxed{0.495 \ (49.5\%)}$$

The answer is (B).

SI Solution

The maximum possible thermal efficiency is given by the Carnot cycle. From Eq. 28.8,

$$\eta_{th} = \frac{T_{high} - T_{low}}{T_{high}}$$

The absolute temperatures are

$$T_{high} = 340°C + 273 = 613K$$
$$T_{low} = 38°C + 273 = 311K$$
$$\eta_{th} = \frac{613K - 311K}{613K} = \boxed{0.493 \ (49.3\%)}$$

The answer is (B).

2. *Customary U.S. Solution*

Refer to the Carnot cycle (Fig. 28.2).

At point A, from App. 24.A, for saturated liquid at $T_A = 650°F$,

$$h_A = 696.0 \ Btu/lbm$$
$$s_A = 0.8833 \ Btu/lbm\text{-}°F$$

At point B, from App. 24.A, for saturated vapor at $T_B = 650°F$,

$$h_B = 1119.7 \ Btu/lbm$$
$$s_B = 1.2651 \ Btu/lbm\text{-}°F$$

At point C,

$$T_C = 100°F$$
$$s_C = s_B = 1.2651 \ Btu/lbm\text{-}°F$$

From App. 24.A,

$$s_f = 0.1296 \text{ Btu/lbm-°F}$$
$$s_g = 1.9819 \text{ Btu/lbm-°F}$$
$$h_f = 68.03 \text{ Btu/lbm}$$
$$h_{fg} = 1036.7 \text{ Btu/lbm}$$

$$x_C = \frac{s_C - s_f}{s_g - s_f} = \frac{1.2651 \frac{\text{Btu}}{\text{lbm-°F}} - 0.1296 \frac{\text{Btu}}{\text{lbm-°F}}}{1.9819 \frac{\text{Btu}}{\text{lbm-°F}} - 0.1296 \frac{\text{Btu}}{\text{lbm-°F}}}$$
$$= 0.613$$

$$h_C = h_f + x_C h_{fg}$$
$$= 68.03 \frac{\text{Btu}}{\text{lbm}} + (0.613)\left(1036.7 \frac{\text{Btu}}{\text{lbm}}\right)$$
$$= 703.5 \text{ Btu/lbm}$$

At point D,

$$T_D = 100°F$$
$$s_D = s_A = 0.8833 \text{ Btu/lbm-°F}$$

$$x_D = \frac{s_D - s_f}{s_g - s_f} = \frac{0.8833 \frac{\text{Btu}}{\text{lbm-°F}} - 0.1296 \frac{\text{Btu}}{\text{lbm-°F}}}{1.9819 \frac{\text{Btu}}{\text{lbm-°F}} - 0.1296 \frac{\text{Btu}}{\text{lbm-°F}}}$$
$$= 0.407$$

$$h_D = h_f + x_D h_{fg}$$
$$= 68.03 \frac{\text{Btu}}{\text{lbm}} + (0.407)\left(1036.7 \frac{\text{Btu}}{\text{lbm}}\right)$$
$$= 489.8 \text{ Btu/lbm}$$

From Eq. 28.9, due to the inefficiency of the turbine,

$$h'_C = h_B - \eta_{s,\text{turbine}}(h_B - h_C)$$
$$= 1119.7 \frac{\text{Btu}}{\text{lbm}} - (0.9)\left(1119.7 \frac{\text{Btu}}{\text{lbm}} - 703.5 \frac{\text{Btu}}{\text{lbm}}\right)$$
$$= 745.1 \text{ Btu/lbm}$$

From Eq. 28.10, due to the inefficiency of the pump,

$$h'_A = h_D + \frac{h_A - h_D}{\eta_{s,\text{pump}}}$$
$$= 489.8 \frac{\text{Btu}}{\text{lbm}} + \frac{696.0 \frac{\text{Btu}}{\text{lbm}} - 489.8 \frac{\text{Btu}}{\text{lbm}}}{0.8}$$
$$= 747.6 \text{ Btu/lbm}$$

From Eq. 28.8, the thermal efficiency of the entire cycle is

$$\eta_{\text{th}} = \frac{(h_B - h'_C) - (h'_A - h_D)}{h_B - h'_A}$$
$$= \frac{\left(1119.7 \frac{\text{Btu}}{\text{lbm}} - 745.1 \frac{\text{Btu}}{\text{lbm}}\right) - \left(747.6 \frac{\text{Btu}}{\text{lbm}} - 489.8 \frac{\text{Btu}}{\text{lbm}}\right)}{1119.7 \frac{\text{Btu}}{\text{lbm}} - 747.6 \frac{\text{Btu}}{\text{lbm}}}$$
$$= \boxed{0.314 \quad (31.4\%)}$$

The answer is (A).

SI Solution

At point A, $T_A = 340°C$.

From App. 24.N, for saturated liquid,

$$h_A = 1594.5 \text{ kJ/kg}$$
$$s_A = 3.6601 \text{ kJ/kg·K}$$

At point B, $T_B = 340°C$.

From App. 24.N, for saturated vapor,

$$h_B = 2621.9 \text{ kJ/kg}$$
$$s_B = 5.3356 \text{ kJ/kg·K}$$

At point C,

$$T_C = 38°C$$
$$s_C = s_B = 5.3356 \text{ kJ/kg·K}$$

From App. 24.N,

$$s_f = 0.5456 \text{ kJ/kg·K}$$
$$s_g = 8.2935 \text{ kJ/kg·K}$$
$$h_f = 159.17 \text{ kJ/kg}$$
$$h_{fg} = 2410.7 \text{ kJ/kg}$$

$$x_C = \frac{s_C - s_f}{s_g - s_f} = \frac{5.3356 \frac{\text{kJ}}{\text{kg·K}} - 0.5456 \frac{\text{kJ}}{\text{kg·K}}}{8.2935 \frac{\text{kJ}}{\text{kg·K}} - 0.5456 \frac{\text{kJ}}{\text{kg·K}}}$$
$$= 0.618$$

$$h_C = h_f + x_C h_{fg}$$
$$= 159.17 \frac{\text{kJ}}{\text{kg}} + (0.618)\left(2410.7 \frac{\text{kJ}}{\text{kg}}\right)$$
$$= 1649.0 \text{ kJ/kg}$$

At point D,

$$T_D = 38°C$$

$$s_D = s_A = 3.6601 \text{ kJ/kg·K}$$

$$x_D = \frac{s_D - s_f}{s_g - s_f}$$

$$= \frac{3.6601 \dfrac{\text{kJ}}{\text{kg·K}} - 0.5456 \dfrac{\text{kJ}}{\text{kg·K}}}{8.2935 \dfrac{\text{kJ}}{\text{kg·K}} - 0.5456 \dfrac{\text{kJ}}{\text{kg·K}}}$$

$$= 0.402$$

$$h_D = h_f + x_D h_{fg}$$

$$= 159.17 \frac{\text{kJ}}{\text{kg}} + (0.402)\left(2410.7 \frac{\text{kJ}}{\text{kg}}\right)$$

$$= 1128.3 \text{ kJ/kg}$$

From Eq. 28.9, due to the inefficiency of the turbine,

$$h_C' = h_B - \eta_{s,\text{turbine}}(h_B - h_C)$$

$$= 2621.9 \frac{\text{kJ}}{\text{kg}} - (0.9)\left(2621.9 \frac{\text{kJ}}{\text{kg}} - 1649.0 \frac{\text{kJ}}{\text{kg}}\right)$$

$$= 1746.3 \text{ kJ/kg}$$

From Eq. 28.10, due to the inefficiency of the pump,

$$h_A' = h_D + \frac{h_A - h_D}{\eta_{s,\text{pump}}}$$

$$= 1128.3 \frac{\text{kJ}}{\text{kg}} + \frac{1594.5 \dfrac{\text{kJ}}{\text{kg}} - 1128.3 \dfrac{\text{kJ}}{\text{kg}}}{0.8}$$

$$= 1711.1 \text{ kJ/kg}$$

From Eq. 28.8, the thermal efficiency of the entire cycle is

$$\eta_{\text{th}} = \frac{(h_B - h_C') - (h_A' - h_D)}{h_B - h_A'}$$

$$= \frac{\left(2621.9 \dfrac{\text{kJ}}{\text{kg}} - 1746.3 \dfrac{\text{kJ}}{\text{kg}}\right) - \left(1711.1 \dfrac{\text{kJ}}{\text{kg}} - 1128.3 \dfrac{\text{kJ}}{\text{kg}}\right)}{2621.9 \dfrac{\text{kJ}}{\text{kg}} - 1711.1 \dfrac{\text{kJ}}{\text{kg}}}$$

$$= \boxed{0.321 \quad (31.1\%)}$$

The answer is (A).

3. *Customary U.S. Solution*

From Eq. 28.1, the thermal efficiency is

$$\eta_{\text{th}} = \frac{Q_{\text{in}} - Q_{\text{out}}}{Q_{\text{in}}}$$

The condenser load is $Q_{\text{out}} = 3.07 \times 10^9$ Btu/hr.

The net work is

$$W_{\text{net}} = Q_{\text{in}} - Q_{\text{out}}$$

The boiler load is

$$Q_{\text{in}} = W_{\text{net}} + Q_{\text{out}}$$

$$= (600 \text{ MW})\left(1000 \frac{\text{kW}}{\text{MW}}\right)\left(3412 \frac{\text{Btu}}{\text{hr}}\frac{}{\text{kW}}\right)$$

$$+ 3.07 \times 10^9 \frac{\text{Btu}}{\text{hr}}$$

$$= 5.12 \times 10^9 \text{ Btu/hr}$$

$$\eta_{\text{th}} = \frac{5.12 \times 10^9 \dfrac{\text{Btu}}{\text{hr}} - 3.07 \times 10^9 \dfrac{\text{Btu}}{\text{hr}}}{5.12 \times 10^9 \dfrac{\text{Btu}}{\text{hr}}}$$

$$= \boxed{0.400 \quad (40\%)}$$

The answer is (C).

SI Solution

From Eq. 28.1, the thermal efficiency is

$$\eta_{\text{th}} = \frac{Q_{\text{in}} - Q_{\text{out}}}{Q_{\text{in}}}$$

The condenser load is $Q_{\text{out}} = 900$ MW.

The boiler load is

$$Q_{\text{in}} = W_{\text{net}} + Q_{\text{out}}$$

$$= 600 \text{ MW} + 900 \text{ MW}$$

$$= 1500 \text{ MW}$$

$$\eta_{\text{th}} = \frac{1500 \text{ MW} - 900 \text{ MW}}{1500 \text{ MW}}$$

$$= \boxed{0.400 \quad (40\%)}$$

The answer is (C).

4. *Customary U.S. Solution*

At point A, $p_A = 100$ psia.

From App. 24.B, the enthalpy of saturated liquid is h_A = 298.5 Btu/lbm.

At point B, $p_B = 100$ psia.

From App.24.B, the enthalpy and entropy of saturated vapor are

$$h_B = 1187.5 \text{ Btu/lbm}$$

$$s_B = 1.6032 \text{ Btu/lbm-°R}$$

At point C,

$$p_C = 1 \text{ atm}$$
$$s_C = s_B = 1.6032 \text{ Btu/lbm-}°R$$

From App. 24.B, the entropy and enthalpy of saturated liquid and entropy and enthalpy of vaporization are

$$s_f = 0.3122 \text{ Btu/lbm-}°R$$
$$h_f = 180.2 \text{ Btu/lbm}$$
$$s_{fg} = 1.4445 \text{ Btu/lbm-}°R$$
$$h_{fg} = 970.1 \text{ Btu/lbm}$$

$$x_C = \frac{s_C - s_f}{s_{fg}} = \frac{1.6032 \dfrac{\text{Btu}}{\text{lbm-}°R} - 0.3122 \dfrac{\text{Btu}}{\text{lbm-}°R}}{1.4445 \dfrac{\text{Btu}}{\text{lbm-}°R}}$$

$$= 0.894$$

$$h_C = h_f + x_C h_{fg}$$
$$= 180.2 \frac{\text{Btu}}{\text{lbm}} + (0.894)\left(970.1 \frac{\text{Btu}}{\text{lbm}}\right)$$
$$= 1047.5 \text{ Btu/lbm}$$

At point D, $T = 80°F$ and $p_D = 1$ atm (subcooled). h and v are essentially independent of pressure.

From App. 24.A, the enthalpy and specific volume of saturated liquid are

$$h_D = 48.07 \text{ Btu/lbm}$$
$$v_D = 0.01607 \text{ ft}^3/\text{lbm}$$

At point E, $p_E = p_A = 100$ psia.

From Eq. 28.14,

$$h_E = h_D + v_D(p_E - p_D)$$
$$= 48.07 \frac{\text{Btu}}{\text{lbm}} + \left(0.01607 \frac{\text{ft}^3}{\text{lbm}}\right)$$
$$\times \frac{\left(100 \dfrac{\text{lbf}}{\text{in}^2} - 14.7 \dfrac{\text{lbf}}{\text{in}^2}\right)\left(144 \dfrac{\text{in}^2}{\text{ft}^2}\right)}{778 \dfrac{\text{ft-lbf}}{\text{Btu}}}$$
$$= 48.32 \text{ Btu/lbm}$$

From Eq. 28.18, due to the inefficiency of the turbine,

$$h_C' = h_B - \eta_{s,\text{turbine}}(h_B - h_C)$$
$$= 1187.5 \frac{\text{Btu}}{\text{lbm}} - (0.80)\left(1187.5 \frac{\text{Btu}}{\text{lbm}}\right.$$
$$\left. - 1047.5 \frac{\text{Btu}}{\text{lbm}}\right)$$
$$= 1075.5 \text{ Btu/lbm}$$

From Eq. 28.19, due to the inefficiency of the pump,

$$h_E' = h_D + \frac{h_E - h_D}{\eta_{s,\text{pump}}}$$
$$= 48.07 \frac{\text{Btu}}{\text{lbm}} + \frac{48.32 \dfrac{\text{Btu}}{\text{lbm}} - 48.07 \dfrac{\text{Btu}}{\text{lbm}}}{0.6}$$
$$= 48.49 \text{ Btu/lbm}$$

From Eq. 28.17, the thermal efficiency of the cycle is

$$\eta_{\text{th}} = \frac{(h_B - h_C') - (h_E' - h_D)}{h_B - h_E'}$$

$$= \frac{\left(1187.5 \dfrac{\text{Btu}}{\text{lbm}} - 1075.5 \dfrac{\text{Btu}}{\text{lbm}}\right) - \left(48.49 \dfrac{\text{Btu}}{\text{lbm}} - 48.07 \dfrac{\text{Btu}}{\text{lbm}}\right)}{1187.5 \dfrac{\text{Btu}}{\text{lbm}} - 48.49 \dfrac{\text{Btu}}{\text{lbm}}}$$

$$= \boxed{0.098 \ (9.8\%)}$$

The answer is (A).

SI Solution

At point A, $p_A = 700$ kPa (7 bars).

From App. 24.O, the enthalpy of saturated liquid is $h_A = 697.00$ kJ/kg.

At point B, $p_B = 700$ kPa.

From App. 24.O, the enthalpy and entropy of saturated vapor are

$$h_B = 2762.8 \text{ kJ/kg}$$
$$s_B = 6.7071 \text{ kJ/kg·K}$$

At point C,

$$p_C = 1 \text{ atm}$$
$$s_C = s_B = 6.7071 \text{ kJ/kg·K}$$

From App. 24.O, the entropy and enthalpy of saturated liquid, the entropy of saturated vapor, and the enthalpy of vaporization are

$$s_f = 1.3062 \text{ kJ/kg·K}$$
$$h_f = 418.79 \text{ kJ/kg}$$
$$s_g = 7.3553 \text{ kJ/kg·K}$$
$$h_{fg} = 2256.6 \text{ kJ/kg}$$

$$x_C = \frac{s_C - s_f}{s_g - s_f}$$

$$= \frac{6.7071\ \dfrac{\text{kJ}}{\text{kg·K}} - 1.3062\ \dfrac{\text{kJ}}{\text{kg·K}}}{7.3553\ \dfrac{\text{kJ}}{\text{kg·K}} - 1.3062\ \dfrac{\text{kJ}}{\text{kg·K}}}$$

$$= 0.893$$

$$h_C = h_f + x_C h_{fg}$$

$$= 418.79\ \frac{\text{kJ}}{\text{kg}} + (0.893)\left(2256.6\ \frac{\text{kJ}}{\text{kg}}\right)$$

$$= 2433.9\ \text{kJ/kg}$$

At point D, $T = 27°C$ and $p_D = 1$ atm (subcooled). h and v are essentially independent of pressure.

From App. 24.N, the enthalpy and specific volume of saturated liquid are

$$h_D = 113.19\ \text{kJ/kg}$$
$$v_D = 1.0035\ \text{cm}^3/\text{g}$$

At point E, $p_E = p_A = 700$ kPa.

From Eq. 28.14,

$$h_E = h_D + v_D(p_E - p_D)$$

$$= 113.19\ \frac{\text{kJ}}{\text{kg}} + \left(1.0035\ \frac{\text{cm}^3}{\text{g}}\right)\left(\frac{1\ \text{m}^3}{10^6\ \text{cm}^3}\right)$$

$$\times \left(1000\ \frac{\text{g}}{\text{kg}}\right)(700\ \text{kPa} - 101\ \text{kPa})$$

$$= 113.79\ \text{kJ/kg}$$

From Eq. 28.18, due to the inefficiency of the turbine,

$$h'_C = h_B - \eta_{s,\text{turbine}}(h_B - h_C)$$

$$= 2762.8\ \frac{\text{kJ}}{\text{kg}} - (0.80)\left(2762.8\ \frac{\text{kJ}}{\text{kg}} - 2433.9\ \frac{\text{kJ}}{\text{kg}}\right)$$

$$= 2499.7\ \text{kJ/kg}$$

From Eq. 28.19, due to the inefficiency of the pump,

$$h'_E = h_D + \frac{h_E - h_D}{\eta_{s,\text{pump}}}$$

$$= 113.19\ \frac{\text{kJ}}{\text{kg}} + \frac{113.79\ \dfrac{\text{kJ}}{\text{kg}} - 113.19\ \dfrac{\text{kJ}}{\text{kg}}}{0.6}$$

$$= 114.19\ \text{kJ/kg}$$

From Eq. 28.17, the thermal efficiency of the cycle is

$$\eta_{th} = \frac{(h_B - h'_C) - (h'_E - h_D)}{h_B - h'_E}$$

$$= \frac{\left(2762.8\ \dfrac{\text{kJ}}{\text{kg}} - 2499.7\ \dfrac{\text{kJ}}{\text{kg}}\right)}{\quad}$$

$$\frac{- \left(114.19\ \dfrac{\text{kJ}}{\text{kg}} - 113.19\ \dfrac{\text{kJ}}{\text{kg}}\right)}{2762.8\ \dfrac{\text{kJ}}{\text{kg}} - 114.19\ \dfrac{\text{kJ}}{\text{kg}}}$$

$$= \boxed{0.10\ \ (10\%)}$$

The answer is (A).

5. *Customary U.S. Solution*

Refer to the reheat cycle (Fig. 28.8).

At point B, $p_B = 600$ psia.

From App. 24.B, the enthalpy of the saturated liquid is $h_B = 471.7$ Btu/lbm.

At point C, $p_C = 600$ psia.

From App. 24.B, the enthalpy of the saturated vapor is $h_C = 1203.8$ Btu/lbm.

At point D,
$$T_D = 600°\text{F}$$
$$p_D = 600\ \text{psia}$$

From App. 24.C, the enthalpy and entropy of superheated vapor are

$$h_D = 1289.9\ \text{Btu/lbm}$$
$$s_D = 1.5326\ \text{Btu/lbm-}°\text{R}$$

At point E,
$$p_E = 20\ \text{psia}$$
$$s_E = s_D = 1.5326\ \text{Btu/lbm-}°\text{R}$$

From App. 24.B, the various saturation properties are

$$s_f = 0.3358\ \text{Btu/lbm-}°\text{R}$$
$$s_{fg} = 1.3961\ \text{Btu/lbm-}°\text{R}$$
$$h_f = 196.3\ \text{Btu/lbm}$$
$$h_{fg} = 959.9\ \text{Btu/lbm}$$

$$x_E = \frac{s_E - s_f}{s_g - s_f} = \frac{1.5326\ \dfrac{\text{Btu}}{\text{lbm-}°\text{R}} - 0.3358\ \dfrac{\text{Btu}}{\text{lbm-}°\text{R}}}{1.3961\ \dfrac{\text{Btu}}{\text{lbm-}°\text{R}}}$$

$$= 0.857$$

$$h_E = h_f + x_E h_{fg}$$

$$= 196.3\ \frac{\text{Btu}}{\text{lbm}} + (0.857)\left(959.9\ \frac{\text{Btu}}{\text{lbm}}\right)$$

$$= 1018.9\ \text{Btu/lbm}$$

From Eq. 28.38, due to the inefficiency of the turbine,

$$h'_E = h_D - \eta_{s,\text{turbine}}(h_D - h_E)$$

$$= 1289.9 \; \frac{\text{Btu}}{\text{lbm}} - (0.88) \left(1289.9 \; \frac{\text{Btu}}{\text{lbm}} \right.$$

$$\left. - 1018.9 \; \frac{\text{Btu}}{\text{lbm}} \right)$$

$$= 1051.4 \; \text{Btu/lbm}$$

At point F, the temperature has been returned to 600°F, but the pressure stays at the expansion pressure, p_E.

$$p_F = 20 \; \text{psia}$$

$$T_F = 600°\text{F}$$

From App. 24.C, the enthalpy and entropy of the superheated vapor is

$$h_F = 1334.9 \; \text{Btu/lbm}$$

$$s_F = 1.9399 \; \text{Btu/lbm-}°\text{R}$$

At point G,

$$T_G = 60°\text{F}$$

$$s_G = s_F = 1.9399 \; \text{Btu/lbm-}°\text{R}$$

From App. 24.A, various saturation properties are

$$s_f = 0.05554 \; \text{Btu/lbm-}°\text{R}$$

$$s_g = 2.0940 \; \text{Btu/lbm-}°\text{R}$$

$$h_f = 28.08 \; \text{Btu/lbm}$$

$$h_{fg} = 1059.3 \; \text{Btu/lbm}$$

$$x_G = \frac{s_G - s_f}{s_g - s_f} = \frac{1.9399 \; \frac{\text{Btu}}{\text{lbm-}°\text{R}} - 0.05554 \; \frac{\text{Btu}}{\text{lbm-}°\text{R}}}{2.0940 \; \frac{\text{Btu}}{\text{lbm-}°\text{R}} - 0.05554 \; \frac{\text{Btu}}{\text{lbm-}°\text{R}}}$$

$$= 0.924$$

$$h_G = h_f + x_G h_{fg}$$

$$= 28.08 \; \frac{\text{Btu}}{\text{lbm}} + (0.924) \left(1059.3 \; \frac{\text{Btu}}{\text{lbm}} \right)$$

$$= 1006.9 \; \text{Btu/lbm}$$

From Eq. 28.39, due to the inefficiency of the turbine,

$$h'_G = h_F - \eta_{s,\text{turbine}}(h_F - h_G)$$

$$= 1334.9 \; \frac{\text{Btu}}{\text{lbm}} - (0.88) \left(1334.9 \; \frac{\text{Btu}}{\text{lbm}} \right.$$

$$\left. - 1006.9 \; \frac{\text{Btu}}{\text{lbm}} \right)$$

$$= 1046.3 \; \text{Btu/lbm}$$

At point H, $T_H = 60°$F.

From App. 24.A, the saturation pressure, enthalpy, and specific volume of the saturated liquid are

$$p_H = 0.2564 \; \text{psia}$$

$$h_H = 28.08 \; \text{Btu/lbm}$$

$$v_H = 0.01604 \; \text{ft}^3/\text{lbm}$$

At point A, $p_A = 600$ psia.

From Eq. 28.14,

$$h_A = h_H + v_H(p_A - p_H)$$

$$= 28.08 \; \frac{\text{Btu}}{\text{lbm}}$$

$$+ \frac{\left(0.01604 \; \frac{\text{ft}^3}{\text{lbm}} \right) \times \left(600 \; \frac{\text{lbf}}{\text{in}^2} - 0.2564 \; \frac{\text{lbf}}{\text{in}^2} \right) \left(144 \; \frac{\text{in}^2}{\text{ft}^2} \right)}{778 \; \frac{\text{lbf-ft}}{\text{Btu}}}$$

$$= 29.86 \; \text{Btu/lbm}$$

From Eq. 28.40, due to the inefficiency of the pump,

$$h'_A = h_H + \frac{h_A - h_H}{\eta_{s,\text{pump}}}$$

$$= 28.08 \; \frac{\text{Btu}}{\text{lbm}} + \frac{29.86 \; \frac{\text{Btu}}{\text{lbm}} - 28.08 \; \frac{\text{Btu}}{\text{lbm}}}{0.96}$$

$$= 29.93 \; \text{Btu/lbm}$$

From Eq. 28.37, the thermal efficiency of the cycle for a non-isentropic process for the turbine and the pump is

$$\eta_{\text{th}} = \frac{(h_D - h'_A) + (h_F - h'_E) - (h'_G - h_H)}{(h_D - h'_A) + (h_F - h'_E)}$$

$$= \frac{\begin{array}{c} \left(1289.9 \; \frac{\text{Btu}}{\text{lbm}} - 29.93 \; \frac{\text{Btu}}{\text{lbm}} \right) \\ + \left(1334.9 \; \frac{\text{Btu}}{\text{lbm}} - 1051.4 \; \frac{\text{Btu}}{\text{lbm}} \right) \\ - \left(1046.3 \; \frac{\text{Btu}}{\text{lbm}} - 28.08 \; \frac{\text{Btu}}{\text{lbm}} \right) \end{array}}{\begin{array}{c} \left(1289.9 \; \frac{\text{Btu}}{\text{lbm}} - 29.93 \; \frac{\text{Btu}}{\text{lbm}} \right) \\ + \left(1334.9 \; \frac{\text{Btu}}{\text{lbm}} - 1051.4 \; \frac{\text{Btu}}{\text{lbm}} \right) \end{array}}$$

$$= \boxed{0.340 \; (34.0\%)}$$

The answer is (B).

SI Solution

Refer to the reheat cycle (Fig. 28.8).

At point B, $p_B = 4$ MPa.

From App. 24.O, the enthalpy of saturated liquid is $h_B = 1087.5$ kJ/kg.

At point C, $p_C = 4$ MPa.

From App. 24.O, the enthalpy of saturated vapor is $h_C = 2800.8$ kJ/kg.

At point D, $p_D = 4$ MPa and $T_D = 300°$C.

From the Mollier diagram, the enthalpy of superheated vapor is $h_D = 2980$ kJ/kg.

At point E, from the Mollier diagram, assuming isentropic expansion, $h_E = 2395$ kJ/kg.

From Eq. 28.38, due to the inefficiency of the turbine,

$$h'_E = h_D - \eta_{s,\text{turbine}}(h_D - h_E)$$
$$= 2980 \, \frac{\text{kJ}}{\text{kg}} - (0.88)\left(2980 \, \frac{\text{kJ}}{\text{kg}} - 2395 \, \frac{\text{kJ}}{\text{kg}}\right)$$
$$= 2465.2 \text{ kJ/kg}$$

At point F,
$$p_F = 150 \text{ kPa}$$
$$T_F = 300°\text{C}$$

From App. 24.P, the enthalpy and entropy of the superheated vapor is
$$h_F = 3073.3 \text{ kJ/kg}$$
$$s_F = 8.0284 \text{ kJ/kg·K}$$

At point G,
$$T_G = 16°\text{C}$$
$$s_G = s_F = 8.0284 \text{ kJ/kg·K}$$

From App. 24.N, the various saturation properties are
$$s_f = 0.2390 \text{ kJ/kg·K}$$
$$s_g = 8.7570 \text{ kJ/kg·K}$$
$$h_f = 67.17 \text{ kJ/kg}$$
$$h_{fg} = 2463.0 \text{ kJ/kg}$$

$$x_G = \frac{s_G - s_f}{s_g - s_f} = \frac{8.0284 \, \frac{\text{kJ}}{\text{kg·K}} - 0.2390 \, \frac{\text{kJ}}{\text{kg·K}}}{8.7570 \, \frac{\text{kJ}}{\text{kg·K}} - 0.2390 \, \frac{\text{kJ}}{\text{kg·K}}}$$
$$= 0.914$$

$$h_G = h_f + x_G h_{fg}$$
$$= 67.17 \, \frac{\text{kJ}}{\text{kg}} + (0.914)\left(2463.0 \, \frac{\text{kJ}}{\text{kg}}\right)$$
$$= 2318.4 \text{ kJ/kg}$$

From Eq. 28.39, due to the inefficiency of the turbine,

$$h'_G = h_F - \eta_{s,\text{turbine}}(h_F - h_G)$$
$$= 3073.3 \, \frac{\text{kJ}}{\text{kg}} - (0.88)\left(3073.3 \, \frac{\text{kJ}}{\text{kg}} - 2318.4 \, \frac{\text{kJ}}{\text{kg}}\right)$$
$$= 2409.0 \text{ kJ/kg}$$

At point H, $T_H = 16°$C.

From App. 24.N, the saturation pressure, enthalpy, and specific volume of the saturated liquid are

$$p_H = (0.01819 \text{ bar})\left(100 \, \frac{\text{kPa}}{\text{bar}}\right) = 1.819 \text{ kPa}$$
$$h_H = 67.17 \text{ kJ/kg}$$
$$v_H = \left(1.0011 \, \frac{\text{cm}}{\text{g}^3}\right)\left(1000 \, \frac{\text{g}}{\text{kg}}\right)\left(\frac{1 \text{ m}}{100 \text{ cm}}\right)^3$$
$$= 1.0011 \times 10^{-3} \text{ m}^3/\text{kg}$$

At point A,

$$p_A = (4 \text{ MPa})\left(1000 \, \frac{\text{kPa}}{\text{MPa}}\right)$$
$$= 4000 \text{ kPa}$$

From Eq. 28.14,

$$h_A = h_H + v_H(p_A - p_H)$$
$$= 67.17 \, \frac{\text{kJ}}{\text{kg}} + \left(1.0011 \times 10^{-3} \, \frac{\text{m}^3}{\text{kg}}\right)$$
$$\times (4000 \text{ kPa} - 1.819 \text{ kPa})$$
$$= 71.17 \text{ kJ/kg}$$

From Eq. 28.40, due to the inefficiency of the pump,

$$h'_A = h_H + \frac{h_A - h_H}{\eta_{s,\text{pump}}}$$
$$= 67.17 \, \frac{\text{kJ}}{\text{kg}} + \frac{71.17 \, \frac{\text{kJ}}{\text{kg}} - 67.17 \, \frac{\text{kJ}}{\text{kg}}}{0.96}$$
$$= 71.34 \text{ kJ/kg}$$

From Eq. 28.37, the thermal efficiency of the cycle for a non-isentropic process for the turbine and the pump is

$$\eta_{th} = \frac{(h_D - h'_A) + (h_F - h'_E) - (h'_G - h_H)}{(h_D - h'_A) + (h_F - h'_E)}$$

$$= \frac{\left(2980 \ \frac{kJ}{kg} - 71.34 \ \frac{kJ}{kg}\right) + \left(3073.3 \ \frac{kJ}{kg} - 2465.2 \ \frac{kJ}{kg}\right) - \left(2409.0 \ \frac{kJ}{kg} - 67.17 \ \frac{kJ}{kg}\right)}{\left(2980 \ \frac{kJ}{kg} - 71.34 \ \frac{kJ}{kg}\right) + \left(3073.3 \ \frac{kJ}{kg} - 2465.2 \ \frac{kJ}{kg}\right)}$$

$$= \boxed{0.334 \quad (33.4\%)}$$

The answer is (B).

6. *Customary U.S. Solution*

Refer to the following diagram.

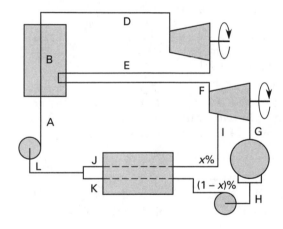

From Prob. 5,

$$h_B = 471.7 \ \text{Btu/lbm}$$
$$h_D = 1289.5 \ \text{Btu/lbm}$$
$$h'_E = 1051.5 \ \text{Btu/lbm}$$
$$h_F = 1334.8 \ \text{Btu/lbm}$$
$$s_F = 1.9395 \ \text{Btu/lbm-}^\circ\text{R}$$
$$h'_G = 1046.5 \ \text{Btu/lbm}$$
$$h_H = 28.08 \ \text{Btu/lbm}$$
$$p_H = 0.2563 \ \text{psia}$$
$$T_H = 60^\circ\text{F}$$

At point I,

$$p_I = 4.5 \ \text{psia}$$
$$s_I = s_F = 1.9395 \ \text{Btu/lbm-}^\circ\text{R}$$

From App. 24.C,

$$h_I = 1181.0 \ \text{Btu/lbm}$$

From Eq. 28.39, due to the inefficiency of the turbine,

$$h'_I = h_F - \eta_{s,\text{turbine}}(h_F - h_I)$$
$$= 1334.8 \ \frac{\text{Btu}}{\text{lbm}} - (0.88)\left(1334.8 \ \frac{\text{Btu}}{\text{lbm}} - 1181.0 \ \frac{\text{Btu}}{\text{lbm}}\right)$$
$$= 1199.5 \ \text{Btu/lbm}$$

At point J, from App. 24.B, the saturated liquid enthalpy and temperature at 4.5 psia are approximately $h_J = 125.6$ Btu/lbm and 157.5°F, respectively.

At point K, the temperature is

$$157.5^\circ\text{F} - 6^\circ\text{F} \approx 152^\circ\text{F}$$

Since the water is subcooled, enthalpy is a function of temperature only.

$$h_K = 119.99 \ \text{Btu/lbm}$$

The condensate pump will match the pressure at J: $p_K = 4.5$ psia.

From the energy balance in the heater,

$$(1-x)(h_K - h_H) = x(h'_I - h_J)$$
$$(1-x)\left(119.99 \ \frac{\text{Btu}}{\text{lbm}} - 28.08 \ \frac{\text{Btu}}{\text{lbm}}\right) = x\left(1199.5 \ \frac{\text{Btu}}{\text{lbm}} - 125.6 \ \frac{\text{Btu}}{\text{lbm}}\right)$$
$$1165.8x = 91.91$$
$$x = 0.0788$$

At point L, $p_L = 4.5$ psia.

$$h_L = xh_J + (1-x)h_K$$
$$= (0.0788)\left(125.6 \ \frac{\text{Btu}}{\text{lbm}}\right)$$
$$\quad + (1 - 0.0788)\left(119.99 \ \frac{\text{Btu}}{\text{lbm}}\right)$$
$$= 120.4 \ \text{Btu/lbm}$$

Since this is a subcooled liquid, enthalpy is a function of temperature only, and from App. 24.A,

$$T_L = 152.4°F$$

$$v_L = 0.01635 \text{ ft}^3/\text{lbm}$$

At point A, $p_A = 600$ psia.

From Eq. 28.14,

$$h_A = h_L + v_L(p_A - p_L)$$

$$= 120.4 \frac{\text{Btu}}{\text{lbm}}$$

$$+ \frac{\left(0.01635 \frac{\text{ft}^3}{\text{lbm}}\right)}{778 \frac{\text{lbf-ft}}{\text{lbm}}} \times \left(600 \frac{\text{lbf}}{\text{in}^2} - 4.5 \frac{\text{lbf}}{\text{in}^2}\right)\left(144 \frac{\text{in}^2}{\text{ft}^2}\right)$$

$$= 122.2 \text{ Btu/lbm}$$

From Eq. 28.40, due to the inefficiency of the pump,

$$h'_A = h_L + \frac{h_A - h_L}{\eta_{s,\text{pump}}}$$

$$= 120.4 \frac{\text{Btu}}{\text{lbm}} + \frac{122.2 \frac{\text{Btu}}{\text{lbm}} - 120.4 \frac{\text{Btu}}{\text{lbm}}}{0.96}$$

$$= 122.3 \text{ Btu/lbm}$$

From Eq. 28.37, the thermal efficiency of the entire cycle neglecting condensation and drip pump, is

$$\eta_{\text{th}} = \frac{W_{\text{turbines}} - W_{\text{pump}}}{Q_{\text{in}}}$$

$$= \frac{\begin{array}{c}(h_D - h'_E) + (h_F - h'_I) \\ + (1 - x)(h'_I - h'_G) - (h'_A - h_L)\end{array}}{(h_D - h'_A) + (h_F - h'_E)}$$

$$= \frac{\begin{array}{c}\left(1289.5 \frac{\text{Btu}}{\text{lbm}} - 1051.5 \frac{\text{Btu}}{\text{lbm}}\right) \\ + \left(1334.8 \frac{\text{Btu}}{\text{lbm}} - 1199.5 \frac{\text{Btu}}{\text{lbm}}\right) \\ + (1 - 0.0788)\left(1199.5 \frac{\text{Btu}}{\text{lbm}} - 1046.5 \frac{\text{Btu}}{\text{lbm}}\right) \\ - \left(122.3 \frac{\text{Btu}}{\text{lbm}} - 120.4 \frac{\text{Btu}}{\text{lbm}}\right)\end{array}}{\begin{array}{c}\left(1289.5 \frac{\text{Btu}}{\text{lbm}} - 122.3 \frac{\text{Btu}}{\text{lbm}}\right) \\ + \left(1334.8 \frac{\text{Btu}}{\text{lbm}} - 1051.5 \frac{\text{Btu}}{\text{lbm}}\right)\end{array}}$$

$$= \boxed{0.353 \text{ (35.3\%)}}$$

The answer is (C).

SI Solution

From Prob. 5,

$$h_B = 1087.3 \text{ kJ/kg}$$

$$h_D = 2980 \text{ kJ/kg}$$

$$h'_E = 2465.2 \text{ kJ/kg}$$

$$h_F = 3073.1 \text{ kJ/kg}$$

$$s_F = 8.0720 \text{ kJ/kg·K}$$

$$h'_G = 2420.2 \text{ kJ/kg}$$

$$h_H = 67.19 \text{ kJ/kg}$$

At point I,

$$p_I = 37.76 \text{ kPa} \quad [0.3776 \text{ bars}]$$

$$s_I = s_F$$

$$= 8.0720 \text{ kJ/kg·K}$$

From App. 24.P,

$$h_I = 2742.2 \text{ kJ/kg}$$

From Eq. 28.39, due to inefficiency of the turbine,

$$h'_I = h_F - \eta_{s,\text{turbine}}(h_F - h_I)$$

$$= 3073.1 \frac{\text{kJ}}{\text{kg}} - (0.88)\left(3073.1 \frac{\text{kJ}}{\text{kg}} - 2742.2 \frac{\text{kJ}}{\text{kg}}\right)$$

$$= 2781.9 \text{ kJ/kg}$$

At point J, from App. 24.O, the saturated liquid enthalpy and temperature at 37.73 kPa are $h_J = 311.3$ kJ/kg and 74.3°C, respectively.

At point K, the temperature is

$$74.3°C - 3°C = 71.3°C$$

Since the water is subcooled, enthalpy is a function of temperature only.

$$h_K = 298.5 \text{ kJ/kg}$$

The condensate pump will match the pressure at J: $p_k = 37.73$ kPa.

From an energy balance in the heater,

$$(1 - x)(h_K - h_H) = x(h'_I - h_J)$$

$$(1 - x)\left(298.5 \frac{\text{kJ}}{\text{kg}}\right.$$

$$\left. - 67.19 \frac{\text{kJ}}{\text{kg}}\right) = x\left(2781.9 \frac{\text{kJ}}{\text{kg}} - 311.3 \frac{\text{kJ}}{\text{kg}}\right)$$

$$2701.9x = 231.3$$

$$x = 0.0856$$

At point L, $p_L = 37.73$ kPa.

$$h_L = xh_J + (1-x)h_K$$
$$= (0.0856)\left(311.3 \ \frac{kJ}{kg}\right) + (1-0.0856)\left(298.5 \ \frac{kJ}{kg}\right)$$
$$= 299.6 \ kJ/kg$$

Since this is a subcooled liquid, enthalpy is a function of temperature only, and from App. 24.N,

$$T_L = 71.6°C$$
$$v_L = 1.0237 \ cm^3/g$$

At point A, $p_A = 4$ MPa.

From Eq. 28.14,

$$h_A = h_L + v_L(p_A - p_L)$$
$$h_A = 299.6 \ \frac{kJ}{kg} + \left(1.0237 \ \frac{cm^3}{g}\right)\left(1000 \ \frac{g}{kg}\right)$$
$$\times \left(\frac{1 \ m}{100 \ cm}\right)^3 \left((4 \ MPa)\left(1000 \ \frac{kPa}{MPa}\right)\right.$$
$$\left. -(0.3776 \ bar)\left(100 \ \frac{kPa}{bar}\right)\right)$$
$$= 303.7 \ kJ/kg$$

From Eq. 28.40, due to the inefficiency in the pump,

$$h'_A = h_L + \frac{h_A - h_L}{\eta_{s,pump}}$$
$$= 299.6 \ \frac{kJ}{kg} + \frac{303.7 \ \frac{kJ}{kg} - 299.6 \ \frac{kJ}{kg}}{0.96}$$
$$= 303.9 \ kJ/kg$$

From Eq. 28.37, the thermal efficiency of the entire cycle, neglecting condensation and drip pump, is

$$\eta_{th} = \frac{W_{turbine} - W_{pump}}{Q_{in}}$$

$$= \frac{\begin{array}{c}((h_D - h'_E) + (h_F - h'_I) \\ + (1-x)(h'_I - h'_G)) - (h'_A - h_L)\end{array}}{(h_D - h'_A) + (h_F - h'_E)}$$

$$= \frac{\begin{array}{c}\left(\left(2980 \ \frac{kJ}{kg} - 2465.2 \ \frac{kJ}{kg}\right)\right. \\ + \left(3073.1 \ \frac{kJ}{kg} - 2781.9 \ \frac{kJ}{kg}\right) \\ + (1-0.0856)\left(2781.9 \ \frac{kJ}{kg} - 2420.2 \ \frac{kJ}{kg}\right)\right) \\ - \left(303.9 \ \frac{kJ}{kg} - 299.6 \ \frac{kJ}{kg}\right)\end{array}}{\begin{array}{c}\left(2980 \ \frac{kJ}{kg} - 303.9 \ \frac{kJ}{kg}\right) \\ + \left(3073.1 \ \frac{kJ}{kg} - 2465.2 \ \frac{kJ}{kg}\right)\end{array}}$$

$$= \boxed{0.345 \ \ (34.5\%)}$$

The answer is (C).

7. *Customary U.S. Solution*

(a) Refer to the given illustration for Prob. 7 and to the following diagram.

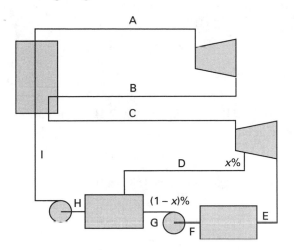

At point A,
$$p_A = 900 \ psia$$
$$T_A = 800°F$$

Using the superheated steam table, App. 24.C,
$$h_A = 1393.9 \ Btu/lbm$$
$$s_A = 1.5816 \ Btu/lbm\text{-}°R$$

At point B, $h'_B = 1270$ Btu/lbm.

From the Mollier diagram, assuming isentropic expansion to 200 psia, $h_B = 1230$ Btu/lbm.

Isentropic efficiency of the high pressure turbine is

$$\eta_{s,\text{turbine}} = \frac{h_A - h'_B}{h_A - h_B}$$

$$= \left(\frac{1393.9 \, \frac{\text{Btu}}{\text{lbm}} - 1270 \, \frac{\text{Btu}}{\text{lbm}}}{1393.9 \, \frac{\text{Btu}}{\text{lbm}} - 1230 \, \frac{\text{Btu}}{\text{lbm}}} \right) (100\%)$$

$$= \boxed{75.59\%}$$

The answer is (D).

(b) At point C,

$$p_C = 190 \text{ psia}$$
$$T_C = 800°\text{F}$$

Using the superheated steam table from App. 24.C, $h_C = 1426.0$ Btu/lbm.

At point D: $\quad h'_D = 1280$ Btu/lbm
At point E: $\quad h'_E = 1075$ Btu/lbm
At point F: $\quad h_F = 69.73$ Btu/lbm
At point G: $\quad W_{\text{pump}} = 0.15$ Btu/lbm
$$\quad W_{\text{pump}} = h'_G - h_F$$
$$\quad h'_G = W_{\text{pump}} + h_F$$
$$\quad = 0.15 \, \frac{\text{Btu}}{\text{lbm}} + 69.73 \, \frac{\text{Btu}}{\text{lbm}}$$
$$\quad = 69.88 \text{ Btu/lbm}$$
At point H: $\quad h_H = 250.2$ Btu/lbm
At point I: $\quad h'_I = 253.1$ Btu/lbm

From an energy balance in the heater,

$$xh'_D + (1-x)h'_G = h_H$$

$$x(h'_D - h'_G) = h_H - h'_G$$

$$x = \frac{h_H - h'_G}{h'_D - h'_G}$$

$$= \frac{250.2 \, \frac{\text{Btu}}{\text{lbm}} - 69.88 \, \frac{\text{Btu}}{\text{lbm}}}{1280 \, \frac{\text{Btu}}{\text{lbm}} - 69.88 \, \frac{\text{Btu}}{\text{lbm}}}$$

$$= 0.149$$

The thermal efficiency of the cycle is

$$\eta_{th} = \frac{W_{\text{out}} - W_{\text{in}}}{Q_{\text{in}}}$$

$$= \frac{\begin{array}{c}(h_A - h'_B) + (h_C - h'_D) + (1-x) \\ \times (h'_D - h'_E) - (h'_I - h_H) - (1-x)(h'_G - h_F)\end{array}}{(h_A - h'_I) + (h_C - h'_B)}$$

$$= \frac{\begin{array}{c}\left(1393.9 \, \frac{\text{Btu}}{\text{lbm}} - 1270 \, \frac{\text{Btu}}{\text{lbm}}\right) \\ + \left(1426.0 \, \frac{\text{Btu}}{\text{lbm}} - 1280 \, \frac{\text{Btu}}{\text{lbm}}\right) \\ + (1 - 0.149)\left(1280 \, \frac{\text{Btu}}{\text{lbm}} - 1075 \, \frac{\text{Btu}}{\text{lbm}}\right) \\ - \left(253.1 \, \frac{\text{Btu}}{\text{lbm}} - 250.2 \, \frac{\text{Btu}}{\text{lbm}}\right) \\ - (1 - 0.149)\left(69.88 \, \frac{\text{Btu}}{\text{lbm}} - 69.73 \, \frac{\text{Btu}}{\text{lbm}}\right)\end{array}}{\begin{array}{c}\left(1393.9 \, \frac{\text{Btu}}{\text{lbm}} - 253.1 \, \frac{\text{Btu}}{\text{lbm}}\right) \\ + \left(1426.0 \, \frac{\text{Btu}}{\text{lbm}} - 1270 \, \frac{\text{Btu}}{\text{lbm}}\right)\end{array}}$$

$$= \boxed{0.340 \ (34.0\%)}$$

The answer is (C).

SI Solution

(a) Refer to the illustration for Prob. 7 and to the diagram for the customary U.S. solution.

At point A,

$$p_A = 6.2 \text{ MPa}$$
$$T_A = 420°\text{C}$$

Using the Mollier diagram for steam, App. 24.R,

$$h_A = 3235.0 \text{ kJ/kg}$$
$$s_A = 6.65 \text{ kJ/kg·K}$$

At point B, $h'_B = 2960$ kJ/kg.

From the Mollier diagram, assuming isentropic expansion to 1.5 MPa, $h_B = 2860$ kJ/kg.

Isentropic efficiency of the high pressure turbine is

$$\eta_{s,\text{turbine}} = \frac{h_\text{A} - h'_\text{B}}{h_\text{A} - h_\text{B}}$$

$$= \frac{3235.0\ \dfrac{\text{kJ}}{\text{kg}} - 2960\ \dfrac{\text{kJ}}{\text{kg}}}{3235.0\ \dfrac{\text{kJ}}{\text{kg}} - 2860\ \dfrac{\text{kJ}}{\text{kg}}}$$

$$= \boxed{0.733\ (73.3\%)}$$

The answer is (D).

(b) At point C,

$$p_\text{C} = 1.4\ \text{MPa}$$
$$T_\text{C} = 420°\text{C}$$

Using the superheated steam table from App. 24.P, $h_\text{C} = 3301.3\ \text{kJ/kg}$.

At point D, $h'_\text{D} = 2980\ \text{kJ/kg}$.

At point E, $h'_\text{E} = 2500\ \text{kJ/kg}$.

At point F, $h_\text{F} = 162.5\ \text{kJ/kg}$.

At point G,

$$W_\text{pump} = 0.3\ \text{kJ/kg}$$
$$= h'_\text{G} - h_\text{F}$$
$$h'_\text{G} = W_\text{pump} + h_\text{F}$$
$$= 0.3\ \frac{\text{kJ}}{\text{kg}} + 162.5\ \frac{\text{kJ}}{\text{kg}}$$
$$= 162.8\ \text{kJ/kg}$$

At point H, $h_\text{H} = 583.0\ \text{kJ/kg}$.

At point I, $h'_\text{I} = 589.7\ \text{kJ/kg}$.

From an energy balance in the heater,

$$xh'_\text{D} + (1-x)h'_\text{G} = h_\text{H}$$
$$x(h'_\text{D} - h'_\text{G}) = h_\text{H} - h'_\text{G}$$
$$x = \frac{h_\text{H} - h_\text{G}}{h'_\text{D} - h'_\text{G}}$$

$$= \frac{583.0\ \dfrac{\text{kJ}}{\text{kg}} - 162.8\ \dfrac{\text{kJ}}{\text{kg}}}{2980\ \dfrac{\text{kJ}}{\text{kg}} - 162.8\ \dfrac{\text{kJ}}{\text{kg}}}$$

$$= 0.149$$

The thermal efficiency of the cycle is

$$\eta_\text{th} = \frac{W_\text{out} - W_\text{in}}{Q_\text{in}}$$

$$= \frac{\begin{array}{c}(h_\text{A} - h'_\text{B}) + (h_\text{C} - h'_\text{D}) + (1-x) \\ \times\,(h'_\text{D} - h'_\text{E}) - (h'_\text{I} - h_\text{H}) - (1-x)(h'_\text{G} - h_\text{F})\end{array}}{(h_\text{A} - h'_\text{I}) + (h_\text{C} - h'_\text{B})}$$

$$= \frac{\begin{array}{c}\left(3235.0\ \dfrac{\text{kJ}}{\text{kg}} - 2960\ \dfrac{\text{kJ}}{\text{kg}}\right) \\[4pt] + \left(3301.3\ \dfrac{\text{kJ}}{\text{kg}} - 2980\ \dfrac{\text{kJ}}{\text{kg}}\right) \\[4pt] + (1 - 0.149)\left(2980\ \dfrac{\text{kJ}}{\text{kg}} - 2500\ \dfrac{\text{kJ}}{\text{kg}}\right) \\[4pt] - \left(598.7\ \dfrac{\text{kJ}}{\text{kg}} - 583.0\ \dfrac{\text{kJ}}{\text{kg}}\right) \\[4pt] - (1 - 0.149)\left(162.8\ \dfrac{\text{kJ}}{\text{kg}} - 162.5\ \dfrac{\text{kJ}}{\text{kg}}\right)\end{array}}{\begin{array}{c}\left(3235.0\ \dfrac{\text{kJ}}{\text{kg}} - 598.7\ \dfrac{\text{kJ}}{\text{kg}}\right) \\[4pt] + \left(3301.3\ \dfrac{\text{kJ}}{\text{kg}} - 2960\ \dfrac{\text{kJ}}{\text{kg}}\right)\end{array}}$$

$$= \boxed{0.332\ \ (33.2\%)}$$

The answer is (C).

8.

15 lbm/hr (1.9 g/s)
140°F (60°C)

0.25 hp (0.19 kW)

15 psia (103.5 kPa)

Customary U.S. Solution

The drill power is

$$P = 0.25\ \text{hp}\quad[\text{given}]$$

$$= (0.25\ \text{hp})\left(2545\ \frac{\text{Btu}}{\text{hp-hr}}\right) = 636.25\ \text{Btu/hr}$$

The absolute inlet temperature is

$$T_\text{I} = 140°\text{F} + 460 = 600°\text{R}$$

From App. 24.F, the properties of air entering the turbine are

$$h_\text{I} = 143.47\ \text{Btu/lbm}$$
$$p_{r,1} = 2.005$$
$$\phi_\text{I} = 0.62607\ \text{Btu/lbm-}°\text{R}$$

From Eq. 27.18,

$$P = \dot{m}(h_1 - h_2')$$

$$\eta_{s,\text{turbine}} = \frac{h_1 - h_2'}{h_1 - h_2}$$

So,

$$P = \dot{m}\eta_{s,\text{turbine}}(h_1 - h_2)$$

$$h_2 = h_1 - \frac{P}{\dot{m}\eta_{s,\text{turbine}}}$$

$$= 143.47 \ \frac{\text{Btu}}{\text{lbm}} - \frac{636.25 \ \frac{\text{Btu}}{\text{hr}}}{\left(15 \ \frac{\text{lbm}}{\text{hr}}\right)(0.60)}$$

$$= 72.776 \ \text{Btu/lbm}$$

Appendix 24.F doesn't go low enough. From the Keenan and Kayes *Gas Tables*, for $h = 72.776$ Btu/lbm,

$$T_2 = 305°\text{R}$$

$$p_{r,2} = 0.18851$$

(a) Due to the irreversibility of the expansion from Eq. 27.19,

$$h_2' = h_1 - \eta_s(h_1 - h_2)$$

$$= 143.47 \ \frac{\text{Btu}}{\text{lbm}}$$

$$\quad - (0.60)\left(143.47 \ \frac{\text{Btu}}{\text{lbm}} - 72.776 \ \frac{\text{Btu}}{\text{lbm}}\right)$$

$$= 101.05 \ \text{Btu/lbm}$$

From App. 24.F for $h = 101.05$ Btu/lbm,

$$T_2' = \boxed{423°\text{R}}$$

$$\phi_2 = 0.54225 \ \text{Btu/lbm-°R}$$

The answer is (B).

(b) The isentropic efficiency does not change the entrance and exit pressures, so the air tables can be used assuming $s_1 = s_2$. Since $p_1/p_2 = p_{r,1}/p_{r,2}$,

$$p_1 = p_2\left(\frac{p_{r,1}}{p_{r,2}}\right) = (15 \ \text{psia})\left(\frac{2.005}{0.18851}\right)$$

$$= \boxed{159.5 \ \text{psia}}$$

The answer is (A).

(c) From Eq. 24.39, the entropy change is

$$s_2 - s_1 = \phi_2 - \phi_1 - R\ln\left(\frac{p_2}{p_1}\right)$$

From Table 24.7, $R = 53.35$ ft-lbf/lbm-°R.

$$s_2 - s_1 = 0.54225 \ \frac{\text{Btu}}{\text{lbm-°R}} - 0.62607 \ \frac{\text{Btu}}{\text{lbm-°R}}$$

$$\quad - \left(\frac{53.35 \ \frac{\text{ft-lbf}}{\text{lbm-°R}}}{778 \ \frac{\text{ft-lbf}}{\text{Btu}}}\right)\ln\left(\frac{15 \ \text{psia}}{159.5 \ \text{psia}}\right)$$

$$= \boxed{0.07829 \ \text{Btu/lbm-°R}}$$

The answer is (A).

SI Solution

The absolute inlet temperature is

$$T_I = 60°\text{C} + 273 = 333\text{K}$$

From App. 24.S, the properties of air entering the turbine are

$$h_1 = 333.70 \ \text{kJ/kg}$$

$$p_{r,1} = 2.0064$$

$$\phi_1 = 1.80784 \ \text{kJ/kg·K}$$

From Eq. 27.18,

$$P = \dot{m}(h_1 - h_2')$$

$$\eta_{s,\text{turbine}} = \frac{h_1 - h_2'}{h_1 - h_2}$$

$$P = \dot{m}\eta_{s,\text{turbine}}(h_1 - h_2)$$

$$h_2 = h_1 - \frac{P}{\dot{m}\eta_{s,\text{turbine}}}$$

$$= 333.70 \ \frac{\text{kJ}}{\text{kg}} - \frac{0.19 \ \text{kW}}{\left(1.9 \ \frac{\text{g}}{\text{s}}\right)\left(\frac{1 \ \text{kg}}{1000 \ \text{g}}\right)(0.60)}$$

$$= 167.03 \ \text{kJ/kg}$$

Appendix 24.S doesn't go low enough. From gas tables, for $h = 167.03$ kJ/kg,

$$T_2 = 164.4\text{K}$$

$$p_{r,2} = 0.16980$$

(a) Due to the irreversibility of the expansion from Eq. 27.19,

$$h_2' = h_1 - \eta_s(h_1 - h_2)$$

$$= 333.70 \ \frac{\text{kJ}}{\text{kg}} - (0.60)\left(330.70 \ \frac{\text{kJ}}{\text{kg}} - 167.03 \ \frac{\text{kJ}}{\text{kg}}\right)$$

$$= 235.50 \ \text{kJ/kg}$$

From App. 24.S, for $h = 235.50$ kJ/kg,

$$T_2' = \boxed{235.3\text{K}}$$

$$\phi_2 = 1.45819 \text{ kJ/kg·K}$$

The answer is (B).

(b) Since $p_1/p_2 = p_{r,1}/p_{r,2}$,

$$p_1 = (103.5 \text{ kPa}) \left(\frac{2.0064}{0.16980} \right)$$

$$= \boxed{1223 \text{ kPa}}$$

The answer is (A).

(c) From Table 24.7, $R = 287.03$ kJ/kg·K. From Eq. 24.39, the entropy change is

$$s_2 - s_1 = \phi_2 - \phi_1 - R \ln \left(\frac{p_2}{p_1} \right)$$

$$= 1.45819 \frac{\text{kJ}}{\text{kg·K}} - 1.80784 \frac{\text{kJ}}{\text{kg·K}}$$

$$- \left(287.03 \frac{\text{kJ}}{\text{kg·K}} \right) \left(\frac{1 \text{ kJ}}{1000 \text{ J}} \right)$$

$$\times \ln \left(\frac{103.5 \text{ kPa}}{1223 \text{ kPa}} \right)$$

$$= \boxed{0.3592 \text{ kJ/kg·K}}$$

The answer is (A).

Power Cycles

29 Combustion Power Cycles

PRACTICE PROBLEMS

1. Air expands isentropically at the rate of 10 ft³/sec (280 L/s) from 200 psia and 1500°F (1.4 MPa and 820°C) to 50 psia (350 kPa).

(a) What is the air's final temperature?
- (A) 1275°R (710K)
- (B) 1325°R (740K)
- (C) 1375°R (770K)
- (D) 1425°R (790K)

(b) What is the air's final volumetric flow rate?
- (A) 28 ft³/sec (780 L/s)
- (B) 31 ft³/sec (860 L/s)
- (C) 39 ft³/sec (1100 L/s)
- (D) 45 ft³/sec (1300 L/s)

(c) What is the air's enthalpy change?
- (A) −110 Btu/lbm (−250 kJ/kg)
- (B) −160 Btu/lbm (−370 kJ/kg)
- (C) −230 Btu/lbm (−530 kJ/kg)
- (D) −350 Btu/lbm (−770 kJ/kg)

2. A 10 in (250 mm) bore, 18 in (460 mm) stroke, two-cylinder, four-stroke internal combustion engine operates with a mean effective pressure of 95 psig (660 kPa) at 200 rpm. The actual torque developed is 600 ft-lbf (820 N·m). What is the friction horsepower?
- (A) 17 hp (13 kW)
- (B) 22 hp (17 kW)
- (C) 45 hp (33 kW)
- (D) 78 hp (60 kW)

3. A four-cycle internal combustion engine has a bore of 3.1 in (80 mm) and a stroke of 3.8 in (97 mm). When the engine is running at 4000 rpm, the air-fuel mixture enters the cylinders during the intake stroke with an average velocity of 100 ft/sec (30 m/s). The intake valve opens at top dead center (TDC) and closes at 40° past bottom dead center (BDC). The volumetric efficiency is 65%. What is the effective area of the intake valve?
- (A) 0.012 ft² (0.0012 m²)
- (B) 0.023 ft² (0.0023 m²)
- (C) 0.034 ft² (0.0034 m²)
- (D) 0.058 ft² (0.0058 m²)

4. An engine runs on the Otto cycle. The compression ratio is 10:1. The total intake volume of the cylinders is 11 ft³ (0.3 m³). Air enters at 14.2 psia and 80°F (98 kPa and 27°C). In two revolutions of the crank, after the compression strokes, 160 Btu (179 kJ) of energy is added.

(a) What is the temperature after the heat addition?
- (A) 1290°R (720K)
- (B) 1550°R (860K)
- (C) 1750°R (970K)
- (D) 2320°R (1340K)

(b) What is the thermal efficiency of the cycle?
- (A) 45%
- (B) 52%
- (C) 57%
- (D) 64%

5. A 4.25 in × 6 in (110 mm × 150 mm), six-cylinder four-stroke diesel engine runs at 1200 rpm while consuming 28 lbm of fuel per hour (0.0035 kg/s). The actual volumetric fraction of carbon dioxide in the exhaust is 9% (dry basis). At another throttle setting, when the air-fuel ratio is 15:1, there is 13.7% (dry basis) carbon dioxide in the exhaust. The atmospheric conditions are 14.7 psia and 70°F (101.3 kPa and 21°C). What is the volumetric efficiency at 1200 rpm?
- (A) 59%
- (B) 65%
- (C) 72%
- (D) 80%

6. At standard atmospheric conditions, a fully loaded diesel engine runs at 2000 rpm and develops 1000 bhp (750 kW). The brake specific fuel consumption is 0.45 lbm/bhp-hr (76 kg/GJ). The mechanical efficiency is 80%, independent of the altitude.

(a) At an altitude of 5000 ft (1500 m), what will be the brake horsepower?
- (A) 810 hp (630 kW)
- (B) 860 hp (650 kW)
- (C) 890 hp (690 kW)
- (D) 950 hp (740 kW)

(b) At an altitude of 5000 ft (1500 m), what will be the brake specific fuel consumption?

(A) 0.26 lbm/hp-hr (44 kg/GJ)
(B) 0.52 lbm/hp-hr (88 kg/GJ)
(C) 1.4 lbm/hp-hr (240 kg/GJ)
(D) 2.2 lbm/hp-hr (370 kg/GJ)

7. In an air-standard gas turbine, air at 14.7 psia and 60°F (101.3 kPa and 16°C) enters a compressor and is compressed through a volume ratio of 5:1. The compressor efficiency is 83%. Air enters the turbine at 1500°F (820°C) and expands to 14.7 psia (101.3 kPa). The turbine efficiency is 92%. What is the thermal efficiency of the cycle?

(A) 33%
(B) 39%
(C) 44%
(D) 51%

8. A 65% efficient regenerator is added to the gas turbine described in Prob. 7. Assume the specific heat remains constant. What is the new thermal efficiency?

(A) 24%
(B) 28%
(C) 34%
(D) 41%

9. (*Time limit: one hour*) A gasoline-fueled internal combustion engine runs at 4600 rpm. The engine is four-stroke and V-8 in configuration with a displacement of 265 in³ (4.3 L). The indicated work required to compress the air-fuel mixture is 1200 ft-lbf (1.6 kJ) per cycle. The indicated work done by the exhaust gases in expansion is 1500 ft-lbf (2.0 kJ) per cycle. The input energy from fuel combustion is 1.27 Btu (1.33 kJ) per cycle. Atmospheric air is at 14.7 psia and 70°F (101.3 kPa and 21°C). The air-fuel ratio is 20:1. The heating value of gasoline is 18,900 Btu/lbm (44 MJ/kg). Neglect the effects of friction.

(a) What is the indicated horsepower?

(A) 140 hp (110 kW)
(B) 170 hp (120 kW)
(C) 200 hp (160 kW)
(D) 240 hp (190 kW)

(b) What is the thermal efficiency?

(A) 30%
(B) 35%
(C) 39%
(D) 46%

(c) What is the mass per hour of gasoline consumed?

(A) 37 lbm/hr (0.0046 kg/s)
(B) 52 lbm/hr (0.0065 kg/s)
(C) 74 lbm/hr (0.0093 kg/s)
(D) 99 lbm/hr (0.012 kg/s)

(d) What is the specific fuel consumption?

(A) 0.055 lbm/hp-hr (10 kg/GJ)
(B) 0.11 lbm/hp-hr (19 kg/GJ)
(C) 0.22 lbm/hp-hr (38 kg/GJ)
(D) 0.44 lbm/hp-hr (76 kg/GJ)

10. (*Time limit: one hour*) A mixture of carbon dioxide and helium in an engine undergoes the following processes in a cycle.

A to B: compression and heat removal

B to C: constant volume heating

C to A: isentropic expansion

point	temperature	pressure
A	520°R (290K)	14.7 psia (101.3 kPa)
B	1240°R (690K)	unknown
C	1600°R (890K)	568.6 psia (3.920 MPa)

Both gases are ideal gases.

(a) What are the gravimetric fractions of the carbon dioxide and helium in the mixture?

(A) CO_2, 0.86; He, 0.14
(B) CO_2, 0.77; He, 0.23
(C) CO_2, 0.65; He, 0.35
(D) CO_2, 0.54; He, 0.46

(b) What work is done during the isentropic expansion process?

(A) 190 Btu/lbm (440 kJ/kg)
(B) 260 Btu/lbm (610 kJ/kg)
(C) 330 Btu/lbm (760 kJ/kg)
(D) 450 Btu/lbm (1000 kJ/kg)

(c) Draw the temperature-entropy and pressure-volume diagrams for the cycle.

11. (*Time limit: one hour*) When the atmospheric conditions are 14.7 psia and 80°F (101.3 kPa and 27°C), a diesel engine with metered fuel injection has the following operating characteristics.

brake horsepower:	200 bhp (150 kW)
brake specific fuel consumption:	0.48 lbm/hp-hr (81 kg/GJ)
air-fuel ratio:	22:1
mechanical efficiency:	86%

The engine is moved to an altitude where the atmospheric conditions are 12.2 psia and 60°F (84 kPa and 16°C). The running speed is unchanged. What are the corresponding operating characteristics?

12. (*Time limit: one hour*) A gas turbine operating on the Brayton cycle with an 8:1 pressure ratio is located at 7000 ft (2100 m) altitude. The conditions at that altitude are 12 psia and 35°F (82 kPa and 2°C). While consuming 0.609 lbm/hp-hr (100 kg/GJ) of fuel and 50,000 cfm (23 500 L/s) of air, the turbine develops 6000 bhp (4.5 MW). The turbine efficiency is 80%, and the compressor efficiency is 85%. The fuel has a lower heating value of 19,000 Btu/lbm (44 MJ/kg). The fuel mass is small compared to the air mass. The turbine receives combustor gases at 1800°F (980°C). The air inlet filter area is 254 ft² (22.9 m²). The turbine is moved to sea level where the conditions are 14.7 psia and 70°F (101.3 kPa and 21°C). The combustion efficiency and combustor temperature remain the same.

(a) What is the new brake horsepower?

    (A) 2800 hp (2.1 MW)

    (B) 4400 hp (3.4 MW)

    (C) 5100 hp (4.0 MW)

    (D) 6300 hp (4.8 MW)

(b) What is the new brake specific fuel consumption?

    (A) 0.42 lbm/hp-hr (67 kg/GJ)

    (B) 0.51 lbm/hp-hr (82 kg/GJ)

    (C) 0.63 lbm/hp-hr (100 kg/GJ)

    (D) 0.87 lbm/hp-hr (140 kg/GJ)

SOLUTIONS

1. *Customary U.S. Solution*

(a) The absolute temperature is

$$T_1 = 1500°F + 460 = 1960°R$$

From air tables (App. 24.F) at 1960°R,

$$h_1 = 493.64 \text{ Btu/lbm}$$
$$p_{r,1} = 160.48$$
$$v_{r,1} = 4.53$$

After expansion,

$$p_{r,2} = p_{r,1}\left(\frac{p_2}{p_1}\right)$$
$$= (160.48)\left(\frac{50 \text{ psia}}{200 \text{ psia}}\right)$$
$$= 40.12$$

From air tables (App. 24.F) at $p_{r,2} = 40.12$,

$$T_2 = \boxed{1375°R}$$
$$h_2 = 336.39 \text{ Btu/lbm}$$
$$v_{r,2} = 12.721$$

The answer is (C).

(b)
$$\dot{V}_2 = \dot{V}_1\left(\frac{v_{r,2}}{v_{r,1}}\right)$$
$$= \left(10\ \frac{\text{ft}^3}{\text{sec}}\right)\left(\frac{12.721}{4.53}\right)$$
$$= \boxed{28.1 \text{ ft}^3/\text{sec}}$$

The answer is (A).

(c) The enthalpy change is

$$\Delta h = h_2 - h_1$$
$$= 336.39\ \frac{\text{Btu}}{\text{lbm}} - 493.64\ \frac{\text{Btu}}{\text{lbm}}$$
$$= \boxed{-157.25 \text{ Btu/lbm} \quad [\text{decrease}]}$$

The answer is (B).

SI Solution

(a) The absolute temperature is

$$T_1 = 820°C + 273 = 1093\text{K}$$

From air tables (App. 24.S) at 1093K,

$$h_1 = 1153 \text{ kJ/kg}$$
$$p_{r,1} = 162.94$$
$$v_{r,1} = 19.275$$

After expansion,

$$p_{r,2} = p_{r,1} \left(\frac{p_2}{p_1} \right)$$

$$= (162.94) \left(\frac{350 \text{ kPa}}{(1.4 \text{ MPa}) \left(1000 \frac{\text{kPa}}{\text{MPa}} \right)} \right)$$

$$= 40.74$$

From air tables (App. 24.S) at $p_{r,2} = 40.74$,

$$T_2 = \boxed{767.2 \text{K}}$$

$$h_2 = 786.05 \text{ kJ/kg}$$
$$v_{r,2} = 53.99$$

The answer is (C).

(b) $\qquad \dot{V}_2 = \dot{V}_1 \left(\frac{v_{r,2}}{v_{r,1}} \right)$

$$= \left(280 \frac{\text{L}}{\text{s}} \right) \left(\frac{53.99}{19.275} \right)$$

$$= \boxed{784.3 \text{ L/s}}$$

The answer is (A).

(c) The enthalpy change is

$$\Delta h = h_2 - h_1$$

$$= 786.05 \frac{\text{kJ}}{\text{kg}} - 1153 \frac{\text{kJ}}{\text{kg}}$$

$$= \boxed{-366.95 \text{ kJ/kg} \quad [\text{decrease}]}$$

The answer is (B).

2. *Customary U.S. Solution*

The actual brake horsepower from Eq. 29.10(b) is

$$\text{BHP} = \frac{nT}{5252} = \frac{(200 \text{ rpm})(600 \text{ ft-lbf})}{5252}$$

$$= 22.85 \text{ hp}$$

From Eq. 29.41, the number of power strokes per minute is

$$N = \frac{(2n)(\text{no. cylinders})}{\text{no. strokes per cycle}}$$

$$= \frac{(2)(200 \text{ rpm})(2)}{4} = 200 \text{ power strokes/min}$$

The stroke is

$$L = (18 \text{ in}) \left(\frac{1 \text{ ft}}{12 \text{ in}} \right) = 1.5 \text{ ft}$$

The bore area is

$$\left(\frac{\pi}{4} \right) (10 \text{ in})^2 = 78.54 \text{ in}^2$$

From Eq. 29.40(b), the ideal horsepower is

$$\text{hp} = \frac{pLAN}{33,000}$$

$$= \frac{\left(95 \frac{\text{lbf}}{\text{in}^2} \right) (1.5 \text{ ft})(78.54 \text{ in}^2) \left(200 \frac{\text{strokes}}{\text{min}} \right)}{33,000 \frac{\text{ft-lbf}}{\text{hp-min}}}$$

$$= 67.83 \text{ hp}$$

The friction horsepower is

$$\text{ideal hp} - \text{actual BHP} = 67.83 \text{ hp} - 22.85 \text{ hp}$$

$$= \boxed{44.98 \text{ hp}}$$

The answer is (C).

SI Solution

The actual brake power from Eq. 29.10(a) is

$$\text{BkW} = \frac{nT}{5252} = \frac{(200 \text{ rpm})(820 \text{ N·m})}{9549}$$

$$= 17.17 \text{ kW}$$

From Eq. 29.41, the number of power strokes per minute is

$$N = \frac{(2n)(\text{no. cylinders})}{\text{no. strokes per cycle}}$$

$$= \frac{(2)(200 \text{ rpm})(2)}{4} = 200 \text{ power strokes/min}$$

The stroke is

$$\frac{460 \text{ mm}}{1000 \frac{\text{mm}}{\text{m}}} = 0.46 \text{ m}$$

The bore area is

$$\left(\frac{\pi}{4} \right) \left(\frac{250 \text{ mm}}{1000 \frac{\text{mm}}{\text{m}}} \right)^2 = 4.909 \times 10^{-2} \text{ m}^2$$

From Eq. 29.40(a), the ideal power is

$$kW = pLAN$$

$$= \frac{(660 \text{ kPa})(0.46 \text{ m})}{60 \frac{\text{s}}{\text{min}}} \times (4.909 \times 10^{-2} \text{ m}^2)\left(200 \frac{\text{strokes}}{\text{min}}\right)$$

$$= 49.68 \text{ kW}$$

The friction power is

$$\text{ideal kW} - \text{actual kW} = 49.68 \text{ kW} - 17.17 \text{ kW}$$

$$= \boxed{32.51 \text{ kW}}$$

The answer is (C).

3. *Customary U.S. Solution*

The number of degrees that the valve is open is

$$180° + 40° = 220°$$

The time that the valve is open is

$$\left(\frac{220°}{360°}\right)\left(\frac{\text{time}}{\text{rev}}\right) = \left(\frac{220°}{360°}\right)\left(\frac{60 \frac{\text{sec}}{\text{min}}}{4000 \text{ rpm}}\right)$$

$$= 9.167 \times 10^{-3} \text{ sec/rev}$$

The displacement is

$$\left(\frac{\pi}{4}\right)(\text{bore})^2(\text{stroke}) = \left(\frac{\pi}{4}\right)(3.1 \text{ in})^2\left(\frac{1 \text{ ft}}{12 \text{ in}}\right)^2$$

$$\times (3.8 \text{ in})\left(\frac{1 \text{ ft}}{12 \text{ in}}\right)$$

$$= 0.0166 \text{ ft}^3$$

The actual incoming volume per intake stroke is

$$V = (\text{volumetric efficiency})(\text{displacement})$$

$$= (0.65)(0.0166 \text{ ft}^3) = 0.01079 \text{ ft}^3$$

The area is

$$A = \frac{V}{vt} = \frac{0.01079 \text{ ft}^3}{\left(100 \frac{\text{ft}}{\text{sec}}\right)(9.167 \times 10^{-3} \text{ sec})}$$

$$= \boxed{0.0118 \text{ ft}^2 \quad (1.69 \text{ in}^2)}$$

The answer is (A).

SI Solution

The number of degrees that the valve is open is

$$180° + 40° = 220°$$

The time that the valve is open is

$$\left(\frac{220°}{360°}\right)\left(\frac{\text{time}}{\text{rev}}\right) = \left(\frac{220°}{360°}\right)\left(\frac{60 \frac{\text{s}}{\text{min}}}{4000 \text{ rpm}}\right)$$

$$= 9.167 \times 10^{-3} \text{ s/rev}$$

The displacement is

$$\left(\frac{\pi}{4}\right)(\text{bore})^2(\text{stroke}) = \left(\frac{\pi}{4}\right)(0.08 \text{ m})^2(0.097 \text{ m})$$

$$= 4.876 \times 10^{-4} \text{ m}^3$$

The actual incoming volume per intake stroke is

$$V = (\text{volumetric efficiency})(\text{displacement})$$

$$= (0.65)(4.876 \times 10^{-4} \text{ m}^3) = 3.169 \times 10^{-4} \text{ m}^3$$

The area is

$$A = \frac{V}{vt} = \frac{3.169 \times 10^{-4} \text{ m}^3}{\left(30 \frac{\text{m}}{\text{s}}\right)(9.167 \times 10^{-3} \text{ s})}$$

$$= \boxed{1.152 \times 10^{-3} \text{ m}^2}$$

The answer is (A).

4. *Customary U.S. Solution*

Refer to the air-standard Otto cycle diagram (Fig. 29.3).

(a) At A:

$$V_A = 11 \text{ ft}^3$$

The absolute temperature is

$$T_A = 80°F + 460 = 540°R$$

For an ideal gas, the mass of the air in the intake volume is

$$m = \frac{pV}{RT} = \frac{\left(14.2 \frac{\text{lbf}}{\text{in}^2}\right)\left(144 \frac{\text{in}^2}{\text{ft}^2}\right)(11 \text{ ft}^3)}{\left(53.3 \frac{\text{ft-lbf}}{\text{lbm-°R}}\right)(540°R)}$$

$$= 0.781 \text{ lbm}$$

From air tables (App. 24.F) at 540°R,

$$v_{r,A} = 144.32$$

$$u_A = 92.04 \text{ Btu/lbm}$$

At B:

The compression ratio is a ratio of volumes.

$$V_B = \left(\frac{1}{10}\right) V_A = \left(\frac{1}{10}\right) (11 \text{ ft}^3)$$
$$= 1.1 \text{ ft}^3$$

Since the compression from A to B is isentropic,

$$v_{r,B} = \frac{v_{r,A}}{10} = \frac{144.32}{10}$$
$$= 14.432$$

From the air tables (App. 24.F) for $v_r = 14.432$,

$$T_B \approx 1314°R$$
$$u_B \approx 230.5 \text{ Btu/lbm}$$

At C:

$$u_C = u_B + \frac{Q_{in,B\text{-}C}}{m}$$
$$= 230.5 \frac{\text{Btu}}{\text{lbm}} + \frac{160 \text{ Btu}}{0.781 \text{ lbm}}$$
$$= 435.4 \text{ Btu/lbm}$$

From air tables (App. 24.F) at $u = 435.4$ Btu/lbm,

$$T_C = \boxed{2319°R}$$
$$v_{r,C} = 2.694$$

At D:

Since expansion is isentropic and the ratio of volumes is the same,

$$v_{r,D} = 10v_{r,C} = (10)(2.694)$$
$$= 26.94$$

From air tables (App. 24.F), at $v_r = 26.94$,

$$T_D = 1044°R$$
$$u_D = 180.38 \text{ Btu/lbm}$$

The answer is (D).

(b) The heat input is

$$q_{in} = \frac{Q}{m} = \frac{160 \text{ Btu}}{0.781 \text{ lbm}}$$
$$= 204.9 \text{ Btu/lbm}$$

The heat rejected during a constant volume process is
$$q_{out} = \Delta u.$$

Heat is rejected between D and A. Therefore,

$$q_{out} = u_D - u_A = 180.38 \frac{\text{Btu}}{\text{lbm}} - 92.04 \frac{\text{Btu}}{\text{lbm}}$$
$$= 88.34 \text{ Btu/lbm}$$

From Eq. 29.38, the thermal efficiency is

$$\eta_{th} = \frac{q_{in} - q_{out}}{q_{in}} = \frac{204.9 \frac{\text{Btu}}{\text{lbm}} - 88.34 \frac{\text{Btu}}{\text{lbm}}}{204.9 \frac{\text{Btu}}{\text{lbm}}}$$
$$= \boxed{0.569 \quad (56.9\%)}$$

The answer is (C).

SI Solution

(a) At A:

The absolute temperature is

$$T_A = 27°C + 273 = 300K$$

From the ideal gas law, the mass of air in the intake volume is

$$m = \frac{pV}{RT} = \frac{(98 \text{ kPa})\left(1000 \frac{\text{Pa}}{\text{kPa}}\right)(0.3 \text{ m}^3)}{\left(287 \frac{\text{J}}{\text{kg·K}}\right)(300K)}$$
$$= 0.3415 \text{ kg}$$

From air tables (App. 24.S) at 300K,

$$v_{r,A} = 621.2$$
$$u_A = 214.07 \text{ kJ/kg}$$

At B:

The compression ratio is a ratio of volumes.

$$V_B = \left(\frac{1}{10}\right) V_A = \frac{0.3 \text{ m}^3}{10}$$
$$= 0.03 \text{ m}^3$$

Since the compression from A to B is isentropic,

$$v_{r,B} = \frac{v_{r,A}}{10} = \frac{621.2}{10}$$
$$= 62.12$$

From air tables (App. 24.S) for this value of $v_{r,B}$,

$$T_B = 730K$$
$$u_B = 536.0 \text{ kJ/kg}$$

At C:

$$u_C = u_B + \frac{Q_{in,B\text{-}C}}{m}$$

$$= 536.0 \ \frac{kJ}{kg} + \frac{179 \ kJ}{0.3415 \ kg}$$

$$= 1060.2 \ kJ/kg$$

From air tables (App. 24.S) at $u = 1060.2$ kJ/kg,

$$T_C = \boxed{1341.4 \ K}$$

$$v_{r,C} = 10.215$$

At D:

Since expansion is isentropic and the ratio of volumes is the same,

$$v_{r,D} = 10v_{r,C} = (10)(10.215)$$

$$= 102.15$$

From air tables (App. 24.S), at $v_r = 102.15$,

$$T_D = 607.9 \ K$$

$$u_D = 440.8 \ kJ/kg$$

The answer is (D).

(b) The heat input is

$$q_{in} = \frac{Q}{m} = \frac{179 \ kJ}{0.3415 \ kg}$$

$$= 524.2 \ kJ/kg$$

The heat rejected during a constant volume process is

$$q_{out} = \Delta u$$

Heat is rejected between D and A. Therefore,

$$q_{out} = u_D - u_A = 440.8 \ \frac{kJ}{kg} - 214.07 \ \frac{kJ}{kg}$$

$$= 226.7 \ kJ/kg$$

From Eq. 29.38, the thermal efficiency is

$$\eta_{th} = \frac{q_{in} - q_{out}}{q_{in}} = \frac{524.2 \ \dfrac{kJ}{kg} - 226.7 \ \dfrac{kJ}{kg}}{524.2 \ \dfrac{kJ}{kg}}$$

$$= \boxed{0.568 \ (56.8\%)}$$

The answer is (C).

5. *Customary U.S. Solution*

step 1: Find the ideal mass of air ingested.

From Eq. 29.41, the number of power strokes per second is

$$N = \frac{(2n)(\text{no. cylinders})}{\text{no. strokes per cycle}}$$

$$= \frac{(2) \left(1200 \ \dfrac{rev}{min}\right) \left(\dfrac{1 \ min}{60 \ sec}\right)(6 \ \text{cylinders})}{4 \ \text{strokes}}$$

$$= 60/\ sec$$

The swept volume is

$$V_s = \left(\frac{\pi}{4}\right)(\text{bore})^2(\text{stroke})$$

$$= \left(\frac{\pi}{4}\right)(4.25 \ in)^2 \left(\frac{1 \ ft^2}{144 \ in^2}\right)(6 \ in)\left(\frac{1 \ ft}{12 \ in}\right)$$

$$= 0.04926 \ ft^3$$

The ideal volume of air taken in per second is

$$\dot{V}_i = (\text{swept volume})\left(\frac{\text{intake strokes}}{sec}\right)$$

$$= V_s N$$

$$= (0.04926 \ ft^3)\left(60 \ \frac{1}{sec}\right) = 2.956 \ ft^3/\ sec$$

The absolute temperature is

$$70°F + 460 = 530°R$$

From the ideal gas law, the ideal mass of air in the swept volume is

$$\dot{m} = \frac{p\dot{V}}{RT}$$

$$= \frac{\left(14.7 \ \dfrac{lbf}{in^2}\right)\left(144 \ \dfrac{in^2}{ft^2}\right)\left(2.956 \ \dfrac{ft^3}{sec}\right)}{\left(53.35 \ \dfrac{ft\text{-}lbf}{lbm\text{-}°R}\right)(530°R)}$$

$$= 0.2213 \ lbm/\ sec$$

step 2: Find the carbon dioxide volume in the exhaust assuming complete combustion when the air/fuel ratio is 15 (%CO_2 = 13.7%, dry basis).

Air is 76.85% (by weight) nitrogen, so the nitrogen/fuel ratio is

$$(0.7685)(15) = 11.528 \text{ lbm N}_2/\text{lbm fuel}$$

From the ideal gas law, the nitrogen per pound of fuel burned is

$$V_{N_2} = \frac{mRT}{p}$$

$$= \frac{\left(11.528 \ \frac{\text{lbm N}_2}{\text{lbm fuel}}\right)\left(55.16 \ \frac{\text{ft-lbf}}{\text{lbm-}^\circ\text{R}}\right)(530^\circ\text{R})}{\left(14.7 \ \frac{\text{lbf}}{\text{in}^2}\right)\left(144 \ \frac{\text{in}^2}{\text{ft}^2}\right)}$$

$$= 159.2 \text{ ft}^3/\text{lbm fuel}$$

Similarly, air is 23.15% oxygen by weight, so the oxygen/fuel ratio is

$$(0.2315)(15) = 3.472 \text{ lbm O}_2/\text{lbm fuel}$$

From the ideal gas law, the oxygen per pound of fuel burned is

$$V_{O_2} = \frac{mRT}{p}$$

$$= \frac{\left(3.472 \ \frac{\text{lbm O}_2}{\text{lbm fuel}}\right)\left(48.29 \ \frac{\text{ft-lbf}}{\text{lbm-}^\circ\text{R}}\right)(530^\circ\text{R})}{\left(14.7 \ \frac{\text{lbf}}{\text{in}^2}\right)\left(144 \ \frac{\text{in}^2}{\text{ft}^2}\right)}$$

$$= 41.98 \text{ ft}^3/\text{lbm fuel}$$

When oxygen forms carbon dioxide, the chemical equation is $C + O_2 \longrightarrow CO_2$.

It takes one volume of oxygen to form one volume of carbon dioxide. Considering nitrogen and excess oxygen in the exhaust, the percentage of carbon dioxide in the exhaust is found from

$$\%CO_2 = \frac{\text{vol CO}_2}{\text{vol CO}_2 + \text{vol O}_2 + \text{vol N}_2}$$

$$\text{vol CO}_2 = x \text{ [unknown], in ft}^3$$

$$\text{vol O}_2 = 41.98 \text{ ft}^3 - \text{oxygen used to make CO}_2$$

$$= 41.98 \text{ ft}^3 - x$$

$$\%CO_2 = 0.137 \text{ [given]}$$

$$0.137 = \frac{x}{x + (41.98 \text{ ft}^3 - x) + 159.3 \text{ ft}^3}$$

$$x = 27.58 \text{ ft}^3/\text{lbm fuel}$$

Assuming complete combustion, the volume of CO_2 will be constant regardless of the amount of air used.

step 3: Calculate the excess air if the percentage of carbon dioxide in the exhaust is 9%.

$$\%CO_2 = \frac{\text{vol CO}_2}{\begin{array}{c}\text{vol CO}_2 + \text{vol O}_2 \\ + \text{vol N}_2 + \text{vol excess air}\end{array}}$$

$$0.09 = \frac{27.58 \text{ ft}^3}{\begin{array}{c}27.58 \text{ ft}^3 + (41.99 \text{ ft}^3 - 27.58 \text{ ft}^3) \\ + 159.3 \text{ ft}^3 + \text{vol excess air}\end{array}}$$

$$\begin{array}{c}\text{vol} \\ \text{excess} \\ \text{air}\end{array} = 105.2 \text{ ft}^3/\text{lbm fuel}$$

From the ideal gas law, the mass of excess air is

$$m_{\text{excess}} = \frac{\left(14.7 \ \frac{\text{lbf}}{\text{in}^2}\right)\left(144 \ \frac{\text{in}^2}{\text{ft}^2}\right)(105.2 \text{ ft}^3)}{\left(53.35 \ \frac{\text{lbf-ft}}{\text{lbm-}^\circ\text{R}}\right)(530^\circ\text{R})}$$

$$= 7.876 \text{ lbm/lbm fuel}$$

step 4: The actual air/fuel ratio is

$$15 \ \frac{\text{lbm air}}{\text{lbm fuel}} + 7.876 \ \frac{\text{lbm air}}{\text{lbm fuel}}$$

$$= 22.876 \text{ lbm air/lbm fuel}$$

The actual air mass per second is

$$\left(22.876 \ \frac{\text{lbm air}}{\text{lbm fuel}}\right)\left(28 \ \frac{\text{lbm fuel}}{\text{hr}}\right)\left(\frac{1 \text{ hr}}{3600 \text{ sec}}\right)$$

$$= 0.178 \text{ lbm/sec}$$

step 5: The volumetric efficiency is

$$\eta_v = \frac{0.178 \ \frac{\text{lbm}}{\text{sec}}}{0.2213 \ \frac{\text{lbm}}{\text{sec}}} = \boxed{0.804 \ (80.4\%)}$$

The answer is (D).

SI Solution

step 1: Find the ideal mass of air ingested.

From Eq. 29.41, the number of power strokes per second is

$$N = \frac{(2n)(\text{no. cylinders})}{\text{no. strokes per cycle}}$$

$$= \frac{(2)\left(1200 \ \frac{\text{rev}}{\text{min}}\right)\left(\frac{1 \text{ min}}{60 \text{ s}}\right)(6 \text{ cylinders})}{4 \text{ strokes}}$$

$$= 60/\text{s}$$

The swept volume is

$$V_s = \left(\frac{\pi}{4}\right)(\text{bore})^2(\text{stroke})$$
$$= \left(\frac{\pi}{4}\right)(0.110 \text{ m})^2(0.150 \text{ m})$$
$$= 1.425 \times 10^{-3} \text{ m}^3$$

The ideal volume of air taken in per second is

$$\dot{V}_i = (\text{swept volume})\left(\frac{\text{intake strokes}}{\text{s}}\right)$$
$$= V_s N$$
$$= (1.425 \times 10^{-3})\left(60 \frac{1}{\text{s}}\right) = 0.0855 \text{ m}^3/\text{s}$$

The absolute temperature is

$$21°C + 273 = 294K$$

From the ideal gas law, the ideal mass of air in the swept volume is

$$\dot{m} = \frac{p\dot{V}}{RT}$$
$$= \frac{(101.3 \text{ kPa})\left(1000 \frac{\text{Pa}}{\text{kPa}}\right)\left(0.0855 \frac{\text{m}^3}{\text{s}}\right)}{\left(287.03 \frac{\text{J}}{\text{kg·K}}\right)(294K)}$$
$$= 0.1026 \text{ kg/s}$$

step 2: Find the carbon dioxide volume in the exhaust assuming complete combustion when the air/fuel ratio is 15 ($\%CO_2 = 13.7\%$ dry basis).

Air is 76.85% nitrogen by weight, so the nitrogen/fuel ratio is

$$(0.7685)(15) = 11.528 \text{ kg N}_2/\text{kg fuel}$$

From the ideal gas law, the nitrogen per kg of fuel burned is

$$V_{N_2} = \frac{mRT}{p}$$
$$= \frac{\left(11.528 \frac{\text{kg N}_2}{\text{kg fuel}}\right)\left(296.77 \frac{\text{J}}{\text{kg·K}}\right)(294K)}{(101.3 \text{ kPa})\left(1000 \frac{\text{Pa}}{\text{kPa}}\right)}$$
$$= 9.929 \text{ m}^3/\text{kg fuel}$$

Similarly, air is 23.15% oxygen by weight, so the oxygen/fuel ratio is

$$(0.2315)(15) = 3.473 \text{ kg O}_2/\text{kg fuel}$$

From the ideal gas law, the oxygen per kg of fuel burned is

$$V_{O_2} = \frac{mRT}{p}$$
$$= \frac{\left(3.473 \frac{\text{kg O}_2}{\text{kg fuel}}\right)\left(259.82 \frac{\text{J}}{\text{kg·K}}\right)(294K)}{(101.3 \text{ kPa})\left(1000 \frac{\text{Pa}}{\text{kPa}}\right)}$$
$$= 2.619 \text{ m}^3/\text{kg fuel}$$

From the customary U.S. solution,

$$\%CO_2 = \frac{\text{vol CO}_2}{\text{vol CO}_2 + \text{vol O}_2 + \text{vol N}_2}$$

$$\text{vol CO}_2 = x \text{ [unknown], in m}^3$$
$$\text{vol O}_2 = 2.619 \text{ m}^3 - \text{O}_2 \text{ used to make CO}_2$$
$$= 2.619 \text{ m}^3 - x$$
$$\%CO_2 = 0.137 \text{ [given]}$$
$$0.137 = \frac{x}{x + (2.619 \text{ m}^3 - x) + 9.929 \text{ m}^3}$$
$$x = 1.719 \text{ m}^3/\text{kg fuel}$$

Assuming complete combustion, the volume of carbon dioxide will be constant regardless of the amount of air used.

step 3: Calculate the excess air if the percentage of carbon dioxide in the exhaust is 9%. Therefore,

$$\%CO_2 = \frac{\text{vol CO}_2}{\text{vol CO}_2 + \text{vol O}_2} \atop {+ \text{ vol N}_2 + \text{vol excess air}}$$

$$0.09 = \frac{1.719 \text{ m}^3}{1.719 \text{ m}^3 + (2.619 \text{ m}^3 - 1.719 \text{ m}^3)} \atop {+ 9.929 \text{ m}^3 + \text{vol excess air}}$$

$$\text{vol} \atop {\text{cxcc33} \atop \text{air}} = 6.552 \text{ m}^3/\text{kg fuel}$$

From the ideal gas law, the mass of the excess air is

$$m_{\text{excess}} = \frac{(101.3 \text{ kPa})\left(1000 \frac{\text{Pa}}{\text{kPa}}\right)(6.552 \text{ m}^3)}{\left(287.03 \frac{\text{J}}{\text{kg·K}}\right)(294K)}$$
$$= 7.865 \text{ kg/kg fuel}$$

Power Cycles

step 4: The actual air/fuel ratio is

$$15 \, \frac{\text{kg air}}{\text{kg fuel}} + 7.865 \, \frac{\text{kg air}}{\text{kg fuel}} = 22.865 \text{ kg air/kg fuel}$$

The actual air mass per second is

$$\left(22.865 \, \frac{\text{kg air}}{\text{kg fuel}}\right)\left(0.0035 \, \frac{\text{kg}}{\text{s}}\right) = 0.0800 \text{ kg/s}$$

step 5: The volumetric efficiency is

$$\eta_v = \frac{0.0800 \, \frac{\text{kg}}{\text{s}}}{0.1026 \, \frac{\text{kg}}{\text{s}}} = \boxed{0.772 \;\; (77.2\%)}$$

The answer is (D).

6. *Customary U.S. Solution*

(a) *step 1:* From App. 26.E,

Altitude 1: standard atmospheric condition, 14.696 psia, 518.7°R

Altitude 2: $z = 5000$ ft, 12.225 psia, 500.9°R

step 2: Calculate the friction power (Eq. 29.68).

$$\text{IHP}_1 = \frac{\text{BHP}_1}{\eta_{m,1}} = \frac{1000 \text{ hp}}{0.80} = 1250 \text{ hp}$$

step 3: Not needed since η_m is constant with altitude.

step 4: From the ideal gas law,

$$\rho_{a1} = \frac{p}{RT} = \frac{\left(14.696 \, \frac{\text{lbf}}{\text{in}^2}\right)\left(144 \, \frac{\text{in}^2}{\text{ft}^2}\right)}{\left(53.35 \, \frac{\text{lbf-ft}}{\text{lbm-°R}}\right)(518.7°\text{R})}$$

$$= 0.0765 \text{ lbm/ft}^3$$

Similarly,

$$\rho_{a2} = \frac{p}{RT} = \frac{\left(12.225 \, \frac{\text{lbf}}{\text{in}^2}\right)\left(144 \, \frac{\text{in}^2}{\text{ft}^2}\right)}{\left(53.35 \, \frac{\text{ft-lbf}}{\text{lbm-°R}}\right)(500.9°\text{R})}$$

$$= 0.0659 \text{ lbm/ft}^3$$

step 5: Calculate the new frictionless power (Eq. 29.70).

$$\text{IHP}_2 = \text{IHP}_1 \left(\frac{\rho_{a2}}{\rho_{a1}}\right)$$

$$= (1250 \text{ hp})\left(\frac{0.0659 \, \frac{\text{lbm}}{\text{ft}^3}}{0.0765 \, \frac{\text{lbm}}{\text{ft}^3}}\right)$$

$$= 1076.8 \text{ hp}$$

steps 6 and 7: Calculate the new net power using Eq. 29.72.

$$\text{BHP}_2 = \eta_{m,2}(\text{IHP}_2) = (0.80)(1076.8 \text{ hp})$$

$$= \boxed{861.4 \text{ hp}}$$

The answer is (B).

(b) *step 8:* Not needed.

step 9: The original fuel rate (Eq. 29.74) is

$$\dot{m}_{f1} = (\text{BSFC}_1)(\text{BHP}_1)$$

$$= \left(0.45 \, \frac{\text{lbm}}{\text{bhp-hr}}\right)(1000 \text{ bhp})$$

$$= 450 \text{ lbm/hr}$$

step 10: Not needed.

step 11: $\dot{m}_{f2} = \dot{m}_{f1} = 450$ lbm/hr

step 12: The new fuel consumption (Eq. 29.79) is

$$\text{BSFC}_2 = \frac{\dot{m}_{f2}}{\text{BHP}_2} = \frac{450 \, \frac{\text{lbm}}{\text{hr}}}{861.4 \text{ hp}}$$

$$= \boxed{0.522 \text{ lbm/bhp-hr}}$$

The answer is (B).

SI Solution

(a) *step 1:* From App. 26.E,

Altitude 1: standard atmospheric condition, 1.01325 bar, 288.15K

Altitude 2: 0.8456 bar, 278.4K

$$z = 1500 \text{ m}$$

step 2: Calculate the friction power (Eq. 29.68).

$$\text{IHP}_1 = \frac{\text{BHP}_1}{\eta_{m,1}} = \frac{750 \text{ kW}}{0.80} = 937.5 \text{ kW}$$

step 3: Not needed since η_m is constant with altitude.

step 4: From the ideal gas law, the air densities are

$$\rho_{a1} = \frac{p}{RI} = \frac{(1.01325 \text{ bar})\left(10^5 \, \frac{\text{Pa}}{\text{bar}}\right)}{\left(287.03 \, \frac{\text{J}}{\text{kg·K}}\right)(288.15\text{K})}$$

$$= 1.225 \text{ kg/m}^3$$

Similarly,

$$\rho_{a2} = \frac{(0.8456 \text{ bar})\left(10^5 \frac{\text{Pa}}{\text{bar}}\right)}{\left(287.03 \frac{\text{J}}{\text{kg·K}}\right)(278.4\text{K})}$$

$$= 1.058 \text{ kg/m}^3$$

step 5: Calculate the new frictionless power (Eq. 29.70).

$$\text{IHP}_2 = \text{IHP}_1\left(\frac{\rho_{a2}}{\rho_{a1}}\right)$$

$$= (937.5 \text{ kW})\left(\frac{1.058 \frac{\text{kg}}{\text{m}^3}}{1.225 \frac{\text{kg}}{\text{m}^3}}\right)$$

$$= 809.7 \text{ kW}$$

steps 6 and 7: Calculate the new net power (Eq. 29.72).

$$\text{BHP}_2 = \eta_{m,2}(\text{IHP}_2) = (0.80)(809.7 \text{ kW})$$

$$= \boxed{647.8 \text{ kW}}$$

The answer is (B).

(b) *step 8:* Not needed.

step 9: The original fuel rate (Eq. 29.74) is

$$\dot{m}_{f1} = (\text{BSFC}_1)(\text{BHP}_1)$$

$$= \left(76 \frac{\text{kg}}{\text{GJ}}\right)(750 \text{ kW})\left(\frac{\text{GJ}}{10^6 \text{ kJ}}\right)$$

$$= 0.057 \text{ kg/s}$$

step 10: Not needed.

step 11: $\dot{m}_{f2} = \dot{m}_{f1} = 0.057 \text{ kg/s}$

step 12: The new fuel consumption (Eq. 29.79) is

$$\text{BSFC}_2 = \frac{m_{f2}}{\text{BHP}_2}$$

$$= \frac{0.057 \frac{\text{kg}}{\text{s}}}{(647.8 \text{ kW})\left(\frac{\text{GW}}{10^6 \text{ kW}}\right)}$$

$$= \boxed{88.0 \text{ kg/GJ}}$$

The answer is (B).

7. *Customary U.S. Solution*

(Use an air table. The SI solution assumes an ideal gas.) Refer to Fig. 29.11.

At A:

$$T_A = 60°\text{F} + 460 = 520°\text{R} \quad \text{[given]}$$
$$p_A = 14.7 \text{ psia} \quad \text{[given]}$$

From the air table (App. 24.F),

$$v_{r,A} = 158.58$$
$$p_{r,A} = 1.2147$$
$$h_A = 124.27 \text{ Btu/lbm}$$

At B:

The process from A to B is isentropic.

$$v_{r,B} = v_{r,A}\left(\frac{V_B}{V_A}\right) = (158.58)\left(\frac{1}{5}\right) = 31.716$$

Locate this volume ratio in the air table (App. 24.F).

$$T_B \approx 980°\text{R}$$
$$h_B \approx 236.02 \text{ Btu/lbm}$$
$$p_{r,B} = 11.430$$

Since process A-B is isentropic,

$$p_B = \left(\frac{p_{r,B}}{p_{r,A}}\right)p_A = \left(\frac{11.430}{1.2147}\right)(14.7 \text{ psia}) = 138.3 \text{ psia}$$

At C:

$$T_C = 1500°\text{F} + 460 = 1960°\text{R} \quad \text{[given]}$$
$$p_C = p_B = 138.3 \text{ psia}$$

Locate the temperature in the air table.

$$h_C = 493.64 \text{ Btu/lbm}$$
$$p_{r,C} = 100.48$$

At D:

$$p_D = 14.7 \text{ psia}$$

Since process C-D is isentropic,

$$p_{r,D} = p_{r,C}\left(\frac{p_D}{p_C}\right) = (160.48)\left(\frac{14.7 \text{ psia}}{138.3 \text{ psia}}\right)$$

$$= 17.057$$

Locate this pressure ratio in the air table.

$$T_D = 1094°R$$

$$h_D = 264.49 \text{ Btu/lbm}$$

Since the efficiency of compression is 83%, from Eq. 29.98,

$$h'_B = h_A + \frac{h_B - h_A}{\eta_{s,\text{compressor}}}$$

$$= 124.27 \frac{\text{Btu}}{\text{lbm}} + \frac{236.02 \frac{\text{Btu}}{\text{lbm}} - 124.27 \frac{\text{Btu}}{\text{lbm}}}{0.83}$$

$$= 258.9 \text{ Btu/lbm}$$

Since the efficiency of the expansion process is 92%, from Eq. 29.100,

$$h'_D = h_C - \eta_{s,\text{turbine}}(h_C - h_D)$$

$$= 493.64 \frac{\text{Btu}}{\text{lbm}}$$

$$- (0.92)\left(493.64 \frac{\text{Btu}}{\text{lbm}} - 264.49 \frac{\text{Btu}}{\text{lbm}}\right)$$

$$= 282.8 \text{ Btu/lbm}$$

From Eq. 26.96, the thermal efficiency is

$$\eta_{\text{th}} = \frac{(h_C - h'_B) - (h'_D - h_A)}{h_C - h'_B}$$

$$= \frac{\left(493.64 \frac{\text{Btu}}{\text{lbm}} - 258.9 \frac{\text{Btu}}{\text{lbm}}\right) - \left(282.8 \frac{\text{Btu}}{\text{lbm}} - 124.27 \frac{\text{Btu}}{\text{lbm}}\right)}{493.64 \frac{\text{Btu}}{\text{lbm}} - 258.9 \frac{\text{Btu}}{\text{lbm}}}$$

$$= \boxed{0.325 \ (32.5\%)}$$

The answer is (A).

SI Solution

Refer to Fig. 29.11.

At A:
$$T_A = 16°C + 273 = 289K \quad [\text{given}]$$

$$p_A = 101.3 \text{ kPa} \quad [\text{given}]$$

At B:

$$T_B = T_A \left(\frac{v_A}{v_B}\right)^{k-1} = (289K)(5)^{1.4-1} = 550.2K$$

$$p_B = p_A \left(\frac{v_A}{v_B}\right)^k = (101.3 \text{ kPa})(5)^{1.4} = 964.2 \text{ kPa}$$

At C:
$$T_C = 820°C + 273 = 1093K \quad [\text{given}]$$

$$p_C = p_B = 964.2 \text{ kPa}$$

At D:
$$p_D = 101.3 \text{ kPa} \quad [\text{given}]$$

$$T_D = T_C \left(\frac{p_D}{p_C}\right)^{\frac{k-1}{k}} = (1093K)\left(\frac{101.3 \text{ kPa}}{964.2 \text{ kPa}}\right)^{\frac{1.4-1}{1.4}}$$

$$T_D = 574.2K$$

For ideal gases, the specific heats are constant. Therefore, the change in internal energy (and enthalpy, approximately) is proportional to the change in temperature. The actual temperature (Eq. 29.99) is

$$T'_B = T_A + \frac{T_B - T_A}{\eta_{s,\text{compressor}}}$$

$$= 289K + \frac{550.2K - 289K}{0.83}$$

$$= 603.7K$$

From Eq. 29.101,

$$T'_D = T_C - \eta_{s,\text{turbine}}(T_C - T_D)$$

$$= 1093K - (0.92)(1093K - 574.2K)$$

$$= 615.7K$$

From Eq. 29.97, the thermal efficiency is

$$\eta_{\text{th}} = \frac{(T_C - T'_B) - (T'_D - T_A)}{T_C - T'_B}$$

$$= \frac{(1093K - 603.7K) - (615.7K - 289K)}{1093K - 603.7K}$$

$$= \boxed{0.332 \ (33.2\%)}$$

The answer is (A).

8. *Customary U.S. Solution*

Since specific heats are constant, use ideal gas rather than air tables.

Refer to Fig. 29.12.

At A:
$$T_A = 60°F + 460 = 520°R \quad [\text{given}]$$

$$p_A = 14.7 \text{ psia} \quad [\text{given}]$$

At B:

$$T_B = T_A \left(\frac{v_A}{v_B}\right)^{k-1} = (520°R)(5)^{1.4-1} = 989.9°R$$

$$p_B = p_A \left(\frac{v_A}{v_B}\right)^k = (14.7 \text{ psia})(5)^{1.4}$$

$$= 139.9 \text{ psia}$$

At D:

$$T_D = 1500°F + 460 = 1960°R \quad \text{[given]}$$
$$p_D = p_B = 139.9 \text{ psia}$$

At E:

$$p_E = 14.7 \text{ psia} \quad \text{[given]}$$
$$T_E = T_D \left(\frac{p_E}{p_D}\right)^{\frac{k-1}{k}} = (1960°R) \left(\frac{14.7 \text{ psia}}{139.9 \text{ psia}}\right)^{\frac{1.4-1}{1.4}}$$
$$= 1029.6°R$$

From Eq. 29.99,

$$T_B' = T_A + \frac{T_B - T_A}{\eta_{s,\text{compressor}}} = 520°R + \frac{989.9°R - 520°R}{0.83}$$
$$= 1086.1°R$$

From Eq. 29.101,

$$T_E' = T_D - \eta_{s,\text{turbine}}(T_D - T_E)$$
$$= 1960°R - (0.92)(1960°R - 1029.6°R)$$
$$= 1104.0°R$$

From Eq. 29.102,

$$\eta_{\text{regenerator}} = \frac{h_C - h_B'}{h_E' - h_B'}$$

For $c_p \approx$ constant,

$$\eta_{\text{regenerator}} = \frac{T_C - T_B'}{T_E' - T_B'}$$
$$0.65 = \frac{T_C - 1086.1°R}{1104.0°R - 1086.1°R}$$
$$T_C = 1097.7°R$$

From Eq. 29.103 with constant specific heats,

$$\eta_{\text{th}} = \frac{(T_D - T_E') - (T_B' - T_A)}{T_D - T_C}$$
$$= \frac{(1960°R - 1104.0°R) - (1086.1°R - 520°R)}{1960°R - 1097.7°R}$$
$$= \boxed{0.336 \quad (33.6\%)}$$

The answer is (C).

SI Solution

Refer to Fig. 29.12. From Prob. 7,

$$T_A = 289K$$
$$T_B' = 603.7K$$
$$T_D = 1093K$$
$$T_E' = 615.7K$$

For constant specific heats and from Eq. 29.102,

$$\eta_{\text{regenerator}} = \frac{T_C - T_B'}{T_E' - T_B'}$$
$$0.65 = \frac{T_C - 603.7K}{615.7K - 603.7K}$$
$$T_C = 611.5K$$

From Eq. 29.103 with constant specific heats,

$$\eta_{\text{th}} = \frac{(T_D - T_E') - (T_B' - T_A)}{T_D - T_C}$$
$$= \frac{(1093K - 615.7K) - (603.7K - 289K)}{1093K - 611.5K}$$
$$= \boxed{0.338 \quad (33.8\%)}$$

The answer is (C).

9. *Customary U.S. Solution*

(a) From Eq. 29.41, the number of power strokes per minute is

$$N = \frac{(2n)(\text{no. cylinders})}{\text{no. strokes per cycle}}$$
$$= \frac{\left(2 \dfrac{\text{strokes}}{\text{rev}}\right)\left(4600 \dfrac{\text{rev}}{\text{min}}\right)(8 \text{ cylinders})}{4 \dfrac{\text{strokes}}{\text{power stroke}}}$$
$$= 18,400 \text{ power strokes/min}$$

The net work per cycle is

$$W_{\text{net}} = W_{\text{out}} - W_{\text{in}} = 1500 \text{ ft-lbf} - 1200 \text{ ft-lbf}$$
$$= 300 \text{ ft-lbf/cycle}$$

The indicated horsepower is

$$\text{IHP} = \frac{\left(18,400 \dfrac{\text{power strokes}}{\text{min}}\right)(300 \text{ ft-lbf})}{33,000 \dfrac{\text{ft-lbf}}{\text{hp-min}}}$$
$$= \boxed{167.27 \text{ hp}}$$

The answer is (B).

(b) From Eq. 29.3, the thermal efficiency is

$$\eta_{th} = \frac{W_{out} - W_{in}}{Q_{in}} = \frac{W_{net}}{Q_{in}}$$

$$= \frac{300 \, \dfrac{\text{ft-lbf}}{\text{cycle}}}{\left(1.27 \, \dfrac{\text{Btu}}{\text{cycle}}\right)\left(778 \, \dfrac{\text{ft-lbf}}{\text{Btu}}\right)}$$

$$= \boxed{0.304 \quad (30.4\%)}$$

The answer is (A).

(c) The lower heating value of gasoline is LHV = 18,900 Btu/lbm.

The fuel consumption is

$$\dot{m}_F = \frac{\left(1.27 \, \dfrac{\text{Btu}}{\text{power stroke}}\right)\left(18,400 \, \dfrac{\text{power strokes}}{\text{min}}\right)\left(60 \, \dfrac{\text{min}}{\text{hr}}\right)}{18,900 \, \dfrac{\text{Btu}}{\text{lbm}}}$$

$$= \boxed{74.18 \, \text{lbm/hr}}$$

The answer is (C).

(d) Specific fuel consumption is given by Eq. 29.8.

$$\text{SFC} = \frac{\text{fuel usage rate}}{\text{power generated}} = \frac{74.18 \, \dfrac{\text{lbm}}{\text{hr}}}{167.27 \, \text{hp}}$$

$$= \boxed{0.443 \, \text{lbm/hp-hr}}$$

The answer is (D).

SI Solution

(a) From Eq. 29.41, the number of power strokes per minute is

$$N = \frac{(2n)(\text{no. cylinders})}{\text{no. strokes per cycle}}$$

$$= \frac{\left(2 \, \dfrac{\text{strokes}}{\text{rev}}\right)\left(4600 \, \dfrac{\text{rev}}{\text{min}}\right)(8 \, \text{cylinders})}{4 \, \text{strokes per power stroke}}$$

$$= 18\,400 \, \text{power strokes/min}$$

The net work per cycle is

$$W_{net} = W_{out} - W_{in} = 2.0 \, \text{kJ} - 1.6 \, \text{kJ}$$

$$= 0.4 \, \text{kJ/cycle}$$

The indicated power is

$$\text{IkW} = \left(18\,400 \, \frac{\text{power strokes}}{\text{min}}\right)\left(\frac{1 \, \text{min}}{60 \, \text{sec}}\right)(0.4 \, \text{kJ})$$

$$= \boxed{122.7 \, \text{kW}}$$

The answer is (B).

(b) From Eq. 29.3, the thermal efficiency is

$$\eta_{th} = \frac{W_{out} - W_{in}}{Q_{in}} = \frac{W_{net}}{Q_{in}}$$

$$= \frac{0.4 \, \dfrac{\text{kJ}}{\text{cycle}}}{1.33 \, \dfrac{\text{kJ}}{\text{cycle}}} = \boxed{0.300 \quad (30\%)}$$

The answer is (A).

(c) The heating value of gasoline is LHV = 44 MJ/kg.

The fuel consumption is

$$\dot{m}_F = \frac{\left(1.33 \, \dfrac{\text{kJ}}{\text{power stroke}}\right)\left(18\,400 \, \dfrac{\text{power strokes}}{\text{min}}\right)\left(\dfrac{1 \, \text{min}}{60 \, \text{s}}\right)}{\left(44 \, \dfrac{\text{MJ}}{\text{kg}}\right)\left(1000 \, \dfrac{\text{kJ}}{\text{MJ}}\right)}$$

$$= \boxed{0.00927 \, \text{kg/s}}$$

The answer is (C).

(d) The specific fuel consumption (Eq. 29.8) is

$$\text{SFC} = \frac{\text{fuel usage rate}}{\text{power generated}} = \frac{0.00927 \, \dfrac{\text{kg}}{\text{s}}}{122.7 \, \text{kW}}$$

$$= \boxed{7.555 \times 10^{-5} \, \text{kg/kJ}}$$

The answer is (D).

10. *Customary U.S. Solution*

(a) At A:

$$T_A = 520°\text{R}$$

$$p_A = 14.7 \, \text{psia}$$

At C:

$$T_C = 1600°$$

$$p_C = 568.6 \, \text{psia}$$

For isentropic process C-A, from Eq. 25.91,

$$p_A = p_C \left(\frac{T_A}{T_C}\right)^{\frac{k}{k-1}}$$

Therefore,

$$14.7 \text{ psia} = (568.6 \text{ psia}) \left(\frac{520°\text{R}}{1600°\text{R}} \right)^{\frac{k}{k-1}}$$

$$0.02585 = (0.325)^{\frac{k}{k-1}}$$

$$\log(0.02585) = \left(\frac{k}{k-1} \right) \log(0.325)$$

$$\frac{k}{k-1} = 3.252$$

$$k = 1.444$$

From Eq. 24.54, the molar specific heat of the mixture is

$$C_{p,\text{mixture}} = \frac{R^* k}{k-1}$$

$$= \frac{\left(1545 \, \frac{\text{ft-lbf}}{\text{lbmol-}°\text{R}} \right) (1.444)}{\left(778 \, \frac{\text{ft-lbf}}{\text{Btu}} \right) (1.444 - 1)}$$

$$= 6.459 \text{ Btu/lbmol-}°\text{R}$$

From Table 24.7,

$$(c_p)_{\text{He}} = 1.240 \text{ Btu/lbm-}°\text{R}$$

$$(\text{MW})_{\text{He}} = 4.003 \text{ lbm/lbmol}$$

$$(c_p)_{\text{CO}_2} = 0.207 \text{ Btu/lbm-}°\text{R}$$

$$(\text{MW})_{\text{CO}_2} = 44.011 \text{ lbm/lbmol}$$

$$C_{p,\text{He}} = (\text{MW})_{\text{He}}(c_p)_{\text{He}}$$

$$= \left(4.003 \, \frac{\text{lbm}}{\text{lbmol}} \right) \left(1.240 \, \frac{\text{Btu}}{\text{lbm-}°\text{R}} \right)$$

$$= 4.96 \text{ Btu/lbmol-}°\text{R}$$

$$C_{p,\text{CO}_2} = (\text{MW})_{\text{CO}_2}(c_p)_{\text{CO}_2}$$

$$= \left(44.011 \, \frac{\text{lbm}}{\text{lbmol}} \right) \left(0.207 \, \frac{\text{Btu}}{\text{lbm-}°\text{R}} \right)$$

$$= 9.11 \text{ Btu/lbmol-}°\text{R}$$

Let x be the mole fraction of helium in the mixture.

$$x = \frac{n_{\text{He}}}{n_{\text{He}} + n_{\text{CO}_2}}$$

From Eq. 24.82, on a mole basis,

$$c_{p,\text{mixture}} = x(c_p)_{\text{He}} + (1-x)(c_p)_{\text{CO}_2}$$

$$6.459 \, \frac{\text{Btu}}{\text{lbmol-}°\text{R}} = x \left(4.96 \, \frac{\text{Btu}}{\text{lbmol-}°\text{R}} \right)$$

$$+ (1-x) \left(9.11 \, \frac{\text{Btu}}{\text{lbmol-}°\text{R}} \right)$$

$$x = 0.639$$

On a per mole basis, the mass of helium would be

$$m_{\text{He}} = x(\text{MW})_{\text{He}}$$

$$= (0.639) \left(4.003 \, \frac{\text{lbm}}{\text{lbmol}} \right)$$

$$= 2.558 \text{ lbm}$$

Similarly, the mass of carbon dioxide on a per mole basis would be

$$m_{\text{CO}_2} = (1-x)(\text{MW})_{\text{CO}_2}$$

$$= (1 \text{ lbmol} - 0.639 \text{ lbmol}) \left(44.011 \, \frac{\text{lbm}}{\text{lbmol}} \right)$$

$$= 15.888 \text{ lbm}$$

The molecular weight of the mixture is

$$(\text{MW})_{\text{mixture}} = 2.558 \text{ lbm} + 15.888 \text{ lbm}$$

$$= 18.446 \text{ lbm/lbmol}$$

The gravimetric (mass) fraction of the gases is

$$G_{\text{He}} = \frac{m_{\text{He}}}{m_{\text{He}} + m_{\text{CO}_2}}$$

$$= \frac{2.558 \text{ lbm}}{2.558 \text{ lbm} + 15.888 \text{ lbm}} = \boxed{0.139}$$

$$G_{\text{CO}_2} = 1 - G_{\text{He}} = 1 - 0.139 = \boxed{0.861}$$

The answer is (A).

(b) From Eq. 24.52,

$$C_{v,\text{mixture}} = C_{p,\text{mixture}} - R^*$$

$$= 6.459 \, \frac{\text{Btu}}{\text{lbmol-}°\text{R}} - \frac{1545 \, \frac{\text{ft-lbf}}{\text{lbmol-}°\text{R}}}{778 \, \frac{\text{ft-lbf}}{\text{Btu}}}$$

$$= 4.473 \text{ Btu/lbmol-}°\text{R}$$

From Eq. 25.102, the work done during the isentropic expansion process is

$$W = c_v(T_1 - T_2) = \left(\frac{C_v}{\text{MW}} \right)_{\text{mixture}} \times (T_{\text{C}} - T_{\text{A}})$$

$$= \left(\frac{4.473 \, \frac{\text{Btu}}{\text{lbmol-}°\text{R}}}{18.446 \, \frac{\text{lbm}}{\text{lbmol}}} \right) (1600°\text{R} - 520°\text{R})$$

$$= \boxed{261.9 \text{ Btu/lbm}}$$

The answer is (B).

(c)

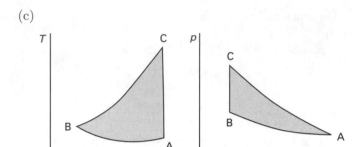

SI Solution

(a) At A:

$$T_A = 290K$$
$$p_A = 101.3 \text{ kPa}$$

At C:

$$T_C = 890K$$
$$p_C = 3.920 \text{ MPa}$$

For isentropic process C-A from Eq. 25.91,

$$p_A = p_C \left(\frac{T_A}{T_C}\right)^{\frac{k}{k-1}}$$

Therefore,

$$101.3 \text{ kPa} = (3.920 \text{ MPa}) \left(1000 \frac{\text{kPa}}{\text{MPa}}\right) \left(\frac{290K}{890K}\right)^{\frac{k}{k-1}}$$

$$0.02584 = (0.3258)^{\frac{k}{k-1}}$$

$$\log(0.02584) = \left(\frac{k}{k-1}\right)\log(0.3258)$$

$$\frac{k}{k-1} = 3.2599$$

$$k = 1.442$$

From Eq. 24.54, the molar specific heat of the mixture is

$$C_{p,\text{mixture}} = \frac{R^*k}{k-1}$$

$$= \frac{\left(8.3143 \frac{\text{kJ}}{\text{kmol·K}}\right)(1.442)}{1.442 - 1}$$

$$= 27.125 \text{ kJ/kmol·K}$$

From Table 24.7,

$$(c_p)_{\text{He}} = \left(5192 \frac{\text{J}}{\text{kg·K}}\right)\left(\frac{1 \text{ kJ}}{1000 \text{ J}}\right)$$

$$= 5.192 \text{ kJ/kg·K}$$

$$(\text{MW})_{\text{He}} = 4.003 \text{ kg/kmol}$$

$$(c_p)_{\text{CO}_2} = \left(867 \frac{\text{J}}{\text{kg·K}}\right)\left(\frac{1 \text{ kJ}}{1000 \text{ J}}\right)$$

$$= 0.867 \text{ kJ/kg·K}$$

$$(\text{MW})_{\text{CO}_2} = 44.011 \text{ kg/kmol}$$

$$C_{p,\text{He}} = (\text{MW})_{\text{He}}(c_p)_{\text{He}}$$

$$= \left(4.003 \frac{\text{kg}}{\text{kmol}}\right)\left(5.192 \frac{\text{kJ}}{\text{kmol}}\right)$$

$$= 20.784 \text{ kJ/kmol·K}$$

$$C_{p,\text{CO}_2} = (\text{MW})_{\text{CO}_2}(c_p)_{\text{CO}_2}$$

$$= \left(44.011 \frac{\text{kg}}{\text{kmol}}\right)\left(0.867 \frac{\text{kJ}}{\text{kg·K}}\right)$$

$$= 38.158 \text{ kJ/kmol·K}$$

Let x be the mole fraction of helium in the mixture.

$$x = \frac{n_{\text{He}}}{n_{\text{He}} + n_{\text{CO}_2}}$$

From Eq. 24.82 on a mole basis,

$$c_{p,\text{mixture}} = x(c_p)_{\text{He}} + (1-x)(c_p)_{\text{CO}_2}$$

$$27.125 \frac{\text{kJ}}{\text{kmol·K}} = x\left(20.784 \frac{\text{kJ}}{\text{kmol·K}}\right)$$

$$+ (1-x)\left(38.158 \frac{\text{kJ}}{\text{kmol·K}}\right)$$

$$x = 0.635 \text{ kmol}$$

On a per mole basis, the mass of helium would be

$$m_{\text{He}} = x(\text{MW})_{\text{He}}$$

$$= (0.635 \text{ kmol})\left(4.003 \frac{\text{kg}}{\text{kmol}}\right)$$

$$= 2.542 \text{ kg}$$

Similarly, the mass of CO_2 on a per mole basis would be

$$m_{\text{CO}_2} = (1-x)(\text{MW})_{\text{CO}_2}$$

$$= (1 \text{ kmol} - 0.635 \text{ kmol})\left(44.011 \frac{\text{kg}}{\text{kmol}}\right)$$

$$= 16.064 \text{ kg}$$

The molecular weight of the mixture is

$$(\text{MW})_{\text{mixture}} = 2.542 \text{ kg} + 16.064 \text{ kg}$$

$$= 18.606 \text{ kg/kmol}$$

The gravimetric (mass) fraction of the gases is

$$G_{He} = \frac{m_{He}}{m_{He} + m_{CO_2}}$$

$$= \frac{2.542 \text{ kg}}{2.542 \text{ kg} + 16.064 \text{ kg}} = \boxed{0.137}$$

$$G_{CO_2} = 1 - G_{He} = 1 - 0.137 = \boxed{0.863}$$

The answer is (A).

(b) From Eq. 24.52,

$$C_{v,\text{mixture}} = C_{p,\text{mixture}} - R^*$$

$$= 27.125 \frac{\text{kJ}}{\text{kmol·K}} - 8.3143 \frac{\text{kJ}}{\text{kmol·K}}$$

$$= 18.811 \text{ kJ/kmol·K}$$

From Eq. 25.102, the work done during the isentropic expansion process is

$$W = c_v(T_1 - T_2) = \left(\frac{C_v}{MW}\right)_{\text{mixture}} \times (T_C - T_A)$$

$$= \left(\frac{18.811 \frac{\text{kJ}}{\text{kmol·K}}}{18.606 \frac{\text{kg}}{\text{kmol}}}\right)(890\text{K} - 290\text{K})$$

$$= \boxed{606.6 \text{ kJ/kg}}$$

The answer is (B).

(c)

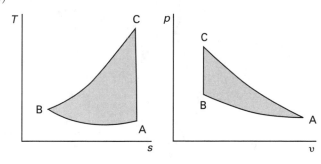

11. *Customary U.S. Solution*

step 1: Not needed.

step 2: From Eq. 29.68, calculate the frictionless power.

$$\text{IHP}_1 = \frac{\text{BHP}_1}{\eta_{m,1}} = \frac{200 \text{ hp}}{0.86} = 232.6 \text{ hp}$$

step 3: From Eq. 29.69, calculate the friction power, which is assumed to be constant at constant speed.

$$\text{FHP} = \text{IHP}_1 - \text{BHP}_1$$

$$= 232.6 \text{ hp} - 200 \text{ hp} = 32.6 \text{ hp}$$

step 4: Calculate the air densities ρ_{a1} and ρ_{a2} from the ideal gas law. The absolute temperatures are

$$T_1 = 80°\text{F} + 460 = 540°\text{R}$$

$$T_2 = 60°\text{F} + 460 = 520°\text{R}$$

$$\rho_{a1} = \frac{p_1}{RT_1} = \frac{\left(14.7 \frac{\text{lbf}}{\text{in}^2}\right)\left(144 \frac{\text{in}^2}{\text{ft}^2}\right)}{\left(53.3 \frac{\text{lbf-ft}}{\text{lbm-}°\text{R}}\right)(540°\text{R})}$$

$$= 0.0735 \text{ lbm/ft}^3$$

$$\rho_{a2} = \frac{p_2}{RT_1} = \frac{\left(12.2 \frac{\text{lbf}}{\text{in}^2}\right)\left(144 \frac{\text{in}^2}{\text{ft}^2}\right)}{\left(53.3 \frac{\text{lbf-ft}}{\text{lbm-}°\text{R}}\right)(520°\text{R})}$$

$$= 0.0634 \text{ lbm/ft}^3$$

step 5: From Eq. 29.70, calculate the new frictionless power.

$$\text{IHP}_2 = \text{IHP}_1 \left(\frac{\rho_{a2}}{\rho_{a1}}\right)$$

$$= (232.6 \text{ hp})\left(\frac{0.0634 \frac{\text{lbm}}{\text{ft}^3}}{0.0735 \frac{\text{lbm}}{\text{ft}^3}}\right)$$

$$= 200.6 \text{ hp}$$

step 6: From Eq. 29.71, calculate the new net power.

$$\text{BHP}_2 = \text{IHP}_2 - \text{FHP}$$

$$= 200.6 \text{ hp} - 32.6 \text{ hp}$$

$$= \boxed{168.0 \text{ hp}}$$

step 7: From Eq. 29.72, calculate the new efficiency.

$$\eta_{m,2} = \frac{\text{BHP}_2}{\text{IHP}_2}$$

$$= \frac{168.0 \text{ hp}}{200.6 \text{ hp}}$$

$$= \boxed{0.837}$$

step 8: From Eq. 29.73, the volumetric air flow rates are the same since the engine speed is constant.

$$\dot{V}_{a2} = \dot{V}_{a1}$$

step 9: The original air and fuel rates from Eqs. 29.74 through 29.76 are

$$\dot{m}_{f1} = (\text{BSFC}_1)(\text{BHP}_1)$$

$$= \left(0.48 \ \frac{\text{lbm}}{\text{hp-hr}}\right)(200 \ \text{hp}) = 96 \ \text{lbm/hr}$$

$$\dot{m}_{a1} = (\text{AFR})(\dot{m}_{f1})$$

$$= (22)\left(96 \ \frac{\text{lbm}}{\text{hr}}\right) = 2112 \ \text{lbm/hr}$$

$$\dot{V}_{a1} = \frac{\dot{m}_{a1}}{\rho_{a1}}$$

$$= \frac{2112 \ \dfrac{\text{lbm}}{\text{hr}}}{0.0735 \ \dfrac{\text{lbm}}{\text{ft}^3}} = 28{,}735 \ \text{ft}^3/\text{hr}$$

$$\dot{V}_{a2} = \dot{V}_{a1} = 28{,}735 \ \text{ft}^3/\text{hr}$$

step 10: From Eq. 29.77, the new air mass flow rate is

$$\dot{m}_{a2} = \dot{V}_{a2}\rho_{a2}$$

$$= \left(28{,}735 \ \frac{\text{ft}^3}{\text{hr}}\right)\left(0.0634 \ \frac{\text{lbm}}{\text{ft}^3}\right)$$

$$= 1821.8 \ \text{lbm/hr}$$

step 11: For engines with metered injection,

$$\dot{m}_{f2} = \dot{m}_{f1} = 96 \ \text{lbm/hr}$$

step 12: From Eq. 29.79, the new fuel consumption is

$$\text{BSFC}_2 = \frac{\dot{m}_{f2}}{\text{BHP}_2} = \frac{96 \ \dfrac{\text{lbm}}{\text{hr}}}{168.0 \ \text{hp}}$$

$$= \boxed{0.571 \ \text{lbm/hp-hr}}$$

From Eq. 29.75, the new air/fuel ratio is

$$\text{AFR}_2 = \frac{\dot{m}_{a2}}{\dot{m}_{f2}} = \frac{1821.8 \ \dfrac{\text{lbm}}{\text{hr}}}{96 \ \dfrac{\text{lbm}}{\text{hr}}} = \boxed{18.98}$$

SI Solution

step 1: Not needed.

step 2: From Eq. 29.68, calculate the frictionless power.

$$\text{IkW}_1 = \frac{\text{BkW}_1}{\eta_{m,1}} = \frac{150 \ \text{kW}}{0.86} = 174.4 \ \text{kW}$$

step 3: From Eq. 29.69, calculate the friction power, which is assumed to be constant at constant speed.

$$\text{FkW} = \text{IkW}_1 - \text{BkW}_1$$

$$= 174.4 \ \text{kW} - 150 \ \text{kW} = 24.4 \ \text{kW}$$

step 4: Calculate the air densities ρ_{a1} and ρ_{a2} from the ideal gas law. The absolute temperatures are

$$T_1 = 27°\text{C} + 273 = 300\text{K}$$
$$T_2 = 16°\text{C} + 273 = 289\text{K}$$

$$\rho_{a1} = \frac{p_1}{RT_1} = \frac{(101.3 \ \text{kPa})\left(1000 \ \dfrac{\text{Pa}}{\text{kPa}}\right)}{\left(287.03 \ \dfrac{\text{kJ}}{\text{kg·K}}\right)(300\text{K})}$$

$$= 1.176 \ \text{kg/m}^3$$

$$\rho_{a2} = \frac{p_2}{RT_2} = \frac{(84 \ \text{kPa})\left(1000 \ \dfrac{\text{Pa}}{\text{kPa}}\right)}{\left(287.03 \ \dfrac{\text{kJ}}{\text{kg·K}}\right)(289\text{K})}$$

$$= 1.013 \ \text{kg/m}^3$$

step 5: From Eq. 29.70, calculate the new frictionless power.

$$\text{IkW}_2 = \text{IkW}_1\left(\frac{\rho_{a2}}{\rho_{a1}}\right)$$

$$= (174.4 \ \text{kW})\left(\frac{1.013 \ \dfrac{\text{kg}}{\text{m}^3}}{1.176 \ \dfrac{\text{kg}}{\text{m}^3}}\right) = 150.2 \ \text{kW}$$

step 6: From Eq. 29.71, calculate the new net power.

$$\text{BkW}_2 = \text{IkW}_2 - \text{FkW} = 150.2 \ \text{kW} - 24.4 \ \text{kW}$$

$$= \boxed{125.8 \ \text{kW}}$$

step 7: From Eq. 29.72, calculate the new efficiency.

$$\eta_{m,2} = \frac{\text{BkW}_2}{\text{IkW}_2} = \frac{125.8 \text{ kW}}{150.2 \text{ kW}} = \boxed{0.838}$$

step 8: From Eq. 29.73, the volumetric air flow rates are the same since the engine speed is constant.

$$\dot{V}_{a2} = \dot{V}_{a1}$$

step 9: The original air and fuel rates from Eqs. 29.74 through 29.76 are

$$\dot{m}_{f1} = (\text{BSFC}_1)(\text{BkW}_1)$$
$$= \left(81 \frac{\text{kg}}{\text{GJ}}\right)\left(\frac{1 \text{ GJ}}{10^6 \text{ J}}\right)(150 \text{ kW})$$
$$= 0.01215 \text{ kg/s}$$

$$\dot{m}_{a1} = (\text{AFR})\dot{m}_{f1}$$
$$= (22)\left(0.01215 \frac{\text{kg}}{\text{s}}\right) = 0.2673 \text{ kg/s}$$

$$\dot{V}_{a1} = \frac{\dot{m}_{a1}}{\rho_{a1}} = \frac{0.2673 \frac{\text{kg}}{\text{s}}}{1.176 \frac{\text{kg}}{\text{m}^3}} = 0.2273 \text{ m}^3/\text{s}$$

$$\dot{V}_{a2} = \dot{V}_{a1} = 0.2273 \text{ m}^3/\text{s}$$

step 10: From Eq. 29.77, the new air mass flow rate is

$$\dot{m}_{a2} = \dot{V}_{a2}\rho_{a2}$$
$$= \left(0.2273 \frac{\text{m}^3}{\text{s}}\right)\left(1.013 \frac{\text{kg}}{\text{m}^3}\right) = 0.2303 \text{ kg/s}$$

step 11: For engines with metered injection,

$$\dot{m}_{f2} = \dot{m}_{f1} = 0.01215 \text{ kg/s}$$

step 12: From Eq. 29.79, the new fuel consumption is

$$\text{BSFC}_2 = \frac{\dot{m}_{f2}}{\text{BkW}_2}$$
$$= \frac{0.01215 \frac{\text{kg}}{\text{s}}}{(125.8 \text{ kW})\left(\frac{1 \text{ GJ}}{10^6 \text{ kJ}}\right)}$$
$$= \boxed{96.58 \text{ kg/GJ}}$$

From Eq. 29.75, the new air/fuel ratio is

$$\text{AFR}_2 = \frac{\dot{m}_{a2}}{\dot{m}_{f2}} = \frac{0.2303 \frac{\text{kg}}{\text{s}}}{0.01215 \frac{\text{kg}}{\text{s}}}$$
$$= \boxed{18.95}$$

12.

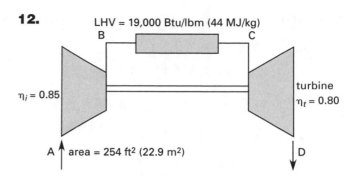

LHV = 19,000 Btu/lbm (44 MJ/kg)

$\eta_i = 0.85$

turbine $\eta_t = 0.80$

A ↑ area = 254 ft² (22.9 m²)   D ↓

Customary U.S. Solution

(a) *At 7000 ft altitude:*

$$\dot{V}_{a1} = 50{,}000 \text{ ft}^3/\text{min}$$
$$\text{BHP}_1 = 6000 \text{ hp}$$
$$\text{BSCF}_1 = 0.609 \text{ lbm/hp-hr}$$

At A:

$$p_A = 12 \text{ psia}$$
$$T_A = 35°F$$

The absolute temperature at A is

$$T_A = 35°F + 460 = 495°R$$

Interpolating from App. 24.F (air table),

$$h_A = 118.28 \text{ Btu/lbm}$$
$$p_{r,a} = 1.0238$$

From the ideal gas law, air density is

$$\rho_{a1} = \frac{p_A}{RT_A} = \frac{\left(12 \frac{\text{lbf}}{\text{in}^2}\right)\left(144 \frac{\text{in}^2}{\text{ft}^2}\right)}{\left(53.35 \frac{\text{ft-lbf}}{\text{lbm-°R}}\right)(495°R)}$$
$$= 0.0654 \text{ lbm/ft}^3$$

From Eq. 29.76,

$$\dot{m}_{a1} = \dot{V}_{a1}\rho_{a1}$$
$$= \left(50{,}000 \frac{\text{ft}^3}{\text{min}}\right)\left(0.0654 \frac{\text{lbm}}{\text{ft}^3}\right)$$
$$= 3270 \text{ lbm/min}$$

At B:

$$p_B = 8p_A = (8)(12 \text{ psia}) = 96 \text{ psia}$$

Assuming isentropic compression,

$$p_{r,B} = 8p_{r,A} = (8)(1.0238)$$
$$= 8.1904$$

From App. 24.F (air table), this $p_{r,\text{B}}$ corresponds to

$$T_{\text{B}} = 893.3°\text{R}$$
$$h_{\text{B}} = 214.62 \text{ Btu/lbm}$$

Due to the inefficiency of the compressor, from Eq. 29.99,

$$h'_{\text{B}} = h_{\text{A}} + \frac{h_{\text{B}} - h_{\text{A}}}{\eta_{s,\text{compression}}}$$

$$= 118.28 \frac{\text{Btu}}{\text{lbm}} + \frac{214.62 \frac{\text{Btu}}{\text{lbm}} - 118.28 \frac{\text{Btu}}{\text{lbm}}}{0.85}$$

$$= 231.62 \text{ Btu/lbm}$$

From App. 24.F, this corresponds to $T'_{\text{B}} = 962.3°\text{R}$.

$$W_{\text{compression}} = h'_{\text{B}} - h_{\text{A}}$$

$$= 231.62 \frac{\text{Btu}}{\text{lbm}} - 118.28 \frac{\text{Btu}}{\text{lbm}}$$

$$= 113.34 \text{ Btu/lbm}$$

At C:

The absolute temperature is

$$T_{\text{C}} = 1800°\text{F} + 460 = 2260°\text{R} \quad [\text{no change if moved}]$$

Assuming there is no pressure drop across the combustor, $p_{\text{C}} = 96$ psia.

From App. 24.F (air table),

$$h_{\text{C}} = 577.52 \text{ Btu/lbm}$$
$$p_{r,\text{C}} = 286.7$$

The energy requirement from the fuel is

$$\dot{m}\Delta h = \dot{m}_{a1}(h_{\text{C}} - h'_{\text{B}})$$

$$= \left(3270 \frac{\text{lbm}}{\text{min}}\right)\left(577.52 \frac{\text{Btu}}{\text{lbm}} - 231.62 \frac{\text{Btu}}{\text{lbm}}\right)$$

$$= 1.133 \times 10^6 \text{ Btu/min}$$

The ideal fuel rate is

$$\frac{\dot{m}\Delta h}{\text{LHV}} = \frac{\left(1.133 \times 10^6 \frac{\text{Btu}}{\text{min}}\right)\left(60 \frac{\text{min}}{\text{hr}}\right)}{19,000 \frac{\text{Btu}}{\text{lbm}}}$$

$$= 3577.9 \text{ lbm/hr}$$

From Eq. 29.74, the ideal BSFC is

$$(\text{BSFC})_{\text{ideal}} = \frac{3577.9 \frac{\text{lbm}}{\text{hr}}}{6000 \text{ hp}} = 0.596 \text{ lbm/hp-hr}$$

The combustor efficiency is

$$\eta_{\text{combustor}} = \frac{(\text{BSFC})_{\text{ideal}}}{(\text{BSFC})_{\text{actual}}} = \frac{0.596 \frac{\text{lbm}}{\text{hp-hr}}}{0.609 \frac{\text{lbm}}{\text{hp-hr}}}$$

$$= 0.979 \quad (97.9\%)$$

At D:

As determined by atmospheric conditions, $p_{\text{D}} = 12$ psia. If expansion is isentropic,

$$p_{r,\text{D}} = \frac{p_{r,\text{C}}}{8} = \frac{286.7}{8} = 35.8375$$

From App. 24.F (air table) this $p_{r,\text{D}}$ corresponds to

$$T_{\text{D}} = 1334.8°\text{R}$$
$$h_{\text{D}} = 325.95 \text{ Btu/lbm}$$

Due to the inefficiency of the turbine, from Eq. 29.101,

$$h'_{\text{D}} = h_{\text{C}} - \eta_{s,\text{turbine}}(h_{\text{C}} - h_{\text{D}})$$

$$= 577.52 \frac{\text{Btu}}{\text{lbm}} - (0.80)\left(577.52 \frac{\text{Btu}}{\text{lbm}}\right.$$

$$\left. - 325.95 \frac{\text{Btu}}{\text{lbm}}\right)$$

$$= 376.26 \frac{\text{Btu}}{\text{lbm}}$$

$$W_{\text{turbine}} = h_{\text{C}} - h'_{\text{D}} = 577.52 \frac{\text{Btu}}{\text{lbm}} - 376.26 \frac{\text{Btu}}{\text{lbm}}$$

$$= 201.26 \text{ Btu/lbm}$$

The theoretical net horsepower is

$$\text{IHP} = \dot{m}_{a1}(W_{\text{turbine}} - W_{\text{compression}})$$

$$= \frac{\left(3270 \frac{\text{lbm}}{\text{min}}\right)\left(201.26 \frac{\text{Btu}}{\text{lbm}} - 113.34 \frac{\text{Btu}}{\text{lbm}}\right)\left(778 \frac{\text{ft-lbf}}{\text{Btu}}\right)}{33,000 \frac{\text{ft-lbf}}{\text{hp-min}}}$$

$$= 6778 \text{ hp}$$

From Eq. 29.69, the friction horsepower is

$$\text{FHP} = \text{IHP} - \text{BHP} = 6778 \text{ hp} - 6000 \text{ hp}$$

$$= 778 \text{ hp}$$

At sea level (zero altitude):

At A:

The absolute temperature is

$$T_A = 70°F + 460 = 530°R$$
$$p_A = 14.7 \text{ psia}$$

From App. 24.F (air table),

$$h_A = 126.86 \text{ Btu/lbm}$$
$$p_{r,A} = 1.2998$$

From the ideal gas law, the air density is

$$\rho_{a2} = \frac{p_A}{RT_A} = \frac{\left(14.7 \ \frac{\text{lbf}}{\text{in}^2}\right)\left(144 \ \frac{\text{in}^2}{\text{ft}^2}\right)}{\left(53.35 \ \frac{\text{lbf-ft}}{\text{lbm-°R}}\right)(530°R)}$$
$$= 0.0749 \text{ lbm/ft}^3$$

From Eq. 29.76, the air mass flow rate is

$$\dot{m}_{a2} = \dot{V}_{a2}\rho_{a2}$$
$$= \left(50,000 \ \frac{\text{ft}^3}{\text{min}}\right)\left(0.0749 \ \frac{\text{lbm}}{\text{ft}^3}\right)$$
$$= 3745 \text{ lbm/min}$$

At B:

$$p_B = 8p_A = (8)(14.7 \text{ psia}) = 117.6 \text{ psia}$$

Assuming isentropic compression,

$$p_{r,B} = 8p_{r,A} = (8)(1.2998) = 10.3984$$

From App. 24.F (air table), this $p_{r,B}$ corresponds to

$$T_B = 954.5°R$$
$$h_B = 229.70 \text{ Btu/lbm}$$

Due to the inefficiency of the compressor, from Eq. 29.99,

$$h'_B = h_A + \frac{h_B - h_A}{\eta_{s,\text{compression}}}$$
$$= 126.86 \ \frac{\text{Btu}}{\text{lbm}} + \frac{229.70 \ \frac{\text{Btu}}{\text{lbm}} - 126.86 \ \frac{\text{Btu}}{\text{lbm}}}{0.85}$$
$$= 247.85 \text{ Btu/lbm}$$

From App. 24.F (air table), $h'_B = 247.85$ Btu/lbm corresponds to $T'_B = 1027.6°R$.

$$W_{\text{compression}} = h'_B - h_A$$
$$= 247.85 \ \frac{\text{Btu}}{\text{lbm}} - 126.86 \ \frac{\text{Btu}}{\text{lbm}}$$
$$= 120.99 \text{ Btu/lbm}$$

At C:

The absolute temperature is

$$T_C = 2260°R \quad [\text{no change}]$$

Assuming there is no pressure drop across the combustor, $p_C = 117.6$ psia.

From App. 24.F (air table),

$$h_C = 577.52 \text{ Btu/lbm}$$
$$p_{r,C} = 286.7$$

The energy requirement from the fuel is $\dot{m}\Delta h$.

$$\dot{m}_{a2}(h_C - h'_B)$$
$$= \left(3745 \ \frac{\text{lbm}}{\text{min}}\right)\left(577.52 \ \frac{\text{Btu}}{\text{lbm}} - 247.85 \ \frac{\text{Btu}}{\text{lbm}}\right)$$
$$= 1.235 \times 10^6 \text{ Btu/min}$$

Assuming a constant combustion efficiency of 97.9%, the ideal fuel rate is

$$\frac{\dot{m}\Delta h}{\text{LHV}} = \frac{\left(1.235 \times 10^6 \ \frac{\text{Btu}}{\text{min}}\right)\left(60 \ \frac{\text{min}}{\text{hr}}\right)}{\left(19,000 \ \frac{\text{Btu}}{\text{lbm}}\right)(0.979)}$$
$$= 3983.7 \text{ lbm/hr}$$

At D:

$$p_D = 14.7 \text{ psia}$$

If expansion is isentropic,

$$p_{r,D} = \frac{p_{r,C}}{8} = \frac{286.7}{8} = 35.8375 \quad [\text{no change}]$$

From App. 24.F (air table), this corresponds to

$$T_D = 1334.8°R$$
$$h_D = 325.95 \text{ Btu/lbm}$$
$$h'_D = 376.26 \text{ Btu/lbm} \quad [\text{no change}]$$
$$W_{\text{turbine}} = 201.26 \text{ Btu/lbm} \quad [\text{no change}]$$

The theoretical net horsepower is

$$\text{IHP} = \dot{m}_{a2}(W_{\text{turbine}} - W_{\text{compression}})$$

$$= \frac{\left(3745 \, \dfrac{\text{lbm}}{\text{min}}\right)\left(201.26 \, \dfrac{\text{Btu}}{\text{lbm}} \right.}{33{,}000 \, \dfrac{\text{ft-lbf}}{\text{hp-min}}}$$

$$\left. - \, 120.99 \, \dfrac{\text{Btu}}{\text{lbm}}\right)\left(778 \, \dfrac{\text{ft-lbf}}{\text{Btu}}\right)$$

$$= 7087 \text{ hp}$$

Assuming the frictional horsepower is constant,

$$\text{BHP} = \text{IHP} - \text{FHP}$$

$$= 7087 \text{ hp} - 778 \text{ hp}$$

$$= \boxed{6309 \text{ hp}}$$

The answer is (D).

(b) From Eq. 29.74, BSFC is

$$\text{BSFC} = \frac{3983.7 \, \dfrac{\text{lbm}}{\text{hr}}}{6309 \text{ hp}} = \boxed{0.631 \text{ lbm/hp-hr}}$$

The answer is (C).

SI Solution

(a) *At 2100 m altitude:*

At A:

$$p_A = 82 \text{ kPa} \quad \text{[given]}$$

The absolute temperature is

$$T_A = 2°C + 273 = 275\text{K} \quad \text{[given]}$$

From App. 24.S (air table),

$$h_A = 275.12 \text{ kJ/kg}$$

$$p_{r,A} = 1.0240$$

From the ideal gas law, the air density is

$$\rho_{a1} = \frac{p_A}{RT_A} = \frac{(82 \text{ kPa})\left(1000 \, \dfrac{\text{Pa}}{\text{kPa}}\right)}{\left(287.03 \, \dfrac{\text{J}}{\text{kg·K}}\right)(275\text{K})}$$

$$= 1.0389 \text{ kg/m}^3$$

From Eq. 29.76, the air mass flow rate is

$$\dot{m}_{a1} = \dot{V}_{a1}\rho_{a1}$$

$$= \left(23{,}500 \, \dfrac{\text{L}}{\text{s}}\right)\left(\dfrac{1 \text{ m}^3}{1000 \text{ L}}\right)\left(1.0389 \, \dfrac{\text{kg}}{\text{m}^3}\right)$$

$$= 24.41 \text{ kg/s}$$

At B:

$$p_C = 8p_A = (8)(82 \text{ kPa}) = 656 \text{ kPa}$$

Assuming isentropic compression,

$$p_{r,B} = 8p_{r,A} = (8)(1.0240) = 8.192$$

From App. 24.S (air table), this $p_{r,B}$ corresponds to

$$T_B = 496.3\text{K}$$

$$h_B = 499.33 \text{ kJ/kg}$$

Due to the inefficiency of the compressor, from Eq. 29.99,

$$h'_B = h_A + \frac{h_B - h_A}{\eta_{s,\text{compression}}}$$

$$= 275.12 \, \frac{\text{kJ}}{\text{kg}} + \frac{499.33 \, \dfrac{\text{kJ}}{\text{kg}} - 275.12 \, \dfrac{\text{kJ}}{\text{kg}}}{0.85}$$

$$= 538.90 \text{ kJ/kg}$$

From App. 24.S (air table), this value of $h = 538.90$ kJ/kg corresponds to $T'_B = 534.7\text{K}$.

$$W_{\text{compression}} = h'_B - h_A$$

$$= 538.90 \, \frac{\text{kJ}}{\text{kg}} - 275.12 \, \frac{\text{kJ}}{\text{kg}} = 263.78 \text{ kJ/kg}$$

At C:

The absolute temperature is

$$T_C = 980°C + 273 = 1253\text{K}$$

Assuming there is no pressure drop across the combustor, $p_C = 656$ kPa.

From App. 24.S (air table),

$$h_C = 1340.28 \text{ kJ/kg}$$

$$p_{r,C} = 284.3$$

The energy requirement from the fuel is

$$\dot{m}\Delta h = \dot{m}_{a1}(h_C - h'_B)$$

$$= \left(24.41 \, \frac{\text{kg}}{\text{s}}\right)\left(1340.28 \, \frac{\text{kJ}}{\text{kg}} - 538.90 \, \frac{\text{kJ}}{\text{kg}}\right)$$

$$= 19\,562 \text{ kJ/s}$$

The ideal fuel rate is

$$\frac{\dot{m}\Delta h}{\text{LHV}} = \frac{19\,562 \, \dfrac{\text{kJ}}{\text{s}}}{\left(44 \, \dfrac{\text{MJ}}{\text{kg}}\right)\left(1000 \, \dfrac{\text{kJ}}{\text{MJ}}\right)} = 0.4446 \text{ kg/s}$$

From Eq. 29.74, the ideal BSFC is

$$(\text{BSFC})_{\text{ideal}} = \left(\frac{0.4446 \, \frac{\text{kg}}{\text{s}}}{4.5 \, \text{MW}} \right) \left(1000 \, \frac{\text{MW}}{\text{GW}} \right)$$

$$= 98.8 \, \text{kg/GJ}$$

The combustor efficiency is

$$\eta_{\text{combustor}} = \frac{(\text{BSFC})_{\text{ideal}}}{(\text{BSFC})_{\text{actual}}} = \frac{98.8 \, \frac{\text{kg}}{\text{GJ}}}{100 \, \frac{\text{kg}}{\text{GJ}}}$$

$$= 0.988 \quad (98.8\%)$$

At D:

Determined by atmospheric conditions, $p_{\text{D}} = 82 \, \text{kPa}$.
If expansion is isentropic,

$$p_{r,\text{D}} = \frac{p_{r,\text{C}}}{8} = \frac{284.3}{8} = 35.54$$

From App. 24.S (air table), this $p_{r,\text{D}}$ corresponds to

$$T_{\text{D}} = 740\text{K}$$
$$h_{\text{D}} = 756.44 \, \text{kJ/kg}$$

From Eq. 29.101, due to the inefficiency of the turbine,

$$h'_{\text{D}} = h_{\text{C}} - \eta_{s,\text{turbine}}(h_{\text{C}} - h_{\text{D}})$$

$$= 1340.28 \, \frac{\text{kJ}}{\text{kg}} - (0.80) \left(1340.28 \, \frac{\text{kJ}}{\text{kg}} \right.$$

$$\left. - 756.44 \, \frac{\text{kJ}}{\text{kg}} \right)$$

$$= 873.21 \, \text{kJ/kg}$$

$$W_{\text{turbine}} = h_{\text{C}} - h'_{\text{D}} = 1340.28 \, \frac{\text{kJ}}{\text{kg}} - 873.21 \, \frac{\text{kJ}}{\text{kg}}$$

$$= 467.07 \, \text{kJ/kg}$$

The theoretical net power is

$$\text{IkW} = \dot{m}_{a1}(W_{\text{turbine}} - W_{\text{compression}})$$

$$= \left(24.41 \, \frac{\text{kg}}{\text{s}} \right) \left(467.07 \, \frac{\text{kJ}}{\text{kg}} - 263.78 \, \frac{\text{kJ}}{\text{kg}} \right)$$

$$\times \left(\frac{1 \, \text{MW}}{1000 \, \text{kW}} \right) = 4.962 \, \text{MW}$$

From Eq. 29.69, the friction horsepower is

$$\text{FkW} = \text{IkW} - \text{BkW} = 4.962 \, \text{MW} - 4.5 \, \text{MW}$$

$$= 0.462 \, \text{MW}$$

At sea level (zero altitude):

At A:

The absolute temperature is

$$T_{\text{A}} = 21°\text{C} + 273 = 294\text{K}$$
$$p_{\text{A}} = 101.3 \, \text{kPa}$$

From App. 24.S,

$$h_{\text{A}} = 294.17 \, \text{kJ/kg}$$
$$p_{r,\text{A}} = 1.2917$$

From the ideal gas law, the air density is

$$\rho_{a2} = \frac{p_{\text{A}}}{RT_{\text{A}}} = \frac{(101.3 \, \text{kPa}) \left(1000 \, \frac{\text{Pa}}{\text{kPa}} \right)}{\left(287.03 \, \frac{\text{J}}{\text{kg·K}} \right) (294\text{K})}$$

$$= 1.2004 \, \text{kg/m}^3$$

From Eq. 29.76, the air mass flow rate is

$$\dot{m}_{a2} = \dot{V}_{a2}\rho_{a2}$$

$$= \left(23\,500 \, \frac{\text{L}}{\text{s}} \right) \left(\frac{1 \, \text{m}^3}{1000 \, \text{L}} \right) \left(1.2004 \, \frac{\text{kg}}{\text{m}^3} \right)$$

$$= 28.21 \, \text{kg/s}$$

At B:

$$p_{\text{B}} = 8p_{\text{A}} = (8)(101.3 \, \text{kPa}) = 810.4 \, \text{kPa}$$

Assuming isentropic expansion,

$$p_{r,\text{B}} = 8p_{r,\text{A}} = (8)(1.2917) = 10.3336$$

From App. 24.S (air table), this $p_{r,\text{B}}$ corresponds to

$$T_{\text{B}} = 529.5\text{K}$$
$$h_{\text{B}} = 533.46 \, \text{kJ/kg}$$

Due to the inefficiency of the compressor, from Eq. 29.99,

$$h'_{\text{B}} = h_{\text{A}} + \frac{h_{\text{B}} - h_{\text{A}}}{\eta_{s,\text{compression}}}$$

$$= 294.17 \, \frac{\text{kJ}}{\text{kg}} + \frac{533.46 \, \frac{\text{kJ}}{\text{kg}} - 294.17 \, \frac{\text{kJ}}{\text{kg}}}{0.85}$$

$$= 575.69 \, \text{kJ/kg}$$

From App. 24.S (air table), this value of h_B corresponds to

$$T'_B = 570\text{K}$$

$$\begin{aligned} W_{\text{compression}} &= h'_B - h_A \\ &= 575.69 \,\frac{\text{kJ}}{\text{kg}} - 294.17 \,\frac{\text{kJ}}{\text{kg}} \\ &= 281.52 \text{ kJ/kg} \end{aligned}$$

At C:

The absolute temperature is

$$T_C = 1253\text{K} \quad \text{[no change]}$$

Assuming there is no pressure drop across the combustor, $p_C = 810.4$ kPa.

From App. 24.S (air table),

$$h_C = 1340.28 \text{ kJ/kg}$$

$$p_{r,C} = 284.3$$

The energy requirement from the fuel is

$$\begin{aligned} \dot{m}\Delta h &= \dot{m}_{a2}(h_C - h'_B) \\ &= \left(28.21 \,\frac{\text{kg}}{\text{s}}\right)\left(1340.28 \,\frac{\text{kJ}}{\text{kg}} - 575.69 \,\frac{\text{kJ}}{\text{kg}}\right) \\ &= 21\,569 \text{ kJ/s} \end{aligned}$$

Assuming a constant combustion efficiency of 91.9%, the ideal fuel rate is

$$\frac{\dot{m}\Delta h}{\text{LHV}} = \frac{\left(21\,569\,\frac{\text{kJ}}{\text{s}}\right)\left(\frac{1 \text{ MJ}}{1000 \text{ kJ}}\right)}{44 \,\frac{\text{MJ}}{\text{kg}}} = 0.490 \text{ kg/s}$$

At D:

$$p_D = 101.3 \text{ kPa}$$

If expansion is isentropic,

$$p_{r,D} = \frac{p_{r,C}}{8} = \frac{284.3}{8} = 35.538 \quad \text{[no change]}$$

From App. 24.S (air table), this $p_{r,D}$ corresponds to

$$T_D = 740\text{K}$$

$$h_D = 756.44 \text{ kJ/kg}$$

$$h'_D = 873.21 \text{ kJ/kg} \quad \text{[no change]}$$

$$W_{\text{turbine}} = 467.07 \text{ kJ/kg}$$

The theoretical net power is

$$\begin{aligned} \text{IkW} &= \dot{m}_{a2}(W_{\text{turbine}} - W_{\text{compression}}) \\ &= \left(28.21 \,\frac{\text{kg}}{\text{s}}\right)\left(467.07 \,\frac{\text{kJ}}{\text{kg}} - 281.52 \,\frac{\text{kJ}}{\text{kg}}\right) \\ &\quad \times \left(\frac{1 \text{ MW}}{1000 \text{ kW}}\right) = 5.234 \text{ MW} \end{aligned}$$

Assuming the frictional horsepower is constant,

$$\text{BkW} = \text{IkW} - \text{FkW} = 5.234 \text{ MW} - 0.462 \text{ MW}$$

$$= \boxed{4.772 \text{ MW}}$$

The answer is (D).

(b) From Eq. 29.74,

$$\text{BSFC} = \frac{0.490 \,\frac{\text{kg}}{\text{s}}}{(4.772 \text{ MW})\left(\frac{1 \text{ GW}}{1000 \text{ MW}}\right)}$$

$$= \boxed{102.7 \text{ kg/GJ}}$$

The answer is (C).

30 Nuclear Power Cycles

PRACTICE PROBLEMS

1. Radioactive sodium emits 2.75 MeV gamma rays to initiate the photodisintegration of deuterium.

$$\lambda + {}_1D^2 \longrightarrow {}_1H^1 + {}_0n^1$$

What is the kinetic energy of the neutron?
- (A) 0.15 MeV
- (B) 0.26 MeV
- (C) 0.31 MeV
- (D) 0.55 MeV

2. The half-life of Cs-132 is approximately 6.47 days. How long will it take to reduce the activity of a Cs-132 sample to 5% of the original value?
- (A) 28 days
- (B) 39 days
- (C) 64 days
- (D) 81 days

3. A source with an activity of 2 curies emits 2 MeV gamma rays. For this intensity, the linear attenuation coefficient and build-up factor for a lead shield are 0.5182 cm^{-1} and 2.78, respectively. How thick must a lead shield be to reduce the activity to 1%?
- (A) 5 cm
- (B) 8 cm
- (C) 11 cm
- (D) 25 cm

4. A 10 cm thick lead plate shields a 2 MeV gamma source. The source flux density is 1,000,000 λ/cm^2·s. For this intensity, the linear attenuation coefficient and build-up factor for a lead shield are approximately 0.5182 cm^{-1} and 2.78, respectively. Use 0.0238 cm^2/g as the energy absorption coefficient in air.

(a) What is the approximate total exit flux?
- (A) 5.6×10^3 λ/cm^2·s
- (B) 9.9×10^3 λ/cm^2·s
- (C) 1.1×10^4 λ/cm^2·s
- (D) 1.6×10^4 λ/cm^2·s

(b) What is the approximate dose rate for exposure in air?
- (A) 20 mR/hr
- (B) 30 mR/hr
- (C) 40 mR/hr
- (D) 50 mR/hr

5. A neutron flux of 1×10^8 neutrons/cm^2·s irradiates a 50°C gold-foil target for 24 hours. At 20°C, gold's absorption cross section is 98 b and its density is 19.32 g/cm^3. The half-life of activated gold, Au-198, is 2.7 days. If the target is a thin disk with a diameter of 25 mm and a volume of 0.4909 cm^3, what is the removal activity?
- (A) 1.1×10^{-3} curie
- (B) 1.5×10^{-3} curie
- (C) 3.2×10^{-3} curie
- (D) 8.4×10^{-3} curie

6. A neutron point source surrounded on all sides by 20°C water emits 1×10^7 neutrons/s. The average diffusion coefficient for water is approximately 0.16, and its diffusion length is approximately 2.85 cm. What is the neutron flux 20 cm from the point source?
- (A) 92 n/cm^2·s
- (B) 150 n/cm^2·s
- (C) 220 n/cm^2·s
- (D) 350 n/cm^2·s

7. The thermal fission and absorption cross sections for natural uranium are 4.18 b and 7.68 b, respectively. What is the probability that a 20°C neutron that has been absorbed will cause fission in natural uranium?
- (A) 8%
- (B) 35%
- (C) 45%
- (D) 54%

8. A source of 1 MeV gamma rays has an activity of 1×10^8 λ/s. For this intensity, the linear attenuation coefficient and build-up factor for an iron shield are approximately 0.4683 cm^{-1} and 14.93, respectively. 0.0280 cm^2/g is the energy absorption coefficient in air.

What thickness of spherical iron shield will reduce the exposure rate to 1 mR/hr outside the shield?

- (A) 8 cm
- (B) 15 cm
- (C) 37 cm
- (D) 64 cm

9. A 1 GW breeder reactor is being designed to process 2000 kg of a 20% Pu-239 and 80% U-238 mixture. For fast neutrons, Pu-239 has an absorption cross section of 1.95 b, has a fission cross section of 1.8 b, and releases 2.95 neutrons per fission. For fast neutrons, U-238 has an absorption cross section of 0.59 b, has a fission cross section of 0.5 b, and releases 2.45 neutrons per fission. The fast fission factor is 1.05. What is the exponential doubling time?

- (A) 1400 days
- (B) 2100 days
- (C) 3600 days
- (D) 4500 days

10. The maximum neutron flux in a bare spherical reactor is 4.5×10^{15} n/cm^2·s. The radius of the reactor is 40 cm, and the macroscopic fission cross section of the fuel is 0.005 cm^{-1}. It takes 3.1×10^{10} fissions to generate one watt of power. What power is generated in the reactor?

- (A) 18 MW
- (B) 31 MW
- (C) 59 MW
- (D) 140 MW

11. A nuclear reactor produces 500,000 kW thermal. The cylindrical core is 10 ft in diameter and 10 ft high. The core contains 100,000 lbm of uranium enriched to 2% U-235. The U-235 fission cross section is 547 barns. What is the average thermal neutron flux?

- (A) 1.2×10^{13} n/cm^2·s
- (B) 7.5×10^{13} n/cm^2·s
- (C) 3.4×10^{14} n/cm^2·s
- (D) 6.9×10^{14} n/cm^2·s

SOLUTIONS

1. Use carbon-based atomic masses. The mass increase is

$$\Delta m = m_H + m_n - m_D$$
$$= 1.007825 \text{ amu} + 1.008665 \text{ amu} - 2.01410 \text{ amu}$$
$$= 0.00239 \text{ amu}$$

Express the mass increase as energy. Convert to electron volts.

$$\Delta E = (0.00239 \text{ amu})\left(931.481 \ \frac{\text{MeV}}{\text{amu}}\right)$$
$$= 2.226 \text{ MeV}$$

Thus, 2.75 MeV $-$ 2.226 MeV $=$ 0.524 MeV are shared by the neutron and the hydrogen atom. Since the neutron and hydrogen atom have roughly the same mass, the neutron receives half.

$$\Delta E_n = \frac{0.524 \text{ MeV}}{2}$$
$$= \boxed{0.262 \text{ MeV}}$$

The answer is (B).

2. The decay constant is

$$\lambda = \frac{0.693}{6.47 \text{ days}} = 0.1071 \text{ day}^{-1}$$

The activity can be predicted by

$$\frac{A}{A_0} = e^{-\lambda t}$$
$$0.05 = e^{\left(-0.1071 \frac{1}{\text{day}}\right)t}$$

Take the natural logarithm of both sides to determine t.

$$\boxed{t = 27.97 \text{ days}}$$

The answer is (A).

3. The linear attenuation coefficient and build-up factor for 2 MeV gamma radiation are, respectively, approximately

$$\mu_l = 0.5182 \text{ cm}^{-1}; \ B = 2.78$$

(If necessary, an initial estimate, say 10 cm, of shield thickness can be used with a table of mass attenuation coefficients. In that case, the solution will be iterative.)

$$\frac{\phi}{\phi_0} = Be^{-\mu_l x}$$
$$0.01 = 2.78 e^{-\left(0.5182 \frac{1}{\text{cm}}\right)x}$$

Solve for x by taking the natural logarithm of both sides.

$$x = 10.86 \text{ cm}$$

The answer is (C).

4. (a) The uncollided exit flux is

$$\phi = \phi_0 e^{-\mu x} = \left(10^6 \frac{1}{\text{cm}^2 \cdot \text{s}}\right) e^{-\left(0.5182 \frac{1}{\text{cm}}\right)(10 \text{ cm})}$$

$$= 5.62 \times 10^3 \frac{1}{\text{cm}^2 \cdot \text{s}}$$

The build-up flux is

$$\phi_B = BI = (2.78)\left(5.62 \times 10^3 \frac{1}{\text{cm}^2 \cdot \text{s}}\right)$$

$$= \boxed{1.56 \times 10^4 \frac{1}{\text{cm}^2 \cdot \text{s}}}$$

The answer is (D).

(b) There are many approximate correlations between dose rate and intensity in air. One of them is

$$D' = \left(0.0659 \frac{\text{g} \cdot \text{s} \cdot \text{mR}}{\text{MeV} \cdot \text{hr}}\right) E_0 \phi_B \mu_{a,\text{air}}$$

$$= (0.0659E)(2 \text{ MeV})\left(1.56 \times 10^4 \frac{1}{\text{cm}^2 \cdot \text{s}}\right)$$

$$\times \left(0.0238 \frac{\text{cm}^2}{\text{g}}\right)$$

$$= \boxed{48.9 \text{ mR/hr}}$$

The answer is (D).

5. The absorption cross section at 50°C is

$$\bar{\sigma} = \frac{\sigma_{a,\text{thermal}}}{1.128}\sqrt{\frac{273 + 20°}{273 + T_{°C}}}$$

$$= \frac{98 \text{ b}}{1.128}\sqrt{\frac{273 + 20°\text{C}}{273 + 50°\text{C}}}$$

$$= 82.75 \text{ b}$$

The number of atoms in the target foil is

$$N = \frac{mN_o}{A} = \frac{\rho V N_o}{A}$$

$$= \frac{\left(19.32 \frac{\text{g}}{\text{cm}^3}\right)(0.4909 \text{ cm}^3)}{197 \frac{\text{g}}{\text{mole}}}$$

$$\times \left(6.022 \times 10^{23} \frac{\text{atoms}}{\text{mole}}\right)$$

$$= 2.9 \times 10^{22} \text{ atoms}$$

The decay constant for the activated gold is

$$\lambda = \frac{0.693}{2.7 \text{ days}} = 0.2567 \text{ day}^{-1}$$

The removal activity is

$$A_R = J\bar{\sigma}_a N(1 - e^{-\lambda t})$$

$$= \frac{\left(10^8 \frac{1}{\text{cm}^2 \times \text{s}}\right)(82.75 \text{ b})\left(1 \times 10^{-24} \frac{\text{cm}^2}{\text{b}}\right)}{3.7 \times 10^{10} \frac{\text{curie}}{\text{s}}}$$

$$\times (2.9 \times 10^{22} \text{ atoms})\left(1 - e^{\left(-0.2567 \frac{1}{\text{day}}\right)(1\text{day})}\right)$$

$$= 1.47 \times 10^{-3} \text{ curie}$$

The answer is (B).

6. For an isotropic point source, the diffused flux is given by

$$\phi = \frac{A_{0,\text{isotropic}} e^{-\frac{r}{L}}}{4\pi \overline{D} r}$$

$$= \frac{\left(1 \times 10^7 \frac{1}{\text{s}}\right) e^{-\frac{20 \text{ cm}}{2.85 \text{ cm}}}}{4\pi(0.16 \text{ cm})(20 \text{ cm})}$$

$$= 223 \frac{1}{\text{cm}^2 \times \text{s}}$$

The answer is (C).

7.
$$p\{f\} = \frac{\sigma_f}{\sigma_a} = \frac{4.18 \text{ b}}{7.68 \text{ b}}$$
$$= 0.544 \quad (54.4\%)$$

The answer is (D).

8. The build-up flux in a spherical shield from an isotropic source is

$$\phi_B = B\frac{\phi_{0,\text{isotropic}} e^{-\mu_l r}}{4\pi r^2}$$

$$= (14.93)\frac{\left(1 \times 10^8 \frac{1}{\text{s}}\right) e^{-\left(0.4683 \frac{1}{\text{cm}}\right)r}}{4\pi r^2}$$

$$= \frac{\left(1.19 \times 10^8 \frac{1}{\text{s}}\right) e^{-\left(0.4683 \frac{1}{\text{cm}}\right)r}}{r^2}$$

Power Cycles

The approximate dose (exposure) rate equation used is

$$D' = \left(0.0659 \, \frac{\text{g} \cdot \text{s} \cdot \text{mR}}{\text{MeV} \cdot \text{hr}}\right) E_0 \phi_B \mu_{a,\text{air}}$$

$$1 \, \frac{\text{mR}}{\text{hr}} = \frac{\begin{array}{c}\left(0.0659 \, \frac{\text{g} \cdot \text{s} \cdot \text{mR}}{\text{MeV} \cdot \text{hr}}\right)(1 \text{ MeV}) \\[6pt] \times \left(1.19 \times 10^8 \, \frac{1}{\text{cm}^2 \cdot \text{s}}\right) e^{-\left(0.4683 \, \frac{1}{\text{cm}}\right)r} \\[6pt] \times \left(0.0280 \, \frac{\text{cm}^2}{\text{g}}\right)\end{array}}{r^2}$$

By trial and error, $r = 14.8$ cm.

The answer is (B).

9. Calculate the average number of neutrons generated per fast neutron released.

$$\eta = G_{\text{Pu-239}} n_{\text{Pu-239}}\left(\frac{\sigma_{f,\text{Pu-239}}}{\sigma_{a,\text{Pu-239}}}\right)$$
$$\quad + G_{\text{U-238}} n_{\text{U-238}}\left(\frac{\sigma_{f,\text{U-238}}}{\sigma_{a,\text{U-238}}}\right)$$
$$= (0.2)(2.95)\left(\frac{1.8 \text{ b}}{1.95 \text{ b}}\right) + (0.8)(2.45)\left(\frac{0.5 \text{ b}}{0.59 \text{ b}}\right)$$
$$= 2.206$$

The conversion ratio is

$$\text{CR} = \eta\varepsilon - 1 = (2.206)(1.05) - 1 = 1.316$$

The linear doubling time is

$$t_{d,\text{linear}} = \frac{m}{(\text{CR} - 1)m'P}$$
$$= \frac{(2000 \text{ kg})\left(1000 \, \frac{\text{g}}{\text{kg}}\right)}{(1.316 - 1)\left(1.23 \, \frac{\text{g}}{\text{MW} \cdot \text{day}}\right)(1000 \text{ MW})}$$
$$= 5146 \text{ days}$$

The exponential doubling time is

$$t_{d,\text{exponential}} = 0.693 t_{d,\text{linear}}$$
$$= (0.693)(5146 \text{ days})$$
$$= 3566 \text{ days}$$

The answer is (C).

10. In a spherical reactor, the ratio of maximum to average neutron flux is 3.29.

Reactor power is predicted by

$$P = \frac{\overline{\phi} \Sigma_f V}{3.1 \times 10^{10} \, \frac{1}{\text{W}}}$$

$$= \frac{\begin{array}{c}\left(\dfrac{4.5 \times 10^{15} \, \frac{1}{\text{cm}^2 \cdot \text{s}}}{3.29}\right)\left(0.005 \, \frac{1}{\text{cm}}\right) \\[10pt] \times \left(\dfrac{4\pi}{3}\right)(40 \text{ cm})^3\end{array}}{3.1 \times 10^{10} \, \frac{1}{\text{W}}}$$

$$= 5.9 \times 10^7 \text{ W}$$

The answer is (C).

11. The mass of uranium is

$$m = (100{,}000 \text{ lbm})\left(0.4536 \, \frac{\text{kg}}{\text{lbm}}\right)\left(1000 \, \frac{\text{g}}{\text{kg}}\right)$$
$$= 4.54 \times 10^7 \text{ g}$$

The number of fuel molecules per unit volume (cm^3) is

$$N_f = \frac{G_f m N_o}{V(MW)}$$
$$= \frac{(0.02)(4.54 \times 10^7 \text{ g})\left(6.022 \times 10^{23} \, \frac{\text{atoms}}{\text{mol}}\right)}{V\left(235 \, \frac{\text{g}}{\text{mol}}\right)}$$
$$= \frac{2.33 \times 10^{27}}{V}$$

From the thermal power equation,

$$\overline{\phi} = \frac{\left(3.1 \times 10^{10} \, \frac{1}{\text{W}}\right) P_{\text{thermal}}}{\Sigma_f V}$$
$$= \frac{\left(3.1 \times 10^{10} \, \frac{1}{\text{W}}\right) P_{\text{thermal}}}{\sigma_f N_f V}$$
$$= \frac{\left(3.1 \times 10^{10} \, \frac{\text{fissions}}{\text{s}}\right)(5 \times 10^8 E)}{(547 \text{ b})\left(1 \times 10^{-24} \, \frac{\text{cm}^2}{\text{b}}\right)(2.33 \times 10^{27})}$$
$$= 1.22 \times 10^{13} \, \frac{\text{fissions}}{\text{cm}^2 \cdot \text{s}}$$

The answer is (A).

31 Advanced and Alternative Power-Generating Systems

PRACTICE PROBLEMS

1. An ocean thermal energy conversion plant draws 40°F (4.4°C) water from a depth of 1200 ft (360 m). The water temperature at the surface is 82°F (27.8°C). What is the maximum achievable thermal efficiency?

- (A) 7.8%
- (B) 11%
- (C) 15%
- (D) 21%

SOLUTIONS

1. *Customary U.S. Solution*

The maximum achievable thermal efficiency is achieved with the Carnot cycle and is given by Eq. 28.8.

$$\eta_{th} = \frac{T_{high} - T_{low}}{T_{high}}$$

$$T_{high} = 82°F + 460 = 542°R$$

$$T_{low} = 40°F + 460 = 500°R$$

$$\eta_{th} = \frac{542°R - 500°R}{542°R} = \boxed{0.0775 \ (7.75\%)}$$

The answer is (A).

SI Solution

The maximum achievable thermal efficiency is achieved with the Carnot cycle and is given by Eq. 28.8.

$$\eta_{th} = \frac{T_{high} - T_{low}}{T_{high}}$$

$$T_{high} = 27.8°C + 273 = 300.8K$$

$$T_{low} = 4.4°C + 273 = 277.4K$$

$$\eta_{th} = \frac{300.8K - 277.4K}{300.8K} = \boxed{0.0778 \ (7.78\%)}$$

The answer is (A).

32 Gas Compression Cycles

PRACTICE PROBLEMS

1. What is the isentropic efficiency of a compressor that takes in air at 8 psia and $-10°F$ (55 kPa and $-20°C$) and discharges it at 40 psia and $315°F$ (275 kPa and $160°C$)?

(A) 61%
(B) 66%
(C) 74%
(D) 81%

2. A reciprocating air compressor has a 7% clearance and discharges 65 psia (450 kPa) air at the rate of 48 lbm/min (0.36 kg/s). Ambient air is at 14.7 psia and $65°F$. The polytropic exponent is 1.33. What mass of air is compressed each minute?

(A) 42 lbm/min (0.32 kg/s)
(B) 51 lbm/min (0.38 kg/s)
(C) 56 lbm/min (0.42 kg/s)
(D) 74 lbm/min (0.55 kg/s)

3. Air at 14.7 psia and $500°F$ (101.3 kPa and $260°C$) is compressed in a centrifugal compressor to 6 atm. The isentropic efficiency of the compression process is 65%.

(a) What is the compression work?
(A) 140 Btu/lbm (320 kJ/kg)
(B) 190 Btu/lbm (440 kJ/kg)
(C) 230 Btu/lbm (540 kJ/kg)
(D) 350 Btu/lbm (810 kJ/kg)

(b) What is the final temperature?
(A) $1550°R$ (860K)
(B) $1650°R$ (920K)
(C) $1750°R$ (970K)
(D) $1850°R$ (1030K)

(c) What is the increase in entropy?
(A) 0.048 Btu/lbm-$°R$ (0.21 kJ/kg·K)
(B) 0.099 Btu/lbm-$°R$ (0.47 kJ/kg·K)
(C) 0.23 Btu/lbm-$°R$ (1.0 kJ/kg·K)
(D) 0.44 Btu/lbm-$°R$ (2.1 kJ/kg·K)

4. (*Time limit: one hour*) Compressors A and B both discharge into a common tank. The storage tank contains 100 psia, $90°F$ air (700 kPa, $32°C$). Both compressors receive air at 14.7 psia and $80°F$ (101.3 kPa and $27°C$). The flow rate through compressor A is 600 cfm (300 L/s). Process C uses 100 cfm of 80 psia, $85°F$ (50 L/s of 550 kPa, $29°C$) air. Process D uses 120 cfm of 85 psia, $80°F$ (60 L/s of 590 kPa, $27°C$) air. Process E uses 8 lbm/min (0.06 kg/s) of $85°F$ ($29°C$) air. Air is an ideal gas, and compressibility effects are to be disregarded. Calculate the flow rate through compressor B.

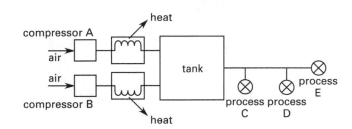

(A) 620 ft³/min (320 L/s)
(B) 740 ft³/min (370 L/s)
(C) 900 ft³/min (450 L/s)
(D) 1100 ft³/min (550 L/s)

5. (*Time limit: one hour*) 300 cfm (150 L/s) of air at 14.7 psia and $90°F$ (101.3 kPa and $32°C$) enter a compressor. The compressor discharges air into a water-cooled heat exchanger. The compressed air is stored at 300 psig and $90°F$ (2.07 MPa and $32°C$) in a 1000 ft³ (27 m³) tank. The tank feeds three air-driven tools with the flow rates and properties listed. The pressures to the air-driven tools are regulated and remain constant as the tank pressure changes. Air is an ideal gas, and compressibility effects are to be disregarded. How long can the system run?

	tool 1	tool 2	tool 3
flow rate			
(cfm)	40	15	unknown
(L/s)	19	7	unknown
flow rate			
(lbm/min)	unknown	unknown	6
(kg/s)	unknown	unknown	0.045
minimum pressure			
(psig)	90	50	80
(kPa)	620	350	550
temperature			
(°F)	90	85	80
(°C)	32	29	27

(A) 0.6 hr (1.2 hr)

(B) 1.8 hr (3.6 hr)

(C) 3.4 hr (6.8 hr)

(D) 6.6 hr (13 hr)

SOLUTIONS

1.

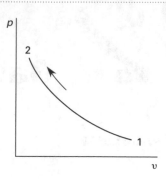

Customary U.S. Solution

Assume air is an ideal gas.

From Eq. 29.18, the isentropic temperature at point 2 is

$$T_2 = T_1 \left(\frac{p_2}{p_1}\right)^{\frac{k-1}{k}}$$

The absolute temperature at point 1 is

$$T_1 = -10°F + 460 = 450°R$$

$$T_2 = (450°R)\left(\frac{40 \text{ psia}}{8 \text{ psia}}\right)^{\frac{1.4-1}{1.4}}$$

$$= 712.7°R$$

$$T_2 = 712.7°R - 460 = 252.7°F$$

The efficiency of the compressor is given by Eq. 29.99.

$$\eta_{s,\text{compressor}} = \frac{T_2 - T_1}{T_2' - T_1}$$

$$= \frac{252.7°F - (-10°F)}{315°F - (-10°F)}$$

$$= \boxed{0.808 \ (80.8\%)}$$

The answer is (D).

SI Solution

Assume air is an ideal gas.

From Eq. 29.18, the isentropic temperature at point 2 is

$$T_2 = T_1 \left(\frac{p_2}{p}\right)^{\frac{k-1}{k}}$$

The absolute temperature at point 1 is

$$T_1 = -20°C + 273 = 253K$$

$$T_2 = (253K)\left(\frac{275 \text{ kPa}}{55 \text{ kPa}}\right)^{\frac{1.4-1}{1.4}}$$

$$= 400.7K$$

$$T_2 = 400.7K - 273 = 127.7°C$$

The efficiency of the compressor is given by Eq. 29.99.

$$\eta_{s,\text{compressor}} = \frac{T_2 - T_1}{T_2' - T_1}$$

$$= \frac{127.7°\text{C} - (-20°\text{C})}{160°\text{C} - (-20°\text{C})}$$

$$= \boxed{0.821 \quad (82.1\%)}$$

The answer is (D).

2.

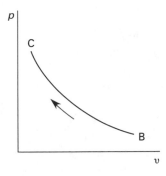

Customary U.S. Solution

The compression ratio for a reciprocating compressor is defined by Eq. 32.4.

$$r_p = \frac{p_C}{p_B}$$

$$= \frac{65 \text{ psia}}{14.7 \text{ psia}} = 4.42$$

The volumetric efficiency is given by Eq. 32.6.

$$\eta_v = 1 - \left(r_p^{\frac{1}{n}} - 1 \right) \left(\frac{c}{100} \right)$$

$$= 1 - \left(4.42^{\frac{1}{1.33}} - 1 \right) \left(\frac{7\%}{100} \right)$$

$$= 0.856 \quad (85.6\%)$$

The mass of air compressed per minute from Eq. 32.8 is

$$\dot{m} = \frac{\dot{m}_{\text{actual}}}{\eta_v} = \frac{48 \dfrac{\text{lbm}}{\text{min}}}{0.856}$$

$$= \boxed{56.07 \text{ lbm/min}}$$

The answer is (C).

SI Solution

The compression ratio for the reciprocating compressor is defined by Eq. 32.4.

$$r_p = \frac{p_C}{p_B} = \frac{450 \text{ kPa}}{101 \text{ kPa}} = 4.46$$

The volumetric efficiency is given by Eq. 32.6.

$$\eta_v = 1 - \left(r_p^{\frac{1}{n}} - 1 \right) \left(\frac{c}{100} \right)$$

$$= 1 - \left(4.46^{\frac{1}{1.33}} - 1 \right) \left(\frac{7\%}{100} \right)$$

$$= 0.855 \quad (86\%)$$

The mass of air compressed per minute from Eq. 32.8 is

$$\dot{m} = \frac{\dot{m}_{\text{actual}}}{\eta_v} = \frac{0.36 \dfrac{\text{kg}}{\text{s}}}{0.855}$$

$$= \boxed{0.4211 \text{ kg/s}}$$

The answer is (C).

3. *Customary U.S. Solution*

((a) and (b)) Although the ideal gas laws can be used, it is more expedient to use air tables for the low pressures.

The absolute inlet temperature is

$$T_1 = 500°\text{F} + 460 = 960°\text{R}$$

From App. 24.F,

$$h_1 = 231.06 \text{ Btu/lbm}$$
$$p_{r,1} = 10.61$$
$$\phi_1 = 0.74030 \text{ Btu/lbm-°R}$$

Since $p_1/p_2 = p_{r,1}/p_{r,2}$ and the compression ratio is 6, for isentropic compression,

$$p_{r,2} = 6 p_{r,1}$$
$$= (6)(10.61) = 63.66$$

From App. 24, at $p_{r,2} = 63.66$,

$$T_2 = 1552°\text{R}$$
$$h_2 = 382.95 \text{ Btu/lbm}$$

The actual enthalpy from the definition of isentropic efficiency is

$$h_2' = h_1 + \frac{h_2 - h_1}{\eta_s}$$

$$= 231.06 \frac{\text{Btu}}{\text{lbm}} + \frac{382.95 \dfrac{\text{Btu}}{\text{lbm}} - 231.06 \dfrac{\text{Btu}}{\text{lbm}}}{0.65}$$

$$= 464.74 \text{ Btu/lbm}$$

From App. 24.F, this corresponds to $\boxed{T_2' = 1855°R}$ and $\phi_2' = 0.91129$ Btu/lbm-°R.

The answer is (D) for Prob. 3(b).

The compression work is

$$W = h_2' - h_1 = 464.74 \frac{\text{Btu}}{\text{lbm}} - 231.06 \frac{\text{Btu}}{\text{lbm}}$$

$$\boxed{= 233.68 \text{ Btu/lbm}}$$

The answer is (C) for Prob. 3(a).

(c) From Eq. 24.39, the increase in entropy is

$$\Delta s = \phi_2' - \phi_1 - R \ln\left(\frac{p_2}{p_1}\right)$$

$$= 0.91129 \frac{\text{Btu}}{\text{lbm-°R}} - 0.74030 \frac{\text{Btu}}{\text{lbm-°R}}$$

$$- \left(\frac{53.3 \frac{\text{ft-lbf}}{\text{lbm-°R}}}{778 \frac{\text{ft-lbf}}{\text{Btu}}}\right) \ln\left(\frac{6 \text{ atm}}{1 \text{ atm}}\right)$$

$$= \boxed{0.04824 \text{ Btu/lbm-°R}}$$

The answer is (A).

SI Solution

((a) and (b)) Although the ideal gas laws can be used, it is more expedient to use air tables for the lower pressures.

The absolute inlet temperature is

$$T_1 = 260°C + 273 = 533K$$

From App. 24.S,

$$h_1 = 537.09 \text{ kJ/kg}$$

$$p_{r,1} = 10.59$$

$$\phi_1 = 2.28161 \text{ kJ/kg·K}$$

Since $p_1/p_2 = p_{r,1}/p_{r,2}$ and the compression ratio is 6, for isentropic compression,

$$p_{r,2} = 6p_{r,1}$$

$$= (6)(10.59) = 63.54$$

From App. 24.S, at $p_{r,2} = 63.54$,

$$T_2 = 861.5K$$

$$h_2 = 889.9 \text{ kJ/kg}$$

The actual enthalpy from the definition of isentropic efficiency is

$$h_2' = h_1 + \frac{h_2 - h_1}{\eta_s}$$

$$= 537.09 \frac{\text{kJ}}{\text{kg}} + \frac{889.9 \frac{\text{kJ}}{\text{kg}} - 537.09 \frac{\text{kJ}}{\text{kg}}}{0.65}$$

$$= 1079.87 \text{ kJ/kg}$$

From App. 24.S, this corresponds to $\boxed{T_2' = 1029.6K}$ and $\phi_2' = 3.00099$ kJ/kg·K.

The answer is (D) for Prob. 3(b).

The compression work is

$$W = h_2' - h_1 = 1079.87 \frac{\text{kJ}}{\text{kg}} - 537.09 \frac{\text{kJ}}{\text{kg}}$$

$$\boxed{= 542.78 \text{ kJ/kg}}$$

The answer is (C) for Prob. 3(a).

(c) From Eq. 24.39, the increase in entropy is

$$\Delta s = \phi_2' - \phi_1 - R \ln\left(\frac{p_2}{p_1}\right)$$

$$= 3.00099 \frac{\text{kJ}}{\text{kg·K}} - 2.28161 \frac{\text{kJ}}{\text{kg·K}}$$

$$- \left(287.03 \frac{\text{J}}{\text{kg·K}}\right)\left(\frac{1 \text{ kJ}}{1000 \text{ J}}\right) \ln\left(\frac{6 \text{ atm}}{1 \text{ atm}}\right)$$

$$= \boxed{0.20509 \text{ kJ/kg·K}}$$

The answer is (A).

4. *Customary U.S. Solution*

The absolute temperature for process C is $85°F + 460 = 545°R$. Using ideal gas laws, the mass flow rate for process C is

$$\dot{m}_C = \frac{p_C \dot{V}_C}{RT_C} = \frac{\left(80 \frac{\text{lbf}}{\text{in}^2}\right)\left(144 \frac{\text{in}^2}{\text{ft}^2}\right)\left(100 \frac{\text{ft}^3}{\text{min}}\right)}{\left(53.3 \frac{\text{ft-lbf}}{\text{lbm-°R}}\right)(545°R)}$$

$$= 39.66 \text{ lbm/min}$$

Similarly, the mass flow rate for process D with an absolute temperature of $80°F + 460 = 540°R$ is

$$\dot{m}_D = \frac{p_D \dot{V}_D}{RT_D} = \frac{\left(85 \frac{\text{lbf}}{\text{in}^2}\right)\left(144 \frac{\text{in}^2}{\text{ft}^2}\right)\left(120 \frac{\text{ft}^3}{\text{min}}\right)}{\left(53.3 \frac{\text{ft-lbf}}{\text{lbm-°R}}\right)(540°R)}$$

$$= 51.03 \text{ lbm/min}$$

For process E, the mass flow rate is given as $\dot{m}_E = 8$ lbm/min.

The total mass flow rate for all three processes is

$$\dot{m}_{total} = \dot{m}_C + \dot{m}_D + \dot{m}_E$$
$$= 39.66 \frac{lbm}{min} + 51.03 \frac{lbm}{min} + 8 \frac{lbm}{min}$$
$$= 98.69 \text{ lbm/min}$$

The mass flow rate into compressor A is with an absolute temperature of $80°F + 460 = 540°R$.

$$\dot{m}_A = \frac{p_A \dot{V}_A}{RT_A} = \frac{\left(14.7 \frac{lbf}{in^2}\right)\left(144 \frac{in^2}{ft^2}\right)\left(600 \frac{ft^3}{min}\right)}{\left(53.3 \frac{ft\text{-}lbf}{lbm\text{-}°R}\right)(540°R)}$$
$$= 44.13 \text{ lbm/min}$$

The required input for compressor B is

$$\dot{m}_B = \dot{m}_{total} - \dot{m}_A$$
$$= 98.69 \frac{lbm}{min} - 44.13 \frac{lbm}{min}$$
$$= 54.56 \text{ lbm/min}$$

The volumetric flow rate for compressor B is

$$\dot{V}_B = \frac{\dot{m}_B RT_B}{p_B}$$
$$= \frac{\left(54.56 \frac{lbm}{min}\right)\left(53.3 \frac{ft\text{-}lbf}{lbm\text{-}°R}\right)(540°R)}{\left(14.7 \frac{lbf}{in^2}\right)\left(144 \frac{in^2}{ft^2}\right)}$$
$$= \boxed{742 \text{ cfm}}$$

The answer is (B).

SI Solution

The absolute temperature for process C is $29°C + 273 = 302K$. Using ideal gas laws, the mass flow rate for process C is

$$\dot{m}_C = \frac{p_C \dot{V}_C}{RT_C}$$
$$= \frac{(550 \text{ kPa})\left(1000 \frac{Pa}{kPa}\right)\left(50 \frac{L}{s}\right)\left(\frac{1 \text{ m}^3}{1000 \text{ L}}\right)}{\left(287 \frac{J}{kg\cdot K}\right)(302K)}$$
$$= 0.3173 \text{ kg/s}$$

The absolute temperature for process D is $27°C + 273 = 300K$. Using ideal gas laws, the mass flow rate for process D is

$$\dot{m}_D = \frac{p_D \dot{V}_D}{RT_D}$$
$$= \frac{(590 \text{ kPa})\left(1000 \frac{Pa}{kPa}\right)\left(60 \frac{L}{s}\right)\left(\frac{1 \text{ m}^3}{1000 \text{ L}}\right)}{\left(287 \frac{J}{kg\cdot K}\right)(300K)}$$
$$= 0.4111 \text{ kg/s}$$

For process E, the mass flow rate is given as $\dot{m}_E = 0.06$ kg/s.

The total mass flow rate for all three processes is

$$\dot{m}_{total} = \dot{m}_C + \dot{m}_D + \dot{m}_E$$
$$= 0.3173 \frac{kg}{s} + 0.4111 \frac{kg}{s} + 0.06 \frac{kg}{s}$$
$$= 0.7884 \text{ kg/s}$$

The absolute temperature for air into compressors A and B is $27°C + 273 = 300K$.

The mass flow rate into compressor A is

$$\dot{m}_A = \frac{p_A \dot{V}_A}{RT_A}$$
$$= \frac{(101.3 \text{ kPa})\left(1000 \frac{Pa}{kPa}\right)\left(300 \frac{L}{s}\right)\left(\frac{1 \text{ m}^3}{1000 \text{ L}}\right)}{\left(287 \frac{J}{kg\cdot K}\right)(300K)}$$
$$= 0.3530 \text{ kg/s}$$

The required input for compressor B is

$$\dot{m}_B = \dot{m}_{total} - \dot{m}_A$$
$$= 0.7884 \frac{kg}{s} - 0.3530 \frac{kg}{s}$$
$$= 0.4354 \text{ kg/s}$$

The volumetric flow rate for compressor B is

$$\dot{V}_B = \frac{\dot{m}_B RT_B}{p_B}$$
$$= \frac{\left(0.4354 \frac{kg}{s}\right)\left(287 \frac{J}{kg\cdot K}\right)(300K)\left(1000 \frac{L}{m^3}\right)}{(101.3 \text{ kPa})\left(1000 \frac{Pa}{kPa}\right)}$$
$$= \boxed{370 \text{ L/s}}$$

The answer is (B).

Power Cycles

5.

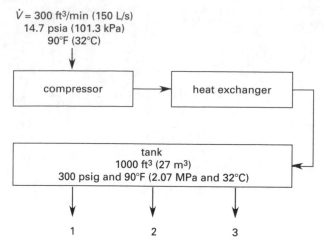

$\dot{V} = 300$ ft³/min (150 L/s)
14.7 psia (101.3 kPa)
90°F (32°C)

Customary U.S. Solution

Assume steady flow and constant properties.

The absolute temperature for compressor air is $90°\text{F} + 460 = 550°\text{R}$.

Using ideal gas laws, the mass flow rate of air into the compressor is

$$\dot{m} = \frac{p\dot{V}}{RT}$$

$$= \frac{\left(14.7 \ \frac{\text{lbf}}{\text{in}^2}\right)\left(144 \ \frac{\text{in}^2}{\text{ft}^2}\right)\left(300 \ \frac{\text{ft}^3}{\text{min}}\right)}{\left(53.3 \ \frac{\text{ft-lbf}}{\text{lbm-}°\text{R}}\right)(550°\text{R})}$$

$$= 21.7 \ \text{lbm/min}$$

The absolute temperature and absolute pressure of stored compressed air are

$$T = 90°\text{F} + 460$$
$$= 550°\text{R}$$
$$p = 300 \ \text{psig} + 14.7$$
$$= 314.7 \ \text{psia}$$

The mass of stored compressed air in a 1000 ft³ tank is

$$m_{\text{tank}} = \frac{pV}{RT}$$

$$= \frac{\left(314.7 \ \frac{\text{lbf}}{\text{in}^2}\right)\left(144 \ \frac{\text{in}^2}{\text{ft}^2}\right)(1000 \ \text{ft}^3)}{\left(53.3 \ \frac{\text{ft-lbf}}{\text{lbm-}°\text{R}}\right)(550°\text{R})}$$

$$= 1545.9 \ \text{lbm}$$

Assuming that each tool operates at its minimum pressure, the mass leaving the system can be calculated as follows.

Tool 1:

The absolute pressure is

$$90 \ \text{psig} + 14.7 = 104.7 \ \text{psia}$$

The absolute temperature is

$$90°\text{F} + 460 = 550°\text{R}$$

$$\dot{m}_{\text{tool 1}} = \frac{p\dot{V}}{RT}$$

$$= \frac{\left(104.7 \ \frac{\text{lbf}}{\text{in}^2}\right)\left(144 \ \frac{\text{in}^2}{\text{ft}^2}\right)\left(40 \ \frac{\text{ft}^3}{\text{min}}\right)}{\left(53.3 \ \frac{\text{ft-lbf}}{\text{lbm-}°\text{R}}\right)(550°\text{R})}$$

$$= 20.57 \ \text{lbm/min}$$

Tool 2:

The absolute pressure is

$$50 \ \text{psig} + 14.7 = 64.7 \ \text{psia}$$

The absolute temperature is

$$85°\text{F} + 460 = 545°\text{R}$$

$$\dot{m}_{\text{tool 2}} = \frac{p\dot{V}}{RT}$$

$$= \frac{\left(64.7 \ \frac{\text{lbf}}{\text{in}^2}\right)\left(144 \ \frac{\text{in}^2}{\text{ft}^2}\right)\left(15 \ \frac{\text{ft}^3}{\text{min}}\right)}{\left(53.3 \ \frac{\text{ft-lbf}}{\text{lbm-}°\text{R}}\right)(545°\text{R})}$$

$$= 4.81 \ \text{lbm/min}$$

Tool 3:

$$\dot{m}_{\text{tool 3}} = 6 \ \text{lbm/min} \quad \text{[given]}$$

The total mass flow leaving the system is

$$\dot{m}_{\text{total}} = \dot{m}_{\text{tool 1}} + \dot{m}_{\text{tool 2}} + \dot{m}_{\text{tool 3}}$$

$$= 20.57 \ \frac{\text{lbm}}{\text{min}} + 4.81 \ \frac{\text{lbm}}{\text{min}} + 6 \ \frac{\text{lbm}}{\text{min}}$$

$$= 31.38 \ \text{lbm/min}$$

The critical pressure of 104.7 psia is required for tool 1 operation. The mass in the tank when critical pressure is achieved is

$$m_{\text{tank,critical}} = \frac{pV}{RT}$$

$$= \frac{\left(104.7 \ \frac{\text{lbf}}{\text{in}^2}\right)\left(144 \ \frac{\text{in}^2}{\text{ft}^2}\right)(1000 \ \text{ft}^3)}{\left(53.3 \ \frac{\text{ft-lbf}}{\text{lbm-}°\text{R}}\right)(550°\text{R})}$$

$$= 514.3 \ \text{lbm}$$

The amount in the tank to be depleted is

$$m_{depleted} = m_{tank} - m_{tank,critical}$$
$$= 1545.9 \text{ lbm} - 514.3 \text{ lbm} = 1031.6 \text{ lbm}$$

The net flow rate of air to the tank is

$$\dot{m}_{net} = \dot{m}_{in} - \dot{m}_{out} = 21.7 \frac{\text{lbm}}{\text{min}} - 31.38 \frac{\text{lbm}}{\text{min}}$$
$$= -9.68 \text{ lbm/min}$$

The time the system can run is

$$\frac{m_{depleted}}{\dot{m}_{net}} = \frac{1031.6 \text{ lbm}}{9.68 \dfrac{\text{lbm}}{\text{min}}} = \boxed{106.6 \text{ min} \quad (1.78 \text{ hr})}$$

The answer is (B).

SI Solution

The absolute temperature of the compressor air is

$$32°C + 273 = 305K$$

Using ideal gas laws, the mass flow rate of air into the compressor is

$$\dot{m} = \frac{p\dot{V}}{RT}$$
$$= \frac{(101.3 \text{ kPa})\left(1000 \dfrac{\text{Pa}}{\text{kPa}}\right)\left(150 \dfrac{\text{L}}{\text{s}}\right)\left(\dfrac{1 \text{ m}^3}{1000 \text{ L}}\right)}{\left(287 \dfrac{\text{J}}{\text{kg·K}}\right)(305K)}$$
$$= 0.1736 \text{ kg/s}$$

The absolute temperature and pressure of stored compressed air are

$$T = 32°C + 273 = 305K$$
$$p = (2.07 \text{ MPa})\left(10^6 \dfrac{\text{Pa}}{\text{MPa}}\right)$$
$$= 2.07 \times 10^6 \text{ Pa}$$

The mass of stored compressed air in a 1000 ft^3 tank is

$$m_{tank} = \frac{pV}{RT}$$
$$= \frac{(2.07 \times 10^6 \text{ Pa})(27 \text{ m}^3)}{\left(287 \dfrac{\text{J}}{\text{kg·K}}\right)(305K)}$$
$$= 638.5 \text{ kg}$$

Assuming that each tool operates at its minimum pressure, the mass leaving the system can be calculated as follows.

Tool 1:

The absolute temperature is $32°C + 273 = 305K$.

$$\dot{m}_{tool\ 1} = \frac{p\dot{V}}{RT}$$
$$= \frac{(620 \text{ kPa})\left(1000 \dfrac{\text{Pa}}{\text{kPa}}\right)\left(19 \dfrac{\text{L}}{\text{s}}\right)\left(\dfrac{1 \text{ m}^3}{1000 \text{ L}}\right)}{\left(287 \dfrac{\text{J}}{\text{kg·K}}\right)(305K)}$$
$$= 0.1346 \text{ kg/s}$$

Tool 2:

The absolute temperature is $29°C + 273 = 302K$.

$$\dot{m}_{tool\ 2} = \frac{p\dot{V}}{RT}$$
$$= \frac{(350 \text{ kPa})\left(1000 \dfrac{\text{Pa}}{\text{kPa}}\right)\left(7 \dfrac{\text{L}}{\text{s}}\right)\left(\dfrac{1 \text{ m}^3}{1000 \text{ L}}\right)}{\left(287 \dfrac{\text{J}}{\text{kg·K}}\right)(302K)}$$
$$= 0.02827 \text{ kg/s}$$

Tool 3:

$$\dot{m}_{tool\ 3} = 0.045 \text{ kg/s} \quad [\text{given}]$$

The total mass flow rate leaving the system is

$$\dot{m}_{total} = \dot{m}_{tool\ 1} + \dot{m}_{tool\ 2} + \dot{m}_{tool\ 3}$$
$$= 0.1346 \frac{\text{kg}}{\text{s}} + 0.02827 \frac{\text{kg}}{\text{s}} + 0.045 \frac{\text{kg}}{\text{s}}$$
$$= 0.2079 \text{ kg/s}$$

The critical pressure of 620 kPa is required for tool 1 operation. The mass in the tank when critical pressure is achieved is

$$m_{tank,critical} = \frac{pV}{RT}$$
$$= \frac{(620 \text{ kPa})\left(1000 \dfrac{\text{Pa}}{\text{kPa}}\right)(27 \text{ m}^3)}{\left(287 \dfrac{\text{J}}{\text{kg·K}}\right)(305K)}$$
$$= 191.2 \text{ kg}$$

The amount in the tank to be depleted is

$$m_{depleted} = m_{tank} - m_{tank,critical}$$
$$= 638.5 \text{ kg} - 191.2 \text{ kg} = 447.3 \text{ kg}$$

The net flow rate of air into the tank is

$$\dot{m}_{net} = \dot{m}_{in} - \dot{m}_{out} = 0.1736 \frac{\text{kg}}{\text{s}} - 0.2079 \frac{\text{kg}}{\text{s}}$$
$$= -0.0343 \text{ kg/s}$$

Power Cycles

The time the system can run is

$$\frac{m_{\text{depleted}}}{\dot{m}_{\text{net}}} = \frac{447.3 \text{ kg}}{0.0343 \, \dfrac{\text{kg}}{\text{s}}} = \boxed{13\,041 \text{ s} \quad (3.62 \text{ h})}$$

The answer is (B).

33 Refrigeration Cycles

PRACTICE PROBLEMS

Carnot Refrigeration Cycle

1. A heat pump operates on the Carnot cycle between 40°F and 700°F (4°C and 370°C). What is the coefficient of performance?

- (A) 1.5
- (B) 1.8
- (C) 2.2
- (D) 2.7

2. A heat pump using refrigerant R-12 operates on the Carnot cycle. The refrigerant evaporates and is compressed at 35.7 psia and 172.4 psia (246 kPa and 1190 kPa), respectively. What is the coefficient of performance?

- (A) 3.2
- (B) 3.9
- (C) 4.3
- (D) 5.8

3. A refrigerator uses refrigerant R-12. The input power is 585 W. Heat absorbed from the cooled space is 450 Btu/hr (0.13 kW). What is the coefficient of performance?

- (A) 0.2
- (B) 0.4
- (C) 0.7
- (D) 0.9

4. Ammonia is used in a reversed Carnot cycle refrigerator with reservoirs at 110°F (45°C) and 10°F (−10°C). 1000 Btu/hr (1000 kJ/h) are to be removed.

(a) Find the coefficient of performance.

- (A) 2.3
- (B) 2.9
- (C) 4.7
- (D) 5.2

(b) Find the work input.

- (A) 210 Btu/hr (210 kJ/h)
- (B) 340 Btu/hr (340 kJ/h)
- (C) 400 Btu/hr (400 kJ/h)
- (D) 630 Btu/hr (630 kJ/h)

(c) Find the rejected heat.

- (A) 1000 Btu/hr (1000 kJ/h)
- (B) 1200 Btu/hr (1200 kJ/h)
- (C) 1400 Btu/hr (1400 kJ/h)
- (D) 1600 Btu/hr (1600 kJ/h)

5. A refrigerator cools a continuous aqueous solution (c_p = 1 Btu/lbm-°F; 4.19 kJ/kg·°C) flow of 100 gal/min (0.4 m³/min) from 80°F (25°C) to 20°F (−5°C) in an 80°F (25°C) environment. What is the minimum power requirement?

- (A) 82 hp (63 kW)
- (B) 100 hp (74 kW)
- (C) 130 hp (90 kW)
- (D) 150 hp (94 kW)

Vapor Compression Cycle

6. An ammonia compressor is used in a heat pump cycle. Suction pressure is 30 psia (200 kPa). Discharge pressure is 160 psia (1.0 MPa). Saturated liquid ammonia enters the throttle valve. The refrigeration effect is 500 Btu/lbm (1200 kJ/kg) ammonia. Find the coefficient of performance as a heat pump.

- (A) 2.1
- (B) 4.5
- (C) 5.3
- (D) 5.9

7. A refrigeration cycle uses Freon-12 as a refrigerant between a 70°F (20°C) environment and a −30°F (−30°C) heat source. If the vapor leaving the evaporator and liquid leaving the condenser are both saturated, find the volume flow of refrigerant leaving the evaporator per ton (kW) of refrigeration. Assume isentropic compression.

- (A) 9.8 ft³/min-ton (0.00081 m³/min-kW)
- (B) 12 ft³/min-ton (0.00099 m³/min-kW)
- (C) 25 ft³/min-ton (0.0021 m³/min-kW)
- (D) 44 ft³/min-ton (0.0036 m³/min-kW)

Air Refrigeration Cycle

8. An air refrigeration cycle compresses air from 70°F (20°C) and 14.7 psia (101 kPa) to 60 psia (400 kPa) in a 70% efficient compressor. The air is cooled to 25°F (−4.0°C) in a constant pressure process before entering a turbine with isentropic efficiency of 0.80.

(a) What is the temperature of the air leaving the compressor?

- (A) 720°R (400K)
- (B) 790°R (440K)
- (C) 860°R (470K)
- (D) 900°R (490K)

(b) What is the temperature of the air leaving the turbine?

- (A) 360°R (200K)
- (B) 430°R (240K)
- (C) 490°R (270K)
- (D) 560°R (310K)

(c) What is the coefficient of performance of the cycle?

- (A) 0.7
- (B) 0.8
- (C) 0.9
- (D) 1.1

SOLUTIONS

1. *Customary U.S. Solution*

The coefficient of performance for a heat pump operating on the Carnot cycle is given by Eq. 33.9.

$$\text{COP}_{\text{heat pump}} = \frac{T_{\text{high}}}{T_{\text{high}} - T_{\text{low}}}$$

The absolute temperatures are

$$T_{\text{high}} = 700°F + 460 = 1160°R$$
$$T_{\text{low}} = 40°F + 460 = 500°R$$
$$\text{COP}_{\text{heat pump}} = \frac{1160°R}{1160°R - 500°R} = \boxed{1.76}$$

The answer is (B).

SI Solution

The coefficient of performance for a heat pump operating on the Carnot cycle is given by Eq. 33.9.

$$\text{COP}_{\text{heat pump}} = \frac{T_{\text{high}}}{T_{\text{high}} - T_{\text{low}}}$$

The absolute temperatures are

$$T_{\text{high}} = 370°C + 273 = 643K$$
$$T_{\text{low}} = 4°C + 273 = 277K$$
$$\text{COP}_{\text{heat pump}} = \frac{643K}{643K - 277K} = \boxed{1.76}$$

The answer is (B).

2. *Customary U.S. Solution*

The coefficient of performance for a heat pump operating on the Carnot cycle is given by Eq. 33.9.

$$\text{COP}_{\text{heat pump}} = \frac{T_{\text{high}}}{T_{\text{high}} - T_{\text{low}}}$$

The temperature T_{high} is the saturation temperature at 172.4 psia, and the temperature T_{low} is the saturation temperature at 35.7 psia. From App. 24.H,

$$T_{\text{high}} = 120.2°F + 460 = 580.2°R$$
$$T_{\text{low}} = 19.5°F + 460 = 479.5°R$$
$$\text{COP}_{\text{heat pump}} = \frac{580.2°R}{580.2°R - 479.5°R} = \boxed{5.76}$$

The answer is (D).

SI Solution

The coefficient of performance for a heat pump operating on the Carnot cycle is given by Eq. 33.9.

$$\text{COP}_{\text{heat pump}} = \frac{T_{\text{high}}}{T_{\text{high}} - T_{\text{low}}}$$

The temperature T_{high} is the saturation temperature at 1190 kPa, and the temperature T_{low} is the saturation temperature at 246 kPa. From App. 24.T,

$$T_{\text{high}} = 49°\text{C} + 273 = 322\text{K}$$

$$T_{\text{low}} = -6.8°\text{C} + 273 = 266.2\text{K}$$

$$\text{COP}_{\text{heat pump}} = \frac{322\text{K}}{322\text{K} - 266.2\text{K}} = \boxed{5.77}$$

The answer is (D).

3. *Customary U.S. Solution*

The coefficient of performance for a refrigerator is given by Eq. 33.1.

$$\text{COP}_{\text{refrigerator}} = \frac{Q_{\text{in}}}{W_{\text{in}}}$$

$$= \frac{450 \frac{\text{Btu}}{\text{hr}}}{(585\text{W})\left(3.4121 \frac{\frac{\text{Btu}}{\text{hr}}}{\text{W}}\right)}$$

$$= \boxed{0.225}$$

The answer is (A).

SI Solution

From Eq. 33.1, the coefficient of performance for a refrigerator is

$$\text{COP}_{\text{refrigerator}} = \frac{Q_{\text{in}}}{W_{\text{in}}}$$

$$= \frac{(0.13 \text{ kW})\left(1000 \frac{\text{W}}{\text{kW}}\right)}{585\text{W}}$$

$$= \boxed{0.222}$$

The answer is (A).

4. *Customary U.S. Solution*

(a) $$\text{COP} = \frac{T_{\text{low}}}{T_{\text{high}} - T_{\text{low}}} = \frac{460 + 10°\text{F}}{110°\text{F} - 10°\text{F}}$$

$$= \boxed{4.7}$$

The answer is (C).

(b) $$W_{\text{in}} = \frac{Q_{\text{in}}}{\text{COP}} = \frac{1000 \frac{\text{Btu}}{\text{hr}}}{4.7}$$

$$= \boxed{212.77 \text{ Btu/hr}}$$

The answer is (A).

(c) $$Q_{\text{out}} = Q_{\text{in}} + W_{\text{in}} = 1000 \frac{\text{Btu}}{\text{hr}} + 212.77 \frac{\text{Btu}}{\text{hr}}$$

$$= \boxed{1212.77 \text{ Btu/hr}}$$

The answer is (B).

SI Solution

(a) $$\text{COP} = \frac{T_{\text{low}}}{T_{\text{high}} - T_{\text{low}}}$$

$$= \frac{-10°\text{C} + 273.15}{45°\text{C} - (-10°\text{C})}$$

$$= \boxed{4.785}$$

The answer is (C).

(b) $$W_{\text{in}} = \frac{Q_{\text{in}}}{\text{COP}} = \frac{1000 \frac{\text{kJ}}{\text{h}}}{4.785}$$

$$= \boxed{208.99 \text{ kJ/h}}$$

The answer is (A).

(c) $$Q_{\text{out}} = Q_{\text{in}} + W_{\text{in}} = 1000 \frac{\text{kJ}}{\text{h}} + 208.99 \frac{\text{kJ}}{\text{h}}$$

$$= \boxed{1208.99 \text{ kJ/h}}$$

The answer is (B).

5. *Customary U.S. Solution*

The density of an aqueous solution is essentially the same as water.

$$\dot{m} = \dot{V}\rho = \left(100 \frac{\text{gal}}{\text{min}}\right)\left(0.1337 \frac{\text{ft}^3}{\text{gal}}\right)$$

$$\times \left(62.4 \frac{\text{lbm}}{\text{ft}^3}\right)$$

$$= 834.5 \text{ lbm/min}$$

Power Cycles

$$\dot{Q}_{in} = \dot{m}c_p \Delta T$$

$$= \left(834.5 \frac{\text{lbm}}{\text{min}}\right) \left(1 \frac{\text{Btu}}{\text{lbm -°F}}\right) (80°\text{F} - 20°\text{F})$$

$$= 50{,}070 \text{ Btu/min}$$

$$\text{COP} = \frac{T_{low}}{T_{high} - T_{low}} = \frac{460 + 20°\text{F}}{80°\text{F} - 20°\text{F}}$$

$$= 8$$

$$W_{in,hp} = \frac{4.715 \, Q_{in,tons}}{\text{COP}}$$

$$= \frac{(4.715) \left(\dfrac{50{,}070 \dfrac{\text{Btu}}{\text{min}}}{200 \dfrac{\text{Btu}}{\text{min-ton}}}\right)}{8}$$

$$= \boxed{147.5 \text{ hp}}$$

The answer is (D).

SI Solution

The density of an aqueous solution is essentially the same as water.

$$\dot{m} = \dot{V}\rho = \left(0.4 \frac{\text{m}^3}{\text{min}}\right) \left(1000 \frac{\text{kg}}{\text{m}^3}\right)$$

$$= 400 \text{ kg/min}$$

$$\dot{Q}_{in} = \dot{m}c_p \Delta T$$

$$= \left(400 \frac{\text{kg}}{\text{min}}\right) \left(4.19 \frac{\text{kJ}}{\text{kg·°C}}\right) (25°\text{C} - (-5°\text{C}))$$

$$= 50\,280 \text{ kJ/min}$$

$$\text{COP} = \frac{T_{low}}{T_{high} - T_{low}} = \frac{-5°\text{C} + 273.15}{25°\text{C} - (-5°\text{C})}$$

$$= 8.938$$

$$\dot{W}_{in} = \frac{\dot{Q}_{in}}{\text{COP}} = \left(\frac{50\,280 \dfrac{\text{kJ}}{\text{min}}}{8.938}\right) \left(\frac{\text{min}}{60 \text{ s}}\right)$$

$$= \boxed{93.76 \text{ kW}}$$

The answer is (D).

6. *Customary U.S. Solution*

At a: $p_a = 160$ psia

Interpolating between 140 psia and 170 psia,

$$\frac{h_a - 126 \dfrac{\text{Btu}}{\text{lbm}}}{139.3 \dfrac{\text{Btu}}{\text{lbm}} - 126 \dfrac{\text{Btu}}{\text{lbm}}} = \frac{160 \text{ psia} - 140 \text{ psia}}{170 \text{ psia} - 140 \text{ psia}}$$

$$h_a = 134.9 \text{ Btu/lbm}$$

At b: $h_b = h_a = 134.9$ Btu/lbm

At c: $h_c = h_b + q_{in} = 134.9 \dfrac{\text{Btu}}{\text{lbm}} + 500 \dfrac{\text{Btu}}{\text{lbm}}$

$$= 634.9 \text{ Btu/lbm}$$

From a superheated ammonia table at 30 psia,

$$s_c = 1.3845 \text{ Btu/lbm-°F}$$

At d: $s_d = s_c = 1.3845$ Btu/lbm-°F

$$p_d = p_a = 160 \text{ psia}$$

Interpolating between 250°F and 300°F,

$$\frac{h_d - 737.6 \dfrac{\text{Btu}}{\text{lbm}}}{1.3845 \dfrac{\text{Btu}}{\text{lbm-°F}} - 1.3675 \dfrac{\text{Btu}}{\text{lbm-°F}}}$$

$$= \frac{767.1 \dfrac{\text{Btu}}{\text{lbm}} - 737.6 \dfrac{\text{Btu}}{\text{lbm}}}{1.4076 \dfrac{\text{Btu}}{\text{lbm-°F}} - 1.3675 \dfrac{\text{Btu}}{\text{lbm-°F}}}$$

$$h_d = 750.1 \text{ Btu/lbm}$$

$$\text{COP} = \frac{q_{out}}{W_{in}} = \frac{h_c - h_b}{h_d - h_c} + 1$$

$$= \frac{634.9 \dfrac{\text{Btu}}{\text{lbm}} - 134.9 \dfrac{\text{Btu}}{\text{lbm}}}{750.1 \dfrac{\text{Btu}}{\text{lbm}} - 634.9 \dfrac{\text{Btu}}{\text{lbm}}} + 1$$

$$= \boxed{5.34}$$

The answer is (C).

SI Solution

At a: $p_a = 1.0$ MPa

From a saturated ammonia table,

$$h_a = 443.65 \text{ kJ/kg}$$

At b: $h_b = h_a = 443.65$ kJ/kg

At c: $h_c = h_b + q_{in} = 443.65 \dfrac{\text{kJ}}{\text{kg}} + 1200 \dfrac{\text{kJ}}{\text{kg}}$

$$= 1643.65 \text{ kJ/kg (superheated)}$$

$$p_c = 200 \text{ kPa}$$

Interpolating from a superheated ammonia table at 200 kPa (0.20 MPa),

$$s_c = 6.4960 \; \frac{\text{kJ}}{\text{kg·K}} + \left(\frac{1643.65 \; \frac{\text{kJ}}{\text{kg}} - 1625.18 \; \frac{\text{kJ}}{\text{kg}}}{1647.94 \; \frac{\text{kJ}}{\text{kg}} - 1625.18 \; \frac{\text{kJ}}{\text{kg}}} \right)$$

$$\times \left(6.5759 \; \frac{\text{kJ}}{\text{kg·K}} - 6.4960 \; \frac{\text{kJ}}{\text{kg·K}} \right)$$

$$= 6.5608 \; \text{kJ/kg·K}$$

$$\text{At d: } s_d = s_c = 6.5608 \; \text{kJ/kg·K}$$

$$p_d = p_a = 1.0 \; \text{MPa}$$

Interpolating from a superheated ammonia table at 1 MPa,

$$h_d = 1900.02 \; \frac{\text{kJ}}{\text{kg}} + \left(\frac{6.5608 \; \frac{\text{kJ}}{\text{kg·K}} - 6.5376 \; \frac{\text{kJ}}{\text{kg·K}}}{6.5965 \; \frac{\text{kJ}}{\text{kg·K}} - 6.5376 \; \frac{\text{kJ}}{\text{kg·K}}} \right)$$

$$\times \left(1924.43 \; \frac{\text{kJ}}{\text{kg}} - 1900.02 \; \frac{\text{kJ}}{\text{kg}} \right)$$

$$= 1909.63 \; \text{kJ/kg}$$

$$\text{COP} = \frac{q_{\text{out}}}{W_{\text{in}}} = \frac{h_c - h_b}{h_d - h_c} + 1$$

$$= \frac{1643.65 \; \frac{\text{kJ}}{\text{kg}} - 443.65 \; \frac{\text{kJ}}{\text{kg}}}{1909.63 \; \frac{\text{kJ}}{\text{kg}} - 1643.65 \; \frac{\text{kJ}}{\text{kg}}} + 1$$

$$= \boxed{4.512}$$

The answer is (B).

7. *Customary U.S. Solution*

$$\text{At a: } T_a = 70°\text{F from saturated Freon-12 table}$$
$$h_a = 23.9 \; \text{Btu/lbm}$$
$$\text{At b: } h_b = h_a = 23.9 \; \text{Btu/lbm}$$
$$\text{At c: } T_c = T_b = -30°\text{F}$$
$$h_c = 74.7 \; \text{Btu/lbm}$$
$$v_c = 3.088 \; \text{ft}^3/\text{lbm}$$

$$q_{\text{in}} = h_c - h_b = 74.7 \; \frac{\text{Btu}}{\text{lbm}} - 23.9 \; \frac{\text{Btu}}{\text{lbm}}$$
$$= 50.8 \; \text{Btu/lbm}$$

$$\dot{m} = \frac{200 \; \frac{\text{Btu}}{\text{min-ton}}}{50.8 \; \frac{\text{Btu}}{\text{lbm}}}$$
$$= 3.94 \; \text{lbm/min-ton}$$

$$\dot{V} = \dot{m} v_c = \left(3.94 \; \frac{\text{lbm}}{\text{min-ton}} \right) \left(3.088 \; \frac{\text{ft}^3}{\text{lbm}} \right)$$

$$= \boxed{12.16 \; \text{ft}^3/\text{min-ton}}$$

The answer is (B).

SI Solution

$$\text{At a: } T_a = 20°\text{C}$$
$$= 293\text{K}$$
From the saturated Freon-12 table,

$$h_a = 83.57 \; \text{kJ/kg}$$

$$\text{At b: } h_b = h_a = 83.57 \; \text{kJ/kg}$$

$$\text{At c: } T_c = T_b = -30°\text{C} = 243\text{K}$$

$$h_c \approx 243 \; \text{kJ/kg}$$

$$v_c \approx 0.1580 \; \text{m}^3/\text{kg}$$

$$q_{\text{in}} = h_c - h_b = 243 \; \frac{\text{kJ}}{\text{kg}} - 83.57 \; \frac{\text{kJ}}{\text{kg}}$$

$$= 159.43 \; \text{kJ/kg}$$

$$\dot{m} = \frac{Q}{q_{\text{in}}} = \frac{1 \; \text{kW}}{159.43 \; \frac{\text{kJ}}{\text{kg}}} = 0.006272 \; \text{kg/s}$$

$$\dot{V} = \dot{m} v_c = \left(0.006272 \; \frac{\text{kg}}{\text{s}} \right) \left(0.1580 \; \frac{\text{m}^3}{\text{kg}} \right)$$

$$= \boxed{0.000991 \; \text{m}^3/\text{s}}$$

The answer is (B).

8. *Customary U.S. Solution*

Assuming ideal gas,

$$\frac{T_d}{T_c} = \left(\frac{p_{\text{high}}}{p_{\text{low}}} \right)^{\frac{k-1}{k}}$$

For air, $k = 1.4$.

$$T_d = T_c \left(\frac{60 \; \text{psia}}{14.7 \; \text{psia}} \right)^{\frac{1.4-1}{1.4}}$$

$$= (460 + 70°\text{F})(1.495)$$

$$= 792.1°\text{R}$$

Power Cycles

(a) Temperature leaving compressor if process is not isentropic:

$$T'_d = T_c + \frac{T_d - T_c}{\eta_{\text{compressor}}}$$

$$= 530\,°\text{R} + \frac{792.1\,°\text{R} - 530\,°\text{R}}{0.7}$$

$$= \boxed{904.5\,°\text{R} \ (444.5\,°\text{F})}$$

The answer is (D).

(b)
$$\frac{T_a}{T_b} = \left(\frac{p_{\text{high}}}{p_{\text{low}}}\right)^{\frac{k-1}{k}}$$

$$T_b = \frac{T_a}{\left(\dfrac{60\ \text{psia}}{14.7\ \text{psia}}\right)^{\frac{1.4-1}{1.4}}} = \frac{460 + 25\,°\text{F}}{1.495}$$

$$= 324.4\,°\text{R}$$

Temperature leaving the turbine if process is not isentropic:

$$T'_b = T_a - \eta_{\text{turbine}}(T_a - T_b)$$

$$= 485\,°\text{R} - (0.80)(485\,°\text{R} - 324.4\,°\text{R})$$

$$= \boxed{356.5\,°\text{R} \ (-103.5\,°\text{F})}$$

The answer is (A).

(c) $\text{COP} = \dfrac{T_c - T'_b}{(T'_d - T_a) - (T_c - T'_b)}$

$$= \frac{530\,°\text{R} - 356.5\,°\text{R}}{(904.5\,°\text{R} - 485\,°\text{R}) - (530\,°\text{R} - 356.5\,°\text{R})}$$

$$= \boxed{0.705}$$

The answer is (A).

SI Solution

(a) Assuming air is an ideal gas with $k = 1.4$,

$$\frac{T_d}{T_c} = \left(\frac{p_{\text{high}}}{p_{\text{low}}}\right)^{\frac{k-1}{k}}$$

$$T_c = 20\text{K} + 273.15 = 293.15$$

$$T_d = T_c\left(\frac{p_{\text{high}}}{p_{\text{low}}}\right)^{\frac{k-1}{k}} = (293.15\text{K})\left(\frac{400\ \text{kPa}}{101\ \text{kPa}}\right)^{\frac{1.4-1}{1.4}}$$

$$= 434.4\text{K}$$

The temperature leaving the compressor is

$$T'_d = T_c + \frac{T_d - T_c}{\eta_{\text{compressor}}}$$

$$= 293.15\text{K} + \frac{434.4\text{K} - 293.15\text{K}}{0.7}$$

$$= \boxed{494.9\text{K} \ (221.8\,°\text{C})}$$

The answer is (D).

(b) $T_a = -4\,°\text{C} + 273.15 = 269.15\text{K}$

$$T_b = \frac{T_a}{\left(\dfrac{p_{\text{high}}}{p_{\text{low}}}\right)^{\frac{k-1}{k}}} = \frac{269.15\text{K}}{\left(\dfrac{400\ \text{kPa}}{101\ \text{kPa}}\right)^{\frac{1.4-1}{1.4}}}$$

$$= 181.6\text{K}$$

The temperature leaving the turbine is

$$T'_b = T_a - \eta_{\text{turbine}}(T_a - T_b)$$

$$= 269.15\text{K} - (0.80)(269.15\text{K} - 181.6\text{K})$$

$$= \boxed{199.1\text{K} \ (-74.0\,°\text{C})}$$

The answer is (A).

(c) $\text{COP} = \dfrac{T_c - T'_b}{(T'_d - T_a) - (T_c - T'_b)}$

$$= \frac{293.15\text{K} - 199.1\text{K}}{(494.9\text{K} - 269.15\text{K}) - (293.15\text{K} - 199.1\text{K})}$$

$$= \boxed{0.714}$$

The answer is (A).

34 Fundamental Heat Transfer

PRACTICE PROBLEMS

Thermal Conductivity

1. Experiments have shown that the thermal conductivity, k, of a particular material varies with temperature, T, according to the following relationship.

$$k_T = (0.030)(1 + 0.0015T)$$

What is the value of k that should be used for a transfer of heat through the material if the hot-side temperature is 350° (°F or °C) and the cold-side temperature is 150° (°F or °C)?

   (A) 0.041
   (B) 0.055
   (C) 0.13
   (D) 0.22

Conduction

2. What is the heat flow through insulating brick, 1.0 ft (30 cm) thick, in an oven wall with a thermal conductivity of 0.038 Btu-ft/hr-ft²-°F (0.066 W/m·K)? The thermal gradient is 350°F (195°C).

   (A) 9.4 Btu/hr-ft² (31 W/m²)
   (B) 13 Btu/hr-ft² (43 W/m²)
   (C) 19 Btu/hr-ft² (63 W/m²)
   (D) 31 Btu/hr-ft² (100 W/m²)

3. A composite wall is made up of 3.0 in (7.6 cm) of material A exposed to 1000°F (540°C), 5.0 in (13 cm) of material B, and 6.0 in (15 cm) of material C exposed to 200°F (90°C). The mean thermal conductivities for materials A, B, and C are 0.06, 0.5, and 0.8 Btu-ft/hr-ft²-°F (0.1, 0.9, and 1.4 W/m·K), respectively. What are the temperatures at the A-B and B-C material interfaces?

Unsteady Heat Flow

4. The heat supply of a large building is turned off at 5:00 P.M. when the interior temperature is 70°F (21°C). The outdoor temperature is constant at 40°F (4°C). The thermal capacity of the building and its contents is 100,000 Btu/°F (60 MJ/K), and the conductance is 6500 Btu/hr-°F (1.1 kW/K). What is the interior temperature at 1:00 A.M.?

   (A) 45°F (7.2°C)
   (B) 51°F (11°C)
   (C) 58°F (14°C)
   (D) 64°F (18°C)

5. Steel ball bearings varying in diameter from $^{1}/_{4}$ in to $1^{1}/_{2}$ in (6.35 to 38.1 mm) are quenched from 1800°F (980°C) in an oil bath that remains at 110°F (43°C). The ball bearings are removed when their internal (center) temperature reaches 250°F (120°C). The average film coefficient is 56 Btu/hr-ft²-°F (320 W/m²·K).

(a) What is the time constant for the cooling process?

(b) Derive a linear equation for the time the ball bearings remain in the oil bath as a function of diameter.

Internal Heat Generation

6. A 0.4 in (1.0 cm) diameter uranium dioxide fuel rod with a thermal conductivity of 1.1 Btu-ft/hr-ft²-°F (1.9 W/m·K) is clad with 0.020 in (0.5 mm) of stainless steel. The fuel rod generates 4×10^7 Btu/hr-ft³ (4.1×10^8 W/m³) internally. A coolant at 500°F (260°C) circulates around the clad rod. The outside film coefficient is 10,000 Btu/hr-ft²-°F (57 kW/m²·K). What is the temperature at the longitudinal centerline of the rod?

   (A) 2100°F (1150°C)
   (B) 2400°F (1320°C)
   (C) 2700°F (1480°C)
   (D) 3100°F (1650°C)

Finned Radiators

7. Two long pieces of 1/16 in (1.6 mm) copper wire are connected end-to-end with a hot soldering iron. The minimum melting temperature of the solder is 450°F (230°C). The surrounding air temperature is 80°F (27°C). The unit film coefficient is 3 Btu/hr-ft²-°F (17 W/m²·K). Disregard radiation losses. What is the minimum rate of heat application to keep the solder molten?

   (A) 4.3 Btu/hr (1.3 W)
   (B) 11 Btu/hr (3.3 W)
   (C) 20 Btu/hr (6.0 W)
   (D) 47 Btu/hr (14 W)

SOLUTIONS

1. Use the value of k at an average temperature of $(1/2)(T_1 + T_2)$.

$$T = \left(\tfrac{1}{2}\right)(150° + 350°) = 250°$$
$$k = (0.030)(1 + 0.0015T)$$
$$= (0.030)\big(1 + (0.0015)(250°)\big) = \boxed{0.04125}$$

The answer is (A).

2. *Customary U.S. Solution*

From Fourier's law of heat conduction (Eq. 34.16),

$$\frac{Q_{1-2}}{A} = \frac{k\Delta T}{L}$$
$$= \frac{\left(0.038 \dfrac{\text{Btu-ft}}{\text{hr-ft}^2\text{-}°\text{F}}\right)(350°\text{F})}{1.0 \text{ ft}}$$
$$= \boxed{13.3 \text{ Btu/hr-ft}^2}$$

The answer is (B).

SI Solution

From Fourier's law of heat conduction (Eq. 34.16),

$$\frac{Q_{1-2}}{A} = \frac{k\Delta T}{L}$$
$$= \frac{\left(0.066 \dfrac{\text{W}}{\text{m·K}}\right)(195\text{K})}{(30 \text{ cm})\left(\dfrac{1 \text{ m}}{100 \text{ cm}}\right)}$$
$$= \boxed{42.9 \text{ W/m}^2}$$

The answer is (B).

3.

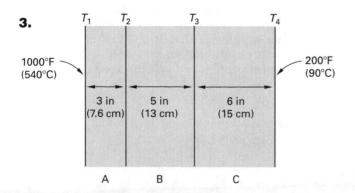

Customary U.S. Solution

Since the wall temperatures are given, it is not necessary to consider films.

From Eq. 34.21, the heat flow through the composite wall is

$$Q = \frac{A(T_1 - T_4)}{\sum\limits_{i=1}^{n} \dfrac{L_i}{k_i}}$$

On a per unit area basis,

$$\frac{Q}{A} = \frac{1000°\text{F} - 200°\text{F}}{\dfrac{(3 \text{ in})\left(\dfrac{1 \text{ ft}}{12 \text{ in}}\right)}{0.06 \dfrac{\text{Btu-ft}}{\text{hr-ft}^2\text{-}°\text{F}}} + \dfrac{(5 \text{ in})\left(\dfrac{1 \text{ ft}}{12 \text{ in}}\right)}{0.5 \dfrac{\text{Btu-ft}}{\text{hr-ft}^2\text{-}°\text{F}}} + \dfrac{(6 \text{ in})\left(\dfrac{1 \text{ ft}}{12 \text{ in}}\right)}{0.8 \dfrac{\text{Btu-ft}}{\text{hr-ft}^2\text{-}°\text{F}}}}$$

$$= 142.2 \text{ Btu/hr-ft}^2$$

$$= \frac{T_1 - T_4}{\sum\limits_{i=1}^{n} \dfrac{L_i}{k_i}} = \frac{T_1 - T_2}{\dfrac{L_A}{k_A}} = \frac{T_2 - T_3}{\dfrac{L_B}{k_B}}$$

To find the temperature at the A-B interface (T_2), use

$$\frac{Q}{A} = \frac{T_1 - T_2}{\dfrac{L_A}{k_A}}$$

$$T_2 = T_1 - \left(\frac{Q}{A}\right)\left(\frac{L_A}{k_A}\right)$$

$$= 1000°\text{F} - \frac{\left(142.2 \dfrac{\text{Btu}}{\text{hr-ft}^2}\right)(3 \text{ in})\left(\dfrac{1 \text{ ft}}{12 \text{ in}}\right)}{0.06 \dfrac{\text{Btu-ft}}{\text{hr-ft}^2\text{-}°\text{F}}}$$

$$= \boxed{407.5°\text{F}}$$

To find the temperature at the B-C interface (T_3), use

$$\frac{Q}{A} = \frac{T_2 - T_3}{\dfrac{L_B}{k_B}}$$

$$T_3 = T_2 - \left(\frac{Q}{A}\right)\left(\frac{L_B}{K_B}\right)$$

$$= 407.5°\text{F} - \left(142.2 \dfrac{\text{Btu}}{\text{hr-ft}^2}\right)\left(\frac{(5 \text{ in})\left(\dfrac{1 \text{ ft}}{12 \text{ in}}\right)}{0.5 \dfrac{\text{Btu-ft}}{\text{hr-ft}^2\text{-}°\text{F}}}\right)$$

$$= \boxed{289°\text{F}}$$

SI Solution

On a per unit area basis,

$$\frac{Q}{A} = \frac{540°C - 90°C}{\dfrac{(7.6\text{ cm})\left(\dfrac{1\text{ m}}{100\text{ cm}}\right)}{0.1\ \dfrac{W}{m\cdot K}} + \dfrac{(13\text{ cm})\left(\dfrac{1\text{ m}}{100\text{ cm}}\right)}{0.9\ \dfrac{W}{m\cdot K}} + \dfrac{(15\text{ cm})\left(\dfrac{1\text{ m}}{100\text{ cm}}\right)}{1.4\ \dfrac{W}{m\cdot K}}}$$

$$= 444.8\text{ W/m}^2$$

$$T_2 = T_1 - \left(\frac{Q}{A}\right)\left(\frac{L_A}{k_A}\right)$$

$$= 540°C - \left(444.8\ \frac{W}{m^2}\right)\left(\frac{(7.6\text{ cm})\left(\dfrac{1\text{ m}}{100\text{ cm}}\right)}{0.1\ \dfrac{W}{m\cdot K}}\right)$$

$$= \boxed{202.0°C}$$

$$T_3 = T_2 - \left(\frac{Q}{A}\right)\left(\frac{L_B}{k_B}\right)$$

$$= 202.0°C - \left(444.8\ \frac{W}{m^2}\right)\left(\frac{(13\text{ cm})\left(\dfrac{1\text{ m}}{100\text{ cm}}\right)}{0.9\ \dfrac{W}{m\cdot K}}\right)$$

$$= \boxed{137.8°C}$$

4. *Customary U.S. Solution*

This is a transient problem. The total time is from 5 P.M. to 1 A.M., which is 8 hr.

The thermal capacitance (capacity), C_e, is given as 100,000 Btu/°F.

The thermal resistance is

$$R_e = \frac{1}{\text{thermal conductance}}$$

$$= \frac{1}{6500\ \dfrac{\text{Btu}}{\text{hr-}°F}} = 0.0001538\ \text{hr-}°F/\text{Btu}$$

From Eq. 34.50,

$$T_t = T_\infty + (T_0 - T_\infty)e^{-\frac{t}{R_e C_e}}$$
$$T_{8\text{ hr}} = 40°F + (70°F - 40°F)$$

$$\times \exp\left(\frac{-8\text{ hr}}{\left(0.0001538\ \dfrac{\text{hr-}°F}{\text{Btu}}\right)\left(100{,}000\ \dfrac{\text{Btu}}{°F}\right)}\right)$$

$$= \boxed{57.8°F}$$

The answer is (C).

SI Solution

The thermal capacitance (capacity) is

$$C_e = \left(60\ \frac{MJ}{K}\right)\left(1000\ \frac{kJ}{MJ}\right)$$

$$= 60{,}000\ \text{kJ/K}\quad[\text{given}]$$

The thermal resistance is

$$R_e = \frac{1}{\text{thermal conductance}}$$

$$= \frac{1}{1.1\ \dfrac{kW}{K}} = 0.909\ \text{K/kW}$$

From Eq. 34.50,

$$T_t = T_\infty + (T_0 - T_\infty)e^{-\frac{t}{R_e C_e}}$$
$$= 4°C + (21°C - 4°C)$$

$$\times \exp\left(\frac{(-8\text{ h})\left(3600\ \dfrac{s}{h}\right)}{\left(0.909\ \dfrac{K}{kW}\right)\left(60\,000\ \dfrac{kJ}{K}\right)}\right)$$

$$= \boxed{14.0°C}$$

The answer is (C).

5. *Customary U.S. Solution*

(a) This is a transient problem. Check the Biot number to see if the lumped parameter method can used.

The characteristic site length from Eq. 34.8 is

$$L_c = \frac{V}{A_s} = \frac{\left(\dfrac{\pi}{6}\right)d^3}{\pi d^2} = \frac{d}{6}$$

For the largest ball, $d = 1.5$ in.

$$L_c = \frac{d}{6} = \frac{(1.5\text{ in})\left(\dfrac{1\text{ ft}}{12\text{ in}}\right)}{6} = 0.0208\text{ ft}$$

Evaluate the thermal conductivity, k, of steel at

$$\left(\tfrac{1}{2}\right)(1800°F + 250°F) = 1025°F$$

From App. 34.B, for steel, $k \approx 22.0$ Btu-ft/hr-ft^2-°F.
From Eq. 34.10, the Biot number is

$$\text{Bi} = \frac{hL_c}{k} = \frac{\left(56\ \dfrac{\text{Btu}}{\text{hr-ft}^2\text{-}°F}\right)(0.0208\text{ ft})}{22.0\ \dfrac{\text{Btu-ft}}{\text{hr-ft}^2\text{-}°F}}$$

$$= 0.053$$

For small balls, Bi will be even smaller.

Since Bi < 0.10, the lumped parameter method can be used.

The assumptions are

- homogeneous body temperature

- minimal radiation losses

- oil bath temperature remains constant

- h remains constant

From Eqs. 34.48 and 34.49, the time constant is

$$C_e R_e = c_p \rho V \left(\frac{1}{hA_s} \right) = \left(\frac{c_p \rho}{h} \right) \left(\frac{V}{A_s} \right)$$

$$= \left(\frac{c_p \rho}{h} \right) L_c = \left(\frac{c_p \rho}{h} \right) \left(\frac{d}{6} \right)$$

From App. 34.B, $\rho = 490$ lbm/ft^3 and $c_p = 0.11$ Btu/lbm-°F, even though those values are for 32°F.

The time constant is (measuring d in inches),

$$C_e R_e = \left(\frac{\left(0.11 \ \dfrac{\text{Btu}}{\text{lbm-°F}} \right) \left(490 \ \dfrac{\text{lbm}}{\text{ft}^3} \right)}{56 \ \dfrac{\text{Btu}}{\text{hr-ft}^2\text{-°F}}} \right)$$

$$\times \left(\frac{d}{6} \right) \left(\frac{1 \ \text{ft}}{12 \ \text{in}} \right)$$

$$= \boxed{0.01337d}$$

(b) Taking the natural log of the transient equation (Eq. 34.51),

$$\ln(T_t - T_\infty) = \ln \left(\Delta T e^{-\frac{t}{R_e C_e}} \right)$$

$$= \ln \Delta T + \ln \left(e^{-\frac{t}{R_e C_e}} \right)$$

$$= \ln \Delta T - \frac{t}{R_e C_e}$$

$$T_t = 250°\text{F}$$
$$T_\infty = 110°\text{F}$$
$$\Delta T = 1800°\text{F} - 110°\text{F} = 1690°\text{F}$$

$$\ln(250°\text{F} - 110°\text{F}) = \ln(1690°\text{F}) - \frac{t}{0.01337d}$$

$$4.942 = 7.432 - \frac{t}{0.01337d}$$

$$t = \boxed{0.0333d}$$

SI Solution

For the largest ball, the characteristic length is

$$L_c = \frac{d}{6} = \frac{(38.1 \ \text{mm}) \left(\dfrac{1 \ \text{m}}{1000 \ \text{mm}} \right)}{6} = 6.35 \times 10^{-3} \ \text{m}$$

Evaluate the thermal conductivity, k, of steel at

$$\left(\tfrac{1}{2} \right) (980°\text{C} + 120°\text{C}) = 550°\text{C}$$

From App. 34.B and its footnote,

$$k \approx (22.0) \left(\frac{1.7307 \ \text{W·hr·ft}^2\text{·°F}}{\text{m·K·Btu·ft}} \right)$$

$$= 38.08 \ \text{W/m·K}$$

From Eq. 34.10, the Biot number is

$$\text{Bi} = \frac{hL_c}{k} = \frac{\left(320 \ \dfrac{\text{W}}{\text{m}^2\text{·K}} \right) (6.35 \times 10^{-3} \ \text{m})}{38.08 \ \dfrac{\text{W}}{\text{m·K}}}$$

$$= 0.053$$

(a) For small balls, Bi will be even smaller.

Since Bi < 0.10, the lumped method can be used. The assumptions are given in the customary U.S. solution. From Eqs. 34.48 and 34.49, the time constant is

$$C_e R_e = c_p \rho V \left(\frac{1}{hA_s} \right) = \left(\frac{c_p \rho}{h} \right) \left(\frac{V}{A_s} \right)$$

$$= \left(\frac{c_p \rho}{h} \right) L_c = \left(\frac{c_p \rho}{h} \right) \left(\frac{d}{6} \right)$$

From App. 34.B and its footnote,

$$\rho = \left(490 \ \frac{\text{lbm}}{\text{ft}^3} \right) \left(16.0185 \ \frac{\text{kg·ft}^3}{\text{m}^3\text{·lbm}} \right)$$

$$= 7849.1 \ \text{kg/m}^3$$

$$c_p = (0.11) \left(4186.8 \ \frac{\text{J·lbm·°F}}{\text{kg·K·Btu}} \right)$$

$$= 460.5 \ \text{J/kg·K}$$

$$C_e R_e = \left(\frac{\left(460.5 \ \dfrac{\text{J}}{\text{kg·K}} \right) \left(7849.1 \ \dfrac{\text{kg}}{\text{m}^3} \right)}{\left(320 \ \dfrac{\text{W}}{\text{m}^2\text{·K}} \right)} \right) \left(\frac{d}{6} \right)$$

$$= \boxed{1882.6d}$$

(b) From the customary U.S. solution,

$$\ln(T_t - T_\infty) = \ln \Delta T - \frac{t}{R_e C_e}$$

$$T_t = 120°C$$

$$T_\infty = 43°C$$

$$\Delta T = 980°C - 43°C - 937°C$$

$$\ln(120°C - 43°C) = \ln(937°C) - \frac{t}{R_e C_e}$$

$$4.344 = 6.843 - \frac{t}{R_e C_e}$$

$$t = (2.499) R_e C_e$$

$$= (2.499)(1882.6 \text{ d})$$

$$= \boxed{4704.6 \text{ d}}$$

6. *Customary U.S. Solution*

The volume of the rod is

$$V = \frac{\pi}{4} d^2 L$$

$$\frac{V}{L} = \frac{\pi}{4} d^2 = \left(\frac{\pi}{4}\right)(0.4 \text{ in})^2 \left(\frac{1 \text{ ft}}{12 \text{ in}}\right)^2$$

$$= 8.727 \times 10^{-4} \text{ ft}^3/\text{ft}$$

The heat output per unit length of rod is

$$\frac{Q}{L} = \left(\frac{V}{L}\right) G$$

$$= \left(8.727 \times 10^{-4} \frac{\text{ft}^3}{\text{ft}}\right)\left(4 \times 10^7 \frac{\text{Btu}}{\text{hr-ft}^3}\right)$$

$$= 3.491 \times 10^4 \text{ Btu/hr-ft}$$

The diameter of the cladding is

$$d_o = 0.4 \text{ in} + (2)(0.020 \text{ in}) = 0.44 \text{ in}$$

The surface area per unit length of cladding is

$$A = \pi d_o = \pi(0.44 \text{ in})\left(\frac{1 \text{ ft}}{12 \text{ in}}\right) = 0.1152 \text{ ft}^2/\text{ft}$$

From Eq. 34.23, the surface temperature of the cladding is

$$T_s = \frac{Q}{hA} + T_\infty$$

$$= \frac{3.491 \times 10^4 \frac{\text{Btu}}{\text{hr-ft}}}{\left(10{,}000 \frac{\text{Btu}}{\text{hr-ft}^2\text{-}°F}\right)\left(0.1152 \frac{\text{ft}^2}{\text{ft}}\right)} + 500°F$$

$$= 530.3°F$$

For the cladding,

$$r_o = \frac{d_o}{2} = \left(\frac{0.44 \text{ in}}{2}\right)\left(\frac{1 \text{ ft}}{12 \text{ in}}\right) = 0.01833 \text{ ft}$$

$$r_i = \frac{d}{2} = \left(\frac{0.4 \text{ in}}{2}\right)\left(\frac{1 \text{ ft}}{12 \text{ in}}\right) = 0.01667 \text{ ft}$$

From App. 34.B, k for stainless steel (at 572°F) is 11 Btu-ft/hr-ft²-°F. This is reasonable because the inside cladding temperature is greater than the surface temperature (530.3°F).

For a cylinder from Eqs. 34.5 and 34.19,

$$T_{\text{inside}} - T_{\text{outside}} = \frac{Q \ln\left(\frac{r_o}{r_i}\right)}{2\pi k L} = \frac{\left(\frac{Q}{L}\right) \ln\left(\frac{r_o}{r_i}\right)}{2\pi k}$$

$$= \frac{\left(3.491 \times 10^4 \frac{\text{Btu}}{\text{hr-ft}}\right) \ln\left(\frac{0.01833 \text{ ft}}{0.01667 \text{ ft}}\right)}{2\pi \left(11 \frac{\text{Btu-ft}}{\text{hr-ft}^2\text{-}°F}\right)}$$

$$= 47.9°F$$

$$\underset{\text{cladding}}{T_{\text{inside}}} = \underset{\text{fuel rod}}{T_{\text{outside}}} + 47.9°F$$

$$= 530.3°F + 47.9°F = 578.2°F$$

From Eq. 34.56,

$$T_{\text{center}} = T_o + \frac{G r_o^2}{4k}$$

$$= 578.2°F + \frac{\left(4 \times 10^7 \frac{\text{Btu}}{\text{hr-ft}^3}\right)(0.01667 \text{ ft})^2}{(4)\left(1.1 \frac{\text{Btu-ft}}{\text{hr-ft}^2\text{-}°F}\right)}$$

$$= \boxed{3104°F}$$

The answer is (D).

SI Solution

$$\frac{V}{L} = \frac{\pi}{4} d = \left(\frac{\pi}{4}\right)(1.0 \text{ cm})^2 \left(\frac{1 \text{ m}}{100 \text{ cm}}\right)^2$$

$$= 7.854 \times 10^{-5} \text{ m}^3/\text{m}$$

The heat output per unit length of rod is

$$\frac{Q}{L} = \left(\frac{V}{L}\right) G = \left(7.854 \times 10^{-5} \frac{\text{m}^3}{\text{m}}\right)\left(4.1 \times 10^8 \frac{\text{W}}{\text{m}^3}\right)$$

$$= 32{,}201.4 \text{ W/m}$$

From Eq. 34.23, the surface temperature of the cladding is

$$T_s = \frac{Q}{hA} + T_\infty$$

The diameter of the cladding is

$$d_o = \left(1.0 \text{ cm} + (2)(0.5 \text{ mm})\left(\frac{1 \text{ cm}}{10 \text{ mm}}\right)\right)\left(\frac{1 \text{ m}}{100 \text{ cm}}\right)$$

$$= 0.011 \text{ m}$$

The surface area per unit length of cladding is

$$A = \pi d_o = \pi(0.011 \text{ m}) = 0.0346 \text{ m}^2/\text{m}$$

$$T_s = \frac{32\,201.4 \dfrac{\text{W}}{\text{m}}}{\left(57 \dfrac{\text{kW}}{\text{m}^2 \cdot \text{K}}\right)\left(1000 \dfrac{\text{W}}{\text{kW}}\right)\left(0.0346 \dfrac{\text{m}^2}{\text{m}}\right)} + 260°\text{C}$$

$$= 276.3°\text{C}$$

For the cladding,

$$r_o = \frac{d_o}{2} = \frac{0.011 \text{ m}}{2} = 0.0055 \text{ m}$$

$$r_i = \frac{d_i}{2} = \frac{0.01 \text{ m}}{2} = 0.0050 \text{ m}$$

From App. 34.B and the table's footnote, k for stainless steel at 300°C is

$$k \approx \left(11 \frac{\text{Btu-ft}}{\text{hr-ft}^2\text{-°F}}\right)\left(1.7307 \frac{\text{W}\cdot\text{hr}\cdot\text{ft}^2\cdot\text{°F}}{\text{m}\cdot\text{K}\cdot\text{Btu}\cdot\text{ft}}\right)$$

$$= 19.038 \text{ W/m}\cdot\text{K}$$

This is reasonable because the inside cladding temperature is greater than the surface temperature (276.3°C).

For a cylinder from Eqs. 34.5 and 34.19,

$$T_{\text{inside}} - T_{\text{outside}} = \frac{Q \ln\left(\dfrac{r_o}{r_i}\right)}{2\pi k L} = \left(\frac{Q}{L}\right)\left(\frac{\ln\left(\dfrac{r_o}{r_i}\right)}{2\pi k}\right)$$

$$= \left(32\,201 \frac{\text{W}}{\text{m}}\right)\left(\frac{\ln\left(\dfrac{0.0055 \text{ m}}{0.0050 \text{ m}}\right)}{2\pi\left(19.038 \dfrac{\text{W}}{\text{m}\cdot\text{K}}\right)}\right)$$

$$= 25.7°\text{C}$$

$$T_{\substack{\text{inside} \\ \text{cladding}}} = T_{\substack{\text{outside} \\ \text{fuel rod}}} + 25.7°\text{C}$$

$$= 276.3°\text{C} + 25.7°\text{C} = 302°\text{C}$$

From Eq. 34.56,

$$T_{\text{center}} = T_o + \frac{Gr_o^2}{4k}$$

$$= 302°\text{C} + \frac{\left(4.1 \times 10^8 \dfrac{\text{W}}{\text{m}^3}\right)(0.0050 \text{ m})^2}{(4)\left(1.9 \dfrac{\text{W}}{\text{m}\cdot\text{K}}\right)}$$

$$= \boxed{1651°\text{C}}$$

The answer is (D).

---

7. *Customary U.S. Solution*

Consider this an infinite cylindrical fin with

$$T_b = 450°\text{F}$$
$$T_\infty = 80°\text{F}$$
$$h = 3 \text{ Btu/hr-ft}^2\text{-°F}$$

From Eq. 34.66, the perimeter length is

$$P = \pi d = \pi\left(\frac{1}{16} \text{ in}\right)\left(\frac{1 \text{ ft}}{12 \text{ in}}\right) = 0.01636 \text{ ft}$$

From App. 34.B, k at 450°F is approximately 215 Btu-ft/hr-ft²-°F.

From Eq. 34.65, the cross-sectional area of the fin at its base is

$$A_b = \pi r^2 = \pi\left(\frac{d}{2}\right)^2 = \frac{\pi}{4}d^2$$

$$= \left(\frac{\pi}{4}\right)\left(\frac{1}{16} \text{ in}\right)^2\left(\frac{1 \text{ ft}}{12 \text{ in}}\right)^2$$

$$= 2.131 \times 10^{-5} \text{ ft}^2$$

From Eq. 34.64, with two fins joined at the middle,

$$Q = 2\sqrt{hPkA_b}(T_b - T_\infty)$$

$$= (2)\sqrt{\begin{aligned}&\left(3 \frac{\text{Btu}}{\text{hr-ft}^2\text{-°F}}\right)(0.01636 \text{ ft}) \\ &\times \left(215 \frac{\text{Btu-ft}}{\text{hr-ft}^2\text{-°F}}\right)(2.131 \times 10^{-5} \text{ ft}^2)\end{aligned}}$$

$$\times (450°\text{F} - 80°\text{F})$$

$$= \boxed{11.1 \text{ Btu/hr}} \quad \text{[This disregards radiation.]}$$

The answer is (B).

SI Solution

Consider this an infinite cylindrical fin with

$$T_b = 230°\text{C}$$
$$T_\infty = 27°\text{C}$$
$$h = 17 \text{ W/m}^2\cdot\text{K}$$

From Eq. 34.66, the perimeter length is

$$P = \pi d = \pi(1.6 \text{ mm})\left(\frac{1 \text{ m}}{1000 \text{ mm}}\right) = 5.027 \times 10^{-3} \text{ m}$$

From App. 34.B and the table footnote, k at 230°C is

$$k \approx \left(215 \frac{\text{Btu-ft}}{\text{hr-ft}^2\text{-°F}}\right)\left(1.7307 \frac{\text{W}\cdot\text{hr}\cdot\text{ft}^2\cdot\text{°F}}{\text{m}\cdot\text{K}\cdot\text{Btu}\cdot\text{ft}}\right)$$

$$= 372.1 \text{ W/m}\cdot\text{K}$$

From Eq. 34.65, the cross-sectional area of the fin at its base is

$$A_b = \pi r^2 = \pi \left(\frac{d}{2}\right)^2 = \frac{\pi}{4} d^2$$

$$= \left(\frac{\pi}{4}\right)(1.6 \text{ mm})^2 \left(\frac{1 \text{ m}}{1000 \text{ mm}}\right)^2$$

$$= 2.011 \times 10^{-6} \text{ m}^2$$

From Eq. 34.64, with two fins joined at the middle,

$$Q = 2\sqrt{hPkA_b}(T_b - T_\infty)$$

$$= (2)\sqrt{\begin{aligned}&\left(17 \frac{\text{W}}{\text{m}^2 \cdot \text{K}}\right)(5.027 \times 10^{-3} \text{ m})\\ &\times \left(372.1 \frac{\text{W}}{\text{m} \cdot \text{K}}\right)(2.011 \times 10^{-6} \text{ m}^2)\end{aligned}}$$

$$\times (230°\text{C} - 27°\text{C})$$

$$= \boxed{3.25 \text{ W}} \quad \text{[This disregards radiation.]}$$

The answer is (B).

35 Natural Convection, Evaporation, and Condensation

PRACTICE PROBLEMS

1. What is the density of 87% wet steam at 50 psia (350 kPa)?
- (A) 0.75 lbm/ft^3 (12 kg/m^3)
- (B) 0.89 lbm/ft^3 (14 kg/m^3)
- (C) 0.94 lbm/ft^3 (15 kg/m^3)
- (D) 1.07 lbm/ft^3 (17 kg/m^3)

2. What is the viscosity of 100°F (38°C) water in units of lbm/hr-ft (kg/s·m)?
- (A) 1.2 lbm/ft-hr (0.00052 kg/m·s)
- (B) 1.4 lbm/ft-hr (0.00060 kg/m·s)
- (C) 1.6 lbm/ft-hr (0.00068 kg/m·s)
- (D) 1.8 lbm/ft-hr (0.00077 kg/m·s)

3. A fluid in a tank is maintained at 85°F (29°C) by a copper tube carrying hot water. The water decreases in temperature from 190°F (88°C) to 160°F (71°C) as it flows through the tube. At what temperature should the fluid's film coefficient be evaluated?
- (A) 130°F (54°C)
- (B) 160°F (71°C)
- (C) 175°F (79°C)
- (D) 190°F (88°C)

4. A bare, horizontal conductor with circular cross section with an outside diameter of 0.6 in (1.5 cm) dissipates 8.0 W/ft (25 W/m). The conductor is cooled by free convection, and the surrounding air temperature is 60°F (15°C). Assume the film temperature is 100°F (38°C). What is the conductor's surface temperature?
- (A) 85°F (29°C)
- (B) 110°F (43°C)
- (C) 130°F (54°C)
- (D) 160°F (67°C)

SOLUTIONS

1. *Customary U.S. Solution*

If the steam is 87% wet, the quality is $x = 0.13$.

From App. 24.B at 50 psia,

$$v_f = 0.01727 \text{ ft}^3/\text{lbm}$$
$$v_g = 8.517 \text{ ft}^3/\text{lbm}$$

From Eq. 24.43, the specific volume of steam is

$$v = v_f + x v_{fg}$$
$$= 0.01727 \frac{\text{ft}^3}{\text{lbm}} + (0.13)\left(8.517 \frac{\text{ft}^3}{\text{lbm}} - 0.01727 \frac{\text{ft}^3}{\text{lbm}}\right)$$
$$= 1.122 \text{ ft}^3/\text{lbm}$$

The density is

$$\rho = \frac{1}{v} = \frac{1}{1.122 \dfrac{\text{ft}^3}{\text{lbm}}} = \boxed{0.891 \text{ lbm/ft}^3}$$

The answer is (B).

SI Solution

If the steam is 87% wet, the quality is $x = 0.13$.

From App. 24.O at 350 kPa,

$$v_f = 1.0786 \text{ cm}^3/\text{g}$$
$$v_g = 524.2 \text{ cm}^3/\text{g}$$

From Eq. 24.43, the specific volume of steam is

$$v = v_f + x v_{fg}$$
$$= 1.0786 \frac{\text{cm}^3}{\text{g}} + (0.13)\left(524.2 \frac{\text{cm}^3}{\text{g}}\right)$$
$$= 69.22 \text{ cm}^3/\text{g}$$

The density is

$$\rho = \frac{1}{v} = \left(\frac{1}{69.22 \dfrac{\text{cm}^3}{\text{g}}}\right)\left(100 \frac{\text{cm}}{\text{m}}\right)^3 \left(\frac{1 \text{ kg}}{1000 \text{ g}}\right)$$
$$= \boxed{14.45 \text{ kg/m}^3}$$

The answer is (B).

2. *Customary U.S. Solution*

From App. 35.A, the viscosity of water at 100°F is

$$\mu = \left(0.458 \times 10^{-3} \ \frac{\text{lbm}}{\text{ft-sec}}\right)\left(3600 \ \frac{\text{sec}}{\text{hr}}\right)$$

$$= \boxed{1.6488 \ \text{lbm/ft-hr}}$$

The answer is (C).

SI Solution

From App. 35.B, the viscosity of water at 38°C is

$$\mu = \boxed{0.682 \times 10^{-3} \ \text{kg/m·s}}$$

The answer is (C).

3. *Customary U.S. Solution*

The midpoint tube temperature is

$$T_s = \left(\tfrac{1}{2}\right)(190°\text{F} + 160°\text{F}) = 175°\text{F}$$
$$T_\infty = 85°\text{F} \quad \text{[given]}$$

From Eq. 35.11, h should be evaluated at $(1/2)(T_s + T_\infty)$.

The film temperature is

$$T_h = \left(\tfrac{1}{2}\right)(175°\text{F} + 85°\text{F}) = \boxed{130°\text{F}}$$

The answer is (A).

SI Solution

The midpoint tube temperature is

$$T_2 = \left(\tfrac{1}{2}\right)(88°\text{C} + 71°\text{C}) = 79.5°\text{C}$$
$$T_\infty = 29°\text{C} \quad \text{[given]}$$

From Eq. 35.11, h should be evaluated at $(1/2)(T_s + T_\infty)$.

The film temperature is

$$T_h = \left(\tfrac{1}{2}\right)(79.5°\text{C} + 29°\text{C}) = \boxed{54.3°\text{C}}$$

The answer is (A).

4. *Customary U.S. Solution*

The heat loss per unit length is

$$\frac{Q}{L} = \left(8 \ \frac{\text{W}}{\text{ft}}\right)\left(3.413 \ \frac{\text{Btu}}{\text{hr-W}}\right) = 27.3 \ \text{Btu/hr-ft}$$

From App. 35.C, the air properties are at 100°F film.

$$\text{Pr} = 0.72$$

$$\frac{g\beta\rho^2}{\mu^2} = 1.76 \times 10^6 \ \frac{1}{\text{ft}^3\text{-°F}}$$

From Eq. 35.4, the Grashof number is

$$\text{Gr} = L^3 \Delta T \left(\frac{g\beta\rho^2}{\mu^2}\right)$$

The characteristic length is the wire diameter.

$$L = (0.6 \ \text{in})\left(\frac{1 \ \text{ft}}{12 \ \text{in}}\right) = 0.05 \ \text{ft}$$

The temperature gradient is

$$\Delta T = T_s - T_\infty = T_{\text{wire}} - 60°\text{F}$$

T_{wire} is unknown, so assume $T_{\text{wire}} = 150°\text{F}$.

$$\text{Gr} = (0.05 \ \text{ft})^3(150°\text{F} - 60°\text{F})\left(1.76 \times 10^6 \ \frac{1}{\text{ft}^3\text{-°F}}\right)$$

$$= 19{,}800$$

$$\text{Pr Gr} = (0.72)(19{,}800) = 14{,}256$$

From Table 35.3,

$$h \approx (0.27)\left(\frac{T_{\text{wire}} - T_\infty}{d}\right)^{0.25}$$

$$= (0.27)\left(\frac{150°\text{F} - 60°\text{F}}{0.05 \ \text{ft}}\right)^{0.25}$$

$$= 1.76 \ \text{Btu/hr-ft}^2\text{-°F}$$

The heat transfer from the wire is given by Eq. 35.1.

$$Q = Aq = \pi dLh(T_{\text{wire}} - T_\infty)$$

$$\frac{Q}{L} = \pi dh(T_{\text{wire}} - T_\infty)$$

$$T_{\text{wire}} = \frac{\frac{Q}{L}}{\pi dh} + T_\infty$$

$$= \frac{27.3 \ \frac{\text{Btu}}{\text{hr-ft}}}{\pi(0.05 \ \text{ft})\left(1.76 \ \frac{\text{Btu}}{\text{hr-ft}^2\text{-°F}}\right)} + 60°\text{F}$$

$$= 158.7°\text{F}$$

Perform one more iteration with $T_{\text{wire}} = 158°\text{F}$.

$$\text{Gr} = (0.05)^3(158°\text{F} - 60°\text{F})\left(1.76 \times 10^6 \ \frac{1}{\text{ft}^3\text{-°F}}\right)$$

$$= 21{,}560$$

$$\text{Pr Gr} = (0.72)(21{,}560) = 15{,}523$$

From Table 35.3,

$$h = (0.27)\left(\frac{T_{\text{wire}} - T_\infty}{d}\right)^{0.25}$$

$$= (0.27)\left(\frac{158°F - 60°F}{0.05\text{ ft}}\right)^{0.25}$$

$$= 1.80\text{ Btu/hr-ft}^2\text{-}°F$$

$$T_{\text{wire}} = \frac{\frac{Q}{L}}{\pi dh} + T_\infty$$

$$= \frac{27.3\,\dfrac{\text{Btu}}{\text{hr-ft}}}{\pi(0.05\text{ ft})\left(1.80\,\dfrac{\text{Btu}}{\text{hr-ft}^2\text{-}°F}\right)} + 60°F$$

$$= \boxed{156.6°F}$$

There is no need to repeat iterations since the assumed temperature and the calculated temperature are about the same.

The answer is (D).

SI Solution

From App. 35.D, for 38°C film, the air properties are

$$\text{Pr} = 0.705$$

$$\frac{g\beta\rho^2}{\mu^2} = 1.12 \times 10^8\,\frac{1}{\text{K·m}^3}$$

From Eq. 35.4, the Grashof number is

$$\text{Gr} = L^3 \Delta T\left(\frac{g\beta\rho^2}{\mu^2}\right)$$

The characteristic length is the wire diameter.

$$L = (1.5\text{ cm})\left(\frac{1\text{ m}}{100\text{ cm}}\right) = 0.015\text{ m}$$

The temperature gradient is

$$\Delta T = T_s - T_\infty = T_{\text{wire}} - 15°C$$

$$\text{Gr} = (0.015\text{ m})^3(T_{\text{wire}} - 15°C)\left(1.12 \times 10^8\,\frac{1}{\text{K·m}^3}\right)$$

Assume $T_{\text{wire}} = 70°C$.

$$\text{Gr} = (0.015\text{ m})^3(70°C - 15°C)\left(1.12 \times 10^8\,\frac{1}{\text{K·m}^3}\right)$$

$$= 20\,790$$

$$\text{Pr}\,\text{Gr} = (0.705)(20\,790) = 14\,657$$

From Table 35.3,

$$h \approx (1.32)\left(\frac{T_{\text{wire}} - T_\infty}{d}\right)^{0.25}$$

$$= (1.32)\left(\frac{70°C - 15°C}{0.015\text{ m}}\right)^{0.25}$$

$$= 10.27\text{ W/m}^2\text{·K}$$

From the customary U.S. solution,

$$T_{\text{wire}} = \frac{\frac{Q}{L}}{\pi dh} + T_\infty$$

$$= \frac{25\,\dfrac{\text{W}}{\text{m}}}{\pi(0.015\text{ m})\left(10.27\,\dfrac{\text{W}}{\text{m}^2\text{·K}}\right)} + 15°C$$

$$= \boxed{66.7°C}$$

There is no need to perform another iteration since the assumed temperature and the calculated temperature are about the same.

The answer is (D).

36 Forced Convection and Heat Exchangers

PRACTICE PROBLEMS

Logarithmic Mean Temperature Difference

1. Water is heated from 55°F to 87°F (15°C to 30°C) by stack gases that are cooled from 350°F to 270°F (175°C to 130°C). What is the logarithmic mean temperature difference?

- (A) 190°F (105°C)
- (B) 210°F (117°C)
- (C) 235°F (130°C)
- (D) 270°F (150°C)

2. A fluid in a tank is maintained at 85°F (30°C) by an immersed hot water coil. The hot water enters at 190°F (90°C) and leaves at 160°F (70°C). What is the logarithmic mean temperature difference?

- (A) 89°F (49°C)
- (B) 96°F (53°C)
- (C) 111°F (62°C)
- (D) 127°F (71°C)

Film Temperature

3. What film temperature should be used for the heated fluid in Prob. 2?

- (A) 110°F (43°C)
- (B) 130°F (55°C)
- (C) 140°F (60°C)
- (D) 150°F (66°C)

4. If the fluid in Prob. 2 is gradually heated from 85°F (30°C) to 110°F (45°C), what initial film temperature should be used?

- (A) 110°F (43°C)
- (B) 130°F (55°C)
- (C) 140°F (60°C)
- (D) 150°F (66°C)

Reynolds Number

5. Light No. 10 oil is heated from 95°F to 105°F (35°C to 41°C) in a 0.6 in (1.52 cm) inside diameter tube. The average velocity of the oil is 2.0 ft/sec (0.6 m/s). The viscosity of the oil at 100°F (38°C) is 45 centistokes. What is the Reynolds number?

- (A) 200
- (B) 2500
- (C) 4600
- (D) 19,000

Convective Heat Transfer

6. A white, uninsulated, rectangular duct passes through a 50 ft (15 m) wide room. The duct is 18 in (45 cm) wide and 12 in (30 cm) high. The room and its contents are at 70°F (21°C). Air at 100°F (40°C) enters the duct flowing at 800 ft/min (4.0 m/s). Considering both radiation and convection, the film coefficient for the outside of the duct is 2.0 Btu/hr-ft^2-°F (11 W/m^2·K).

(a) What is the heat transfer to the room?

- (A) 4300 Btu/hr (1.4 kW)
- (B) 5800 Btu/hr (1.9 kW)
- (C) 7400 Btu/hr (2.4 kW)
- (D) 9100 Btu/hr (2.9 kW)

(b) What is the temperature of the air after it has traveled in the duct the full 50 ft (15 m)?

- (A) 85°F (29°C)
- (B) 88°F (31°C)
- (C) 91°F (33°C)
- (D) 94°F (36°C)

(c) What is the pressure drop in in (cm) of water due to friction in the duct?

- (A) 0.020 in wg (4.6 Pa)
- (B) 0.033 in wg (7.5 Pa)
- (C) 0.041 in wg (9.4 Pa)
- (D) 0.055 in wg (13 Pa)

Laminar Flow in Tubes

7. A steel pipe carrying 350°F (175°C) air is 100 ft (30 m) long. The outside and inside diameters are 4.00 in and 3.50 in (10 cm and 9.0 cm), respectively. The pipe is covered with 2.0 in (5.0 cm) of insulation with a thermal conductivity of 0.05 Btu-ft/hr-ft^2-°F (0.086 W/m·K). The pipe passes through a 50°F (10°C) basement. Flow is laminar and fully developed. What is the heat loss?

- (A) 3100 Btu/hr (0.93 kW)
- (B) 3500 Btu/hr (1.1 kW)
- (C) 4100 Btu/hr (1.2 kW)
- (D) 8700 Btu/hr (2.0 kW)

Turbulent Flow in Tubes

8. An uninsulated horizontal pipe with 4.00 in (10 cm) outside diameter carries saturated 300 psia (2.1 MPa) steam through a 70°F (21°C) room. The steam flow rate is 5000 lbm/hr (0.63 kg/s). The pipe emissivity is 0.80. What decrease in quality will occur in the first 50 ft (15 m) of length?

- (A) 1.6%
- (B) 2.2%
- (C) 2.8%
- (D) 4.3%

Heat Exchangers

9. A crossflow tubular feedwater heater is being designed to heat 2940 lbm/hr (0.368 kg/s) of water from 70°F (21°C) to 190°F (90°C). Saturated steam at 134 psia (923 kPa) is condensing on the outside of the tubes. The tubes are a copper alloy containing 70% Cu and 30% Ni. Each tube has a 1 in (2.54 cm) outside diameter and a 0.9 in (2.29 cm) inside diameter. The water velocity inside the tubes is 3 ft/sec (0.9 m/s). What outside tube surface area is required?

- (A) 3.0 ft^2 (0.30 m^2)
- (B) 4.0 ft^2 (0.39 m^2)
- (C) 10 ft^2 (0.97 m^2)
- (D) 18 ft^2 (1.7 m^2)

10. A U-tube surface feedwater heater with one shell pass and two tube passes is being designed to heat 500,000 lbm/hr (60 kg/s) of water from 200°F to 390°F (100°C to 200°C). The water flows at 5 ft/sec (1.5 m/s) through the tubes. Dry, saturated steam at 400°F (205°C) is to be used as the heating medium. The heater is to operate straight condensing (i.e., the condensed steam will not be mixed with the heated water). Saturated water at 400°F (205°C) is removed from the heater. The tubes in the heater are 7/8 in (2.2 cm) outside diameter with 1/16 in (1.6 mm) walls. The overall heat transfer coefficient is estimated as 700 Btu/hr-ft^2-°F.

(a) How many tubes are required?

- (A) 80
- (B) 110
- (C) 140
- (D) 170

(b) What should be the length of the tube bundle?

- (A) 30 ft (9 m)
- (B) 40 ft (12 m)
- (C) 60 ft (17 m)
- (D) 80 ft (24 m)

11. A single-pass heat exchanger is tested in a clean condition and is found to heat 100 gal/min (6.3 L/s) of 70°F (21°C) water to 140°F (60°C). The hot side uses 230°F (110°C) steam. The tube's inner surface area is 50 ft^2 (4.7 m^2). After being used in the field for several months, the exchanger heats 100 gal/min (6.3 L/s) of 70°F (21°C) water to 122°F (50°C). What is the fouling factor?

- (A) 0.00044 hr-ft^2-°F/Btu (0.000075 m^2·K/W)
- (B) 0.00081 hr-ft^2-°F/Btu (0.00014 m^2·K/W)
- (C) 0.0011 hr-ft^2-°F/Btu (0.00019 m^2·K/W)
- (D) 0.0023 hr-ft^2-°F/Btu (0.00039 m^2·K/W)

Tubes in Crossflow

12. A heat-conducting rod has an outside diameter of 0.35 in (8.9 mm). Its uniform temperature is 100°F (38°C). The rod is inserted perpendicularly into a 100 ft/sec (30 m/s) airflow. The air temperature is 150°F (66°C). What is the film coefficient on the outside of the rod?

- (A) 36 Btu/hr-ft^2-°F (210 W/m^2·K)
- (B) 45 Btu/hr-ft^2-°F (260 W/m^2·K)
- (C) 66 Btu/hr-ft^2-°F (380 W/m^2·K)
- (D) 91 Btu/hr-ft^2-°F (530 W/m^2·K)

SOLUTIONS

1. *Customary U.S. Solution*

The logarithmic mean temperature difference will be different for different types of flow.

Parallel flow:

55°F (15°C) ⟶ 87°F (30°C)

A   B

350°F (175°C) ⟶ 270°F (130°C)

$$\Delta T_A = 350°\text{F} - 55°\text{F} = 295°\text{F}$$
$$\Delta T_B = 270°\text{F} - 87°\text{F} = 183°\text{F}$$

From Eq. 36.67, the logarithmic mean temperature difference is

$$\Delta T_{lm} = \frac{\Delta T_A - \Delta T_B}{\ln\left(\frac{\Delta T_A}{\Delta T_B}\right)} = \frac{295°\text{F} - 183°\text{F}}{\ln\left(\frac{295°\text{F}}{183°\text{F}}\right)} = \boxed{234.6°\text{F}}$$

Counterflow:

55°F (15°C) ⟶ 87°F (30°C)

A   B

270°F (130°C) ⟵ 350°F (175°C)

$$\Delta T_A = 270°\text{F} - 55°\text{F} = 215°\text{F}$$
$$\Delta T_B = 350°\text{F} - 87°\text{F} = 263°\text{F}$$

From Eq. 36.67, the logarithmic mean temperature difference is

$$\Delta T_{lm} = \frac{215°\text{F} - 263°\text{F}}{\ln\left(\frac{215°\text{F}}{263°\text{F}}\right)} = \boxed{238.2°\text{F}}$$

The answer is (C).

SI Solution

Parallel flow:

$$\Delta T_A = 175°\text{C} - 15°\text{C} = 160°\text{C}$$
$$\Delta T_B = 130°\text{C} - 30°\text{C} = 100°\text{C}$$

From Eq. 36.67, the logarithmic mean temperature difference is

$$\Delta T_{lm} = \frac{160°\text{C} - 100°\text{C}}{\ln\left(\frac{160°\text{C}}{100°\text{C}}\right)} = \boxed{127.7°\text{C}}$$

Counterflow:

$$\Delta T_A = 130°\text{C} - 15°\text{C} = 115°\text{C}$$
$$\Delta T_B = 175°\text{C} - 30°\text{C} = 145°\text{C}$$

From Eq. 36.67, the logarithmic mean temperature difference is

$$\Delta T_{lm} = \frac{115°\text{C} - 145°\text{C}}{\ln\left(\frac{115°\text{C}}{145°\text{C}}\right)} = \boxed{129.4°\text{C}}$$

The answer is (C).

2. *Customary U.S. Solution*

$$\Delta T_A = 190°\text{F} - 85°\text{F} = 105°\text{F}$$
$$\Delta T_B = 160°\text{F} - 85°\text{F} = 75°\text{F}$$

From Eq. 36.67, the logarithmic mean temperature difference is

$$\Delta T_{lm} = \frac{\Delta T_A - \Delta T_B}{\ln\left(\frac{\Delta T_A}{\Delta T_B}\right)} = \frac{105°\text{F} - 75°\text{F}}{\ln\left(\frac{105°\text{F}}{75°\text{F}}\right)} = \boxed{89.2°\text{F}}$$

The answer is (A).

SI Solution

$$\Delta T_A = 90°\text{C} - 30°\text{C} = 60°\text{C}$$
$$\Delta T_B = 70°\text{C} - 30°\text{C} = 40°\text{C}$$

From Eq. 36.67, the logarithmic mean temperature difference is

$$\Delta T_{lm} = \frac{60°\text{C} - 40°\text{C}}{\ln\left(\frac{60°\text{C}}{40°\text{C}}\right)} = \boxed{49.3°\text{C}}$$

The answer is (A).

3. *Customary U.S. Solution*

$$T_{\text{tank}} = 85°\text{F}$$
$$T_{\text{coil}} = \left(\tfrac{1}{2}\right)(190°\text{F} + 160°\text{F}) = 175°\text{F}$$

The film temperature is

$$T_f = \left(\tfrac{1}{2}\right)(T_{\text{tank}} + T_{\text{coil}})$$
$$= \left(\tfrac{1}{2}\right)(85°\text{F} + 175°\text{F}) = \boxed{130°\text{F}}$$

The answer is (B).

SI Solution

$$T_{\text{tank}} = 30°C$$
$$T_{\text{coil}} = \left(\tfrac{1}{2}\right)(90°C + 70°C) = 80°C$$

The film temperature is

$$T_f = \left(\tfrac{1}{2}\right)(T_{\text{tank}} + T_{\text{coil}})$$
$$= \left(\tfrac{1}{2}\right)(30°C + 80°C) = \boxed{55°C}$$

The answer is (B).

4. *Customary U.S. Solution*

$$T_{\text{coil,initial}} = \left(\tfrac{1}{2}\right)(190°F + 160°F) = 175°F$$

The initial film temperature is

$$T_{f,\text{initial}} = \left(\tfrac{1}{2}\right)(T_{\text{tank,initial}} + T_{\text{coil,initial}})$$
$$= \left(\tfrac{1}{2}\right)(85°F + 175°F) = \boxed{130°F}$$

The answer is (B).

SI Solution

$$T_{\text{coil,initial}} = \left(\tfrac{1}{2}\right)(90°C + 70°C) = 80°C$$

The initial film temperature is

$$T_{f,\text{initial}} = \left(\tfrac{1}{2}\right)(T_{\text{tank,initial}} + T_{\text{coil,initial}})$$
$$= \left(\tfrac{1}{2}\right)(30°C + 80°C) = \boxed{55°C}$$

The answer is (B).

5. *Customary U.S. Solution*

The Reynolds number is

$$Re_d = \frac{vD}{\nu}$$
$$D = (0.6 \text{ in})\left(\frac{1 \text{ ft}}{12 \text{ in}}\right) = 0.05 \text{ ft}$$
$$\nu = (45 \text{ cS})\left(\frac{1 \text{ S}}{100 \text{ cS}}\right)\left(\frac{1 \text{ ft}^2}{929 \text{ sec-stoke}}\right)$$
$$= 4.84 \times 10^{-4} \text{ ft}^2/\text{sec}$$
$$v = 2 \text{ ft/sec}$$
$$Re = \frac{\left(2 \frac{\text{ft}}{\text{sec}}\right)(0.05 \text{ ft})}{4.84 \times 10^{-4} \frac{\text{ft}^2}{\text{sec}}} = \boxed{206.6}$$

The answer is (A).

SI Solution

The Reynolds number is

$$Re_d = \frac{vD}{\nu}$$
$$D = (1.52 \text{ cm})\left(\frac{1 \text{ m}}{100 \text{ cm}}\right) = 0.0152 \text{ m}$$
$$v = 0.6 \text{ m/s}$$
$$\nu = (45 \text{ cS})\left(1 \frac{\frac{\mu m^2}{s}}{cS \cdot s}\right)\left(\frac{1 \text{ m}^2}{10^6 \ \mu m^2}\right)$$
$$= 45 \times 10^{-6} \text{ m}^2/\text{s}$$
$$Re_d = \frac{\left(0.6 \frac{m}{s}\right)(0.0152 \text{ m})}{45 \times 10^{-6} \frac{m^2}{s}} = \boxed{202.7}$$

The answer is (A).

6. *Customary U.S. Solution*

(a) The exposed duct area is

$$A = (2W + 2H)L$$
$$= \big((2)(18 \text{ in}) + (2)(12 \text{ in})\big)\left(\frac{1 \text{ ft}}{12 \text{ in}}\right)(50 \text{ ft})$$
$$= 250 \text{ ft}^2$$

The duct is noncircular; therefore, the hydraulic diameter of the duct will be used. From Eq. 36.48, the hydraulic diameter is

$$d_H = (4)\left(\frac{\text{area in flow}}{\text{wetted perimeter}}\right) = (4)\left(\frac{WH}{(2)(W+H)}\right)$$
$$= (4)\left(\frac{(18 \text{ in})(12 \text{ in})}{(2)(18 \text{ in} + 12 \text{ in})}\right)\left(\frac{1 \text{ ft}}{12 \text{ in}}\right)$$
$$= 1.2 \text{ ft}$$

From App. 35.C for air at 100°F,

$$\nu = 18.0 \times 10^{-5} \text{ ft}^2/\text{sec}$$
$$\rho = 0.0710 \text{ lbm/ft}^3$$

The Reynolds number is

$$Re = \frac{vD}{\nu} = \frac{\left(800 \frac{\text{ft}}{\text{min}}\right)(1.2 \text{ ft})\left(\frac{1 \text{ min}}{60 \text{ sec}}\right)}{18.0 \times 10^{-5} \frac{\text{ft}^2}{\text{sec}}}$$
$$= 8.90 \times 10^4$$

This is a turbulent flow. From Eq. 36.39,

$$h_i \approx (0.00351 + 0.000001583 T_{°F}) \left(\frac{(G_{\text{lbm/hr-ft}^2})^{0.8}}{(d_{\text{ft}})^{0.2}} \right)$$

$$G = \rho \text{v} = \left(0.0710 \, \frac{\text{lbm}}{\text{ft}^3} \right) \left(800 \, \frac{\text{ft}}{\text{min}} \right) \left(60 \, \frac{\text{min}}{\text{hr}} \right)$$

$$= 3408.0 \, \text{lbm/hr-ft}^2$$

$$h_i = (0.00351 + (0.000001583)(100°F))$$

$$\times \left(\frac{\left(3408.0 \, \frac{\text{lbm}}{\text{hr-ft}^2} \right)^{0.8}}{(1.2 \, \text{ft})^{0.2}} \right)$$

$$= 2.37 \, \text{Btu/hr-ft}^2\text{-°F}$$

Disregarding the duct thermal resistance, the overall heat transfer coefficient (from Eq. 36.72) is

$$\frac{1}{U} \approx \frac{1}{h_i} + \frac{1}{h_o}$$

$$= \frac{1}{2.37 \, \frac{\text{Btu}}{\text{hr-ft}^2\text{-°F}}} + \frac{1}{2.0 \, \frac{\text{Btu}}{\text{hr-ft}^2\text{-°F}}}$$

$$= 0.922 \, \text{hr-ft}^2\text{-°F/Btu}$$

$$U = 1.08 \, \text{Btu/hr-ft}^2\text{-°F}$$

The heat transfer to the room is

$$Q = UA(T_{\text{ave}} - T_\infty) = UA \left(\left(\tfrac{1}{2} \right) (T_{\text{in}} + T_{\text{out}}) - T_\infty \right)$$

Since T_{out} is unknown, an iteration procedure may be required. Assume $T_{\text{out}} \approx 95°F$.

$$Q = \left(1.08 \, \frac{\text{Btu}}{\text{hr-ft}^2\text{-°F}} \right) (250 \, \text{ft}^2)$$

$$\times \left(\left(\tfrac{1}{2} \right) (100°F + 95°F) - 70°F \right)$$

$$= \boxed{7425 \, \text{Btu/hr}}$$

The answer is (C).

Notice that ΔT (not ΔT_{lm}) is used in accordance with standard conventions in the HVAC industry.

(b) Temperature T_{out} can be verified by using

$$Q = \dot{m} c_p (T_{\text{in}} - T_{\text{out}})$$

$$\dot{m} = GA_{\text{flow}} = GWH$$

$$= \left(3408.0 \, \frac{\text{lbm}}{\text{hr-ft}^2} \right) (18 \, \text{in})(12 \, \text{in})$$

$$\times \left(\frac{1 \, \text{ft}^2}{144 \, \text{in}^2} \right)$$

$$= 5112 \, \text{lbm/hr}$$

$$c_p = 0.240 \, \text{Btu/lbm-°F}$$

[from App. 35.C at 100°F]

$$7425 \, \frac{\text{Btu}}{\text{hr}} = \left(5112 \, \frac{\text{lbm}}{\text{hr}} \right) \left(0.240 \, \frac{\text{Btu}}{\text{lbm-°F}} \right)$$

$$\times (100°F - T_{\text{out}})$$

$$T_{\text{out}} = \boxed{94°F} \quad \text{[close enough]}$$

The answer is (D).

(c) Assume clean galvanized ductwork with $\epsilon = 0.0005$ ft and about 25 joints per 100 ft. From Eq. 20.33, the equivalent diameter of a rectangular duct is

$$D_e = \frac{(1.3)(\text{short side} \times \text{long side})^{\frac{5}{8}}}{(\text{short side} + \text{long side})^{\frac{1}{4}}}$$

$$= \frac{(1.3)\left((12 \, \text{in})(18 \, \text{in})\right)^{\frac{5}{8}}}{(12 \, \text{in} + 18 \, \text{in})^{\frac{1}{4}}} = 16 \, \text{in}$$

The flow rate is

$$\dot{V} = \text{v}A = \frac{\left(800 \, \frac{\text{ft}}{\text{min}} \right) (12 \, \text{in})(18 \, \text{in})}{144 \, \frac{\text{in}^2}{\text{ft}^2}}$$

$$= 1200 \, \text{cfm}$$

From Fig. 20.4, the friction loss is 0.066 in wg/100 ft. Therefore,

$$\Delta p = (0.066 \, \text{in wg}) \left(\frac{50 \, \text{ft}}{100 \, \text{ft}} \right) = \boxed{0.033 \, \text{in wg}}$$

The answer is (B).

(Notice that the flow rate and not the flow velocity must be used with D_e in Fig. 20.4.)

SI Solution

(a) The exposed duct area is

$$A = (2W + 2H)L$$

$$= ((2)(45 \, \text{cm}) + (2)(30 \, \text{cm})) \left(\frac{1 \, \text{m}}{100 \, \text{cm}} \right) (15 \, \text{m})$$

$$= 22.5 \, \text{m}^2$$

The duct is noncircular; therefore, the hydraulic diameter of the duct will be used. From Eq. 36.48, the hydraulic diameter is

$$d_H = (4) \left(\frac{\text{area in flow}}{\text{wetted perimeter}} \right)$$
$$= (4) \left(\frac{WH}{(2)(W+H)} \right)$$
$$= (4) \left(\frac{(45 \text{ cm})(30 \text{ cm})}{(2)(45 \text{ cm} + 30 \text{ cm})} \right) \left(\frac{1 \text{ m}}{100 \text{ cm}} \right)$$
$$= 0.36 \text{ m}$$

From App. 35.D, for air at $40°C$,

$$\mu = 1.91 \times 10^{-5} \text{ kg/m·s}$$
$$\rho = 1.130 \text{ kg/m}^3$$
$$c_p = 1.0051 \text{ kJ/kg·K}$$
$$k = 0.02718 \text{ W/m·K}$$

The Reynolds number is

$$\text{Re} = \frac{\rho v D}{\mu} = \frac{\left(1.130 \frac{\text{kg}}{\text{m}^3} \right) \left(4.0 \frac{\text{m}}{\text{s}} \right) (0.36 \text{ m})}{1.91 \times 10^{-5} \frac{\text{kg}}{\text{m·s}}}$$
$$= 8.52 \times 10^4$$

This is a turbulent flow. From Eq. 36.34, the Nusselt number is

$$\text{Nu} = (0.023)(\text{Re}^{0.8})$$

The film coefficient is

$$h = (0.023)(\text{Re}^{0.8}) \left(\frac{k}{d} \right)$$
$$= (0.023)(8.52 \times 10^4)^{0.8} \left(\frac{0.02718 \frac{\text{W}}{\text{m·K}}}{0.36 \text{ m}} \right)$$
$$= 15.28 \text{ W/m}^2\text{·K}$$

Disregarding the duct thermal resistance, the overall heat transfer coefficient from Eq. 36.72 is

$$\frac{1}{U} = \frac{1}{h_i} + \frac{1}{h_o}$$
$$= \frac{1}{11 \frac{\text{W}}{\text{m}^2\text{·K}}} + \frac{1}{15.28 \frac{\text{W}}{\text{m}^2\text{·K}}} = 0.1564 \text{ m}^2\text{·K/W}$$
$$U = 6.39 \text{ W/m}^2\text{·K}$$

The heat transfer to the room is

$$Q = UA(T_{\text{ave}} - T_\infty) = UA \left(\left(\tfrac{1}{2} \right) (T_{\text{in}} + T_{\text{out}}) - T_\infty \right)$$

Since T_{out} is unknown, an iterative procedure may be required. Assume $T_{\text{out}} = 36°C$.

$$Q = \left(6.39 \frac{\text{W}}{\text{m}^2\text{·K}} \right) (22.5 \text{ m}^2)$$
$$\times \left(\left(\tfrac{1}{2} \right) (40°C + 36°C) - 21°C \right)$$
$$= \boxed{2444.2 \text{ W}}$$

The answer is (C).

Notice that ΔT (not ΔT_{lm}) is used in accordance with standard conventions in the HVAC industry.

(b) Verify temperature T_{out}.

$$Q = \dot{m} c_p (T_{\text{in}} - T_{\text{out}})$$
$$\dot{m} = \rho A_{\text{flow}} v$$
$$= \left(1.130 \frac{\text{kg}}{\text{m}^3} \right) (45 \text{ cm})(30 \text{ cm}) \left(\frac{1 \text{ m}}{100 \text{ cm}} \right)^2$$
$$\times \left(4 \frac{\text{m}}{\text{s}} \right)$$
$$= 0.6102 \text{ kg/s}$$

$$2444.2 \text{ W} = \left(0.6102 \frac{\text{kg}}{\text{s}} \right) \left(1.0051 \frac{\text{kJ}}{\text{kg·K}} \right) \left(1000 \frac{\text{J}}{\text{kJ}} \right)$$
$$\times (40°C - T_{\text{out}})$$

$$T_{\text{out}} = \boxed{36°C} \quad \text{[same as assumed]}$$

The answer is (D).

(c) Assume clean galvanized ductwork with $\epsilon = 0.15$ mm and about 1 joint per meter. From Eq. 20.33, the equivalent diameter of a rectangular duct is

$$D_e = \frac{(1.3)(\text{short side} \times \text{long side})^{\frac{5}{8}}}{(\text{short side} + \text{long side})^{\frac{1}{4}}}$$
$$= \left(\frac{(1.3)((30 \text{ cm})(45 \text{ cm}))^{\frac{5}{8}}}{(30 \text{ cm} + 45 \text{ cm})^{\frac{1}{4}}} \right) \left(\frac{10 \text{ mm}}{1 \text{ cm}} \right)$$
$$= 400 \text{ mm}$$

The flow rate is

$$\dot{V} = vA = \left(4 \frac{\text{m}}{\text{s}} \right) (0.45 \text{ m})(0.30 \text{ m}) \left(1000 \frac{\text{L}}{\text{m}^3} \right)$$
$$= 540 \text{ L/s}$$

From Fig. 20.5, the friction loss is 0.5 Pa/m. For 15 m,

$$\Delta p = \left(0.5 \; \frac{\text{Pa}}{\text{m}}\right)(15 \text{ m}) = \boxed{7.5 \text{ Pa}}$$

The answer is (B).

(Notice that the flow rate and not the flow velocity must be used with D_e in Fig. 20.4.)

7. *Customary U.S. Solution*

Refer to Fig. 34.3. The corresponding radii are

$$r_a = \frac{d_i}{2} = \frac{(3.5 \text{ in})\left(\frac{1 \text{ ft}}{12 \text{ in}}\right)}{2} = 0.1458 \text{ ft}$$

$$r_b = \frac{d_o}{2} = \frac{(4 \text{ in})\left(\frac{1 \text{ ft}}{12 \text{ in}}\right)}{2} = 0.1667 \text{ ft}$$

$$r_c = r_b + t_{\text{insulation}} = 0.1667 \text{ ft} + (2 \text{ in})\left(\frac{1 \text{ ft}}{12 \text{ in}}\right)$$

$$= 0.3334 \text{ ft}$$

From App. 34.B, for steel at 350°F, $k_{\text{pipe}} \approx 25.6$ Btu-ft/hr-ft²-°F.

Initially assume a typical value of $h_c = 1.5$ Btu/hr-ft²-°F.

For fully developed laminar flow from Eq. 36.28, $\text{Nu}_d = 3.658$.

From App. 35.C, for air at 350°F,

$$k_{\text{air}} \approx 0.0203 \text{ Btu/hr-ft-°F}$$

$$\text{Nu}_d = \frac{h_a d_i}{k_{\text{air}}} = 3.658$$

$$h_a = \frac{(3.658)k_{\text{air}}}{d_i} = \frac{(3.658)\left(0.0203 \; \frac{\text{Btu}}{\text{hr-ft-°F}}\right)}{(3.5 \text{ in})\left(\frac{1 \text{ ft}}{12 \text{ in}}\right)}$$

$$= 0.255 \text{ Btu/hr-ft}^2\text{-°F}$$

Neglect thermal resistance between pipe and insulation. From Eq. 34.34, the heat transfer is

$$Q = \frac{2\pi L(T_i - T_\infty)}{\dfrac{1}{r_a h_a} + \dfrac{\ln\left(\dfrac{r_b}{r_a}\right)}{k_{\text{pipe}}} + \dfrac{\ln\left(\dfrac{r_c}{r_b}\right)}{k_{\text{insulation}}} + \dfrac{1}{r_c h_c}}$$

$$= \frac{2\pi(100 \text{ ft})(350\text{°F} - 50\text{°F})}{\dfrac{1}{(0.1458 \text{ ft})\left(0.255 \; \dfrac{\text{Btu}}{\text{hr-ft}^2\text{-°F}}\right)} + \dfrac{\ln\left(\dfrac{0.1667 \text{ ft}}{0.1458 \text{ ft}}\right)}{25.6 \; \dfrac{\text{Btu-ft}}{\text{hr-ft}^2\text{-°F}}}}$$

$$+ \frac{\ln\left(\dfrac{0.3334 \text{ ft}}{0.1667 \text{ ft}}\right)}{0.05 \; \dfrac{\text{Btu-ft}}{\text{hr-ft}^2\text{-°F}}} + \dfrac{1}{(0.3334 \text{ ft})\left(1.5 \; \dfrac{\text{Btu}}{\text{hr-ft}^2\text{-°F}}\right)}$$

$$= \frac{188{,}496 \text{ ft-°F}}{26.90 \; \dfrac{\text{hr-ft-°F}}{\text{Btu}} + 0.00523 \; \dfrac{\text{hr-ft-°F}}{\text{Btu}}}$$

$$+ 13.863 \; \dfrac{\text{hr-ft-°F}}{\text{Btu}} + 2.00 \; \dfrac{\text{hr-ft-°F}}{\text{Btu}}$$

$$= 4407 \text{ Btu/hr}$$

Using Eq. 34.34 to find T_2, use all resistances except the outer $(T_2 - T_\infty)$ resistance.

$$\left(\frac{4407 \; \dfrac{\text{Btu}}{\text{hr}}}{2\pi(100 \text{ ft})}\right)\left(26.9 \; \dfrac{\text{hr-ft-°F}}{\text{Btu}}\right)$$

$$+ 0.00523 \; \frac{\text{hr-ft-°F}}{\text{Btu}} + 13.863 \; \frac{\text{hr-ft-°F}}{\text{Btu}}$$

$$= 285.9\text{°F}$$

$$T_2 = T_i - 285.9\text{°F} = 350\text{°F} - 285.9\text{°F} = 64.1\text{°F}$$

To evaluate h_c, use film temperature.

$$T_{\text{film}} = \left(\tfrac{1}{2}\right)(T_2 + T_\infty) = \left(\tfrac{1}{2}\right)(64.1\text{°F} + 50\text{°F})$$

$$= 57\text{°F}$$

From App. 35.C air at 57°F,

$$\text{Pr} = 0.72$$

$$\frac{g\beta\rho^2}{\mu^2} = 2.645 \times 10^6 \; \frac{1}{\text{ft}^3\text{-°F}}$$

From Eq. 35.4, the Grashof number is

$$\text{Gr} = \frac{L^3 g\beta\rho^2(T_2 - T_\infty)}{\mu^2}$$

Heat Transfer

For pipe,

$$L = d_c = 2r_c = (2)(0.3334 \text{ ft}) = 0.6668 \text{ ft}$$

$$\text{Gr} = (0.6668 \text{ ft})^3 \left(2.645 \times 10^6 \; \frac{1}{\text{ft}^3\text{-}°\text{F}} \right)$$
$$\times (64.1°\text{F} - 50°\text{F})$$
$$= 1.1 \times 10^7$$
$$\text{Gr Pr} = (1.1 \times 10^7)(0.72)$$
$$= 7.9 \times 10^6$$

From Table 35.3,

$$h_c \approx (0.27) \left(\frac{T_2 - T_\infty}{d_c} \right)^{\frac{1}{4}}$$

$$= (0.27) \left(\frac{64.1°\text{F} - 50°\text{F}}{0.6668 \text{ ft}} \right)^{\frac{1}{4}}$$

$$= 0.579 \text{ Btu/hr-ft}^2\text{-}°\text{F}$$

At the second iteration, the heat transfer is

$$Q = \cfrac{188{,}496 \text{ ft-}°\text{F}}{26.90 \; \frac{\text{hr-ft-}°\text{F}}{\text{Btu}} + 0.00523 \; \frac{\text{hr-ft-}°\text{F}}{\text{Btu}}}$$
$$+ 13.863 \; \frac{\text{hr-ft-}°\text{F}}{\text{Btu}}$$
$$+ \cfrac{1}{(0.3334 \text{ ft}) \left(0.579 \; \frac{\text{Btu}}{\text{hr-ft-}°\text{F}} \right)}$$

$$= \boxed{4102 \text{ Btu/hr}}$$

The answer is (C).

Additional iterations will improve the accuracy further.

SI Solution

Refer to Fig. 34.3. The corresponding radii are

$$r_a = \frac{d_i}{2} = \frac{(9.0 \text{ cm}) \left(\dfrac{1 \text{ m}}{100 \text{ cm}} \right)}{2} = 0.045 \text{ m}$$

$$r_b = \frac{d_o}{2} = \frac{(10 \text{ cm}) \left(\dfrac{1 \text{ m}}{100 \text{ cm}} \right)}{2} = 0.050 \text{ m}$$

$$r_c = r_b + t_{\text{insulation}} = 0.050 \text{ m} + (5.0 \text{ cm}) \left(\frac{1 \text{ m}}{100 \text{ cm}} \right)$$

$$= 0.100 \text{ m}$$

From App. 34.B and the table footnote, for steel at 175°C (~347°F),

$$k_{\text{pipe}} \approx \left(25.6 \; \frac{\text{Btu}}{\text{hr-ft-}°\text{F}} \right) \left(1.7307 \; \frac{\text{W·hr·ft·}°\text{F}}{\text{m·K·Btu}} \right)$$

$$= 44.31 \text{ W/m·K}$$

Initially assume a typical value of $h_c = 3.5 \text{ W/m}^2\text{·K}$. For fully developed laminar flow from Eq. 36.28,

$$\text{Nu}_d = \frac{h_a d_i}{k_{\text{air}}} = 3.658$$

From App. 35.D for air at 175°C,

$$k_{\text{air}} \approx 0.03709 \text{ W/m·K}$$

$$h_a = \frac{3.658 k_{\text{air}}}{d_i}$$

$$= \frac{(3.658) \left(0.03709 \; \dfrac{\text{W}}{\text{m·K}} \right)}{(9.0 \text{ cm}) \left(\dfrac{1 \text{ m}}{100 \text{ cm}} \right)}$$

$$= 1.508 \text{ W/m}^2\text{·K}$$

Neglect thermal resistance between pipe and insulation. From Eq. 34.34, the heat transfer is

$$Q = \cfrac{2\pi L (T_i - T_\infty)}{\dfrac{1}{r_a h_a} + \dfrac{\ln \left(\dfrac{r_b}{r_a} \right)}{k_{\text{pipe}}} + \dfrac{\ln \left(\dfrac{r_c}{r_b} \right)}{k_{\text{insulation}}} + \dfrac{1}{r_c h_c}}$$

$$= \cfrac{2\pi (30 \text{ m})(175°\text{C} - 10°\text{C})}{\dfrac{1}{(0.045 \text{ m}) \left(1.508 \; \dfrac{\text{W}}{\text{m}^2\text{·K}} \right)} + \dfrac{\ln \left(\dfrac{0.050 \text{ m}}{0.045 \text{ m}} \right)}{44.31 \; \dfrac{\text{W}}{\text{m·K}}}}$$

$$+ \dfrac{\ln \left(\dfrac{0.10 \text{ m}}{0.050 \text{ m}} \right)}{0.086 \; \dfrac{\text{W}}{\text{m·K}}} + \dfrac{1}{(0.10 \text{ m}) \left(3.5 \; \dfrac{\text{W}}{\text{m}^2\text{·K}} \right)}$$

$$= \cfrac{31\,101.8 \text{ m·}°\text{C}}{14.74 \; \frac{\text{m·K}}{\text{W}} + 0.00238 \; \frac{\text{m·K}}{\text{W}}}$$
$$+ 8.06 \; \frac{\text{m·K}}{\text{W}} + 2.86 \; \frac{\text{m·K}}{\text{W}}$$

$$= 1212 \text{ W}$$

Use Eq. 34.34 to find T_2 by using all resistances except the outer $(T_2 - T_\infty)$ resistance.

$$T_i - T_2 = \left(\frac{1212 \text{ W}}{2\pi (30 \text{ m})} \right)$$

$$\times \left(14.74 \; \frac{\text{m·K}}{\text{W}} + 0.00238 \; \frac{\text{m·K}}{\text{W}} + 8.06 \; \frac{\text{m·K}}{\text{W}} \right)$$

$$= 146.6°\text{C}$$

$$T_2 = T_i - 146.6°\text{C} = 175°\text{C} - 146.6°\text{C}$$

$$= 28.4°\text{C}$$

To evaluate h_c, use film temperature.

$$T_{\text{film}} = \left(\tfrac{1}{2}\right)(T_2 - T_\infty) = \left(\tfrac{1}{2}\right)(28.4°C + 10°C)$$
$$= 19.2°C$$

From App. 35.D for air at 19.2°C,

$$\text{Pr} = 0.710$$

$$\frac{g\beta\rho^2}{\mu^2} = 1.52 \times 10^8 \ \frac{1}{\text{K·m}^3}$$

From Eq. 35.4, the Grashof number is

$$\text{Gr} = \frac{L^3 g\beta\rho^2(T_2 - T_\infty)}{\mu^2}$$

For pipe,

$$L = d_c = 2r_c = (2)(0.10 \text{ m}) = 0.20 \text{ m}$$
$$\text{Gr} = (0.20 \text{ m})^3 \left(1.52 \times 10^8 \ \frac{1}{\text{K·m}^3}\right)(28.4°C - 10°C)$$
$$= 2.24 \times 10^7$$
$$\text{Gr Pr} = (2.24 \times 10^7)(0.710)$$
$$= 1.59 \times 10^7$$

From Table 35.3,

$$h_c \approx (1.37)\left(\frac{T_2 - T_\infty}{d_c}\right)^{\frac{1}{4}}$$

$$= (1.37)\left(\frac{28.4°C - 10°C}{0.20 \text{ m}}\right)^{\frac{1}{4}}$$

$$= 4.24 \text{ W/m}^2\text{·K} \quad \left[\begin{array}{c}\text{versus assumed value of} \\ 3.5 \text{ W/m}^2\text{·K}\end{array}\right]$$

Further iteration will improve accuracy.

$$Q = \cfrac{31\,101.8 \text{ m·°C}}{14.74 \ \dfrac{\text{m·K}}{\text{W}} + 0.00238 \ \dfrac{\text{m·K}}{\text{W}}}$$
$$+ 8.06 \ \dfrac{\text{m·K}}{\text{W}} + \cfrac{1}{(0.10 \text{ m})\left(4.24 \ \dfrac{\text{W}}{\text{m}^2\text{·K}}\right)}$$

$$= \boxed{1236 \text{ W}}$$

The answer is (C).

8. *Customary U.S. Solution*

Neglecting pipe resistance (no information for pipe is given), $T_{\text{pipe}} = T_{\text{sat}}$.

From App. 24.B for 300 lbf/in² steam, $T_{\text{sat}} = 417.35°F$. When a vapor condenses, the vapor and condensed liquid are at the same temperature. Therefore, the entire

pipe is assumed to be at 417.35°F. The outside film coefficient should be evaluated from Eq. 36.11.

$$T_{\text{film}} = \left(\tfrac{1}{2}\right)(T_s + T_\infty)$$
$$= \left(\tfrac{1}{2}\right)(417.35°F + 70°F)$$
$$= 243.7°F$$

From App. 35.C for air at 243.7°F,

$$\text{Pr} = 0.715$$

$$\frac{g\beta\rho^2}{\mu^2} = 0.673 \times 10^6 \ \frac{1}{\text{ft}^3\text{-°F}}$$

From Eq. 35.4, the Grashof number is

$$\text{Gr} = \frac{L^3 g\beta\rho^2(T_s - T_\infty)}{\mu^2}$$

$$L = d_{\text{outside}} = (4 \text{ in})\left(\frac{1 \text{ ft}}{12 \text{ in}}\right) = 0.3333 \text{ ft}$$

$$\text{Gr} = (0.3333 \text{ ft})^3 \left(0.673 \times 10^6 \ \frac{1}{\text{ft}^3\text{-°F}}\right)$$
$$\times (417.35°F - 70°F)$$
$$= 8.66 \times 10^6$$
$$\text{Gr Pr} = (8.66 \times 10^6)(0.715)$$
$$= 6.19 \times 10^6$$

From Table 35.3,

$$h_c \approx (0.27)\left(\frac{T_s - T_\infty}{d_{\text{outside}}}\right)^{\frac{1}{4}}$$

$$= (0.27)\left(\frac{417.35°F - 70°F}{0.3333 \text{ ft}}\right)^{\frac{1}{4}}$$

$$= 1.53 \text{ Btu/hr-ft}^2\text{-°F}$$

From Eq. 35.1, the heat transfer for the first 50 ft due to convection is

$$Q = h_c A_{\text{outside}}(T_s - T_\infty)$$
$$= h_c(\pi d_{\text{outside}} L)(T_s - T_\infty)$$
$$= \left(1.53 \ \frac{\text{Btu}}{\text{hr-ft}^2\text{-°F}}\right)\pi(0.3333 \text{ ft})(50 \text{ ft})$$
$$\times (417.35°F - 70°F)$$
$$Q_{\text{convection}} = 27{,}824 \text{ Btu/hr}$$

To determine heat transfer due to radiation, assume oxidized steel pipe, completely enclosed.

$$F_a = 1$$

The absolute temperatures are

$$T_1 = 417.35°F + 460 = 877.35°R$$
$$T_2 = 70°F + 460 = 530°R$$
$$F_e = \epsilon_{\text{pipe}} = 0.80$$

The radiation heat transfer is

$$E_{net} = \sigma F_a F_e \left(T_1^4 - T_2^4 \right)$$
$$= \left(0.1713 \times 10^{-8} \ \frac{\text{Btu}}{\text{hr-ft}^2\text{-}^\circ\text{R}^4} \right)$$
$$\times (1)(0.80) \left((877.35^\circ\text{R})^4 - (530^\circ\text{R})^4 \right)$$
$$= 704 \ \text{Btu/hr-ft}^2$$
$$Q_{radiation} = E_{net} A = E_{net} (\pi d_{outside} L)$$
$$= \left(704 \ \frac{\text{Btu}}{\text{hr-ft}^2} \right) \pi (0.3333 \ \text{ft})(50 \ \text{ft})$$
$$= 36,858 \ \text{Btu/hr}$$

The total heat loss is

$$Q_{total} = Q_{convection} + Q_{radiation}$$
$$= 27,824 \ \frac{\text{Btu}}{\text{hr}} + 36,858 \ \frac{\text{Btu}}{\text{hr}}$$
$$= 64,682 \ \text{Btu/hr}$$
$$Q_{total} = \dot{m} \Delta h$$

The enthalpy decrease per pound is

$$\Delta h = \frac{Q_{total}}{\dot{m}_{steam}} = \frac{64,682 \ \frac{\text{Btu}}{\text{hr}}}{5000 \ \frac{\text{lbm}}{\text{hr}}} = 12.94 \ \text{Btu/lbm}$$

This is a quality loss of

$$\Delta x = \frac{\Delta h}{h_{fg}} = \frac{12.94 \ \frac{\text{Btu}}{\text{lbm}}}{809.4 \ \frac{\text{Btu}}{\text{lbm}}}$$
$$= \boxed{0.0160 \ (1.6\%)}$$

The answer is (A).

SI Solution

From the customary U.S. solution, T_{pipe} is the same for the entire length.

From App. 24.O for 2.1 MPa, $T_{sat} = 214.72^\circ\text{C}$. The outside film coefficient should be evaluated from Eq. 36.11 as

$$T_{film} = \left(\tfrac{1}{2} \right) (T_s + T_\infty) = \left(\tfrac{1}{2} \right) (214.72^\circ\text{C} + 21^\circ\text{C})$$
$$= 117.9^\circ\text{C}$$

From App. 35.D for air at 117.9°C,

$$\text{Pr} = 0.692$$
$$\frac{g\beta\rho^2}{\mu^2} = 0.403 \times 10^8 \ \frac{1}{\text{K·m}^3}$$

From Eq. 35.4, the Grashof number is

$$\text{Gr} = \frac{L^3 g\beta\rho^2 (T_s - T_\infty)}{\mu^2}$$
$$L = d_{outside} = (10 \ \text{cm}) \left(\frac{1 \ \text{m}}{100 \ \text{cm}} \right) = 0.10 \ \text{m}$$
$$\text{Gr} = (0.10 \ \text{m})^3 \left(0.403 \times 10^8 \ \frac{1}{\text{K·m}^3} \right)$$
$$\times (214.72^\circ\text{C} - 21^\circ\text{C})$$
$$= 7.807 \times 10^6$$
$$\text{Gr Pr} = (7.807 \times 10^6)(0.692)$$
$$= 5.40 \times 10^6$$

From Table 35.3,

$$h_c \approx (1.37) \left(\frac{T_s - T_\infty}{d_{outside}} \right)^{\frac{1}{4}}$$
$$= (1.37) \left(\frac{214.72^\circ\text{C} - 21^\circ\text{C}}{0.10 \ \text{m}} \right)^{\frac{1}{4}}$$
$$= 9.09 \ \text{W/m·K}$$

From Eq. 35.1, the heat transfer for the first 15 m due to convection is

$$Q_{convection} = h_c (\pi d_{outside} L)(T_s - T_\infty)$$
$$= \left(9.09 \ \frac{\text{W}}{\text{m·K}} \right) \pi (0.10 \ \text{m})(15 \ \text{m})$$
$$\times (214.72^\circ\text{C} - 21^\circ\text{C})$$
$$= 8298 \ \text{W}$$

To determine heat transfer due to radiation, assume oxidized steel pipe, completely enclosed.

$$F_a = 1$$

The absolute temperatures are

$$T_1 = 214.72^\circ\text{C} + 273 = 487.72\text{K}$$
$$T_2 = 21^\circ\text{C} + 273 = 294\text{K}$$
$$F_e = \epsilon_{pipe} = 0.80$$

The radiation heat transfer is

$$E_{net} = \sigma F_a F_e \left(T_1^4 - T_2^4 \right)$$
$$= \left(5.67 \times 10^{-8} \ \frac{\text{W}}{\text{m}^2\text{·K}^4} \right)$$
$$\times (1)(0.80) \left((487.72\text{K})^4 - (294\text{K})^4 \right)$$
$$= 2228 \ \text{W/m}^2$$
$$Q_{radiation} = E_{net} (\pi d_{outside} L)$$
$$= \left(2228 \ \frac{\text{W}}{\text{m}^2} \right) \pi (0.10 \ \text{m})(15 \ \text{m})$$
$$= 10\,499 \ \text{W}$$

The total heat loss is

$$Q_{\text{total}} = Q_{\text{convection}} + Q_{\text{radiation}}$$
$$= 8298 \text{ W} + 10\,499 \text{ W}$$
$$= 18\,797 \text{ W}$$
$$Q_{\text{total}} = \dot{m}\Delta h$$

The enthalpy decrease per kilogram is

$$\Delta h = \frac{Q_{\text{total}}}{\dot{m}_{\text{steam}}} = \frac{(18\,797 \text{ W})\left(\dfrac{1 \text{ kJ}}{1000 \text{ J}}\right)}{0.63 \dfrac{\text{kg}}{\text{s}}}$$
$$= 29.84 \text{ kJ/kg}$$

This is a quality loss of

$$\Delta x = \frac{\Delta h}{h_{fg}} = \frac{29.84 \dfrac{\text{kJ}}{\text{kg}}}{1879.8 \dfrac{\text{kJ}}{\text{kg}}}$$
$$= \boxed{0.0159 \quad (1.59\%)}$$

The answer is (A).

9. *Customary U.S. Solution*

The bulk temperature of the water is

$$T_b = \left(\tfrac{1}{2}\right)(T_{\text{in}} + T_{\text{out}})$$
$$= \left(\tfrac{1}{2}\right)(70°\text{F} + 190°\text{F}) = 130°\text{F}$$

From App. 35.A, the properties of water at 130°F are

$$c_p = 0.999 \text{ Btu/lbm-°F}$$
$$\nu = 0.582 \times 10^{-5} \text{ ft}^2/\text{sec}$$
$$\text{Pr} = 3.45$$
$$k = 0.376 \text{ Btu/hr-ft-°F}$$

The heat transfer is found from the temperature gain of the water.

$$Q = \dot{m}c_p\Delta T$$
$$= \left(2940 \frac{\text{lbm}}{\text{hr}}\right)\left(0.999 \frac{\text{Btu}}{\text{lbm °F}}\right)(190°\text{F} - 70°\text{F})$$
$$= 352{,}447 \text{ Btu/hr}$$

The Reynolds number is

$$\text{Re} = \frac{\text{v}D}{\nu} = \frac{\left(3 \dfrac{\text{ft}}{\text{sec}}\right)(0.9 \text{ in})\left(\dfrac{1 \text{ ft}}{12 \text{ in}}\right)}{0.582 \times 10^{-5} \dfrac{\text{ft}^2}{\text{sec}}}$$
$$= 3.87 \times 10^4$$

From Eq. 36.33, the film coefficient is

$$h = (0.023)(\text{Re})^{0.8}(\text{Pr})^n\left(\frac{k}{d}\right)$$
$$= (0.023)(3.87 \times 10^4)^{0.8}(3.45)^{0.4}$$
$$\times \left(\frac{0.376 \dfrac{\text{Btu}}{\text{hr-ft-°F}}}{(0.9 \text{ in})\left(\dfrac{1 \text{ ft}}{12 \text{ in}}\right)}\right)$$
$$= 885 \text{ Btu/hr-ft}^2\text{-°F}$$

The saturation temperature for 134 psia steam is ≈ 350°F. Assume the wall is 20°F lower (≈ 330°F). The film properties are evaluated at the average of the wall and saturation temperatures.

$$T_h = \left(\tfrac{1}{2}\right)(T_{\text{sat},v} + T_s) = \left(\tfrac{1}{2}\right)(350°\text{F} + 330°\text{F})$$
$$= 340°\text{F}$$

Film properties are obtained from App. 35.A for liquid water and App. 24.A for vapor.

$$k_{340°\text{F}} = 0.392 \text{ Btu/hr-ft-°F}$$
$$\mu_{340°\text{F}} = \left(0.109 \times 10^{-3} \frac{\text{lbm}}{\text{sec-ft}}\right)\left(3600 \frac{\text{sec}}{\text{hr}}\right)$$
$$= 0.392 \text{ lbm/ft-hr}$$
$$\rho_{l,340°\text{F}} = \frac{1}{v_{f,340°\text{F}}} = \frac{1}{0.01787 \dfrac{\text{ft}^3}{\text{lbm}}}$$
$$= 55.96 \text{ lbm/ft}^3$$
$$\rho_{v,340°\text{F}} = \frac{1}{v_{g,340°\text{F}}} = \frac{1}{3.788 \dfrac{\text{ft}^3}{\text{lbm}}}$$
$$= 0.2640 \text{ lbm/ft}^3$$
$$h_{fg,134 \text{ psia}} = 871.3 \text{ Btu/lbm}$$
$$d = (1.0 \text{ in})\left(\frac{1 \text{ ft}}{12 \text{ in}}\right) = 0.083 \text{ ft}$$
$$g = \left(32.2 \frac{\text{ft}}{\text{sec}^2}\right)\left(3600 \frac{\text{sec}}{\text{hr}}\right)^2$$
$$= 4.17 \times 10^8 \text{ ft/hr}^2$$

Heat Transfer

From Eq. 35.27, the film coefficient is

$$h_o = (0.725)\left(\frac{\rho_l(\rho_l - \rho_v)gh_{fg}(k_l)^3}{d\mu_l(T_{sat,v} - T_s)}\right)^{\frac{1}{4}}$$

$$= (0.725)$$

$$\times \left(\frac{\left(55.96\ \frac{lbm}{ft^3}\right)\left(55.96\ \frac{lbm}{ft^3} - 0.2640\ \frac{lbm}{ft^3}\right)}{(0.083\ ft)\left(0.392\ \frac{lbm}{ft\text{-}hr}\right)(350°F - 330°F)}\right)^{\frac{1}{4}}$$

$$= 2320\ Btu/hr\text{-}ft^2\text{-}°F$$

From Table 36.7, for copper alloy (70% Cu, 30% Ni), at 330°F, $k = 19.6$ Btu-ft/hr-ft²-°F.

From Eq. 36.69, the overall heat transfer coefficient based on the outside area is

$$\frac{1}{U_o} = \frac{1}{h_o} + \left(\frac{r_o}{k_{tube}}\right)\ln\left(\frac{r_o}{r_i}\right) + \frac{r_o}{r_i h_i}$$

$$r_o = \frac{d_o}{2} = \frac{(1\ in)\left(\frac{1\ ft}{12\ in}\right)}{2} = 0.0417\ ft$$

$$r_i = \frac{d_i}{2} = \frac{(0.9\ in)\left(\frac{1\ ft}{12\ in}\right)}{2} = 0.0375\ ft$$

$$U_o = \cfrac{1}{\cfrac{1}{2320\ \frac{Btu}{hr\text{-}ft^2\text{-}°F}} + \left(\cfrac{0.0417\ ft}{19.6\ \frac{Btu\text{-}ft}{hr\text{-}ft^2\text{-}°F}}\right)\ln\left(\cfrac{0.0417\ ft}{0.0375\ ft}\right) + \cfrac{0.0417\ ft}{(0.0375\ ft)\left(885\ \frac{Btu}{hr\text{-}ft^2\text{-}°F}\right)}}$$

$$= 523\ Btu/hr\text{-}ft^2\text{-}°F$$

For crossflow operation,

70°F ⟶ 190°F
A   B
350°F ⟵ 350°F

$$\Delta T_A = 350°F - 70°F = 280°F$$
$$\Delta T_B = 350°F - 190°F = 160°F$$

From Eq. 36.67, the logarithmic mean temperature difference is

$$\Delta T_{lm} = \frac{\Delta T_A - \Delta T_B}{\ln\left(\frac{\Delta T_A}{\Delta T_B}\right)} = \frac{280°F - 160°F}{\ln\left(\frac{280°F}{160°F}\right)} = 214.4°F$$

The heat transfer is known; therefore, the outside area can be calculated from Eq. 36.68.

$$Q = U_o A_o F_c \Delta T_{lm}$$

For steam condensation, the temperature of steam remains constant. Therefore, $F_c = 1$.

$$A_o = \frac{352{,}447\ \frac{Btu}{hr}}{\left(523\ \frac{Btu}{hr\text{-}ft^2\text{-}°F}\right)(1)(214.4°F)} = \boxed{3.14\ ft^2}$$

At this point the assumption that $T_{sat,v} - T_s = 20°F$ could be checked using $Q = U_{partial}A_o\Delta T_{partial}$, working from the inside (at 130°F) to the outside (at T_s).

The answer is (A).

SI Solution

The bulk temperature of the water is

$$T_b = \left(\tfrac{1}{2}\right)(T_{in} + T_{out})$$
$$= \left(\tfrac{1}{2}\right)(21°C + 90°C) = 55.5°C$$

From App. 35.B, the properties of water at 55.5°C are

$$c_p = 4.186\ kJ/kg\text{·}K$$
$$\rho = 986.6\ kg/m^3$$
$$\mu = 0.523 \times 10^{-3}\ kg/m\text{·}s$$
$$k = 0.6503\ W/m\text{·}K$$
$$Pr = 3.37$$

The heat transfer is found from the temperature gain of the water.

$$Q = \dot{m}c_p\Delta T$$
$$= \left(0.368\ \frac{kg}{s}\right)\left(4.186\ \frac{kJ}{kg\text{·}K}\right)\left(1000\ \frac{J}{kJ}\right)$$
$$\times (90°C - 21°C)$$
$$= 106\,291\ W$$

The Reynolds number is

$$\text{Re} = \frac{\rho v D}{\mu}$$

$$= \frac{\left(986.6 \ \frac{\text{kg}}{\text{m}^3}\right)\left(0.9 \ \frac{\text{m}}{\text{s}}\right)(2.29 \ \text{cm})\left(\frac{1 \ \text{m}}{100 \ \text{cm}}\right)}{0.523 \times 10^{-3} \ \frac{\text{kg}}{\text{m·s}}}$$

$$= 3.89 \times 10^4$$

From Eq. 36.33, the film coefficient is

$$h = (0.023)(\text{Re})^{0.8}(\text{Pr})^n \left(\frac{k}{d}\right)$$

$$= (0.023)(3.89 \times 10^4)^{0.8}(3.37)^{0.4}$$

$$\times \left(\frac{0.6503 \ \frac{\text{W}}{\text{m·K}}}{(2.29 \ \text{cm})\left(\frac{1 \ \text{m}}{100 \ \text{cm}}\right)}\right)$$

$$= 4989 \ \text{W/m}^2\text{·K}$$

The saturation temperature for 923 kPa steam is 176.4°C. Assume the wall to be at 10°C lower (or 166.4°C). The film properties are evaluated at the average of the wall and saturation temperatures.

$$T_h = \left(\tfrac{1}{2}\right)(T_{\text{sat},v} + T_s) = \left(\tfrac{1}{2}\right)(176.4°\text{C} + 166.4°\text{C})$$

$$= 171.4°\text{C}$$

Film properties are obtained from App. 35.B for liquid water and App. 24.N for vapor.

$$k_{171.4°\text{C}} = 0.6745 \ \text{W/m·K}$$

$$\mu_{171.4°\text{C}} = 0.1712 \times 10^{-3} \ \text{kg/m·s}$$

$$\rho_{l,171.4°\text{C}} = \frac{1}{v_{f,171.4°\text{C}}}$$

$$= \frac{1}{\left(1.1161 \ \frac{\text{cm}^3}{\text{g}}\right)\left(1000 \ \frac{\text{g}}{\text{kg}}\right)\left(\frac{1 \ \text{m}}{100 \ \text{cm}}\right)^3}$$

$$= 896.0 \ \text{kg/m}^3$$

$$\rho_{v,171.4°\text{C}} = \frac{1}{v_{g,171.4°\text{C}}}$$

$$= \frac{1}{\left(235.3 \ \frac{\text{cm}^3}{\text{g}}\right)\left(1000 \ \frac{\text{g}}{\text{kg}}\right)\left(\frac{1 \ \text{m}}{100 \ \text{cm}}\right)^3}$$

$$= 4.25 \ \text{kg/m}^3$$

$$h_{fg,923 \ \text{kPa}} = \left(2026.8 \ \frac{\text{kJ}}{\text{kg}}\right)\left(1000 \ \frac{\text{J}}{\text{kJ}}\right)$$

$$= 2.0268 \times 10^6 \ \text{J/kg}$$

$$d = (2.29 \ \text{cm})\left(\frac{1 \ \text{m}}{100 \ \text{cm}}\right) = 0.0229 \ \text{m}$$

From Eq. 35.27, the film coefficient is

$$h_o = (0.725)\left(\frac{\rho_l(\rho_l - \rho_v)gh_{fg}(k_l)^3}{d\mu_l(T_{\text{sat},v} - T_s)}\right)^{\frac{1}{4}}$$

$$= (0.725)$$

$$\times \left(\frac{\begin{array}{c}\left(896 \ \frac{\text{kg}}{\text{m}^3}\right)\left(896 \ \frac{\text{kg}}{\text{m}^3} - 4.25 \ \frac{\text{kg}}{\text{m}^3}\right)\left(9.81 \ \frac{\text{m}}{\text{s}^2}\right) \\ \times \left(2.0268 \times 10^6 \ \frac{\text{J}}{\text{kg}}\right)\left(0.6745 \ \frac{\text{W}}{\text{m·K}}\right)^3 \end{array}}{\begin{array}{c}(0.0254 \ \text{m})\left(0.1712 \times 10^{-3} \ \frac{\text{kg}}{\text{m·s}}\right) \\ \times (176.4°\text{C} - 166.4°\text{C})\end{array}}\right)^{\frac{1}{4}}$$

$$= 13\,266 \ \text{W/m}^2\text{·K}$$

From Table 36.7 and the table footnote, for copper alloy (70% Cu, 30% Ni) at 166.4°C,

$$k \approx \left(19.6 \ \frac{\text{Btu}}{\text{hr·ft·°F}}\right)\left(1.731 \ \frac{\text{W·hr·ft·°F}}{\text{m·K·Btu}}\right)$$

$$= 33.93 \ \text{W/m·K}$$

From Eq. 36.69, the overall heat transfer coefficient based on outside area is

$$\frac{1}{U_o} = \frac{1}{h_o} + \left(\frac{r_o}{k_{\text{tube}}}\right)\ln\left(\frac{r_o}{r_i}\right) + \frac{r_o}{r_i h_i}$$

$$r_o = \frac{d_o}{2} = \frac{(2.54 \ \text{cm})\left(\frac{1 \ \text{m}}{100 \ \text{cm}}\right)}{2} = 0.0127 \ \text{m}$$

$$r_i = \frac{d_i}{2} = \frac{0.0229 \ \text{m}}{2} = 0.0115 \ \text{m}$$

$$U_o = \frac{1}{\begin{array}{c}\dfrac{1}{13\,266 \ \frac{\text{W}}{\text{m}^2\text{·K}}} + \left(\dfrac{0.0127 \ \text{m}}{33.93 \ \frac{\text{W}}{\text{m·K}}}\right)\ln\left(\dfrac{0.0127 \ \text{m}}{0.0115 \ \text{m}}\right) \\ + \dfrac{0.0127 \ \text{m}}{(0.0115 \ \text{m})\left(4989 \ \frac{\text{W}}{\text{m}^2\text{·K}}\right)}\end{array}}$$

$$= 2995 \ \text{W/m}^2\text{·K}$$

For crossflow operation,

```
21°C ─────────────────────→ 90°C
      A                  B
176.4°C ←───────────────── 176.4°C
```

$$\Delta T_A = 176.4°\text{C} - 21°\text{C} = 155.4°\text{C}$$

$$\Delta T_B = 176.4°\text{C} - 90°\text{C} = 86.4°\text{C}$$

Heat Transfer

From Eq. 36.67, the logarithmic mean temperature difference is

$$\Delta T_{lm} = \frac{\Delta T_A - \Delta T_B}{\ln\left(\frac{\Delta T_A}{\Delta T_B}\right)} = \frac{155.4°C - 86.4°C}{\ln\left(\frac{155.4°C}{86.4°C}\right)}$$

$$= 117.5°C$$

The heat transfer is known; therefore, the outside area can be calculated from Eq. 36.68.

$$Q = U_o A_o F_c \Delta T_{lm}$$

For steam condensation, the temperature of steam remains constant. Therefore, $F_c = 1$.

$$A_o = \frac{106\,291 \text{ W}}{\left(2995 \frac{\text{W}}{\text{m}^2 \cdot \text{K}}\right)(1)(117.5°C)}$$

$$= \boxed{0.302 \text{ m}^2}$$

At this point the assumption that $T_{\text{sub,v}} - T_s = 10°C$ could be checked using $Q = U_{\text{partial}} A_o \Delta T_{\text{partial}}$.

The answer is (A).

10.

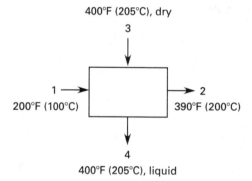

400°F (205°C), dry
3
1 →
200°F (100°C)
→ 2
390°F (200°C)
4
400°F (205°C), liquid

Customary U.S. Solution

(a) From App. 24.A, the enthalpy of each point is

$$h_1 = 168.13 \text{ Btu/lbm}$$
$$h_2 = 364.3 \text{ Btu/lbm}$$
$$h_3 = 1201.4 \text{ Btu/lbm}$$
$$h_4 = 375.1 \text{ Btu/lbm}$$

The heat transfer is due to the temperature gain of water.

$$Q = \dot{m}(h_2 - h_1)$$

$$= \left(500,000 \frac{\text{lbm}}{\text{hr}}\right)\left(364.3 \frac{\text{Btu}}{\text{lbm}} - 168.13 \frac{\text{Btu}}{\text{lbm}}\right)$$

$$= 9.809 \times 10^7 \text{ Btu/hr}$$

The mass flow rate per tube is

$$\dot{m}_{\text{tube}} = \rho A_{\text{tube}} \text{v}$$

Select ρ where the specific volume is greatest (at 390°F). From App. 24.A,

$$v_f = 0.01850 \text{ ft}^3/\text{lbm}$$

$$\rho = \frac{1}{v_f} = \frac{1}{0.01850 \frac{\text{ft}^3}{\text{lbm}}}$$

$$= 54.05 \text{ lbm/ft}^3$$

The inside diameter of the tube is

$$d_i = d_o - (2)(\text{wall})$$

$$= \frac{7}{8} \text{ in} - (2)\left(\frac{1}{16} \text{ in}\right)$$

$$= 0.750 \text{ in}$$

The area per tube is

$$A_{\text{tube}} = \left(\frac{\pi}{4}\right)(d_i)^2 = \left(\frac{\pi}{4}\right)(0.750 \text{ in})^2 \left(\frac{1 \text{ ft}}{12 \text{ in}}\right)^2$$

$$= 0.003068 \text{ ft}^2$$

$$\dot{m}_{\text{tube}} = \left(54.05 \frac{\text{lbm}}{\text{ft}^3}\right)(0.003068 \text{ ft}^2)$$

$$\times \left(5 \frac{\text{ft}}{\text{sec}}\right)\left(3600 \frac{\text{sec}}{\text{hr}}\right)$$

$$= 2985 \text{ lbm/hr}$$

The required number of tubes is

$$N = \frac{500,000 \frac{\text{lbm}}{\text{hr}}}{2985 \frac{\text{lbm}}{\text{hr}}} = \boxed{167.5} \quad [\text{say } 168]$$

The answer is (D).

(b) Consider counterflow or parallel flow. Since one fluid temperature remains constant, it will not make a difference. Also, from App. 36.A, $F_c = 1$.

$$\Delta T_A = 400°F - 200°F = 200°F$$
$$\Delta T_B = 400°F - 390°F = 10°F$$

From Eq. 36.67, the logarithmic mean temperature difference is

$$\Delta T_{lm} = \frac{\Delta T_A - \Delta T_B}{\ln\left(\frac{\Delta T_A}{\Delta T_B}\right)} = \frac{200°F - 10°F}{\ln\left(\frac{200°F}{10°F}\right)}$$

$$= 63.4°F$$

The heat transfer is known. So,

$$Q = UA\Delta T_{lm}F_c$$

Let L be the length of the tube bundle, approximately one-half the length of a single tube. The surface area is

$$A = (\pi D_o L)N \times 2 \text{ passes}$$

$$Q = U(\pi D_o L)N\Delta T_{lm}F_c \times 2 \text{ passes}$$

The length of the bundle takes into consideration both passes (i.e., uses the total surface area).

$$L = \frac{Q}{2U\pi D_o \Delta T_{lm}F_c N}$$

$$= \frac{9.812 \times 10^7 \ \dfrac{\text{Btu}}{\text{hr}}}{(2 \text{ passes})\left(700 \ \dfrac{\text{Btu}}{\text{hr-ft}^2\text{-}°\text{F}}\right)\pi\left(\dfrac{7}{8} \text{ in}\right)}$$

$$\times \left(\frac{1 \text{ ft}}{12 \text{ in}}\right)(63.4°\text{F})(1)(168)$$

$$= \boxed{28.7 \text{ ft}}$$

The answer is (C).

This is the approximate length of the tube bundle and one half of the total straight length of the bent tubes.

SI Solution

(a) From App. 24.N, the enthalpy of each point is

$$h_1 = 419.17 \text{ kJ/kg}$$
$$h_2 = 852.27 \text{ kJ/kg}$$
$$h_3 = 2794.8 \text{ kJ/kg}$$
$$h_4 = 874.88 \text{ kJ/kg}$$

The heat transfer is due to the temperature gain of water.

$$Q = \dot{m}(h_2 - h_1)$$

$$= \left(60 \ \frac{\text{kg}}{\text{s}}\right)\left(852.27 \ \frac{\text{kJ}}{\text{kg}} - 419.17 \ \frac{\text{kJ}}{\text{kg}}\right)$$

$$= 25\,986 \text{ kW}$$

The mass flow rate per tube is

$$\dot{m}_{\text{tube}} = \rho A_{\text{tube}}\text{v}$$

Select ρ where the specific volume is greatest (at 200°C). From App. 24.N,

$$v_f = \left(1.1565 \ \frac{\text{cm}^3}{\text{g}}\right)\left(1000 \ \frac{\text{g}}{\text{kg}}\right)\left(\frac{1 \text{ m}}{100 \text{ cm}}\right)^3$$

$$= 1.1565 \times 10^{-3} \text{ m}^3/\text{kg}$$

$$\rho = \frac{1}{v_f} = \frac{1}{1.1565 \times 10^{-3} \ \dfrac{\text{m}^3}{\text{kg}}}$$

$$= 864.7 \text{ kg/m}^3$$

The inside diameter of the tube is

$$d_i = d_o - (2)(\text{wall})$$

$$= \left(2.2 \text{ cm} - (2)(1.6 \text{ mm})\left(\frac{1 \text{ cm}}{10 \text{ mm}}\right)\right)\left(\frac{1 \text{ m}}{100 \text{ cm}}\right)$$

$$= 0.0188 \text{ m}$$

The area per tube is

$$A_{\text{tube}} = \left(\frac{\pi}{4}\right)(d_i)^2 = \left(\frac{\pi}{4}\right)(0.0188 \text{ m})^2$$

$$= 2.776 \times 10^{-4} \text{ m}^2$$

$$\dot{m}_{\text{tube}} = \left(864.7 \ \frac{\text{kg}}{\text{m}^3}\right)(2.776 \times 10^{-4} \text{ m}^2)\left(1.5 \ \frac{\text{m}}{\text{s}}\right)$$

$$= 0.360 \text{ kg/s}$$

The required number of tubes is

$$N = \frac{60 \ \dfrac{\text{kg}}{\text{s}}}{0.360 \ \dfrac{\text{kg}}{\text{s}}} = \boxed{166.7} \quad \text{[say 167]}$$

The answer is (D).

(b) Since one fluid remains at constant temperature, from App. 36.A, $F_c = 1$. Also, it will not make a difference whether counterflow or parallel flow is considered.

$$\Delta T_A = 205°\text{C} - 100°\text{C} = 105°\text{C}$$

$$\Delta T_B = 205°\text{C} - 200°\text{C} = 5°\text{C}$$

From Eq. 36.67, the logarithmic mean temperature difference is

$$\Delta T_{lm} = \frac{\Delta T_A - \Delta T_B}{\ln\left(\dfrac{\Delta T_A}{\Delta T_B}\right)} = \frac{105°\text{C} - 5°\text{C}}{\ln\left(\dfrac{105°\text{C}}{5°\text{C}}\right)}$$

$$= 32.8°\text{C}$$

The heat transfer is known. So,

$$Q = UA\Delta T_{lm}F_c$$

The surface area is

$$A = (\pi D_o L)N \times 2 \text{ passes}$$

$$Q = 2U(\pi D_o L)N\Delta T_{lm}F_c$$

The length of the bundle takes into consideration both passes (i.e., uses the total surface area).

$$L = \frac{Q}{2U\pi D_o N\Delta T_{lm}F_c}$$

$$= \frac{(26\,005 \text{ kW})\left(1000 \ \dfrac{\text{W}}{\text{kW}}\right)}{(2)\left(700 \ \dfrac{\text{Btu}}{\text{hr-ft}^2\text{-}°\text{F}}\right)\left(5.6783 \ \dfrac{\text{W}}{\text{m}^2\text{·}°\text{C}}\right)\pi}$$

$$\times (2.2 \text{ cm})\left(\frac{1 \text{ m}}{100 \text{ cm}}\right)(167)(32.8°\text{C})(1)$$

$$= \boxed{17.3 \text{ m}}$$

This is the approximate length of the tube bundle and one half of the total straight length of the bent tubes.

The answer is (C).

11. *Customary U.S. Solution*

The water's bulk temperature is

$$T_{b,\text{water}} = \left(\tfrac{1}{2}\right)(70°F + 140°F) = 105°F$$

The fluid properties at 105°F are obtained from App. 35.A.

$$\rho_{105°F} = 61.92 \text{ lbm/ft}^3$$

$$c_{p,105°F} = 0.998 \text{ Btu/lbm-°F}$$

The mass flow rate of water is

$$\dot{m}_{\text{water}} = \dot{V}\rho$$

$$= \frac{\left(100 \, \dfrac{\text{gal}}{\text{min}}\right)\left(60 \, \dfrac{\text{min}}{\text{hr}}\right)\left(61.92 \, \dfrac{\text{lbm}}{\text{ft}^3}\right)}{7.48 \, \dfrac{\text{gal}}{\text{ft}^3}}$$

$$= 49{,}668 \text{ lbm/hr}$$

The heat transfer is found from the temperature gain of the water.

$$Q_{\text{clean}} = \dot{m}c_p\Delta T$$

$$= \left(49{,}668 \, \dfrac{\text{lbm}}{\text{hr}}\right)\left(0.998 \, \dfrac{\text{Btu}}{\text{lbm-°F}}\right)$$

$$\times (140°F - 70°F)$$

$$= 3.470 \times 10^6 \text{ Btu/hr}$$

$$\Delta T_A = 230°F - 70°F = 160°F$$

$$\Delta T_B = 230°F - 140°F = 90°F$$

From Eq. 36.67, the logarithmic mean temperature difference is

$$\Delta T_{lm} = \frac{\Delta T_A - \Delta T_B}{\ln\left(\dfrac{\Delta T_A}{\Delta T_B}\right)} = \frac{160°F - 90°F}{\ln\left(\dfrac{160°F}{90°F}\right)} = 121.66°F$$

The heat transfer is known. Therefore, the overall heat transfer coefficient can be calculated from Eq. 36.68.

$$Q_{\text{clean}} = U_{\text{clean}}AF_c\Delta T_{lm}$$

For steam condensation, the temperature of steam remains constant. Therefore, $F_c = 1$.

$$U_{\text{clean}} = \frac{Q_{\text{clean}}}{AF_c\Delta T_{lm}}$$

$$= \frac{3.470 \times 10^6 \, \dfrac{\text{Btu}}{\text{hr}}}{(50 \text{ ft}^2)(1)(121.66°F)}$$

$$= 570.4 \text{ Btu/hr-ft}^2\text{-°F}$$

After fouling,

$$Q_{\text{fouled}} = \left(49{,}668 \, \dfrac{\text{lbm}}{\text{hr}}\right)\left(0.998 \, \dfrac{\text{Btu}}{\text{lbm-°F}}\right)$$

$$\times (122°F - 70°F)$$

$$= 2.578 \times 10^6 \text{ Btu/hr}$$

$$\Delta T_B = 230°F - 122°F = 108°F$$

$$\Delta T_A = 230°F - 70°F = 160°F$$

From Eq. 36.67, the logarithmic mean temperature difference is

$$\Delta T_{lm} = \frac{160°F - 108°F}{\ln\left(\dfrac{160°F}{108°F}\right)} = 132.3°F$$

$$U_{\text{fouled}} = \frac{2.578 \times 10^6 \, \dfrac{\text{Btu}}{\text{hr}}}{(50 \text{ ft}^2)(1)(132.3°F)} = 389.7 \text{ Btu/hr-ft}^2\text{-°F}$$

From Eq. 36.74, the fouling factor is

$$R_f = \frac{1}{U_{\text{fouled}}} - \frac{1}{U_{\text{clean}}}$$

$$= \frac{1}{389.7 \, \dfrac{\text{Btu}}{\text{hr-ft}^2\text{-°F}}} - \frac{1}{570.4 \, \dfrac{\text{Btu}}{\text{hr-ft}^2\text{-°F}}}$$

$$= \boxed{0.000813 \text{ hr-ft}^2\text{-°F/Btu}}$$

The answer is (B).

SI Solution

The water's bulk temperature is

$$T_{b,\text{water}} = \left(\tfrac{1}{2}\right)(21°C + 60°C) = 40.5°C$$

The fluid properties at 40.5°C are obtained from App. 35.B.

$$\rho_{40.5°C} = 993.5 \text{ kg/m}^3$$

$$c_{p,40.5°C} = 4.183 \text{ kJ/kg·K}$$

The mass flow rate of water is

$$\dot{m}_{\text{water}} = \dot{V}\rho$$

$$= \left(6.3 \, \dfrac{\text{L}}{\text{s}}\right)\left(\dfrac{1 \text{ m}^3}{1000 \text{ L}}\right)\left(993.5 \, \dfrac{\text{kg}}{\text{m}^3}\right)$$

$$= 6.26 \text{ kg/s}$$

The heat transfer is found from the temperature gain of the water.

$$Q_{\text{clean}} = \dot{m}c_p\Delta T$$

$$= \left(6.26 \, \dfrac{\text{kg}}{\text{s}}\right)\left(4.183 \, \dfrac{\text{kJ}}{\text{kg·K}}\right)\left(1000 \, \dfrac{\text{J}}{\text{kg}}\right)$$

$$\times (60°C - 21°C)$$

$$= 1.021 \times 10^6 \text{ W}$$

$$\Delta T_A = 110°C - 21°C = 89°C$$

$$\Delta T_B = 110°C - 60°C = 50°C$$

From Eq. 36.67, the logarithmic mean temperature difference is

$$\Delta T_{lm} = \frac{\Delta T_A - \Delta T_B}{\ln\left(\frac{\Delta T_A}{\Delta T_B}\right)} = \frac{89°\text{C} - 50°\text{C}}{\ln\left(\frac{89°\text{C}}{50°\text{C}}\right)} = 67.64°\text{C}$$

The heat transfer is known. Therefore, the overall heat transfer coefficient can be calculated from Eq. 36.68.

$$Q_{\text{clean}} = U_{\text{clean}} A F_c \Delta T_{lm}$$

For steam condensation, the temperature of steam remains constant. Therefore, $F_c = 1$.

$$\begin{aligned} U_{\text{clean}} &= \frac{Q_{\text{clean}}}{A F_c \Delta T_{lm}} \\ &= \frac{1.021 \times 10^6 \text{ W}}{(4.7 \text{ m}^2)(1)(67.64°\text{C})} \\ &= 3211.6 \text{ W/m}^2\text{·K} \end{aligned}$$

After fouling,

$$\begin{aligned} Q_{\text{fouled}} &= \left(6.26 \ \frac{\text{kg}}{\text{s}}\right)\left(4.183 \ \frac{\text{kJ}}{\text{kg·K}}\right)\left(1000 \ \frac{\text{J}}{\text{kJ}}\right) \\ &\quad \times (50°\text{C} - 21°\text{C}) \\ &= 7.594 \times 10^5 \text{ W} \\ \Delta T_A &= 110°\text{C} - 21°\text{C} = 89°\text{C} \\ \Delta T_B &= 110°\text{C} - 50°\text{C} = 60°\text{C} \end{aligned}$$

From Eq. 36.67, the logarithmic mean temperature difference is

$$\Delta T_{lm} = \frac{89°\text{C} - 60°\text{C}}{\ln\left(\frac{89°\text{C}}{60°\text{C}}\right)} = 73.55°\text{C}$$

$$U_{\text{fouled}} = \frac{7.594 \times 10^5 \text{ W}}{(4.7 \text{ m}^2)(1)(73.55°\text{C})} = 2196.8 \text{ W/m}^2\text{·K}$$

From Eq. 36.74, the fouling factor is

$$\begin{aligned} R_f &= \frac{1}{U_{\text{fouled}}} - \frac{1}{U_{\text{clean}}} \\ &= \frac{1}{2196.8 \ \dfrac{\text{W}}{\text{m}^2\text{·K}}} - \frac{1}{3211.6 \ \dfrac{\text{W}}{\text{m}^2\text{·K}}} \\ &= \boxed{0.000144 \text{ m}^2\text{·K/W}} \end{aligned}$$

The answer is (B).

12. *Customary U.S. Solution*

The film temperature of the air from Eq. 35.11 is

$$T_h = \left(\tfrac{1}{2}\right)(T_s + T_\infty) = \left(\tfrac{1}{2}\right)(100°\text{F} + 150°\text{F}) = 125°\text{F}$$

From App. 35.C, the air properties at 125°F are

$$\begin{aligned} \nu &= 0.195 \times 10^{-3} \text{ ft}^2/\text{sec} \\ k &= 0.0159 \text{ Btu/hr-ft-°F} \\ \text{Pr} &= 0.72 \end{aligned}$$

The Reynolds number based on the diameter, d, is

$$\begin{aligned} \text{Re}_d = \frac{\text{v}d}{\nu} &= \frac{\left(100 \ \dfrac{\text{ft}}{\text{sec}}\right)(0.35 \text{ in})\left(\dfrac{1 \text{ ft}}{12 \text{ in}}\right)}{0.195 \times 10^{-3} \ \dfrac{\text{ft}^2}{\text{sec}}} \\ &= 1.50 \times 10^4 \end{aligned}$$

From Eq. 36.50, the film coefficient is

$$h = C_1(\text{Re}_d)^n(\text{Pr})^{\frac{1}{3}}\left(\frac{k}{d}\right)$$

From Table 36.3, $C_1 = 0.193$ and $n = 0.618$.

$$\begin{aligned} h &= (0.193)(1.50 \times 10^4)^{0.618}(0.72)^{\frac{1}{3}} \\ &\quad \times \left(\frac{0.0159 \ \dfrac{\text{Btu}}{\text{hr-ft-°F}}}{(0.35 \text{ in})\left(\dfrac{1 \text{ ft}}{12 \text{ in}}\right)}\right) \\ &= \boxed{35.9 \text{ Btu/hr-ft}^2\text{-°F}} \end{aligned}$$

The answer is (A).

SI Solution

From Eq. 35.11, the film temperature of the air is

$$T_h = \left(\tfrac{1}{2}\right)(T_s + T_\infty) = \left(\tfrac{1}{2}\right)(38°\text{C} + 66°\text{C}) = 52°\text{C}$$

From App. 35.D, the air properties at 52°C are

$$\begin{aligned} \rho &= 1.09 \text{ kg/m}^3 \\ \mu &= 1.966 \times 10^{-5} \text{ kg/m·s} \\ k &= 0.02815 \text{ W/m·K} \\ \text{Pr} &= 0.703 \end{aligned}$$

The Reynolds number is

$$\begin{aligned} \text{Re}_d &= \frac{\rho \text{v} d}{\mu} \\ &= \frac{\left(1.09 \ \dfrac{\text{kg}}{\text{m}^3}\right)\left(30 \ \dfrac{\text{m}}{\text{s}}\right)(8.9 \text{ mm})\left(\dfrac{1 \text{ m}}{1000 \text{ mm}}\right)}{1.966 \times 10^{-5} \ \dfrac{\text{kg}}{\text{m·s}}} \\ &= 1.48 \times 10^4 \end{aligned}$$

From Eq. 36.50, the film coefficient is

$$h = C_1(\text{Re}_d)^h(\text{Pr})^{\frac{1}{3}}\left(\frac{k}{d}\right)$$

From Table 36.3, $C_1 = 0.193$ and $n = 0.618$.

$$h = (0.193)(1.48 \times 10^4)^{0.618}(0.703)^{\frac{1}{3}}$$
$$\times \left(\frac{0.02815 \ \frac{\text{W}}{\text{m·K}}}{(8.9 \ \text{mm})\left(\frac{1 \ \text{m}}{1000 \ \text{mm}}\right)}\right)$$
$$= \boxed{205.0 \ \text{W/m}^2\text{·K}}$$

The answer is (A).

37 Radiation and Combined Heat Transfer

PRACTICE PROBLEMS

Arrangement Factors

1. A 6 in (15 cm) thick furnace wall has a 3 in (8 cm) square peephole. The interior of the furnace is at 2200°F (1200°C). The surrounding air temperature is 70°F (20°C). What is the heat loss due to radiation when the peephole is open?

(A) 450 Btu/hr (150 W)
(B) 1300 Btu/hr (440 W)
(C) 2000 Btu/hr (680 W)
(D) 7900 Btu/hr (2.7 kW)

Combined Heat Transfer

2. A 9 in (23 cm) diameter duct is painted with white enamel. The surface of the duct is at 200°F (95°C). The duct carries hot air through a room whose walls are 70°F (20°C). The air in the room is at 80°F (27°C). What is the unit heat transfer?

(A) 650 Btu/hr-ft length (650 W/m length)
(B) 900 Btu/hr-ft length (900 W/m length)
(C) 1100 Btu/hr-ft length (1100 W/m length)
(D) 1500 Btu/hr-ft length (1500 W/m length)

3. The walls of a cold storage unit have the cross section shown.

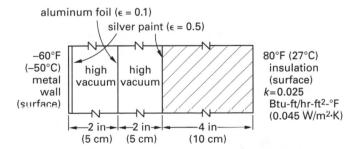

(a) What is the heat transfer per unit area of wall?

(A) 3 Btu/hr-ft² (10 W/m²)
(B) 7 Btu/hr-ft² (23 W/m²)
(C) 14 Btu/hr-ft² (46 W/m²)
(D) 53 Btu/hr-ft² (180 W/m²)

(b) What is the temperature of the aluminum foil?

(A) 420°R (230K)
(B) 460°R (260K)
(C) 475°R (264K)
(D) 490°R (270K)

4. Dry air at 1 atmospheric pressure flows at 500 ft³/min (0.25 m³/s) through 50 ft (15 m) of 12 in (30 cm) diameter uninsulated duct. The emissivity of the duct surface is 0.28. Air enters the duct at 45°F (7°C). The walls, air, and contents of the room through which the duct passes are at 80°F (27°C). An engineer states that the air leaving the duct will be at 50°F (10°C). Consider both convection and radiation to prove or disprove the engineer's statement.

5. A steel pipe is painted on the outside with dull gray (oil-based) paint. The pipe is 35 ft (10 m) long. The pipe is 4.00 in (10.2 cm) inside diameter and 4.25 in (10.8 cm) outside diameter. The pipe carries 200 ft³/min (0.1 m³/s) of 500°F (260°C), 25 psig (170 kPa) air through a 70°F (20°C) room. The conditions of the air at the end of the pipe are 350°F (180°C) and 15 psig (100 kPa).

(a) What is the overall coefficient of heat transfer?

(A) 1.0 Btu/hr-ft²-°F (6.1 W/m²·K)
(B) 1.8 Btu/hr-ft²-°F (11 W/m²·K)
(C) 3.6 Btu/hr-ft²-°F (22 W/m²·K)
(D) 49 Btu/hr-ft²-°F (300 W/m²·K)

(b) Using theoretical methods or empirical correlations, what is the calculated overall coefficient of heat transfer?

(A) 1.0 Btu/hr-ft²-°F (6.1 W/m²·K)
(B) 1.8 Btu/hr-ft²-°F (11 W/m²·K)
(C) 3.8 Btu/hr-ft²-°F (22 W/m²·K)
(D) 54 Btu/hr-ft²-°F (330 W/m²·K)

(c) Explain the possible reasons for differences between the actual and calculated overall coefficients of heat transfer.

6. A semiconductor device is modeled as a circular cylinder 0.75 in (19 mm) in diameter and 1.5 in (38 mm) high standing on one end. The device emits 5.0 W and is cooled by a combination of natural convection and radiation. The surface emissivity is 0.65. The base is insulated and transmits no heat. The air and surroundings are at 14.7 psia (101 kPa) and 75°F (24°C).

Heat Transfer

(a) What is the surface temperature of the device?

(b) What percentages of heat are lost through convection and radiation?

7. The temperature of a gas in a duct with 600°F (315°C) walls is evaluated with a 0.5 in (13 mm) diameter thermocouple probe. The emissivity of the probe is 0.8. The gas flow rate is 3480 lbm/hr-ft^2 (4.7 kg/s·m^2). The gas velocity is 400 ft/min (2 m/s). The film coefficient on the probe is given empirically as

$$h = \frac{0.024 G^{0.8}}{D^{0.4}}$$

h in Btu/hr-ft^2-°F

G in lbm/hr-ft^2

D in ft

$$h = \frac{2.9 G^{0.8}}{D^{0.4}}$$

h in W/m^2·K

G in kg/s·m^2

D in m

(a) If the actual gas temperature is 300°F (150°C), what is the probe's reading?

   (A) 700°R (390K)
   (B) 740°R (410K)
   (C) 760°R (420K)
   (D) 780°R (480K)

(b) If the probe reading indicates that the gas temperature is 300°F (150°C), what is the actual gas temperature?

   (A) 650°R (360K)
   (B) 740°R (350K)
   (C) 770°R (430K)
   (D) 810°R (450K)

SOLUTIONS

1. *Customary U.S. Solution*

The absolute temperatures are

$$T_{\text{furnace}} = 2200°\text{F} + 460 = 2660°\text{R}$$
$$T_\infty = 70°\text{F} + 460 = 530°\text{R}$$

Assuming that the walls are reradiating, nonconducting, and varying in temperature from 2200°F at the inside to 70°F at the outside, Fig. 37.3, curve 6, can be used to find F_{1-2} using $x = 3$ in/6 in $= 0.5$; $F_{1-2} = 0.38$.

The radiation heat loss is

$$
\begin{aligned}
Q &= A E_{\text{net}} \\
&= A \sigma F_{1-2}(T_{\text{furnace}}^4 - T_\infty^4) \\
&= (3 \text{ in})^2 \left(\frac{1 \text{ ft}^2}{144 \text{ in}^2}\right)\left(0.1713 \times 10^{-8} \frac{\text{Btu}}{\text{hr-ft}^2\text{-}°\text{R}^4}\right) \\
&\quad \times (0.38)\left((2660°\text{R})^4 - (530°\text{R})^4\right) \\
&= \boxed{2033.6 \text{ Btu/hr}}
\end{aligned}
$$

The answer is (C).

SI Solution

The absolute temperatures are

$$T_{\text{furnace}} = 1200°\text{C} + 273 = 1473\text{K}$$
$$T_\infty = 20°\text{C} + 273 = 293\text{K}$$

Making the same assumptions as for the customary U.S. solution, Fig. 37.3, curve 6, can be used to find F_{1-2} using $x = 8$ cm/15 cm $= 0.533$; $F_{1-2} = 0.4$.

The radiation heat loss is

$$
\begin{aligned}
Q &= A E_{\text{net}} \\
&= A \sigma F_{1-2}(T_{\text{furnace}}^4 - T_\infty^4) \\
&= (8 \text{ cm})^2 \left(\frac{1 \text{ m}}{100 \text{ cm}}\right)^2 \left(5.67 \times 10^{-8} \frac{\text{W}}{\text{m}^2\text{·K}^4}\right) \\
&\quad \times (0.4)\left((1473\text{K})^4 - (293\text{K})^4\right) \\
&= \boxed{682.3 \text{ W}}
\end{aligned}
$$

The answer is (C).

2. *Customary U.S. Solution*

The absolute temperatures are

$$T_\infty = 80°\text{F} + 460 = 540°\text{R}$$
$$T_{\text{duct}} = 200°\text{F} + 460 = 660°\text{R}$$
$$T_{\text{wall}} = 70°\text{F} + 460 = 530°\text{R}$$

Assume laminar flow. From Table 35.3, the convective film coefficient on the outside of the duct is approximately

$$h_{\text{convective}} = (0.27) \left(\frac{T_{\text{duct}} - T_{\infty}}{L} \right)^{0.25}$$

$$= (0.27) \left(\frac{660°\text{R} \quad 540°\text{R}}{(9 \text{ in}) \left(\frac{1 \text{ ft}}{12 \text{ in}} \right)} \right)^{0.25}$$

$$= 0.96 \text{ Btu/hr-ft}^2\text{-}°\text{F}$$

The duct area per unit length is

$$\frac{A}{L} = \pi D = \pi(9 \text{ in}) \left(\frac{1 \text{ ft}}{12 \text{ in}} \right) = 2.356 \text{ ft}^2/\text{ft}$$

The convection losses (per unit length) are

$$\frac{Q_{\text{convection}}}{L} = h \left(\frac{A}{L} \right) \Delta T = h \left(\frac{A}{L} \right) (T_{\text{duct}} - T_{\infty})$$

$$= \left(0.96 \frac{\text{Btu}}{\text{hr-ft}^2\text{-}°\text{F}} \right) \left(2.356 \frac{\text{ft}^2}{\text{ft}} \right)$$

$$\times (660°\text{R} - 540°\text{R})$$

$$= 271.4 \text{ Btu/hr-ft}$$

Assume $\epsilon_{\text{duct}} \approx 0.90$. Then $F_e = \epsilon_{\text{duct}} = 0.90$. $F_a = 1$ since the duct is enclosed. The radiation losses (per unit length) are

$$\frac{E_{\text{net}}}{L} = \left(\frac{A}{L} \right) \sigma F_a F_e \left((T_{\text{duct}})^4 - (T_{\text{wall}})^4 \right)$$

$$= (2.356 \text{ ft}^2) \left(0.1713 \times 10^{-8} \frac{\text{Btu}}{\text{hr-ft}^2\text{-}°\text{R}^4} \right)$$

$$\times (1)(0.90) \left((660°\text{R})^4 - (530°\text{R})^4 \right)$$

$$= 402.6 \text{ Btu/hr-ft}$$

The total heat transfer per unit length is

$$\frac{Q_{\text{total}}}{L} = \frac{Q_{\text{convection}}}{L} + \frac{E_{\text{net}}}{L}$$

$$= 271.4 \frac{\text{Btu}}{\text{hr-ft}} + 402.6 \frac{\text{Btu}}{\text{hr-ft}}$$

$$= \boxed{674 \text{ Btu/hr-ft length}}$$

The answer is (A).

SI Solution

The absolute temperatures are

$$T_{\text{duct}} = 95°\text{C} + 273 = 368\text{K}$$

$$T_{\text{wall}} = 20°\text{C} + 273 = 293\text{K}$$

$$T_{\infty} = 27°\text{C} + 273 = 300\text{K}$$

Assume laminar flow. From Table 35.3, the convective film coefficient on the outside of the duct is approximately

$$h_{\text{convective}} = (1.32) \left(\frac{T_{\text{duct}} - T_{\infty}}{L} \right)^{0.25}$$

$$= (1.32) \left(\frac{368\text{K} - 300\text{K}}{(23 \text{ cm}) \left(\frac{1 \text{ m}}{100 \text{ cm}} \right)} \right)^{0.25}$$

$$= 5.47 \text{ W/m}^2\text{·K}$$

The duct area per unit length is

$$\frac{A}{L} = \pi D = \pi(23 \text{ cm}) \left(\frac{1 \text{ m}}{100 \text{ cm}} \right) = 0.723 \text{ m}^2/\text{m}$$

The convective losses per unit length are

$$\frac{Q_{\text{convective}}}{L} = h \left(\frac{A}{L} \right) (T_{\text{duct}} - T_{\infty})$$

$$= \left(5.47 \frac{\text{W}}{\text{m}^2\text{·K}} \right) \left(0.723 \frac{\text{m}^2}{\text{m}} \right) (368\text{K} - 300\text{K})$$

$$= 268.9 \text{ W/m}$$

Assuming $\epsilon_{\text{duct}} \approx 0.90$, $F_e = \epsilon_{\text{duct}} = 0.90$. $F_a = 1$ since the duct is enclosed. The radiation losses per unit length are

$$\frac{E_{\text{net}}}{L} = \left(\frac{A}{L} \right) \sigma F_a F_e \left((T_{\text{duct}})^4 - (T_{\text{wall}})^4 \right)$$

$$= \left(0.723 \frac{\text{m}^2}{\text{m}} \right) \left(5.67 \times 10^{-8} \frac{\text{W}}{\text{m}^2\text{·K}^4} \right)$$

$$\times (1)(0.90) \left((368\text{K})^4 - (293\text{K})^4 \right)$$

$$= 404.7 \text{ W/m}$$

The total heat transfer per unit length is

$$\frac{Q_{\text{total}}}{L} = \frac{Q_{\text{convection}}}{L} + \frac{E_{\text{net}}}{L}$$

$$= 268.9 \frac{\text{W}}{\text{m}} + 404.7 \frac{\text{W}}{\text{m}} = \boxed{673.6 \text{ W/m length}}$$

The answer is (A).

3.

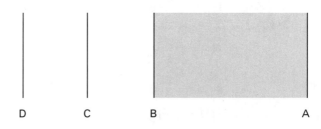

Customary U.S. Solution

(a) The absolute temperatures are

$$T_D = -60°F + 460 = 400°R$$
$$T_A = 80°F + 460 = 540°R$$

The conductive heat transfer from A to B per unit area is

$$\frac{Q_{A-B}}{A} = \frac{k(T_A - T_B)}{L}$$
$$= \frac{\left(0.025 \ \frac{\text{Btu-ft}}{\text{hr-ft}^2\text{-}°F}\right)(540°R - T_B)}{(4 \text{ in})\left(\frac{1 \text{ ft}}{12 \text{ in}}\right)}$$
$$= 40.5 - 0.075T_B \quad [\text{Eq. 1}]$$

Since the spaces are evacuated, only radiation should be considered from B to C and from C to D.

The radiation heat transfer per unit area from B to C is

$$\frac{E_{B-C}}{A} = \sigma F_e F_a (T_B^4 - T_C^4)$$

Since the freezer is assumed to be large, $F_a = 1$.

From Table 37.1 for infinite parallel planes,

$$F_e = \frac{1}{\dfrac{1}{\epsilon_1} + \dfrac{1}{\epsilon_2} - 1} = \frac{1}{\dfrac{1}{0.5} + \dfrac{1}{0.1} - 1}$$
$$= 0.0909$$
$$\frac{E_{B-C}}{A} = \left(0.1713 \times 10^{-8} \ \frac{\text{Btu}}{\text{hr-ft}^2\text{-}°R^4}\right)$$
$$\times (0.0909)(1)(T_B^4 - T_C^4)$$
$$= (1.56 \times 10^{-10})T_B^4 - (1.56 \times 10^{-10})T_C^4 \quad [\text{Eq. 2}]$$

Similarly, radiation heat transfer per unit area from C to D is

$$\frac{E_{C-D}}{A} = \left(0.1713 \times 10^{-8} \ \frac{\text{Btu}}{\text{hr-ft}^2\text{-}°R^4}\right)$$
$$\times (0.0909)(1)\left(T_C^4 - (400°R)^4\right)$$
$$= (1.56 \times 10^{-10})T_C^4 - 3.99 \quad [\text{Eq. 3}]$$

The heat transfer from B to C is equal to the heat transfer from C to D.

$$\frac{E_{B-C}}{A} = \frac{E_{C-D}}{A}$$
$$(1.56 \times 10^{-10})T_B^4$$
$$- (1.56 \times 10^{-10})T_C^4 = (1.56 \times 10^{-10})T_C^4 - 3.99$$
$$T_B^4 - T_C^4 = T_C^4 - 2.56 \times 10^{10}$$
$$T_B^4 + 2.56 \times 10^{10} = 2T_C^4$$
$$T_C^4 = \tfrac{1}{2}T_B^4 + 1.28 \times 10^{10} \quad [\text{Eq. 4}]$$

The heat transfer from A to B is equal to the heat transfer from B to C.

$$\frac{Q_{A-B}}{A} = \frac{E_{B-C}}{A}$$
$$40.5 - 0.075T_B = (1.56 \times 10^{-10})T_B^4$$
$$- (1.56 \times 10^{-10})T_C^4$$
$$T_C^4 = T_B^4 + (4.81 \times 10^8)T_B$$
$$- 2.60 \times 10^{11} \quad [\text{Eq. 5}]$$

Since Eq. 4 = Eq. 5,

$$\tfrac{1}{2}T_B^4 + 1.28 \times 10^{10} = T_B^4 + (4.81 \times 10^8)T_B$$
$$- 2.60 \times 10^{11}$$
$$T_B^4 + (9.62 \times 10^8)T_B = 5.456 \times 10^{11}$$

By trial and error, $T_B = 501.4°R$.

From Eq. 1,

$$\frac{Q_{A-B}}{A} = 40.5 - (0.075)(501.4°R)$$
$$= \boxed{2.895 \text{ Btu/hr-ft}^2}$$

The answer is (A).

(b) From Eq. 4,

$$T_C = \left(\left(\tfrac{1}{2}\right)(501.4)^4 + 1.28 \times 10^{10}\right)^{\frac{1}{4}}$$
$$= \boxed{459°R}$$

The answer is (B).

SI Solution

(a) The absolute temperatures are

$$T_D = -50°C + 273 = 223K$$
$$T_A = 27°C + 273 = 300K$$

The conductive heat transfer per unit area from A to B is

$$\frac{Q_{A-B}}{A} = \frac{k(T_A - T_B)}{L} = \frac{\left(0.045 \ \frac{\text{W}}{\text{m·K}}\right)(300K - T_B)}{(10 \text{ cm})\left(\frac{1 \text{ m}}{100 \text{ cm}}\right)}$$
$$= 135 - 0.45T_B \quad [\text{Eq. 1}]$$

Since spaces are evacuated, only radiation should be considered from B to C and from C to D.

The radiation heat transfer per unit area from B to C is

$$\frac{E_{\text{B-C}}}{A} = \sigma F_e F_a (T_{\text{B}}^4 - T_{\text{C}}^4)$$

From the customary U.S. solution, $F_e = 0.0909$ and $F_a = 1$.

$$\frac{E_{\text{B-C}}}{A} = \left(5.67 \times 10^{-8} \ \frac{\text{W}}{\text{m}^2 \cdot \text{K}^4} \right)$$
$$\times (0.0909)(1)(T_{\text{B}}^4 - T_{\text{C}}^4)$$
$$= (5.15 \times 10^{-9})T_{\text{B}}^4 - (5.15 \times 10^9)T_{\text{C}}^4 \quad \text{[Eq. 2]}$$

Similarly, the radiation heat transfer per unit area from C to D is

$$\frac{E_{\text{C-D}}}{A} = \left(5.67 \times 10^{-8} \ \frac{\text{W}}{\text{m}^2 \cdot \text{K}^4} \right)$$
$$\times (0.0909)(1)\left(T_{\text{C}}^4 - (223\text{K})^4\right)$$
$$= (5.15 \times 10^{-9})T_{\text{C}}^4 - 12.746 \quad \text{[Eq. 3]}$$

The heat transfer from B to C is equal to the heat transfer from C to D.

$$\frac{E_{\text{B-C}}}{A} = \frac{E_{\text{C-D}}}{A}$$
$$(5.15 \times 10^{-9})T_{\text{B}}^4$$
$$- (5.15 \times 10^{-9})T_{\text{C}}^4 = (5.15 \times 10^{-9})T_{\text{C}}^4 - 12.746$$
$$T_{\text{B}}^4 - T_{\text{C}}^4 = T_{\text{C}}^4 - 2.475 \times 10^9$$
$$T_{\text{B}}^4 + 2.475 \times 10^9 = 2T_{\text{C}}^4$$
$$T_{\text{C}}^4 = \tfrac{1}{2}T_{\text{B}}^4 + 1.237 \times 10^9 \quad \text{[Eq. 4]}$$

The heat transfer from A to B is equal to the heat transfer from B to C.

$$\frac{Q_{\text{A-B}}}{A} = \frac{E_{\text{B-C}}}{A}$$
$$135 - 0.45T_{\text{B}} = (5.15 \times 10^{-9})T_{\text{B}}^4 - (5.15 \times 10^{-9})T_{\text{C}}^4$$
$$T_{\text{C}}^4 = T_{\text{B}}^4 + 8.74 \times 10^7 T_{\text{B}} - 2.62 \times 10^{10}$$
$$\text{[Eq. 5]}$$

Since Eq. 4 = Eq. 5,

$$\tfrac{1}{2}T_{\text{B}}^4 + 1.237 \times 10^9 = T_{\text{B}}^4 + 8.74 \times 10^7 T_{\text{B}}$$
$$- 2.62 \times 10^{10}$$
$$T_{\text{B}}^4 + 17.48 \times 10^7 T_{\text{B}} = 5.49 \times 10^{10}$$

By trial and error, $T_{\text{B}} \approx 278\text{K}$.

From Eq. 1,

$$\frac{Q_{\text{A-B}}}{A} = 135 - (0.45)(278\text{K}) = \boxed{9.9 \ \text{W/m}^2}$$

The answer is (A).

(b) From Eq. 4,

$$T_{\text{C}} = \left(\left(\tfrac{1}{2}\right)(278\text{K})^4 + 1.237 \times 10^9 \right)^{\frac{1}{4}} = \boxed{254.9\text{K}}$$

The answer is (B).

4. *Customary U.S. Solution*

The absolute temperature of air entering the duct is

$$45°\text{F} + 460 = 505°\text{R}$$

From the ideal gas law, the density of air entering the duct is

$$\rho = \frac{p}{RT}$$
$$= \frac{\left(14.7 \ \frac{\text{lbf}}{\text{in}^2} \right)\left(144 \ \frac{\text{in}^2}{\text{ft}^2} \right)}{\left(53.35 \ \frac{\text{ft-lbf}}{\text{lbm-°R}} \right)(505°\text{R})} = 0.07857 \ \text{lbm/ft}^3$$

The mass flow rate of entering air is

$$\dot{m} = \rho \dot{V} = \left(0.07857 \ \frac{\text{lbm}}{\text{ft}^3} \right)\left(500 \ \frac{\text{ft}^3}{\text{min}} \right)\left(60 \ \frac{\text{min}}{\text{hr}} \right)$$
$$= 2357.1 \ \text{lbm/hr}$$

The mass velocity is

$$G = \frac{\dot{m}}{A_{\text{flow}}} = \frac{2357.1 \ \frac{\text{lbm}}{\text{hr}}}{\left(\frac{\pi}{4}\right)(12 \ \text{in})^2 \left(\frac{1 \ \text{ft}}{12 \ \text{in}} \right)^2}$$
$$= 3001.1 \ \text{lbm/ft}^2\text{-hr}$$

To calculate the initial film coefficients, estimate the temperature based on the claim. The film coefficients are not highly sensitive to small temperature differences.

$$T_{\text{bulk,air}} = \left(\tfrac{1}{2}\right)(T_{\text{air,in}} + T_{\text{air,out}})$$
$$= \left(\tfrac{1}{2}\right)(45°\text{F} + 50°\text{F}) = 47.5°\text{F}$$

$$T_{\text{surface}} = 70°\text{F} \quad \text{[estimate]}$$

The film coefficient for air flowing inside the duct is given by Eq. 36.39 as

$$h_i \approx (0.00351 + 0.000001583T°_{\text{F}})$$
$$\times \left(\frac{(G'_{\text{lbm/hr-ft}^2})^{0.8}}{(d_{\text{ft}})^{0.2}} \right)$$
$$= (0.00351 + (0.000001583)(47.5°\text{F}))$$
$$\times \left(\frac{\left(3001.1 \ \frac{\text{lbm}}{\text{hr-ft}^2} \right)^{0.8}}{\left((12 \ \text{in}) \left(\frac{1 \ \text{ft}}{12 \ \text{in}} \right) \right)^{0.2}} \right)$$

$$h_i = 2.17 \ \text{Btu/hr-ft}^2\text{-°F}$$

If Eq. 36.40(b) is used instead, the value of h_i is approximately 2.01 Btu/hr-ft^2-°F, indicating that the temperature is not important.

For natural convection on the outside of the duct, estimate the film temperature.

$$T_{\text{film}} = \left(\tfrac{1}{2}\right)(T_{\text{surface}} + T_\infty) = \left(\tfrac{1}{2}\right)(70°F + 80°F)$$
$$= 75°F$$

From App. 35.C, the properties of air at 75°F are

$$\text{Pr} \approx 0.72$$

$$\frac{g\beta\rho^2}{\mu^2} = 2.27 \times 10^6 \; \frac{1}{\text{ft}^3\text{-°F}}$$

The characteristic length is the diameter of the duct.

$$L = (12 \text{ in})\left(\frac{1 \text{ ft}}{12 \text{ in}}\right) = 1 \text{ ft}$$

The Grashof number is

$$\text{Gr} = L^3 \left(\frac{\rho^2 \beta g}{\mu^2}\right)(T_\infty - T_s)$$
$$= (1 \text{ ft})^3 \left(2.27 \times 10^6 \; \frac{1}{\text{ft}^3\text{-°F}}\right)(80°F - 70°F)$$
$$= 2.27 \times 10^7$$
$$\text{Pr Gr} = (0.72)(2.27 \times 10^7) = 1.63 \times 10^7$$

From Table 35.3, the film coefficient for a horizontal cylinder is

$$h_o = (0.27)\left(\frac{T_\infty - T_s}{d}\right)^{0.25}$$
$$= (0.27)\left(\frac{80°F - 70°F}{1 \text{ ft}}\right)^{0.25}$$
$$= 0.48 \text{ Btu/hr-ft}^2\text{-°F}$$

Neglecting the wall resistance, the overall film coefficient from Eq. 36.71 is

$$\frac{1}{U} = \frac{1}{h_o} + \frac{1}{h_i}$$
$$= \frac{1}{2.17 \; \dfrac{\text{Btu}}{\text{hr-ft}^2\text{-°F}}} + \frac{1}{0.48 \; \dfrac{\text{Btu}}{\text{hr-ft}^2\text{-°F}}}$$
$$= 2.544 \text{ hr-ft}^2\text{-°F/Btu}$$

$$U = \frac{1}{2.544 \; \dfrac{\text{hr-ft}^2\text{-°F}}{\text{Btu}}} = 0.393 \text{ Btu/hr-ft}^2\text{-°F}$$

The heat transfer due to convection is

$$Q_{\text{convection}} = UA_{\text{surface}}(T_\infty - T_{\text{bulk,air}})$$
$$= U(\pi d L)(T_\infty - T_{\text{bulk,air}})$$
$$= \left(0.393 \; \frac{\text{Btu}}{\text{hr-ft}^2\text{-°F}}\right)$$
$$\times \pi(1 \text{ ft})(50 \text{ ft})(80°F - 47.5°F)$$
$$= 2006.3 \text{ Btu/hr}$$

The heat transfer due to radiation is

$$Q_{\text{radiation}} = \sigma F_e F_a A_{\text{surface}}(T_\infty^4 - T_{\text{surface}}^4)$$

Assume the room and duct have an unobstructed view of each other. Then $F_a = 1.0$ and $F_e = \epsilon = 0.28$. The absolute temperatures are

$$T_\infty = 80°F + 460 = 540°R$$
$$T_{\text{surface}} = 70°F + 460 = 530°R$$
$$Q_{\text{radiation}} = \left(0.1713 \times 10^{-8} \; \frac{\text{Btu}}{\text{hr-ft}^2\text{-°R}^4}\right)(0.28)(1.0)$$
$$\times \pi(1 \text{ ft})(50 \text{ ft})\left((540°R)^4 - (530°R)^4\right)$$
$$= 461.5 \text{ Btu/hr}$$

The total heat transfer to the air is

$$Q_{\text{total}} = Q_{\text{convection}} + Q_{\text{radiation}}$$
$$= 2006.3 \; \frac{\text{Btu}}{\text{hr}} + 461.5 \; \frac{\text{Btu}}{\text{hr}}$$
$$= 2467.8 \text{ Btu/hr}$$

(A second iteration could begin by using Eq. 36.2 to calculate a more accurate surface temperature for the duct.)

At 47.5°F, the specific heat of air is approximately 0.240 Btu/lbm-°F. Since the heat transfer is known, the temperature of air leaving the duct can be calculated from

$$Q_{\text{total}} = \dot{m}c_p(T_{\text{air,out}} - T_{\text{air,in}})$$
$$T_{\text{air,out}} = T_{\text{air,in}} + \frac{Q_{\text{total}}}{\dot{m}c_p}$$
$$= 45°F + \frac{2467.8 \; \dfrac{\text{Btu}}{\text{hr}}}{\left(2357.1 \; \dfrac{\text{lbm}}{\text{hr}}\right)\left(0.24 \; \dfrac{\text{Btu}}{\text{lbm-°F}}\right)}$$
$$= \boxed{49.4°F}$$

This agrees with the engineer's estimate.

SI Solution

The absolute temperature of air entering the duct is 7°C + 273 = 280K. From the ideal gas law, the density of air entering the duct is

$$\rho = \frac{p}{RT}$$
$$= \frac{(101.3 \text{ kPa})\left(1000 \; \dfrac{\text{Pa}}{\text{kPa}}\right)}{\left(287.03 \; \dfrac{\text{J}}{\text{kg·K}}\right)(280\text{K})}$$
$$= 1.2604 \text{ kg/m}^3$$

The mass flow rate of air entering the duct is

$$\dot{m} = \rho\dot{V} = \left(1.2604 \; \frac{\text{kg}}{\text{m}^3}\right)\left(0.25 \; \frac{\text{m}^3}{\text{s}}\right) = 0.3151 \text{ kg/s}$$

The diameter of the duct is

$$d = (30 \text{ cm}) \left(\frac{1 \text{ m}}{100 \text{ cm}} \right) = 0.30 \text{ m}$$

The velocity of air entering the duct is

$$\text{v} = \frac{\dot{V}}{A_{\text{flow}}} = \frac{0.25 \frac{\text{m}^3}{\text{s}}}{\left(\frac{\pi}{4} \right) (0.30 \text{ m})^2} = 3.537 \text{ m/s}$$

To calculate the initial film coefficients, estimate the temperatures based on the claim. The film coefficients are not highly sensitive to small temperature differences.

$$T_{\text{bulk,air}} = \left(\tfrac{1}{2} \right) (T_{\text{air,in}} + T_{\text{air,out}})$$
$$= \left(\tfrac{1}{2} \right) (7°C + 10°C) = 8.5°C$$
$$T_{\text{surface}} = 20°C \quad [\text{estimate}]$$

The film coefficient for air flowing inside the duct is given from Eq. 36.40(a) (independent of temperature).

$$h_i \approx \frac{(3.52)(\text{v}_{\text{m/s}})^{0.8}}{(d_{\text{m}})^{0.2}}$$
$$= \frac{(3.52) \left(3.537 \frac{\text{m}}{\text{s}} \right)^{0.8}}{(0.30 \text{ m})^{0.2}} = 12.3 \text{ W/m}^2 \cdot °C$$

For natural convection on the outside of the duct, estimate the film coefficient.

$$T_{\text{film}} = \left(\tfrac{1}{2} \right) (T_{\text{surface}} + T_\infty)$$
$$= \left(\tfrac{1}{2} \right) (20°C + 27°C) = 23.5°C$$

From App. 35.D, the properties of air at 23.5°C are

$$\text{Pr} = 0.709$$
$$\frac{g\beta\rho^2}{\mu^2} = 1.43 \times 10^8 \frac{1}{\text{K} \cdot \text{m}^3}$$

The characteristic length, L, is the diameter of the duct, which is 0.30 m. The Grashof number is

$$\text{Gr} = L^3 \left(\frac{\rho^2 \beta g}{\mu^2} \right) (T_\infty - T_{\text{surface}})$$
$$= (0.30 \text{ m})^3 \left(1.43 \times 10^8 \frac{1}{\text{K} \cdot \text{m}^3} \right) (27°C - 20°C)$$
$$= 2.70 \times 10^7$$
$$\text{Pr} \, \text{Gr} = (0.709)(2.70 \times 10^7) = 1.9 \times 10^7$$

From Table 35.3, the film coefficient for a horizontal cylinder is

$$h_o \approx (1.32) \left(\frac{T_\infty - T_{\text{surface}}}{d} \right)^{0.25}$$
$$= (1.32) \left(\frac{27°C - 20°C}{0.30 \text{ m}} \right)^{0.25} = 2.90 \text{ W/m}^2 \cdot \text{K}$$

Neglecting the wall resistance, the overall film coefficient from Eq. 37.63 is

$$\frac{1}{U} = \frac{1}{h_o} + \frac{1}{h_i}$$
$$= \frac{1}{12.3 \frac{\text{W}}{\text{m}^2 \cdot \text{K}}} + \frac{1}{2.90 \frac{\text{W}}{\text{m}^2 \cdot \text{K}}} = 0.426 \text{ m}^2 \cdot \text{K/W}$$
$$U = \frac{1}{0.426 \frac{\text{m}^2 \cdot \text{K}}{\text{W}}} = 2.35 \text{ W/m}^2 \cdot \text{K}$$

The heat transfer due to convection is

$$Q_{\text{convection}} = U A_{\text{surface}} (T_\infty - T_{\text{bulk,air}})$$
$$= U \pi d L (T_\infty - T_{\text{bulk,air}})$$
$$= \left(2.35 \frac{\text{W}}{\text{m}^2 \cdot \text{K}} \right) \pi (0.30 \text{ m})(15 \text{ m})$$
$$\times (27°C - 8.5°C)$$
$$= 614.6 \text{ W}$$

The heat transfer due to radiation is

$$Q_{\text{radiation}} = \sigma F_e F_a A_{\text{surface}} \left(T_\infty^4 - T_{\text{surface}}^4 \right)$$

Assume the room and duct have an unobstructed view of each other. Then $F_a = 1.0$ and $F_e = \epsilon = 0.28$. The absolute temperatures are

$$T_\infty = 27°C + 273 = 300\text{K}$$
$$T_{\text{surface}} = 20°C + 273 = 293\text{K}$$
$$Q_{\text{radiation}} = \left(5.67 \times 10^{-8} \frac{\text{W}}{\text{m}^2 \cdot \text{K}^4} \right) (0.28)(1)\pi$$
$$\times (0.30 \text{ m})(15 \text{ m}) \left((300\text{K})^4 - (293\text{K})^4 \right)$$
$$= 163.8 \text{ W}$$

The total heat transfer to the air is

$$Q_{\text{total}} = Q_{\text{convection}} + Q_{\text{radiation}}$$
$$= 614.6 \text{ W} + 163.8 \text{ W} = 778.4 \text{ W}$$

(A second iteration could begin by using Eq. 36.2 to calculate a more accurate surface temperature for the duct.)

At 8.5°C, the specific heat of air is

$$\left(1.0048 \frac{\text{kJ}}{\text{kg} \cdot \text{K}} \right) \left(1000 \frac{\text{J}}{\text{kJ}} \right) = 1004.8 \text{ J/kg} \cdot \text{K}$$

Since the heat transfer is known, the temperature of air leaving the duct can be calculated by

$$Q_{\text{total}} = \dot{m} c_p (T_{\text{air,out}} - T_{\text{air,in}})$$
$$T_{\text{air,out}} = T_{\text{air,in}} + \frac{Q_{\text{total}}}{\dot{m} c_p}$$
$$= 7°C + \frac{778.4 \text{ W}}{\left(0.3151 \frac{\text{kg}}{\text{s}} \right) \left(1004.8 \frac{\text{J}}{\text{kg} \cdot \text{K}} \right)}$$
$$= \boxed{9.5°C}$$

This agrees with the engineer's estimate.

5. *Customary U.S. Solution*

(a) The absolute temperature of entering air is

$$500°F + 460 = 960°R$$

The absolute pressure of entering air is

$$25 \text{ psig} + 14.7 \text{ psi} = 39.7 \text{ psia}$$

The density of air entering, from the ideal gas law, is

$$\rho = \frac{p}{RT}$$

$$= \frac{\left(39.7 \dfrac{\text{lbf}}{\text{in}^2}\right)\left(144 \dfrac{\text{in}^2}{\text{ft}^2}\right)}{\left(53.35 \dfrac{\text{ft-lbf}}{\text{lbm-°R}}\right)(960°R)} = 0.1116 \text{ lbm/ft}^3$$

The mass flow rate of entering air is

$$\dot{m} = \rho\dot{V} = \left(0.1116 \dfrac{\text{lbm}}{\text{ft}^3}\right)\left(200 \dfrac{\text{ft}^3}{\text{min}}\right)\left(60 \dfrac{\text{min}}{\text{hr}}\right)$$

$$= 1339.2 \text{ lbm/hr}$$

At low pressures, the air enthalpy is found from air tables (App. 24.F).

The absolute temperature of leaving air is

$$350°F + 460 = 810°R$$

From App. 24.F,

$$h_1 = 231.06 \text{ Btu/lbm at } 960°R$$

$$h_2 = 194.25 \text{ Btu/lbm at } 810°R$$

The heat loss is

$$Q = \dot{m}(h_1 - h_2)$$

$$= \left(1339.2 \dfrac{\text{lbm}}{\text{hr}}\right)\left(231.06 \dfrac{\text{Btu}}{\text{lbm}} - 194.25 \dfrac{\text{Btu}}{\text{lbm}}\right)$$

$$= 49{,}296 \text{ Btu/hr}$$

Assuming midpoint pipe surface temperature,

$$\left(\tfrac{1}{2}\right)(T_{\text{in}} + T_{\text{out}}) = \left(\tfrac{1}{2}\right)(500°F + 350°F) = 425°F$$

Since the heat loss is known, the overall heat transfer coefficient can be determined from

$$Q = UA\Delta T = U(\pi dL)\Delta T$$

$$U = \frac{Q}{(\pi dL)\Delta T}$$

$$= \frac{49{,}296 \dfrac{\text{Btu}}{\text{hr}}}{\pi(4.25 \text{ in})\left(\dfrac{1 \text{ ft}}{12 \text{ in}}\right)(35 \text{ ft})(425°F - 70°F)}$$

$$= \boxed{3.57 \text{ Btu/hr-ft}^2\text{-°F}}$$

The answer is (C).

(b) To calculate the overall heat transfer coefficient, disregard the pipe thermal resistance and the inside film coefficient (small compared with outside film and radiation).

Work with the midpoint pipe temperature of 425°F.

The absolute temperatures are

$$T_1 = 425°F + 460 = 885°R$$

$$T_2 = 70°F + 460 = 530°R$$

For radiation heat loss, assume $F_a = 1$. For 500°F enamel paint of any color,

$$F_e = \epsilon \approx 0.9$$

$$\frac{Q_{\text{net}}}{A} = E_{\text{net}} = \sigma F_e F_a \left(T_1^4 - T_2^4\right)$$

$$= \left(0.1713 \times 10^{-8} \dfrac{\text{Btu}}{\text{hr-ft}^2\text{-°R}^4}\right)$$

$$\times (0.9)(1.0)\left((885°R)^4 - (530°R)^4\right)$$

$$= 824.1 \text{ Btu/hr-ft}^2$$

From Eq. 37.21, the radiant heat transfer coefficient is

$$h_{\text{radiation}} = \frac{E_{\text{net}}}{T_1 - T_2} = \frac{824.1 \dfrac{\text{Btu}}{\text{hr}}}{885°R - 530°R}$$

$$= 2.32 \text{ Btu/hr-ft}^2\text{-°F}$$

For the outside film coefficient, evaluate the film at the pipe midpoint. The film temperature is

$$T_f = \left(\tfrac{1}{2}\right)(425°F + 70°F) = 247.5°F$$

From App. 35.C,

$$\text{Pr} = 0.72$$

$$\frac{g\beta\rho^2}{\mu^2} = 0.657 \times 10^6 \dfrac{1}{\text{ft}^3\text{-°F}}$$

The characteristic length, L, is $d_o = 4.25$ in. The Grashof number is

$$\text{Gr} = L^3\left(\frac{\rho^2\beta g}{\mu^2}\right)\Delta T$$

$$= (4.25 \text{ in})^3\left(\frac{1 \text{ ft}}{12 \text{ in}}\right)^3\left(0.657 \times 10^6 \dfrac{1}{\text{ft}^3\text{-°F}}\right)$$

$$\times (425°F - 70°F)$$

$$= 1.04 \times 10^7$$

$$\text{Pr Gr} = (0.72)(1.04 \times 10^7) = 7.5 \times 10^6$$

From Table 35.3, the film coefficient for a horizontal cylinder is

$$h_o = (0.27)\left(\frac{\Delta T}{d}\right)^{0.25}$$

$$= (0.27)\left(\frac{425°F - 70°F}{(4.25 \text{ in})\left(\dfrac{1 \text{ ft}}{12 \text{ in}}\right)}\right)^{0.25}$$

$$= 1.52 \text{ Btu/hr-ft}^2\text{-°F}$$

The overall film coefficient is

$$U = h_{\text{total}} = h_{\text{radiation}} + h_o$$

$$= 2.32 \ \frac{\text{Btu}}{\text{hr-ft}^2\text{-}°\text{F}} + 1.52 \ \frac{\text{Btu}}{\text{hr-ft}^2\text{-}°\text{F}}$$

$$= \boxed{3.84 \ \text{Btu/hr-ft}^2\text{-}°\text{F}}$$

The answer is (C).

This is not too far from the actual value.

(c) • The internal film coefficient was disregarded.

   • The emissivity could be lower due to dirt on the outside of the duct.

   • h_r and h_o are not really additive.

   • Pipe thermal resistance was disregarded.

   • The midpoint calculations should be replaced with integration along the length.

SI Solution

(a) The absolute temperature of entering air is

$$260°\text{C} + 273 = 533\text{K}$$

The absolute pressure of entering air is

$$170 \ \text{kPa} + 101.3 \ \text{kPa} = 271.3 \ \text{kPa}$$

The density of air entering, from the ideal gas law, is

$$\rho = \frac{p}{RT} = \frac{(271.3 \ \text{kPa})\left(1000 \ \dfrac{\text{Pa}}{\text{kPa}}\right)}{\left(287.03 \ \dfrac{\text{J}}{\text{kg·K}}\right)(533\text{K})}$$

$$= 1.773 \ \text{kg/m}^3$$

The mass flow rate of entering air is

$$\dot{m} = \rho \dot{V} = \left(1.773 \ \frac{\text{kg}}{\text{m}^3}\right)\left(0.1 \ \frac{\text{m}^3}{\text{s}}\right) = 0.1773 \ \text{kg/s}$$

At low pressure, the air enthalpy is found from air tables. The absolute temperature of leaving air is

$$180°\text{C} + 273 = 453\text{K}$$

From App. 24.S,

$$h_1 = 537.09 \ \text{kJ/kg at } 533\text{K}$$

$$h_2 = 454.87 \ \text{kJ/kg at } 453\text{K}$$

The heat loss is

$$Q = \dot{m}(h_1 - h_2)$$

$$= \left(0.1773 \ \frac{\text{kg}}{\text{s}}\right)\left(537.09 \ \frac{\text{kJ}}{\text{kg}} - 454.87 \ \frac{\text{kJ}}{\text{kg}}\right)$$

$$\times \left(1000 \ \frac{\text{W}}{\text{kW}}\right)$$

$$= 14\,578 \ \text{W}$$

Assume the midpoint pipe surface temperature.

$$\left(\tfrac{1}{2}\right)(T_{\text{in}} - T_{\text{out}}) = \left(\tfrac{1}{2}\right)(260°\text{C} + 180°\text{C}) = 220°\text{C}$$

Since the heat loss is known, the overall heat transfer coefficient can be determined from

$$Q = UA\Delta T = U(\pi dL)\Delta T$$

$$U = \frac{Q}{(\pi dL)\Delta T}$$

$$= \frac{14\,578 \ \text{W}}{\pi(10.8 \ \text{cm})\left(\dfrac{1 \ \text{m}}{100 \ \text{cm}}\right)(10 \ \text{m})(220°\text{C} - 20°\text{C})}$$

$$= \boxed{21.5 \ \text{W/m}^2\text{·K}}$$

The answer is (C).

(b) From the customary U.S. solution, work with the midpoint pipe temperature of 220°C.

The absolute temperatures are

$$T_1 = 220°\text{C} + 273 = 493\text{K}$$

$$T_2 = 20°\text{C} + 273 = 293\text{K}$$

For radiation heat loss, assume $F_a = 1$. For 260°C enamel paint of any color,

$$F_e = \epsilon \approx 0.9$$

From Eq. 37.21, the radiant heat transfer coefficient is

$$h_r = \frac{\sigma F_a F_e \left(T_1^4 - T_2^4\right)}{T_1 - T_2}$$

$$= \frac{\left(5.67 \times 10^{-8} \ \dfrac{\text{W}}{\text{m}^2\text{·K}^4}\right)(1)(0.9)}{493\text{K} - 293\text{K}}$$

$$\qquad \times \left((493\text{K})^4 - (293\text{K})^4\right)$$

$$= 13.2 \ \text{W/m}^2\text{·K}$$

For the outside film coefficient, evaluate the film at the pipe midpoint. The film temperature is

$$T_f = \left(\tfrac{1}{2}\right)(220°\text{C} + 20°\text{C}) = 120°\text{C}$$

From App. 35.D, at 120°C,

$$Pr \approx 0.692$$

$$\frac{g\beta\rho^2}{\mu^2} = 0.528 \times 10^8 \; \frac{1}{K \cdot m^3}$$

The characteristic length is

$$L = \text{outside diameter} = (10.8 \text{ cm})\left(\frac{1 \text{ m}}{100 \text{ cm}}\right)$$

$$= 0.108 \text{ m}$$

The Grashof number is

$$Gr = L^3 \left(\frac{\rho^2 g\beta}{\mu^2}\right) \Delta T$$

$$= (0.108 \text{ m})^3 \left(0.528 \times 10^8 \; \frac{1}{K \cdot m^3}\right)$$

$$\times (220°C - 20°C)$$

$$= 1.33 \times 10^7$$

$$Pr\,Gr = (0.692)(1.33 \times 10^7) = 9.20 \times 10^6$$

From Table 35.3, the film coefficient for a horizontal cylinder is

$$h_o = (1.32)\left(\frac{\Delta T}{d}\right)^{0.25}$$

$$= (1.32)\left(\frac{220°C - 20°C}{0.108 \text{ m}}\right)^{0.25}$$

$$= 8.66 \text{ W/m}^2 \cdot \text{K}$$

The overall film coefficient is

$$U = h_{\text{total}} = h_r + h_o = 13.2 \; \frac{W}{m^2 \cdot K} + 8.66 \; \frac{W}{m^2 \cdot K}$$

$$= \boxed{21.86 \text{ W/m}^2 \cdot \text{K}}$$

The answer is (C).

This solution is almost the same as the actual value.

(c) See the customary U.S. solution.

6. *Customary U.S. Solution*

(a) Heat is lost from the top and sides by radiation and convection. The absolute temperature of the surroundings is

$$75°F + 460 = 535°R$$

$$A_{\text{sides}} = \pi dL = \pi(0.75 \text{ in})(1.5 \text{ in})\left(\frac{1 \text{ ft}}{12 \text{ in}}\right)^2$$

$$= 0.0245 \text{ ft}^2$$

$$A_{\text{top}} = \frac{\pi}{4}d^2 = \left(\frac{\pi}{4}\right)(0.75 \text{ in})^2\left(\frac{1 \text{ ft}}{12 \text{ in}}\right)^2$$

$$= 0.003068 \text{ ft}^2$$

$$Q_{\text{total}} = Q_{\text{convection}} + Q_{\text{radiation}}$$

$$= h_{\text{sides}}A_{\text{sides}}(T_s - T_\infty) + h_{\text{top}}A_{\text{top}}(T_s - T_\infty)$$

$$+ \sigma F_e F_a (A_{\text{sides}} + A_{\text{top}})(T_s^4 - T_\infty^4) \quad \text{[Eq. 1]}$$

For the first approximation of T_s, assume $h_{\text{sides}} = h_{\text{top}} = 1.65$ Btu/hr-ft²-°F, $F_a = 1$, and $F_e = \epsilon = 0.65$.

$$(5.0 \text{ W})\left(\frac{3.412 \; \frac{\text{Btu}}{\text{hr}}}{1 \text{ W}}\right) = \left(1.65 \; \frac{\text{Btu}}{\text{hr-ft}^2\text{-°F}}\right)$$

$$\times (0.0245 \text{ ft}^2)(T_s - 535°R)$$

$$+ \left(1.65 \; \frac{\text{Btu}}{\text{hr-ft}^2\text{-°F}}\right)$$

$$\times (0.003068 \text{ ft}^2)(T_s - 535°R)$$

$$+ \left(0.1713 \times 10^{-8} \; \frac{\text{Btu}}{\text{hr-ft}^2\text{-°R}^4}\right)$$

$$\times (0.65)(1)(0.0245 \text{ ft}^2$$

$$+ 0.003068 \text{ ft}^2)$$

$$\times \left(T_s^4 - (535°R)^4\right)$$

By trial and error, $T_s \approx 750°R$.

For natural convection on the outside, estimate the film temperature.

$$T_{\text{film}} = \left(\tfrac{1}{2}\right)(T_s + T_\infty) = \left(\tfrac{1}{2}\right)(750°R + 535°R)$$

$$= 642.5°R \quad (182.5°F)$$

From App. 35.C, the properties of air at 182.5°F are

$$Pr \approx 0.72$$

$$\frac{g\beta\rho^2}{\mu^2} = 1.01 \times 10^6 \; \frac{1}{\text{ft}^3\text{-°F}}$$

For the sides, the characteristic length is 1.5 in. The Grashof number is

$$Gr = L^3 \left(\frac{\rho^2 \beta g}{\mu^2}\right)(T_s - T_\infty)$$

$$= (1.5 \text{ in})^3 \left(\frac{1 \text{ ft}}{12 \text{ in}}\right)^3 \left(1.01 \times 10^6 \; \frac{1}{\text{ft}^3\text{-°F}}\right)$$

$$\times (750°R - 535°R)$$

$$= 4.24 \times 10^5$$

$$Pr\,Gr = (0.72)(4.24 \times 10^5) = 3.05 \times 10^5$$

From Table 35.3, the film coefficient for a vertical surface is

$$h_{\text{sides}} = (0.29)\left(\frac{T_s - T_\infty}{L}\right)^{0.25}$$

$$= (0.29)\left(\frac{750°R - 535°R}{(1.5 \text{ in})\left(\frac{1 \text{ ft}}{12 \text{ in}}\right)}\right)^{0.25}$$

$$= 1.87 \text{ Btu/hr-ft}^2\text{-°F}$$

(This application is slightly outside the useful range of this correlation.)

For the top, the characteristic length is 0.75 in, so

$$Gr = (0.75 \text{ in})^3 \left(\frac{1 \text{ ft}}{12 \text{ in}}\right)^3 \left(1.01 \times 10^6 \frac{1}{\text{ft}^3\text{-}°F}\right)$$
$$\times (750°R - 535°R)$$
$$= 5.30 \times 10^4$$
$$Pr\, Gr = (0.72)(5.30 \times 10^4) = 3.8 \times 10^4$$

From Table 35.3, the film coefficient for a horizontal surface is

$$h_{\text{top}} = (0.27)\left(\frac{T_s - T_\infty}{L}\right)^{0.25}$$

$$= (0.27)\left(\frac{750°R - 535°R}{(0.9)(0.75 \text{ in})\left(\frac{1 \text{ ft}}{12 \text{ in}}\right)}\right)^{0.25}$$

$$= 2.12 \text{ Btu/hr-ft}^2\text{-}°F$$

Substituting the calculated values of h_{top} and h_{sides} into Eq. 1 gives

$$Q_{\text{total}} = (5.0 \text{ W})\left(\frac{3.412 \frac{\text{Btu}}{\text{hr}}}{1 \text{ W}}\right)$$

$$= \left(1.87 \frac{\text{Btu}}{\text{hr-ft}^2\text{-}°F}\right)(0.0245 \text{ ft}^2)(T_s - 535°R)$$

$$+ \left(2.12 \frac{\text{Btu}}{\text{hr-ft}^2\text{-}°F}\right)(0.003068 \text{ ft}^2)(T_s - 535°R)$$

$$+ \left(0.1713 \times 10^{-8} \frac{\text{Btu}}{\text{hr-ft}^2\text{-}°R^4}\right)(0.65)(1)$$

$$\times (0.0245 \text{ ft}^2 + 0.003068 \text{ ft}^2)\left(T_s^4 - (535°R)^4\right)$$

By trial and error, $T_s \approx \boxed{736°R.}$

(b) Substituting $T_s = 736°R$ in the preceding equation,

$$Q_{\text{total}} = 9.209 \frac{\text{Btu}}{\text{hr}} + 1.277 \frac{\text{Btu}}{\text{hr}} + 6.492 \frac{\text{Btu}}{\text{hr}}$$

$$= 16.98 \text{ Btu/hr}$$

$$\frac{Q_{\text{convection}}}{Q_{\text{total}}} = \frac{9.209 \frac{\text{Btu}}{\text{hr}} + 1.277 \frac{\text{Btu}}{\text{hr}}}{16.98 \frac{\text{Btu}}{\text{hr}}}$$

$$= \boxed{0.618 \quad (61.8\%)}$$

$$\frac{Q_{\text{radiation}}}{Q_{\text{total}}} = \frac{6.492 \frac{\text{Btu}}{\text{hr}}}{16.98 \frac{\text{Btu}}{\text{hr}}} = \boxed{0.382 \quad (38.2\%)}$$

SI Solution

(a) Heat is lost from the top and sides by radiation and convection.

The absolute temperature of the surrounding is

$$24°C + 273 = 297K$$

$$A_{\text{sides}} = \pi dL = \pi (19 \text{ mm})(38 \text{ mm})\left(\frac{1 \text{ m}}{1000 \text{ mm}}\right)^2$$
$$= 0.00227 \text{ m}^2$$

$$A_{\text{top}} = \frac{\pi}{4}d^2 = \left(\frac{\pi}{4}\right)(19 \text{ mm})^2\left(\frac{1 \text{ m}}{1000 \text{ mm}}\right)^2$$
$$= 0.000284 \text{ m}^2$$

$$Q_{\text{total}} = Q_{\text{convection}} + Q_{\text{radiation}}$$
$$= h_{\text{sides}}A_{\text{sides}}(T_s - T_\infty) + h_{\text{top}}A_{\text{top}}(T_s - T_\infty)$$
$$+ \sigma F_e F_a (A_{\text{sides}} + A_{\text{top}})(T_s^4 - T_\infty^4) \quad [\text{Eq. 1}]$$

For a first approximation of T_s, assume $h_{\text{sides}} = h_{\text{top}} = 9.4 \text{ W/m}^2\text{·K}$, $F_a = 1$, and $F_e = \epsilon = 0.65$.

$$5.0 \text{ W} = \left(9.4 \frac{\text{W}}{\text{m}^2\text{·K}}\right)(0.00227 \text{ m}^2)(T_s - 297K)$$

$$+ \left(9.4 \frac{\text{W}}{\text{m}^2\text{·K}}\right)(0.000284 \text{ m}^2)(T_s - 297K)$$

$$+ \left(5.67 \times 10^{-8} \frac{\text{W}}{\text{m}^2\text{·K}^4}\right)(0.65)(1)$$

$$\times (0.00227 \text{ m}^2 + 0.000284 \text{ m}^2)$$

$$\times \left(T_s^4 - (297K)^4\right)$$

By trial and error, $T_s = 416.75K$.

For natural convection on the outside, estimate the film temperature.

$$T_{\text{film}} = \left(\tfrac{1}{2}\right)(T_s + T_\infty) = \left(\tfrac{1}{2}\right)(416.75K + 297K)$$
$$= 356.9K \quad (83.9°C)$$

From App. 35.D, the properties of air at 83.9°C are

$$Pr \approx 0.697$$
$$\frac{g\beta\rho^2}{\mu^2} \approx 0.616 \times 10^8 \frac{1}{\text{K·m}^3}$$

For the sides, the characteristic length is 38 mm.

$$Gr = (38 \text{ mm})^3 \left(\frac{1 \text{ m}}{1000 \text{ mm}}\right)^3 \left(0.616 \times 10^8 \frac{1}{\text{K·m}^3}\right)$$
$$\times (416.75K - 297K)$$
$$= 4.048 \times 10^5$$
$$Pr\, Gr = (0.697)(4.048 \times 10^5) = 2.82 \times 10^5$$

Heat Transfer

For the top, the characteristic length is 19 mm.

$$\text{Gr} = (19 \text{ mm})^3 \left(\frac{1 \text{ m}}{1000 \text{ mm}}\right)^3 \left(0.616 \times 10^8 \frac{1}{\text{K·m}^3}\right)$$
$$\times (416.75\text{K} - 297\text{K})$$
$$= 5.06 \times 10^4$$
$$\text{Pr Gr} = (0.697)(5.06 \times 10^4) = 3.53 \times 10^4$$

From Table 35.2, the film coefficient for a horizontal surface is

$$h_{\text{top}} = (1.32) \left(\frac{T_s - T_\infty}{L}\right)^{0.25}$$

$$= (1.32) \left(\frac{416.75\text{K} - 297\text{K}}{(0.9)(19 \text{ mm})\left(\frac{1 \text{ m}}{1000 \text{ mm}}\right)}\right)^{0.25}$$

$$= 12.08 \text{ W/m}^2\text{·K}$$

From Table 35.3, the film coefficient for a vertical surface is

$$h_{\text{sides}} = (1.37) \left(\frac{416.75\text{K} - 297\text{K}}{(38 \text{ mm})\left(\frac{1 \text{ m}}{1000 \text{ mm}}\right)}\right)^{0.25}$$

$$= 10.26 \text{ W/m}^2\text{·K}$$

(This application is slightly outside the useful range of this correlation.)

Using the calculated values of h_{top} and h_{sides} into Eq. 1,

$$Q_{\text{total}} = 5 \text{ W}$$
$$= \left(10.26 \frac{\text{W}}{\text{m}^2\text{·K}}\right)(0.00227 \text{ m}^2)(T_s - 297\text{K})$$
$$+ \left(12.08 \frac{\text{W}}{\text{m}^2\text{·K}}\right)(0.000284 \text{ m}^2)(T_s - 297\text{K})$$
$$+ \left(5.67 \times 10^{-8} \frac{\text{W}}{\text{m}^2\text{·K}^4}\right)(0.65)(1)$$
$$\times (0.00227 \text{ m}^2 + 0.000284 \text{ m}^2)$$
$$\times (T_s^4 - (297\text{K})^4)$$

By trial and error, $T_s = \boxed{411\text{K.}}$

(b) Substitute $T_s = 411\text{K}$ into the preceding equation.

$$Q_{\text{total}} = 2.655 \text{ W} + 0.381 \text{ W} + 1.953 \text{ W}$$
$$= 4.989 \text{ W}$$

$$\frac{Q_{\text{convection}}}{Q_{\text{total}}} = \frac{2.655 \text{ W} + 0.381 \text{ W}}{4.989 \text{ W}} = \boxed{0.609 \quad (60.9\%)}$$

$$\frac{Q_{\text{radiation}}}{Q_{\text{total}}} = \frac{1.953 \text{ W}}{4.989 \text{ W}} = \boxed{0.391 \quad (39.1\%)}$$

7. *Customary U.S. Solution*

The velocity is relatively low, so incompressible flow can be assumed.

The film coefficient on the probe is

$$h = \frac{0.024 G^{0.8}}{D^{0.4}}$$

$$= \frac{(0.024)\left(3480 \frac{\text{lbm}}{\text{hr-ft}^2}\right)^{0.8}}{\left((0.5 \text{ in})\left(\frac{1 \text{ ft}}{12 \text{ in}}\right)\right)^{0.4}} = 58.3 \text{ Btu/hr-ft}^2\text{-}^\circ\text{F}$$

The absolute temperature of the walls is

$$T_{\text{walls}} = 600^\circ\text{F} + 460 = 1060^\circ\text{R}$$

Neglect conduction and the insignificant kinetic energy loss. The thermocouple gains heat through radiation from the walls and loses heat through convection to the gas.

$$Q_{\text{convection}} = A E_{\text{radiation}}$$
$$hA(T_{\text{probe}} - T_{\text{gas}}) = A\sigma\epsilon(T_{\text{walls}}^4 - T_{\text{probe}}^4)$$
$$h(T_{\text{probe}} - T_{\text{gas}}) = \sigma\epsilon(T_{\text{walls}}^4 - T_{\text{probe}}^4)$$

(a) If the actual gas temperature is $300^\circ\text{F} + 460 = 760^\circ\text{R}$,

$$\left(58.3 \frac{\text{Btu}}{\text{hr-ft}^2\text{-}^\circ\text{F}}\right)$$
$$\times (T_{\text{probe}} - 760^\circ\text{R}) = \left(0.1713 \times 10^{-8} \frac{\text{Btu}}{\text{hr-ft}^2\text{-}^\circ\text{R}^4}\right)$$
$$\times (0.8)\left((1060^\circ\text{R})^4 - T_{\text{probe}}^4\right)$$
$$(1.37 \times 10^{-9})T_{\text{probe}}^4$$
$$+ (58.3)T_{\text{probe}} = 46{,}038$$

By trial and error, $T_{\text{probe}} = \boxed{781^\circ\text{R.}}$

The answer is (D).

(b) If $T_{\text{probe}} = 300^\circ\text{F} + 460 = 760^\circ\text{R}$,

$$\left(58.3 \frac{\text{Btu}}{\text{hr-ft}^2\text{-}^\circ\text{F}}\right)$$
$$\times (760^\circ\text{R} - T_{\text{gas}}) = \left(0.1713 \times 10^{-8} \frac{\text{Btu}}{\text{hr-ft}^2\text{-}^\circ\text{R}^4}\right)$$
$$\times (0.8)\left((1060^\circ\text{R})^4 - (760^\circ\text{R})^4\right)$$
$$T_{\text{gas}} = \boxed{738.2^\circ\text{R}}$$

The answer is (B).

SI Solution

The velocity is relatively low, so incompressible flow can be assumed.

The film coefficient on the probe is

$$h = \frac{2.9G^{0.8}}{D^{0.4}}$$

$$= \frac{(2.9)\left(4.7 \ \dfrac{\text{kg}}{\text{s·m}^2}\right)^{0.8}}{(13 \text{ mm})^{0.4}\left(\dfrac{1 \text{ m}}{1000 \text{ mm}}\right)^{0.4}} = 56.82 \text{ W/m}^2\text{·K}$$

The absolute temperature of the walls is

$$T_{\text{walls}} = 315°\text{C} + 273 = 588\text{K}$$

Neglecting conduction and the insignificant kinetic energy loss, the thermocouple gains heat through radiation from the walls and loses heat through convection to the gas.

$$\frac{Q_{\text{convection}}}{A} = E_{\text{radiation}}$$
$$h(T_{\text{probe}} - T_{\text{gas}}) = \sigma\epsilon(T_{\text{walls}}^4 - T_{\text{probe}}^4)$$

(a) If the actual gas temperature is $150°\text{C} + 273 = 423\text{K}$,

$$\left(56.82 \ \frac{\text{W}}{\text{m}^2\text{·K}}\right)$$
$$\times (T_{\text{probe}} - 423\text{K}) = \left(5.67 \times 10^{-8} \ \frac{\text{W}}{\text{m}^2\text{·K}^4}\right)(0.8)$$
$$\times \left((588\text{K})^4 - T_{\text{probe}}^4\right)$$

$$(4.536 \times 10^{-8})T_{\text{probe}}^4$$
$$+ (56.82)T_{\text{probe}} = 29\,457$$

By trial and error, $T_{\text{probe}} = \boxed{477\text{K.}}$

The answer is (D).

(b) If $T_{\text{probe}} = 150°\text{C} + 273 = 423\text{K}$,

$$\left(56.82 \ \frac{\text{W}}{\text{m}^2\text{·K}}\right)$$
$$\times (423\text{K} - T_{\text{gas}}) = \left(5.67 \times 10^{-8} \ \frac{\text{W}}{\text{m}^2\text{·K}^4}\right)(0.8)$$
$$\times \left((588\text{K})^4 - (423\text{K})^4\right)$$
$$T_{\text{gas}} = \boxed{353.1\text{K}}$$

The answer is (B).

Heat Transfer

38 Psychrometrics

PRACTICE PROBLEMS

1. A room contains air at 80°F (27°C) dry-bulb and 67°F (19°C) wet-bulb. The total pressure is 1 atm.

(a) What is the humidity ratio?
- (A) 0.009 lbm/lbm (0.009 kg/kg)
- (B) 0.011 lbm/lbm (0.011 kg/kg)
- (C) 0.014 lbm/lbm (0.014 kg/kg)
- (D) 0.018 lbm/lbm (0.018 kg/kg)

(b) What is the enthalpy?
- (A) 30.2 Btu/lbm (51.3 kJ/kg)
- (B) 30.8 Btu/lbm (52.4 kJ/kg)
- (C) 31.5 Btu/lbm (53.9 kJ/kg)
- (D) 31.9 Btu/lbm (54.2 kJ/kg)

(c) What is the specific heat?
- (A) 0.234 Btu/lbm-°F (0.979 kJ/kg·K)
- (B) 0.237 Btu/lbm-°F (0.991 kJ/kg·K)
- (C) 0.239 Btu/lbm-°F (0.999 kJ/kg·K)
- (D) 0.242 Btu/lbm-°F (1.012 kJ/kg·K)

2. If one layer of cooling coils effectively bypasses one-third of the air passing through it, what is the theoretical bypass factor for four layers of identical cooling coils in series?
- (A) 0.01
- (B) 0.09
- (C) 0.33
- (D) 0.67

3. 1000 ft³/min (0.5 m³/s) of air at 50°F (10°C) dry-bulb and 95% relative humidity are mixed with 1500 ft³/min (0.75 m³/s) of air at 76°F (24°C) dry-bulb and 45% relative humidity.

(a) What is the dry-bulb temperature of the mixture?
- (A) 55°F (13°C)
- (B) 65°F (18°C)
- (C) 68°F (20°C)
- (D) 70°F (21°C)

(b) What is the specific humidity of the mixture?
- (A) 0.008 lbm/lbm (0.008 kg/kg)
- (B) 0.009 lbm/lbm (0.009 kg/kg)
- (C) 0.010 lbm/lbm (0.010 kg/kg)
- (D) 0.011 lbm/lbm (0.011 kg/kg)

(c) What is the dew point of the mixture?
- (A) 45°F (7.2°C)
- (B) 48°F (8.9°C)
- (C) 51°F (11°C)
- (D) 54°F (12°C)

4. Air at 60°F (16°C) dry-bulb and 45°F (7°C) wet-bulb passes through an air washer with a humidifying efficiency of 70%.

(a) What is the effective bypass factor of the system?
- (A) 0.30
- (B) 0.50
- (C) 0.67
- (D) 0.70

(b) What is the dry-bulb temperature of the air leaving the washer?
- (A) 45°F (7.2°C)
- (B) 50°F (9.7°C)
- (C) 54°F (12°C)
- (D) 57°F (14°C)

5. 95°F (35°C) dry-bulb, 75°F (24°C) wet-bulb air passes through a cooling tower and leaves at 85°F (29°C) dry-bulb and 90% relative humidity.

(a) What is the enthalpy change per cubic foot (meter) of air?
- (A) 0.57 Btu/ft³ (18 kJ/m³)
- (B) 1.3 Btu/ft³ (40 kJ/m³)
- (C) 3.2 Btu/ft³ (99 kJ/m³)
- (D) 7.6 Btu/ft³ (240 kJ/m³)

(b) What is the change in moisture content per cubic foot (meter) of air?
- (A) 1.8×10^{-4} lbm/ft³ (2.7×10^{-3} kg/m³)
- (B) 3.3×10^{-4} lbm/ft³ (4.5×10^{-3} kg/m³)
- (C) 6.7×10^{-4} lbm/ft³ (9.9×10^{-3} kg/m³)
- (D) 9.2×10^{-4} lbm/ft³ (14×10^{-3} kg/m³)

6. An air washer receives 1800 ft³/min (0.85 m³/s) of air at 70°F (21°C) and 40% relative humidity and discharges the air at 75% relative humidity. A recirculating water spray with a constant temperature of 50°F (10°C) is used.

(a) What will be the condition of the discharged air?

(b) What mass of makeup water is required per minute?

7. Repeat Probs. 6(a) and 6(b) using saturated steam at atmospheric pressure in place of the 50°F (10°C) water spray.

8. During performances, a theater experiences a sensible heat load of 500,000 Btu/hr (150 kW) and a moisture load of 175 lbm/hr (80 kg/h). Air enters the theater at 65°F (18°C) and 55% relative humidity and is removed when it reaches 75°F (24°C) or 60% relative humidity, whichever comes first.

(a) What is the ventilation rate in mass of air per hour?

(b) What are the conditions of the air leaving the theater?

9. 500 ft³/min (0.25 m³/s) of air at 80°F (27°C) dry-bulb and 70% relative humidity are removed from a room. 150 ft³/min (0.075 m³/s) pass through an air conditioner and leave saturated at 50°F (10°C). The remaining 350 ft³/min (0.175 m³/s) bypass the air conditioner and mix with the conditioned air at 1 atm.

(a) What is the mixture's temperature?
  (A) 55°F (12.8°C)
  (B) 66°F (18.8°C)
  (C) 71°F (21.9°C)
  (D) 74°F (23.3°C)

(b) What is the mixture's humidity ratio?
  (A) 0.010 lbm/lbm (1.0 g/kg)
  (B) 0.013 lbm/lbm (1.3 g/kg)
  (C) 0.017 lbm/lbm (1.7 g/kg)
  (D) 0.021 lbm/lbm (2.1 g/kg)

(c) What is the mixture's relative humidity?
  (A) 45%
  (B) 57%
  (C) 73%
  (D) 81%

(d) What is the heat load (in tons) of the air conditioner?
  (A) 0.9 ton
  (B) 1.3 ton
  (C) 2.4 tons
  (D) 2.9 tons

10. (*Time limit: one hour*) A dehumidifier takes 5000 ft³/min (2.36 m³/s) of air at 95°F (35°C) dry-bulb and 70% relative humidity and discharges it at 60°F (16°C) dry-bulb and 95% relative humidity. The dehumidifier uses a wet R-12 refrigeration cycle operating between 100°F (saturated) (38°C) and 50°F (10°C).

(a) Locate the air entering and leaving points on the psychrometric chart.

(b) Find the quantity of water removed from the air.

(c) Find the quantity of heat removed from the air.

(d) Draw the temperature-entropy and enthalpy-entropy diagrams for the refrigeration cycle.

(e) Find the temperature, pressure, enthalpy, entropy, and specific volume for each endpoint of the refrigeration cycle.

11. (*Time limit: one hour*) 1500 ft³/min (0.71 m³/s) of saturated 25 psia (170 kPa) air is heated from 200°F to 400°F (93°C to 204°C) in a constant pressure, constant moisture drying process.

(a) What is the final relative humidity?
  (A) 4.7%
  (B) 9.2%
  (C) 13%
  (D) 23%

(b) What is the final specific humidity?
  (A) 0.41 lbm/lbm (0.41 kg/kg)
  (B) 0.53 lbm/lbm (0.53 kg/kg)
  (C) 0.66 lbm/lbm (0.66 kg/kg)
  (D) 0.79 lbm/lbm (0.79 kg/kg)

(c) How much heat is required per unit mass of dry air?
  (A) 31 Btu/lbm (71 kJ/kg)
  (B) 57 Btu/lbm (130 kJ/kg)
  (C) 99 Btu/lbm (230 kJ/kg)
  (D) 120 Btu/lbm (280 kJ/kg)

(d) What is the final dew point?
  (A) 180°F (82°C)
  (B) 200°F (93°C)
  (C) 220°F (104°C)
  (D) 240°F (115°C)

12. (*Time limit: one hour*) 410 lbm/hr (0.052 kg/s) of dry 800°F (427°C) air pass through a scrubber to reduce particulate emissions. To protect the elastomeric seals in the scrubber, the air temperature is reduced to 350°F (177°C) by passing the air through a spray of 80°F (27°C) water. The pressure in the spray chamber is 20 psia (140 kPa).

(a) How much water is evaporated per hour?
  (A) 18 lbm/hr (0.0023 kg/s)
  (B) 27 lbm/hr (0.0035 kg/s)
  (C) 31 lbm/hr (0.0040 kg/s)
  (D) 39 lbm/hr (0.0050 kg/s)

(b) What will be the relative humidity of the air leaving the spray chamber?
  (A) 1.3%
  (B) 2.0%
  (C) 3.1%
  (D) 4.4%

13. (*Time limit: one hour*) An evaporative counter-flow air cooling tower removes 1×10^6 Btu/hr (290 kW) from a water flow. The temperature of the water is reduced from 120°F to 110°F (49°C to 43°C). Air enters the cooling tower at 91°F (33°C) and 60% relative humidity, and air leaves at 100°F (38°C) and 82% relative humidity.

(a) Calculate the air flow rate.

(b) Calculate the quantity of makeup water.

SOLUTIONS

1. *Customary U.S. Solution*

((a) and (b)) Locate the intersection of 80°F dry bulb and 67°F wet bulb on the psychrometric chart (App. 38.A). Read the value of humidity and enthalpy.

$$\omega = 0.0112 \text{ lbm moisture/lbm dry air}$$
$$h = 31.5 \text{ Btu/lbm dry air}$$

The answer is (B) for Prob. 1(a).

The answer is (C) for Prob. 1(b).

(c) c_p is gravimetrically weighted. c_p for air is 0.240 Btu/lbm-°F, and c_p for steam is approximately 0.40 Btu/lbm-°F.

$$G_{\text{air}} = \frac{1}{1 + 0.0112} = 0.989$$

$$G_{\text{steam}} = \frac{0.0112}{1 + 0.0112} = 0.011$$

$$c_{p,\text{mixture}} = G_{\text{air}}c_{p,\text{air}} + G_{\text{steam}}c_{p,\text{steam}}$$

$$= (0.989)\left(0.240 \,\frac{\text{Btu}}{\text{lbm-°F}}\right)$$

$$+ (0.011)\left(0.40 \,\frac{\text{Btu}}{\text{lbm-°F}}\right)$$

$$= \boxed{0.242 \text{ Btu/lbm-°F}}$$

The answer is (D).

SI Solution

((a) and (b)) Locate the intersection of 27°C dry bulb and 19°C wet bulb on the psychrometric chart (App. 38.B). Read the value of humidity and enthalpy.

$$\omega = \left(10.5 \,\frac{\text{g}}{\text{kg dry air}}\right)\left(\frac{1 \text{ kg}}{1000 \text{ g}}\right)$$

$$\boxed{\begin{array}{l} = 0.0105 \text{ kg/kg dry air} \\ h = 53.9 \text{ kJ/kg dry air} \end{array}}$$

The answer is (B) for Prob. 1(a).

The answer is (C) for Prob. 1(b).

(c) c_p is gravimetrically weighted. c_p for air is 1.0048 kJ/kg·K, and c_p for steam is approximately 1.675 kJ/kg·K.

$$G_{\text{air}} = \frac{1}{1 + 0.0105} = 0.990$$

$$G_{\text{steam}} = \frac{0.0105}{1 + 0.0105} = 0.010$$

$$c_{p,\text{mixture}} = G_{\text{air}}c_{p,\text{air}} + G_{\text{air}}c_{p,\text{steam}}$$

$$= (0.990)\left(1.0048 \ \frac{\text{kJ}}{\text{kg·K}}\right)$$

$$+ (0.010)\left(1.675 \ \frac{\text{kJ}}{\text{kg·K}}\right)$$

$$= \boxed{1.0115 \ \text{kJ/kg·K}}$$

The answer is (D).

2.

$$\text{BF}_{n \text{ layers}} = (\text{BF}_{1 \text{ layer}})^n$$

$$= \left(\tfrac{1}{3}\right)^4 = \boxed{0.0123}$$

Only 1% of the air will be untreated.

The answer is (A).

3.

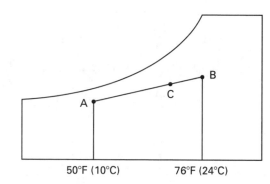

50°F (10°C)   76°F (24°C)

Customary U.S. Solution

(a) Locate the two points on the psychrometric chart and draw a line between them.

Reading from the chart (specific volumes),

$$v_A = 13.0 \ \text{ft}^3/\text{lbm}$$

$$v_B = 13.7 \ \text{ft}^3/\text{lbm}$$

The density at each point is

$$\rho_A = \frac{1}{v_A} = \frac{1}{13.0 \ \dfrac{\text{ft}^3}{\text{lbm}}} = 0.0769 \ \text{lbm/ft}^3$$

$$\rho_B = \frac{1}{v_B} = \frac{1}{13.7 \ \dfrac{\text{ft}^3}{\text{lbm}}} = 0.0730 \ \text{lbm/ft}^3$$

The mass flow at each point is

$$\dot{m}_A = \rho_A \dot{V}_A = \left(0.0769 \ \frac{\text{lbm}}{\text{ft}^3}\right)\left(1000 \ \frac{\text{ft}^3}{\text{min}}\right)$$

$$= 76.9 \ \text{lbm/min}$$

$$\dot{m}_B = \rho_B \dot{V}_B = \left(0.0730 \ \frac{\text{lbm}}{\text{ft}^3}\right)\left(1500 \ \frac{\text{ft}^3}{\text{min}}\right)$$

$$= 109.5 \ \text{lbm/min}$$

The gravimetric fraction of flow A is

$$\frac{76.9 \ \dfrac{\text{lbm}}{\text{min}}}{76.9 \ \dfrac{\text{lbm}}{\text{min}} + 109.5 \ \dfrac{\text{lbm}}{\text{min}}} = 0.413$$

Since the scales are all linear,

$$0.413 = \frac{T_B - T_C}{T_B - T_A}$$

$$T_C = T_B - (0.413)(T_B - T_A)$$

$$= 76°\text{F} - (0.413)(76°\text{F} - 50°\text{F})$$

$$= \boxed{65.3°\text{F}}$$

The answer is (B).

(b) $\omega = \boxed{0.0082 \ \text{lbm moisture/lbm dry air}}$

The answer is (A).

(c) $T_{\text{dp}} = \boxed{51°\text{F}}$

The answer is (C).

SI Solution

(a) Locate the two points on the psychrometric chart and draw a line between them.

Reading from the chart (specific volumes),

$$v_A = 0.813 \ \text{m}^3/\text{kg dry air}$$

$$v_B = 0.856 \ \text{m}^3/\text{kg dry air}$$

The density at each point is

$$\rho_A = \frac{1}{v_A} = \frac{1}{0.813 \ \dfrac{\text{m}^3}{\text{kg}}} = 1.23 \ \text{kg/m}^3$$

$$\rho_B = \frac{1}{v_B} = \frac{1}{0.856 \ \dfrac{\text{m}^3}{\text{kg}}} = 1.17 \ \text{kg/m}^3$$

The mass flow at each point is

$$\dot{m}_A = \rho_A \dot{V}_A = \left(1.23 \ \frac{\text{kg}}{\text{m}^3}\right)\left(0.5 \ \frac{\text{m}^3}{\text{s}}\right) = 0.615 \ \text{kg/s}$$

$$\dot{m}_B = \rho_B \dot{V}_B = \left(1.17 \ \frac{\text{kg}}{\text{m}^3}\right)\left(0.75 \ \frac{\text{m}^3}{\text{s}}\right) = 0.878 \ \text{kg/s}$$

The gravimetric fraction of flow A is

$$\frac{0.615 \, \dfrac{\text{kg}}{\text{s}}}{0.615 \, \dfrac{\text{kg}}{\text{s}} + 0.878 \, \dfrac{\text{kg}}{\text{s}}} = 0.412$$

Since the scales are linear,

$$0.412 = \frac{T_B - T_C}{T_B - T_A}$$
$$T_C = T_B - (0.412)(T_B - T_A)$$
$$= 24°\text{C} - (0.412)(24°\text{C} - 10°\text{C})$$
$$= \boxed{18.2°\text{C}}$$

The answer is (B).

(b) $\qquad \omega = \left(8.0 \, \dfrac{\text{g}}{\text{kg dry air}}\right)\left(\dfrac{1 \, \text{kg}}{1000 \, \text{g}}\right)$

$$= \boxed{0.008 \, \text{kg/kg dry air}}$$

The answer is (A).

(c) $\qquad T_{\text{dp}} = \boxed{10.6°\text{C}}$

The answer is (C).

4. *Customary U.S. Solution*

(a) From Eq. 38.24, the bypass factor is

$$\text{BF} = 1 - \eta_{\text{sat}}$$
$$= 1 - 0.70 = \boxed{0.30}$$

The answer is (A).

(b) From Eq. 38.34, the dry-bulb temperature of air leaving the washer can be determined.

$$\eta_{\text{sat}} = \frac{T_{\text{db,in}} - T_{\text{db,out}}}{T_{\text{db,in}} - T_{\text{wb,in}}}$$
$$0.70 = \frac{60°\text{F} - T_{\text{db,out}}}{60°\text{F} - 45°\text{F}}$$
$$T_{\text{db,out}} = \boxed{49.5°\text{F}}$$

The answer is (B).

SI Solution

(a) See the customary U.S. solution.

(b) From Eq. 38.34, the dry-bulb temperature of air leaving the washer can be determined.

$$\eta_{\text{sat}} = \frac{T_{\text{db,in}} - T_{\text{db,out}}}{T_{\text{db,in}} - T_{\text{wb,in}}}$$
$$0.70 = \frac{16°\text{C} - T_{\text{db,out}}}{16°\text{C} - 7°\text{C}}$$
$$T_{\text{db,out}} = \boxed{9.7°\text{C}}$$

The answer is (B).

5. *Customary U.S. Solution*

(a) Refer to the psychrometric chart (App. 38.A).

At point 1, properties of air at $T_{\text{db}} = 95°\text{F}$ and $T_{\text{wb}} = 75°\text{F}$ are

$$\omega_1 = 0.0141 \, \text{lbm moisture/lbm air}$$
$$h_1 = 38.4 \, \text{Btu/lbm air}$$
$$v_1 = 14.3 \, \text{ft}^3/\text{lbm air}$$

At point 2, properties of air at $T_{\text{db}} = 85°\text{F}$ and 90% relative humidity are

$$\omega_2 = 0.0237 \, \text{lbm moisture/lbm air}$$
$$h_2 = 46.6 \, \text{Btu/lbm air}$$

The enthalpy change is

$$\frac{h_2 - h_1}{v_1} = \frac{46.6 \, \dfrac{\text{Btu}}{\text{lbm air}} - 38.4 \, \dfrac{\text{Btu}}{\text{lbm air}}}{14.3 \, \dfrac{\text{ft}^3}{\text{lbm air}}}$$

$$= \boxed{0.573 \, \text{Btu/ft}^3 \, \text{air}}$$

The answer is (A).

(b) The moisture added is

$$\frac{\omega_2 - \omega_1}{v_1} = \frac{0.0237 \, \dfrac{\text{lbm moisture}}{\text{lbm air}} - 0.0141 \, \dfrac{\text{lbm moisture}}{\text{lbm air}}}{14.3 \, \dfrac{\text{ft}^3}{\text{lbm air}}}$$

$$= \boxed{6.71 \times 10^{-4} \, \text{lbm/ft}^3 \, \text{air}}$$

The answer is (C).

HVAC

SI Solution

(a) Refer to the psychrometric chart (App. 38.B).

At point 1, properties of air at $T_{db} = 35°C$ and $T_{wb} = 24°C$ are

$$\omega_1 = 14.3 \text{ g/kg air}$$
$$h_1 = 71.8 \text{ kJ/kg air}$$
$$v_1 = 0.8893 \text{ m}^3\text{/kg air}$$

At point 2, properties of air at $T_{db} = 29°C$ and 90% relative humidity are

$$\omega_2 = 23.1 \text{ g/kg air}$$
$$h_2 = 88 \text{ kJ/kg air}$$

The enthalpy change is

$$\frac{h_2 - h_1}{v_1} = \frac{88 \dfrac{\text{kJ}}{\text{kg air}} - 71.8 \dfrac{\text{kJ}}{\text{kg air}}}{0.8893 \dfrac{\text{m}^3}{\text{kg air}}}$$

$$= \boxed{18.2 \text{ kJ/m}^3 \text{ air}}$$

The answer is (A).

(b) The moisture added is

$$\frac{\omega_2 - \omega_1}{v_1} = \frac{\left(\left(23.1 \dfrac{\text{g}}{\text{kg air}}\right) - \left(14.3 \dfrac{\text{kg}}{\text{kg air}}\right)\right) \times \left(\dfrac{1 \text{ kg}}{1000 \text{ g}}\right)}{0.8893 \dfrac{\text{m}^3}{\text{kg air}}}$$

$$= \boxed{9.90 \times 10^{-3} \text{ kg/m}^3 \text{ air}}$$

The answer is (C).

6.

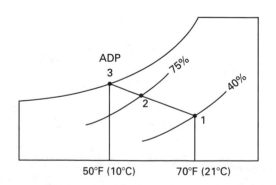

Customary U.S. Solution

(a) Refer to the psychrometric chart (App. 38.A).

At point 1, properties of air at $T_{db} = 70°F$ and $\phi = 40\%$ are

$$h_1 = 23.6 \text{ Btu/lbm air}$$
$$\omega_1 = 0.00623 \text{ lbm moisture/lbm air}$$
$$v_1 = 13.48 \text{ ft}^3\text{/lbm air}$$

The mass flow rate of incoming air is

$$\dot{m}_{a,1} = \frac{\dot{V}_1}{v_1} = \frac{1800 \dfrac{\text{ft}^3}{\text{min}}}{13.48 \dfrac{\text{ft}^3}{\text{lbm air}}} = 133.53 \text{ lbm air/min}$$

Locate point 1 on the psychrometric chart.

Notice that the temperature of the recirculating water is constant but not equal to the air's entering wet-bulb temperature. Therefore, this is not an adiabatic process.

Locate point 3 as 50°F saturated condition (water being sprayed) on the psychrometric chart.

Draw a line from point 1 to point 3. The intersection of this line with 75% relative humidity defines point 2 as

$$\boxed{\begin{aligned} h_2 &= 21.4 \text{ Btu/lbm air} \\ \omega_2 &= 0.0072 \text{ lbm moisture/lbm air} \\ T_{db,2} &= 56°F \\ T_{wb,2} &= 51.8°F \end{aligned}}$$

(b) The moisture (water) added is

$$\dot{m}_w = \dot{m}_{a,1}(\omega_2 - \omega_1)$$
$$= \left(133.53 \dfrac{\text{lbm air}}{\text{min}}\right)\left(0.0072 \dfrac{\text{lbm moisture}}{\text{lbm air}}\right.$$
$$\left. - 0.00625 \dfrac{\text{lbm moisture}}{\text{lbm air}}\right)$$
$$= \boxed{0.127 \text{ lbm/min}}$$

SI Solution

(a) Refer to the psychrometric chart (App. 38.B).

At point 1, properties of air at $T_{db} = 21°C$ and $\phi = 40\%$ are

$$h_1 = 36.75 \text{ kJ/kg air}$$
$$\omega_1 = 6.2 \text{ g moisture/kg air}$$
$$v_1 = 0.842 \text{ m}^3\text{/kg air}$$

The mass flow rate of incoming air is

$$\dot{m}_{a,1} = \frac{\dot{V}_1}{v_1} = \frac{0.85 \dfrac{\text{m}^3}{\text{s}}}{0.842 \dfrac{\text{m}^3}{\text{kg air}}} = 1.010 \text{ kg air/s}$$

Locate point 1 on the psychrometric chart.

Notice that the temperature of the recirculating water is constant but not equal to the air's entering wet-bulb temperature. Therefore, this is not an adiabatic process.

Locate point 3 as 10°C saturated condition (water being sprayed) on the psychrometric chart.

Draw a line from point 1 to point 3. The intersection of this line with 75% relative humidity defines point 2 as

$$
\begin{array}{rcl}
h_2 &=& 34.1 \text{ kJ/kg air} \\
\omega_2 &=& 7.6 \text{ g moisture/kg air} \\
T_{\text{db},2} &=& 14.7°C \\
T_{\omega b,2} &=& 12.0°C
\end{array}
$$

(b) The water added is

$$
\begin{aligned}
\dot{m}_w &= \dot{m}_{a,1}(\omega_2 - \omega_1) \\
&= \left(1.010 \ \frac{\text{kg air}}{\text{s}}\right)\left(7.6 \ \frac{\text{g moisture}}{\text{kg air}} - 6.2 \ \frac{\text{g moisture}}{\text{kg air}}\right) \\
&\quad \times \left(\frac{1 \text{ kg}}{1000 \text{ g}}\right) \\
&= \boxed{0.00141 \text{ kg/s}}
\end{aligned}
$$

7. *Customary U.S. Solution*

((a) and (b)) From Prob. 6,

$$
\begin{aligned}
\omega_1 &= 0.00623 \text{ lbm moisture/lbm air} \\
h_1 &= 23.6 \text{ Btu/lbm air} \\
\dot{m}_{a,1} &= 133.53 \text{ lbm air/min}
\end{aligned}
$$

From the steam table (App. 24.B) for 1 atm steam, $h_{\text{steam}} = 1150.3$ Btu/lbm.

From the conservation of energy equation (Eq. 38.32),

$$
\dot{m}_{a,1}h_1 + \dot{m}_{\text{steam}}h_{\text{steam}} = \dot{m}_{a,1}h_2
$$

$$
\left(133.53 \ \frac{\text{lbm air}}{\text{min}}\right)\left(23.6 \ \frac{\text{Btu}}{\text{lbm air}}\right)
$$

$$
+ \dot{m}_{\text{steam}}\left(1150.3 \ \frac{\text{Btu}}{\text{lbm}}\right) = \left(133.53 \ \frac{\text{lbm air}}{\text{min}}\right)h_2
$$

[Eq. 1]

From conservation of mass for the water (Eq. 38.33),

$$
\dot{m}_{a,1}\omega_1 + \dot{m}_{\text{steam}} = \dot{m}_{a,2}\omega_2
$$

$$
\left(133.53 \ \frac{\text{lbm air}}{\text{min}}\right)\left(0.00623 \ \frac{\text{lbm moisture}}{\text{lbm air}}\right)
$$

$$
+ \dot{m}_{\text{steam}} = \left(133.53 \ \frac{\text{lbm air}}{\text{min}}\right)\omega_2 \quad \textit{[Eq. 2]}
$$

Since no single relationship exists between ω_2, $\dot{m}_{\text{steam}}$, and h_2, a trial-and-error solution is required. Once $\dot{m}_{\text{steam}}$ is selected, ω_2 and h_2 can be found from Eq. 1 and Eq. 2 as

$$
h_2 = 8.615\dot{m}_{\text{steam}} + 23.6
$$

$$
\omega_2 = 0.00623 + 0.00749\dot{m}_{\text{steam}}
$$

Once h_2 and ω_2 are known, the relative humidity can be determined from the psychrometric chart. Continue the process until a relative humidity of 75% is achieved.

$\dot{m}_{\text{steam}}$ $\left(\dfrac{\text{lbm}}{\text{min}}\right)$	ω_2 $\left(\dfrac{\text{lbm moisture}}{\text{lbm air}}\right)$	h_2 $\left(\dfrac{\text{Btu}}{\text{lbm air}}\right)$	ϕ_2 (%)
0.3	0.00848	26.18	53
0.4	0.00923	27.05	58
0.5	0.00998	27.91	62
0.6	0.01072	28.77	66
0.7	0.01147	29.63	70
0.8	0.01222	30.49	74
0.82	0.01237	30.66	74.8
0.83	0.01245	30.75	75.3
0.825	0.01241	30.71	75.0

$$
\begin{aligned}
\dot{m}_{\text{steam}} &= 0.825 \text{ lbm/min} \\
\omega_2 &= 0.01241 \text{ lbm moisture/lbm air} \\
h_2 &= 30.71 \text{ Btu/lbm air} \\
T_{\text{db}} &= 71.5°F \\
T_{\text{wb}} &= 65.9°F
\end{aligned}
$$

SI Solution

((a) and (b)) From Prob. 6,

$$
\begin{aligned}
\omega_1 &= 6.2 \text{ g moisture/kg air} \\
h_1 &= 36.75 \text{ kJ/kg air} \\
\dot{m}_{a,1} &= 1.010 \text{ kg air/s}
\end{aligned}
$$

From the steam table (App. 24.O), for 1 atm steam, $h_{\text{steam}} = 2675.4$ kJ/kg.

From the conservation of energy equation (Eq. 38.32),

$$
\dot{m}_{a,1}h_1 + \dot{m}_{\text{steam}}h_{\text{steam}} = \dot{m}_{a,1}h_2
$$

$$
\left(1.010 \ \frac{\text{kg air}}{\text{s}}\right)\left(36.75 \ \frac{\text{kJ}}{\text{kg air}}\right)
$$

$$
+ \dot{m}_{\text{steam}}\left(2675.4 \ \frac{\text{kJ}}{\text{kg}}\right) = \left(1.010 \ \frac{\text{kg air}}{\text{s}}\right)h_2
$$

[Eq. 1]

From conservation of mass of water (Eq. 38.33),

$$\dot{m}_{a,1}\omega_1 + \dot{m}_{steam} = \dot{m}_{a,2}\omega_2$$

$$\left(1.010 \ \frac{\text{kg air}}{\text{s}}\right)\left(6.2 \ \frac{\text{g moisture}}{\text{kg air}}\right)\left(\frac{1 \ \text{kg}}{1000 \ \text{kg}}\right)$$

$$+ m_{steam} = \left(1.010 \ \frac{\text{kg air}}{\text{s}}\right)\omega_2$$

[Eq. 2]

Since no single relationship exists between ω_2, $\dot{m}_{steam}$, and h_2, a trial-and-error solution is required. Once $\dot{m}_{steam}$ is selected, ω_2 and h_2 can be found from Eq. 1 and Eq. 2 as

$$h_2 = 36.75 + 2648.9\dot{m}_{steam}$$

$$\omega_2 = 0.0062 + 0.99\dot{m}_{steam}$$

Once h_2 and ω_2 are known, the relative humidity can be determined from the psychrometric chart. Continue the process until a relative humidity of 75% is achieved.

$\dot{m}_{steam}$ $\left(\dfrac{\text{kg}}{\text{s}}\right)$	ω_2 $\left(\dfrac{\text{kg moisture}}{\text{kg air}}\right)$	h_2 $\left(\dfrac{\text{kJ}}{\text{kg air}}\right)$	ϕ_2 (%)
0.005	0.0112	49.99	69.5
0.0055	0.0116	51.32	74.5
0.0056	0.0117	51.58	75.0

$$\dot{m}_{steam} = 0.0056 \ \text{kg/s}$$
$$\omega_2 = \left(0.0117 \ \frac{\text{kg moisture}}{\text{kg air}}\right)\left(1000 \ \frac{\text{g}}{\text{kg}}\right)$$
$$= 11.7 \ \text{g moisture/kg air}$$
$$T_{db} = 21.3°C$$
$$T_{wb} = 18.2°C$$

8. *Customary U.S. Solution*

(a) From the psychrometric chart (App. 38.A), for incoming air at 65°F and 55% relative humidity, $\omega_1 = 0.0072$ lbm moisture/lbm air.

With sensible heating as a limiting factor, calculate the mass flow rate of air entering the theater from Eq. 38.26 (ventilation rate).

$$\dot{q} = \dot{m}_a(c_{p,air} + \omega c_{p,moisture})(T_2 - T_1)$$

$$500,000 \ \frac{\text{Btu}}{\text{hr}} = \dot{m}_a\left(0.240 \ \frac{\text{Btu}}{\text{lbm-°F}}\right.$$

$$+ \left(0.0072 \ \frac{\text{lbm moisture}}{\text{lbm air}}\right)$$

$$\left. \times \left(0.444 \ \frac{\text{Btu}}{\text{lbm-°F}}\right)\right)(75°F - 65°F)$$

$$\dot{m}_a = \boxed{2.056 \times 10^5 \ \text{lbm air/hr}}$$

(b) Assume that this air absorbs all the moisture. Then, the final humidity ratio is given by

$$\dot{m}_w = \dot{m}_a(\omega_2 - \omega_1)$$

$$\omega_2 = \left(\frac{\dot{m}_w}{\dot{m}_a}\right) + \omega_1$$

$$= \frac{175 \ \dfrac{\text{lbm moisture}}{\text{hr}}}{2.056 \times 10^5 \ \dfrac{\text{lbm air}}{\text{hr}}} + 0.0072 \ \frac{\text{lbm moisture}}{\text{lbm air}}$$

$$= 0.00805 \ \text{lbm moisture/lbm air}$$

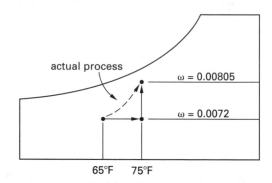

The final conditions are

$$\boxed{\begin{array}{l} T_{db} = 75°F \quad \text{[given]} \\ \omega_2 = 0.00805 \ \text{lbm moisture/lbm air} \end{array}}$$

From the psychrometric chart (App. 38.A), the relative humidity is 44%. This is below 60%.

SI Solution

(a) From the psychrometric chart (App. 38.B), for incoming air at 18°C and 55% relative humidity, $\omega_1 = 7.1$ g moisture/kg air.

With sensible heating as a limiting factor, calculate the mass flow rate of air entering the theater from Eq. 38.26 (ventilation rate).

$$\dot{q} = \dot{m}_a(c_{p,air} + \omega c_{p,moisture})$$
$$\times (T_2 - T_1)$$

$$(150 \ \text{kW})\left(1000 \ \frac{\text{W}}{\text{kW}}\right) = \dot{m}_a\left(\left(1.005 \ \frac{\text{kJ}}{\text{kg·°C}}\right)\left(1000 \ \frac{\text{J}}{\text{kJ}}\right)\right.$$

$$+ \left(1.805 \ \frac{\text{kJ}}{\text{kg·°C}}\right)\left(1000 \ \frac{\text{J}}{\text{kJ}}\right)$$

$$\times \left(7.1 \ \frac{\text{g moisture}}{\text{kg air}}\right)\left(\frac{1 \ \text{kg}}{1000 \ \text{g}}\right)\right)$$

$$\times (24°C - 18°C)$$

$$\dot{m}_a = \boxed{24.56 \ \text{kg/s}}$$

(b) Assume that this air absorbs all the moisture. Then, the final humidity ratio is given by

$$\dot{m}_w = \dot{m}_a(\omega_2 - \omega_1)$$

$$\omega_2 = \frac{\dot{m}_w}{\dot{m}_a} + \omega_1$$

$$= \frac{\left(80 \dfrac{\text{kg}}{\text{h}}\right)\left(\dfrac{1 \text{ h}}{3600 \text{ s}}\right)}{24.56 \dfrac{\text{kg}}{\text{s}}}$$

$$+ \left(7.0 \dfrac{\text{g moisture}}{\text{kg air}}\right)\left(\dfrac{1 \text{ kg}}{1000 \text{ g}}\right)$$

$$= 0.00790 \text{ kg moisture/kg air}$$

The final conditions are

$$\boxed{\begin{aligned} T_{db} &= 24°\text{C} \\ \omega_2 &= \left(0.00790 \dfrac{\text{kg moisture}}{\text{kg air}}\right)\left(1000 \dfrac{\text{g}}{\text{kg}}\right) \\ &= 7.9 \text{ g moisture/kg air} \end{aligned}}$$

From the psychrometric chart (App. 38.B), the relative humidity is 44%. This is below 60%.

9.

$T_{db} = 80°\text{F} (27°\text{C})$
$\phi = 70\%$
$\dot{V} = 500 \text{ ft}^3/\text{min} (0.25 \text{ m}^3/\text{s})$

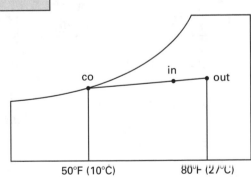

Customary U.S. Solution

Locate point "out" ($T_{db} = 80°\text{F}$ and $\phi = 70\%$) and point "co" (saturated at 50°F) on the psychrometric chart. At point "out" from App. 38.A,

$$v_{out} = 13.95 \text{ ft}^3/\text{lbm air}$$

$$h_{out} = 36.2 \text{ Btu/lbm air}$$

At point "co," $h_{co} = 20.3$ Btu/lbm air.

The air mass flow rate through the air conditioner is

$$\dot{m}_1 = \frac{\dot{V}_1}{v_1} = \frac{\dot{V}_1}{v_{out}} = \frac{150 \dfrac{\text{ft}^3}{\text{min}}}{13.95 \dfrac{\text{ft}^3}{\text{lbm air}}}$$

$$= 10.75 \text{ lbm air/min}$$

The mass flow rate of the bypass air is

$$\dot{m}_2 = \frac{\dot{V}_2}{v} = \frac{350 \dfrac{\text{ft}^3}{\text{min}}}{13.95 \dfrac{\text{ft}^3}{\text{lbm air}}} = 25.09 \text{ lbm air/min}$$

The percentage of bypass air is

$$x = \frac{25.09 \dfrac{\text{lbm air}}{\text{min}}}{10.75 \dfrac{\text{lbm air}}{\text{min}} + 25.09 \dfrac{\text{lbm air}}{\text{min}}} = 0.70 \quad (70\%)$$

Using the lever rule and the fact that all of the temperature scales are linear,

$$\begin{aligned} T_{db,in} &= T_{co} + (0.70)(T_{out} - T_{co}) \\ &= 50°\text{F} + (0.70)(80°\text{F} - 50°\text{F}) \\ &= 71°\text{F} \end{aligned}$$

At that point,

(a) $$\boxed{T_{db,in} = 71°\text{F}}$$

The answer is (C) for Prob. 9(a).

(b) $$\boxed{\omega_{in} = 0.0132 \text{ lbm moisture/lbm air}}$$

The answer is (B) for Prob. 9(b).

(c) $$\boxed{\phi_{in} = 81\%}$$

The answer is (D) for Prob. 9(c).

(d) The air conditioner capacity is given by

$$\dot{Q} = \dot{m}_{air}(h_{t,2} - h_{t,1}) = \dot{m}_1(h_{out} - h_{co})$$

$$= \left(10.75 \dfrac{\text{lbm air}}{\text{min}}\right)\left(36.2 \dfrac{\text{Btu}}{\text{lbm air}} - 20.3 \dfrac{\text{Btu}}{\text{lbm air}}\right)$$

$$\times \left(\dfrac{1 \text{ ton}}{200 \dfrac{\text{Btu}}{\text{min}}}\right)$$

$$= \boxed{0.85 \text{ ton}}$$

The answer is (A).

SI Solution

Locate point "out" ($T_{db} = 27°C$, $\phi = 70\%$) and point "co" (saturated at $10°C$) on the psychrometric chart. At point "out" from App. 38.B,

$$v_{out} = 0.872 \text{ m}^3/\text{kg air}$$

$$h_{out} = 67.3 \text{ kJ/kg air}$$

At point "co" from App. 38.B, $h_{co} = 29.26$ kJ/kg air.

At mass flow rate through the air conditioner,

$$\dot{m}_1 = \frac{\dot{V}_1}{v} = \frac{0.075 \frac{\text{m}^3}{\text{s}}}{0.872 \frac{\text{m}^3}{\text{kg air}}} = 0.0860 \text{ kg air/s}$$

The flow rate of bypass air is

$$\dot{m}_2 = \frac{\dot{V}_2}{v} = \frac{0.175 \frac{\text{m}^3}{\text{s}}}{0.872 \frac{\text{m}^3}{\text{kg air}}} - 0.2007 \text{ kg air/s}$$

The percentage bypass air is

$$x = \frac{0.2007 \frac{\text{kg air}}{\text{s}}}{0.0860 \frac{\text{kg air}}{\text{s}} + 0.2007 \frac{\text{kg air}}{\text{s}}}$$

$$= 0.70 \quad (70\%)$$

Using the lever rule and the fact that all of the temperature scales are linear,

$$T_{db,in} = T_{co} + (0.70)(T_{out} - T_{co})$$
$$= 10°C + (0.70)(27°C - 10°C)$$
$$= 21.9°C$$

At that point,

(a)
$$\boxed{T_{db,in} = 21.9°C}$$

The answer is (C) for Prob. 9(a).

(b)
$$\boxed{\omega_{in} = 13.4 \text{ g moisture/kg air}}$$

The answer is (B) for Prob. 9(b).

(c)
$$\boxed{\phi_{in} = 81\%}$$

The answer is (D) for Prob. 9(c).

(d) The air conditioner capacity is given by

$$\dot{Q} = \dot{m}_{air}(h_{t,2} - h_{t,1}) = \dot{m}_1(h_{out} - h_{co})$$
$$= \left(0.0860 \frac{\text{kg air}}{\text{s}}\right)\left(67.3 \frac{\text{kJ}}{\text{kg air}} - 29.26 \frac{\text{kJ}}{\text{kg air}}\right)$$
$$\times \left(0.2843 \frac{\text{ton}}{\text{kW}}\right)$$
$$= \boxed{0.93 \text{ ton}}$$

The answer is (A).

10.

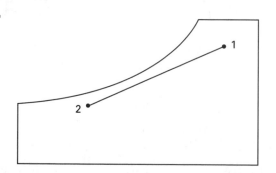

Customary U.S. Solution

(a) At point 1, from the psychrometric chart (App. 38.A) at $T_{db} = 95°F$ and $\phi = 70\%$,

$$\boxed{\begin{array}{l} h_1 = 50.7 \text{ Btu/lbm air} \\ v_1 = 14.56 \text{ ft}^3/\text{lbm air} \\ \omega_1 = 0.0253 \text{ lbm water/lbm air} \end{array}}$$

At point 2, from the psychrometric chart (App. 38.A) at $T_{db} = 60°F$ and $\phi = 95\%$,

$$\boxed{\begin{array}{l} h_2 = 25.8 \text{ Btu/lbm air} \\ \omega_2 = 0.0105 \text{ lbm water/lbm air} \end{array}}$$

(b) The air mass flow rate is

$$\dot{m}_a = \frac{\dot{V}}{v_1} = \frac{5000 \frac{\text{ft}^3}{\text{min}}}{14.56 \frac{\text{ft}^3}{\text{lbm air}}} = 343.4 \text{ lbm air/min}$$

From Eq. 38.27, the water removed is

$$\dot{m}_w = \dot{m}_a(\omega_1 - \omega_2)$$
$$= \left(343.4 \frac{\text{lbm air}}{\text{min}}\right)$$
$$\times \left(0.0253 \frac{\text{lbm water}}{\text{lbm air}} - 0.0105 \frac{\text{lbm water}}{\text{lbm air}}\right)$$
$$= \boxed{5.08 \text{ lbm water/min}}$$

(c) From Eq. 38.28, the quantity of heat removed is

$$\dot{q} = \dot{m}_a(h_1 - h_2)$$

$$= \left(343.4 \; \frac{\text{lbm air}}{\text{min}}\right)\left(50.7 \; \frac{\text{Btu}}{\text{lbm air}} - 25.8 \; \frac{\text{Btu}}{\text{lbm air}}\right)$$

$$= \boxed{8551 \; \text{Btu/min}}$$

(d) Considering an R-12 refrigeration cycle operating at saturated condition at 100°F, the T-s and h-s diagrams are as follows.

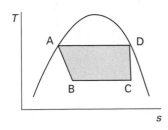 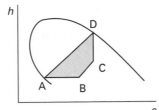

Use App. 24.G for saturated conditions.

(e) At A,

$$T = 100°\text{F} \quad [\text{given}]$$

$$p = p_{\text{sat}} \text{ at } 100°\text{F} = 131.6 \text{ psia}$$

$$h_\text{A} = h_f \text{ at } 100°\text{F} = 31.16 \text{ Btu/lbm}$$

$$s_\text{A} = s_f \text{ at } 100°\text{F} = 0.06316 \text{ Btu/lbm-°R}$$

$$v_\text{A} = v_f \text{ at } 100°\text{F} = 0.0127 \text{ ft}^3/\text{lbm}$$

At point B,

$$T = 50°\text{F} \quad [\text{given}]$$

$$p = p_{\text{sat}} \text{ at } 50°\text{F} = 61.39 \text{ psia}$$

$$h_\text{B} = h_a = 31.16 \text{ Btu/lbm}$$

$$h_{f,\text{B}} = 19.27 \text{ Btu/lbm}$$

$$h_{fg,\text{B}} = 64.51 \text{ Btu/lbm}$$

$$x = \frac{h_\text{B} - h_{f,\text{B}}}{h_{fg,\text{B}}}$$

$$= \frac{31.16 \; \frac{\text{Btu}}{\text{lbm}} - 19.27 \; \frac{\text{Btu}}{\text{lbm}}}{64.51 \; \frac{\text{Btu}}{\text{lbm}}}$$

$$= 0.184$$

$$s_\text{B} = s_{f,\text{B}} + x s_{fg,\text{B}}$$

$$= 0.04126 \; \frac{\text{Btu}}{\text{lbm-°R}}$$

$$+ (0.184)\left(0.12659 \; \frac{\text{Btu}}{\text{lbm-°R}}\right)$$

$$= 0.06455 \text{ Btu/lbm-°R}$$

$$v_\text{B} = v_{f,\text{B}} + x(v_g - v_f)$$

$$= 0.0118 \; \frac{\text{ft}^3}{\text{lbm}}$$

$$+ (0.184)\left(0.673 \; \frac{\text{ft}^3}{\text{lbm}} - 0.0118 \; \frac{\text{ft}^3}{\text{lbm}}\right)$$

$$= 0.1335 \text{ ft}^3/\text{lbm}$$

At point D,

$$T = 100°\text{F} \quad [\text{given}]$$

$$p = p_{\text{sat}} \text{ at } 100°\text{F} = 131.6 \text{ psia}$$

$$h_\text{D} = h_g \text{ at } 100°\text{F} = 88.62 \text{ Btu/lbm}$$

$$s_\text{D} = s_g \text{ at } 100°\text{F} = 0.16584 \text{ Btu/lbm-°R}$$

$$v_\text{D} = v_g \text{ at } 100°\text{F} = 0.319 \text{ ft}^3/\text{lbm}$$

At point C,

$$T = 50°\text{F} \quad [\text{given}]$$

$$p = p_{\text{sat}} \text{ at } 50°\text{F} = 61.39 \text{ psia}$$

$$s_\text{C} = s_\text{D} = 0.16584 \text{ Btu/lbm-°R}$$

$$s_{f,\text{C}} = 0.04126 \text{ Btu/lbm-°R}$$

$$s_{fg,\text{C}} = 0.12659 \text{ Btu/lbm-°R}$$

$$x = \frac{s_\text{C} - s_{f,\text{C}}}{s_{fg,\text{C}}}$$

$$= \frac{0.16584 \; \frac{\text{Btu}}{\text{lbm-°R}} - 0.04126 \; \frac{\text{Btu}}{\text{lbm-°R}}}{0.12659 \; \frac{\text{Btu}}{\text{lbm-°R}}}$$

$$= 0.984$$

$$h_\text{C} = h_{f,\text{C}} + x h_{fg,\text{C}}$$

$$= 19.27 \; \frac{\text{Btu}}{\text{lbm}} + (0.984)\left(64.51 \; \frac{\text{Btu}}{\text{lbm}}\right)$$

$$= 82.75 \text{ Btu/lbm}$$

$$v_\text{C} = v_{f,\text{C}} + x(v_{g,\text{C}} - v_{f,\text{C}})$$

$$= 0.0118 \; \frac{\text{ft}^3}{\text{lbm}}$$

$$+ (0.984)\left(0.673 \; \frac{\text{ft}^3}{\text{lbm}} - 0.0118 \; \frac{\text{ft}^3}{\text{lbm}}\right)$$

$$= 0.662 \text{ ft}^3/\text{lbm}$$

SI Solution

(a) The psychrometric chart is shown in the customary U.S. solution. At point 1, from the psychrometric chart (App. 38.B) at $T_{\text{db}} = 35°\text{C}$ and $\phi = 70\%$,

$$\begin{aligned} h_1 &= 99.9 \text{ kJ/kg air} \\ v_1 &= 0.91 \text{ m}^3/\text{kg air} \\ \omega_1 &= \left(25.3 \; \frac{\text{g moisture}}{\text{kg air}}\right)\left(\frac{1 \text{ kg}}{1000 \text{ g}}\right) \\ &= 0.0253 \text{ kg moisture/kg air} \end{aligned}$$

At point 2, from the psychrometric chart (App. 38.B) at $T_{db} = 16°C$ and $\phi = 95\%$,

$$h_2 = 43.4 \text{ kJ/kg air}$$
$$\omega_2 = \left(10.8 \frac{\text{g moisture}}{\text{kg air}}\right)\left(\frac{1 \text{ kg}}{1000 \text{ g}}\right)$$
$$= 0.0108 \text{ kg moisture/kg air}$$

(b) The air mass flow rate is

$$\dot{m}_a = \frac{\dot{V}_1}{v_1} = \frac{\dot{V}_1}{v_{out}} = \frac{2.36 \frac{\text{m}^3}{\text{s}}}{0.91 \frac{\text{m}^3}{\text{kg air}}} = 2.593 \text{ kg air/s}$$

The water removed from Eq. 38.27 is

$$\dot{m}_w = \dot{m}_a(\omega_1 - \omega_2)$$
$$= \left(2.593 \frac{\text{kg air}}{\text{s}}\right)\left(0.0253 \frac{\text{kg moisture}}{\text{kg air}}\right.$$
$$\left. - 0.0108 \frac{\text{kg moisture}}{\text{kg air}}\right)$$
$$= \boxed{0.0376 \text{ kg moisture/s}}$$

(c) From Eq. 38.28, the quantity of heat removed is

$$\dot{q} = \dot{m}_a(h_1 - h_2)$$
$$= \left(2.593 \frac{\text{kg air}}{\text{s}}\right)\left(99.9 \frac{\text{kJ}}{\text{kg air}} - 43.4 \frac{\text{kJ}}{\text{kg air}}\right)$$
$$= \boxed{146.5 \text{ kW}}$$

(d) The T-s and h-s diagrams are shown in the customary U.S. solution.

(e) Use App. 24.T for saturated conditions.

At point A,

$$T = 38°C \quad \text{[given]}$$
$$p = p_{sat} \text{ at } 38°C = 0.91324 \text{ MPa}$$
$$h_A = h_f \text{ at } 38°C = 237.23 \text{ kJ/kg}$$
$$s_A = s_f \text{ at } 38°C = 1.1259 \text{ kJ/kg·°C}$$
$$v_A = v_f \text{ at } 38°C = \frac{1}{\rho_f} = \frac{1}{1261.9 \frac{\text{kg}}{\text{m}^3}}$$
$$= 0.0007925 \text{ m}^3/\text{kg}$$

At point B,

$$T = 10°C \quad \text{[given]}$$
$$p = p_{sat} \text{ at } 10°C = 0.42356 \text{ MPa}$$
$$h_B = h_A = 237.23 \text{ kJ/kg}$$
$$h_{f,B} = 209.48 \text{ kJ/kg}$$
$$h_{g,B} = 356.79 \text{ kJ/kg}$$

$$x = \frac{h_B - h_{f,B}}{h_{g,B} - h_{f,B}} = \frac{237.23 \frac{\text{kJ}}{\text{kg}} - 209.48 \frac{\text{kJ}}{\text{kg}}}{356.79 \frac{\text{kJ}}{\text{kg}} - 209.48 \frac{\text{kJ}}{\text{kg}}}$$
$$= 0.188$$
$$s_B = s_{f,B} + x(s_{g,B} - s_{f,B})$$
$$= 1.0338 \frac{\text{kJ}}{\text{kg·K}}$$
$$+ (0.188)\left(1.5541 \frac{\text{kJ}}{\text{kg·K}} - 1.0338 \frac{\text{kJ}}{\text{kg·K}}\right)$$
$$= 1.1316 \text{ kJ/kg·K}$$
$$v_B = v_{f,B} + x(v_{g,B} - v_{f,B})$$
$$v_{f,B} = \frac{1}{\rho_{f,B}} = \frac{1}{1363.0 \frac{\text{kg}}{\text{m}^3}} = 0.0007337 \text{ m}^3/\text{kg}$$
$$v_{g,B} = 0.04119 \text{ m}^3/\text{kg}$$
$$v_B = 0.0007337 \frac{\text{m}^3}{\text{kg}}$$
$$+ (0.188)\left(0.04119 \frac{\text{m}^3}{\text{kg}} - 0.0007337 \frac{\text{m}^3}{\text{kg}}\right)$$
$$= 0.008339 \text{ m}^3/\text{kg}$$

At point D,

$$T = 38°C \quad \text{[given]}$$
$$p = p_{sat} \text{ at } 38°C = 0.42356 \text{ MPa}$$
$$h_D = h_g \text{ at } 38°C = 367.95 \text{ kJ/kg}$$
$$s_D = s_g \text{ at } 38°C = 1.5461 \text{ kJ/kg·K}$$
$$v_D = v_g \text{ at } 38°C = 0.01931 \text{ m}^3/\text{kg}$$

At point C,

$$T = 10°C$$
$$p = p_{sat} \text{ at } 10°C = 0.42356 \text{ MPa}$$
$$s_C = s_D = 1.5461 \text{ kJ/kg·K}$$
$$s_{f,C} = 1.0338 \text{ kJ/kg·K}$$
$$s_{g,C} = 1.5541 \text{ kJ/kg·K}$$

$$x = \frac{s_C - s_{f,C}}{s_{g,C} - s_{f,C}} = \frac{1.5461 \frac{\text{kJ}}{\text{kg·K}} - 1.0338 \frac{\text{kJ}}{\text{kg·K}}}{1.5541 \frac{\text{kJ}}{\text{kg·K}} - 1.0338 \frac{\text{kJ}}{\text{kg·K}}}$$
$$= 0.985$$

$$h_C = h_{f,C} + x(h_{g,C} - h_{f,C})$$

$$= 209.48 \ \frac{\text{kJ}}{\text{kg}} + (0.985)\left(356.79 \ \frac{\text{kJ}}{\text{kg}} - 209.48 \ \frac{\text{kJ}}{\text{kg}}\right)$$

$$= 354.58 \ \text{kJ/kg}$$

$$v_C = v_{f,C} + x(v_{g,C} - v_{f,C})$$

$$v_{g,C} = 0.04119 \ \text{m}^3/\text{kg}$$

$$v_C = 0.0007337 \ \frac{\text{m}^3}{\text{kg}}$$

$$+ (0.985)\left(0.04119 \ \frac{\text{m}^3}{\text{kg}} - 0.0007337 \ \frac{\text{m}^3}{\text{kg}}\right)$$

$$= 0.04058 \ \text{m}^3/\text{kg}$$

11. Customary U.S. Solution

(a) The saturation pressure at 200°F from App. 24.A is $p_{\text{sat},1} = 11.538$ psia.

Since air is saturated (100% relative humidity), the water vapor pressure is equal to the saturation pressure.

$$p_{w,1} = p_{\text{sat},1} = 11.538 \ \text{psia}$$

The partial pressure of the air is

$$p_{a,1} = p_1 - p_{w,1} = 25 \ \text{psia} - 11.538 \ \text{psia}$$
$$= 13.462 \ \text{psia}$$

From Table 24.7, the specific gas constants are

$$R_w = 85.78 \ \text{ft-lbf/lbm-°R}$$
$$R_{\text{air}} = 53.35 \ \text{ft-lbf/lbm-°R}$$

The mass of water vapor from the ideal gas law is

$$\dot{m}_{w,1} = \frac{p_{w,1}\dot{V}}{R_w T}$$

$$= \frac{\left(11.538 \ \frac{\text{lbf}}{\text{in}^2}\right)\left(144 \ \frac{\text{in}^2}{\text{ft}^2}\right)\left(1500 \ \frac{\text{ft}^3}{\text{min}}\right)}{\left(85.78 \ \frac{\text{ft-lbf}}{\text{lbm-°R}}\right)(200°\text{F} + 460)}$$

$$= 44.02 \ \text{lbm/min water}$$

The mass of air from the ideal gas law is

$$\dot{m}_{a,1} = \frac{p_{a,1}\dot{V}}{R_{\text{air}} T}$$

$$= \frac{\left(13.471 \ \frac{\text{lbf}}{\text{in}^2}\right)\left(144 \ \frac{\text{in}^2}{\text{ft}^2}\right)\left(1500 \ \frac{\text{ft}^3}{\text{min}}\right)}{\left(53.35 \ \frac{\text{ft-lbf}}{\text{lbm-°R}}\right)(200°\text{F} + 460)}$$

$$= 82.64 \ \text{lbm/min air}$$

The humidity ratio is

$$\omega_1 = \frac{\dot{m}_{w,1}}{\dot{m}_{a,1}} = \frac{44.02 \ \frac{\text{lbm water}}{\text{min}}}{82.64 \ \frac{\text{lbm air}}{\text{min}}}$$

$$= 0.533 \ \text{lbm water/lbm air}$$

Since it is a constant pressure, constant moisture drying process, mole fractions and partial pressures do not change.

$$p_{w,2} = p_{w,1} = 11.538 \ \text{psia}$$

The saturation pressure at 400°F from App. 24.A is $p_{\text{sat},2} = 247.3$ psia.

The relative humidity at state 2 is

$$\phi_2 = \frac{p_{w,2}}{p_{\text{sat},2}} = \frac{11.538 \ \text{psia}}{247.3 \ \text{psia}} = \boxed{0.0467 \ (4.67\%)}$$

The answer is (A).

(b) Although the volume may change, the mass does not. The specific humidity remains constant.

$$\omega_2 = \omega_1 = \boxed{0.533 \ \text{lbm water/lbm air}}$$

The answer is (B).

(c) The heat required consists of two parts.

Obtain enthalpy for air from App. 24.F.

The absolute temperatures are

$$T_1 = 200°\text{F} + 460 = 660°\text{R}$$
$$T_2 = 400°\text{F} + 460 = 860°\text{R}$$
$$h_1 = 157.92 \ \text{Btu/lbm}$$
$$h_2 = 206.46 \ \text{Btu/lbm}$$

The heat absorbed by the air is

$$q_1 = h_2 - h_1 = 206.46 \ \frac{\text{Btu}}{\text{lbm}} - 157.92 \ \frac{\text{Btu}}{\text{lbm}}$$

$$= 48.54 \ \text{Btu/lbm dry air}$$

(There will be a small error if constant specific heat is used instead.)

For water, use the Mollier diagram. From App. 24.E, h_1 at 200°F and 11.529 psia is 1146 Btu/lbm (almost saturated).

Follow a constant 11.529 psia pressure curve up to 400°F.

$$h_2 = 1240 \ \text{Btu/lbm}$$

(There will be a small error if Eq. 38.19(b) is used instead.)

The heat absorbed by the steam is

$$q_2 = \omega(h_2 - h_1)$$
$$= \left(0.532 \ \frac{\text{lbm water}}{\text{lbm air}}\right)\left(1240 \ \frac{\text{Btu}}{\text{lbm}} - 1146 \ \frac{\text{Btu}}{\text{lbm}}\right)$$
$$= 50.01 \ \text{Btu/lbm air}$$

The total heat absorbed is

$$q_{\text{total}} = q_1 + q_2 = 48.54 \ \frac{\text{Btu}}{\text{lbm air}} + 50.01 \ \frac{\text{Btu}}{\text{lbm air}}$$
$$\boxed{= 98.55 \ \text{Btu/lbm air}}$$

The answer is (C).

(d) The dew point is the temperature at which water starts to condense out in a constant pressure process. Following the constant 11.538 psia pressure line back to the saturation line, $\boxed{T_{\text{dp}} = 200°\text{F}.}$

The answer is (B).

SI Solution

(a) From App. 24.N, the saturation pressure at 93°C is

$$p_{\text{sat},1} = (0.7884 \ \text{bar})\left(100 \ \frac{\text{kPa}}{\text{bar}}\right) = 78.84 \ \text{kPa}$$

Since air is saturated (100% relative humidity), water vapor pressure is equal to saturation pressure.

$$p_{w,1} = p_{\text{sat},1} = 78.84 \ \text{kPa}$$

The partial pressure of air is

$$p_{a,1} = p_1 - p_{w,1} = 170 \ \text{kPa} - 78.84 \ \text{kPa} = 91.16 \ \text{kPa}$$

From Table 24.7, the specific gas constants are

$$R_w = 461.50 \ \text{J/kg·K}$$
$$R_{\text{air}} = 287.03 \ \text{J/kg·K}$$

The mass of water vapor from the ideal gas law is

$$\dot{m}_{w,1} = \frac{p_{w,1}\dot{V}}{R_w T}$$
$$= \frac{(78.84 \ \text{kPa})\left(1000 \ \frac{\text{Pa}}{\text{kPa}}\right)\left(0.71 \ \frac{\text{m}^3}{\text{s}}\right)}{\left(461.50 \ \frac{\text{J}}{\text{kg·K}}\right)(93°\text{C} + 273)}$$
$$= 0.3314 \ \text{kg/s water}$$

The mass of air from the ideal gas law is

$$\dot{m}_{a,1} = \frac{p_{a,1}\dot{V}}{R_{\text{air}} T}$$
$$= \frac{(91.21 \ \text{kPa})\left(1000 \ \frac{\text{Pa}}{\text{kPa}}\right)\left(0.71 \ \frac{\text{m}^3}{\text{s}}\right)}{\left(287.03 \ \frac{\text{J}}{\text{kg·K}}\right)(93°\text{C} + 273)}$$
$$= 0.6164 \ \text{kg/s air}$$

The humidity ratio is

$$\omega_1 = \frac{\dot{m}_{w,1}}{\dot{m}_{a,1}} = \frac{0.3314 \ \frac{\text{kg water}}{\text{s}}}{0.6164 \ \frac{\text{kg air}}{\text{s}}}$$
$$= 0.538 \ \text{kg water/kg air}$$

Since it is a constant pressure, constant moisture drying process, mole fractions and partial pressure do not change.

$$p_{w,2} = p_{w,1} = 78.84 \ \text{kPa}$$

The saturation pressure at 204°C from App. 24.N is

$$p_{\text{sat},2} = (16.90 \ \text{bar})\left(100 \ \frac{\text{kPa}}{\text{bar}}\right) = 1690 \ \text{kPa}$$

The relative humidity at state 2 is

$$\phi_2 = \frac{p_{w,2}}{p_{\text{sat},2}} = \frac{78.84 \ \text{kPa}}{1690 \ \text{kPa}} = \boxed{0.0467 \ \ (4.67\%)}$$

The answer is (A).

(b) Although the volume may change, the mass does not. The specific humidity remains constant.

$$\omega_2 = \omega_1 = \boxed{0.538 \ \text{kg water/kg air}}$$

The answer is (B).

(c) The heat required consists of two parts.

Obtain enthalpy for air from App. 24.S.

The absolute temperatures are

$$T_1 = 93°\text{C} + 273 = 366\text{K}$$
$$T_2 = 204°\text{C} + 273 = 477\text{K}$$
$$h_1 = 366.63 \ \text{kJ/kg}$$
$$h_2 = 479.42 \ \text{kJ/kg}$$

The heat absorbed by air is

$$q_1 = h_2 - h_1 = 479.42 \ \frac{\text{kJ}}{\text{kg}} - 366.63 \ \frac{\text{kJ}}{\text{kg}}$$
$$= 112.79 \ \text{kJ/kg air}$$

(There will be a small error if constant specific heat is used instead.)

For water use the Mollier diagram. From App. 24.R, h_1 at 93°C and 78.79 kPa is 2670 kJ/kg (almost saturated).

Follow a constant 78.79 kPa pressure curve up to 204°C.

$$h_2 = 2890 \ \text{kJ/kg}$$

(There will be a small error if Eq. 38.19(a) is used instead.)

The heat absorbed by steam is

$$q_2 = \omega(h_2 - h_1)$$
$$= \left(0.537 \ \frac{\text{kg water}}{\text{kg air}}\right)\left(2890 \ \frac{\text{kJ}}{\text{kg}} - 2670 \ \frac{\text{kJ}}{\text{kg}}\right)$$
$$= 118.14 \ \text{kJ/kg air}$$

The total heat absorbed is

$$q_{\text{total}} = q_1 + q_2$$
$$= 112.79 \ \frac{\text{kJ}}{\text{kg air}} + 118.14 \ \frac{\text{kJ}}{\text{kg air}}$$
$$= \boxed{230.93 \ \text{kJ/kg air}}$$

The answer is (C).

(d) The dew point is the temperature at which water starts to condense out in a constant pressure process. Following the constant 78.84 kPa pressure line back to the saturation line, $\boxed{T_{\text{dp}} \approx 93°\text{C}.}$

The answer is (B).

12. *Customary U.S. Solution*

(a) The absolute air temperatures are

$$T_1 = 800°\text{F} + 460 = 1260°\text{R}$$
$$T_2 = 350°\text{F} + 460 = 810°\text{R}$$

At low pressures, use air tables. From App. 24.F,

$$h_1 = 306.65 \ \text{Btu/lbm}$$
$$h_2 = 194.25 \ \text{Btu/lbm}$$

From App. 24.A, the enthalpy of water at 80°F is $h_{w,1}$ = 48.07 Btu/lbm.

From Eq. 38.19(b), the enthalpy of steam at 350°F is

$$h_{w,2} \approx \left(0.444 \ \frac{\text{Btu}}{\text{lbm-°F}}\right)(350°\text{F}) + 1061 \ \frac{\text{Btu}}{\text{lbm}}$$
$$= 1216.4 \ \text{Btu/lbm}$$

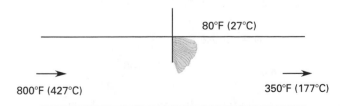

Air temperature is reduced from 800°F to 350°F, and this energy is used to change water at 80°F to steam at 350°F. From the energy balance equation,

$$\dot{m}_w(h_{w,2} - h_{w,1}) = \dot{m}_a(h_1 - h_2)$$
$$\dot{m}_w = \frac{\dot{m}_a(h_1 - h_2)}{h_{w,2} - h_{w,1}}$$
$$= \frac{\left(410 \ \frac{\text{lbm}}{\text{hr}}\right)\left(306.65 \ \frac{\text{Btu}}{\text{lbm}} - 194.25 \ \frac{\text{Btu}}{\text{lbm}}\right)}{1216.4 \ \frac{\text{Btu}}{\text{lbm}} - 48.07 \ \frac{\text{Btu}}{\text{lbm}}}$$
$$= \boxed{39.4 \ \text{lbm/hr water}}$$

The answer is (D).

(b) The number of moles of water evaporated (in the air mixture) is

$$\dot{n}_w = \frac{39.4 \ \frac{\text{lbm}}{\text{hr}}}{18.016 \ \frac{\text{lbm}}{\text{lbmol}}} = 2.19 \ \text{lbmol/hr}$$

The number of moles of air at the exit in the mixture is

$$\dot{n}_a = \frac{410 \ \frac{\text{lbm}}{\text{hr}}}{28.967 \ \frac{\text{lbm}}{\text{lbmol}}} = 14.15 \ \text{lbmol/hr}$$

The mole fraction of water in the mixture is

$$x_w = \frac{\dot{n}_w}{\dot{n}_a + \dot{n}_w}$$
$$= \frac{2.19 \ \frac{\text{lbmol}}{\text{hr}}}{14.15 \ \frac{\text{lbmol}}{\text{hr}} + 2.19 \ \frac{\text{lbmol}}{\text{hr}}} = 0.134$$

Partial pressure of water vapor is

$$p_w = x(p_{\text{chamber}}) = (0.134)(20 \ \text{psia}) = 2.68 \ \text{psia}$$

From App. 24.A, the saturation pressure at 350°F is $p_{\text{sat}} = 134.63 \ \text{psia}$.

The relative humidity is

$$\phi = \frac{p_w}{p_{\text{sat}}} = \frac{2.68 \ \text{psia}}{134.63 \ \text{psia}} = \boxed{0.020 \ \ (2.0\%)}$$

The answer is (B).

SI Solution

(a) The absolute temperatures are

$$T_1 = 427°C + 273 = 700K$$
$$T_2 = 177°C + 273 = 450K$$

Air tables can be used at low pressures. From App. 24.S,

$$h_1 = 713.27 \text{ kJ/kg}$$
$$h_2 = 451.80 \text{ kJ/kg}$$

From App. 24.N, the enthalpy of water at 27°C is $h_{w,1}$ = 113.19 kJ/kg.

From Eq. 38.19(a), the enthalpy of steam at 177°C is

$$h_{w,2} = \left(1.805 \frac{\text{kJ}}{\text{kg}\cdot°\text{C}}\right)(177°C) + 2501 \frac{\text{kJ}}{\text{kg}}$$
$$= 2820.5 \text{ kJ/kg}$$

Air temperature is reduced from 427°C to 177°C, and this energy is used to change water at 27°C to steam at 177°C. From the energy balance equation,

$$\dot{m}_w = \frac{\dot{m}_a(h_1 - h_2)}{h_{w,2} - h_{w,1}}$$

$$= \frac{\left(0.052 \frac{\text{kg}}{\text{s}}\right)\left(713.27 \frac{\text{kJ}}{\text{kg}} - 451.80 \frac{\text{kJ}}{\text{kg}}\right)}{2820.5 \frac{\text{kJ}}{\text{kg}} - 113.19 \frac{\text{kJ}}{\text{kg}}}$$

$$= \boxed{0.00502 \text{ kg/s}}$$

The answer is (D).

(b) The number of moles of water evaporated (in the air mixture) is

$$\dot{n}_w = \frac{0.00502 \frac{\text{kg}}{\text{s}}}{18.016 \frac{\text{kg}}{\text{kmol}}} = 2.79 \times 10^{-4} \text{ kmol/s}$$

The number of moles of air in the mixture at the exit is

$$\dot{n}_a = \frac{0.052 \frac{\text{kg}}{\text{s}}}{28.967 \frac{\text{kg}}{\text{kmol}}} = 1.80 \times 10^{-3} \text{ kmol/s}$$

The mole fraction of water in the mixture is

$$x_w = \frac{\dot{n}_w}{\dot{n}_a + \dot{n}_w}$$

$$= \frac{2.79 \times 10^{-4} \frac{\text{kmol}}{\text{s}}}{1.80 \times 10^{-3} \frac{\text{kmol}}{\text{s}} + 2.79 \times 10^{-4} \frac{\text{kmol}}{\text{s}}}$$

$$= 0.134$$

The partial pressure of water vapor is

$$p_w = x p_{\text{chamber}} = (0.134)(140 \text{ kPa})$$
$$= 18.76 \text{ kPa}$$

From App. 24.N, the saturation pressure at 177°C is

$$p_{\text{sat}} = (9.368 \text{ bar})\left(100 \frac{\text{kPa}}{\text{bar}}\right) = 936.8 \text{ kPa}$$

The relative humidity is

$$\phi = \frac{p_w}{p_{\text{sat}}} = \frac{18.76 \text{ kPa}}{936.8 \text{ kPa}} = \boxed{0.020 \ (2.0\%)}$$

The answer is (B).

13. *Customary U.S. Solution*

(a) The cooled water flow rate is given by

$$Q = \dot{m}_w c_p \Delta T$$
$$1 \times 10^6 \text{ Btu} = \dot{m}_w \left(1.0 \frac{\text{Btu}}{\text{lbm-°F}}\right)(120°F - 110°F)$$
$$\dot{m}_w = 1 \times 10^5 \text{ lbm/hr}$$

From the psychrometric chart (App. 38.A), for air in at $T_{\text{db}} = 91°F$ and $\phi = 60\%$,

$$h_{\text{in}} \approx 42.7 \text{ Btu/lbm}$$
$$\omega_{\text{in}} = 0.0190 \text{ lbm moisture/lbm air}$$

For air out, the normal psychrometric chart (offscale) cannot be used, so use Eq. 38.11 to find the humidity ratio and Eqs. 38.17, 38.18(b), and 38.19(b) to calculate the enthalpy of air. (App. 38.D could also be used as a simpler solution.)

From App. 24.A, the saturated steam pressure at 100°F is 0.9505 psia.

$$p_w = \phi p_{\text{sat}}$$
$$= (0.82)(0.9505 \text{ psia})$$
$$= 0.7794 \text{ psia}$$

From Eq. 38.1,

$$p_a = p - p_w = 14.696 \text{ psia} - 0.7794 \text{ psia}$$
$$= 13.9166 \text{ psia}$$

From Eq. 38.11, the humidity ratio for air out is

$$\phi = 1.608 \omega_{\text{out}}\left(\frac{p_a}{p_{\text{sat}}}\right)$$
$$0.82 = 1.608 \omega_{\text{out}}\left(\frac{13.9166 \text{ psia}}{0.9505 \text{ psia}}\right)$$
$$\omega_{\text{out}} = 0.0348 \text{ lbm moisture/lbm air}$$

From Eqs. 38.17, 38.18(b), and 38.19(b), the enthalpy of air out is

$$h_2 = h_a + \omega_2 h_w$$
$$= \left(0.240 \; \frac{\text{Btu}}{\text{lbm-}°\text{F}} \right) T_{2,°\text{F}} + \omega_2$$
$$\times \left(\left(0.444 \; \frac{\text{Btu}}{\text{lbm-}°\text{F}} \right) T_{2,°\text{F}} + 1061 \; \frac{\text{Btu}}{\text{lbm}} \right)$$
$$= \left(0.240 \; \frac{\text{Btu}}{\text{lbm-}°\text{F}} \right) (100°\text{F})$$
$$+ \left(0.0348 \; \frac{\text{lbm moisture}}{\text{lbm air}} \right) \left(\left(0.444 \; \frac{\text{Btu}}{\text{lbm-}°\text{F}} \right) \right.$$
$$\times (100°\text{F}) + 1061 \; \frac{\text{Btu}}{\text{lbm}} \right)$$
$$= 62.47 \; \text{Btu/lbm air}$$

The mass flow rate of air can be determined from Eq. 38.26.

$$q = \dot{m}_a (h_2 - h_1)$$
$$\dot{m}_{\text{air}} = \frac{q}{h_2 - h_1} = \frac{1 \times 10^6 \; \dfrac{\text{Btu}}{\text{hr}}}{62.47 \; \dfrac{\text{Btu}}{\text{lbm air}} - 42.7 \; \dfrac{\text{Btu}}{\text{lbm air}}}$$
$$= \boxed{5.058 \times 10^4 \; \text{lbm air/hr}}$$

(b) From conservation of water vapor,

$$\omega_1 \dot{m}_{\text{air}} + \dot{m}_{\text{make-up}} = \omega_2 \dot{m}_{\text{air}}$$
$$\dot{m}_{\text{make-up}} = \dot{m}_{\text{air}} (\omega_{\text{out}} - \omega_{\text{in}})$$
$$= \left(5.058 \times 10^4 \; \frac{\text{lbm air}}{\text{hr}} \right)$$
$$\times \left(\begin{array}{c} 0.0348 \; \dfrac{\text{lbm moisture}}{\text{lbm air}} \\ - \; 0.0191 \; \dfrac{\text{lbm moisture}}{\text{lbm air}} \end{array} \right)$$
$$= \boxed{794 \; \text{lbm water/hr}}$$

SI Solution

(a) The cooled water flow rate is given by

$$Q = \dot{m}_w c_p \Delta T$$
$$290 \; \text{kW} = \dot{m}_w \left(4.187 \; \frac{\text{kJ}}{\text{kg-}°\text{C}} \right) (49°\text{C} - 43°\text{C})$$
$$\dot{m}_w = 11.54 \; \text{kg/s}$$

From the psychrometric chart (App. 38.B), for air in at $T_{\text{db}} = 33°\text{C}$ and $\phi = 60\%$,

$$h_{\text{in}} = 82.3 \; \text{kJ/kg air}$$
$$\omega_{\text{in}} = \left(19.2 \; \frac{\text{g moisture}}{\text{kg air}} \right) \left(\frac{1 \; \text{kg}}{1000 \; \text{g}} \right)$$
$$= 0.0192 \; \text{kg moisture/kg air}$$

For air out, the psychrometric chart (off scale) cannot be used, so use Eq. 38.11 to find the humidity ratio and Eqs. 38.17, 38.18(a), and 38.19(a) to calculate enthalpy of air.

From App. 24.N, the saturated steam pressure at $38°\text{C}$ is 0.06633 bars.

$$p_w = \phi p_{\text{sat}}$$
$$= (0.82)(0.06633 \; \text{bar})$$
$$= 0.05439 \; \text{bar}$$

From Eq. 38.1,

$$p_a = p - p_w = 1 \; \text{bar} - 0.05439 \; \text{bar}$$
$$= 0.94561 \; \text{bar}$$

From Eq. 38.11, the humidity ratio for air out is

$$\phi = 1.608 \omega_{\text{out}} \left(\frac{p_a}{p_{\text{sat}}} \right)$$
$$0.82 = 1.608 \omega_{\text{out}} \left(\frac{0.94561 \; \text{bar}}{0.06633 \; \text{bar}} \right)$$
$$\omega_{\text{out}} = 0.0358 \; \text{kg moisture/kg air}$$

From Eqs. 38.17, 38.18(a), and 38.19(a), the enthalpy of air out is

$$h_2 = h_a + \omega_2 h_w$$
$$= \left(1.005 \; \frac{\text{kJ}}{\text{kg-}°\text{C}} \right) T_{°\text{C}}$$
$$+ \omega_{\text{out}} \left(\left(1.805 \; \frac{\text{kJ}}{\text{kg-}°\text{C}} \right) T_{°\text{C}} + 2501 \; \frac{\text{kJ}}{\text{kg}} \right)$$
$$= \left(1.005 \; \frac{\text{kJ}}{\text{kg-}°\text{C}} \right) (38°\text{C}) + \left(0.0358 \; \frac{\text{kg moisture}}{\text{kg air}} \right)$$
$$\times \left(\left(1.805 \; \frac{\text{kJ}}{\text{kg-}°\text{C}} \right) (38°\text{C}) + 2501 \; \frac{\text{kJ}}{\text{kg}} \right)$$
$$= 130.2 \; \text{kJ/kg air}$$

The mass flow rate of air can be determined from Eq. 38.26.

$$q = \dot{m}_a (h_2 - h_1)$$
$$\dot{m}_{\text{air}} = \frac{q}{h_2 - h_1} = \frac{290 \; \text{kW}}{130.2 \; \dfrac{\text{kJ}}{\text{kg air}} - 82.3 \; \dfrac{\text{kJ}}{\text{kg air}}}$$
$$= \boxed{6.054 \; \text{kg air/s}}$$

HVAC

(b) From conservation of water vapor,

$$\omega_1 \dot{m}_{\text{air}} + \dot{m}_{\text{make-up}} = \omega_2 \dot{m}_{\text{air}}$$

$$\dot{m}_{\text{make-up}} = \dot{m}_{\text{air}}(\omega_{\text{out}} - \omega_{\text{in}})$$

$$= \left(6.054 \, \frac{\text{kg air}}{\text{s}}\right)$$

$$\times \left(0.0358 \, \frac{\text{kg moisture}}{\text{kg air}} - 0.0192 \, \frac{\text{kg moisture}}{\text{kg air}}\right)$$

$$= \boxed{0.100 \text{ kg water/s}}$$

39 Ventilation

PRACTICE PROBLEMS

1. An office room has floor dimensions of 60 ft by 95 ft (18 m by 29 m) and a ceiling height of 10 ft (3 m). 45 people occupy the office, and half of them smoke.

(a) Calculate the ventilation rate based on six air changes per hour.

(b) State your assumptions and determine the ventilation rate based on occupancy.

2. 150 ppm of methanol (TLV = 200 ppm; MW = 32.04; SG = 0.792) and 285 ppm of methylene chloride (TLV = 500 ppm; MW = 84.94; SG = 1.336) are found in the air in a plating booth. Two pints of each are evaporated per hour. Use a mixing safety factor (i.e., a K value) of 6. What ventilation rate is required?

    (A) 2500 ft^3/min
    (B) 5200 ft^3/min
    (C) 10,000 ft^3/min
    (D) 13,000 ft^3/min

3. An auditorium is designed to seat 4500 people. The ventilation rate is 60 ft^3/min (1.68 m^3/min) per person of outside air. The outside temperature is 0°F (−18°C) dry-bulb, and the outside pressure is 14.6 psia (100.6 kPa). Air leaves the auditorium at 70°F (21°C) dry-bulb. There is no recirculation. The furnace has a capacity of 1,250,000 Btu/hr (370 kW).

(a) At what temperature should the air enter the auditorium?

    (A) 52°F (11.1°C)
    (B) 55°F (12.8°C)
    (C) 62°F (16.7°C)
    (D) 67°F (19.3°C)

(b) How much heat should be supplied to the ventilation air?

    (A) 1.5×10^7 Btu/hr (4.5 MW)
    (B) 2.2×10^7 Btu/hr (6.5 MW)
    (C) 3.9×10^7 Btu/hr (12 MW)
    (D) 5.1×10^7 Btu/hr (16 MW)

(c) Has the furnace been sized properly?

    (A) The furnace is less than half the required capacity.
    (B) The furnace is more than half the required capacity.
    (C) The furnace is less than twice the required capacity.
    (D) The furnace is more than twice the required capacity.

4. A room is maintained at design conditions of 75°F (23.9°C) dry-bulb and 50% relative humidity. The air outside is at 95°F (35°C) dry-bulb and 75°F (23.9°C) wet-bulb. The outside air is conditioned and mixed with some room exhaust air. The mixed, conditioned air enters the room and increases 20°F (11.1°C) in temperature before being removed from the room. The sensible and latent loads are 200,000 Btu/hr (60 kW) and 50,000 Btu/hr (15 kW), respectively. Air leaves the coil at 50.8°F (10°C). What is the volume of air flowing through the coil?

    (A) 4200 ft^3/min (120 m^3/min)
    (B) 5800 ft^3/min (160 m^3/min)
    (C) 7600 ft^3/min (220 m^3/min)
    (D) 9300 ft^3/min (260 m^3/min)

SOLUTIONS

1. *Customary U.S. Solution*

(a) The office volume is

$$V = (60 \text{ ft})(95 \text{ ft})(10 \text{ ft}) = 57{,}000 \text{ ft}^3$$

Based on six air changes per hour, the flow rate is

$$\dot{V} = \left(57{,}000 \; \frac{\text{ft}^3}{\text{air change}}\right) \left(6 \; \frac{\text{air changes}}{\text{hr}}\right) \left(\frac{1 \text{ hr}}{60 \text{ min}}\right)$$

$$= \boxed{5700 \text{ ft}^3/\text{min}}$$

(b) For preferred ventilation based on Sec. 39-2, "Ventilation Standards," assume the following.

- The ventilation rate for a nonsmoking area is 20 ft³/min.

- The ventilation rate for a smoking area ranges from 30 to 60 ft³/min, with an average value of 45 ft³/min.

The preferred ventilation rate is

$$\dot{V} = \left(\tfrac{1}{2}\right)(45 \text{ persons}) \left(\frac{20 \; \frac{\text{ft}^3}{\text{min}}}{1 \text{ person}}\right)$$

$$+ \left(\tfrac{1}{2}\right)(45 \text{ persons}) \left(\frac{45 \; \frac{\text{ft}^3}{\text{min}}}{1 \text{ person}}\right)$$

$$= \boxed{1463 \text{ ft}^3/\text{min}}$$

SI Solution

(a) The office volume is

$$(18 \text{ m})(29 \text{ m})(3 \text{ m}) = 1566 \text{ m}^3$$

Based on six air changes per hour, the flow rate is

$$\dot{V} = \left(1566 \; \frac{\text{m}^3}{\text{air change}}\right) \left(6 \; \frac{\text{air changes}}{\text{h}}\right) \left(\frac{1 \text{ h}}{60 \text{ min}}\right)$$

$$= \boxed{156.6 \text{ m}^3/\text{min}}$$

(b) For preferred ventilation based on Sec. 39-2, "Ventilation Standards," assume the following.

- The ventilation rate for a nonsmoking area is 0.57 m³/min.

- The ventilation rate for a smoking area ranges from 0.84 to 1.68 m³/min, with an average value of 1.26 m³/min.

The preferred ventilation rate is

$$\dot{V} = \left(\tfrac{1}{2}\right)(45 \text{ persons}) \left(\frac{0.57 \; \frac{\text{m}^3}{\text{min}}}{1 \text{ person}}\right)$$

$$+ \left(\tfrac{1}{2}\right)(45 \text{ persons}) \left(\frac{1.26 \; \frac{\text{m}^3}{\text{min}}}{1 \text{ person}}\right)$$

$$= \boxed{41.2 \text{ m}^3/\text{min}}$$

2. From Eq. 39.11(b), the ventilation rate required is

$$\dot{V}_{\text{cfm}} = \frac{(4.03 \times 10^8)K(\text{SG})R_{\text{pints/min}}}{(\text{MW})\text{TLV}_{\text{ppm}}}$$

For the methanol,

$$\dot{V}_{\text{cfm}} = \frac{(4.03 \times 10^8)(6)(0.792)\left(2 \; \frac{\text{pints}}{\text{h}}\right)\left(\frac{1 \text{ h}}{60 \text{ min}}\right)}{(32.04)(200 \text{ ppm})}$$

$$= 9962 \text{ ft}^3/\text{min}$$

For the methylene chloride,

$$\dot{V}_{\text{cfm}} = \frac{(4.03 \times 10^8)(6)(1.336)\left(2 \; \frac{\text{pints}}{\text{hr}}\right)\left(\frac{1 \text{ hr}}{60 \text{ min}}\right)}{(84.94)(500 \text{ ppm})}$$

$$= 2535 \text{ ft}^3/\text{min}$$

The total ventilation rate is

$$9962 \; \frac{\text{ft}^3}{\text{min}} + 2535 \; \frac{\text{ft}^3}{\text{min}} = \boxed{12{,}497 \text{ ft}^3/\text{min}}$$

The answer is (D).

3. *Customary U.S. Solution*

(a) The total ventilation rate is

$$\dot{V} = \left(\frac{60 \ \frac{\text{ft}^3}{\text{min}}}{1 \ \text{person}} \right) (4500 \ \text{persons}) \left(60 \ \frac{\text{min}}{\text{hr}} \right)$$

$$= 1.62 \times 10^7 \ \text{ft}^3/\text{hr}$$

The absolute temperature of outside air is

$$T = 0°\text{F} + 460 = 460°\text{R}$$

From the ideal gas law, the density of outside air is

$$\rho = \frac{p}{RT} = \frac{\left(14.6 \ \frac{\text{lbf}}{\text{in}^2} \right) \left(144 \ \frac{\text{in}^2}{\text{ft}^2} \right)}{\left(53.35 \ \frac{\text{ft-lbf}}{\text{lbm-°R}} \right) (460°\text{R})}$$

$$= 0.08567 \ \text{lbm/ft}^3$$

The mass flow rate is

$$\dot{m} = \dot{V}\rho = \left(1.62 \times 10^7 \ \frac{\text{ft}^3}{\text{hr}} \right) \left(0.08567 \ \frac{\text{lbm}}{\text{ft}^3} \right)$$

$$= 1.388 \times 10^6 \ \text{lbm/hr}$$

Assume no latent heat (no moisture) at 0°F.

From Table 39.1, the sensible heat generated by each person seated in the theater is 225 Btu/hr. Therefore,

$$Q_{\text{in from people}} = \left(225 \ \frac{\frac{\text{Btu}}{\text{hr}}}{\text{person}} \right) (4500 \ \text{persons})$$

$$= 1.01 \times 10^6 \ \text{Btu/hr}$$

The air leaves the auditorium at 70°F. From App. 35.C, the specific heat is 0.240 Btu/lbm-°F (remains fairly constant). Since Q is known, the air temperature entering the auditorium can be calculated by

$$Q = \dot{m}c_p(T_{\text{out,air}} - T_{\text{in,air}})$$

$$T_{\text{in,air}} = T_{\text{out,air}} - \frac{Q}{\dot{m}c_p}$$

$$= 70°\text{F} - \frac{1.01 \times 10^6 \ \frac{\text{Btu}}{\text{hr}}}{\left(1.388 \times 10^6 \ \frac{\text{lbm}}{\text{hr}} \right) \left(0.240 \ \frac{\text{Btu}}{\text{lbm-°F}} \right)}$$

$$= \boxed{67.0°\text{F}}$$

The answer is (D).

(b) The heat needed to heat dry ventilation air from 0 to 67°F is

$$Q = \dot{m}c_p\Delta T$$

$$= \left(1.388 \times 10^6 \ \frac{\text{lbm}}{\text{hr}} \right) \left(0.240 \ \frac{\text{Btu}}{\text{lbm-°F}} \right) (67°\text{F} - 0°\text{F})$$

$$= \boxed{2.23 \times 10^7 \ \text{Btu/hr}}$$

The answer is (B).

(c) Since $1.25 \times 10^6 \ \text{Btu/hr} < 2.23 \times 10^7 \ \text{Btu/hr}$, the furnace is too small.

The answer is (A).

SI Solution

(a) The total ventilation rate is

$$\dot{V} = \left(\frac{1.68 \ \frac{\text{m}^3}{\text{min}}}{\text{person}} \right) (4500 \ \text{persons}) \left(60 \ \frac{\text{min}}{\text{h}} \right)$$

$$= 4.54 \times 10^5 \ \text{m}^3/\text{h}$$

The absolute temperature of outside air is

$$T = -18°\text{C} + 273 = 255\text{K}$$

From the ideal gas law, the density of outside air is

$$\rho = \frac{p}{RT} = \frac{(100.6 \ \text{kPa}) \left(1000 \ \frac{\text{Pa}}{\text{kPa}} \right)}{\left(287.03 \ \frac{\text{J}}{\text{kg·K}} \right) (255\text{K})}$$

$$= 1.374 \ \text{kg/m}^3$$

The mass flow rate is

$$\dot{m} = \rho\dot{V} = \left(1.374 \ \frac{\text{kg}}{\text{m}^3} \right) \left(4.54 \times 10^5 \ \frac{\text{m}^3}{\text{h}} \right)$$

$$= 6.238 \times 10^5 \ \text{kg/h}$$

Assume no latent heat (no moisture) at −18°C.

HVAC

From Table 39.1 and the table footnote, the sensible heat generated by people seated in the theater is

$$\left(225 \; \frac{\text{Btu}}{\text{h}}\right)\left(0.293 \; \frac{\text{W}}{\frac{\text{Btu}}{\text{h}}}\right) = 65.93 \; \text{W}$$

Therefore,

$$Q_{\text{in from people}} = \left(65.93 \; \frac{\text{W}}{\text{persons}}\right)(4500 \; \text{persons})$$
$$\times \left(\frac{1 \; \text{kW}}{1000 \; \text{W}}\right)$$
$$= 296.7 \; \text{kW}$$

The air leaves the auditorium at 21°C. From App. 35.D, the specific heat ≈ 1.0048 kJ/kg·K (remains fairly constant). The air temperature entering the auditorium can be found from known Q as

$$Q = \dot{m}c_p(T_{\text{out,air}} - T_{\text{in,air}})$$
$$T_{\text{in,air}} = T_{\text{out,air}} - \frac{Q}{\dot{m}c_p}$$
$$= 21°\text{C} - \frac{296.7 \; \text{kW}}{\left(6.238 \times 10^5 \; \frac{\text{kg}}{\text{h}}\right)}$$
$$\times \left(\frac{1 \; \text{h}}{3600 \; \text{s}}\right)\left(1.0048 \; \frac{\text{kJ}}{\text{kg·K}}\right)$$
$$= \boxed{19.3°\text{C}}$$

The answer is (D).

(b) The heat needed to heat dry ventilation air from −18°C to 19.3°C is

$$Q = \dot{m}c_p\Delta T$$
$$= \left(6.238 \times 10^5 \; \frac{\text{kg}}{\text{h}}\right)\left(\frac{1 \; \text{h}}{3600 \; \text{s}}\right)\left(1.0048 \; \frac{\text{kJ}}{\text{kg·K}}\right)$$
$$\times (19.3°\text{C} - (-18°\text{C}))$$
$$= \boxed{6494 \; \text{kW}}$$

The answer is (B).

(c) Since 370 kW < 6494 kW, the furnace is too small.

The answer is (A).

4. $\qquad T_{\text{db,in}} = 75°\text{F} - 20°\text{F} = 55°\text{F}$

From Eq. 39.6(b), the volumetric flow rate of air entering the room is

$$\dot{V}_{\text{in,cfm}} = \frac{\dot{q}_{\text{s,Btu/hr}}}{\left(1.08 \; \frac{\text{Btu-min}}{\text{ft}^3\text{-hr-°F}}\right)(T_{\text{i,°F}} - T_{\text{in,°F}})}$$
$$= \frac{200{,}000 \; \frac{\text{Btu}}{\text{hr}}}{\left(1.08 \; \frac{\text{Btu-min}}{\text{ft}^3\text{-hr-°F}}\right)(75°\text{F} - 55°\text{F})}$$
$$= 9259 \; \text{ft}^3/\text{min}$$

This is a mixing problem.

(i)

75°F $\quad \dot{V}_2$

$\dot{V}_1$ $\qquad$ (in)

(co) $\qquad \dot{V}_{\text{in}}$ $\quad$ 55°F
$\qquad\qquad$ 9259 ft³/min

50.8°F

Using the lever rule, and since the temperature scales are all linear, the fraction of air passing through the coil is

$$\frac{T_i - T_{\text{in}}}{T_i - T_{\text{co}}} = \frac{75°\text{F} - 55°\text{F}}{75°\text{F} - 50.8°\text{F}} = 0.826$$
$$\dot{V}_1 = (0.826)\left(9259 \; \frac{\text{ft}^3}{\text{min}}\right) = \boxed{7648 \; \text{ft}^3/\text{min}}$$

The answer is (C).

SI Solution

$$T_{\text{db,in}} = 23.9°\text{C} - 11.1°\text{C} = 12.8°\text{C}$$

From Eq. 39.6(a),

$$\dot{V}_{\text{in,m}^3/\text{min}} = \frac{\dot{q}_{\text{s,kW}}}{\left(0.02 \; \frac{\text{kJ·min}}{\text{m}^3\text{·s·°C}}\right)(T_{\text{i,°C}} - T_{\text{in,°C}})}$$
$$= \frac{60 \; \text{kW}}{\left(0.02 \; \frac{\text{kJ-min}}{\text{m}^3\text{·s·°C}}\right)(23.9°\text{C} - 12.8°\text{C})}$$
$$= 270.3 \; \text{m}^3/\text{min}$$

This is a mixing problem.

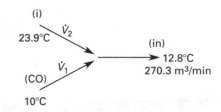

(i)

23.9°C $\quad \dot{V}_2$

$\dot{V}_1$ $\qquad$ (in)

(CO) $\qquad$ 12.8°C
$\qquad\qquad$ 270.3 m³/min

10°C

Using the lever rule, and since the temperature scales are all linear, the fraction of air passing through the coil is

$$\frac{T_i - T_{\text{in}}}{T_i - T_{\text{co}}} = \frac{23.9°\text{C} - 12.8°\text{C}}{23.9°\text{C} - 10°\text{C}}$$

$$= 0.799$$

$$\dot{V}_1 = (0.799)\left(270.3 \ \frac{\text{m}^3}{\text{min}}\right)$$

$$= \boxed{216.0 \ \text{m}^3/\text{min}}$$

The answer is (C).

40 Heating Load

PRACTICE PROBLEMS

1. A conditioned room contains 12,000 W of fluorescent lights (with older coil ballasts) and twelve 90% efficient, 10 hp motors operating at 80% of their rated capacities. The lights are pendant-mounted on chains from the ceiling. The motors drive various pieces of machinery located in the conditioned space. What is the internal heat gain?

(A) 1.9×10^5 Btu/hr (55 kW)

(B) 3.2×10^5 Btu/hr (94 kW)

(C) 3.9×10^5 Btu/hr (110 kW)

(D) 4.6×10^5 Btu/hr (130 kW)

2. A building is located in the city of New York. There are 4772 (2651 in SI) degree days during the October 15 to May 15 period for this area. The building is heated by fuel oil whose heating value is 153,600 Btu/gal ($42\,800$ MJ/m^3) and which costs \$0.15/gal (\$0.034/L). The calculated design heat loss is 3.5×10^6 Btu/hr (1 MW) based on 70°F (21.1°C) inside and 0°F (−17.8°C) outside design temperatures. The furnace has an efficiency of 70%. What is the approximate cost for heating this building during the winter (October 15 through May 15)?

(A) \$7000

(B) \$8000

(C) \$9000

(D) \$10,000

3. A flat roof consists of $1^1/_2$ in (38 mm) of insulation installed over 3 in (76 mm) of soft pine sheathing, and a $^3/_4$ in (19 mm) acoustical ceiling suspended 4 in (100 mm) below the pine. The outside wind velocity is 15 mi/hr (24 km/h). The interior design temperature is 80°F (26.7°C). The exterior design temperature is 95°F (35°C). What is the overall coefficient of heat transfer?

(A) 0.08 Btu/ft^2-hr-°F (0.5 W/m^2-°C)

(B) 1 Btu/ft^2-hr-°F (6 W/m^2-°C)

(C) 4.08 Btu/ft^2-hr-°F (24 W/m^2-°C)

(D) 18 Btu/ft^2-hr-°F (110 W/m^2-°C)

4. A 12 ft × 12 ft (3.6 m × 3.6 m) floor is constructed as a concrete slab with two exposed edges. The slab edge heat loss coefficient for these two edges is 0.55 Btu/ft-hr-°F (0.95 W/m·°C). The other two edges form part of a basement wall exposed to 70°F air. The inside design temperature is 70°F (21.1°C); the outdoor design temperature is −10°F (−23.3°C). Use the slab edge method to determine the heat loss from the slab.

(A) 500 Btu/hr (140 W)

(B) 800 Btu/hr (240 W)

(C) 1100 Btu/hr (300 W)

(D) 4200 Btu/hr (1300 W)

5. A first-floor office in a remodeled historic building has floor dimensions of 100 ft × 40 ft (30 m × 12 m) and a ceiling height of 10 ft (3 m). One of the 40 ft (12 m) walls is shared with an adjacent heated space. The three remaining walls have one 4 ft (1.2 m) wide × 6 ft (1.8 m) high, double glass with $^1/_4$ in (6.4 mm) air space, weatherstripped, double-hung window per 10 ft (3 m). The crack coefficient for the windows is 32 ft^3/hr-ft (3.0 m^3/h·m). One wall of 10 windows is exfiltrating. The basement and second floor are heated to 70°F (21.1°C). The wall coefficient (exclusive of film resistance) is 0.2 Btu/ft^2-hr-°F (1.1 W/m^2·°C), and the outside wind velocity is 15 mi/hr (24 km/h). The inside design temperature is 70°F (21.1°C); the outside design temperature is −10°F (−23.3°C). Including infiltration but disregarding ventilation air, what is the heating load?

6. A building is located in Newark, New Jersey. The heating season lasts for 245 days; the degree days are 5252 (2918 in SI); the outside design temperature is 0°F (−17.8°C). The temperature in the building is maintained at 70°F (21.1°C) between the hours of 8:30 A.M. and 5:30 P.M. During the rest of the day and the night, the temperature is allowed to drop to 50°F (10°C). The building is heated with coal that has a heating value of 13,000 Btu/lbm (30.2 MJ/kg). A heat loss of 650,000 Btu/hr (0.19 MW) has been calculated based on 70°F (21.1°C) inside and 0°F (−17.8°C) outside temperatures. The furnace has an efficiency of 70%. What mass of coal is required each year?

(A) 18,000 lbm/yr (4000 kg/yr)

(B) 43,000 lbm/yr (9500 kg/yr)

(C) 69,000 lbm/yr (15,000 kg/yr)

(D) 84,000 lbm/yr (40,000 kg/yr)

7. The design heat loss of a building is 200,000 Btu/hr (60 kW) originally based on 70°F (21.1°C) inside and 0°F (−17.8°C) outside design temperatures. At that location, there are 4200 (2333 in SI) degree days over the 210-day heating season. The building is occupied

24 hr/day. What is the percentage reduction in heating fuel if the thermostat is lowered from 70°F to 68°F (21.1°C to 20°C)?

  (A) 2%
  (B) 4%
  (C) 8%
  (D) 14%

8. (*Time limit: one hour*) A building with the characteristics listed is located where the annual heating season lasts 21 weeks. The inside design temperature is 70°F (21.1°C). The average outside air temperature is 30°F(−1°C). The gas furnace has an efficiency of 75%. Fuel costs $0.25 per therm. The building is occupied only from 8 A.M. until 6 P.M., Monday through Friday. In the past, the building was maintained 70°F (21.1°C) at all times and ventilated with one air change per hour (based on inside conditions). The thermostat is now being set back 12°F (6.7°C) and the ventilation reduced 50% during the unoccupied times. Infiltration through cracks and humidity changes are disregarded. What is the annual savings?

internal volume:	801,000 ft^3 (22 700 m^3)
wall area:	11,040 ft^2 (993 m^2)
wall overall heat transfer coefficient:	0.15 Btu/ft^2-hr-°F (0.85 W/m^2·°C)
window area:	2760 ft^2 (260 m^2)
window overall heat transfer coefficient:	1.13 Btu/ft^2-hr-°F (6.42 W/m^2·°C)
roof area:	26,700 ft^2 (2480 m^2)
roof overall heat transfer coefficient:	0.05 Btu/ft^2-hr-°F (0.3 W/m^2·°C)
slab on grade:	690 lineal ft (210 m) of exposed slab edge
slab edge coefficient:	1.5 Btu/ft-hr-°F (2.6 W/m·°C)

  (A) $3100/yr
  (B) $3800/yr
  (C) $4600/yr
  (D) $6200/yr

SOLUTIONS

1. *Customary U.S. Solution*

Based on Sec. 40-10, for fluorescent lights, the rated wattage should be increased by 20–25% to account for ballast heating. Since the lights are pendant-mounted on chain from the ceiling, most of this heat enters the conditioned space. Assume rated wattage increases by 20%. From Eq. 40.7(b), the internal heat gain due to lights (SF = 1.2) is

$$\dot{q}_{lights} = \left(3413 \ \frac{Btu}{kW\text{-}hr} \right) (SF) \left(\frac{P}{\eta} \right)$$

$$\eta = 1 \quad \text{[given]}$$

$$\dot{q}_{lights} = \left(3413 \ \frac{Btu}{kW\text{-}hr} \right) (1.2)$$

$$\times \left(\frac{(12{,}000 \ W) \left(\frac{1 \ kW}{1000 \ W} \right)}{1} \right)$$

$$= 4.9147 \times 10^4 \ Btu/hr$$

From Eq. 40.7(b), the internal heat gain due to motors (SF = 0.8) is

$$\dot{q}_{motors} = \left(2545 \ \frac{Btu}{hr\text{-}hp} \right) (SF) \left(\frac{P}{\eta} \right)$$

$$\eta = 0.90 \quad \text{[given]}$$

$$\dot{q}_{motors} = (12 \ \text{motors}) \left(2545 \ \frac{Btu}{hr\text{-}hp} \right) (0.8) \left(\frac{10 \ hp}{0.90} \right)$$

$$= 2.7147 \times 10^5 \ Btu/hr$$

$$\dot{q}_{total} = \dot{q}_{lights} + \dot{q}_{motors}$$

$$= 4.9147 \times 10^4 \ \frac{Btu}{hr} + 2.7147 \times 10^5 \ \frac{Btu}{hr}$$

$$= \boxed{3.206 \times 10^5 \ Btu/hr}$$

The answer is (B).

SI Solution

Based on the customary U.S. solution, SF = 1.2 for lights. From Eq. 40.7(a), the internal heat gain due to lights is

$$\eta = 1 \quad \text{[given]}$$

$$\dot{q}_{lights} = (SF) \left(\frac{P}{\eta} \right)$$

$$= (1.2) \left(\frac{(12\,000 \ W) \left(\frac{1 \ kW}{1000 \ W} \right)}{1} \right)$$

$$= 14.4 \ kW$$

From Eq. 40.7(a), the internal heat gain due to motors (12 motors) is

$$\dot{q}_{\text{motors}} = \left(0.7457 \; \frac{\text{kW}}{\text{hp}}\right)(\text{SF})\left(\frac{P}{\eta}\right)$$

$$\text{SF} = 0.8 \quad \text{[given]}$$

$$\eta = 0.90 \quad \text{[given]}$$

$$\dot{q}_{\text{motors}} = (12 \text{ motors})\left(0.7457 \; \frac{\text{kW}}{\text{hp}}\right)(0.8)\left(\frac{10 \text{ hp}}{0.90}\right)$$

$$= 79.5 \text{ kW}$$

$$\dot{q}_{\text{total}} = \dot{q}_{\text{lights}} + \dot{q}_{\text{motors}}$$

$$= 14.4 \text{ kW} + 79.5 \text{ kW}$$

$$= \boxed{93.9 \text{ kW}}$$

The answer is (B).

2. *Customary U.S. Solution*

From Eq. 40.10(b), the fuel consumption (in gal/heating season) is

$$\frac{\left(24 \; \frac{\text{hr}}{\text{day}}\right)\dot{q}_{\text{Btu/hr}}(\text{DD})}{(T_i - T_o)(\text{HV}_{\text{Btu/gal}})\eta_{\text{furnace}}}$$

$$= \frac{\left(24 \; \frac{\text{hr}}{\text{day}}\right)\left(3.5 \times 10^6 \; \frac{\text{Btu}}{\text{hr}}\right)(4772^\circ\text{F-days})}{(70^\circ\text{F} - 0^\circ\text{F})\left(153{,}600 \; \frac{\text{Btu}}{\text{gal}}\right)(0.70)}$$

$$= 53{,}260 \text{ gal}$$

The total cost of 53,260 gal fuel at \$0.15/gal is

$$(53{,}260 \text{ gal})\left(\frac{\$0.15}{\text{gal}}\right) = \boxed{\$7989}$$

The answer is (B).

SI Solution

$$\text{HV}_{\text{kJ/L}} = \left(42\,800 \; \frac{\text{MJ}}{\text{m}^3}\right)\left(\frac{1000 \text{ kJ}}{1 \text{ MJ}}\right)\left(\frac{1 \text{ m}^3}{1000 \text{ L}}\right)$$

$$= 42\,800 \text{ kJ/L}$$

From Eq. 40.10(a), fuel consumption (in L/heating season) is

$$\frac{\left(86\,400 \; \frac{\text{s}}{\text{day}}\right)\dot{q}(\text{DD})}{(T_c - T_o)(\text{HV})\eta_{\text{furnace}}}$$

$$= \frac{\left(86\,400 \; \frac{\text{s}}{\text{day}}\right)(1 \text{ MW})\left(\frac{1000 \text{ kW}}{1 \text{ MW}}\right)}{(21.1^\circ\text{C} - (-17.8^\circ\text{C}))\left(42\,800 \; \frac{\text{kJ}}{\text{L}}\right)(0.70)}$$

$$\qquad \times (2651^\circ\text{C-days})$$

$$= 196\,531 \text{ L}$$

The total cost at \$0.034/L is

$$(196\,531 \text{ L})\left(\frac{\$0.034}{\text{L}}\right) = \boxed{\$6682}$$

The answer is (A).

3. *Customary U.S. Solution*

From Table 40.3, the surface film coefficient for outside air (horizontal, heat flow down, 15 mph) is $h_o = 6$ Btu/ft²-hr-°F. The film thermal resistance is

$$R_1 = \frac{1}{h_o} = \frac{1}{6 \; \dfrac{\text{Btu}}{\text{ft}^2\text{-hr-}^\circ\text{F}}} = 0.167 \text{ ft}^2\text{-hr-}^\circ\text{F/Btu}$$

From App. 40.A, for roof insulation, the thermal resistance per inch is 2.78 ft²-°F-hr/Btu-in.

$$R_2 = \left(2.78 \; \frac{\text{ft}^2\text{-}^\circ\text{F-hr}}{\text{Btu-in}}\right)(1.5 \text{ in}) = 4.17 \text{ ft}^2\text{-hr-}^\circ\text{F/Btu}$$

From App. 40.A, the thermal resistance per inch for soft wood is 1.25 ft²-°F-hr/Btu-in.

$$R_3 = \left(1.25 \; \frac{\text{ft}^2\text{-}^\circ\text{F-hr}}{\text{Btu-in}}\right)(3 \text{ in}) = 3.75 \text{ ft}^2\text{-}^\circ\text{F-hr/Btu}$$

For the space between soft wood and an acoustic ceiling, the effective surface emissivity from Eq. 40.4 is

$$E = \frac{1}{\dfrac{1}{\epsilon_1} + \dfrac{1}{\epsilon_2} - 1}$$

$$\epsilon_1 = \epsilon_{\text{wood}} = 0.90 \quad \text{[Sec. 6]}$$

$$\epsilon_2 = \epsilon_{\text{paper}} = 0.95 \quad \text{[used for acoustic tile]}$$

$$E = \frac{1}{\dfrac{1}{0.90} + \dfrac{1}{0.95} - 1} = 0.86$$

From Table 40.2, the thermal conductance of 4 in horizontal planar air space at $E = 0.86$ with heat flow down is approximately 0.81 Btu/hr-ft²-°F. The thermal resistance is

$$R_4 = \frac{1}{0.81 \; \dfrac{\text{Btu}}{\text{ft}^2\text{-hr-}^\circ\text{F}}} = 1.235 \text{ ft}^2\text{-hr-}^\circ\text{F/Btu}$$

From App. 40.A, the thermal resistance for ³/₄ in acoustical tile is $R_5 = 1.78$ ft²-°F-hr/Btu.

From Table 40.3, the surface film coefficient for inside air ($v = 0$) (horizontal, heat flow down) is $h_i = 1.08$ Btu/ft²-hr-°F. The film thermal resistance is

$$R_6 = \frac{1}{h_i} = \frac{1}{1.08 \; \dfrac{\text{Btu}}{\text{ft}^2\text{-hr-}^\circ\text{F}}} = 0.926 \text{ ft}^2\text{-hr-}^\circ\text{F/Btu}$$

HVAC

The total resistance is

$$R_{\text{total}} = R_1 + R_2 + R_3 + R_4 + R_5 + R_6$$

$$= 0.167 \, \frac{\text{ft}^2\text{-hr-}°\text{F}}{\text{Btu}} + 4.17 \, \frac{\text{ft}^2\text{-hr-}°\text{F}}{\text{Btu}}$$

$$+ 3.75 \, \frac{\text{ft}^2\text{-hr-}°\text{F}}{\text{Btu}} + 1.235 \, \frac{\text{ft}^2\text{-hr-}°\text{F}}{\text{Btu}}$$

$$+ 1.78 \, \frac{\text{ft}^2\text{-hr-}°\text{F}}{\text{Btu}} + 0.926 \, \frac{\text{ft}^2\text{-hr-}°\text{F}}{\text{Btu}}$$

$$= 12.03 \, \text{ft}^2\text{-hr-}°\text{F/Btu}$$

The overall coefficient of heat transfer is

$$U = \frac{1}{R_{\text{total}}} = \frac{1}{12.03 \, \dfrac{\text{ft}^2\text{-hr-}°\text{F}}{\text{Btu}}}$$

$$= \boxed{0.0831 \, \text{Btu/ft}^2\text{-hr-}°\text{F}}$$

The answer is (A).

SI Solution

From Table 40.3, the surface film coefficient for outside air (horizntal, heat flow down) at 15 mph is $h_o = 34.1 \, \text{W/m}^2\text{-}°\text{C}$. The film thermal resistance is

$$R_1 = \frac{1}{h_o} = \frac{1}{34.1 \, \dfrac{\text{W}}{\text{m}^2\text{-}°\text{C}}} = 0.0293 \, \text{m}^2\text{-}°\text{C/W}$$

From App. 40.A for roof insulation, the thermal resistance per inch is 2.78 $\text{ft}^2\text{-}°\text{F-hr/Btu-in}$.

$$R_2 = \left(2.78 \, \frac{\text{ft}^2\text{-}°\text{F-hr}}{\text{Btu-in}} \right) \left(6.93 \, \frac{\dfrac{\text{m-}°\text{C}}{\text{W}}}{\dfrac{\text{ft}^2\text{-}°\text{F-hr}}{\text{Btu-in}}} \right)$$

$$\times (38 \, \text{mm}) \left(\frac{1 \, \text{m}}{1000 \, \text{mm}} \right)$$

$$= 0.732 \, \text{m}^2\text{-}°\text{C/W}$$

From App. 40.A, the thermal resistance per inch for soft wood is 1.25 $\text{ft}^2\text{-}°\text{F-hr/Btu-in}$.

$$R_3 = \left(1.25 \, \frac{\text{ft}^2\text{-}°\text{F-hr}}{\text{Btu-in}} \right) \left(6.93 \, \frac{\dfrac{\text{m-}°\text{C}}{\text{W}}}{\dfrac{\text{ft}^2\text{-}°\text{F-hr}}{\text{Btu-in}}} \right)$$

$$\times (76 \, \text{mm}) \left(\frac{1 \, \text{m}}{1000 \, \text{mm}} \right)$$

$$= 0.658 \, \text{m}^2\text{-}°\text{C/W}$$

From the customary U.S. solution, the effective surface emissivity between soft wood and an acoustic ceiling is $E = 0.86$.

From Table 40.2, the thermal conductance of 100 mm horizontal planar air space at $E = 0.86$ with heat flow down is approximately 4.6 $\text{W/m}^2\text{-}°\text{C}$. The thermal resistance is

$$R_4 = \frac{1}{4.6 \, \dfrac{\text{W}}{\text{m}^2\text{-}°\text{C}}} = 0.217 \, \text{m}^2\text{-}°\text{C/W}$$

From App. 40.A, the thermal resistance for 19 mm ($^3/_4$ in) acoustical tile is

$$R_5 = \left(1.78 \, \frac{\text{ft}^2\text{-}°\text{F-hr}}{\text{Btu-in}} \right) \left(0.176 \, \frac{\dfrac{\text{m}^2\text{-}°\text{C}}{\text{W}}}{\dfrac{\text{ft}^2\text{-}°\text{F-hr}}{\text{Btu}}} \right)$$

$$= 0.313 \, \text{m}^2\text{-}°\text{C/W}$$

From Table 40.3, the surface film coefficient for inside air (v = 0) (horizontal, heat flow down) is $h_i = 6.13 \, \text{W/m}^2\text{-}°\text{C}$. The film thermal resistance is

$$R_6 = \frac{1}{h_i} = \frac{1}{6.13 \, \dfrac{\text{W}}{\text{m}^2\text{-}°\text{C}}} = 0.163 \, \text{m}^2\text{-}°\text{C/W}$$

The total resistance is

$$R_{\text{total}} = R_1 + R_2 + R_3 + R_4 + R_5 + R_6$$

$$= 0.0293 \, \frac{\text{m}^2\text{-}°\text{C}}{\text{W}} + 0.732 \, \frac{\text{m}^2\text{-}°\text{C}}{\text{W}}$$

$$+ 0.658 \, \frac{\text{m}^2\text{-}°\text{C}}{\text{W}} + 0.217 \, \frac{\text{m}^2\text{-}°\text{C}}{\text{W}}$$

$$+ 0.313 \, \frac{\text{m}^2\text{-}°\text{C}}{\text{W}} + 0.163 \, \frac{\text{m}^2\text{-}°\text{C}}{\text{W}}$$

$$= 2.112 \, \text{m}^2\text{-}°\text{C/W}$$

The overall coefficient of heat transfer is

$$U = \frac{1}{R_{\text{total}}} = \frac{1}{2.112 \, \dfrac{\text{m}^2\text{-}°\text{C}}{\text{W}}} = \boxed{0.473 \, \text{W/m}^2\text{-}°\text{C}}$$

The answer is (A).

4. *Customary U.S. Solution*

From Eq. 40.5, the heat loss from the slab is

$$\dot{q} = pF(T_i - T_o)$$

The slab edge coefficient is given as $F = 0.55$ Btu/ft-hr-°F.

The perimeter is

$$p = 12 \, \text{ft} + 12 \, \text{ft} \quad \text{[2 edges only]}$$

$$= 24 \, \text{ft}$$

$$\dot{q} = (24 \, \text{ft}) \left(0.55 \, \frac{\text{Btu}}{\text{ft-hr-}°\text{F}} \right) (70°\text{F} - (-10°\text{F}))$$

$$= \boxed{1056 \, \text{Btu/hr}}$$

The answer is (C).

SI Solution

From Eq. 40.5, the heat loss from the slab is

$$\dot{q} = pF(T_i - T_o)$$

The slab edge coefficient is given as $F = 0.95$ W/m·°C.

The perimeter is

$$p = 3.6 \text{ m} + 3.6 \text{ m} \quad [2 \text{ edges only}]$$
$$= 7.2 \text{ m}$$

$$\dot{q} = (7.2 \text{ m})\left(0.95 \frac{\text{W}}{\text{m·°C}}\right)(21.1°\text{C} - (-23.3°\text{C}))$$

$$= \boxed{303.7 \text{ W}}$$

The answer is (C).

5.

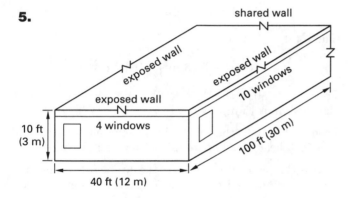

Customary U.S. Solution

For three unshared walls, the total exposed (walls + windows) area is

$$(10 \text{ ft})(40 \text{ ft} + 100 \text{ ft} + 100 \text{ ft}) = 2400 \text{ ft}^2$$

- There are two windows per 20 ft.
- The number of windows for a 40 ft wall is 4.
- The number of windows for each 100 ft wall is 10.
- The total number of windows is

$$4 + 10 + 10 = 24$$

- The total window area is

$$(24)(4 \text{ ft})(6 \text{ ft}) = 576 \text{ ft}^2$$

- The total wall area is

$$2400 \text{ ft}^2 - 576 \text{ ft}^2 = 1824 \text{ ft}^2$$

From Table 40.3, the outside film coefficient for a vertical wall in a 15 mph wind is $h_o = 6.00$ Btu/hr-ft²-°F.

From Table 40.3, the inside film coefficient for still air (vertical, heat flow horizontal) is $h_i = 1.46$ Btu/ft²-hr-°F.

Notice that L/k was given in the problem statement, not k. The overall coefficient of heat transfer for the wall is

$$U_{\text{wall}} = \cfrac{1}{\cfrac{1}{h_o} + \cfrac{L_{\text{wall}}}{k_{\text{wall}}} + \cfrac{1}{h_i}}$$

$$= \cfrac{1}{\cfrac{1}{6.00 \dfrac{\text{Btu}}{\text{hr-ft}^2\text{-°F}}} + \cfrac{1}{0.2 \dfrac{\text{Btu}}{\text{ft}^2\text{-hr-°F}}} + \cfrac{1}{1.46 \dfrac{\text{Btu}}{\text{ft}^2\text{-hr-°F}}}}$$

$$= 0.1709 \text{ Btu/hr-ft}^2\text{-°F}$$

The heat loss from the wall is

$$\dot{q}_{\text{wall}} = U_{\text{wall}} A_{\text{wall}} \Delta T$$

$$= \left(0.1709 \frac{\text{Btu}}{\text{hr-ft}^2\text{-°F}}\right)(1824 \text{ ft}^2)(70°\text{F} - (-10°\text{F}))$$

$$= 24{,}938 \text{ Btu/hr}$$

From App. 40.A, for double, vertical (1/4 in air space) glass windows, the thermal resistance is 1.63 ft²-°F-hr/Btu. From the footnote, this includes the inside and outside film coefficients.

$$U_{\text{windows}} = \cfrac{1}{1.63 \dfrac{\text{ft}^2\text{-°F-hr}}{\text{Btu}}} = 0.613 \text{ Btu/ft}^2\text{-°F-hr}$$

The heat loss from the windows is

$$\dot{q}_{\text{windows}} = U_{\text{windows}} A_{\text{windows}} \Delta T$$

$$= \left(0.613 \frac{\text{Btu}}{\text{ft}^2\text{-°F-hr}}\right)(576 \text{ ft}^2)$$
$$\times (70°\text{F} - (-10°\text{F}))$$

$$= 28{,}247 \text{ Btu/hr}$$

From Eq. 39.1, the infiltration flow rate is

$$\dot{V} = BL$$

The crack coefficient is given as $B = 32$ ft³/hr-ft.

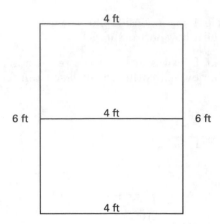

For double-hung windows, the crack length is

$$L = 4 \text{ ft} + 4 \text{ ft} + 4 \text{ ft} + 6 \text{ ft} + 6 \text{ ft}$$
$$= 24 \text{ ft}$$

$$\dot{V} = \left(32 \frac{\text{ft}^3}{\text{hr-ft}}\right)(24 \text{ ft})$$
$$= 768 \text{ ft}^3/\text{hr per window}$$

Although there are 24 windows, one wall of 10 is exfiltrating.

$$\dot{V} = (24 - 10)\left(768 \frac{\text{ft}^3}{\text{hr}}\right) = 10{,}752 \text{ ft}^3/\text{hr}$$

For atmospheric air, $p = 14.7$ psia and $c_p = 0.24$ Btu/lbm-°F.

The air density is

$$\rho = \frac{p}{RT} = \frac{\left(14.7 \frac{\text{lbf}}{\text{in}^2}\right)\left(144 \frac{\text{in}^2}{\text{ft}^2}\right)}{\left(53.35 \frac{\text{ft-lbf}}{\text{lbm-°R}}\right)(-10\text{°F} + 460°)}$$
$$= 0.088 \text{ lbm/ft}^3$$

The mass flow rate of infiltrated air is

$$\dot{m} = \rho\dot{V} = \left(0.088 \frac{\text{lbm}}{\text{ft}^3}\right)\left(10{,}752 \frac{\text{ft}^3}{\text{hr}}\right)$$
$$= 946.2 \text{ lbm/hr}$$

The heat loss due to infiltration is

$$\dot{q}_{\text{infiltration}} = \dot{m}c_p\Delta T$$
$$= \left(946.2 \frac{\text{lbm}}{\text{hr}}\right)\left(0.24 \frac{\text{Btu}}{\text{lbm-°F}}\right)$$
$$\times (70\text{°F} - (-10\text{°F}))$$
$$= 18{,}167 \text{ Btu/hr}$$

The total heat loss is

$$\dot{q}_{\text{total}} = \dot{q}_{\text{wall}} + \dot{q}_{\text{windows}} + \dot{q}_{\text{infiltration}}$$
$$= 24{,}938 \frac{\text{Btu}}{\text{hr}} + 28{,}247 \frac{\text{Btu}}{\text{hr}} + 18{,}167 \frac{\text{Btu}}{\text{hr}}$$
$$= \boxed{71{,}352 \text{ Btu/hr}}$$

SI Solution

For three unshared walls, the total exposed (walls + windows) area is

$$(3 \text{ m})(12 \text{ m} + 30 \text{ m} + 30 \text{ m}) = 216 \text{ m}^2$$

For 24 windows (from the customary U.S. solution), the total window area is

$$(24)(1.2 \text{ m})(1.8 \text{ m}) = 51.8 \text{ m}^2$$

The total wall area is

$$216 \text{ m}^2 - 51.8 \text{ m}^2 = 164.2 \text{ m}^2$$

From Table 40.3, the outside film coefficient for a vertical wall in a 15 mph wind is $h_o = 34.1$ W/m²·°C.

From Table 40.3, the inside film coefficient for still air (vertical, heat flow horizontal) is $h_i = 8.29$ W/m²·°C.

Notice that L/k was given in the problem statement, not k. The overall coefficient of heat transfer for the wall is

$$U_{\text{wall}} = \frac{1}{\dfrac{1}{h_o} + \dfrac{L_{\text{wall}}}{k_{\text{wall}}} + \dfrac{1}{h_i}}$$

$$= \frac{1}{\dfrac{1}{34.1 \frac{\text{W}}{\text{m}^2\cdot\text{°C}}} + \dfrac{1}{1.1 \frac{\text{W}}{\text{m}^2\cdot\text{°C}}} + \dfrac{1}{8.29 \frac{\text{W}}{\text{m}^2\cdot\text{°C}}}}$$

$$= 0.944 \text{ W/m}^2\cdot\text{°C}$$

The heat loss from the wall is

$$\dot{q}_{\text{wall}} = U_{\text{wall}}A_{\text{wall}}\Delta T$$
$$= \left(0.944 \frac{\text{W}}{\text{m}^2\cdot\text{°C}}\right)(164.2 \text{ m}^2)$$
$$\times (21.1\text{°C} - (-23.3\text{°C}))$$
$$= 6882 \text{ W}$$

From the customary U.S. solution,

$$U_{\text{windows}} = \left(0.613 \frac{\text{Btu}}{\text{ft}^2\text{-°F-hr}}\right)\left(5.68 \frac{\frac{\text{W}}{\text{m}^2\cdot\text{°C}}}{\frac{\text{Btu}}{\text{ft}^2\text{-°F-hr}}}\right)$$
$$= 3.48 \text{ W/m}^2\cdot\text{°C}$$

The heat loss from the windows is

$$\dot{q}_{\text{windows}} = U_{\text{windows}} A_{\text{windows}} \Delta T$$
$$= \left(3.48 \; \frac{\text{W}}{\text{m}^2 \cdot {}^\circ \text{C}} \right) (51.8 \; \text{m}^2)$$
$$\times \left(21.1 {}^\circ \text{C} - (-23.3 {}^\circ \text{C}) \right)$$
$$- 8004 \; \text{W}$$

From Eq. 39.1, the infiltration flow rate is

$$\dot{V} = BL$$

The crack coefficient is given as $B = 3.0 \; \text{m}^3/\text{h} \cdot \text{m}$.

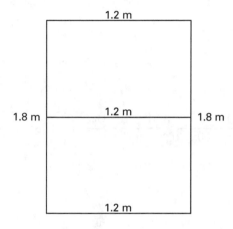

For double-hung windows, the crack length is

$$L = 1.2 \; \text{m} + 1.2 \; \text{m} + 1.2 \; \text{m} + 1.8 \; \text{m} + 1.8 \; \text{m}$$
$$= 7.2 \; \text{m}$$
$$\dot{V} = \left(3.0 \; \frac{\text{m}^3}{\text{h} \cdot \text{m}} \right) (7.2 \; \text{m}) = 21.6 \; \text{m}^2/\text{h per window}$$

Although there are 24 windows, one wall of 10 is exfil-trating.

$$\dot{V} = (24 - 10) \left(21.6 \; \frac{\text{m}^3}{\text{h}} \right) = 302.4 \; \text{m}^3/\text{h}$$

For atmospheric air, $p = 100{,}000 \; \text{Pa}$ and $c_p = 1.005 \; \text{kJ/kg} \cdot {}^\circ \text{C}$.

The air density is

$$\rho = \frac{p}{RT} = \frac{100{,}000 \; \text{Pa}}{\left(287.03 \; \frac{\text{J}}{\text{kg} \cdot \text{K}} \right) (-23.3 {}^\circ \text{C} + 273^\circ)}$$
$$= 1.4 \; \text{kg/m}^3$$

The mass flow rate of infiltrated air is

$$\dot{m} = \rho \dot{V} = \left(1.4 \; \frac{\text{kg}}{\text{m}^3} \right) \left(302.4 \; \frac{\text{m}^3}{\text{h}} \right)$$
$$= 423.4 \; \text{kg/h}$$

The heat loss due to infiltration is

$$\dot{q}_{\text{infiltration}} = \dot{m} c_p \Delta T$$
$$= \left(423.4 \; \frac{\text{kg}}{\text{h}} \right) \left(1.005 \; \frac{\text{kJ}}{\text{kg} \cdot {}^\circ \text{C}} \right) \left(1000 \; \frac{\text{J}}{\text{kg}} \right)$$
$$\times \left(\frac{1 \; \text{h}}{3600 \; \text{s}} \right) (21.1 {}^\circ \text{C} - (-23.3 {}^\circ \text{C}))$$
$$= 5248 \; \text{W}$$

The total heat loss is

$$\dot{q}_{\text{total}} = \dot{q}_{\text{wall}} + \dot{q}_{\text{windows}} + \dot{q}_{\text{infiltration}}$$
$$= 6882 \; \text{W} + 8004 \; \text{W} + 5248 \; \text{W}$$
$$= \boxed{20\,134 \; \text{W} \quad (20.1 \; \text{kW})}$$

6. *Customary U.S. Solution*

The heat loss per degree of temperature difference is

$$\frac{\dot{q}}{\Delta T} = \frac{650{,}000 \; \frac{\text{Btu}}{\text{hr}}}{70 {}^\circ \text{F} - 0 {}^\circ \text{F}}$$
$$= 9286 \; \text{Btu/hr-}{}^\circ \text{F}$$

From Eq. 40.9(b), the average temperature, $\overline{\overline{T}}$, over the entire heating season can be calculated.

$$\text{DD} = N(65 {}^\circ \text{F} - \overline{\overline{T}})$$

The heating degree days, DD, are 5252.

The number of days in the heating season, N, is 245.

$$5252 \; \text{days} = (245 \; \text{days})(65 {}^\circ \text{F} - \overline{\overline{T}})$$
$$\overline{\overline{T}} = 43.56 {}^\circ \text{F}$$

The period between 8:30 A.M. and 5:30 P.M. is 9 hr. For this period, the temperature in the building is 70°F. For the remaining 15 hr, the temperature in the building is 50°F. The total winter heat loss is

$$(245 \; \text{days}) \left(9286 \; \frac{\text{Btu}}{\text{hr-}{}^\circ \text{F}} \right)$$
$$\times \left(\left(9 \; \frac{\text{hr}}{\text{day}} \right) (70 {}^\circ \text{F} - 43.56 {}^\circ \text{F}) \right.$$
$$\left. + \left(15 \; \frac{\text{hr}}{\text{day}} \right) (50 {}^\circ \text{F} - 43.56 {}^\circ \text{F}) \right)$$
$$= 7.61 \times 10^8 \; \text{Btu}$$

The fuel consumption is

$$\frac{7.61 \times 10^8 \; \text{Btu}}{\left(13{,}000 \; \frac{\text{Btu}}{\text{lbm}} \right) (0.70)} = \boxed{83{,}626 \; \text{lbm/yr}}$$

The answer is (D).

SI Solution

The heat loss per degree of temperature difference is

$$\frac{\dot{q}}{\Delta T} = \frac{(0.19 \text{ MW})\left(10^6 \frac{\text{W}}{\text{MW}}\right)}{21.1^{\circ}\text{C} - (-17.8^{\circ}\text{C})}$$

$$= 4884 \text{ W}/^{\circ}\text{C}$$

From Eq. 40.9(a), the average temperature, $\overline{\overline{T}}$, over the entire heating season can be calculated.

$$\text{DD} = N(18^{\circ}\text{C} - \overline{\overline{T}})$$

The heating Kelvin degree days, DD, are 2918 K·days.

The number of days in the entire heating season, N, is 245.

$$2918 \text{ days} = (245 \text{ days})(18^{\circ}\text{C} - \overline{\overline{T}})$$

$$\overline{\overline{T}} = 6.09^{\circ}\text{C}$$

The period between 8:30 A.M. and 5:30 P.M. is 9 h. For this period, the temperature in the building is 21.1°C. For the remaining 15 h, the temperature in the building is 10°C. The total winter heat loss is

$$(245 \text{ days})\left(4884 \frac{\text{W}}{^{\circ}\text{C}}\right)$$

$$\times \left(\left(9 \frac{\text{h}}{\text{day}}\right)\left(3600 \frac{\text{s}}{\text{h}}\right)(21.1^{\circ}\text{C} - 6.09^{\circ}\text{C})\right.$$

$$\left. + \left(15 \frac{\text{h}}{\text{day}}\right)\left(3600 \frac{\text{s}}{\text{h}}\right)(10^{\circ}\text{C} - 6.09^{\circ}\text{C})\right)\left(\frac{1 \text{ MJ}}{10^6 \text{ J}}\right)$$

$$= 8.345 \times 10^5 \text{ MJ}$$

The fuel consumption is

$$\frac{8.345 \times 10^5 \text{ MJ}}{\left(30.2 \frac{\text{MJ}}{\text{kg}}\right)(0.70)} = \boxed{39\,475 \text{ kg/yr}}$$

The answer is (D).

7. *Customary U.S. Solution*

The heat loss per degree of temperature difference is

$$\frac{\dot{q}}{\Delta T} = \frac{200,000 \frac{\text{Btu}}{\text{hr}}}{70^{\circ}\text{F} - 0^{\circ}\text{F}} = 2857 \text{ Btu/hr-}^{\circ}\text{F}$$

From Eq. 40.9(b), the average temperature, $\overline{\overline{T}}$, over the entire heating season can be calculated.

$$\text{DD} = N(65^{\circ}\text{F} - \overline{\overline{T}})$$

The heating degree days, DD, are 4200.

The number of days in the heating season, N, is 210.

$$4200 \text{ days} = (210 \text{ days})(65^{\circ}\text{F} - \overline{\overline{T}})$$

$$\overline{\overline{T}} = 45^{\circ}\text{F}$$

The total original winter heat loss based on 70°F inside is

$$(210 \text{ days})\left(24 \frac{\text{hr}}{\text{day}}\right)\left(2857 \frac{\text{Btu}}{\text{hr-}^{\circ}\text{F}}\right)$$

$$\times (70^{\circ}\text{F} - 45^{\circ}\text{F}) = 3.60 \times 10^8 \text{ Btu}$$

The reduced heat loss based on 68°F inside is

$$(210 \text{ days})\left(24 \frac{\text{hr}}{\text{day}}\right)\left(2857 \frac{\text{Btu}}{\text{hr-}^{\circ}\text{F}}\right)$$

$$\times (68^{\circ}\text{F} - 45^{\circ}\text{F}) = 3.31 \times 10^8 \text{ Btu}$$

The reduction is

$$\frac{3.60 \times 10^8 \text{ Btu} - 3.31 \times 10^8 \text{ Btu}}{3.60 \times 10^8 \text{ Btu}} = \boxed{0.081 \quad (8.1\%)}$$

The answer is (C).

SI Solution

Since the problem only concerns the percentage reduction in heat loss, determine only the average temperature, $\overline{\overline{T}}$, as all other parameters except inside temperature remain constant.

From Eq. 40.9(a), the average temperature, $\overline{\overline{T}}$, over the entire heating season can be calculated.

$$\text{DD} = N(18^{\circ}\text{C} - \overline{\overline{T}})$$

The heating Kelvin degree days, DD, are 2333 K·days.

The number of days in the entire heating season, N, is 210.

$$2333 \text{ days} = (210 \text{ days})(18^{\circ}\text{C} - \overline{\overline{T}})$$

$$\overline{\overline{T}} = 6.89^{\circ}\text{C}$$

The reduction is

$$\frac{\Delta T_{\text{original}} - \Delta T_{\text{reduced}}}{\Delta T_{\text{original}}}$$

$$\Delta T_{\text{original}} = 21.1^{\circ}\text{C} - 6.89^{\circ}\text{C} = 14.21^{\circ}\text{C}$$

$$\Delta T_{\text{reduced}} = 20^{\circ}\text{C} - 6.89^{\circ}\text{C} = 13.11^{\circ}\text{C}$$

The reduction is

$$\frac{14.21^{\circ}\text{C} - 13.11^{\circ}\text{C}}{14.21^{\circ}\text{C}} = \boxed{0.077 \quad (7.7\%)}$$

The answer is (C).

8. *Customary U.S. Solution*

Prior to the thermostat change, the energy requirement for ventilation air is

$$q_{\text{air},1} = m_1 c_p (T_{i,1} - T_o)$$
$$= N_1 V t \rho c_p (T_{i,1} - T_o)$$

After the change, the energy requirement is

$$q_{\text{air},2} = m_2 c_p (T_{i,2} - T_o)$$
$$= N_2 V t \rho c_p (T_{i,2} - T_o)$$

The energy savings is

$$\Delta q_{\text{air}} = q_{\text{air},1} - q_{\text{air},2} = N_1 V t \rho c_p (T_{i,1} - T_o)$$
$$- N_2 V t \rho c_p (T_{i,2} - T_o)$$
$$= V t \rho c_p \big(N_1 (T_{i,1} - T_o) - N_2 (T_{i,2} - T_o) \big)$$

During an entire week, the building is occupied from 8:00 A.M. to 6:00 P.M. (for 10 hr) and is unoccupied 14 hr each day for 5 days, and it is unoccupied 24 hr each day for 2 days (over the weekend).

The unoccupied time for a 21-week period over a year is

$$t = \left(14\ \frac{\text{hr}}{\text{day}}\right)\left(5\ \frac{\text{days}}{\text{wk}}\right)\left(21\ \frac{\text{wk}}{\text{yr}}\right)$$
$$+ \left(24\ \frac{\text{hr}}{\text{day}}\right)\left(2\ \frac{\text{days}}{\text{wk}}\right)\left(21\ \frac{\text{wk}}{\text{yr}}\right)$$
$$= 2478\ \text{hr/yr}$$

With air at 70°F, $c_p \approx 0.24$ Btu/lbm-°F and $\rho \approx 0.075$ lbm/ft^3.

$$\Delta q_{\text{air}} = \left(801{,}000\ \frac{\text{ft}^3}{\text{air change}}\right)\left(2478\ \frac{\text{hr}}{\text{yr}}\right)\left(0.075\ \frac{\text{lbm}}{\text{ft}^3}\right)$$
$$\times \left(0.24\ \frac{\text{Btu}}{\text{lbm-°F}}\right)\left(\left(1\ \frac{\text{air change}}{\text{hr}}\right)\right.$$
$$\times (70°\text{F} - 30°\text{F}) - \left(0.5\ \frac{\text{air change}}{\text{hr}}\right)$$
$$\left. \times \big((70°\text{F} - 12°\text{F}) - 30°\text{F}\big)\right)$$
$$= 9.289 \times 10^8\ \text{Btu/hr}$$

The savings due to reduced heat losses from the roof, floor, and walls (for 12°F set back) is

$$\dot{q} = UA\Delta T$$
$$= \left(\left(0.15\ \frac{\text{Btu}}{\text{ft}^2\text{-hr-°F}}\right)(11{,}040\ \text{ft}^2) + \left(1.13\ \frac{\text{Btu}}{\text{ft}^2\text{-hr-°F}}\right)\right.$$
$$\times\ (2760\ \text{ft}^2) + \left(0.05\ \frac{\text{Btu}}{\text{ft}^2\text{-hr-°F}}\right)(26{,}700\ \text{ft}^2)$$
$$\left. + \left(1.5\ \frac{\text{Btu}}{\text{ft-hr-°F}}\right)(690\ \text{ft})\right)(12°\text{F})$$

$$= 85{,}738\ \text{Btu/hr}$$
$$q = \dot{q}t = \left(85{,}738\ \frac{\text{Btu}}{\text{hr}}\right)\left(2478\ \frac{\text{hr}}{\text{yr}}\right)$$
$$= 2.125 \times 10^8\ \text{Btu/yr}$$

The energy saved per year is

$$q_{\text{actual}} = \frac{q_{\text{ideal}}}{\eta_{\text{furnace}}} = \frac{\Delta q_{\text{air}} + q}{\eta_{\text{furnace}}}$$
$$= \frac{9.289 \times 10^8\ \dfrac{\text{Btu}}{\text{yr}} + 2.125 \times 10^8\ \dfrac{\text{Btu}}{\text{yr}}}{(0.75)\left(100{,}000\ \dfrac{\text{Btu}}{\text{therm}}\right)}$$
$$= 15{,}218\ \text{therm/yr}$$

The cost savings is

$$\left(15{,}218\ \frac{\text{therm}}{\text{yr}}\right)\left(\frac{\$0.25}{\text{therm}}\right) = \boxed{\$3805\ \text{per year}}$$

The answer is (B).

SI Solution

Prior to the thermostat change, the energy requirement for ventilation air is

$$q_{\text{air},1} = m_1 c_p (T_{i,1} - T_o)$$
$$= N_1 V t \rho c_p (T_{i,1} - T_o)$$

After the change, the energy requirement is

$$q_{\text{air},2} = m_2 c_p (T_{i,2} - T_o)$$
$$= N_2 V t \rho c_p (T_{i,2} - T_o)$$

The energy savings is

$$\Delta q_{\text{air}} = q_{\text{air},1} - q_{\text{air},2} = N_1 V t \rho c_p (T_{i,1} - T_o)$$
$$- N_2 V t \rho c_p (T_{i,2} - T_o)$$
$$= V t \rho c_p \big(N_1 (T_{i,1} - T_o) - N_2 (T_{i,2} - T_o) \big)$$

During an entire week, the building is occupied from 8:00 A.M. to 6:00 P.M. (for 10 h) and is unoccupied 14 h

HVAC

each day for 5 days, and it is unoccupied 24 h each day for 2 days (over the weekend).

The unoccupied time for a 21-week period over a year is

$$
\begin{aligned}
t &= \left(14 \ \frac{\text{h}}{\text{day}}\right)\left(5 \ \frac{\text{days}}{\text{wk}}\right)\left(21 \ \frac{\text{wk}}{\text{yr}}\right) \\
&\quad + \left(24 \ \frac{\text{h}}{\text{day}}\right)\left(2 \ \frac{\text{days}}{\text{wk}}\right)\left(21 \ \frac{\text{wk}}{\text{yr}}\right) \\
&= 2478 \ \text{h/yr}
\end{aligned}
$$

With air at 21°C, $c_p \approx 1.005$ kJ/kg·°C and $\rho \approx 1.2$ kg/m³.

$$
\begin{aligned}
\Delta q_{\text{air}} &= \left(22\,700 \ \frac{\text{m}^3}{\text{air change}}\right)\left(2478 \ \frac{\text{h}}{\text{yr}}\right)\left(1.2 \ \frac{\text{kg}}{\text{m}^3}\right) \\
&\quad \times \left(1.005 \ \frac{\text{kJ}}{\text{kg·°C}}\right)\left(\left(1 \ \frac{\text{air change}}{\text{h}}\right)\right. \\
&\quad \times (21.1°\text{C} - (-1°\text{C})) - \left(0.5 \ \frac{\text{air change}}{\text{h}}\right) \\
&\quad \left. \times ((21.1°\text{C} - 6.7°\text{C}) - (-1°\text{C})) \right) \\
&= 9.769 \times 10^8 \ \text{kJ/yr}
\end{aligned}
$$

The savings due to reduced heat losses from the roof, floor, and walls for 6.7°C set back is

$$
\begin{aligned}
\dot{q} &= UA\Delta T \\
&= \left(\left(0.85 \ \frac{\text{W}}{\text{m}^2\cdot°\text{C}}\right)(993 \ \text{m}^2)\right. \\
&\quad + \left(6.42 \ \frac{\text{W}}{\text{m}^2\cdot°\text{C}}\right)(260 \ \text{m}^2) \\
&\quad + \left(0.3 \ \frac{\text{W}}{\text{m}^2\cdot°\text{C}}\right)(2480 \ \text{m}^2) \\
&\quad \left. + \left(2.6 \ \frac{\text{W}}{\text{m}\cdot°\text{C}}\right)(210 \ \text{m})\right)(6.7°\text{C}) \\
&= 25\,482 \ \text{W} \\
q = \dot{q}t &= (25\,482 \ \text{W})\left(1 \ \frac{\text{J}}{\text{s·W}}\right)\left(2478 \ \frac{\text{h}}{\text{yr}}\right) \\
&\quad \times \left(3600 \ \frac{\text{s}}{\text{h}}\right)\left(\frac{1 \ \text{kJ}}{1000 \ \text{J}}\right) \\
&= 2.273 \times 10^8 \ \text{kJ/yr}
\end{aligned}
$$

The energy saved per year is

$$
\begin{aligned}
q_{\text{actual}} &= \frac{q_{\text{ideal}}}{\eta_{\text{furnace}}} = \frac{\Delta q_{\text{air}} + q}{\eta_{\text{furnace}}} \\
&= \frac{9.769 \times 10^8 \ \dfrac{\text{kJ}}{\text{yr}} + 2.273 \times 10^8 \ \dfrac{\text{kJ}}{\text{yr}}}{(0.75)\left(1000 \ \dfrac{\text{kJ}}{\text{MJ}}\right)\left(105.506 \ \dfrac{\text{MJ}}{\text{therm}}\right)} \\
&= 15\,218 \ \text{therm/yr}
\end{aligned}
$$

The cost savings is

$$
\left(15\,218 \ \frac{\text{therm}}{\text{yr}}\right)\left(\frac{\$0.25}{\text{therm}}\right) = \boxed{\$3805 \text{ per year}}
$$

The answer is (B).

41 Cooling Load

PRACTICE PROBLEMS

1. (*Time limit: one hour*) The bypass air conditioning system shown has the following operating characteristics.

> total air pressure: 14.7 psia (101.3 kPa)
> supply temperature: 58°F (14.4°C) dry-bulb
> sensible load: 200,000 Btu/hr (58.6 kW)
> latent load: 450,000 grains/hr (29 kg/h)
> outside air: 90°F (32.2°C) dry-bulb,
>   76°F (24.4°C) wet-bulb
> make-up air: 2000 ft³/min (940 L/s)
> condition of air leaving washer: saturated

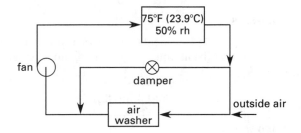

(a) Find the temperature of the air leaving the washer.
- (A) 34°F (1°C)
- (B) 38°F (3°C)
- (C) 49°F (9°C)
- (D) 55°F (13°C)

(b) Find the rate of supply air.
- (A) 7400 ft³/min (3400 L/s)
- (B) 11,000 ft³/min (5100 L/s)
- (C) 14,000 ft³/min (6400 L/s)
- (D) 19,000 ft³/min (8700 L/s)

(c) Find the moisture content of the supply air.
- (A) 0.0081 lbm/lbm (8.1 g/kg)
- (B) 0.0097 lbm/lbm (9.7 g/kg)
- (C) 0.0110 lbm/lbm (11 g/kg)
- (D) 0.0130 lbm/lbm (13 g/kg)

(d) What is the bypass factor of the washing process?
- (A) 0
- (B) 0.35
- (C) 0.65
- (D) 1.0

(e) What is the bypass factor of the system?
- (A) 0
- (B) 0.35
- (C) 0.65
- (D) 1.0

2. (*Time limit: one hour*) A building is located at 32°N latitude. The inside design temperature is 78°F (25.6°C). The outside design temperature is 95°F (35°C). The daily temperature range is 22°F (12.2°C). Energy gains through the floor, from lights, and from occupants are insignificant. The building's construction is as follows.

walls: 1600 ft² (144 m²) facing north
           1400 ft² (126 m²) facing south
           1500 ft² (135 m²) facing east
           1400 ft² (126 m²) facing west
           4 in (100 mm) brick facing
           3 in (75 mm) concrete block
           1 in (25 mm) mineral wool
              (emissivity = 0.96)
           2 in (50 mm) furring
           ³/₈ in (9.5 mm) drywall gypsum
              (emissivity = 0.95)
           ¹/₂ in (12 mm) plaster

roof: 6000 ft² (540 m²)
         4 in concrete (100 mm)
         2 in insulation (50 mm)
         insulation conductivity:
            2.28 ft²-°F-hr/Btu-in
         single felt layer
         1 in (25 mm) air gap
         ¹/₂ in (12 mm) acoustical ceiling tile

windows: 100 ft² (9 m²) facing east
             ¹/₄ in (6.4 mm) thick, single glazing
             cream-colored Venetian shades
             no exterior shading

(a) What is the cooling load for mid-July at 4:00 P.M. sun time?

(b) Is this the peak cooling load? Explain.

SOLUTIONS

1. *Customary U.S. Solution*

(a) This is a standard bypass problem.

The indoor conditions are

$$T_i = 75°F$$

$$\phi_i = 50\%$$

The outdoor conditions are

$$T_{o,db} = 90°F$$

$$T_{o,wb} = 76°F$$

The ventilation rate is 2000 ft³/min.

The loads are

$$q_s = 200{,}000 \text{ Btu/hr}$$

$$q_l = 450{,}000 \text{ gr/hr}$$

To find the sensible heat ratio, q_l must be expressed in Btu/hr.

From the psychrometric chart (App. 38.A), for the room conditions, $\omega = 0.0095$ lbm moisture/lbm dry air.

From Eq. 38.7,

$$\omega = (0.622)\left(\frac{p_w}{p_a}\right)$$

$$= (0.622)\left(\frac{p_w}{p - p_w}\right)$$

$$0.0095 \,\frac{\text{lbm moisture}}{\text{lbm dry air}} = (0.622)\left(\frac{p_{w,\text{psia}}}{14.7 \text{ psia} - p_{w,\text{psia}}}\right)$$

$$p_w = 0.22 \text{ psia}$$

From steam tables (App. 24.A), for 0.22 psia,

$$h_{fg} = 1061.7 \text{ Btu/lbm}$$

$$q_l = \left(450{,}000 \,\frac{\text{gr}}{\text{hr}}\right)\left(\frac{1 \text{ lbm}}{7000 \text{ gr}}\right)\left(1061.7 \,\frac{\text{Btu}}{\text{lbm}}\right)$$

$$= 68{,}255 \text{ Btu/hr}$$

From Eq. 41.7, the room sensible heat ratio is

$$\text{RSHR} = \frac{q_s}{q_s + q_l} = \frac{200{,}000 \,\dfrac{\text{Btu}}{\text{hr}}}{200{,}000 \,\dfrac{\text{Btu}}{\text{hr}} + 68{,}255 \,\dfrac{\text{Btu}}{\text{hr}}}$$

$$= 0.75$$

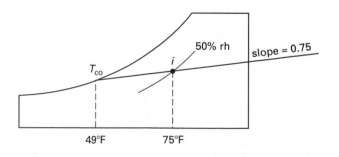

Locate 75°F and 50% relative humidity on the psychrometric chart (App. 38.A). Draw a line with a slope of 0.75 through this point.

The left-hand intersection shows $T_{co} = 49°F$. Since the air leaves the conditioner saturated,

$$\text{BF}_{\text{coil}} = 0$$

$$T_{co} = \boxed{49°F}$$

The answer is (C).

(b) Calculate the air flow through the room from Eq. 41.9(b).

$$\dot{V}_{\text{in}} = \frac{\dot{q}_{s,\text{Btu/hr}}}{\left(1.08 \,\dfrac{\text{Btu-min}}{\text{ft}^3\text{-hr}}\right)(T_i - T_{\text{in}})}$$

$$\dot{V}_{\text{in,cfm}} = \frac{200{,}000 \,\dfrac{\text{Btu}}{\text{hr}}}{\left(1.08 \,\dfrac{\text{Btu-min}}{\text{ft}^3\text{-hr}}\right)(75°F - 58°F)}$$

$$= \boxed{10{,}893 \text{ ft}^3/\text{min}}$$

The answer is (B).

(c) Since $T_{\text{in}} = 58°F$, locate this dry-bulb temperature on the condition line. Reading from the chart (App. 38.A),

$$\omega_{\text{in}} = \boxed{0.0081 \text{ lbm/lbm}}$$

The answer is (A).

(d) Since the air leaves the air washer saturated, the efficiency of the air washer is 100%. No air escapes being saturated, so the process bypass factor is 0.

The answer is (A).

(e) The bypass factor of the system can be calculated from the dry-bulb temperatures.

$$\text{BF} = \frac{T_{\text{supply}} - T_{\text{air washer}}}{T_{\text{bypass}} - T_{\text{air washer}}}$$

$$= \frac{58°F - 49°F}{75°F - 49°F} = 0.35$$

The answer is (B).

SI Solution

(a) The indoor conditions are

$$T_i = 23.9°C$$

$$\phi_i = 50\%$$

The outdoor conditions are

$$T_{o,db} = 32.2°C$$

$$T_{o,wb} = 24.4°C$$

The ventilation rate is 940 L/s.

The loads are
$$q_s = 58.6 \text{ kW}$$
$$q_l = 29 \text{ kg/h}$$

To find the sensible heat ratio, q_l must be expressed in kW.

From the psychrometric chart (App. 38.B), for the room conditions, $\omega \approx 9.3$ g moisture/kg dry air (9.3×10^{-3} kg moisture/kg dry air).

From Eq. 38.7,
$$\omega = (0.622)\left(\frac{p_w}{p_a}\right)$$
$$= (0.622)\left(\frac{p_w}{p - p_w}\right)$$
$$9.3 \times 10^{-3} \; \frac{\text{kg moisture}}{\text{kg dry air}} = (0.622)\left(\frac{p_w}{101.3 \text{ kPa} - p_w}\right)$$
$$p_w = 1.49 \text{ kPa}$$

From steam tables (App. 24.N), for 1.49 kPa,
$$h_{fg} = 2470.3 \text{ kJ/kg}$$
$$q_l = \left(29 \; \frac{\text{kg}}{\text{h}}\right)\left(\frac{1 \text{ h}}{3600 \text{ s}}\right)\left(2470.3 \; \frac{\text{kJ}}{\text{kg}}\right) = 19.9 \text{ kW}$$

From Eq. 38.22, the sensible heat ratio is
$$\text{SHR} = \frac{q_s}{q_s + q_l} = \frac{58.6 \text{ kW}}{58.6 \text{ kW} + 19.9 \text{ kW}}$$
$$= 0.75$$

Locate 23.9°C and 50% relative humidity on the psychrometric chart (Fig. 38.B). Draw a line with a slope of 0.75 through this point.

The left-hand intersection shows ADP = 9°C. Since the air leaves the conditioner saturated,
$$\text{BF}_{\text{coil}} = 0$$
$$T_{\text{co}} = \boxed{9°\text{C}}$$

The answer is (C).

(b) Calculate the air flow through the room from Eq. 41.9(a).

$$\dot{V}_{\text{in,L/s}} = \frac{\dot{q}_{s,\text{W}}}{\left(1.20 \; \frac{\text{W·s}}{\text{L·°C}}\right)(T_i - T_{\text{in}})}$$

$$= \frac{(58.6 \text{ kW})\left(1000 \; \frac{\text{W}}{\text{kW}}\right)}{\left(1.20 \; \frac{\text{W·s}}{\text{L·°C}}\right)(23.9°\text{C} - 14.4°\text{C})}$$

$$= \boxed{5140 \text{ L/s}}$$

The answer is (B).

(c) Since $T_{\text{in}} = 14.4°\text{C}$, locate this dry-bulb temperature on the condition line. Reading from the chart (Fig. 38.B),

$$\omega_{\text{in}} = \boxed{8.1 \text{ g/kg dry air}}$$

The answer is (A).

(d) Since the air leaves the air washer saturated, the efficiency of the air washer is 100%. No air escapes being saturated, so the process bypass factor is 0.

The answer is (A).

(e) The bypass factor of the system can be calculated from the dry-bulb temperatures.

$$BF = \frac{T_{\text{supply}} - T_{\text{outwasher}}}{T_{\text{bypass}} - T_{\text{outwasher}}}$$
$$= \frac{58° - 49°}{75° - 49°} = 0.35$$

The answer is (B).

2. (a) *step 1:* Determine thermal resistances for walls, roof, and windows.

For walls:

Assume 2 in furring to be 2 in air space.

From Eq. 40.4, the effective space emissivity is

$$\frac{1}{E} = \frac{1}{\epsilon_1} + \frac{1}{\epsilon_2} - 1$$
$$= \frac{1}{0.96} + \frac{1}{0.95} - 1$$
$$E = 0.91$$

From Table 40.2, for vertical air space orientation, the thermal conductance (by extrapolating for $E = 0.91$) is

$$\frac{1}{C} = \frac{1}{1.07 \; \frac{\text{Btu}}{\text{hr-ft}^2\text{-°F}}}$$
$$= 0.93 \text{ hr-ft}^2\text{-°F/Btu}$$

Use App. 40.A.

walls

type of construction	R, $\dfrac{\text{ft}^2\text{-}^\circ\text{F-hr}}{\text{Btu}}$	notes
4 in brick spacing	0.44	
3 in concrete block	0.40	
1 in mineral wool	3.85	value given per in
2 in furring	0.93	see calculation above
$^3/_8$ in drywall gypsum	0.34	0.45 for $^1/_2$ is given
$^1/_2$ in plaster (sand aggregate)	0.09	
surface outside (Table 40.3)	0.25	assume $7^1/_2$ mph air speed
surface inside (Table 40.3)	0.68	still air, $U = 1.46 \dfrac{\text{Btu}}{\text{ft}^2\text{-}^\circ\text{F-hr}}$

roof

type of construction	R, $\dfrac{\text{ft}^2\text{-}^\circ\text{F-hr}}{\text{Btu}}$	notes
4 in concrete	0.32	sand aggregate
2 in insulation	5.56	2.78/in
felt	0.06	
1 in air gap	0.87	$\epsilon = 0.90$ for buildup material
$^1/_2$ in accoustical ceiling tile	1.19	
surface outside (Table 40.3)	0.25	assume $7^1/_2$ mph air speed and downward heat flow $\sim$ upward heat flow
surface inside (Table 40.3)	0.93	still air, $U = 1.08 \dfrac{\text{Btu}}{\text{ft}^2\text{-}^\circ\text{F-hr}}$

windows

type of construction	R, $\dfrac{\text{ft}^2\text{-}^\circ\text{F-hr}}{\text{Btu}}$	notes
$^1/_4$ in thick, single glazing	0.66	without shades, $R = 0.88 \dfrac{\text{ft}^2\text{-}^\circ\text{F-hr}}{\text{Btu}}$
cream colored venetian shades, no exterior shades		assume shading factor of 0.75

step 2: Determine the overall heat transfer coefficient for walls, roof, and windows using Eq. 40.3.

$$U = \frac{1}{\sum R_i}$$

For the walls,

$$U = \frac{1}{\begin{aligned}&0.25\ \tfrac{\text{ft}^2\text{-}^\circ\text{F-hr}}{\text{Btu}} + 0.44\ \tfrac{\text{ft}^2\text{-}^\circ\text{F-hr}}{\text{Btu}}\\&+ 0.40\ \tfrac{\text{ft}^2\text{-}^\circ\text{F-hr}}{\text{Btu}} + 3.85\ \tfrac{\text{ft}^2\text{-}^\circ\text{F-hr}}{\text{Btu}}\\&+ 0.93\ \tfrac{\text{ft}^2\text{-}^\circ\text{F-hr}}{\text{Btu}} + 0.34\ \tfrac{\text{ft}^2\text{-}^\circ\text{F-hr}}{\text{Btu}}\\&+ 0.09\ \tfrac{\text{ft}^2\text{-}^\circ\text{F-hr}}{\text{Btu}} + 0.68\ \tfrac{\text{ft}^2\text{-}^\circ\text{F-hr}}{\text{Btu}}\end{aligned}}$$

$$= 0.143\ \text{Btu/ft}^2\text{-}^\circ\text{F-hr}$$

The cooling load is calculated from the equivalent temperature differences for the walls and roof. Tables of these values are not common; they are now often incorporated directly into HVAC computer application databases. Results from the remainder of this problem will vary with the temperature differences used.

Assume the following equivalent temperature differences for 4 in brick face walls at 4 P.M.

facing east: $\Delta T = 20^\circ\text{F}$

facing west: $\Delta T = 20^\circ\text{F}$

facing south: $\Delta T = 28^\circ\text{F}$

facing north: $\Delta T = 17^\circ\text{F}$

For 4 in brick facing at 4 P.M.,

$$\begin{aligned}Q_{\text{walls}} &= \left(0.143\ \frac{\text{Btu}}{\text{ft}^2\text{-}^\circ\text{F-hr}}\right)\big((1600\ \text{ft}^2)(17^\circ\text{F})\\&\quad + (1400\ \text{ft}^2)(28^\circ\text{F}) + (1500\ \text{ft}^2)(20^\circ\text{F})\\&\quad + (1400\ \text{ft}^2)(20^\circ\text{F})\big)\end{aligned}$$

$$= 17{,}789\ \text{Btu/hr}$$

For the roof,

$$U = \frac{1}{\begin{aligned}&0.25\ \tfrac{\text{ft}^2\text{-}^\circ\text{F-hr}}{\text{Btu}} + 0.32\ \tfrac{\text{ft}^2\text{-}^\circ\text{F-hr}}{\text{Btu}}\\&+ 5.56\ \tfrac{\text{ft}^2\text{-}^\circ\text{F-hr}}{\text{Btu}} + 0.06\ \tfrac{\text{ft}^2\text{-}^\circ\text{F-hr}}{\text{Btu}}\\&+ 0.87\ \tfrac{\text{ft}^2\text{-}^\circ\text{F-hr}}{\text{Btu}} + 1.19\ \tfrac{\text{ft}^2\text{-}^\circ\text{F-hr}}{\text{Btu}}\\&+ 0.93\ \tfrac{\text{ft}^2\text{-}^\circ\text{F-hr}}{\text{Btu}}\end{aligned}}$$

$$= 0.109\ \text{Btu/ft}^2\text{-}^\circ\text{F-hr}$$

Assume an equivalent temperature difference for 4 in concrete roof at 4 P.M. = 74°F. Therefore,

$$Q_{\text{roof}} = \left(0.109 \ \frac{\text{Btu}}{\text{ft}^2\text{-°F-hr}}\right)(6000 \ \text{ft}^2)(74°F)$$

$$= 48{,}396 \ \text{Btu/hr}$$

For the windows:

Since all windows facing east will receive no direct sunlight at 4 P.M., consider $\Delta T = 95°F - 78°F$.

$$Q_{\text{windows}} = \left(\frac{1}{0.66 \ \dfrac{\text{ft}^2\text{-°F-hr}}{\text{Btu}}}\right)(100 \ \text{ft}^2)(95°F - 78°F)$$

$$= 2576 \ \text{Btu/hr}$$

The total sensible transmission load (at 4 P.M.) is

$$Q_{\text{total}} = 17{,}789 \ \frac{\text{Btu}}{\text{hr}} + 48{,}396 \ \frac{\text{Btu}}{\text{hr}} + 2576 \ \frac{\text{Btu}}{\text{hr}}$$

$$= \boxed{68{,}761 \ \text{Btu/hr}}$$

(b) This may not be the peak load. Since Q_{walls} and Q_{roof} are the major contributors and ΔT equivalent temperature differences are maximum between 4 P.M. and 6 P.M., the peak cooling load will be maximum somewhere between 4 P.M. and 6 P.M. Consider Q_{walls} and Q_{roof} for 6 P.M.

At 6 P.M., the equivalent temperature differences for 4 in brick facing are assumed as

$$\text{facing east: } \Delta T = 22°F$$
$$\text{facing west: } \Delta T = 35°F$$
$$\text{facing south: } \Delta T = 28°F$$
$$\text{facing north: } \Delta T = 20°F$$

$$Q_{\text{walls}} = \left(0.143 \ \frac{\text{Btu}}{\text{ft}^2\text{-°F-hr}}\right)((1600 \ \text{ft}^2)(20°F)$$
$$+ \ (1400 \ \text{ft}^2)(28°F) + (1500 \ \text{ft}^2)(22°F)$$
$$+ \ (1400 \ \text{ft}^2)(35°F))$$

$$= 21{,}908 \ \text{Btu/hr}$$

Assume the equivalent temperature difference for a 4 in concrete roof at 6 P.M. is 68°F.

$$Q_{\text{roof}} = \left(0.109 \ \frac{\text{Btu}}{\text{ft}^2\text{-°F-hr}}\right)(6000 \ \text{ft}^2)(68°F)$$

$$= 44{,}472 \ \text{Btu/hr}$$

The total sensible transmission load at 6 P.M. is

$$Q_{\text{total}} = 21{,}908 \ \frac{\text{Btu}}{\text{hr}} + 44{,}472 \ \frac{\text{Btu}}{\text{hr}} + 2576 \ \frac{\text{Btu}}{\text{hr}}$$

$$= \boxed{68{,}956 \ \text{Btu/hr}}$$

This is very close to the 4 P.M. load.

(An SI solution of this problem is not supported by this book.)

HVAC

42 Air Conditioning Systems and Controls

PRACTICE PROBLEMS

1. What is the typical approximate supply air pressure in a pneumatic control system?

(A) 7 psig
(B) 18 psig
(C) 35 psig
(D) 140 psig

2. The control signal in a pneumatic control system is air pressure that falls within which approximate range?

(A) 0–4 psig
(B) 3–15 psig
(C) 12–45 psig
(D) 30–120 psig

3. What kind of device is a pneumatic thermostat?

(A) reverse acting
(B) normally closed
(C) normally open
(D) direct acting

4. What is the term for a device that converts a pneumatic signal of 15 psig to a control pressure of 20 psig?

(A) a pneumatic transformer
(B) a pneumatic amplifier
(C) a pneumatic relay
(D) a pneumatic booster

5. What kind of device is a pneumatic hot water valve?

(A) reverse acting
(B) normally closed
(C) normally open
(D) direct acting

6. What kind of device is a pneumatic steam valve?

(A) reverse acting
(B) normally closed
(C) normally open
(D) direct acting

SOLUTIONS

1. Pneumatic air pressure is typically supplied at 18–20 psig but may be as high as 25 psig.

The answer is (B).

2. A device's pneumatic signal cannot be less than the supply signal, and it must be sufficient for detection. 3–15 psig is a typical range of output pressures.

The answer is (B).

3. Thermostats are direct-acting (DA) because, as the room cools, an increase in control pressure is needed.

The answer is (D).

4. A pneumatic relay has three ports: a signal input, a supply input, and a control output. When the full input signal is received, the control output valve is opened to the full supply input.

The answer is (C).

5. A hot water valve is normally open so that the valve will provide [uncontrolled] heat in the event of a power failure or air pressure outage.

The answer is (C).

6. A steam valve is normally closed so that uncontrolled steam will not be introduced into the HVAC system in the event of a power failure or air pressure outage.

The answer is (B).

HVAC

43 Determinate Statics

PRACTICE PROBLEMS

1. Two towers are located on level ground 100 ft (30 m) apart. They support a transmission line with a mass of 2 lbm/ft (3 kg/m). The midpoint sag is 10 ft (3 m).

(a) What is the midpoint tension?
- (A) 125 lbf (0.55 kN)
- (B) 250 lbf (1.2 kN)
- (C) 375 lbf (1.6 kN)
- (D) 500 lbf (2.2 kN)

(b) What is the maximum tension in the transmission line?
- (A) 170 lbf (0.70 kN)
- (B) 210 lbf (0.86 kN)
- (C) 270 lbf (1.2 kN)
- (D) 330 lbf (1.4 kN)

(c) If the midpoint tension is increased to 500 lbf (2200 N), what is the sag in the cable?
- (A) 1.3 ft (0.4 m)
- (B) 3.2 ft (1.0 m)
- (C) 4.0 ft (1.2 m)
- (D) 5.1 ft (1.5 m)

2. Two legs of a tripod are mounted on a vertical wall. Both legs are horizontal. The apex is 12 distance units from the wall. The right leg is 13.4 units long. The wall mounting points are 10 units apart. A third leg is mounted on the wall 6 units to the left of the right upper leg and and 9 units below the two top legs. A vertical downward load of 200 is supported at the apex. What is the reaction at the lowest mounting point?

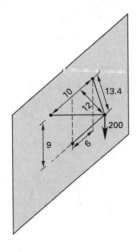

- (A) 120
- (B) 170
- (C) 250
- (D) 330

3. The ideal truss shown is supported by a pinned connection at point D and a roller connection at point C. Loads are applied at points A and F. What is the force in member DE?

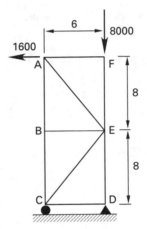

- (A) 1200
- (B) 2700
- (C) 3300
- (D) 3700

4. A pin-connected tripod is loaded at the apex by a horizontal force of 1200 as shown. What is the magnitude of the force in member AD?

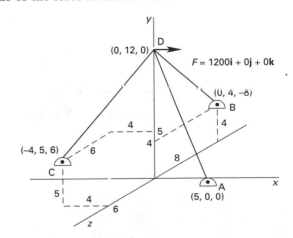

Statics

(A)   1100
(B)   1300
(C)   1800
(D)   2500

5. A truss is loaded by forces of 4000 at each upper connection point and forces of 60,000 at each lower connection point as shown.

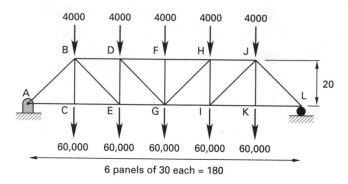

6 panels of 30 each = 180

(a) What is the force in member DE?
(A)   36,000
(B)   45,000
(C)   60,000
(D)   160,000

(b) What is the force in member HJ?
(A)   24,000
(B)   60,000
(C)   160,000
(D)   380,000

6. The rigid rod AO is supported by guy wires BO and CO, as shown. (Points A, B, and C are all in the same vertical plane.) Vertical and horizontal forces are 12,000 and 6000, respectively, as carried at the end of the rod. What are the x-, y-, and z-components of the reactions at point C?

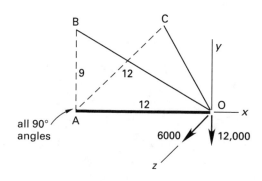

(A) $(C_x, C_y, C_z) = (0, 6000, 0)$
(B) $(C_x, C_y, C_z) = (6000, 0, 6000)$
(C) $(C_x, C_y, C_z) = (6000, 0, 4200)$
(D) $(C_x, C_y, C_z) = (4200, 0, 8500)$

7. When the temperature is 70°F (21.11°C), sections of steel railroad rail are welded end to end to form a continuous, horizontal track exactly 1 mi long (1.6 km). Both ends of the track are constrained by preexisting installed sections of rail. Before the 1 mi section of track can be nailed to the ties, however, the sun warms it to a uniform temperature of 99.14°F (37.30°C). Laborers watch in amazement as the rail pops up in the middle and takes on a parabolic shape. The laborers prop the rail up (so that it does not buckle over) while they take souvenir pictures. How high off the ground is the midpoint of the hot rail?
(A)   0.8 ft (2.4 m)
(B)   2.1 ft (3.6 m)
(C)   17 ft (5.1 m)
(D)   45 ft (14 m)

8. (a) Find the force in member FC.

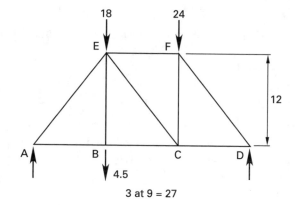

3 at 9 = 27

(A)   0.5
(B)   17
(C)   23
(D)   29

(b) What is the force in member CE?
(A)   0.50
(B)   0.63
(C)   11
(D)   17

SOLUTIONS

1. *Customary U.S. Solution*

Use Eq. 43.67 to relate the midpoint sag S to the constant c.

$$S = c\left(\cosh\left(\frac{a}{c}\right) - 1\right)$$

$$10\text{ ft} = c\left(\cosh\left(\frac{50\text{ ft}}{c}\right) - 1\right)$$

Solve by trial and error.

$$c = 126.6\text{ ft}$$

(a) Use Eq. 43.69 to find the midpoint tension.

$$H = wc = m\left(\frac{g}{g_c}\right)c$$

$$= \left(2\,\frac{\text{lbm}}{\text{ft}}\right)\left(\frac{32.2\,\frac{\text{ft}}{\text{sec}^2}}{32.2\,\frac{\text{ft-lbm}}{\text{lbf-sec}^2}}\right)(126.6\text{ ft})$$

$$= \boxed{253.2\text{ lbf}}$$

The answer is (B).

(b) Use Eq. 43.71 to find the maximum tension.

$$T = wy = w(c + S) = m\left(\frac{g}{g_c}\right)(c + S)$$

$$= \left(2\,\frac{\text{lbm}}{\text{ft}}\right)\left(\frac{32.2\,\frac{\text{ft}}{\text{sec}^2}}{32.2\,\frac{\text{ft-lbm}}{\text{lbf-sec}^2}}\right)(126.6\text{ ft} + 10\text{ ft})$$

$$= \boxed{273.2\text{ lbf}}$$

The answer is (C).

(c) From $T = wy$,

$$y = \frac{T}{w} = \frac{500\text{ lbf}}{2\,\frac{\text{lbf}}{\text{ft}}} = 250\text{ ft}\quad\text{[at right support]}$$

$$250\text{ ft} = c\left(\cosh\left(\frac{50\text{ ft}}{c}\right)\right)$$

By trial and error, $c = 245$ ft.

Substitute into Eq. 43.67.

$$S = c\left(\cosh\left(\frac{a}{c}\right) - 1\right)$$

$$= \left(245\,\frac{\text{ft}}{\text{sec}}\right)\left(\cosh\left(\frac{50\text{ ft}}{245\,\frac{\text{ft}}{\text{sec}}}\right) - 1\right)$$

$$= \boxed{5.12\text{ ft}}$$

The sag decreases when the tension is increased.

The answer is (C).

SI Solution

Use Eq. 43.67 to relate the midpoint sag S to the constant c.

$$S = c\left(\cosh\left(\frac{a}{c}\right) - 1\right)$$

$$3\text{ m} = c\left(\cosh\left(\frac{15\text{ m}}{c}\right) - 1\right)$$

Solve by trial and error.

$$c = 38.0\text{ m}$$

(a) Use Eq. 43.69 to find the midpoint tension.

$$H = wc = mgc = \left(3\,\frac{\text{kg}}{\text{m}}\right)\left(9.81\,\frac{\text{m}}{\text{s}^2}\right)(38.0\text{ m})$$

$$= \boxed{1118.3\text{ N}}$$

The answer is (B).

(b) Use Eq. 43.71 to find the maximum tension.

$$T = wy = w(c + S) = mg(c + S)$$

$$= \left(3\,\frac{\text{kg}}{\text{m}}\right)\left(9.81\,\frac{\text{m}}{\text{s}^2}\right)(38.0\text{ m} + 3.0\text{ m})$$

$$= \boxed{1206.6\text{ N}}$$

The answer is (C).

(c) From $T = wy$,

$$y = \frac{T}{w} = \frac{2200\text{ N}}{\left(3\,\frac{\text{kg}}{\text{m}}\right)\left(9.81\,\frac{\text{m}}{\text{s}^2}\right)}$$

$$= 74.75\text{ m}\quad\text{[at right support]}$$

$$74.75\text{ m} = c\left(\cosh\left(\frac{15\text{ m}}{c}\right)\right)$$

By trial and error, $c = 73.2$ ft.

Substitute into Eq. 43.67

$$S = c\left(\cosh\left(\frac{a}{c}\right) - 1\right)$$

$$= (73.72\text{ m})\left(\cosh\left(\frac{15\text{ m}}{73.2\text{ m·s}}\right) - 1\right)$$

$$= \boxed{1.54\text{ m}}$$

The answer is (D).

Statics

2. *step 1:* Draw the tripod with the origin at the apex.

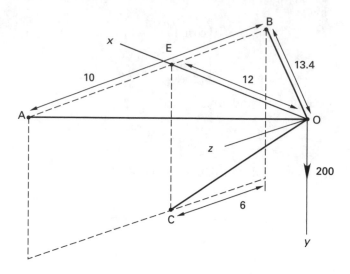

step 2:   By inspection, the force components are $F_x = 0$, $F_y = 200$, and $F_z = 0$.

step 3:   First, from triangle EBO, length BE is

$$BE = \sqrt{(13.4 \text{ units})^2 - (12 \text{ units})^2}$$
$$= 5.96 \text{ units} \quad [\text{use 6 units}]$$

The (x, y, z) coordinates of the three support points are

point A—$(12, 0, 4)$
point B—$(12, 0, -6)$
point C—$(12, 9, 0)$

step 4:   Find the lengths of the legs.

$$AO = \sqrt{x^2 + y^2 + z^2}$$
$$= \sqrt{(12 \text{ units})^2 + (0 \text{ units})^2 + (4 \text{ units})^2}$$
$$= 12.65 \text{ units}$$

$$BO = \sqrt{x^2 + y^2 + z^2}$$
$$= \sqrt{(12 \text{ units})^2 + (0 \text{ units})^2 + (-6 \text{ units})^2}$$
$$= 13.4 \text{ units}$$

$$CO = \sqrt{x^2 + y^2 + z^2}$$
$$= \sqrt{(12 \text{ units})^2 + (9 \text{ units})^2 + (0 \text{ units})^2}$$
$$= 15.0 \text{ units}$$

step 5:   Use Eqs. 43.75, 43.76, and 43.77 to find the direction cosines for each leg.

For leg AO,

$$\cos \theta_{A,x} = \frac{x_A}{AO} = \frac{12 \text{ units}}{12.65 \text{ units}} = 0.949$$

$$\cos \theta_{A,y} = \frac{y_A}{AO} = \frac{0 \text{ units}}{12.65 \text{ units}} = 0$$

$$\cos \theta_{A,z} = \frac{z_A}{AO} = \frac{4 \text{ units}}{12.65 \text{ units}} = 0.316$$

For leg BO,

$$\cos \theta_{B,x} = \frac{x_B}{BO} = \frac{12 \text{ units}}{13.4 \text{ units}} = 0.896$$

$$\cos \theta_{B,y} = \frac{y_B}{BO} = \frac{0 \text{ units}}{13.4 \text{ units}} = 0$$

$$\cos \theta_{B,z} = \frac{z_B}{BO} = \frac{-6 \text{ units}}{13.4 \text{ units}} = -0.448$$

For leg CO,

$$\cos \theta_{C,x} = \frac{x_C}{CO} = \frac{12 \text{ units}}{15.0 \text{ units}} = 0.80$$

$$\cos \theta_{C,y} = \frac{y_C}{CO} = \frac{9 \text{ units}}{15.0 \text{ units}} = 0.60$$

$$\cos \theta_{C,z} = \frac{z_C}{CO} = \frac{0 \text{ units}}{15.0 \text{ units}} = 0$$

steps 6 and 7:   Substitute Eqs. 43.78, 43.79, and 43.80 into equilibrium Eqs. 43.81, 43.82, and 43.83.

$$F_A \cos \theta_{A,x} + F_B \cos \theta_{B,x} + F_C \cos \theta_{C,x} + F_x = 0$$
$$F_A \cos \theta_{A,y} + F_B \cos \theta_{B,y} + F_C \cos \theta_{C,y} + F_y = 0$$
$$F_A \cos \theta_{A,z} + F_B \cos \theta_{B,z} + F_C \cos \theta_{C,z} + F_z = 0$$
$$0.949F_A + 0.896F_B + 0.80F_C + 0 = 0$$
$$0F_A + 0F_B + 0.60F_C + 200 = 0$$
$$0.316F_A - 0.448F_B + 0F_C + 0 = 0$$

Solve the three equations simultaneously.

$$F_A = 168.6 \quad (\text{T})$$
$$F_B = 118.9 \quad (\text{T})$$
$$F_C = \boxed{-333.3 \ (\text{C})}$$

The answer is (D).

3. First, find the vertical reaction at point D.

$$\sum M_C = (CD)D_y - (AF)(8000) + (AC)(1600) = 0$$
$$6D_y - (6)(8000) + (16)(1600) = 0$$

Solve for $D_y = 3733.3$.

The free-body diagram of pin D is as follows.

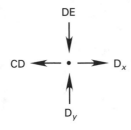

DE

CD $\longleftarrow$ • $\longrightarrow$ D$_x$

D$_y$

$$\sum F_y = D_y - DE = 0$$

Therefore,

$$DE = D_y = \boxed{3733.3 \ \ (C)}$$

The answer is (D).

4. *step 1:* Move the origin to the apex of the tripod. Call this point O.

step 2: By inspection, the force components are $F_x = 1200$, $F_y = 0$, and $F_z = 0$.

step 3: The (x, y, z) coordinates of the three support points are

point A—$(5, -12, 0)$
point B—$(0, -8, -8)$
point C—$(-4, -7, 6)$

step 4: Find the lengths of the legs.

$$AO = \sqrt{x^2 + y^2 + z^2} = \sqrt{(5)^2 + (-12)^2 + (0)^2}$$
$$= 13.0$$

$$BO = \sqrt{x^2 + y^2 + z^2} = \sqrt{(0)^2 + (-8)^2 + (-8)^2}$$
$$= 11.31$$

$$CO = \sqrt{x^2 + y^2 + z^2} = \sqrt{(-4)^2 + (-7)^2 + (6)^2}$$
$$= 10.05$$

step 5: Use Eqs. 43.75, 43.76, and 43.77 to find the direction cosines for each leg.

For leg AO,

$$\cos \theta_{A,x} = \frac{x_A}{AO} = \frac{5}{13.0} = 0.385$$

$$\cos \theta_{A,y} = \frac{y_A}{AO} = \frac{-12}{13.0} = -0.923$$

$$\cos \theta_{A,z} = \frac{z_A}{AO} = \frac{0}{13.0} = 0$$

For leg BO,

$$\cos \theta_{B,x} = \frac{x_B}{BO} = \frac{0}{11.31} = 0$$

$$\cos \theta_{B,y} = \frac{y_B}{BO} = \frac{-8}{11.31} = -0.707$$

$$\cos \theta_{B,z} = \frac{z_B}{BO} = \frac{-8}{11.31} = -0.707$$

For leg CO,

$$\cos \theta_{C,x} = \frac{x_C}{CO} = \frac{-4}{10.05} = -0.398$$

$$\cos \theta_{C,y} = \frac{y_C}{CO} = \frac{-7}{10.05} = -0.697$$

$$\cos \theta_{C,z} = \frac{z_C}{CO} = \frac{6}{10.05} = 0.597$$

steps 6 and 7: Substitute Eqs. 43.78, 43.79, and 43.80 into equilibrium Eqs. 43.81, 43.82, and 43.83.

$$F_A \cos \theta_{A,x} + F_B \cos \theta_{B,x} + F_C \cos \theta_{C,x} + F_x = 0$$
$$F_A \cos \theta_{A,y} + F_B \cos \theta_{B,y} + F_C \cos \theta_{C,y} + F_y = 0$$
$$F_A \cos \theta_{A,z} + F_B \cos \theta_{B,z} + F_C \cos \theta_{C,z} + F_z = 0$$
$$0.385 F_A + 0 F_B - 0.398 F_C + 1200 = 0$$
$$-0.923 F_A - 0.707 F_B - 0.697 F_C + 0 = 0$$
$$0 F_A - 0.707 F_B + 0.597 F_C + 0 = 0$$

Solve the three equations simultaneously.

$$\boxed{F_A = -1793 \ \ (C)}$$

$$F_B = 1080 \ \ (T)$$
$$F_C = 1279 \ \ (T)$$

$$F = \sqrt{F_A^2 + F_B^2 + F_C^2}$$
$$= \sqrt{(-1793)^2 + (1080)^2 + (1279)^2}$$
$$= 2453$$

The answer is (C).

5. First, find the vertical reactions.

$$\sum F_y = A_y + L_y - (5)(4000) - (5)(60{,}000) = 0$$

By symmetry, $A_y = L_y$.

$$2A_y = (5)(4000) + (5)(60{,}000)$$
$$A_y = 160{,}000$$
$$L_y = 160{,}000$$

Statics

(a) For DE, make a cut in members BD and DE.

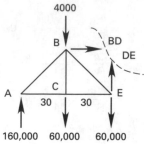

$$\sum F_y = 160{,}000 - 60{,}000 - 60{,}000 + \text{DE} - 4000 = 0$$

$$\text{DE} = \boxed{-36{,}000 \ (\text{C})}$$

The answer is (A).

(b) For HJ, make a cut in members HJ, HI, and GI.

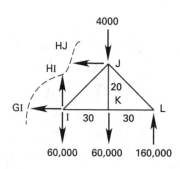

$$\sum M_\text{I} = (160{,}000)(60) - (60{,}000)(30)$$
$$- (4000)(30) + \text{HJ}(20)$$
$$= 0$$

$$\text{HJ} = \boxed{-384{,}000 \quad (\text{C})}$$

The answer is (D).

6. First, consider a free-body diagram at point O in the x-y plane.

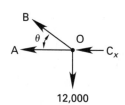

θ is obtained from triangle AOB.

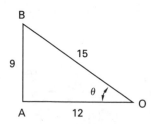

C_x is obtained from triangle AOC.

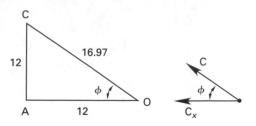

Equilibrium in the x-y plane at point O is $\sum F_y = 0$.

$$\text{B} \sin \theta = 12{,}000$$

$$\text{B} = \frac{12{,}000}{\sin \theta} = \frac{12{,}000}{\dfrac{9}{15}} = 20{,}000$$

$$\text{B}_x = \text{B} \cos \theta = (20{,}000)\left(\frac{12}{15}\right) = 16{,}000$$

$$\text{B}_y = \text{B} \sin \theta = (20{,}000)\left(\frac{9}{15}\right) = 12{,}000$$

Since BO is in the x-y plane, $\text{B}_z = 0$.

Next, consider a free-body diagram at point O in the x-z plane.

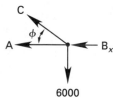

$\sum F_x = 0$:

$$\text{A} + \text{B}_x + \text{C} \cos \phi = 0$$
$$\text{A} + \text{C} \cos \phi = -\text{B}_x = -\text{B} \cos \theta$$
$$= (-20{,}000)\left(\frac{12}{15}\right)$$
$$= -16{,}000$$

$\sum F_z = 0$:

$$\text{C} \sin \phi = 6000$$
$$\text{C} = \frac{6000}{\sin \phi}$$
$$= \frac{6000}{\dfrac{12}{16.97}} = 8485$$

Thus,

$$\text{A} = -\text{C} \cos \phi - 16{,}000$$
$$= (-8485)\left(\frac{12}{16.97}\right) - 16{,}000$$
$$= -22{,}000$$

Since AO is on the x-axis,

$$A_x = -22,000$$
$$A_y = 0$$
$$A_z = 0$$

Solve for C reactions.

$$C_x = \text{C} \cos\phi = (8485)\left(\frac{12}{16.97}\right) = 6000$$

$$C_z = \text{C} \sin\phi = (8485)\left(\frac{12}{16.97}\right) = 6000$$

Since CO is in the x-z plane, $C_y = 0$.

The answer is (B).

7. *Customary U.S. Solution*

First, find the amount of thermal expansion. From Table 49.2, the coefficient of thermal expansion for steel is 6.5×10^{-6} $1/°F$. Use Eq. 49.9.

$$\delta = \alpha L_o(T_2 - T_1)$$
$$= \left(6.5 \times 10^{-6}\ \frac{1}{°F}\right)(1\ \text{mi})\left(5280\ \frac{\text{ft}}{\text{mi}}\right)$$
$$\times (99.14°F - 70°F)$$
$$= 1.000085\ \text{ft} \approx 1\ \text{ft}$$

For thermal expansion, assume the distributed load is uniform along the length of the rail. This resembles the case of a cable under its own weight. From the parabolic cable figure (Fig. 43.18), when distance S is small relative to distance a, the problem can be solved by using the parabolic equation, Eq. 43.57.

$$L \approx a\left(1 + \left(\tfrac{2}{3}\right)\left(\frac{S}{a}\right)^2 - \left(\tfrac{2}{5}\right)\left(\frac{S}{a}\right)^4\right)$$

$$\frac{5280\ \text{ft} + 1\ \text{ft}}{2} \approx (2640\ \text{ft})\left(1 + \left(\tfrac{2}{3}\right)\left(\frac{S}{2640\ \text{ft}}\right)^2\right.$$
$$\left. - \left(\tfrac{2}{5}\right)\left(\frac{S}{2640\ \text{ft}}\right)^4\right)$$

Using trial and error, $S \approx \boxed{44.5\ \text{ft.}}$

The answer is (D).

SI Solution

First, find the amount of thermal expansion. From Table 49.2, the coefficient of thermal expansion for steel is 11.7×10^{-6} $1/°C$. Use Eq. 49.9.

$$\delta = \alpha L_o(T_2 - T_1)$$
$$= \left(11.7 \times 10^{-6}\ \frac{1}{°C}\right)(1.6\ \text{km})\left(1000\ \frac{\text{m}}{\text{km}}\right)$$
$$\times (37.30°C - 21.11°C)$$
$$= 0.30308\ \text{m} \approx 0.30\ \text{m}$$

For thermal expansion, assume the distributed load is uniform along the length of the rail. This resembles the case of a cable under its own weight. From the parabolic cable figure (Fig. 43.18), when distance S is small relative to distance a, the problem can be solved by using the parabolic equation, Eq. 43.57.

$$L \approx a\left(1 + \left(\tfrac{2}{3}\right)\left(\frac{S}{a}\right)^2 - \left(\tfrac{2}{5}\right)\left(\frac{S}{a}\right)^4\right)$$

$$\frac{1600\ \text{m} + 0.3\ \text{m}}{2} \approx (800\ \text{m})\left(1 + \left(\tfrac{2}{3}\right)\left(\frac{S}{800\ \text{m}}\right)^2\right.$$
$$\left. - \left(\tfrac{2}{5}\right)\left(\frac{S}{800\ \text{m}}\right)^4\right)$$

Using trial and error, $S \approx \boxed{13.6\ \text{m.}}$

The answer is (D).

8. (a) First, find the reactions. Take clockwise moments about A as positive.

$$\sum M_A = (27)(-D_y) + (18)(24) + (9)(22.5) = 0$$
$$D_y = 23.5$$
$$A_y = 18 + 24 + 4.5 - 23.5 = 23$$

The general force triangle is

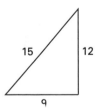

At pin A,

$$AE_y = 23$$
$$AE_x = \left(\frac{9}{12}\right)(23) = 17.25$$
$$AE = \left(\frac{15}{12}\right)(23) = 28.75\ \text{(C)}$$
$$AB = AE_x = 17.25\ \text{(T)}$$

At pin B,

$$BE = 4.5 \text{ (T)}$$
$$BC = AB = 17.25 \text{ (T)}$$

At pin D,

$$DF_y = 23.5$$
$$DF_x = \left(\frac{9}{12}\right)(23.5) = 17.63$$
$$DF = \left(\frac{15}{12}\right)(23.5) = 29.38 \text{ (C)}$$
$$DC = DF_x = 17.63 \text{ (T)}$$

At pin F,

$$FE = DF_x = 17.63 \text{ (C)}$$
$$FC = 24 - DF_y = \boxed{0.5 \text{ (C)}}$$

The answer is (A).

(b) At pin C,

$$CE_y = FC = 0.5$$
$$CE = \left(\frac{15}{12}\right)(0.5) = \boxed{0.625 \text{ (T)}}$$

The answer is (B).

44 Indeterminate Statics

PRACTICE PROBLEMS

Degree of Indeterminacy

1. What is the degree of indeterminacy of the following structure?

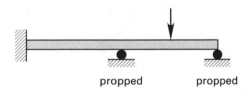

propped   propped

- (A) 1
- (B) 2
- (C) 3
- (D) 4

2. What is the degree of indeterminacy of the following truss?

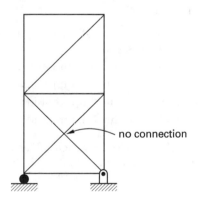

no connection

- (A) 1
- (B) 2
- (C) 3
- (D) 4

Elastic Deformation

3. What is the compressive load if a 1 in × 2 in × 10 in (2 cm × 5 cm × 30 cm) copper tie rod ($E = 18 \times 10^6$ lbf/in^2 (12×10^4 MPa)) experiences a 55°F (30°C) increase from the no-stress temperature?

- (A) 8000 lbf (26 kN)
- (B) 12,000 lbf (38 kN)
- (C) 15,000 lbf (48 kN)
- (D) 18,000 lbf (58 kN)

4. A 2 in diameter (5 cm diameter) and 15 in long (40 cm long) steel rod ($E = 30 \times 10^6$ lbf/in^2 (20×10^4 MPa)) supports a 2250 lbf (10 000 N) compressive load.

(a) What is the stress?
- (A) 570 lbf/in^2 (4.0 MPa)
- (B) 720 lbf/in^2 (5.1 MPa)
- (C) 1100 lbf/in^2 (7.8 MPa)
- (D) 1400 lbf/in^2 (9.9 MPa)

(b) What is the decrease in length?
- (A) 8.7×10^{-5} in (2.3×10^{-6} m)
- (B) 1.1×10^{-4} in (3.0×10^{-6} m)
- (C) 3.6×10^{-4} in (1.0×10^{-5} m)
- (D) 9.3×10^{-4} in (2.6×10^{-5} m)

Consistent Deformation Method

5.

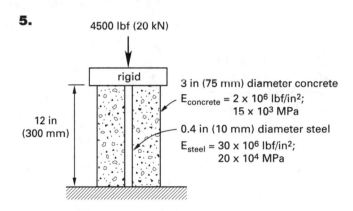

(a) Find the force and stress in each member for the structure shown.

(b) Find the total deflection for the structure shown.

6. The concentric pipe assembly shown is rigidly attached at both ends. The rigid ends are unconstrained. What is the stress generated in each pipe if the temperature of the assembly rises 200°F (110°C)?

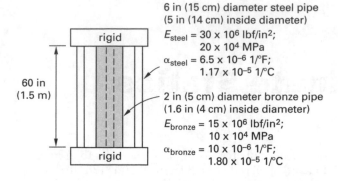

6 in (15 cm) diameter steel pipe
(5 in (14 cm) inside diameter)

$E_{steel} = 30 \times 10^6$ lbf/in²;
20 × 10⁴ MPa

$\alpha_{steel} = 6.5 \times 10^{-6}$ 1/°F;
1.17 × 10⁻⁵ 1/°C

2 in (5 cm) diameter bronze pipe
(1.6 in (4 cm) inside diameter)

$E_{bronze} = 15 \times 10^6$ lbf/in²;
10 × 10⁴ MPa

$\alpha_{bronze} = 10 \times 10^{-6}$ 1/°F;
1.80 × 10⁻⁵ 1/°C

60 in
(1.5 m)

7. For the structure shown, what is the load carried by the steel cable? The beam is made of steel, $I = 10.0$ in⁴ $(4.17 \times 10^6$ mm⁴). The cable cross section is 0.0124 in² (8 mm²). Before the 270 lbf load is applied, the cable is taut but carries no load.

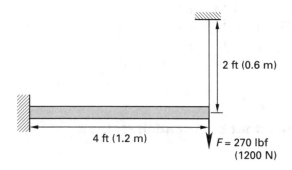

2 ft (0.6 m)

4 ft (1.2 m)

$F = 270$ lbf
(1200 N)

(A)   180 lbf (780 N)
(B)   200 lbf (870 N)
(C)   220 lbf (950 N)
(D)   240 lbf (1.0 kN)

8. A beam is simply supported at its ends and by a column of area A at its center. The beam and column are the same material. If the beam is subject to uniform load, w, what are the reactions and the force in the column?

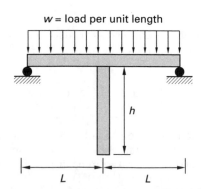

w = load per unit length

h

L L

9.

$a = 6$ ft (2 m)
$b = 3$ ft (1 m)
$c = 3$ ft (1 m)
$d = 12$ ft (4 m)
$P = 4500$ lbf (20 kN)
rod area $= 0.124$ in² (80 mm²)
$E_{aluminum} = 10 \times 10^6$ lbf/in² $(70 \times 10^3$ MPa)

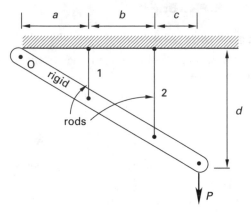

(a) What is the load of each of the aluminum rods for the structure shown?
(A)   3200 lbf (14 kN)
(B)   3600 lbf (16 kN)
(C)   4000 lbf (18 kN)
(D)   4300 lbf (19 kN)

(b) What is the elongation of each of the aluminum rods for the structure shown?
(A)   0.13 in and 0.26 in (3.3 mm and 6.6 mm)
(B)   0.21 in and 0.47 in (5.7 mm and 12 mm)
(C)   0.26 in and 0.62 in (6.6 mm and 16 mm)
(D)   0.21 in and 0.31 in (5.7 mm and 8.6 mm)

10. What are the reactions at each of the supports for the structure? The beams are identical in length, cross-sectional shape and area, and material.

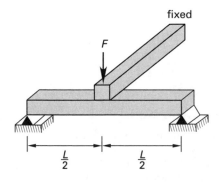

fixed

F

$\frac{L}{2}$ $\frac{L}{2}$

(A) $F/13$
(B) $4F/21$
(C) $8F/17$
(D) $2F/3$

Superposition Method

11. Using the superposition method, find the intermediate support reaction, R_2, for the structure.

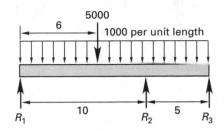

(A) 9000

(B) 11,000

(C) 15,000

(D) 18,000

12. Using the superposition method, determine the reaction, R, at the prop.

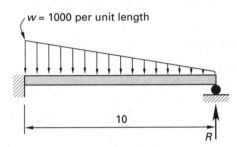

(A) 800

(B) 1000

(C) 1200

(D) 1400

Three-Moment Equation

13. What are the reactions R_1, R_2, and R_3?

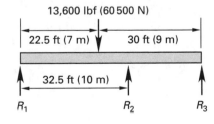

14. What are the reactions R_1, R_2, and R_3?

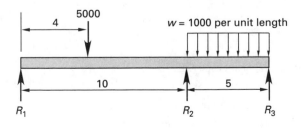

Fixed-End Moments

15. What are the vertical reactions at both ends of the structure?

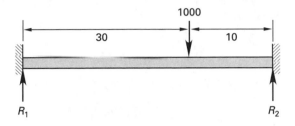

SOLUTIONS

1. The degree of indeterminacy is 2. Remove the two props (vertical reactions) in order to make the structure statically determinate.

The answer is (B).

2. $\dfrac{\text{degree of}}{\text{indeterminacy}} = 3 + \dfrac{\text{no.}}{\text{members}} - (2)\left(\dfrac{\text{no.}}{\text{joints}}\right)$

$= 3 + 10 - (2)(6) = \boxed{1}$

The answer is (A).

3. *Customary U.S. Solution*

The thermal strain is

$$\varepsilon_{\text{th}} = \alpha \Delta T = \left(8.9 \times 10^{-6} \dfrac{1}{^\circ \text{F}}\right)(55^\circ \text{F}) = 0.000490$$

The strain must be counteracted to maintain the rod in its original position.

$$\sigma = E\varepsilon_{\text{th}} = \left(18 \times 10^6 \dfrac{\text{lbf}}{\text{in}^2}\right)(0.000490) = 8820 \, \text{lbf/in}^2$$

$$F = \sigma A = \left(8820 \dfrac{\text{lbf}}{\text{in}^2}\right)(1 \, \text{in})(2 \, \text{in}) = \boxed{17{,}640 \, \text{lbf}}$$

The answer is (D).

SI Solution

The thermal strain is

$$\varepsilon_{\text{th}} = \alpha \Delta T = \left(16.0 \times 10^{-6} \dfrac{1}{^\circ \text{C}}\right)(30^\circ \text{C}) = 0.00048$$

The strain must be counteracted to maintain the rod in its original position.

$$\sigma = E\varepsilon_{\text{th}} = \left(12 \times 10^4 \, \text{MPa}\right)(0.00048)$$
$$= 57.6 \, \text{MPa}$$

$$F = \sigma A = (57.6 \times 10^6 \, \text{Pa})(0.02 \, \text{m})(0.05 \, \text{m})$$
$$= 57\,600 \, \text{N} = \boxed{57.6 \, \text{kN}}$$

The answer is (D).

4. *Customary U.S. Solution*

(a) $\quad \sigma = \dfrac{F}{A} = \dfrac{2250 \, \text{lbf}}{\dfrac{(\pi)(2 \, \text{in})^2}{4}} = \boxed{716.2 \, \text{lbf/in}^2}$

The answer is (B).

(b) $\quad \varepsilon = \dfrac{\sigma}{E} = \dfrac{\delta}{L_o}$

$$\delta = L_o \dfrac{\sigma}{E} = (15 \, \text{in})\left(\dfrac{716.2 \dfrac{\text{lbf}}{\text{in}^2}}{30 \times 10^6 \dfrac{\text{lbf}}{\text{in}^2}}\right)$$

$$= \boxed{3.58 \times 10^{-4} \, \text{in}}$$

The answer is (C).

SI Solution

(a) $\quad \sigma = \dfrac{F}{A} = \dfrac{10\,000 \, \text{N}}{\dfrac{\pi(0.05 \, \text{m})^2}{4}}$

$$= 5.093 \times 10^6 \, \text{Pa} = \boxed{5.093 \, \text{MPa}}$$

The answer is (B).

(b) $\quad \varepsilon = \dfrac{\sigma}{E} = \dfrac{\delta}{L_o}$

$$\delta = L_o\left(\dfrac{\sigma}{E}\right) = (0.4 \, \text{m})\left(\dfrac{5.093 \, \text{MPa}}{20 \times 10^4 \, \text{MPa}}\right)$$

$$= \boxed{1.02 \times 10^{-5} \, \text{m}}$$

The answer is (C).

5. *Customary U.S. Solution*

(a) Let F_c and F_{st} be the loads carried by the concrete and steel, respectively.

$$F = F_c + F_{\text{st}} = 4500 \, \text{lbf} \qquad \text{[Eq. 1]}$$

The deformation of the steel is

$$\delta_{\text{st}} = \dfrac{F_{\text{st}} L}{A_{\text{st}} E_{\text{st}}}$$

The deformation of the concrete is

$$\delta_c = \dfrac{F_c L}{A_c E_c}$$

The geometric constraint is $\delta_{\text{st}} = \delta_c$, or

$$\dfrac{F_{\text{st}} L}{A_{\text{st}} E_{\text{st}}} = \dfrac{F_c L}{A_c E_c} \qquad \text{[Eq. 2]}$$

Solving Eqs. 1 and 2,

$$F_c = \cfrac{F}{1 + \cfrac{A_{st}E_{st}}{A_cE_c}}$$

$$= \cfrac{4500 \text{ lbf}}{1 + \cfrac{\cfrac{(\pi)(0.4 \text{ in})^2}{4}\left(30 \times 10^6 \cfrac{\text{lbf}}{\text{in}^2}\right)}{\cfrac{(\pi)\left((3.0 \text{ in})^2 - (0.4 \text{ in})^2\right)}{4}\left(2 \times 10^6 \cfrac{\text{lbf}}{\text{in}^2}\right)}}$$

$$= \boxed{3539 \text{ lbf}}$$

$$F_{st} = 4500 \text{ lbf} - F_c = 4500 \text{ lbf} - 3539 \text{ lbf} = \boxed{961 \text{ lbf}}$$

$$\sigma_{st} = \frac{F_{st}}{A_{st}} = \frac{961 \text{ lbf}}{\cfrac{(\pi)(0.4 \text{ in})^2}{4}} = \boxed{7647 \text{ lbf/in}^2}$$

$$\sigma_c = \frac{F_c}{A_c} = \frac{3539 \text{ lbf}}{\cfrac{(\pi)\left((3 \text{ in})^2 - (0.4 \text{ in})^2\right)}{4}}$$

$$= \boxed{510 \text{ lbf/in}^2}$$

(b) $\quad \delta = \cfrac{F_{st}L}{A_{st}E_{st}} = \cfrac{(961 \text{ lbf})(12 \text{ in})}{\cfrac{(\pi)(0.4 \text{ in})^2}{4}\left(30 \times 10^6 \cfrac{\text{lbf}}{\text{in}^2}\right)}$

$$= \boxed{3.06 \times 10^{-3} \text{ in}}$$

SI Solution

(a) Let F_c and F_{st} be the loads carried by the concrete and steel, respectively.

$$F = F_c + F_{st} = 20 \text{ kN} \qquad \text{[Eq. 1]}$$

The deformation of the steel is

$$\delta_{st} = \frac{F_{st}L}{A_{st}E_{st}}$$

The deformation of the concrete is

$$\delta_c = \frac{F_cL}{A_cE_c}$$

The geometric constraint is $\delta_{st} = \delta_c$, or

$$\frac{F_{st}L}{A_{st}E_{st}} = \frac{F_cL}{A_cE_c} \qquad \text{[Eq. 2]}$$

Solving Eqs. 1 and 2,

$$F_c = \cfrac{F}{1 + \cfrac{A_{st}E_{st}}{A_cE_c}}$$

$$= \cfrac{20 \text{ kN}}{1 + \cfrac{\left(\cfrac{\pi}{4}\right)(0.01 \text{ m})^2(20 \times 10^4 \text{ MPa})}{\left(\cfrac{\pi}{4}\right)\left((0.075 \text{ m})^2 - (0.01 \text{ m})^2\right)(15 \times 10^3 \text{ MPa})}}$$

$$= \boxed{16.11 \text{ kN}}$$

$$F_{st} = 20 \text{ kN} - F_c = 20 \text{ kN} - 16.11 \text{ kN} = \boxed{3.89 \text{ kN}}$$

$$\sigma_{st} = \frac{F_{st}}{A_{st}} = \frac{3.89 \times 10^{-3} \text{ MN}}{\left(\cfrac{\pi}{4}\right)(0.01 \text{ m})^2} = \boxed{49.5 \text{ MPa}}$$

$$\sigma_c = \frac{F_c}{A_c} = \frac{16.11 \times 10^{-3} \text{ MN}}{\left(\cfrac{\pi}{4}\right)\left((0.075 \text{ m})^2 - (0.01 \text{ m})^2\right)}$$

$$= \boxed{3.71 \text{ MPa}}$$

(b) $\quad \delta = \cfrac{F_{st}L}{A_{st}E_{st}}$

$$= \cfrac{(3.89 \times 10^3 \text{ N})(0.3 \text{ m})}{\left(\cfrac{\pi}{4}\right)(0.01 \text{ m})^2(20 \times 10^{10} \text{ Pa})}$$

$$= 7.43 \times 10^{-5} \text{ m} = \boxed{74.3 \ \mu\text{m}}$$

6. *Customary U.S. Solution*

$$A_{st} = \left(\frac{\pi}{4}\right)\left((6 \text{ in})^2 - (5 \text{ in})^2\right) = 8.639 \text{ in}^2$$

$$A_b = \left(\frac{\pi}{4}\right)\left((2 \text{ in})^2 - (1.6 \text{ in})^2\right) = 1.131 \text{ in}^2$$

The thermal deformation that would occur if the pipes were free is

$$\delta_b = \alpha_b L \Delta T = \left(10 \times 10^{-6} \frac{1}{{}^\circ\text{F}}\right)(60 \text{ in})(200{}^\circ\text{F})$$

$$= 0.120 \text{ in}$$

$$\delta_{st} = \alpha_{st} L \Delta T = \left(6.5 \times 10^{-6} \frac{1}{{}^\circ\text{F}}\right)(60 \text{ in})(200{}^\circ\text{F})$$

$$= 0.078 \text{ in}$$

Statics

The two pipes expand by the same amount, δ, so that $\delta_{st} < \delta < \delta_b$.

$$\delta - \delta_{st} = \frac{F_{st}L}{A_{st}E_{st}} \qquad \text{[Eq. 1]}$$

$$\delta_b - \delta = \frac{F_b L}{A_b E_b} \qquad \text{[Eq. 2]}$$

F_{st} is a tensile force, F_b is a compressive force. Since there is no external force,

$$F_{st} = F_b = F$$

Adding Eq. 1 to Eq. 2,

$$\delta_b - \delta_{st} = \frac{FL}{A_{st}E_{st}} + \frac{FL}{A_b E_b}$$

$$= FL\left(\frac{1}{A_{st}E_{st}} + \frac{1}{A_b E_b}\right)$$

$$F = \frac{\dfrac{\delta_b - \delta_{st}}{L}}{\dfrac{1}{A_{st}E_{st}} + \dfrac{1}{A_b E_b}}$$

$$= \frac{\dfrac{0.120 \text{ in} - 0.078 \text{ in}}{60 \text{ in}}}{\dfrac{1}{(8.639 \text{ in}^2)\left(30 \times 10^6 \, \dfrac{\text{lbf}}{\text{in}^2}\right)} + \dfrac{1}{(1.131 \text{ in}^2)\left(15 \times 10^6 \, \dfrac{\text{lbf}}{\text{in}^2}\right)}}$$

$$= 11{,}146 \text{ lbf}$$

$$\sigma_{st} = \frac{F}{A_{st}} = \frac{11{,}146 \text{ lbf}}{8.639 \text{ in}^2} = \boxed{1290 \text{ psi}}$$

$$\sigma_b = \frac{F}{A_b} = \frac{11{,}146 \text{ lbf}}{1.131 \text{ in}^2} = \boxed{9855 \text{ psi}}$$

SI Solution

$$A_{st} = \left(\frac{\pi}{4}\right)\left((15 \text{ cm})^2 - (14 \text{ cm})^2\right) = 22.78 \text{ cm}^2$$

$$= 0.002278 \text{ m}^2$$

$$A_b = \left(\frac{\pi}{4}\right)\left((5 \text{ cm})^2 - (4 \text{ cm})^2\right) = 7.069 \text{ cm}^2$$

$$= 0.0007069 \text{ m}^2$$

The thermal deformation that would occur if the pipes were free is

$$\delta_b = \alpha_b L \Delta T = \left(1.8 \times 10^{-5} \frac{1}{\text{°C}}\right)(1.5 \text{ m})(110\text{°C})$$

$$= 0.00297 \text{ m}$$

$$\delta_{st} = \alpha_{st} L \Delta T = \left(1.17 \times 10^{-5} \frac{1}{\text{°C}}\right)(1.5 \text{ m})(110\text{°C})$$

$$= 0.00193 \text{ m}$$

The two pipes expand by the same amount, δ, so that $\delta_{st} < \delta < \delta_b$.

$$\delta - \delta_{st} = \frac{F_{st}L}{A_{st}E_{st}} \qquad \text{[Eq. 1]}$$

$$\delta_b - \delta = \frac{F_b L}{A_b E_b} \qquad \text{[Eq. 2]}$$

F_{st} is a tensile force, F_b is a compressive force. Since there is no external force,

$$F_{st} = F_b = F$$

Adding Eq. 1 to Eq. 2,

$$\delta_b - \delta_{st} = \frac{FL}{A_{st}E_{st}} + \frac{FL}{A_b E_b}$$

$$= FL\left(\frac{1}{A_{st}E_{st}} + \frac{1}{A_b E_b}\right)$$

$$F = \frac{\dfrac{\delta_b - \delta_{st}}{L}}{\dfrac{1}{A_{st}E_{st}} + \dfrac{1}{A_b E_b}}$$

$$= \frac{\dfrac{0.00297 \text{ m} - 0.00193 \text{ m}}{1.5 \text{ m}}}{\dfrac{1}{(0.002278 \text{ m}^2)(20 \times 10^{10} \text{ Pa})} + \dfrac{1}{(0.0007069 \text{ m}^2)(10 \times 10^{10} \text{ Pa})}}$$

$$= 4.24 \times 10^4 \text{ N}$$

$$\sigma_{st} = \frac{F}{A_{st}} = \frac{4.24 \times 10^4 \text{ N}}{0.002278 \text{ m}^2} = 1.86 \times 10^7 \text{ Pa}$$

$$= \boxed{18.6 \text{ MPa}}$$

$$\sigma_b = \frac{F}{A_b} = \frac{4.24 \times 10^4 \text{ N}}{0.0007069 \text{ m}^2} = 6.00 \times 10^7 \text{ Pa}$$

$$= \boxed{60.0 \text{ MPa}}$$

7. *Customary U.S. Solution*

The deflection of the beam is $\delta_b = (P_b L^3)/(3EI)$ where P_b is the net load at the beam tip. If P_c is the tension in the cable,

$$\delta_c = \frac{P_c L_c}{AE}$$

$\delta_c = \delta_b$ is the constraint on the deformation. Therefore,

$$\frac{P_b L_b^3}{3EI} = \frac{P_c L_c}{AE} \qquad \text{[Eq. 1]}$$

Another equation is the equilibrium equation.

$$F - P_c = P_b \qquad \text{[Eq. 2]}$$

Solving Eqs. 1 and 2 simultaneously,

$$P_c = \frac{\dfrac{F L_b^3}{3I}}{\dfrac{L_c}{A} + \dfrac{L_b^3}{3I}}$$

$$= \frac{\dfrac{(270 \text{ lbf}) \left((4 \text{ ft}) \left(\dfrac{12 \text{ in}}{\text{ft}} \right) \right)^3}{(3)(10.0 \text{ in}^4)}}{\dfrac{(2 \text{ ft}) \left(\dfrac{12 \text{ in}}{\text{ft}} \right)}{0.0124 \text{ in}^2} + \dfrac{\left((4 \text{ ft}) \left(\dfrac{12 \text{ in}}{\text{ft}} \right) \right)^3}{(3)(10.0 \text{ in}^4)}}$$

$$= \boxed{177 \text{ lbf}}$$

The answer is (A).

SI Solution

The deflection of the beam is $\delta_b = (P_b L^3)/(3EI)$ where P_b is the net load at the beam tip. If P_c is the tension in the cable,

$$\delta_c = \frac{P_c L_c}{AE}$$

$\delta_c = \delta_b$ is the constraint on the deformation. Therefore,

$$\frac{P_b L_b^3}{3EI} = \frac{P_c L_c}{AE} \qquad \text{[Eq. 1]}$$

Another equation is the equilibrium equation.

$$F - P_c = P_b \qquad \text{[Eq. 2]}$$

Solving Eqs. 1 and 2 simultaneously,

$$P_c = \frac{\dfrac{F L_b^3}{3I}}{\dfrac{L_c}{A} + \dfrac{L_b^3}{3I}}$$

$$= \frac{\dfrac{(1200 \text{ N})(1.2 \text{ m})^3}{(3) \left(4.17 \times 10^6 \text{ mm}^4 \right) \left(\dfrac{\text{m}^4}{10^{12} \text{ mm}^4} \right)}}{\dfrac{0.6 \text{ m}}{(8 \text{ mm}^2) \left(\dfrac{\text{m}^2}{10^6 \text{ mm}^2} \right)}}$$

$$+ \frac{(1.2 \text{ m})^3}{(3) \left(4.17 \times 10^6 \text{ mm}^4 \right) \left(\dfrac{\text{m}^4}{10^{12} \text{ mm}^4} \right)}$$

$$= \boxed{778 \text{ N}}$$

The answer is (A).

8. *Customary U.S. Solution*

Let deflection down be positive.

The deflection at the center of the beam is

$$\delta_b = \frac{5w(2L)^4}{384EI} - \frac{F(2L)^3}{48EI}$$

F is the force applied by the column at the beam center. The beam deflection must be equal to the shortening of the column.

$$\delta_c = \frac{Fh}{EA}$$

Since $\delta_b = \delta_c$,

$$\frac{Fh}{EA} = \frac{5w(2L)^4}{384EI} - \frac{F(2L)^3}{48EI}$$

$$F = \boxed{\frac{5AwL^4}{24hI + 4AL^3}} \qquad \text{[Eq. 1]}$$

Another equation is the equilibrium equation. Let R_1 and R_2 be the left and right support reactions, respectively, on the beam. By symmetry,

$$R_1 = R_2 = R$$
$$2R + F - 2wL = 0 \qquad \text{[Eq. 2]}$$

Solving Eqs. 1 and 2 simultaneously,

$$R = \frac{2wL - F}{2}$$

$$= \boxed{wL \left(\frac{48hI + 3AL^3}{48hI + 8AL^3} \right)}$$

Statics

SI Solution

Let deflection down be positive.

The deflection at the center of the beam is

$$\delta_b = \frac{5w(2L)^4}{384EI} - \frac{F(2L)^3}{48EI}$$

F is the force applied by the column at the beam center. The beam deflection must be equal to the shortening of the column.

$$\delta_c = \frac{Fh}{EA}$$

Since $\delta_b = \delta_c$,

$$\frac{Fh}{EA} = \frac{5w(2L)^4}{384EI} - \frac{F(2L)^3}{48EI}$$

$$\boxed{F = \frac{5AwL^4}{24hI + 4AL^3}} \quad \text{[Eq. 1]}$$

Another equation is the equilibrium equation. Let R_1 and R_2 be the left and right support reactions, respectively, on the beam. By symmetry,

$$R_1 = R_2 = R$$
$$2R + F - 2wL = 0 \quad \text{[Eq. 2]}$$

Solving Eqs. 1 and 2 simultaneously,

$$R = \frac{2wL - F}{2}$$

$$= \boxed{wL\left(\frac{48hI + 3AL^3}{48hI + 8AL^3}\right)}$$

9. *Customary U.S. Solution*

(a) Let F_1 and F_2 and δ_1 and δ_2 be the tensions and the deformations in the cables, respectively. The moment equilibrium equation taken at the hinge of the rigid bar is

$$\sum M_O = aF_1 + (a+b)F_2 - (a+b+c)P = 0 \quad \text{[Eq. 1]}$$

The relationship between the elongations is

$$\frac{\delta_1}{a} = \frac{\delta_2}{a+b}$$

This can be rewritten as

$$\frac{F_1 L_1}{AEa} = \frac{F_2 L_2}{AE(a+b)}$$

Since $L_1/a = L_2/(a+b)$,

$$F_1 = F_2 \quad \text{[Eq. 2]}$$

Solving Eqs. 1 and 2,

$$aF + (a+b)F - (a+b+c)P = 0$$

$$F = F_1 = F_2$$

$$= \left(\frac{a+b+c}{2a+b}\right)P$$

$$= \left(\frac{6\,\text{ft} + 3\,\text{ft} + 3\,\text{ft}}{(2)(6\,\text{ft}) + 3\,\text{ft}}\right)(4500\,\text{lbf}) = \boxed{3600\,\text{lbf}}$$

The answer is (B).

(b) Take downward as a positive deflection. The slope of the rigid member is

$$\frac{d}{a+b+c}$$

$$\delta_1 = \left(\frac{F_1}{AE}\right)L_1 = \left(\frac{F_1}{AE}\right)\left(\frac{d}{a+b+c}\right)a$$

$$= \frac{3600\,\text{lbf}}{(0.124\,\text{in}^2)\left(10 \times 10^6\,\dfrac{\text{lbf}}{\text{in}^2}\right)}$$

$$\times \left(\frac{12\,\text{ft}}{6\,\text{ft} + 3\,\text{ft} + 3\,\text{ft}}\right)(6\,\text{ft})$$

$$= 1.74 \times 10^{-2}\,\text{ft} = \boxed{0.209\,\text{in}}$$

$$\delta_2 = \left(\frac{F_2}{AE}\right)L_2 = \left(\frac{F_2}{AE}\right)\left(\frac{d}{a+b+c}\right)(a+b)$$

$$= \frac{3600\,\text{lbf}}{(0.124\,\text{in}^2)\left(10 \times 10^6\,\dfrac{\text{lbf}}{\text{in}^2}\right)}$$

$$\times \left(\frac{12\,\text{ft}}{6\,\text{ft} + 3\,\text{ft} + 3\,\text{ft}}\right)(6\,\text{ft} + 3\,\text{ft})$$

$$= 2.61 \times 10^{-2}\,\text{ft} = \boxed{0.314\,\text{in}}$$

The answer is (D).

SI Solution

(a) Let F_1 and F_2 and δ_1 and δ_2 be the tensions and the deformations in the cables, respectively. The moment equilibrium equation taken at the hinge of the rigid bar is

$$\sum M_O = aF_1 + (a+b)F_2 - (a+b+c)P = 0 \quad \text{[Eq. 1]}$$

The relationship between the elongations is

$$\frac{\delta_1}{a} = \frac{\delta_2}{a+b}$$

This can be rewritten as

$$\frac{F_1 L_1}{AEa} = \frac{F_2 L_2}{AE(a+b)}$$

Since $\dfrac{L_1}{a} = \dfrac{L_2}{a+b}$,

$$F_1 = F_2 \qquad \text{[Eq. 2]}$$

Solving Eqs. 1 and 2,

$$aF + (a+b)F - (a+b+c)P = 0$$

$$
\begin{aligned}
F = F_1 &= F_2 \\
&= \left(\frac{a+b+c}{2a+b}\right)P \\
&= \left(\frac{2\,\text{m}+1\,\text{m}+1\,\text{m}}{4\,\text{m}+1\,\text{m}}\right)(20\,\text{kN}) = \boxed{16\,\text{kN}}
\end{aligned}
$$

The answer is (B).

(b) Take downward as a positive deflection. The slope of the rigid member is

$$\frac{d}{a+b+c}$$

$$
\begin{aligned}
\delta_1 &= \left(\frac{F_1}{AE}\right)L_1 = \left(\frac{F_1}{AE}\right)\left(\frac{d}{a+b+c}\right)a \\
&= \frac{16\,000\,\text{N}}{(80\,\text{mm}^2)\left(\dfrac{\text{m}^2}{10^6\,\text{mm}^2}\right)(70\times10^9\,\text{Pa})} \\
&\quad \times \left(\frac{4\,\text{m}}{2\,\text{m}+1\,\text{m}+1\,\text{m}}\right)(2\,\text{m}) \\
&= 5.71\times10^{-3}\,\text{m} = \boxed{5.71\,\text{mm}}
\end{aligned}
$$

$$
\begin{aligned}
\delta_2 &= \left(\frac{F_2}{AE}\right)L_2 = \left(\frac{F_2}{AE}\right)\left(\frac{d}{a+b+c}\right)(a+b) \\
&= \frac{16\,000\,\text{N}}{(80\,\text{mm}^2)\left(\dfrac{\text{m}^2}{10^6\,\text{mm}^2}\right)(70\times10^9\,\text{Pa})} \\
&\quad \times \left(\frac{4\,\text{m}}{2\,\text{m}+1\,\text{m}+1\,\text{m}}\right)(2\,\text{m}+1\,\text{m}) \\
&= 8.57\times10^{-3}\,\text{m} = \boxed{8.57\,\text{mm}}
\end{aligned}
$$

The answer is (D).

10. *Customary U.S. Solution*

Let subscript s refer to the supported beam and subscript c refer to the cantilever beam. The deflections are equal.

$$\delta_s = \delta_c$$

$$\frac{P_s L^3}{48EI} = \frac{P_c L^3}{3EI}$$

$$P_s = 16P_c \qquad \text{[Eq. 1]}$$

P_s and P_c are the net loads exerted on the supported beam center and the cantilever beam tip, respectively. The equilibrium equation for the supported beam is

$$\sum F_{s,y} = 2R - P_s = 0 \qquad \text{[Eq. 2]}$$

The equilibrium equation for the cantilever beam is

$$\sum F_{c,y} = P_c = F - P_s \qquad \text{[Eq. 3]}$$

Solving Eqs. 1 and 3 simultaneously,

$$P_c = \frac{F}{17}$$

$$P_s = \frac{16}{17}F$$

From Eq. 2,

$$R = \boxed{\frac{8}{17}F}$$

The answer is (C).

SI Solution

Let subscript s refer to the supported beam and subscript c refer to the cantilever beam. The deflections are equal.

$$\delta_s = \delta_c$$

$$\frac{P_s L^3}{48EI} = \frac{P_c L^3}{3EI}$$

$$P_s = 16P_c \qquad \text{[Eq. 1]}$$

P_s and P_c are the net loads exerted on the supported beam center and the cantilever beam tip, respectively. The equilibrium equation for the supported beam is

$$\sum F_{s,y} = 2R - P_s = 0 \qquad \text{[Eq. 2]}$$

The equilibrium equation for the cantilever beam is

$$\sum F_{c,y} = P_c = F - P_s \qquad \text{[Eq. 3]}$$

Solving Eqs. 1 and 3 simultaneously,

$$P_c = \frac{F}{17}$$

$$P_s = \frac{16}{17}F$$

From Eq. 2,

$$R = \boxed{\frac{8}{17}F}$$

The answer is (C).

Statics

11. *Customary U.S. Solution*

Assume deflection downward is positive.

step 1: Remove support 2 to make the structure statically determinate.

step 2: The deflection at the location of (removed) support 2 is the sum of the deflections induced by the discrete load and the distributed load.

From a beam deflection table,

$$\delta_{discrete} = \frac{Pb}{6EIL}\left(\left(\frac{L}{b}\right)(x-a)^3 + (L^2-b^2)x - x^3\right)$$

$$(P = 5000, \ L = 15, \ a = 6, \ b = 9, \ x = 10)$$

From a beam deflection table,

$$\delta_{distributed} = -\frac{w}{24EI}(L^3x - 2Lx^3 + x^4)$$

$$(w = 1000)$$

step 3: The deflection induced by R_2 alone considered as a load is

$$\delta_{R_2} = \frac{-R_2 a^2 b^2}{3EIL} \quad (a = 10, \ b = 5)$$

step 4: The total deflection at the location of support 2 is zero.

$$\delta_{discrete} + \delta_{distributed} + \delta_{R_2} = 0$$

$$\left(\frac{(5000)(9)}{(6)(15)}\right)$$

$$\times\left(\left(\frac{15}{9}\right)(10-6)^3 + \left((15)^2 - (9)^2\right)(10) - (10)^3\right)$$

$$+\left(\frac{1000}{24}\right)\left((15)^3(10) - (2)(15)(10)^3 + (10)^4\right)$$

$$-\frac{R_2(10)^2(5)^2}{(3)(15)} = 0$$

$$R_2 = \boxed{15,233}$$

The answer is (C).

SI Solution

Assume deflection downward is positive.

step 1: Remove support 2 to make the structure statically determinate.

step 2: The deflection at the location of (removed) support 2 is the sum of the deflections induced by the discrete load and the distributed load.

From a beam deflection table,

$$\delta_{discrete} = \frac{Pb}{6EIL}\left(\left(\frac{L}{b}\right)(x-a)^3 + (L^2-b^2)x - x^3\right)$$

$$(P = 5000, \ L = 15, \ a = 6, \ b = 9, \ x = 10)$$

From a beam deflection table,

$$\delta_{distributed} = -\frac{w}{24EI}(L^3x - 2Lx^3 + x^4)$$

$$(w = 1000)$$

step 3: The deflection induced by R_2 alone considered as a load is

$$\delta_{R_2} = \frac{-R_2 a^2 b^2}{3EIL} \quad (a = 10, \ b = 5)$$

step 4: The total deflection at the location of support 2 is zero.

$$\delta_{discrete} + \delta_{distributed} + \delta_{R_2} = 0$$

$$\left(\frac{(5000)(9)}{(6)(15)}\right)$$

$$\times\left(\left(\frac{15}{9}\right)(10-6)^3 + \left((15)^2 - (9)^2\right)(10) - (10)^3\right)$$

$$+\left(\frac{1000}{24}\right)\left((15)^3(10) - (2)(15)(10)^3 + (10)^4\right)$$

$$-\frac{R_2(10)^2(5)^2}{(3)(15)} = 0$$

$$R_2 = \boxed{15\,233}$$

The answer is (C).

12. *Customary U.S. Solution*

First, remove the prop to make the structure statically determinate. The deflection induced by the distributed load at the tip is

$$\delta_{distributed} = \frac{-wL^4}{30EI} \quad (down)$$

The deflection caused by load R_1 is

$$\delta_R = \frac{RL^3}{3EI} \quad (up)$$

Since the deflection is actually zero at the tip,

$$\delta_{\text{distributed}} + \delta_R = 0$$

$$\frac{-wL^4}{30EI} + \frac{RL^3}{3EI} = 0$$

$$R = \frac{wL}{10} = \frac{(1000)(10)}{10} = \boxed{1000}$$

The answer is (B).

SI Solution

First, remove the prop to make the structure statically determinate. The deflection induced by the distributed load at the tip is

$$\delta_{\text{distributed}} = \frac{-wL^4}{30EI} \quad \text{(down)}$$

The deflection caused by load R_1 is

$$\delta_R = \frac{RL^3}{3EI} \quad \text{(up)}$$

Since the deflection is actually zero at the tip,

$$\delta_{\text{distributed}} + \delta_R = 0$$

$$\frac{-wL^4}{30EI} + \frac{RL^3}{3EI} = 0$$

$$R = \frac{wL}{10} = \frac{(1000)(10)}{10} = \boxed{1000}$$

The answer is (B).

13. *Customary U.S. Solution*

Use the three-moment method. The first moment of the area is

$$A_1 a = \tfrac{1}{6}Fc(L^2 - c^2)$$
$$= \left(\tfrac{1}{6}\right)(13,600)(22.5)\left((32.5)^2 - (22.5)^2\right)$$
$$= 28,050,000$$

Since there is no force between R_2 and R_3, $A_2 b = 0$.

The left and right ends of the beam are simply supported; M_1 and M_3 are zero. Therefore, the three-moment equation becomes

$$2M_2(32.5 + 20) = (-6)\left(\frac{28,050,000}{32.5}\right)$$
$$M_2 = -49,318.7$$

M_2 can be written in terms of the load and reactions to the left of support 2.

$$M_2 = (-13,600)(10) + (32.5)(R_1) = -49,318.7$$

$$R_1 = \boxed{2667.1 \text{ lbf}}$$

Now that R_1 is known, moments can be taken about support 3 to the left.

$$\sum M_3 = (2667.1)(52.5) - (13,600)(30) + (R_2)(20) = 0$$

$$R_2 = \boxed{13,398.9 \text{ lbf}}$$

R_3 can be obtained by taking moments about support 1.

$$\sum M_1 = (22.5)(13,600) - (32.5)(13,398.9)$$
$$- (52.5)(R_3) = 0$$

$$R_3 = \boxed{-2466.0 \text{ lbf}}$$

SI Solution

Use the three-moment method. The first moment of the area is

$$A_1 a = \frac{1}{6}Fc(L^2 - c^2)$$
$$= \left(\frac{1}{6}\right)(60\,500 \text{ N})(7 \text{ m})\left((10 \text{ m})^2 - (7 \text{ m})^2\right)$$
$$= 3.60 \times 10^6 \text{ N·m}^3$$

Since there is no force between R_2 and R_3, $A_2 b = 0$.

The left and right ends of the beam are simply supported; M_1 and M_3 are zero. Therefore, the three-moment equation becomes

$$2M_2(10 \text{ m} + 6 \text{ m}) = (-6)\left(\frac{3.6 \times 10^6 \text{ N·m}^3}{10 \text{ m}}\right)$$
$$M_2 = -67\,500 \text{ N·m}$$

M_2 can be written in terms of the load and reactions to the left of support 2.

$$M_2 = (-60\,500 \text{ N})(10 \text{ m} - 7 \text{ m}) + (10 \text{ m})(R_1)$$
$$= -67\,500 \text{ N·m}$$

$$R_1 = \boxed{11\,400 \text{ N}}$$

Now that R_1 is known, moments can be taken about support 3 to the left.

$$\sum M_3 = (11\,400 \text{ N})(16 \text{ m}) - (60\,500 \text{ N})(9 \text{ m})$$
$$+ (R_2)(6 \text{ m}) = 0$$

$$R_2 = \boxed{60\,350 \text{ N}}$$

Statics

R_3 can be obtained by taking moments about support 1.

$$\sum M_1 = (7\,\text{m})(60\,500\,\text{N}) - (10\,\text{m})(60\,350\,\text{N})$$

$$- (16\,\text{m})(R_3) = 0$$

$$R_3 = \boxed{-11\,250\,\text{N}}$$

14. *Customary U.S. Solution*

Use the three-moment method. The first moments of the areas are

$$A_1 a = \frac{1}{6} Fc(L_1^2 - c^2) = \left(\frac{1}{6}\right)(5000)(4)\left((10)^2 - (4)^2\right)$$

$$= 280{,}000$$

$$A_2 b = \frac{wL_2^4}{24} = \frac{(1000)(5)^4}{24} = 26{,}042$$

The two ends of the beam are simply supported; M_1 and M_3 are zero. Therefore, the three moment equation becomes

$$2M_2(10 + 5) = (-6)\left(\frac{280{,}000}{10} + \frac{26{,}042}{5}\right)$$

$$M_2 = -6642$$

M_2 can also be written in terms of the load and reactions to the left of support 2.

$$M_2 = (-5000)(6) + 10R_1 = -6642$$

$$R_1 = \boxed{2336}$$

Sum the moments around support 3.

$$\sum M_3 = (2336)(15) - (5000)(11) + 5R_2$$

$$- (5000)(2.5) = 0$$

$$R_2 = \boxed{6492}$$

Sum the moments around support 1.

$$\sum M_1 = (5000)(4) - (6492)(10) + (5000)(12.5)$$

$$- 15R_3 = 0$$

$$R_3 = \boxed{1172}$$

SI Solution

Use the three-moment method. The first moments of the areas are

$$A_1 a = \frac{1}{6} Fc(L_1^2 - c^2) = \left(\frac{1}{6}\right)(5000)(4)\left((10)^2 - (4)^2\right)$$

$$= 280\,000$$

$$A_2 b = \frac{wL_2^4}{24} = \frac{(1000)(5)^4}{24} = 26\,042$$

The two ends of the beam are simply supported; M_1 and M_3 are zero. Therefore, the three moment equation becomes

$$2M_2(10 + 5) = (-6)\left(\frac{280\,000}{10} + \frac{26\,042}{5}\right)$$

$$M_2 = -6642$$

M_2 can also be written in terms of the load and reactions to the left of support 2.

$$M_2 = (-5000)(6) + 10R_1 = -6642$$

$$R_1 = \boxed{2336}$$

Sum the moments around support 3.

$$\sum M_3 = (2336)(15) - (5000)(11) + 5R_2$$

$$- (5000)(2.5) = 0$$

$$R_2 = \boxed{6492}$$

Sum the moments around support 1.

$$\sum M_1 = (5000)(4) - (6492)(10) + (5000)(12.5)$$

$$- 15R_3 = 0$$

$$R_3 = \boxed{1172}$$

15. *Customary U.S. Solution*

The equilibrium requirement is

$$R_1 + R_2 = 1000$$

The moment equation at support 1 is

$$\sum M_{R_1} = M_1 + M_2 + (1000)(30) - 40R_2 = 0$$

From a table of fixed-end moments,

$$M_1 = \frac{-Fb^2 a}{L^2} = \frac{-(1000)(10)^2(30)}{(40)^2} = -1875$$

$$M_2 = \frac{Fa^2 b}{L^2} = \frac{(1000)(30)^2(10)}{(40)^2} = 5625$$

$$R_2 = \boxed{843.75}$$

The moment equation at support 2 is

$$\sum M_{R_2} = M_1 + M_2 - (1000)(10) + 40R_1 = 0$$

$$R_1 = \boxed{156.25}$$

SI Solution

From a table of fixed-end moments,

$$M_1 = \frac{-Fb^2a}{L^2} = \frac{-(1000)(10)^2(30)}{(40)^2} = -1875$$

$$M_2 = \frac{Fa^2b}{L^2} = \frac{(1000)(30)^2(10)}{(40)^2} = 5625$$

The moment equation at support 1 is

$$\sum M_{R_1} = M_1 + M_2 + (1000)(30) - 40R_2 = 0$$
$$-1875 + 5625 + (1000)(30) - 40R_2 = 0$$

$$R_2 = \boxed{843.75}$$

The moment equation at support 2 is

$$\sum M_{R_2} = M_1 + M_2 - (1000)(10) + 40R_1 = 0$$
$$-1875 + 5625 - (1000)(10) + 40R_1 = 0$$

$$R_1 = \boxed{156.25}$$

The equilibrium requirement is

$$R_1 + R_2 = 1000$$
$$156.25 + 843.75 = 1000 \quad \text{(check)}$$

45 Engineering Materials

PRACTICE PROBLEMS

1. How much (in molecules of HCl per gram of PVC) HCl should be used as an initiator in PVC if the efficiency is 20% and an average molecular weight of 7000 g/mol is desired? The final polymer has the following structure.

H−C−C−C−C− $\cdots$ −C−C−C−C−Cl (with H, Cl substituents)

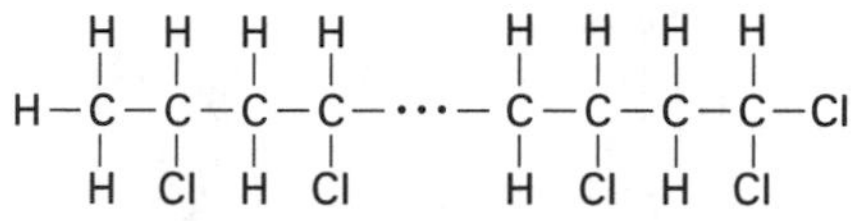

- (A) 7.8×10^{18} molecules HCl per gram PVC
- (B) 4.3×10^{20} molecules HCl per gram PVC
- (C) 9.5×10^{21} molecules HCl per gram PVC
- (D) 3.6×10^{23} molecules HCl per gram PVC

2. 10 ml of a 0.2% solution (by weight) of hydrogen peroxide is added to 12 g of ethylene to stabilize the polymer. What is the average degree of polymerization if the hydrogen peroxide is completely utilized? Assume that hydrogen peroxide breaks down according to

$$H_2O_2 \rightarrow 2(OH^-) + \cdots$$

Assume that the stabilized polymer has the following structure.

OH−C−C−C− $\cdots$ −C−C−C−OH (with H substituents)

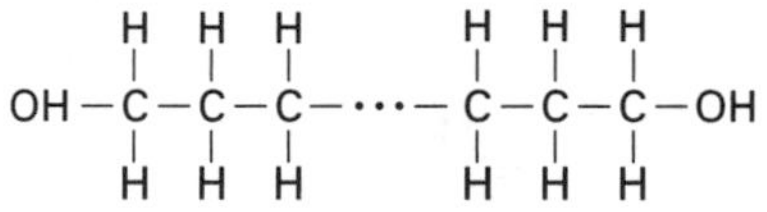

- (A) 180
- (B) 730
- (C) 910
- (D) 1200

SOLUTIONS

1. The vinyl chloride mer is

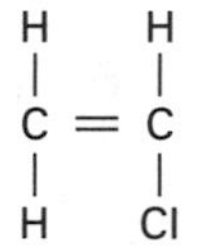

H−C=C−H (with H, Cl substituents)

With 20% efficiency, 5 molecules of HCl per PVC molecule are required to supply each end Cl atom. This is the same as 5 mol HCl per mole PVC. Using Avogadro's number of 6.022×10^{23} molecules per mole, the number of molecules of HCl per gram of PVC is

$$\frac{\left(5 \ \dfrac{\text{mol HCl}}{\text{mol PVC}}\right)\left(6.022 \times 10^{23} \ \dfrac{\text{molecules}}{\text{mol}}\right)}{7000 \ \dfrac{\text{g}}{\text{mol}}}$$

$$= \boxed{4.30 \times 10^{20} \text{ molecules HCl/gram PVC}}$$

The answer is (B).

2. The molecular weight of hydrogen peroxide, H_2O_2, is

$$(2)\left(1 \ \frac{\text{g}}{\text{mol}}\right) + (2)\left(16 \ \frac{\text{g}}{\text{mol}}\right) = 34 \text{ g/mol}$$

The weight of H_2O_2 is

$$(10 \text{ mL})\left(1 \ \frac{\text{g}}{\text{mL}}\right) = 10 \text{ g}$$

The number of H_2O_2 molecules in a 0.2% solution is

$$\frac{(10 \text{ g})\left(\dfrac{0.2\%}{100\%}\right)\left(6.022 \times 10^{23} \ \dfrac{\text{molecules}}{\text{mol}}\right)}{34 \ \dfrac{\text{g}}{\text{mol}}}$$

$$= 3.54 \times 10^{20} \text{ molecules}$$

The molecular weight of ethylene, C_2H_4, is

$$(2)\left(12 \ \frac{\text{g}}{\text{mol}}\right) + (4)\left(1 \ \frac{\text{g}}{\text{mol}}\right) = 28 \text{ g/mol}$$

The number of ethylene molecules is

$$\frac{(12 \text{ g}) \left(6.022 \times 10^{23} \, \frac{\text{molecules}}{\text{mol}}\right)}{28 \, \frac{\text{g}}{\text{mol}}}$$

$$= 2.58 \times 10^{23} \text{ molecules}$$

Since it takes one H_2O_2 molecule (i.e., $2OH^-$ radicals) to stabilize a polyethylene molecule, there are 3.54×10^{20} polymers.

The degree of polymerization is

$$\text{DP} = \frac{2.58 \times 10^{23} \, C_2H_4 \text{ molecules}}{3.54 \times 10^{20} \text{ polymers}} = \boxed{729}$$

The answer is (B).

46 Material Properties and Testing

PRACTICE PROBLEMS

1. The engineering stress and engineering strain for a copper specimen are 20,000 lbf/in² (140 MPa) and 0.0200 in/in (0.0200 mm/mm), respectively. Poisson's ratio for the specimen is 0.3.

(a) What is the true stress?
(A) 14,000 lbf/in² (98 MPa)
(B) 18,000 lbf/in² (130 MPa)
(C) 20,000 lbf/in² (140 MPa)
(D) 22,000 lbf/in² (160 MPa)

(b) What is the true strain?
(A) 0.0182 in/in (0.0182 mm/mm)
(B) 0.0189 in/in (0.0189 mm/mm)
(C) 0.0194 in/in (0.0194 mm/mm)
(D) 0.0198 in/in (0.0198 mm/mm)

2. A graph of engineering stress-strain is shown. Poisson's ratio for the material is 0.3.

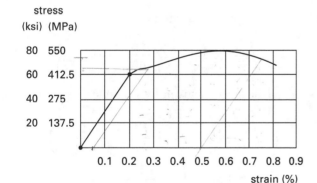

(a) Find the strength at 0.05% strain.
(A) 65,000 lbf/in² (450 MPa)
(B) 76,000 lbf/in² (530 MPa)
(C) 84,000 lbf/in² (590 MPa)
(D) 98,000 lbf/in² (690 MPa)

(b) Find the elastic modulus.
(A) 3.0 × 10⁵ lbf/in² (2.1 GPa)
(B) 2.7 × 10⁶ lbf/in² (19 GPa)
(C) 2.9 × 10⁶ lbf/in² (20 GPa)
(D) 3.0 × 10⁷ lbf/in² (210 GPa)

(c) Find the ultimate strength.
(A) 72,000 lbf/in² (500 MPa)
(B) 76,000 lbf/in² (530 MPa)
(C) 80,000 lbf/in² (550 MPa)
(D) 84,000 lbf/in² (590 MPa)

(d) Find the fracture strength.
(A) 62,000 lbf/in² (430 MPa)
(B) 70,000 lbf/in² (480 MPa)
(C) 76,000 lbf/in² (530 MPa)
(D) 84,000 lbf/in² (590 MPa)

(e) Find the percentage of elongation at fracture.
(A) 0.6%
(B) 0.8%
(C) 1.0%
(D) 1.2%

(f) Find the shear modulus.
(A) 0.8 × 10⁶ lbf/in² (6.3 GPa)
(B) 1.0 × 10⁶ lbf/in² (7.7 GPa)
(C) 1.2 × 10⁷ lbf/in² (79 GPa)
(D) 1.5 × 10⁷ lbf/in² (110 GPa)

(g) Find the toughness.
(A) 500 in-lbf/in³ (3.5 MJ/m³)
(B) 570 in-lbf/in³ (3.8 MJ/m³)
(C) 6300 in-lbf/in³ (42 MJ/m³)
(D) 8900 in-lbf/in³ (60 MJ/m³)

3. A specimen with an unstressed cross-sectional area of 4 in² (25 cm²) necks down to 3.42 in² (22 cm²) before breaking in a standard tensile test. What is the reduction in area of the material?
(A) 0.094 (0.089)
(B) 0.10 (0.094)
(C) 0.13 (0.10)
(D) 0.15 (0.12)

4. A constant 15,000 lbf/in² (100 MPa) tensile stress is applied to a specimen. The stress is known to be less than the material's yield strength. The strain is measured at various times. What is the steady-state creep rate for the material?

time (hr)	strain (in/in)
5	0.018
10	0.022
20	0.026
30	0.031
40	0.035
50	0.040
60	0.046
70	0.058

(A) 0.00037 hr^{-1}

(B) 0.00041 hr^{-1}

(C) 0.00046 hr^{-1}

(D) 0.00049 hr^{-1}

SOLUTIONS

1. *Customary U.S. Solution*

The fractional reduction in diameter is

$$\nu e = (0.3)\left(0.020 \ \frac{\text{in}}{\text{in}}\right) = 0.006$$

(a) The true stress is given by Eq. 46.5.

$$\sigma = \frac{F}{A_o(1-\nu e)^2} = \left(\frac{F_o}{A_o}\right)\left(\frac{1}{(1-\nu e)^2}\right)$$

$$= \left(20{,}000 \ \frac{\text{lbf}}{\text{in}^2}\right)\left(\frac{1}{(1-0.006)^2}\right)$$

$$= \boxed{20{,}242 \ \text{lbf/in}^2}$$

The answer is (C).

(b) The true strain is given by Eq. 46.6.

$$\epsilon = \ln(1+e) = \ln(1+0.020) = \boxed{0.0198 \ \text{in/in}}$$

The answer is (D).

SI Solution

The fractional reduction in diameter is

$$\nu e = (0.3)\left(0.020 \ \frac{\text{mm}}{\text{mm}}\right) = 0.006$$

(a) The true stress is given by Eq. 46.5.

$$\sigma = \frac{F}{A_o(1-\nu e)^2} = \left(\frac{F}{A_o}\right)\left(\frac{1}{(1-\nu e)^2}\right)$$

$$= (140 \ \text{MPa})\left(\frac{1}{(1-0.006)^2}\right)$$

$$= \boxed{141.7 \ \text{MPa}}$$

The answer is (C).

(b) The true strain is

$$\epsilon = \ln(1+e) = \ln(1+0.020) = \boxed{0.0198 \ \text{mm/mm}}$$

The answer is (D).

2. *Customary U.S. Solution*

(a) Extend a line from the 0.5% offset strain value parallel to the linear portion of the curve.

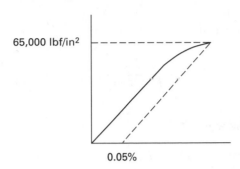

The 0.05% strain strength is $\boxed{65{,}000 \text{ lbf/in}^2.}$

The answer is (A).

(b) At the elastic limit, the stress is 60,000 lbf/in² and the percent strain is 0.2%. The elastic modulus is

$$E = \frac{\text{stress}}{\text{strain}} = \frac{60{,}000 \dfrac{\text{lbf}}{\text{in}^2}}{0.002}$$

$$= \boxed{3 \times 10^7 \text{ lbf/in}^2}$$

The answer is (D).

(c) The ultimate strength is the highest point of the curve. This value is $\boxed{80{,}000 \text{ lbf/in}^2.}$

The answer is (C).

(d) The fracture strength is at the end of the curve. This value is $\boxed{70{,}000 \text{ lbf/in}^2.}$

The answer is (B).

(e) The percentage of elongation at fracture is determined by extending a straight line down from the fracture point parallel to the linear portion of the curve. This gives an approximate value of $\boxed{0.6\%.}$

The answer is (A).

(f) The shear modulus is given by Eq. 46.21.

$$G = \frac{E}{2(1+\nu)} = \frac{3 \times 10^7 \dfrac{\text{lbf}}{\text{in}^2}}{(2)(1+0.3)}$$

$$= \boxed{1.15 \times 10^7 \text{ lbf/in}^2 \quad (1.2 \times 10^7 \text{ lbf/in}^2)}$$

The answer is (C).

(g) The toughness is the area under the stress-strain curve. Divide the area into squares of 20 ksi ×0.1%. There are about 25.5 squares covered.

$$(25.5)\left(20{,}000 \frac{\text{lbf}}{\text{in}^2}\right)\left(0.001 \frac{\text{in}}{\text{in}}\right) = \boxed{510 \text{ in-lbf/in}^3}$$

The answer is (A).

SI Solution

(a) Extend a line from the 0.5% offset strain value parallel to the linear portion of the curve.

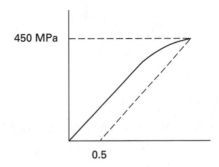

The 0.05% strain strength is $\boxed{450 \text{ MPa.}}$

The answer is (A).

(b) At the elastic limit, the stress is 410 MPa and the percent strain is 0.2%. The elastic modulus is

$$E = \frac{\text{stress}}{\text{strain}} = \frac{410 \text{ MPa}}{(0.002)\left(1000 \dfrac{\text{MPa}}{\text{GPa}}\right)}$$

$$= \boxed{205 \text{ GPa}}$$

The answer is (D).

(c) The ultimate strength is the highest point of the curve. This value is $\boxed{550 \text{ MPa.}}$

The answer is (C).

(d) The fracture strength is at the end of the curve. This value is $\boxed{480 \text{ MPa.}}$

The answer is (B).

(e) The percent elongation at fracture is determined by extending a straight line parallel to the initial strain line from the fracture point. This gives an approximate value of $\boxed{0.6\%.}$

The answer is (A).

(f) The shear modulus is given by Eq. 46.21.

$$G = \frac{E}{2(1+\nu)} = \frac{205 \text{ GPa}}{(2)(1+0.3)}$$

$$= \boxed{78.8 \text{ GPa}}$$

The answer is (C).

(g) The toughness is the area under the stress-strain curve. Divide the area into squares of 137.5 MPa × 0.1%. There are about 25.5 squares covered.

$$(25.5)(137.5 \text{ MPa}) \left(0.001 \, \frac{\text{m}}{\text{m}} \right) = \boxed{3.5 \text{ MJ/m}^3}$$

The answer is (A).

3. *Customary U.S. Solution*

The reduction in area is found from Eq. 46.12.

$$\text{reduction in area} = \frac{A_o - A_f}{A_o}$$

$$= \frac{4.0 \text{ in}^2 - 3.42 \text{ in}^2}{4.0 \text{ in}^2} = \boxed{0.145}$$

The answer is (D).

SI Solution

Use the reduction in area as the measure of ductility. Use Eq. 46.12.

$$\text{ductility} = \text{reduction in area} = \frac{A_o - A_f}{A_o}$$

$$= \frac{25 \text{ cm}^2 - 22 \text{ cm}^2}{25 \text{ cm}^2} = \boxed{0.12}$$

The answer is (D).

4. Plot the data and extrapolate a straight line. Disregard the first and last data points, as these represent primary and tertiary creep, respectively.

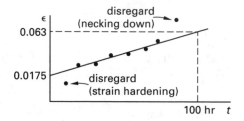

The creep rate is the slope of the line. The extrapolated points are (0, 0.0175) and (100, 0.063). Other points can be used.

$$\text{creep rate} = \frac{\Delta \epsilon}{\Delta t} = \frac{0.063 \, \frac{\text{in}}{\text{in}} - 0.0175 \, \frac{\text{in}}{\text{in}}}{100 \text{ hr}}$$

$$= \boxed{0.000455 \, \frac{1}{\text{hr}}}$$

The answer is (C).

47 Thermal Treatment of Metals

PRACTICE PROBLEMS

1. Refer to the following equilibrium diagram for an alloy of elements A and B.

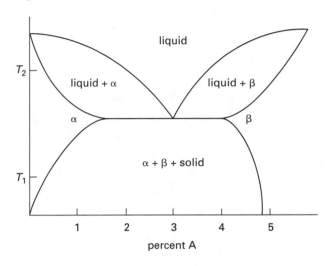

(a) For a 4% A alloy at temperature T_1, what are the compositions of solids α and β?

(b) For a 1% A alloy at temperature T_2, how much liquid and how much solid are present?

2. Write the procedures used with 2011 aluminum for (a) annealing and (b) precipitation hardening.

SOLUTIONS

1. (a) For temperature T_1, the equilibrium diagram is

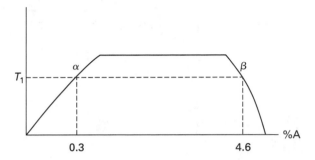

From the diagram, solid α is $\boxed{0.3\%\ \text{A.}}$

The %B for solid α is

$$\%\text{B} = 100\% - 0.3\% = \boxed{99.7\%}$$

For solid β,

$$\%\text{A} = \boxed{4.6\%}$$

$$\%\text{B} = 100\% - 4.6\% = \boxed{95.4\%}$$

(b) For temperature T_2, the equilibrium diagram is

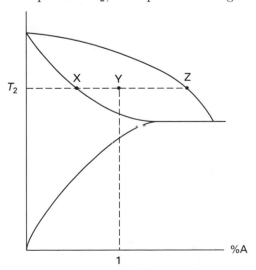

From the diagram, the following distances are measured.

The distance from point X to point Y is 9 mm.

The distance from point X to point Z is 21 mm.

From Eq. 47.2, the percent liquid is

$$\left(\frac{XY}{XZ}\right)(100\%) = \left(\frac{9 \text{ mm}}{21 \text{ mm}}\right)(100\%) = \boxed{42.9\%}$$

From Eq. 47.1, the percent solid is

$$100\% - \text{percent liquid} = 100\% - 42.9\% = \boxed{57.1\%}$$

2. (a) Since 2011 aluminum is a nonferrous substance, it does not readily form allotropes and thus its properties cannot be changed by the controlled cooling in an annealing process.

(b) The procedures for precipitation hardening of 2011 aluminum are

1. precipitation

2. rapid quenching

3. artificial aging

48 Properties of Areas

PRACTICE PROBLEMS

1. Locate the centroid of the area.

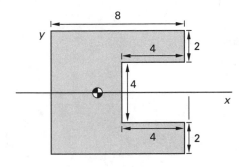

(A) 2.7
(B) 2.9
(C) 3.1
(D) 3.3

2. Replace the distributed load with three concentrated loads, and indicate the points of application.

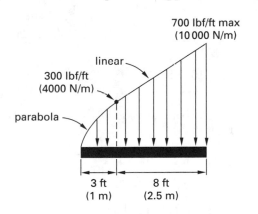

3. Find the centroidal moment of inertia about an axis parallel to the x-axis.

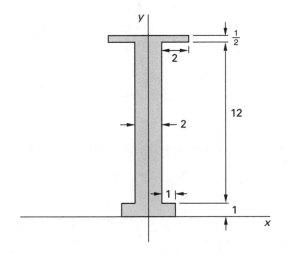

(A) 160 units^4
(B) 290 units^4
(C) 570 units^4
(D) 740 units^4

SOLUTIONS

1. The area is divided into three basic shapes.

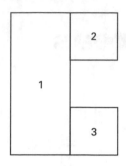

First, calculate the areas of the basic shapes.

$$A_1 = (4)(8) = 32 \text{ units}^2$$
$$A_2 = (4)(2) = 8 \text{ units}^2$$
$$A_3 = (4)(2) = 8 \text{ units}^2$$

Next, find the x-components of the centroids of the basic shapes.

$$x_{c1} = 2 \text{ units}$$
$$x_{c2} = 6 \text{ units}$$
$$x_{c3} = 6 \text{ units}$$

Finally, use Eq. 48.5.

$$x_c = \frac{\sum A_i x_{ci}}{\sum A_i} = \frac{(32)(2) + (8)(6) + (8)(6)}{32 + 8 + 8}$$

$$= \boxed{3.33 \text{ units}}$$

The answer is (D).

2. *Customary U.S. Solution*

The parabolic shape is

$$f(x) = 300\sqrt{\frac{x}{3}}$$

First, use Eq. 48.3 to find the concentrated load given by the area.

$$A = \int f(x)dx = \int_0^3 300\sqrt{\frac{x}{3}}dx = \left(\frac{300}{\sqrt{3}}\right)\left[\frac{x^{\frac{3}{2}}}{\frac{3}{2}}\right]_0^3$$

$$= \left(\frac{300}{\left(\frac{3}{2}\right)\sqrt{3}}\right)\left(3^{\frac{3}{2}} - 0^{\frac{3}{2}}\right) = \boxed{600 \text{ lbf}}$$

From Eq. 48.4,

$$dA = f(x)dx = 300\sqrt{\frac{x}{3}}$$

Finally, use Eq. 48.1 to find the location x_c of the concentrated load from the left end.

$$x_c = \frac{\int x\, dA}{A} = \frac{1}{600 \text{ lbf}} \int_0^3 300x\sqrt{\frac{x}{3}}dx$$

$$= \frac{300}{600\sqrt{3}} \int_0^3 x^{\frac{3}{2}}dx = \left(\frac{300}{600\sqrt{3}}\right)\left[\frac{x^{\frac{5}{2}}}{\frac{5}{2}}\right]_0^3$$

$$= \left(\frac{300}{600\sqrt{3}\left(\frac{5}{2}\right)}\right)\left(3^{\frac{5}{2}} - 0^{\frac{5}{2}}\right) = \boxed{1.8 \text{ ft}}$$

Alternative solution for the parabola:

Use App. 48.A.

$$A = \frac{2bh}{3} = \frac{(2)(300 \text{ lbf})(3 \text{ ft})}{3}$$

$$= \boxed{600 \text{ lbf}}$$

The centroid is located at

$$\frac{3h}{5} = \frac{(3)(3 \text{ ft})}{5} = \boxed{1.8 \text{ ft}}$$

The concentrated load for the triangular shape is the area from App. 48.A.

$$A = \frac{bh}{2} = \frac{\left(700\,\frac{\text{lbf}}{\text{ft}} - 300\,\frac{\text{lbf}}{\text{ft}}\right)(8 \text{ ft})}{2}$$

$$= \boxed{1600 \text{ lbf}}$$

From App. 48.A, the location of the concentrated load from the right end is

$$\frac{h}{3} = \frac{8 \text{ ft}}{3} = \boxed{2.67 \text{ ft}}$$

The concentrated load for the rectangular shape is the area from App. 48.A.

$$A = bh = \left(300\,\frac{\text{lbf}}{\text{ft}}\right)(8 \text{ ft})$$

$$= \boxed{2400 \text{ lbf}}$$

From App. 48.A, the location of the concentrated load from the right end is

$$\frac{h}{2} = \frac{8 \text{ ft}}{2} = \boxed{4 \text{ ft}}$$

SI Solution

The parabolic shape is

$$f(x) = 4000\sqrt{x}$$

First, use Eq. 48.3 to find the concentrated load given by the area.

$$A = \int f(x)dx = \int_0^1 4000\sqrt{x}\,dx = \left[\frac{4000\,x^{\frac{3}{2}}}{\frac{3}{2}}\right]_0^1$$

$$= \left(\frac{4000}{\frac{3}{2}}\right)\left(1^{\frac{3}{2}} - 0^{\frac{3}{2}}\right) = \boxed{2666.7 \text{ N}}$$

[first concentrated load]

From Eq. 48.4,

$$dA = f(x)dx = 4000\sqrt{x}\,dx$$

Finally, use Eq. 48.1 to find the location, x_c, of the concentrated load from the left end.

$$x_c = \frac{\int x\,dA}{A} = \frac{1}{2666.7 \text{ N}}\int_0^1 4000x\sqrt{x}\,dx$$

$$= \frac{4000}{2666.7}\int_0^1 x^{\frac{3}{2}}\,dx = \left(\frac{4000}{2666.7}\right)\left[\frac{x^{\frac{5}{2}}}{\frac{5}{2}}\right]_0^1$$

$$= \left(\frac{4000}{2666.7}\right)\left(\frac{1^{\frac{5}{2}} - 0^{\frac{5}{2}}}{\frac{5}{2}}\right) = \boxed{0.60 \text{ m}} \quad \text{[location]}$$

Alternative solution for the parabola:

Use App. 48.A.

$$A = \frac{2bh}{3} = \frac{(2)\left(4000\,\frac{N}{m}\right)(1 \text{ m})}{3}$$

$$= \boxed{2666.7 \text{ N}}$$

The centroid is located at

$$\frac{3h}{5} = \frac{(3)(1 \text{ m})}{5} = \boxed{0.6 \text{ m}}$$

The concentrated load for the triangular shape is the area from App. 48.A.

$$A = \frac{bh}{2} = \frac{\left(10\,000\,\frac{N}{m} - 4000\,\frac{N}{m}\right)(2.5 \text{ m})}{2}$$

$$= \boxed{7500 \text{ N}} \quad \text{[second concentrated load]}$$

From App. 48.A, the location of the concentrated load from the right end is

$$\frac{h}{3} = \frac{2.5 \text{ m}}{3} = \boxed{0.83 \text{ m}}$$

The concentrated load for the rectangular shape is the area from App. 48.A.

$$A = bh = \left(4000\,\frac{N}{m}\right)(2.5 \text{ m})$$

$$= \boxed{10\,000 \text{ N}} \quad \text{[third concentrated load]}$$

From App. 48.A, the location of the concentrated load from the right end is

$$\frac{h}{2} = \frac{2.5 \text{ m}}{2} = \boxed{1.25 \text{ m}}$$

3. The area is divided into three basic shapes.

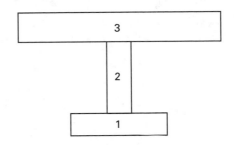

First, calculate the areas of the basic shapes.

$$A_1 = (4)(1) = 4 \text{ units}^2$$
$$A_2 = (2)(12) = 24 \text{ units}^2$$
$$A_3 = (6)(0.5) = 3 \text{ units}^2$$

Next, find the y-components of the centroids of the basic shapes.

$$y_{c1} = 0.5 \text{ units}$$
$$y_{c2} = 7 \text{ units}$$
$$y_{c3} = 13.25 \text{ units}$$

From Eq. 48.6, the centroid of the area is

$$y_c = \frac{\sum A_i y_{ci}}{\sum A_i} = \frac{(4)(0.5) + (24)(7) + (3)(13.25)}{4 + 24 + 3}$$
$$= 6.77 \text{ units}$$

From App. 48.A, the moment of inertia of basic shape 1 about its own centroid is

$$I_{cx1} = \frac{bh^3}{12} = \frac{(4)(1)^3}{12} = 0.33 \text{ units}^4$$

The moment of inertia of basic shape 2 about its own centroid is

$$I_{cx2} = \frac{bh^3}{12} = \frac{(2)(12)^3}{12} = 288 \text{ units}^4$$

The moment of inertia of basic shape 3 about its own centroid is

$$I_{cx3} = \frac{bh^3}{12} = \frac{(6)(0.5)^3}{12} = 0.063 \text{ units}^4$$

From the parallel axis theorem, Eq. 48.20, the moment of inertia of basic shape 1 about the centroidal axis of the section is

$$I_{x1} = I_{cx1} + A_1 d_1^2 = 0.33 + (4)(6.77 - 0.5)^2$$
$$= 157.6 \text{ units}^4$$

The moment of inertia of basic shape 2 about the centroidal axis of the section is

$$I_{x2} = I_{cx2} + A_2 d_2^2 = 288 + (24)(7.0 - 6.77)^2$$
$$= 289.3 \text{ units}^4$$

The moment of inertia of basic shape 3 about the centroidal axis of the section is

$$I_{x3} = I_{cx3} + A_3 d_3^2 = 0.063 + (3)(13.25 - 6.77)^2$$
$$= 126.0 \text{ units}^4$$

The total moment of inertia about the centroidal axis of the section is

$$I_x = I_{x1} + I_{x2} + I_{x3}$$
$$= 157.6 \text{ units}^4 + 289.3 \text{ units}^4 + 126.0 \text{ units}^4$$
$$= \boxed{572.9 \text{ units}^4}$$

The answer is (C).

49 Strength of Materials

PRACTICE PROBLEMS

Shear and Moment Diagrams

1. A beam 14 ft (4.2 m) long is supported at the left end and 2 ft (0.6 m) from the right end. The beam has a mass of 20 lbm/ft (30 kg/m). A 100 lbf (450 N) load is applied 2 ft (0.6 m) from the left end. An 80 lbf (350 N) load is applied at the right end.

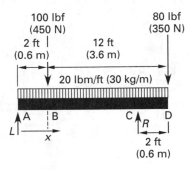

(a) What is the maximum moment?
- (A) 150 ft-lbf (200 N·m)
- (B) 250 ft-lbf (340 N·m)
- (C) 390 ft-lbf (520 N·m)
- (D) 830 ft-lbf (1100 N·m)

(b) What is the maximum shear?
- (A) 80 lbf (360 N)
- (B) 120 lbf (530 N)
- (C) 150 lbf (650 N)
- (D) 190 lbf (830 N)

Beam Deflections

2. A cantilever beam is 6 ft (1.8 m) in length. The cross section is 6 in wide by 4 in high (150 mm wide by 100 mm high). The modulus of elasticity is 1.5×10^6 psi (10 GPa). The beam is loaded by two concentrated forces: 200 lbf (900 N) located 1 ft (0.3 m) from the free end and 120 lbf (530 N) located 2 ft (0.6 m) from the free end. What is the tip deflection?
- (A) 0.29 in (0.0084 m)
- (B) 0.31 in (0.0090 m)
- (C) 0.47 in (0.014 m)
- (D) 0.55 in (0.015 m)

Thermal Deformation

3. A straight steel beam 200 ft (60 m) long is installed when the temperature is 40°F (4°C). It is supported in such a manner as to allow only 0.5 in (12 mm) longitudinal expansion. Lateral support is provided to prevent buckling. If the temperature increases to 110°F (43°C), what will be the compressive stress in the member?
- (A) 3900 lbf/in^2 (28 MPa)
- (B) 7400 lbf/in^2 (51 MPa)
- (C) 9200 lbf/in^2 (67 MPa)
- (D) 12,000 lbf/in^2 (88 MPa)

Elastic Deformation

4. A 1 in (25 mm) diameter soft steel rod carries a tensile load of 15,000 lbf (67 kN). The elongation is 0.158 in (4 mm). The modulus of elasticity is 2.9×10^7 lbf/in^2 (200 GPa). What is the total length of the rod?
- (A) 239.46 in (5.853 m)
- (B) 239.93 in (5.857 m)
- (C) 240.03 in (5.861 m)
- (D) 240.07 in (5.865 m)

Combined Stresses

5. A 3 in (75 mm) diameter horizontal shaft carries a 32 in (80 cm) diameter, 600 lbm (270 kg) pulley on an overhung (cantilever) end. The pulley is 8 in (200 mm) from the face of the outboard bearing. The pulley belt approaches and leaves horizontally. The belt carries upper and lower tensions of 1500 lbf (6.7 kN) and 350 lbf (1.6 kN), respectively. What is the maximum stress in the shaft?

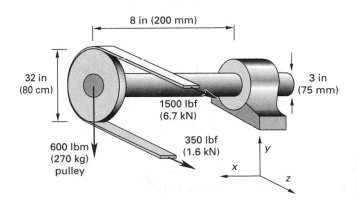

(A) 4500 lbf/in² (32 MPa)

(B) 5500 lbf/in² (39 MPa)

(C) 6500 lbf/in² (46 MPa)

(D) 7700 lbf/in² (55 MPa)

6. A 1.0 in (25 mm) diameter solid rod is held firmly in a chuck. A wrench with a 12 in (300 mm) moment arm applies 60 lbf (270 N) of force 8 in (200 mm) up from the chuck.

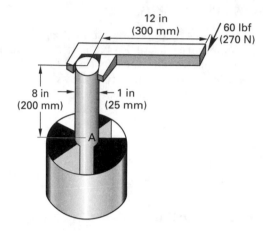

(a) What is the maximum shear stress in the rod?

(A) 2900 lbf/in² (20 MPa)

(B) 3700 lbf/in² (26 MPa)

(C) 4400 lbf/in² (32 MPa)

(D) 6800 lbf/in² (48 MPa)

(b) What is the maximum normal stress in the rod?

(A) 5400 lbf/in² (38 MPa)

(B) 6900 lbf/in² (49 MPa)

(C) 7500 lbf/in² (54 MPa)

(D) 11,000 lbf/in² (77 MPa)

7. A horizontal square bar with a cross-sectional area of 1.5 in² (9.7 cm²) is acted upon by an 18,000 lbf (80 kN) compressive load at each end. The shear stress in the horizontal direction at a particular point is 4000 lbf/in² (28 MPa).

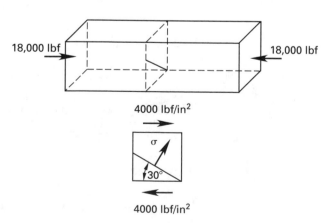

(a) What is the normal stress on a plane inclined +30° from the horizontal?

(A) 460 lbf/in² (3.6 MPa)

(B) 930 lbf/in² (6.5 MPa)

(C) 1700 lbf/in² (12 MPa)

(D) 2400 lbf/in² (17 MPa)

(b) What is the shear stress on a plane inclined +30° from the horizontal?

(A) 1700 lbf/in² (12 MPa)

(B) 2800 lbf/in² (20 MPa)

(C) 3200 lbf/in² (22 MPa)

(D) 4100 lbf/in² (29 MPa)

8. (*Time limit: one hour*) The offset wrench handle shown is constructed from a 5/8 in (16 mm) diameter round bar. The bar's modulus of elasticity is 29.6×10^6 lbf/in² (204 GPa). The 3 in (75 mm) rise is in the plane of the socket.

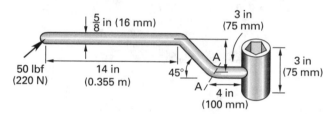

elevation view

view from handle

(a) What are the principal stresses at section A-A?

(b) What is the maximum shear at section A-A?

9. (*Time limit: one hour*) A brass tube 6 ft long, 2.0 in outside diameter, 1.0 in inside diameter, and with a modulus of elasticity of 1.5×10^7 lbf/in² (1.8 m long, 50 mm outside diameter, 25 mm inside diameter, and with a modulus of elasticity of 100 GPa) is used as a cantilever beam. When a concentrated load of 50 lbf (220 N) is applied at the free end, the tip deflection is found to be excessive. To reduce the deflection, it is suggested that a tight-fitting 1 in (25 mm) outside diameter soft steel rod with a modulus of elasticity of 2.9×10^7 lbf/in² (200 GPa) be inserted into the entire length of the brass tube.

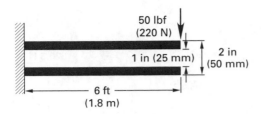

(a) Does the suggestion have merit?

(b) What is the percentage change in tip deflection, if any? (Neglect the self-weights.)

10. A 2500 lbm (1100 kg) stationary flywheel is mounted with a 4 in (100 mm) overhang on a stepped shaft as shown. The step fillet has a 5/16 in (8 mm) radius. Disregarding yielding in the vicinity of the fillet, what is the maximum bending stress in the overhanging section of the shaft?

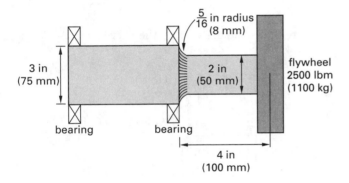

(A) 11,000 lbf/in² (77 MPa)
(B) 13,000 lbf/in² (91 MPa)
(C) 16,000 lbf/in² (110 MPa)
(D) 19,000 lbf/in² (130 MPa)

Composite Structures

11. A bimetallic spring is constructed of an aluminum strip with a ⅛ in by 1½ in (3.2 mm by 38 mm) cross section bonded on top of a 1/16 in by 1½ in (1.6 mm by 38 mm) steel strip. The modulus of elasticity of the aluminum is 10×10^6 lbf/in² (70 GPa). The modulus of elasticity of the steel is 30×10^6 lbf/in² (200 GPa). What is the equivalent, all-aluminum centroidal area moment of inertia of the cross section?

(A) 2.4×10^{-4} in⁴ (190 mm⁴)
(B) 5.2×10^{-4} in⁴ (390 mm⁴)
(C) 1.3×10^{-3} in⁴ (550 mm⁴)
(D) 2.6×10^{-3} in⁴ (1100 mm⁴)

SOLUTIONS

1. *Customary U.S. Solution*

First, determine the reactions. The uniform load can be assumed to be concentrated at the center of the beam.

Sum the moments about A.

$$(100 \text{ lbf})(2 \text{ ft}) + (80 \text{ lbf})(14 \text{ ft})$$

$$+ \left(20 \; \frac{\text{lbm}}{\text{ft}}\right) \left(\frac{32.2 \; \frac{\text{ft}}{\text{sec}^2}}{32.2 \; \frac{\text{lbm-ft}}{\text{lbf-sec}^2}} \right)$$

$$\times (14 \text{ ft})(7 \text{ ft}) - R(12 \text{ ft}) = 0$$

$$R = 273.3 \text{ lbf}$$

Sum the forces in the vertical direction.

$$L + 273.3 \text{ lbf} = 100 \text{ lbf} + 80 \text{ lbf}$$

$$+ \left(20 \; \frac{\text{lbm}}{\text{ft}}\right) \left(\frac{32.2 \; \frac{\text{ft}}{\text{sec}^2}}{32.2 \; \frac{\text{lbm-ft}}{\text{lbf-sec}^2}} \right) (14 \text{ ft})$$

$$L = 186.7 \text{ lbf}$$

The shear diagram starts at +186.7 lbf at the left reaction and decreases linearly at a rate of 20 lbf/ft to 146.7 lbf at point B. The concentrated load reduces the shear to 46.7 lbf. The shear then decreases linearly at a rate of 20 lbf/ft to point C. Measuring x from the left, the shear line goes through zero at

$$x = 2 \text{ ft} + \frac{46.7 \text{ lbf}}{20 \; \frac{\text{lbf}}{\text{ft}}} = 4.3 \text{ ft}$$

The shear at the right of the beam at point D is 80 lbf and increases linearly at a rate of 20 lbf/ft to 120 lbf at point C. The reaction R at point C decreases the shear to −153.3 lbf. This is sufficient to draw the shear diagram.

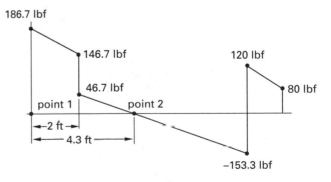

(a) From the shear diagram, the maximum moment occurs when the shear is zero. Call this point 2. The moment at the left reaction is zero. Call this point 1. Use Eq. 49.27.

$$M_2 = M_1 + \int_{x_1}^{x_2} V \, dx$$

The integral is the area under the curve from $x_1 = 0$ to $x_2 = 4.3$ ft.

$$M_2 = 0 + (146.7 \text{ lbf})(2 \text{ ft})$$
$$+ \left(\tfrac{1}{2}\right)(186.7 \text{ lbf} - 146.7 \text{ lbf})(2 \text{ ft})$$
$$+ \left(\tfrac{1}{2}\right)(46.7 \text{ lbf})(4.3 \text{ ft} - 2 \text{ ft})$$
$$= \boxed{387.1 \text{ ft-lbf}}$$

The answer is (C).

(b) From the shear diagram, the maximum shear is

$$\boxed{186.7 \text{ lbf.}}$$

The answer is (D).

SI Solution

First, determine the reactions. The uniform load can be assumed to be concentrated at the center of the beam.

Sum the moments about A.

$$(450 \text{ N})(0.6 \text{ m}) + (350 \text{ N})(4.2 \text{ m})$$
$$+ \left(30 \ \frac{\text{kg}}{\text{m}}\right)(4.2 \text{ m})\left(9.81 \ \frac{\text{N}}{\text{kg}}\right)(2.1 \text{ m})$$
$$- R(3.6 \text{ m}) = 0$$
$$R = 1204.4 \text{ N}$$

Sum the forces in the vertical direction.

$$L + 1204.4 \text{ N} = 450 \text{ N} + 350 \text{ N}$$
$$+ \left(30 \ \frac{\text{kg}}{\text{m}}\right)(4.2 \text{ m})\left(9.81 \ \frac{\text{N}}{\text{kg}}\right)$$
$$L = 831.7 \text{ N}$$

The shear diagram starts at +831.7 N at the left reaction and decreases linearly at a rate of 294.3 N/m to 655.1 N at point B. The concentrated load reduces the shear to 205.1 N. The shear then decreases linearly at a rate of 294.3 N/m to point C. Measuring x from the left, the shear line goes through zero at

$$x = 0.6 \text{ m} + \frac{205.1 \text{ N}}{294.3 \ \dfrac{\text{N}}{\text{m}}} = 1.3 \text{ m}$$

The shear at the right of the beam at point D is 350 N and increases linearly at a rate of 294.3 N/m to 526.3 N at point C. The reaction R at point C decreases the shear to −677.8 N. This is sufficient to draw the shear diagram.

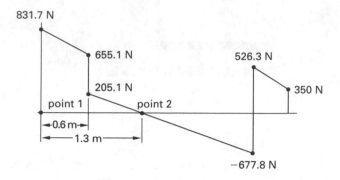

(a) From the shear diagram, the maximum moment occurs when the shear = 0. Call this point 2. The moment at the left reaction = 0. Call this point 1.

Use Eq. 49.27.

$$M_2 = M_1 + \int_{x_1}^{x_2} V \, dx$$

The integral is the area under the curve from $x_1 = 0$ to $x_2 = 1.3$ m.

$$M_2 = 0 + (655.1 \text{ N})(0.6 \text{ m})$$
$$+ \left(\tfrac{1}{2}\right)(831.7 \text{ N} - 655.1 \text{ N})(0.6 \text{ m})$$
$$+ \left(\tfrac{1}{2}\right)(205.1 \text{ N})(1.3 \text{ m} - 0.6 \text{ m})$$
$$= \boxed{517.8 \text{ N·m}}$$

The answer is (C).

(b) From the shear diagram, the maximum shear is

$$\boxed{831.7 \text{ N.}}$$

The answer is (D).

2. *Customary U.S. Solution*

First, the moment of inertia of the beam cross section is

$$I = \frac{bh^3}{12} = \frac{(6 \text{ in})(4 \text{ in})^3}{12} = 32 \text{ in}^4$$

From case 1 in App. 49.A, the deflection of the 200 lbf load is

$$y_1 = \frac{PL^3}{3EI} = \frac{(200 \text{ lbf})(72 \text{ in} - 12 \text{ in})^3}{(3)\left(1.5 \times 10^6 \ \dfrac{\text{lbf}}{\text{in}^2}\right)(32 \text{ in}^4)}$$
$$= 0.30 \text{ in}$$

The slope at the 200 lbf load is

$$\theta = \frac{PL^2}{2EI} = \frac{(200 \text{ lbf})(72 \text{ in} - 12 \text{ in})^2}{(2)\left(1.5 \times 10^6 \ \dfrac{\text{lbf}}{\text{in}^2}\right)(32 \text{ in}^4)}$$
$$= 0.0075 \text{ in/in}$$

The additional deflection at the tip of the beam is

$$y_{1'} = \left(0.0075 \, \frac{\text{in}}{\text{in}}\right)(12 \text{ in}) = 0.09 \text{ in}$$

From case 1 in App. 49.A, the deflection at the 120 lbf load is

$$y_2 = \frac{PL^3}{3EI} = \frac{(120 \text{ lbf})(72 \text{ in} - 24 \text{ in})^3}{(3)\left(1.5 \times 10^6 \, \frac{\text{lbf}}{\text{in}^2}\right)(32 \text{ in}^4)}$$

$$= 0.0922 \text{ in}$$

The slope at the 120 lbf load is

$$\theta_2 = \frac{PL^2}{2EI} = \frac{(120 \text{ lbf})(72 \text{ in} - 24 \text{ in})^2}{(2)\left(1.5 \times 10^6 \, \frac{\text{lbf}}{\text{in}^2}\right)(32 \text{ in}^4)}$$

$$= 0.00288 \text{ in/in}$$

The additional deflection at the tip of the beam is

$$y_{2'} = \left(0.00288 \, \frac{\text{in}}{\text{in}}\right)(24 \text{ in}) = 0.0691 \text{ in}$$

The total deflection is the sum of the preceding four parts.

$$\begin{aligned} y_{\text{tip}} &= y_1 + y_{1'} + y_2 + y_{2'} \\ &= 0.30 \text{ in} + 0.09 \text{ in} + 0.0922 \text{ in} + 0.0691 \text{ in} \\ &= \boxed{0.5513 \text{ in}} \end{aligned}$$

The answer is (D).

SI Solution

First, the moment of inertia of the beam cross section is

$$I = \frac{bh^3}{12} = \frac{(0.15 \text{ m})(0.10 \text{ m})^3}{12} = 1.25 \times 10^{-5} \text{ m}^4$$

From case 1 in App. 49.A, the deflection at the 900 N load is

$$y_1 = \frac{PL^3}{3EI} = \frac{(900 \text{ N})(1.8 \text{ m} - 0.3 \text{ m})^3}{(3)\left(10 \times 10^9 \text{ Pa}\right)(1.25 \times 10^{-5} \text{ m}^4)}$$

$$= 0.0081 \text{ m}$$

The slope at the 900 N load is

$$\theta_1 = \frac{PL^2}{2EI} = \frac{(900 \text{ N})(1.8 \text{ m} - 0.3 \text{ m})^2}{(2)\left(10 \times 10^9 \text{ Pa}\right)(1.25 \times 10^{-5} \text{ m}^4)}$$

$$= 0.0081 \text{ m/m}$$

The additional deflection at the tip of the beam is

$$y_{1'} = \left(0.0081 \, \frac{\text{m}}{\text{m}}\right)(0.3 \text{ m}) = 0.00243 \text{ m}$$

From case 1 in App. 49.A, the deflection at the 530 N load is

$$y_2 = \frac{PL^3}{3EI} = \frac{(530 \text{ N})(1.8 \text{ m} - 0.6 \text{ m})^3}{(3)\left(10 \times 10^9 \text{ Pa}\right)(1.25 \times 10^{-5} \text{ m}^4)}$$

$$= 0.00244 \text{ m}$$

The slope at the 530 N load is

$$\theta_2 = \frac{PL^2}{2EI} = \frac{(530 \text{ N})(1.8 \text{ m} - 0.6 \text{ m})^2}{(2)\left(10 \times 10^9 \text{ Pa}\right)(1.25 \times 10^{-5} \text{ m}^4)}$$

$$= 0.00305 \text{ m/m}$$

The additional deflection at the tip of the beam is

$$y_{2'} = \left(0.00305 \, \frac{\text{m}}{\text{m}}\right)(0.6 \text{ m}) = 0.00183 \text{ m}$$

The total deflection is the sum of the preceding four parts.

$$\begin{aligned} y_{\text{tip}} &= y_1 + y_{1'} + y_2 + y_{2'} \\ &= 0.0081 \text{ m} + 0.00243 \text{ m} + 0.00244 \text{ m} + 0.00183 \text{ m} \\ &= \boxed{0.0148 \text{ m}} \end{aligned}$$

The answer is (D).

3. *Customary U.S. Solution*

Use Eq. 49.9 to find the amount of expansion for an unconstrained beam. Use Table 49.2.

$$\begin{aligned} \Delta L &= \alpha L_o(T_2 - T_1) \\ &= \left(6.5 \times 10^{-6} \, \frac{1}{\text{°F}}\right)(200 \text{ ft})\left(12 \, \frac{\text{in}}{\text{ft}}\right) \\ &\quad \times (110\text{°F} - 40\text{°F}) \\ &= 1.092 \text{ in} \end{aligned}$$

The constrained length is

$$\Delta L_c = 1.092 \text{ in} - 0.5 \text{ in} = 0.592 \text{ in}$$

From Eq. 49.14, the thermal strain is

$$\epsilon_{\text{th}} = \frac{\Delta L_c}{L_o + 0.5 \text{ in}} = \frac{0.592 \text{ in}}{2400 \text{ in} + 0.5 \text{ in}}$$

$$= 2.466 \times 10^{-4} \text{ in/in}$$

From Eq. 49.15, the compressive thermal stress is

$$\sigma_{\text{th}} = E\epsilon_{\text{th}} = \left(30 \times 10^6 \, \frac{\text{lbf}}{\text{in}^2}\right)(2.466 \times 10^{-4})$$

$$= \boxed{7398 \text{ lbf/in}^2}$$

The answer is (B).

SI Solution

Use Eq. 49.9 to find the amount of expansion for an unconstrained beam. Use Table 49.2.

$$\Delta L = \alpha L_o (T_2 - T_1)$$
$$= \left(11.7 \times 10^{-6} \ \frac{1}{°C}\right)(60 \text{ m})(43°C - 4°C)$$
$$= 0.02738 \text{ m}$$

The constrained length is

$$\Delta L_c = 0.02738 \text{ m} - 0.012 \text{ m} = 0.01538 \text{ m}$$

From Eq. 49.14, the thermal strain is

$$\epsilon_{th} = \frac{\Delta L_c}{L_o + 0.012 \text{ m}} = \frac{0.01538 \text{ m}}{60 \text{ m} + 0.012 \text{ m}}$$
$$= 0.00256 \text{ m/m}$$

From Eq. 49.15, the compressive thermal stress is

$$\sigma_{th} = E\epsilon_{th} = (20 \times 10^{10} \text{ Pa})\left(0.000256 \ \frac{m}{m}\right)$$
$$= \boxed{5.12 \times 10^7 \text{ Pa (51.2 MPa)}}$$

The answer is (B).

4. *Customary U.S. Solution*

First, the cross-sectional area of the rod is

$$A = \frac{\pi d^2}{4}$$

From Eq. 49.4, the unstretched length of the rod is

$$L_o = \frac{\delta E A}{F} = \frac{(0.158 \text{ in})\left(2.9 \times 10^7 \ \frac{\text{lbf}}{\text{in}^2}\right)\left(\frac{\pi}{4}\right)(1 \text{ in})^2}{15{,}000 \text{ lbf}}$$
$$= 239.913 \text{ in}$$

The total length of the rod is given by Eq. 49.5.

$$L = L_o + \delta = 239.913 \text{ in} + 0.158 \text{ in} = \boxed{240.071 \text{ in}}$$

The answer is (D).

SI Solution

First, the cross-sectional area of the rod is

$$A = \frac{\pi d^2}{4}$$

From Eq. 49.4, the unstretched length of the rod is

$$L_o = \frac{\delta E A}{F}$$
$$= \frac{(0.004 \text{ m})\left(20 \times 10^{10} \ \frac{N}{m^2}\right)\left(\frac{\pi}{4}\right)(0.025 \text{ m})^2}{67 \times 10^3 \text{ N}}$$
$$= 5.861 \text{ m}$$

The total length of the rod is

$$L = L_o + \delta = 5.861 \text{ m} + 0.004 \text{ m} = \boxed{5.865 \text{ m}}$$

The answer is (D).

5. *Customary U.S. Solution*

First, find the properties of the shaft cross section. The area is

$$A = \frac{\pi d^2}{4} = \frac{\pi(3 \text{ in})^2}{4} = 7.07 \text{ in}^2$$

The moment of inertia is

$$I_c = \frac{\pi r^4}{4} = \frac{\pi \left(\frac{3 \text{ in}}{2}\right)^4}{4} = 3.98 \text{ in}^4$$

The polar moment of inertia is

$$J = \frac{\pi r^4}{2} = \frac{\pi \left(\frac{3 \text{ in}}{2}\right)^4}{2} = 7.95 \text{ in}^4$$

The moment at the bearing face due to the pulley weight is

$$M_y = (600 \text{ lbm})\left(\frac{32.2 \ \frac{\text{ft}}{\text{sec}^2}}{32.2 \ \frac{\text{lbm-ft}}{\text{lbf-sec}^2}}\right)(8 \text{ in}) = 4800 \text{ in-lbf}$$

From Eq. 49.37, the maximum bending stress at the extreme fiber on the y-axis is

$$\sigma_y = \frac{M_y c}{I_c} = \frac{(4800 \text{ in-lbf})\left(\frac{3 \text{ in}}{2}\right)}{3.98 \text{ in}^4} = 1809 \text{ lbf/in}^2$$

The moment at the bearing face due to the belt tensions is

$$M_z = (1500 \text{ lbf} + 350 \text{ lbf})(8 \text{ in}) = 14{,}800 \text{ in-lbf}$$

From Eq. 49.37, the maximum bending stress at the extreme fiber on the z-axis is

$$\sigma_z = \frac{M_z c}{I_c} = \frac{(14{,}800 \text{ in-lbf})\left(\frac{3 \text{ in}}{2}\right)}{3.98 \text{ in}^4}$$
$$= 5578 \text{ lbf/in}^2$$

The net torque from the belt tensions is

$$T = (1500 \text{ lbf} - 350 \text{ lbf})(16 \text{ in}) = 18{,}400 \text{ in-lbf}$$

The maximum torsional shear stress at the extreme fiber is

$$\tau = \frac{Tr}{J} = \frac{(18{,}400 \text{ in-lbf})\left(\frac{3 \text{ in}}{2}\right)}{7.95 \text{ in}^4}$$

$$= 3472 \text{ lbf/in}^2$$

The direct shear stress is zero at the surface of the shaft.

Use Eqs. 49.19 and 49.20 to find the maximum stress in the shaft.

$$\sigma_1 = \frac{\sigma_y + \sigma_z}{2} + \frac{1}{2}\sqrt{(\sigma_y - \sigma_z)^2 + (2\tau)^2}$$

$$= \frac{1809 \dfrac{\text{lbf}}{\text{in}^2} + 5578 \dfrac{\text{lbf}}{\text{in}^2}}{2}$$

$$+ \frac{1}{2}\sqrt{\begin{array}{c}\left(1809 \dfrac{\text{lbf}}{\text{in}^2} - 5578 \dfrac{\text{lbf}}{\text{in}^2}\right)^2 \\ + \left((2)\left(3472 \dfrac{\text{lbf}}{\text{in}^2}\right)\right)^2\end{array}}$$

$$= \boxed{7644 \text{ lbf/in}^2}$$

The answer is (D).

SI Solution

First, find the properties of the shaft cross section. The area is

$$A = \frac{\pi d^2}{4} = \frac{\pi (0.075 \text{ m})^2}{4} = 4.42 \times 10^{-3} \text{ m}^2$$

The moment of inertia is

$$I_c = \frac{\pi r^4}{4} = \frac{\pi\left(\dfrac{0.075 \text{ m}}{2}\right)^4}{4} = 1.55 \times 10^{-6} \text{ m}^4$$

The polar moment of inertia is

$$J = \frac{\pi r^4}{2} = \frac{\pi\left(\dfrac{0.075 \text{ m}}{2}\right)^4}{2} = 3.11 \times 10^{-6} \text{ m}^4$$

The moment at the bearing face due to the pulley weight is

$$M_y = (270 \text{ kg})\left(9.81 \dfrac{\text{m}}{\text{s}^2}\right)(0.2 \text{ m}) = 529.74 \text{ N·m}$$

From Eq. 49.37, the maximum bending stress at the extreme fiber on the y-axis is

$$\sigma_y = \frac{M_y c}{I_c} = \frac{(529.74 \text{ N·m})\left(\dfrac{0.075 \text{ m}}{2}\right)}{1.55 \times 10^{-6} \text{ m}^4}$$

$$= 1.282 \times 10^7 \text{ Pa}$$

The moment at the bearing face due to the belt tensions is

$$M_z = (6.7 \times 10^3 \text{ N} + 1.6 \times 10^3 \text{ N})(0.2 \text{ m}) = 1660 \text{ N·m}$$

From Eq. 49.37, the maximum bending stress at the extreme fiber on the z-axis is

$$\sigma_z = \frac{M_z c}{I_c}$$

$$= \frac{(1660 \text{ N·m})\left(\dfrac{0.075 \text{ m}}{2}\right)}{1.55 \times 10^{-6} \text{ m}^4}$$

$$= 4.016 \times 10^7 \text{ Pa}$$

The net torque from the belt tensions is

$$T = (6.7 \times 10^3 \text{ N} - 1.6 \times 10^3 \text{ N})(0.40 \text{ m}) = 2040 \text{ N·m}$$

The maximum torsional shear stress at the extreme fiber is

$$\tau = \frac{Tr}{J}$$

$$= \frac{(2040 \text{ N·m})\left(\dfrac{0.075 \text{ m}}{2}\right)}{3.11 \times 10^{-6} \text{ m}^4}$$

$$= 2.460 \times 10^7 \text{ Pa}$$

The direct shear stress is zero at the surface of the shaft.

Use Eqs. 49.19 and 49.20 to find the maximum stress in the shaft.

$$\sigma_1 = \frac{\sigma_y + \sigma_z}{2} + \frac{1}{2}\sqrt{(\sigma_y - \sigma_z)^2 + (2\tau)^2}$$

$$= \frac{1.282 \times 10^7 \text{ Pa} + 4.016 \times 10^7 \text{ Pa}}{2}$$

$$+ \frac{1}{2}\sqrt{\begin{array}{c}\left(1.282 \times 10^7 \text{ Pa} - 4.016 \times 10^7 \text{ Pa}\right)^2 \\ + \left((2)(2.460 \times 10^7 \text{ Pa})\right)^2\end{array}}$$

$$= \boxed{5.46 \times 10^7 \text{ Pa} \quad (55 \text{ MPa})}$$

The answer is (D).

Materials

6. *Customary U.S. Solution*

First, find the properties of the rod cross section. The area is

$$A = \frac{\pi d^2}{4} = \frac{\pi (1 \text{ in})^2}{4} = 0.7854 \text{ in}^2$$

The moment of inertia is

$$I_c = \frac{\pi r^4}{4} = \frac{\pi \left(\frac{1.0 \text{ in}}{2}\right)^4}{4} = 0.04909 \text{ in}^4$$

The polar moment of inertia is

$$J = \frac{\pi r^4}{2} = \frac{\pi \left(\frac{1 \text{ in}}{2}\right)^4}{2} = 0.09817 \text{ in}^4$$

The moment at the chuck is

$$M = (60 \text{ lbf})(8 \text{ in}) = 480 \text{ in-lbf}$$

From Eq. 49.37, the maximum bending stress at the extreme fiber of the rod is

$$\sigma = \frac{Mc}{I_c} = \frac{(480 \text{ in-lbf})\left(\frac{1.0 \text{ in}}{2}\right)}{0.04909 \text{ in}^4} = 4889 \text{ lbf/in}^2$$

The torque applied to the rod is

$$T = (60 \text{ lbf})(12 \text{ in}) = 720 \text{ in-lbf}$$

The maximum torsional shear stress at the extreme fiber of the rod is

$$\tau = \frac{Tr}{J} = \frac{(720 \text{ in-lbf})\left(\frac{1.0 \text{ in}}{2}\right)}{0.09817 \text{ in}^4} = 3667 \text{ lbf/in}^2$$

The direct shear stress is zero at the surface of the rod.

(a) Use Eq. 49.20 to find the maximum shear stress in the rod.

$$\tau_1 = \tfrac{1}{2}\sqrt{\sigma^2 + (2\tau)^2}$$

$$= \tfrac{1}{2}\sqrt{\left(4889 \; \frac{\text{lbf}}{\text{in}^2}\right)^2 + \left((2)\left(3667 \; \frac{\text{lbf}}{\text{in}^2}\right)\right)^2}$$

$$= \boxed{4407 \text{ lbf/in}^2}$$

The answer is (C).

(b) Use Eq. 49.19 to find the maximum normal stress in the rod.

$$\sigma_1 = \frac{\sigma}{2} + \tau_1$$

$$= \frac{4889 \; \frac{\text{lbf}}{\text{in}^2}}{2} + 4407 \; \frac{\text{lbf}}{\text{in}^2}$$

$$= \boxed{6852 \text{ lbf/in}^2}$$

The answer is (B).

SI Solution

First, find the properties of the rod cross section. The area is

$$A = \frac{\pi d^2}{4} = \frac{\pi (0.025 \text{ m})^2}{4} = 4.909 \times 10^{-4} \text{ m}^2$$

The moment of inertia is

$$I_c = \frac{\pi r^4}{4} = \frac{\pi \left(\frac{0.025 \text{ m}}{2}\right)^4}{4} = 1.917 \times 10^{-8} \text{ m}^4$$

The polar moment of inertia is

$$J = \frac{\pi r^4}{2} = \frac{\pi \left(\frac{0.025 \text{ m}}{2}\right)^4}{2} = 3.835 \times 10^{-8} \text{ m}^4$$

The moment at the chuck is

$$M = (270 \text{ N})(0.2 \text{ m}) = 54 \text{ N·m}$$

From Eq. 49.37, the maximum bending stress at the extreme fiber of the rod is

$$\sigma = \frac{Mc}{I_c} = \frac{(54 \text{ N·m})\left(\frac{0.025 \text{ m}}{2}\right)}{1.917 \times 10^{-8} \text{ m}^4}$$

$$= 3.52 \times 10^7 \text{ Pa} \quad (35.2 \text{ MPa})$$

The torque applied to the rod is

$$T = (270 \text{ N})(0.3 \text{ m}) = 81 \text{ N·m}$$

The maximum torsional shear stress at the extreme fiber of the rod is

$$\tau = \frac{Tr}{J} = \frac{(81 \text{ N·m})\left(\frac{0.025 \text{ m}}{2}\right)}{3.835 \times 10^{-8} \text{ m}^4}$$

$$= 2.64 \times 10^7 \text{ Pa} \quad (26.4 \text{ MPa})$$

The direct shear stress is zero at the surface of the rod.

(a) Use Eq. 49.20 to find the maximum shear stress in the rod.

$$\tau_1 = \tfrac{1}{2}\sqrt{\sigma^2 + (2\tau)^2}$$

$$= \tfrac{1}{2}\sqrt{(35.2 \text{ MPa})^2 + ((2)(26.4 \text{ MPa}))^2}$$

$$= \boxed{31.7 \text{ MPa}}$$

The answer is (C).

(b) Use Eq. 49.19 to find the maximum normal stress in the rod.

$$\sigma_1 = \frac{\sigma}{2} + \tau_1$$

$$= \frac{35.2 \text{ MPa}}{2} + 31.7 \text{ MPa}$$

$$= \boxed{49.3 \text{ MPa}}$$

The answer is (B).

7. *Customary U.S. Solution*

The normal stress in the bar is compressive (i.e., negative).

$$\sigma_x = \frac{F}{A} = \frac{-18,000 \text{ lbf}}{1.5 \text{ in}^2} = -12,000 \text{ lbf/in}^2$$

The shear stress in the bar is $\tau = 4000 \text{ lbf/in}^2$.

From Fig. 49.7, a plane inclined $+30°$ from the horizontal gives an angle θ of $60°$.

(a) From Eq. 49.17, the normal stress on the plane is

$$\sigma_\theta = \frac{\sigma_x}{2} + \left(\frac{\sigma_x}{2}\right)(\cos 2\theta) + \tau \sin 2\theta$$

$$= \frac{-12,000 \dfrac{\text{lbf}}{\text{in}^2}}{2} + \frac{\left(-12,000 \dfrac{\text{lbf}}{\text{in}^2}\right)(\cos 120°)}{2}$$

$$+ \left(4000 \dfrac{\text{lbf}}{\text{in}^2}\right)(\sin 120°)$$

$$= \boxed{464 \text{ lbf/in}^2}$$

The answer is (A).

(b) From Eq. 49.18, the shear stress on the plane is

$$\tau_\theta = -\frac{\sigma_x \sin 2\theta}{2} + \tau \cos 2\theta$$

$$= -\frac{\left(-12,000 \dfrac{\text{lbf}}{\text{in}^2}\right)(\sin 120°)}{2}$$

$$+ \left(4000 \dfrac{\text{lbf}}{\text{in}^2}\right)(\cos 120°)$$

$$= \boxed{3106 \text{ lbf/in}^2}$$

The answer is (C).

SI Solution

The normal stress in the bar is

$$\sigma_x = \frac{F}{A} = \frac{-80 \times 10^3 \text{ N}}{(9.7 \text{ cm}^2)\left(\dfrac{1 \text{ m}}{100 \text{ cm}}\right)^2}$$

$$= -8.247 \times 10^7 \text{ Pa} \quad (-82.5 \text{ MPa})$$

From Fig. 49.17, a plane inclined $+30°$ from the horizontal gives an angle θ of $60°$.

(a) From Eq. 49.17, the normal stress on the plane is

$$\sigma_\theta = \frac{\sigma_x}{2} + \left(\frac{\sigma_x}{2}\right)(\cos 2\theta) + \tau \sin 2\theta$$

$$= \frac{-82.5 \text{ MPa}}{2} + \frac{(-82.5 \text{ MPa})(\cos 120°)}{2}$$

$$+ (28 \text{ MPa})(\sin 120°)$$

$$= \boxed{3.62 \text{ MPa}}$$

The answer is (A).

(b) From Eq. 49.18, the shear stress on the plane is

$$\tau_\theta = -\frac{\sigma_x \sin 2\theta}{2} + \tau \cos 2\theta$$

$$= -\frac{(-82.4 \text{ MPa})(\sin 120°)}{2}$$

$$+ (28 \text{ MPa})(\cos 120°)$$

$$= \boxed{21.7 \text{ MPa}}$$

The answer is (C).

8. *Customary U.S. Solution*

First, find the properties of the handle cross section. The moment of inertia is

$$I_c = \frac{\pi r^4}{4} = \frac{\pi \left(\dfrac{0.625 \text{ in}}{2}\right)^4}{4} = 0.00749 \text{ in}^4$$

The polar moment of inertia is

$$J = \frac{\pi r^4}{2} = \frac{\pi \left(\dfrac{0.625 \text{ in}}{2}\right)^4}{2} = 0.01498 \text{ in}^4$$

The moment at section A-A is

$$M = (50 \text{ lbf})(14 \text{ in} + 3 \text{ in}) = 850 \text{ in-lbf}$$

From Eq. 49.37, the maximum bending stress at the extreme fiber at section A-A is

$$\sigma = \frac{Mc}{I_c} = \frac{(850 \text{ in-lbf})\left(\dfrac{0.625 \text{ in}}{2}\right)}{0.00749 \text{ in}^4} = 35,464 \text{ lbf/in}^2$$

The torque at section A-A is

$$T = (50 \text{ lbf})(3 \text{ in}) = 150 \text{ in-lbf}$$

The maximum torsional shear stress at the extreme fiber at section A-A is

$$\tau = \frac{Tr}{J} = \frac{(150 \text{ in-lbf})\left(\dfrac{0.625 \text{ in}}{2}\right)}{0.01498 \text{ in}^4} = 3129 \text{ lbf/in}^2$$

Materials

The direct shear at the extreme fiber is assumed to be small enough to be neglected.

(a) Use Eq. 49.20 to find the maximum shear stress at section A-A.

$$\tau_1 = \tfrac{1}{2}\sqrt{\sigma^2 + (2\tau)^2}$$

$$= \tfrac{1}{2}\sqrt{\left(35{,}464 \ \frac{\text{lbf}}{\text{in}^2}\right)^2 + \left((2)\left(3129 \ \frac{\text{lbf}}{\text{in}^2}\right)\right)^2}$$

$$= 18{,}006 \ \text{lbf/in}^2$$

Use Eq. 49.19 to find the principal stresses.

$$\sigma_1 = \frac{\sigma}{2} + \tau_1$$

$$= \frac{35{,}464 \ \dfrac{\text{lbf}}{\text{in}^2}}{2} + 18{,}006 \ \frac{\text{lbf}}{\text{in}^2}$$

$$= \boxed{35{,}738 \ \text{lbf/in}^2}$$

$$\sigma_2 = \frac{\sigma}{2} - \tau_1$$

$$= \frac{35{,}464 \ \dfrac{\text{lbf}}{\text{in}^2}}{2} - 18{,}006 \ \frac{\text{lbf}}{\text{in}^2}$$

$$= \boxed{-274 \ \text{lbf/in}^2}$$

(b) From part (a), the maximum shear at section A-A is $\tau_1 = \boxed{18{,}006 \ \text{lbf/in}^2}.$

SI Solution

First, find the properties of the handle cross section. The moment of inertia is

$$I_c = \frac{\pi r^4}{4} = \frac{\pi \left(\dfrac{0.016 \ \text{m}}{2}\right)^4}{4} = 3.22 \times 10^{-9} \ \text{m}^4$$

The polar moment of inertia is

$$J = \frac{\pi r^4}{2} = \frac{\pi \left(\dfrac{0.016 \ \text{m}}{2}\right)^4}{2} = 6.43 \times 10^{-9} \ \text{m}^4$$

The moment at section A-A is

$$M = (220 \ \text{N})(0.355 \ \text{m} + 0.075 \ \text{m}) = 94.6 \ \text{N·m}$$

From Eq. 49.37, the maximum bending stress at the extreme fiber at section A-A is

$$\sigma = \frac{Mc}{I_c} = \frac{(94.6 \ \text{N·m})\left(\dfrac{0.016 \ \text{m}}{2}\right)}{3.22 \times 10^{-9} \ \text{m}^4}$$

$$= 2.35 \times 10^8 \ \text{Pa} \quad (235 \ \text{MPa})$$

The torque at section A-A is

$$T = (220 \ \text{N})(0.075 \ \text{m}) = 16.5 \ \text{N·m}$$

The maximum torsional shear stress at the extreme fiber at section A-A is

$$\tau = \frac{Tr}{J} = \frac{(16.5 \ \text{N·m})\left(\dfrac{0.016 \ \text{m}}{2}\right)}{6.43 \times 10^{-9} \ \text{m}^4}$$

$$= 2.05 \times 10^7 \ \text{Pa} \quad (20.5 \ \text{MPa})$$

The direct shear at the extreme fiber is assumed to be small enough to be neglected.

(a) Use Eq. 49.20 to find the maximum shear stress at section A-A.

$$\tau_1 = \tfrac{1}{2}\sqrt{\sigma^2 + (2\tau)^2}$$

$$= \tfrac{1}{2}\sqrt{(235 \ \text{MPa})^2 + \left((2)\left((20.5 \ \text{MPa})^2\right)\right)}$$

$$= 119.3 \ \text{MPa}$$

Use Eq. 49.19 to find the principal stresses.

$$\sigma_1 = \frac{\sigma}{2} + \tau_1$$

$$= \frac{235 \ \text{MPa}}{2} + 119.3 \ \text{MPa}$$

$$= \boxed{237 \ \text{MPa}}$$

$$\sigma_2 = \frac{\sigma}{2} - \tau_1$$

$$= \frac{235 \ \text{MPa}}{2} - 119.3 \ \text{MPa}$$

$$= \boxed{-1.8 \ \text{MPa}}$$

(b) From part (a), the maximum shear at section A-A is $\tau_1 = \boxed{119.3 \ \text{MPa}.}$

9. *Customary U.S. Solution*

(a) The moment of inertia of the brass tube annular cross section is

$$I_{\text{brass}} = \left(\frac{\pi}{4}\right)(r_o^4 - r_i^4)$$

$$= \left(\frac{\pi}{4}\right)\left(\left(\frac{2.0 \ \text{in}}{2}\right)^4 - \left(\frac{1.0 \ \text{in}}{2}\right)^4\right)$$

$$= 0.736 \ \text{in}^4$$

The moment of inertia of the steel rod insert circular cross section is

$$I_{\text{steel}} = \frac{\pi r^4}{4} = \left(\frac{\pi}{4}\right)\left(\frac{1.0 \ \text{in}}{2}\right)^4$$

$$= 0.0491 \ \text{in}^4$$

The product EI for the brass tube is

$$E_\text{brass} I_\text{brass} = \left(1.5 \times 10^7 \, \frac{\text{lbf}}{\text{in}^2}\right) (0.736 \text{ in}^4)$$
$$= 1.104 \times 10^7 \text{ lbf-in}^2$$

The product EI for the steel rod insert is

$$E_\text{steel} I_\text{steel} = \left(2.9 \times 10^7 \, \frac{\text{lbf}}{\text{in}^2}\right) (0.0491 \text{ in}^4)$$
$$= 0.142 \times 10^7 \text{ lbf-in}^2$$

The total EI for the composite is

$$E_c I_c = E_\text{brass} I_\text{brass} + E_\text{steel} I_\text{steel}$$
$$= 1.104 \times 10^7 \text{ lbf-in}^2 + 0.142 \times 10^7 \text{ lbf-in}^2$$
$$= 1.246 \times 10^7 \text{ lbf-in}^2$$

From case 1 in App. 49.A, the tip deflection of the tube is

$$y_\text{tip} = \frac{PL^3}{3EI}$$

Since the deflection is inversely proportional to EI, the suggestion of inserting a steel rod does have merit because the insert increases the EI product.

(b) The percent change in tip deflection is

$$\text{percent} = (100\%) \left(\frac{y_\text{tip}_\text{brass} - y_\text{tip}_\text{brass + steel}}{y_\text{tip}_\text{brass}}\right)$$
$$= (100\%) \left(\frac{\dfrac{PL^3}{3E_\text{brass} I_\text{brass}} - \dfrac{PL^3}{3E_c I_c}}{\dfrac{PL^3}{3E_\text{brass} I_\text{brass}}}\right)$$

Simplify.

$$\text{percent} = (100\%) \left(\frac{\dfrac{1}{E_\text{brass} I_\text{brass}} - \dfrac{1}{E_c I_c}}{\dfrac{1}{E_\text{brass} I_\text{brass}}}\right)$$
$$= (100\%) \left(\frac{\dfrac{1}{1.104 \times 10^7 \text{ lbf-in}^2} - \dfrac{1}{1.246 \times 10^7 \text{ lbf-in}^2}}{\dfrac{1}{1.104 \times 10^7 \text{ lbf-in}^2}}\right)$$
$$= \boxed{11.4\%}$$

SI Solution

(a) The moment of inertia of the brass tube annular cross section is

$$I_\text{brass} = \left(\frac{\pi}{4}\right) (r_o^4 - r_i^4)$$
$$= \left(\frac{\pi}{4}\right) \left(\left(\frac{0.050 \text{ m}}{2}\right)^4 - \left(\frac{0.025 \text{ m}}{2}\right)^4\right)$$
$$= 2.876 \times 10^{-7} \text{ m}^4$$

The moment of inertia of the steel rod insert circular cross section is

$$I_\text{steel} = \frac{\pi r^4}{4} = \left(\frac{\pi}{4}\right) \left(\frac{0.025 \text{ m}}{2}\right)^4$$
$$= 1.917 \times 10^{-8} \text{ m}^4$$

The product EI for the brass tube is

$$E_\text{brass} I_\text{brass} = (100 \times 10^9 \text{ Pa}) (2.876 \times 10^{-7} \text{ m}^4)$$
$$= 28\,760 \text{ N·m}^2$$

The product EI for the steel rod insert is

$$E_\text{steel} I_\text{steel} = (200 \times 10^9 \text{ Pa}) (1.917 \times 10^{-8} \text{ m}^4)$$
$$= 3834 \text{ N·m}^2$$

The total EI for the composite is

$$E_c I_c = E_\text{brass} I_\text{brass} + E_\text{steel} I_\text{steel}$$
$$= 28\,760 \text{ N·m}^2 + 3834 \text{ N·m}^2$$
$$= 32\,594 \text{ N·m}^2$$

From case 1 in App. 49.A, the tip deflection of the tube is

$$y_\text{tip} = \frac{PL^3}{3EI}$$

Since the deflection is inversely proportional to EI, the suggestion of inserting a steel rod does have merit because the insert increases the EI product.

(b) The percent change in tip deflection is

$$\text{percent} = (100\%) \left(\frac{y_\text{tip}_\text{brass} - y_\text{tip}_\text{brass + steel}}{y_\text{tip}_\text{brass}}\right)$$
$$= (100\%) \left(\frac{\dfrac{PL^3}{3E_\text{brass} I_\text{brass}} - \dfrac{PL^3}{3E_c I_c}}{\dfrac{PL^3}{3E_\text{brass} I_\text{brass}}}\right)$$

Simplify.

$$\text{percent} = (100\%) \left(\frac{\dfrac{1}{E_\text{brass} I_\text{brass}} - \dfrac{1}{E_c I_c}}{\dfrac{1}{E_\text{brass} I_\text{brass}}}\right)$$
$$= (100\%) \left(\frac{\dfrac{1}{28\,760 \text{ N·m}^2} - \dfrac{1}{32\,594 \text{ N·m}^2}}{\dfrac{1}{28\,760 \text{ N·m}^2}}\right)$$
$$= \boxed{11.8\%}$$

Materials

10. *Customary U.S. Solution*

The centroidal moment of inertia of the shaft is

$$I = \left(\frac{\pi}{4}\right)r^4 = \left(\frac{\pi}{4}\right)\left(\frac{2 \text{ in}}{2}\right)^4 = 0.7854 \text{ in}^4$$

Although the stress is probably greatest at the fillet toe, it is difficult to specify exactly where the greatest stress occurs. By common convention, it is assumed to occur at the shoulder.

The moment at the shoulder is

$$M = Fd$$

$$= m\left(\frac{g}{g_c}\right)d = (2500 \text{ lbm})\left(\frac{32.2 \frac{\text{ft}}{\text{sec}^2}}{32.2 \frac{\text{ft-lbm}}{\text{lbf-sec}^2}}\right)(4 \text{ in})$$

$$= 10{,}000 \text{ in-lbf}$$

Use App. 49.B, Fig. (c).

$$\frac{r}{d} = \frac{\frac{5}{16} \text{ in}}{2 \text{ in}} = 0.156$$

$$\frac{D}{d} = \frac{3 \text{ in}}{2 \text{ in}} = 1.5$$

From App. 49.B, the stress concentration factor is approximately 1.5.

The bending stress is

$$\sigma = K\left(\frac{Mc}{I}\right) = \frac{(1.5)(10{,}000 \text{ in-lbf})\left(\frac{2 \text{ in}}{2}\right)}{0.7854 \text{ in}^4}$$

$$= \boxed{19{,}099 \text{ lbf/in}^2}$$

(Note that shear stress is zero where bending stress is maximum.)

The answer is (D).

SI Solution

The centroidal moment of inertia of the shaft is

$$I = \left(\frac{\pi}{4}\right)r^4 = \left(\frac{\pi}{4}\right)\left(\frac{50 \text{ mm}}{(2)\left(1000 \frac{\text{mm}}{\text{m}}\right)}\right)^4$$

$$= 3.068 \times 10^{-7} \text{ m}^4$$

Although the stress is probably greatest at the fillet toe, it is difficult to specify exactly where the greatest stress occurs. By common convention, it is assumed to occur at the shoulder.

The moment at the shoulder is

$$M = Fd = mgd = (1100 \text{ kg})\left(9.81 \frac{\text{m}}{\text{s}^2}\right)\left(\frac{100 \text{ mm}}{1000 \frac{\text{mm}}{\text{m}}}\right)$$

$$= 1079 \text{ N·m}$$

Use App. 49.B, Fig. (c).

$$\frac{r}{d} = \frac{8 \text{ mm}}{50 \text{ mm}} = 0.16$$

$$\frac{D}{d} = \frac{75 \text{ mm}}{50 \text{ mm}} = 1.5$$

From App. 49.B, the stress concentration factor is approximately 1.5.

The bending stress is

$$\sigma = K\left(\frac{Mc}{I}\right)$$

$$= \frac{(1.5)(1079 \text{ N·m})\left(\frac{50 \text{ mm}}{(2)\left(1000 \frac{\text{mm}}{\text{m}}\right)}\right)}{3.068 \times 10^{-7} \text{ m}^4}$$

$$= \boxed{1.319 \times 10^8 \text{ Pa} \quad (132 \text{ MPa})}$$

The answer is (D).

11. *Customary U.S. Solution*

First, determine an equivalent aluminum area for the steel. The ratio of equivalent aluminum area to steel area is the modular ratio.

$$n = \frac{30 \times 10^6 \frac{\text{lbf}}{\text{in}^2}}{10 \times 10^6 \frac{\text{lbf}}{\text{in}^2}} = 3$$

The equivalent aluminum width to replace the steel is

$$(1.5 \text{ in})(3) = 4.5 \text{ in}$$

The equivalent all-aluminum cross section is

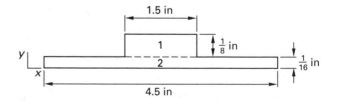

To find the centroid of the section, first calculate the areas of the basic shapes.

$$A_1 = (1.5 \text{ in})\left(\tfrac{1}{8} \text{ in}\right) = 0.1875 \text{ in}^2$$

$$A_2 = (4.5 \text{ in})\left(\tfrac{1}{16} \text{ in}\right) = 0.28125 \text{ in}^2$$

Next, find the y-components of the centroids of the basic shapes.

$$y_{c1} = \tfrac{1}{16} \text{ in} + \left(\tfrac{1}{2}\right)\left(\tfrac{1}{8} \text{ in}\right) = 0.125 \text{ in}$$

$$y_{c2} = \frac{\tfrac{1}{16} \text{ in}}{2} = 0.03125 \text{ in}$$

The centroid of the section is

$$
\begin{aligned}
y_c &= \frac{\displaystyle\sum_i A_i y_{ci}}{\displaystyle\sum_i A_i} \\
&= \frac{(0.1875 \text{ in}^2)(0.125 \text{ in}) + (0.28125 \text{ in}^2)(0.03125 \text{ in})}{0.1875 \text{ in}^2 + 0.28125 \text{ in}^2} \\
&= 0.06875 \text{ in}
\end{aligned}
$$

The moment of inertia of basic shape 1 about its own centroid is

$$I_{cy1} = \frac{bh^3}{12} = \frac{(1.5 \text{ in})(0.125 \text{ in})^3}{12} = 2.441 \times 10^{-4} \text{ in}^4$$

The moment of inertia of basic shape 2 about its own centroid is

$$I_{cy2} = \frac{bh^3}{12} = \frac{(4.5 \text{ in})(0.0625 \text{ in})^3}{12} = 9.155 \times 10^{-5} \text{ in}^4$$

The distance from the centroid of basic shape 1 to the section centroid is

$$d_1 = y_{c1} - y_c = 0.125 \text{ in} - 0.06875 \text{ in} = 0.05625 \text{ in}$$

The distance from the centroid of basic shape 2 to the section centroid is

$$d_2 = y_c - y_{c2} = 0.06875 \text{ in} - 0.03125 \text{ in} = 0.0375 \text{ in}$$

From the parallel axis theorem (Eq. 48.20), the moment of inertia of basic shape 1 about the centroid of the section is

$$
\begin{aligned}
I_{y1} &= I_{yc1} + A_1 d_1^2 \\
&= 2.411 \times 10^{-4} \text{ in}^4 + (0.1875 \text{ in}^2)(0.05625 \text{ in})^2 \\
&= 8.344 \times 10^{-4} \text{ in}^4
\end{aligned}
$$

The moment of inertia of basic shape 2 about the centroid of the section is

$$
\begin{aligned}
I_{y2} &= I_{yc2} + A_2 d_2^2 \\
&= 9.155 \times 10^{-5} \text{ in}^2 + (0.28125 \text{ in}^2)(0.0375 \text{ in})^2 \\
&= 4.871 \times 10^{-4} \text{ in}^2
\end{aligned}
$$

The total moment of inertia of the section about the centroid of the section is

$$I_y = I_{y1} + I_{y2} = 8.344 \times 10^{-4} \text{ in}^4 + 4.871 \times 10^{-4} \text{ in}^4$$

$$\boxed{= 1.322 \times 10^{-3} \text{ in}^4}$$

The answer is (C).

SI Solution

First, determine an equivalent aluminum area for the steel. The ratio of equivalent aluminum area to steel area is the modular ratio.

$$n = \frac{20 \times 10^4 \text{ MPa}}{70 \times 10^3 \text{ MPa}} = 2.86$$

The equivalent aluminum width to replace the steel is

$$(38 \text{ mm})(2.86) = 108.7 \text{ mm}$$

The equivalent all-aluminum cross section is

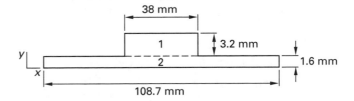

To find the centroid of the section, first calculate the areas of the basic shapes.

$$A_1 = (38 \text{ mm})(3.2 \text{ mm}) = 121.6 \text{ mm}^2$$
$$A_2 = (108.7 \text{ mm})(1.6 \text{ mm}) = 173.9 \text{ mm}^2$$

Next, find the y-components of the centroids of the basic shapes.

$$y_{c1} = 1.6 \text{ mm} + \left(\tfrac{1}{2}\right)(3.2 \text{ mm}) = 3.2 \text{ mm}$$
$$y_{c2} = \frac{1.6 \text{ mm}}{2} = 0.8 \text{ mm}$$

The centroid of the section is

$$
\begin{aligned}
y_c &= \frac{\displaystyle\sum_i A_i y_{ci}}{\displaystyle\sum_i A_i} \\
&= \frac{(121.6 \text{ mm}^2)(3.2 \text{ mm}) + (173.9 \text{ mm}^2)(0.8 \text{ mm})}{121.6 \text{ mm}^2 + 173.9 \text{ mm}^2} \\
&= 1.79 \text{ mm}
\end{aligned}
$$

The moment of inertia of basic shape 1 about its own centroid is

$$I_{cy1} = \frac{bh^3}{12} = \frac{(38 \text{ mm})(3.2 \text{ mm})^3}{12} = 103.77 \text{ mm}^4$$

The moment of inertia of basic shape 2 about its own centroid is

$$I_{cy2} = \frac{bh^3}{12} = \frac{(108.7 \text{ mm})(1.6 \text{ mm})^3}{12} = 37.10 \text{ mm}^4$$

The distance from the centroid of basic shape 1 to the section centroid is

$$d_1 = y_{c1} - y_c = 3.2 \text{ mm} - 1.79 \text{ mm} = 1.41 \text{ mm}$$

The distance from the centroid of basic shape 2 to the section centroid is

$$d_2 = y_c - y_{c2} = 1.79 \text{ mm} - 0.8 \text{ mm} = 0.99 \text{ mm}$$

From the parallel axis theorem, the moment of inertia of basic shape 1 about the centroid of the section is

$$
\begin{aligned}
I_{y1} &= I_{yc1} + A_1 d_1^2 \\
&= 103.77 \text{ mm}^4 + (121.6 \text{ mm}^2)(1.41 \text{ mm})^2 \\
&= 345.52 \text{ mm}^4
\end{aligned}
$$

The moment of inertia of basic shape 2 about the centroid of the section is

$$
\begin{aligned}
I_{y2} &= I_{yc2} + A_2 d_2^2 \\
&= 37.10 \text{ mm}^4 + (173.9 \text{ mm}^2)(0.99 \text{ mm})^2 \\
&= 207.54 \text{ mm}^4
\end{aligned}
$$

The total moment of inertia of the section about the centroid of the section is

$$
\begin{aligned}
I_y &= I_{y1} + I_{y2} = 345.52 \text{ mm}^4 + 207.54 \text{ mm}^4 \\
&= \boxed{553.06 \text{ mm}^4}
\end{aligned}
$$

The answer is (C).

50 Failure Theories

PRACTICE PROBLEMS

1. A shaft with a 1.125 in (28.6 mm) diameter receives 400 in-lbf (45 N·m) of torque through a pinned sleeve. The pin is manufactured from steel with a tensile yield strength of 73.9 ksi (510 MPa). Using a factor of safety of 2.5, what pin diameter is needed?

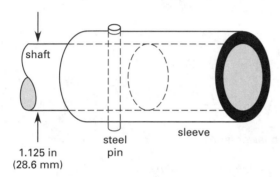

  (A) 0.16 in (0.0041 m)
  (B) 0.19 in (0.0048 m)
  (C) 0.26 in (0.0066 m)
  (D) 0.37 in (0.0094 m)

2. A $^5/_8$ in UNC bolt (with rolled threads) and nut carry a simple tensile load that fluctuates between 1000 lbf and 8000 lbf (4400 N and 35,000 N). Both the bolt and nut are made from medium carbon steel with a yield strength of 57,000 lbf/in^2 (390 MPa) and an endurance limit of 30,000 lbf/in^2 (205 MPa). Using Soderberg theory, and including a stress concentration factor for the threads, what is the factor of safety for the bolt?

  (A) 0.5
  (B) 0.7
  (C) 1.2
  (D) 2.4

3. (*Time limit: one hour*) The spool in the spool valve shown is constructed of aluminum with a yield strength of 19,000 lbf/in^2 (130 MPa). A 500 psi (3.5 MPa) pressure differential exists across the spool shaft. The shaft has a 1.0 in (25 mm) outside diameter and a 0.050 in (1.3 mm) wall thickness. The end disks have a 2.0 in (50 mm) outside diameter and a 0.375 in (9.5 mm) thickness. The end disks are rigidly attached to the tube. Disregard spool collapse. Use the distortion energy theory to calculate the factor of safety for the spool.

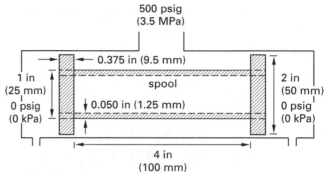

  (A) 0.8
  (B) 1.1
  (C) 1.6
  (D) 2.3

4. (*Time limit: one hour*) The pressure in a small pressure vessel varies continually between the extremes of 50 psig and 350 psig (340 kPa and 2400 kPa). The vessel is closed by a $^1/_2$ in (12 mm) plate. The plate is attached to the vessel flange with six $^3/_8$-24 UNF bolts evenly spaced around a $9^1/_2$ in (240 mm) circle. Each bolt is tightened to an initial preload of 3700 lbf (16.4 kN). The bolts and nuts are constructed of cold-rolled steel with a 90 ksi (620 MPa) yield strength and 110 ksi (760 MPa) ultimate strength. The plate, flange, and vessel are constructed of steel with a 30 ksi (205 MPa) yield strength and a 50 ksi (345 MPa) ultimate strength. The vessel is intended to be used indefinitely. Neglect the effects of the gasket (not shown). Neglect bending of the plate. Using a thread stress concentration factor of 2 and appropriate endurance strength derating factors, what is the factor of safety?

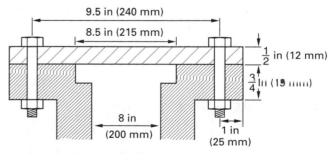

(sealing method not shown)

  (A) 1.2
  (B) 1.8
  (C) 2.1
  (D) 2.5

SOLUTIONS

1. *Customary U.S. Solution*

Use the distortion energy failure theory. From Eq. 50.27, the yield strength in shear is

$$S_{ys} = 0.577 S_{yt} = (0.577) \left(73.9 \times 10^3 \, \frac{\text{lbf}}{\text{in}^2} \right)$$
$$= 42.64 \times 10^3 \, \text{lbf/in}^2$$

For a safety factor of 2.5, the maximum allowable shear stress is

$$\tau_{\text{max}} = \frac{S_{ys}}{\text{FS}} = \frac{42.64 \times 10^3 \, \frac{\text{lbf}}{\text{in}^2}}{2.5}$$
$$= 17.06 \times 10^3 \, \text{lbf/in}^2$$

The total shear at the pin is

$$V = \frac{\text{shaft torque}}{\text{shaft radius}} = \frac{400 \, \text{in-lbf}}{\dfrac{1.125 \, \text{in}}{2}} = 711.1 \, \text{lbf}$$

This is not a case of biaxial stress. There is no bending. The direct shear stress in the pin is

$$\tau_{\text{ave}} = \frac{V}{A}$$

Solve for the required total pin area.

$$A = \frac{V}{\tau_{\text{ave}}} = \frac{711.1 \, \text{lbf}}{17.06 \times 10^3 \, \dfrac{\text{lbf}}{\text{in}^2}} = 0.04168 \, \text{in}^2$$

Since two surfaces of the pin resist the shear,

$$A = (2) \left(\frac{\pi d^2}{4} \right) = \frac{\pi d^2}{2}$$

Solve for the pin diameter.

$$d = \sqrt{\frac{2A}{\pi}} = \sqrt{\frac{(2)(0.04168 \, \text{in}^2)}{\pi}} = \boxed{0.163 \, \text{in}}$$

The answer is (A).

SI Solution

Use the distortion energy failure theory. From Eq. 50.27, the yield strength in shear is

$$S_{ys} = 0.577 S_{yt} = (0.577) \left(510 \times 10^6 \, \text{Pa} \right)$$
$$= 294.3 \times 10^6 \, \text{Pa}$$

For a safety factor of 2.5, the maximum allowable shear stress is

$$\tau_{\text{max}} = \frac{S_{ys}}{\text{FS}} = \frac{294.3 \times 10^6 \, \text{Pa}}{2.5}$$
$$= 117.7 \times 10^6 \, \text{Pa}$$

The total shear at the pin is

$$V = \frac{\text{shaft torque}}{\text{shaft radius}} = \frac{45 \, \text{N·m}}{\dfrac{0.0286 \, \text{m}}{2}} = 3146.9 \, \text{N}$$

This is not a case of biaxial stress. There is no bending. The direct shear stress in the pin is

$$\tau_{\text{ave}} = \frac{V}{A}$$

Solve for the required total pin area.

$$A = \frac{V}{\tau_{\text{ave}}} = \frac{3146.9 \, \text{N}}{117.7 \times 10^6 \, \text{Pa}}$$
$$= 2.674 \times 10^{-5} \, \text{m}^2$$

Since two surfaces of the pin resist the shear, the total pin area is

$$A = (2) \left(\frac{\pi d^2}{4} \right) = \frac{\pi d^2}{2}$$

Solve for the pin diameter.

$$d = \sqrt{\frac{2A}{\pi}} = \sqrt{\frac{(2)(2.674 \times 10^{-5} \, \text{m}^2)}{\pi}} = \boxed{0.00413 \, \text{m}}$$

The answer is (A).

2. *Customary U.S. Solution*

From Table 51.5, the tensile stress area for a $5/8$ in UNC bolt is $A = 0.226 \, \text{in}^2$.

The maximum stress is

$$\sigma_{\text{max}} = \frac{F_{\text{max}}}{A} = \frac{8000 \, \text{lbf}}{0.226 \, \text{in}^2} = 35{,}398 \, \text{lbf/in}^2$$

The minimum stress is

$$\sigma_{\text{min}} = \frac{F_{\text{min}}}{A} = \frac{1000 \, \text{lbf}}{0.226 \, \text{in}^2} = 4425 \, \text{lbf/in}^2$$

From Eq. 50.28, the mean stress is

$$\sigma_m = \frac{\sigma_{\text{max}} + \sigma_{\text{min}}}{2} = \frac{35{,}398 \, \dfrac{\text{lbf}}{\text{in}^2} + 4425 \, \dfrac{\text{lbf}}{\text{in}^2}}{2}$$
$$= 19{,}912 \, \text{lbf/in}^2$$

From Eq. 50.30, the alternating stress is

$$\sigma_{\text{alt}} = \tfrac{1}{2} (\sigma_{\text{max}} - \sigma_{\text{min}})$$
$$= \tfrac{1}{2} \left(35{,}398 \, \frac{\text{lbf}}{\text{in}^2} - 4425 \, \frac{\text{lbf}}{\text{in}^2} \right)$$
$$= 15{,}487 \, \text{lbf/in}^2$$

From Sec. 51.11, the stress concentration factor for rolled threads is 2.2. Thus, the alternating stress is

$$\sigma_{\text{alt}} = \left(15{,}487 \, \frac{\text{lbf}}{\text{in}^2} \right) (2.2) = 34{,}071 \, \text{lbf/in}^2$$

Draw the Soderberg line and locate the (σ_m, σ_{alt}) point.

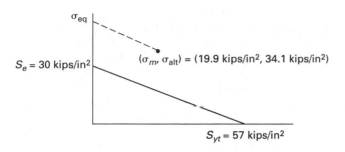

The Soderberg equivalent stress is

$$\sigma_{eq} = \sigma_{alt} + \left(\frac{S_e}{S_{yt}}\right)\sigma_m$$

$$= 34{,}071 \ \frac{\text{lbf}}{\text{in}^2} + \left(\frac{30{,}000 \ \frac{\text{lbf}}{\text{in}^2}}{57{,}000 \ \frac{\text{lbf}}{\text{in}^2}}\right)\left(19{,}912 \ \frac{\text{lbf}}{\text{in}^2}\right)$$

$$= 44{,}551 \ \text{lbf/in}^2$$

The factor of safety is

$$\text{FS} = \frac{S_e}{\sigma_{eq}} = \frac{30{,}000 \ \frac{\text{lbf}}{\text{in}^2}}{44{,}551 \ \frac{\text{lbf}}{\text{in}^2}} = \boxed{0.673}$$

The answer is (B).

SI Solution

From Table 51.5, the tensile stress area for a ⅝ in UNC bolt is

$$A = (0.226 \ \text{in}^2)\left(\frac{1 \ \text{m}}{39.36 \ \text{in}}\right)^2 = 1.459 \times 10^{-4} \ \text{m}^2$$

The maximum stress is

$$\sigma_{max} = \frac{F_{max}}{A} = \frac{35\,000 \ \text{N}}{1.459 \times 10^{-4} \ \text{m}^2} = 239.89 \times 10^6 \ \text{Pa}$$

The minimum stress is

$$\sigma_{min} = \frac{F_{min}}{A} = \frac{4400 \ \text{N}}{1.459 \times 10^{-4} \ \text{m}^2} = 30.16 \times 10^6 \ \text{Pa}$$

From Eq. 50.28, the mean stress is

$$\sigma_m = \frac{\sigma_{max} + \sigma_{min}}{2}$$

$$= \frac{239.89 \times 10^6 \ \text{Pa} + 30.16 \times 10^6 \ \text{Pa}}{2}$$

$$= 135.03 \times 10^6 \ \text{Pa} \quad (135.03 \ \text{MPa})$$

From Eq. 50.30, the alternating stress is

$$\sigma_{alt} = \left(\tfrac{1}{2}\right)(\sigma_{max} - \sigma_{min})$$

$$= \left(\tfrac{1}{2}\right)(239.89 \times 10^6 \ \text{Pa} - 30.16 \times 10^6 \ \text{Pa})$$

$$= 104.87 \times 10^6 \ \text{Pa} \quad (104.87 \ \text{MPa})$$

From Sec. 51.11, the stress concentration factor for rolled threads is 2.2. Thus the alternating stress is

$$\sigma_{alt} = (104.87 \ \text{MPa})(2.2) = 230.71 \ \text{MPa}$$

Draw the Soderberg line and locate the (σ_m, σ_{alt}) point.

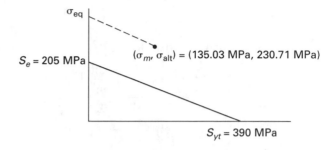

The Soderberg equivalent stress is

$$\sigma_{eq} = \sigma_{alt} + \left(\frac{S_e}{S_{yt}}\right)\sigma_m$$

$$= 230.71 \ \text{MPa} + \left(\frac{205 \ \text{MPa}}{390 \ \text{MPa}}\right)(135.03 \ \text{MPa})$$

$$= 301.7 \ \text{MPa}$$

The factor of safety is

$$\text{FS} = \frac{S_e}{\sigma_{eq}} = \frac{205 \ \text{MPa}}{301.7 \ \text{MPa}} = \boxed{0.679}$$

The answer is (B).

3. *Customary U.S. Solution*

The spool appears to be a thin-walled tube. Only the circumferential stress and the longitudinal stress are significant.

The end disc area exposed to the 500 psi pressure is

$$A_e = \left(\frac{\pi}{4}\right)((2.0 \ \text{in})^2 - (1.0 \ \text{in})^2) = 2.356 \ \text{in}^2$$

The longitudinal force produced by the 500 psig pressure is

$$F = pA_e = \left(500 \ \frac{\text{lbf}}{\text{in}^2}\right)(2.356 \ \text{in}^2) = 1178 \ \text{lbf}$$

The annular area of the spool tube is

$$A = \left(\frac{\pi}{4}\right)(d_o^2 - d_i^2)$$

$$= \left(\frac{\pi}{4}\right)\left((1.0 \ \text{in})^2 - (1 \ \text{in} - (2)(0.05 \ \text{in}))^2\right)$$

$$= 0.149 \ \text{in}^2$$

The longitudinal stress in the spool tube is

$$\sigma_{\text{long}} = \frac{F}{A} = \frac{1178 \text{ lbf}}{0.149 \text{ in}^2} = 7906 \text{ lbf/in}^2$$

Since $t/d = 0.05 \text{ in}/1 \text{ in} = 0.05 < 0.10$, this qualifies as a thin-walled cylinder. However, since the tube is exposed to external pressure (not internal), use Lamé's equation for a thick-walled cylinder. This is in the range for Lamé's solution.

The compressive circumferential stress in the tube is given in Table 51.2.

$$r_o = \frac{1 \text{ in}}{2} = 0.5 \text{ in}$$
$$r_i = 0.5 \text{ in} - 0.050 \text{ in} = 0.45 \text{ in}$$

$$\sigma_{\text{ci}} = \frac{-2r_o^2 p}{r_o^2 - r_i^2} = \frac{(-2)(0.5 \text{ in})^2 \left(500 \, \frac{\text{lbf}}{\text{in}^2}\right)}{(0.5 \text{ in})^2 - (0.45 \text{ in})^2}$$
$$= -5263 \text{ lbf/in}^2 \quad [\text{negative because compression}]$$

The principal stresses are

$$\sigma_1 = \sigma_{\text{long}} = 7906 \text{ lbf/in}^2$$
$$\sigma_2 = \sigma_{\text{ci}} = -5263 \text{ lbf/in}^2$$

For the distortion energy theory, use Eq. 50.23 to find the von Mises stress.

$$\sigma' = \sqrt{\sigma_1^2 + \sigma_2^2 - \sigma_1\sigma_2}$$

$$= \sqrt{\begin{array}{c}\left(7906 \, \frac{\text{lbf}}{\text{in}^2}\right)^2 + \left(-5263 \, \frac{\text{lbf}}{\text{in}^2}\right)^2 \\ - \left(7906 \, \frac{\text{lbf}}{\text{in}^2}\right)\left(-5263 \, \frac{\text{lbf}}{\text{in}^2}\right)\end{array}}$$

$$= 11{,}481 \text{ lbf/in}^2$$

From Eq. 50.26, the factor of safety for the spool is

$$\text{FS} = \frac{S_{yt}}{\sigma'} = \frac{19{,}000 \, \frac{\text{lbf}}{\text{in}^2}}{11{,}481 \, \frac{\text{lbf}}{\text{in}^2}} = \boxed{1.65}$$

The answer is (C).

SI Solution

The spool appears to be a thin-walled tube. Only the circumferential stress and the longitudinal stress are significant.

The end disc area exposed to the 3.5 MPa pressure is

$$A_e = \left(\frac{\pi}{4}\right)\left((0.050 \text{ m})^2 - (0.025 \text{ m})^2\right) = 0.001473 \text{ m}^2$$

The longitudinal force produced by the 3.5 MPa pressure is

$$F = pA_e = (3.5 \times 10^6 \text{ Pa})(0.001473 \text{ m}^2) = 5155.5 \text{ N}$$

The annular area of the spool tube is

$$A = \left(\frac{\pi}{4}\right)(d_o^2 - d_i^2)$$
$$= \left(\frac{\pi}{4}\right)\left((0.025 \text{ m})^2 - (0.025 \text{ m} - (2)(0.00125))^2\right)$$
$$= 9.327 \times 10^{-5} \text{ m}^2$$

The longitudinal stress in the spool tube is

$$\sigma_{\text{long}} = \frac{F}{A} = \frac{5155.5 \text{ N}}{9.327 \times 10^{-5} \text{ m}^2} = 55.28 \times 10^6 \text{ Pa}$$

Since $d/t = 1.3 \text{ mm}/2.5 \text{ mm} = 0.05 < 0.10$, this qualifies as a thin-walled cylinder. However, since the tube is exposed to external pressure (not internal), use Lamé's equation for a thick-walled cylinder. This is in the range for Lamé's solution.

The compressive circumferential stress in the tube is given in Table 51.2.

$$r_o = \frac{25 \text{ mm}}{2} = 12.5 \text{ mm}$$
$$r_i = 12.5 \text{ mm} - 1.25 \text{ mm} = 11.25 \text{ mm}$$
$$\sigma_{\text{ci}} = \frac{-2r_o^2 p}{r_o^2 - r_i^2}$$
$$= \frac{(-2)(12.5 \text{ mm})^2(3.5 \text{ MPa})\left(10^6 \, \frac{\text{Pa}}{\text{MPa}}\right)}{(12.5 \text{ mm})^2 - (11.25 \text{ mm})^2}$$
$$= 36.84 \times 10^6 \text{ Pa}$$

The principal stresses are

$$\sigma_1 = \sigma_{\text{long}} = 55.28 \times 10^6 \text{ Pa} \quad (55.28 \text{ MPa})$$
$$\sigma_2 = \sigma_{\text{ci}} = -36.84 \times 10^6 \text{ Pa} \quad (-36.84 \text{ MPa})$$

For the distortion energy theory, use Eq. 50.23 to find the von Mises stress.

$$\sigma' = \sqrt{\sigma_1^2 + \sigma_2^2 - \sigma_1\sigma_2}$$
$$= \sqrt{\begin{array}{c}(55.28 \text{ MPa})^2 + (-36.84 \text{ MPa})^2 \\ - (55.28 \text{ MPa})(-36.84 \text{ MPa})\end{array}}$$
$$= 80.31 \text{ MPa}$$

From Eq. 50.26, the factor of safety for the spool is

$$\text{FS} = \frac{S_{yt}}{\sigma'} = \frac{130 \text{ MPa}}{80.31 \text{ MPa}} = \boxed{1.62}$$

The answer is (C).

4. *Customary U.S. Solution*

First find the portion of the pressure load, p, taken by the bolts.

The length of the bolt in tension is

$$L = 0.75 \text{ in} + 0.50 \text{ in} = 1.25 \text{ in}$$

The modulus of elasticity of cold-rolled bolts is $E = 30 \times 10^6 \text{ lbf/in}^2$.

The total bolt stiffness is

$$k_{\text{bolt}} = (6)\left(\frac{AE}{L}\right)$$

$$= \frac{(6)\left(\frac{\pi}{4}\right)(0.375 \text{ in})^2\left(30 \times 10^6 \frac{\text{lbf}}{\text{in}^2}\right)}{1.25 \text{ in}}$$

$$= 1.590 \times 10^7 \text{ lbf/in}$$

The plate/vessel contact area is an annulus with an 8.5 in inside diameter and an 11.5 in outside diameter.

$$A = \left(\frac{\pi}{4}\right)\left((11.5 \text{ in})^2 - (8.5 \text{ in})^2\right) = 47.12 \text{ in}^2$$

The modulus of elasticity of the plate/vessel material is $E = 29 \times 10^6 \text{ lbf/in}^2$.

The plate/vessel stiffness is

$$k_{\text{plate/vessel}} = \frac{AE}{L} = \frac{(47.12 \text{ in}^2)\left(29 \times 10^6 \frac{\text{lbf}}{\text{in}^2}\right)}{1.25 \text{ in}}$$

$$= 1.093 \times 10^9 \text{ lbf/in}$$

Let x be the decimal portion of the pressure load, p, taken by the bolts. The increase in bolt length is

$$\delta_{\text{bolt}} = \frac{xp}{k_{\text{bolt}}}$$

The plate/vessel deformation is

$$\delta_{\text{plate/vessel}} = \frac{(1-x)p}{k_{\text{plate/vessel}}}$$

Set $\delta_{\text{bolt}} = \delta_{\text{plate/vessel}}$.

$$\frac{xp}{k_{\text{bolt}}} = \frac{(1-x)p}{k_{\text{plate/vessel}}}$$

Canceling p gives

$$\frac{x}{1.590 \times 10^7 \frac{\text{lbf}}{\text{in}^2}} = \frac{1-x}{1.093 \times 10^9 \frac{\text{lbf}}{\text{in}^2}}$$

$$x = 0.0143$$

Next, find the stresses in the bolts.

The end plate area exposed to pressure is

$$A_{\text{plate}} = \left(\frac{\pi}{4}\right)(8.5 \text{ in})^2 = 56.75 \text{ in}^2$$

The forces on the end plate are

$$F_{\text{max}} = p_{\text{max}}A_{\text{plate}} = \left(350 \frac{\text{lbf}}{\text{in}^2}\right)(56.75 \text{ in}^2)$$

$$= 19{,}863 \text{ lbf}$$

$$F_{\text{min}} = p_{\text{min}}A_{\text{plate}} = \left(50 \frac{\text{lbf}}{\text{in}^2}\right)(56.75 \text{ in}^2)$$

$$= 2838 \text{ lbf}$$

From Table 51.5, the stress area of a $^3/_8$-24 UNF bolt is $A = 0.0878 \text{ in}^2$.

For a preload, F_o, the maximum stress in each bolt is

$$\sigma_{\text{max}} = \frac{F_o + \dfrac{xF_{\text{max}}}{6}}{A}$$

$$= \frac{3700 \text{ lbf} + \dfrac{(0.0143)(19{,}863 \text{ lbf})}{6}}{0.0878}$$

$$= 42{,}680 \text{ lbf/in}^2$$

The minimum stress in each bolt is

$$\sigma_{\text{min}} = \frac{F_o + \dfrac{xF_{\text{min}}}{6}}{A}$$

$$= \frac{3700 \text{ lbf} + \dfrac{(0.0143)(2838 \text{ lbf})}{6}}{0.0878}$$

$$= 42{,}218 \text{ lbf/in}^2$$

From Eq. 50.28, the mean stress is

$$\sigma_m = \frac{\sigma_{\text{max}} + \sigma_{\text{min}}}{2} = \frac{42{,}680 \frac{\text{lbf}}{\text{in}^2} + 42{,}218 \frac{\text{lbf}}{\text{in}^2}}{2}$$

$$= 42{,}449 \text{ lbf/in}^2$$

For a thread stress concentration factor of 2, use Eq. 50.30 to find the alternating stress.

$$\sigma_{\text{alt}} = (2)\left(\frac{\sigma_{\text{max}} - \sigma_{\text{min}}}{2}\right) = \sigma_{\text{max}} - \sigma_{\text{min}}$$

$$= 42{,}680 \frac{\text{lbf}}{\text{in}^2} - 42{,}218 \frac{\text{lbf}}{\text{in}^2} = 462 \text{ lbf/in}^2$$

The endurance strength of the bolt is considered to be one-half the ultimate strength.

$$S'_e = 0.5 S_{ut} = (0.5)\left(110{,}000 \frac{\text{lbf}}{\text{in}^2}\right) = 55{,}000 \text{ lbf/in}^2$$

Materials

Use the following endurance strength derating factors (from Shigley and Mischke, *Mechanical Engineering Design*). (Note: The third edition of Shigley and Mischke gives $K_b = 0.85$ without explanation. By the fifth edition, a quantitative method based on Kugel's theory yields 0.96.)

surface finish: $K_a = 0.72$
size: $K_b = 0.85$
reliability: $K_c = 0.90$

The new endurance strength of the bolt is

$$S_e = K_a K_b K_c S_e' = (0.72)(0.85)(0.90)\left(55,000 \ \frac{\text{lbf}}{\text{in}^2}\right)$$
$$= 30,294 \ \text{lbf/in}^2$$

Draw the Goodman line and locate the $(\sigma_m, \sigma_{\text{alt}})$ point.

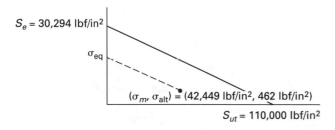

The Goodman equivalent stress is

$$\sigma_{\text{eq}} = \sigma_{\text{alt}} + \left(\frac{S_e}{S_{ut}}\right)\sigma_m$$
$$= 462 \ \frac{\text{lbf}}{\text{in}^2} + \left(\frac{30,294 \ \frac{\text{lbf}}{\text{in}^2}}{110,000 \ \frac{\text{lbf}}{\text{in}^2}}\right)\left(42,449 \ \frac{\text{lbf}}{\text{in}^2}\right)$$
$$= 12,152 \ \text{lbf/in}^2$$

From Eq. 50.32, the factor of safety is

$$\text{FS} = \frac{S_e}{\sigma_{\text{eq}}} = \frac{30,294 \ \frac{\text{lbf}}{\text{in}^2}}{12,152 \ \frac{\text{lbf}}{\text{in}^2}} = \boxed{2.5}$$

The answer is (D).

SI Solution

First find the portion of the pressure load, p, taken by the bolts.

The length of the bolt in tension is

$$L = (19 \ \text{mm} + 12 \ \text{mm})\left(\frac{1 \ \text{m}}{1000 \ \text{mm}}\right) = 0.031 \ \text{m}$$

The modulus of elasticity of cold-rolled bolts is $E = 20 \times 10^4$ MPa.

The total bolt stiffness is

$$k_{\text{bolt}} = (6)\left(\frac{AE}{L}\right)$$
$$= \frac{(6)\left(\frac{\pi}{4}\right)(0.0095 \ \text{m})^2 \left(20 \times 10^{10} \ \text{Pa}\right)}{0.031 \ \text{m}}$$
$$= 2.744 \times 10^9 \ \text{N/m}$$

The plate/vessel contact area is an annulus with a 215 mm inside diameter and a 290 mm outside diameter.

$$A = \left(\frac{\pi}{4}\right)\left((0.290 \ \text{m})^2 - (0.215 \ \text{m})^2\right) = 0.02975 \ \text{m}^2$$

The modulus of elasticity of the plate/vessel material is $E = 19 \times 10^4$ MPa.

The plate/vessel stiffness is

$$k_{\text{plate/vessel}} = \frac{AE}{L} = \frac{(0.02975 \ \text{m}^2)\left(19 \times 10^{10} \ \text{Pa}\right)}{0.031 \ \text{m}}$$
$$= 1.823 \times 10^{11} \ \text{N/m}$$

Let x be the decimal portion of the pressure load, p, taken by the bolts. The increase in bolt length is

$$\delta_{\text{bolt}} = \frac{xp}{k_{\text{bolt}}}$$

The plate/vessel deformation is

$$\delta_{\text{plate/vessel}} = \frac{(1-x)p}{k_{\text{plate/vessel}}}$$

Set $\delta_{\text{bolt}} = \delta_{\text{plate/vessel}}$.

$$\frac{xp}{k_{\text{bolt}}} = \frac{(1-x)p}{k_{\text{plate/vessel}}}$$

Canceling p gives

$$\frac{x}{2.744 \times 10^9 \ \frac{\text{N}}{\text{m}}} = \frac{1-x}{1.823 \times 10^{11} \ \frac{\text{N}}{\text{m}}}$$
$$x = 0.0148$$

Next, find the stresses in the bolts.

The end plate area exposed to pressure is

$$A_{\text{plate}} = \left(\frac{\pi}{4}\right)(0.215 \ \text{m})^2 = 0.0363 \ \text{m}^2$$

The forces on the end plate are

$$F_{\text{max}} = p_{\text{max}} A_{\text{plate}} = \left(2400 \times 10^3 \ \text{Pa}\right)(0.0363 \ \text{m}^2)$$
$$= 87.12 \times 10^3 \ \text{N}$$
$$F_{\text{min}} = p_{\text{min}} A_{\text{plate}} = \left(340 \times 10^3 \ \text{Pa}\right)(0.0363 \ \text{m}^2)$$
$$= 12.34 \times 10^3 \ \text{N}$$

From Table 51.15, the stress area of a $^3/_8$-24 UNF bolt is

$$A = (0.0878 \text{ in}^2)\left(\frac{1 \text{ m}}{39.36 \text{ in}}\right)^2 = 5.667 \times 10^{-5} \text{ m}^2$$

For a preload F_o, the maximum stress in each bolt is

$$\begin{aligned}
\sigma_{\text{max}} &= \frac{F_o + \dfrac{x F_{\text{max}}}{6}}{A} \\
&= \frac{16.4 \times 10^3 \text{ N} + \dfrac{(0.0148)(87.12 \times 10^3 \text{ N})}{6}}{5.667 \times 10^{-5} \text{ m}^2} \\
&= 2.932 \times 10^8 \text{ Pa}
\end{aligned}$$

The minimum stress in each bolt is

$$\begin{aligned}
\sigma_{\text{min}} &= \frac{F_o + \dfrac{x F_{\text{min}}}{6}}{A} \\
&= \frac{16.4 \times 10^3 \text{ N} + \dfrac{(0.0148)(12.34 \times 10^3 \text{ N})}{6}}{5.667 \times 10^{-5} \text{ m}^2} \\
&= 2.899 \times 10^8 \text{ Pa}
\end{aligned}$$

From Eq. 50.28, the mean stress is

$$\begin{aligned}
\sigma_m &= \frac{\sigma_{\text{max}} + \sigma_{\text{min}}}{2} \\
&= \frac{2.932 \times 10^8 \text{ Pa} + 2.899 \times 10^8 \text{ Pa}}{2} \\
&= 2.916 \times 10^8 \text{ Pa} \quad (291.6 \text{ MPa})
\end{aligned}$$

For a thread stress concentration factor of 2, use Eq. 50.36 to find the alternating stress.

$$\begin{aligned}
\sigma_{\text{alt}} &= (2)\left(\frac{\sigma_{\text{max}} - \sigma_{\text{min}}}{2}\right) = \sigma_{\text{max}} - \sigma_{\text{min}} \\
&= 2.932 \times 10^8 \text{ Pa} - 2.899 \times 10^8 \text{ Pa} \\
&= 0.033 \times 10^8 \text{ Pa} \quad (3.5 \text{ MPa})
\end{aligned}$$

The endurance strength of the bolt is considered to be one-half the ultimate strength.

$$S_e' = 0.5 S_{ut} = (0.5)(760 \text{ MPa}) = 380 \text{ MPa}$$

Use the following endurance strength derating factors (from Shigley and Mischke, *Mechanical Engineering Design*). (Note: The third edition of Shigley and Mischke gives $K_b = 0.85$ without explanation. By the fifth edition, a quantitative method based on Kugel's theory yields 0.96.)

> surface finish: $K_a = 0.72$
> size: $K_b = 0.85$
> reliability: $K_c = 0.90$

The new endurance strength of the bolt is

$$\begin{aligned}
S_e &= K_a K_b K_c S_e' = (0.72)(0.85)(0.90)(380 \text{ MPa}) \\
&= 209.3 \text{ MPa}
\end{aligned}$$

Draw the Goodman line and locate the $(\sigma_m, \sigma_{\text{alt}})$ point.

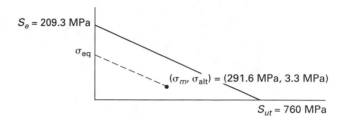

The Goodman equivalent stress is

$$\begin{aligned}
\sigma_{\text{eq}} &= \sigma_{\text{alt}} + \left(\frac{S_e}{S_{ut}}\right)\sigma_m \\
&= 3.3 \text{ MPa} + \left(\frac{209.3 \text{ MPa}}{760 \text{ MPa}}\right)(291.6 \text{ MPa}) \\
&= 83.6 \text{ MPa}
\end{aligned}$$

From Eq. 50.32, the factor of safety is

$$\text{FS} = \frac{S_e}{\sigma_{\text{eq}}} = \frac{209.3 \text{ MPa}}{83.6 \text{ MPa}} = \boxed{2.5}$$

The answer is (D).

51 Basic Machine Design

PRACTICE PROBLEMS

Note: Unless instructed otherwise in a problem, use the following properties:

steel:
$E = 30 \times 10^6$ lbf/in^2 (20×10^4 MPa)
$G = 11.5 \times 10^6$ lbf/in^2 (8.0×10^4 MPa)
$\alpha = 6.5 \times 10^{-6}$ 1/°F (1.2×10^{-5} 1/°C)
$\nu = 0.3$

aluminum: $E = 10 \times 10^6$ lbf/in^2 (70×10^3 MPa)
copper: $E = 17.5 \times 10^6$ lbf/in^2 (12×10^4 MPa)

1. The yield strength of a structural steel member is 36,000 lbf/in^2 (250 MPa). The tensile stress is 8240 lbf/in^2 (57 MPa). What is the factor of safety in tension?

(A) 2.5
(B) 3.1
(C) 3.6
(D) 4.4

2. A structural steel member 50 ft (15 m) long is used as a long column to support 75,000 lbf (330 kN). Both ends are built-in, and there are no intermediate supports. A factor of safety of 2.5 is used. What is the required moment of inertia?

(A) 48 in^4 (2.0×10^{-5} m^4)
(B) 72 in^4 (3.0×10^{-5} m^4)
(C) 96 in^4 (4.0×10^{-5} m^4)
(D) 130 in^4 (5.5×10^{-5} m^4)

3. A long steel column with a yield strength of 36,000 lbf/in^2 (250 MPa) has pinned ends. The column is 25 ft (7.5 m) long, has a cross-sectional area of 25.6 in^2 (165 cm^2), a centroidal moment of inertia of 350 in^4 (14,600 cm^4), and a distance from the neutral axis to the extreme fiber of 7 in (180 mm). It carries an axial concentric load of 100,000 lbf (440 kN) and an eccentric load of 150,000 lbf (660 kN) located 3.33 in (80 mm) from the longitudinal axial axis. Use the secant formula to determine the stress factor of safety.

(A) 1.2
(B) 1.6
(C) 1.8
(D) 2.0

4. A rectangular tank with dimensions of 2 ft by 2 ft by 2 ft (60 cm by 60 cm by 60 cm) is pressurized to 2 lbf/in^2 gage (14 kPa). The steel plate used to construct a tank has a thickness of 0.25 in (6.3 mm) and a yield stress of 36 ksi (250 MPa). Neglecting stress concentration factors at the corners and edges, what is the factor of safety?

(A) 3.1
(B) 3.7
(C) 4.2
(D) 6.3

5. A short section of pipe with an inside diameter of 1.750 in (44.5 mm) is to be produced by turning and boring bar stock. The steel has an allowable normal stress of 20,000 lbf/in^2 (140 MPa). The pipe will be pressurized internally to 2000 lbf/in^2 gage (14 MPa). Disregard longitudinal stress and other end effects. Use a thick-walled analysis to determine the required outside diameter.

(A) 1.9 in (49 mm)
(B) 2.2 in (56 mm)
(C) 2.6 in (66 mm)
(D) 3.9 in (99 mm)

6. A pressurized cylinder has an inside diameter of 0.742 in (18.8 mm) and an outside diameter of 1.486 in (37.7 mm). The cylinder is subjected to an external pressure of 400 lbf/in^2 gage (2.8 MPa). What is the maximum stress developed in the cylinder?

(A) -1100 lbf/in^2 (-7.5 MPa)
(B) -1600 lbf/in^2 (-12 MPa)
(C) -2500 lbf/in^2 (-18 MPa)
(D) -3600 lbf/in^2 (-28 MPa)

7. A shell with an outside diameter of 16 in (406 mm) and a wall thickness of 0.10 in (2.54 mm) is subjected to a 40,000 lbf/in^2 (280 MPa) tensile load and a 400,000 in-lbf torque (45 kN·m).

(a) What are the principal stresses?

(b) What is the maximum shear stress?

8. (*Time limit: one hour*) A projectile launcher is formed by shrinking a jacket over a tube of the same length. The assembly is made with the maximum allowable diametral interference. The tube inside and outside diameters are 4.7 in and 7.75 in (119 mm and 197 mm), respectively. The outside diameter of the jacket is 12 in (300 mm). The jacket and tube are both steel with a Poisson's ratio of 0.3 and modulus of elasticity of 3.0×10^7 lbf/in^2 (204 GPa). The maximum allowable stress at the interface is 18,000 lbf/in^2 (124 MPa).

(a) What is the diametral interference?
   (A)   0.0064 in (0.16 mm)
   (B)   0.0088 in (0.22 mm)
   (C)   0.012 in (0.30 mm)
   (D)   0.022 in (0.56 mm)

(b) What is the maximum circumferential stress in the jacket?
   (A)   12,000 lbf/in^2 (81.6 MPa)
   (B)   14,000 lbf/in^2 (95.2 MPa)
   (C)   16,000 lbf/in^2 (106 MPa)
   (D)   18,000 lbf/in^2 (124 MPa)

(c) What is the minimum circumferential stress in the tube?
   (A)   −12,000 lbf/in^2 (−81.6 MPa)
   (B)   −14,000 lbf/in^2 (−95.2 MPa)
   (C)   −16,000 lbf/in^2 (−106 MPa)
   (D)   −18,000 lbf/in^2 (−124 MPa)

9. (*Time limit: one hour*) Two hollow cylinders of the same length are press-fitted together with a diametral interference of 0.012 in (0.3 mm). The outer cylinder has a wall thickness of 0.079 in (2 mm) and an outside diameter of 4.7 in (120 mm). The inner cylinder has a wall thickness of 0.12 in (3 mm). Both cylinders are the same high-strength material with a Poisson's ratio of 0.3 and modulus of elasticity of 3×10^7 lbf/in^2 (207 GPa). What is the circumferential stress at the interface between the two cylinders?

10. A $^3/_4$-16 UNF steel bolt is used without washers to clamp two rigid steel plates, each 2 in (50 mm) thick. 0.75 in (19 mm) of the threaded section of the bolt remains under the nut. The nut has six threads. The nut is tightened until the stress in the bolt body is 40,000 lbf/in^2 (280 MPa). The bolt's modulus of elasticity is 2.9×10^7 lbf/in^2 (200 GPa). How much does the bolt stretch?

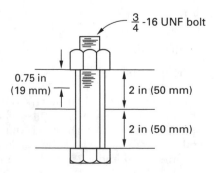

(A)   0.0060 in (0.15 mm)
(B)   0.0087 in (0.22 mm)
(C)   0.011 in (0.28 mm)
(D)   0.019 in (0.48 mm)

11. A class 30 gray cast-iron hub with an outside diameter of 12 in (300 mm) is pressed onto a soft steel solid shaft with an outside diameter of 6 in (150 mm). Poisson's ratio and modulus of elasticity for the cast iron are 0.27 and 1.45×10^7 lbf/in^2 (100 GPa), respectively. Poisson's ratio and modulus of elasticity for the steel are 0.30 and 2.9×10^7 lbf/in^2 (200 GPa), respectively. The stress-strain curve for gray cast iron becomes nonlinear at low stresses. What is the maximum radial interference such that the stress-strain relationship remains linear?

(A)   0.0014 in (0.036 mm)
(B)   0.0060 in (0.15 mm)
(C)   0.0087 in (0.22 mm)
(D)   0.011 in (0.28 mm)

12. The bracket shown is attached to a column with three 0.75 in (19 mm) bolts arranged in an equilateral triangular layout. A force is applied with a moment arm of 20 in (500 mm) measured to the centroid of the bolt group. The maximum shear stress in the bolts is limited to 15,000 lbf/in^2 (100 MPa). What is the maximum force that the connection can support?

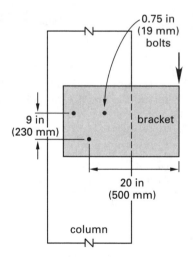

(A)   3300 lbf (15 kN)
(B)   4700 lbf (21 kN)
(C)   5100 lbf (23 kN)
(D)   5800 lbf (26 kN)

13. A fillet weld is used to secure a steel bracket to a column. The bracket supports a 10,000 lbf (44 kN) force applied 12 in (300 mm) from the face of the column, as shown. The maximum shear stress in the weld material is 8000 lbf/in^2 (55 MPa). What size fillet weld is required?

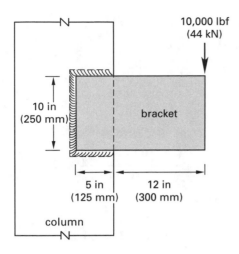

10,000 lbf
(44 kN)

10 in
(250 mm)

bracket

5 in
(125 mm)

12 in
(300 mm)

column

(A) 0.26 in (6.6 mm)
(B) 0.39 in (9.9 mm)
(C) 0.51 in (13 mm)
(D) 0.72 in (18 mm)

14. A 2 in (50 mm) diameter steel solid shaft rotates at 3500 rpm. A 16 in (406 mm) outside diameter, 2 in (50 mm) thick steel disk flywheel is press fitted to the shaft. The contact pressure between the shaft and flywheel is 1250 lbf/in² (8.6 MPa) when the shaft is turning. Centrifugal expansion of the shaft diameter is negligible. For both the shaft and the flywheel, Poisson's ratio is 0.3, modulus of elasticity is 2.9×10^7 lbf/in² (20 GPa), and mass density is 0.283 lbm/in³ (7830 kg/m³). What interference is required when the shaft is not turning?

(A) 1.0×10^{-4} in (2.5×10^{-3} mm)
(B) 1.5×10^{-4} in (3.7×10^{-3} mm)
(C) 2.1×10^{-4} in (5.2×10^{-3} mm)
(D) 2.7×10^{-4} in (6.7×10^{-3} mm)

SOLUTIONS

1. *Customary U.S. Solution*

The factor of safety is

$$FS = \frac{S_{yt}}{\sigma} = \frac{36,000 \; \frac{lbf}{in^2}}{8240 \; \frac{lbf}{in^2}} = \boxed{4.37}$$

The answer is (D).

SI Solution

The factor of safety is

$$FS = \frac{S_{yt}}{\sigma} = \frac{250 \text{ MPa}}{57 \text{ MPa}} = \boxed{4.39}$$

The answer is (D).

2. *Customary U.S. Solution*

The design load for a factor of safety of 2.5 is

$$F = (75,000 \text{ lbf})(2.5) = 187,500 \text{ lbf}$$

From Table 51.1, the theoretical end restraint coefficient for built-in ends is $C = 0.5$, and the minimum design value is 0.65.

From Eq. 51.3, the effective length of the column is

$$L' = CL = (0.65)(50 \text{ ft})\left(12 \; \frac{in}{ft}\right) = 390 \text{ in}$$

Set the design load equal to the Euler load F_e and use Eq. 51.1 to find the required moment of inertia.

$$I = \frac{F_e(L')^2}{\pi^2 E}$$

$$= \frac{(187,500 \text{ lbf})(390 \text{ in})^2}{\pi^2 \left(30 \times 10^6 \; \frac{lbf}{in^2}\right)} = \boxed{96.32 \text{ in}^4}$$

The answer is (C).

SI Solution

The design load for a factor of safety of 2.5 is

$$F = (330 \times 10^3 \text{ N})(2.5) = 825 \times 10^3 \text{ N}$$

From Table 51.1, the theoretical end restraint coefficient for built-in ends is $C = 0.5$, and the minimum design value is 0.65.

From Eq. 51.3, the effective length of the column is

$$L' = CL = (0.65)(15 \text{ m}) = 9.75 \text{ m}$$

Set the design load equal to the Euler load F_e and use Eq. 51.1 to find the required moment of inertia.

$$I = \frac{F_e (L')^2}{\pi^2 E}$$

$$= \frac{(825 \times 10^3 \text{ N})(9.75 \text{ m})^2}{\pi^2 \left(20 \times 10^{10} \dfrac{\text{N}}{\text{m}^2}\right)} = \boxed{3.973 \times 10^{-5} \text{ m}^4}$$

The answer is (C).

3. *Customary U.S. Solution*

The radius of gyration is

$$k = \sqrt{\frac{I}{A}} = \sqrt{\frac{350 \text{ in}^4}{25.6 \text{ in}^2}} = 3.70 \text{ in}$$

The slenderness ratio is

$$\frac{L}{k} = \frac{(25 \text{ ft}) \left(12 \dfrac{\text{in}}{\text{ft}}\right)}{3.70 \text{ in}} = 81.08$$

The total buckling load is

$$F = 100{,}000 \text{ lbf} + 150{,}000 \text{ lbf} = 250{,}000 \text{ lbf}$$

From Eq. 51.9,

$$\phi = \left(\tfrac{1}{2}\right) \left(\frac{L}{k}\right) \sqrt{\frac{F}{AE}}$$

$$= \left(\tfrac{1}{2}\right)(81.08) \sqrt{\frac{250{,}000 \text{ lbf}}{(25.6 \text{ in}^2) \left(30 \times 10^6 \dfrac{\text{lbf}}{\text{in}^2}\right)}}$$

$$= 0.731 \text{ rad}$$

The eccentricity is

$$e = \frac{M}{F} = \frac{(150{,}000 \text{ lbf})(3.33 \text{ in})}{250{,}000 \text{ lbf}} = 2.0 \text{ in}$$

From Eq. 51.8, the critical column stress is

$$\sigma_{\max} = \left(\frac{F}{A}\right) \left(1 + \left(\frac{ec}{k^2}\right)(\sec \phi)\right)$$

$$= \left(\frac{250{,}000 \text{ lbf}}{25.6 \text{ in}^2}\right) \left(1 + \left(\frac{(2.0 \text{ in})(7 \text{ in})}{(3.70 \text{ in})^2}\right.\right.$$

$$\left.\left. \times \left(\sec(0.731 \text{ rad})\right)\right)\right)$$

$$= 23{,}180 \text{ lbf/in}^2$$

The stress factor of safety for the column is

$$\text{FS} = \frac{S_y}{\sigma_{\max}} = \frac{36{,}000 \dfrac{\text{lbf}}{\text{in}^2}}{23{,}180 \dfrac{\text{lbf}}{\text{in}^2}} = \boxed{1.55}$$

The answer is (B).

SI Solution

The radius of gyration is

$$k = \sqrt{\frac{I}{A}} = \sqrt{\frac{14\,600 \text{ cm}^4}{165 \text{ cm}^2}} = 9.41 \text{ cm}$$

The slenderness ratio is

$$\frac{L}{k} = \frac{(7.5 \text{ m}) \left(100 \dfrac{\text{cm}}{\text{m}}\right)}{9.41 \text{ cm}} = 79.70$$

The total buckling load is

$$F = 440 \text{ kN} + 660 \text{ kN} = (1100 \text{ kN}) \left(1000 \dfrac{\text{N}}{\text{kN}}\right)$$

$$= 1.1 \times 10^6 \text{ N}$$

From Eq. 51.9,

$$\phi = \left(\tfrac{1}{2}\right) \left(\frac{L}{k}\right) \sqrt{\frac{F}{AE}}$$

$$= \left(\tfrac{1}{2}\right)(79.70) \sqrt{\frac{1.1 \times 10^6 \text{ N}}{\left(\dfrac{165 \text{ cm}^2}{\left(100 \dfrac{\text{cm}}{\text{m}}\right)^2}\right) \left(20 \times 10^{10} \dfrac{\text{N}}{\text{m}^2}\right)}}$$

$$= 0.728 \text{ rad}$$

The eccentricity is

$$e = \frac{M}{F} = \frac{(660 \text{ kN})(80 \text{ mm})}{1100 \text{ kN}} = 48 \text{ mm}$$

From Eq. 51.8, the critical column stress is

$$\sigma_{\max} = \left(\frac{F}{A}\right) \left(1 + \left(\frac{ec}{k^2}\right)(\sec \phi)\right)$$

$$= \left(\frac{\dfrac{1.1 \times 10^6 \text{ N}}{165 \text{ cm}^2}}{\left(100 \dfrac{\text{cm}}{\text{m}}\right)^2}\right) \left(1 + \left(\frac{(48 \text{ mm})(180 \text{ mm})}{(9.41 \text{ cm})^2 \left(100 \dfrac{\text{mm}^2}{\text{cm}^2}\right)}\right.\right.$$

$$\left.\left. \times \left(\sec(0.728 \text{ rad})\right)\right)\right) \left(\frac{1 \text{ MPa}}{10^6 \text{ Pa}}\right)$$

$$= 153.8 \text{ MPa}$$

The stress factor of safety for the column is

$$\text{FS} = \frac{S_y}{\sigma_{\max}} = \frac{250 \text{ MPa}}{153.8 \text{ MPa}} = \boxed{1.63}$$

The answer is (B).

4. *Customary U.S. Solution*

The dimensions of the face are

$$a = (2 \text{ ft}) \left(12 \, \frac{\text{in}}{\text{ft}} \right) = 24 \text{ in}$$

$$b = (2 \text{ ft}) \left(12 \, \frac{\text{in}}{\text{ft}} \right) = 24 \text{ in}$$

The ratio of the side dimensions is

$$\frac{a}{b} = \frac{24 \text{ in}}{24 \text{ in}} = 1$$

The plate is considered to have built-in edges. From Table 51.7, the maximum stress is

$$\sigma_{\max} = \frac{C_3 p b^2}{t^2} = \frac{(0.308) \left(2 \, \frac{\text{lbf}}{\text{in}^2} \right) (24 \text{ in})^2}{(0.25 \text{ in})^2}$$

$$= 5677 \text{ lbf/in}^2$$

The factor of safety is

$$\text{FS} = \frac{S_y}{\sigma_{\max}} = \frac{36,000 \, \frac{\text{lbf}}{\text{in}^2}}{5677 \, \frac{\text{lbf}}{\text{in}^2}} = \boxed{6.34}$$

The answer is (D).

SI Solution

The dimensions of the face are $a = 60$ cm and $b = 60$ cm.

The ratio of the side dimensions is

$$\frac{a}{b} = \frac{60 \text{ cm}}{60 \text{ cm}} = 1$$

The plate is considered to have built-in edges. From Table 51.7, the maximum stress is

$$\sigma_{\max} = \frac{C_3 p b^2}{t^2}$$

$$= \frac{(0.308)(14 \text{ kPa}) \left((60 \text{ cm}) \left(\frac{1 \text{ m}}{100 \text{ cm}} \right) \right)^2}{\left(6.3 \text{ mm} \left(\frac{1 \text{ m}}{1000 \text{ mm}} \right) \right)^2 \left(1000 \, \frac{\text{kPa}}{\text{MPa}} \right)}$$

$$= 39.11 \text{ MPa}$$

The factor of safety is

$$\text{FS} = \frac{S_y}{\sigma_{\max}} = \frac{250 \text{ MPa}}{39.11 \text{ MPa}} = \boxed{6.39}$$

The answer is (D).

5. *Customary U.S. Solution*

The inside radius is

$$r_i = \frac{d_i}{2} = \frac{1.750 \text{ in}}{2} = 0.875 \text{ in}$$

From Table 51.2, the maximum stress is the circumferential stress at the inside radius. This is a normal stress.

$$\sigma_{ci} = \frac{(r_o^2 + r_i^2)p}{r_o^2 - r_i^2}$$

$$20,000 \, \frac{\text{lbf}}{\text{in}^2} = \frac{\left(r_o^2 + (0.875 \text{ in})^2 \right) \left(2000 \, \frac{\text{lbf}}{\text{in}^2} \right)}{r_o^2 - (0.875 \text{ in})^2}$$

Simplify.

$$(10) \left(r_o^2 - (0.875 \text{ in})^2 \right) = r_o^2 + (0.875 \text{ in})^2$$

$$r_o^2 = \frac{(11)(0.875 \text{ in})^2}{9}$$

$$r_o = 0.9673 \text{ in}$$

The required outside diameter is

$$d_o = 2r_o = (2)(0.9673 \text{ in}) = \boxed{1.935 \text{ in}}$$

The answer is (A).

SI Solution

The inside radius is

$$r_i = \frac{d_i}{2} = \frac{44.5 \text{ mm}}{2} = 22.25 \text{ mm}$$

From Table 51.2, the maximum stress is the circumferential stress at the inside radius. This is a normal stress.

$$\sigma_{ci} = \frac{(r_o^2 + r_i^2)p}{r_o^2 - r_i^2}$$

$$140 \text{ MPa} = \frac{\left(r_o^2 + (22.25 \text{ mm})^2 \right) (14 \text{ MPa})}{r_o^2 - (22.25 \text{ mm})^2}$$

Simplify.

$$(10) \left(r_o^2 - (22.25 \text{ mm})^2 \right) = r_o^2 + (22.25 \text{ mm})^2$$

$$r_o^2 = \frac{(11)(22.25 \text{ mm})^2}{9}$$

$$r_o = 24.598 \text{ mm}$$

The required outside diameter is

$$d_o = 2r_o = (2)(24.598 \text{ mm}) = \boxed{49.20 \text{ mm}}$$

The answer is (A).

6. *Customary U.S. Solution*

The inside radius is

$$r_i = \frac{0.742 \text{ in}}{2} = 0.371 \text{ in}$$

The outside radius is

$$r_o = \frac{1.486 \text{ in}}{2} = 0.743 \text{ in}$$

From Table 51.2, the maximum stress developed in the cylinder is the circumferential stress at the inside radius.

$$\sigma_{ci} = \frac{-2r_o^2 p}{r_o^2 - r_i^2} = \frac{(-2)(0.743 \text{ in})^2 \left(400 \dfrac{\text{lbf}}{\text{in}^2}\right)}{(0.743 \text{ in})^2 - (0.371 \text{ in})^2}$$

$$= \boxed{-1066 \text{ lbf/in}^2 \text{ (C)}}$$

The answer is (A).

SI Solution

The inside radius is

$$r_i = \frac{18.8 \text{ mm}}{2} = 9.4 \text{ mm}$$

The outside radius is

$$r_o = \frac{37.7 \text{ mm}}{2} = 18.85 \text{ mm}$$

From Table 51.2, the maximum stress developed in the cylinder is the circumferential stress at the inside radius.

$$\sigma_{ci} = \frac{-2r_o^2 p}{r_o^2 - r_i^2} = \frac{(-2)(18.85 \text{ mm})^2(2.8 \text{ MPa})}{(18.85 \text{ mm})^2 - (9.4 \text{ mm})^2}$$

$$= \boxed{-7.45 \text{ MPa (C)}}$$

The answer is (A).

7. *Customary U.S. Solution*

The inside diameter of the shell is

$$d_i = d_o - 2t = 16 \text{ in} - (2)(0.10 \text{ in}) = 15.8 \text{ in}$$

The polar moment of inertia is

$$J = \left(\frac{\pi}{32}\right)(d_o^4 - d_i^4) = \left(\frac{\pi}{32}\right)\left((16 \text{ in})^4 - (15.8 \text{ in})^4\right)$$

$$= 315.7 \text{ in}^4$$

The shear stress is

$$\tau = \frac{Tc}{J} = \frac{(400{,}000 \text{ in-lbf})(8 \text{ in})}{315.7 \text{ in}^4} = 10{,}136 \text{ lbf/in}^2$$

(b) The maximum shear stress is

$$\tau_{\max} = \sqrt{\left(\frac{\sigma}{2}\right)^2 + \tau^2}$$

$$= \sqrt{\left(\frac{40{,}000 \dfrac{\text{lbf}}{\text{in}^2}}{2}\right)^2 + \left(10{,}136 \dfrac{\text{lbf}}{\text{in}^2}\right)^2}$$

$$= \boxed{22{,}422 \text{ lbf/in}^2}$$

(a) The principal stresses are

$$\sigma_1 = \frac{\sigma}{2} + \tau_{\max}$$

$$= \frac{40{,}000 \dfrac{\text{lbf}}{\text{in}^2}}{2} + 22{,}422 \dfrac{\text{lbf}}{\text{in}^2}$$

$$= \boxed{42{,}422 \text{ lbf/in}^2}$$

$$\sigma_2 = \frac{\sigma}{2} - \tau_{\max}$$

$$= \frac{40{,}000 \dfrac{\text{lbf}}{\text{in}^2}}{2} - 22{,}422 \dfrac{\text{lbf}}{\text{in}^2}$$

$$= \boxed{-2422 \text{ lbf/in}^2}$$

SI Solution

The inside diameter of the shell is

$$d_i = d_o - 2t = 406 \text{ mm} - (2)(2.54 \text{ mm}) = 400.92 \text{ mm}$$

The polar moment of inertia is

$$J = \left(\frac{\pi}{32}\right)(d_o^4 - d_i^4)$$

$$= \left(\frac{\pi}{32}\right)\left((406 \text{ mm})^4 - (400.92 \text{ mm})^4\right)$$

$$\times \left(\frac{1 \text{ m}}{10^3 \text{ mm}}\right)^4$$

$$= 0.000131 \text{ m}^4$$

The shear stress is

$$\tau = \frac{Tc}{J} = \frac{(45 \text{ kN·m})(0.203 \text{ m})\left(1000 \dfrac{\text{N}}{\text{kN}}\right)}{0.000131 \text{ m}^4}$$

$$= \boxed{6.973 \times 10^6 \text{ Pa} \ \ (69.73 \text{ MPa})}$$

(b) The maximum shear stress is

$$\tau_{\max} = \sqrt{\frac{\sigma}{2}^2 + \tau^2}$$

$$= \sqrt{\left(\frac{280 \text{ MPa}}{2}\right)^2 + (69.73 \text{ MPa})^2}$$

$$= \boxed{156.4 \text{ MPa}}$$

(a) The principal stresses are

$$\sigma_1 = \frac{\sigma}{2} + \tau_{\max}$$

$$= \frac{280 \text{ MPa}}{2} + 156.4 \text{ MPa} = \boxed{296.4 \text{ MPa}}$$

$$\sigma_2 = \frac{\sigma}{2} - \tau_{\max}$$

$$= \frac{280 \text{ MPa}}{2} - 156.4 \text{ MPa} = \boxed{-16.4 \text{ MPa}}$$

8. *Customary U.S. Solution*

For the jacket,

$$r_o = \frac{12 \text{ in}}{2} = 6 \text{ in}$$

$$r_i = \frac{7.75 \text{ in}}{2} = 3.875 \text{ in}$$

From Table 51.2, the highest stress is the circumferential stress at the inside surface.

$$\sigma_{ci} = \frac{(r_o^2 + r_i^2)p}{r_o^2 - r_i^2}$$

Note that the maximum allowable stress is given for the interface, not for the tube. The stress is higher than 18,000 lbf/in² at the inside of the tube. Note, also, that radial and circumferential stresses are principal stresses, and do not combine.

Set σ_{ci} equal to the maximum allowable stress.

$$18,000 \frac{\text{lbf}}{\text{in}^2} = \frac{\left((6 \text{ in})^2 + (3.875 \text{ in})^2\right)p}{(6 \text{ in})^2 - (3.875 \text{ in})^2}$$

$$p = 7404 \text{ lbf/in}^2$$

The radial stress at the inside surface is

$$\sigma_{ri} = -p = -7404 \text{ lbf/in}^2$$

From Eq. 51.18, the diametral strain is

$$\epsilon = \frac{\sigma_c - \nu(\sigma_r + \sigma_l)}{E}$$

$$= \frac{18,000 \frac{\text{lbf}}{\text{in}^2} - (0.3)\left(-7404 \frac{\text{lbf}}{\text{in}^2} + 0\right)}{3 \times 10^7 \frac{\text{lbf}}{\text{in}^2}}$$

$$= 6.7404 \times 10^{-4}$$

The diametral deflection is

$$\Delta d = \epsilon d_i = (6.7404 \times 10^{-4})(7.75 \text{ in}) = 0.00522 \text{ in}$$

For the tube,

$$r_o = \frac{7.75 \text{ in}}{2} = 3.875 \text{ in}$$

$$r_i = \frac{4.7 \text{ in}}{2} = 2.35 \text{ in}$$

The external pressure acting on the tube is the same as the radial stress at the inside surface of the jacket. Thus, $p = 7404 \text{ lbf/in}^2$.

From Table 51.2, the circumferential stress at the outside surface is

$$\sigma_{co} = \frac{-(r_o^2 + r_i^2)p}{r_o^2 - r_i^2}$$

$$= \frac{-\left((3.875 \text{ in})^2 + (2.35 \text{ in})^2\right)\left(7404 \frac{\text{lbf}}{\text{in}^2}\right)}{(3.875 \text{ in})^2 - (2.35 \text{ in})^2}$$

$$= -16,018 \text{ lbf/in}^2$$

The radial stress at the outside surface is

$$\sigma_{ro} = -p = -7404 \text{ lbf/in}^2$$

From Eq. 51.18, the diametral strain is

$$\epsilon = \frac{\sigma_c - \nu(\sigma_r + \sigma_l)}{E}$$

$$= \frac{-16,018 \frac{\text{lbf}}{\text{in}^2} - (0.3)\left(-7404 \frac{\text{lbf}}{\text{in}^2} + 0\right)}{3 \times 10^7 \frac{\text{lbf}}{\text{in}^2}}$$

$$= -4.599 \times 10^{-4}$$

The diametral deflection is

$$\Delta d = \epsilon d_o = (-4.599 \times 10^{-4})(7.75 \text{ in}) = -0.00356 \text{ in}$$

(a) The diametral interference is the sum of the magnitudes of the two deflections. From Eq. 51.20,

$$I_{\text{diametral}} = |\Delta d_{\text{jacket}}| + |\Delta d_{\text{tube}}|$$

$$= |0.00522 \text{ in}| + |-0.00356 \text{ in}|$$

$$= \boxed{0.00878 \text{ in}}$$

The answer is (B).

(b) The maximum circumferential stress in the jacket is at the inside surface.

$$\sigma_{\max} = \sigma_{ci} = \boxed{18,000 \text{ lbf/in}^2}$$

The answer is (D).

(c) The minimum circumferential stress in the tube is at the outside surface.

$$\sigma_{\min} = \sigma_{co} = \boxed{-16,018 \text{ lbf/in}^2}$$

The answer is (C).

SI Solution

For the jacket,

$$r_o = \frac{300 \text{ mm}}{2} = 150 \text{ mm}$$

$$r_i = \frac{197 \text{ mm}}{2} = 98.5 \text{ mm}$$

From Table 51.2, the highest stress is the circumferential stress at the inside surface.

$$\sigma_{ci} = \frac{(r_o^2 + r_i^2)p}{r_o^2 - r_i^2}$$

Note that the maximum allowable stress is given for the interface, not for the tube. The stress is higher than 124 MPa at the inside of the tube. Note, also, that radial and circumferential stresses are principal stresses, and do not combine.

Set σ_{ci} equal to the maximum allowable stress.

$$124 \text{ MPa} = \frac{\left((150 \text{ mm})^2 + (98.5 \text{ mm})^2\right)p}{(150 \text{ mm})^2 - (98.5 \text{ mm})^2}$$

$$p = 49.28 \text{ MPa}$$

The radial stress at the inside surface is

$$\sigma_{ri} = -p = -49.28 \text{ MPa}$$

From Eq. 51.18, the diametral strain is

$$\epsilon = \frac{\sigma_c - \nu(\sigma_r + \sigma_l)}{E}$$

$$= \frac{124 \text{ MPa} - (0.3)(-49.28 \text{ MPa} + 0)}{(204 \text{ GPa})\left(1000 \dfrac{\text{MPa}}{\text{GPa}}\right)}$$

$$= 6.803 \times 10^{-4}$$

The diametral deflection is

$$\Delta d = \epsilon d_i = (6.803 \times 10^{-4})(197 \text{ mm}) = 0.1340 \text{ mm}$$

For the tube,

$$r_o = \frac{197 \text{ mm}}{2} = 98.5 \text{ mm}$$

$$r_i = \frac{119 \text{ mm}}{2} = 59.5 \text{ mm}$$

The external pressure acting on the tube is the same as the radial stress at the inside surface of the jacket. Thus, $p = 49.28$ MPa.

From Table 51.2, the circumferential stress at the outside surface is

$$\sigma_{co} = \frac{-(r_o^2 + r_i^2)p}{r_o^2 - r_i^2}$$

$$= \frac{-\left((98.5 \text{ mm})^2 + (59.5 \text{ mm})^2\right)(49.28 \text{ MPa})}{(98.5 \text{ m})^2 - (59.5 \text{ mm})^2}$$

$$= -105.91 \text{ MPa}$$

The radial stress at the outside surface is

$$\sigma_{ro} = -p = -49.28 \text{ MPa}$$

From Eq. 51.18, the diametral strain is

$$\epsilon = \frac{\sigma_c - \nu(\sigma_r + \sigma_l)}{E}$$

$$= \frac{-105.91 \text{ MPa} - (0.3)(-49.28 \text{ MPa} - 0)}{(204 \text{ GPa})\left(1000 \dfrac{\text{MPa}}{\text{GPa}}\right)}$$

$$= -4.467 \times 10^{-4}$$

The diametral deflection is

$$\Delta d = \epsilon d_o = \left(-4.467 \times 10^{-4}\right)(197 \text{ mm}) = -0.0880 \text{ mm}$$

(a) The diametral interface is the sum of the magnitudes of the two deflections. From Eq. 51.20,

$$I_{\text{diametral}} = |\Delta d_{\text{jacket}}| + |\Delta d_{\text{tube}}|$$

$$= |0.1340 \text{ mm}| + |-0.0880 \text{ mm}|$$

$$= \boxed{0.222 \text{ mm}}$$

The answer is (B).

(b) The maximum circumferential stress in the jacket is at the inside surface.

$$\sigma_{\text{max}} = \sigma_{ci} = \boxed{124 \text{ MPa}}$$

The answer is (D).

(c) The minimum circumferential stress in the tube is at the outside surface.

$$\sigma_{\text{min}} = \sigma_{co} = \boxed{-105.91 \text{ MPa}}$$

The answer is (C).

9. *Customary U.S. Solution*

For the inner cylinder,

$$r_o = \frac{4.7 \text{ in} - (2)(0.079 \text{ in})}{2} = 2.27 \text{ in}$$

Give all of the interference to the outside of the inner cylinder, since it is easier to cut the outside diameter than to ream an inside diameter.

$$r_i = 2.27 \text{ in} - 0.12 \text{ in} = 2.15 \text{ in}$$

From Table 51.2, the circumferential stress at the outside surface due to external pressure, p, is

$$\sigma_{co} = \frac{-(r_o^2 + r_i^2)p}{r_o^2 - r_i^2}$$

$$= \frac{-\left((2.27 \text{ in})^2 + (2.15 \text{ in})^2\right)p}{(2.27 \text{ in})^2 - (2.15 \text{ in})^2}$$

$$= -18.43p \text{ in lbf/in}^2$$

The radial stress in lbf/in² is $\sigma_{ro} = -p$.

From Eq. 51.18, the diametral deflection is

$$\Delta d_{inner} = \left(\frac{d}{E}\right)(\sigma_c - \nu(\sigma_r + \sigma_l))$$

$$= \frac{(4.54 \text{ in})\big(-18.43p - (0.3)(-p + 0)\big)}{3 \times 10^7 \dfrac{\text{lbf}}{\text{in}^2}}$$

$$= -27.437 \times 10^{-7}p \quad \text{[in in]}$$

For the outer cylinder,

$$r_o = \frac{4.7 \text{ in}}{2} = 2.35 \text{ in}$$

$$r_i = 2.35 \text{ in} - 0.079 \text{ in} = 2.27 \text{ in}$$

From Table 51.2, the circumferential stress at the inside surface due to internal pressure, p, is

$$\sigma_{ci} = \frac{(r_o^2 + r_i^2)p}{r_o^2 - r_i^2} = \frac{\big((2.35 \text{ in})^2 + (2.27 \text{ in})^2\big)p}{(2.35 \text{ in})^2 - (2.27 \text{ in})^2}$$

$$= 28.88p \quad \text{[in lbf/in}^2\text{]}$$

The radial stress in lbf/in² is $\sigma_{ri} = -p$.

From Eq. 51.18, the diametral deflection is

$$\Delta d_{outer} = \left(\frac{d}{E}\right)(\sigma_e - \nu(\sigma_r + \sigma_l))$$

$$= \frac{(4.54 \text{ in})\big(28.88p - (0.3)(-p + 0)\big)}{3 \times 10^7 \dfrac{\text{lbf}}{\text{in}^2}}$$

$$= 44.159 \times 10^{-7}p \quad \text{[in in]}$$

From Eq. 51.20, the diametral interference is

$$I_{diametral} = |\Delta d_{inner}| + |\Delta d_{outer}|$$

$$0.012 \text{ in} = |27.437 \times 10^{-7}p| + |44.159 \times 10^{-7}p|$$

$$p = 1676 \text{ lbf/in}^2$$

The circumferential stress at the interface between the two cylinders is as follows.

For the inner cylinder,

$$\sigma_{co} = -18.43p$$

$$= (-18.43)\left(1676 \frac{\text{lbf}}{\text{in}^2}\right) = \boxed{-30{,}889 \text{ lbf/in}^2}$$

For the outer cylinder,

$$\sigma_{ci} = 28.88p$$

$$= (28.88)\left(1676 \frac{\text{lbf}}{\text{in}^2}\right) = \boxed{48{,}403 \text{ lbf/in}^2}$$

SI Solution

For the inner cylinder,

$$r_o = \frac{120 \text{ mm} - (2)(2 \text{ mm})}{2} = 58 \text{ mm}$$

Give all of the interference to the outside of the inner cylinder, since it is easier to cut the outside diameter than to ream an inside diameter.

$$r_i = 58 \text{ mm} - 3 \text{ mm} = 55 \text{ mm}$$

From Table 51.2, the circumferential stress at the outside surface due to external pressure, p, is

$$\sigma_{co} = \frac{-(r_o^2 + r_i^2)p}{r_o^2 - r_i^2}$$

$$= \frac{-\big((58 \text{ mm})^2 + (55 \text{ mm})^2\big)p}{(58 \text{ mm})^2 - (55 \text{ mm})^2}$$

$$= -18.85p \quad \text{[in MPa]}$$

The radial stress in MPa is $\sigma_{ro} = -p$.

From Eq. 51.18, the diametral deflection is

$$\Delta d_{inner} = \left(\frac{d}{E}\right)(\sigma_c - \nu(\sigma_r + \sigma_l))$$

$$= \frac{(116 \text{ mm})\big((-18.85p - (0.3)(-p + 0))\big)}{(207 \text{ GPa})\left(1000 \dfrac{\text{MPa}}{\text{GPa}}\right)}$$

$$= -1.040 \times 10^{-2}p \quad \text{[in mm]}$$

For the outer cylinder,

$$r_o = \frac{120 \text{ mm}}{2} = 60 \text{ mm}$$

$$r_i = 60 \text{ mm} - 2 \text{ mm} = 58 \text{ mm}$$

From Table 51.2, the circumferential stress at the inside surface due to internal pressure, p, is

$$\sigma_{ci} = \frac{(r_o^2 + r_i^2)p}{r_o^2 - r_i^2} = \frac{\big((60 \text{ mm})^2 + (58 \text{ mm})^2\big)p}{(60 \text{ mm})^2 - (58 \text{ mm})^2}$$

$$= 29.51p \quad \text{[in MPa]}$$

The radial stress in MPa is $\sigma_{ri} = -p$.

From Eq. 51.18, the diametral deflection is

$$\Delta d_{outer} = \left(\frac{d}{E}\right)(\sigma_e - \nu(\sigma_r - \sigma_l))$$

$$= \frac{(116 \text{ mm})\big(29.51p - (0.3)(-p + 0)\big)}{(207 \text{ GPa})\left(1000 \dfrac{\text{MPa}}{\text{GPa}}\right)}$$

$$= 1.671 \times 10^{-2}p \quad \text{[in mm]}$$

From Eq. 51.20, the diametral interference is

$$I_{\text{diametral}} = |\Delta d_{\text{inner}}| + |\Delta d_{\text{outer}}|$$

$$0.3 \text{ mm} = |1.040 \times 10^{-2}p| + |1.671 \times 10^{-2}p|$$

$$p = 11.07 \text{ MPa}$$

The circumferential stress at the interface between the two cylinders is as follows.

For the inner cylinder,

$$\sigma_{co} = -18.85p$$

$$= (-18.85)(11.07 \text{ MPa}) = \boxed{-209 \text{ MPa}}$$

For the outer cylinder,

$$\sigma_{ci} = 29.51p$$

$$= (29.51)(11.07 \text{ MPa}) = \boxed{327 \text{ MPa}}$$

10. *Customary U.S. Solution*

The elongation in the unthreaded part of the bolt is

$$\delta_1 = \frac{\sigma_1 L_1}{E} = \frac{\left(40{,}000 \ \frac{\text{lbf}}{\text{in}^2}\right)(4.0 \text{ in} - 0.75 \text{ in})}{2.9 \times 10^7 \ \frac{\text{lbf}}{\text{in}^2}}$$

$$= 0.00448 \text{ in}$$

From Table 51.5, the stress area for a $^3/_4$-16 UNF bolt is $A_2 = 0.373 \text{ in}^2$.

The stress area for the unthreaded part of the bolt is

$$A_1 = \frac{\pi d^2}{4} = \frac{\pi (0.75 \text{ in})^2}{4} = 0.4418 \text{ in}^2$$

The stress in the threaded part of the bolt is

$$\sigma_2 = \sigma_1 \left(\frac{A_1}{A_2}\right) = \left(40{,}000 \ \frac{\text{lbf}}{\text{in}^2}\right)\left(\frac{0.4418 \text{ in}^2}{0.373 \text{ in}^2}\right)$$

$$= 47{,}378 \text{ lbf/in}^2$$

Assume half of the bolt in the nut contributes to elongation. The elongation in the threaded part of the bolt, including three threads in the nut, is

$$\delta_2 = \frac{\sigma_2 L_2}{E}$$

$$= \frac{\left(47{,}378 \ \frac{\text{lbf}}{\text{in}^2}\right)}{2.9 \times 10^7 \ \frac{\text{lbf}}{\text{in}^2}}$$
$$\times \left(0.75 \text{ in} + (3 \text{ threads})\left(\frac{1}{16 \ \frac{\text{threads}}{\text{in}}}\right)\right)$$

$$= 0.00153 \text{ in}$$

The total stretch of the bolt is

$$\delta = \delta_1 + \delta_2 = 0.00448 \text{ in} + 0.00153 \text{ in} = \boxed{0.00601 \text{ in}}$$

The answer is (A).

SI Solution

The elongation in the unthreaded part of the bolt is

$$\delta_1 = \frac{\sigma_1 L_1}{E} = \frac{(280 \text{ MPa})(100 \text{ mm} - 19 \text{ mm})}{(200 \text{ GPa})\left(1000 \ \frac{\text{MPa}}{\text{GPa}}\right)}$$

$$= 0.1134 \text{ mm}$$

From Table 51.5, the stress area for a $^3/_4$-16 UNF bolt is

$$A_1 = (0.373 \text{ in}^2)\left(25.4 \ \frac{\text{mm}}{\text{in}}\right)^2 = 240.64 \text{ mm}^2$$

The stress area for the unthreaded part of the bolt is

$$A_2 = \frac{\pi d^2}{4} = \frac{\pi (19 \text{ mm})^2}{4} = 283.53 \text{ mm}^2$$

The stress in the threaded part of the bolt is

$$\sigma_2 = \sigma_1 \left(\frac{A_2}{A_1}\right) = (280 \text{ MPa})\left(\frac{283.53 \text{ mm}^2}{240.64 \text{ mm}^2}\right)$$

$$= 330 \text{ MPa}$$

Assume half of the bolt in the nut contributes to elongation.

$$\delta_2 = \frac{\sigma_2 L_2}{E}$$

$$= \frac{(330 \text{ MPa})\left(\begin{array}{c} 19 \text{ mm} + (3 \text{ threads}) \\ \times \left(\frac{1}{16 \ \frac{\text{threads}}{\text{in}}}\right)\left(25.4 \ \frac{\text{mm}}{\text{in}}\right) \end{array}\right)}{(200 \text{ GPa})\left(1000 \ \frac{\text{MPa}}{\text{GPa}}\right)}$$

$$= 0.0392 \text{ mm}$$

The total stretch of the bolt is

$$\delta = \delta_1 + \delta_2 = 0.1134 \text{ mm} + 0.0392 \text{ mm} = \boxed{0.1526 \text{ mm}}$$

The answer is (A).

11. *Customary U.S. Solution*

For the cast-iron hub,

$$r_o = \frac{12 \text{ in}}{2} = 6 \text{ in}$$

$$r_i = \frac{6 \text{ in}}{2} = 3 \text{ in}$$

From Table 51.2, the circumferential stress at the inside radius due to internal pressure, p, is

$$\sigma_{ci} = \frac{(r_o^2 + r_i^2)p}{r_o^2 - r_i^2}$$

The ultimate strength of cast iron is $S_{ut} = 30,000 \text{ lbf/in}^2$.

For cast iron, nonlinearity begins at approximately $^1/_6 S_{ut}$.

$$\sigma_{ci} = \frac{S_{ut}}{\text{FS}} = \frac{30,000 \dfrac{\text{lbf}}{\text{in}^2}}{6} = 5000 \text{ lbf/in}^2$$

The σ_{ci} relationship is

$$5000 \frac{\text{lbf}}{\text{in}^2} = \frac{\left((6 \text{ in})^2 + (3 \text{ in})^2\right) p}{(6 \text{ in})^2 - (3 \text{ in})^2}$$

$$p = 3000 \text{ lbf/in}^2$$

The radial stress is

$$\sigma_{ri} = -p = -3000 \text{ lbf/in}^2$$

From Eq. 51.18, the radial deflection at the inside of the cast iron hub is

$$\Delta r_{\text{inner}} = \frac{r\left(\sigma_c - \nu(\sigma_r + \sigma_l)\right)}{E}$$

$$= \frac{(3 \text{ in})\left(5000 \dfrac{\text{lbf}}{\text{in}^2} - (0.27)\left(-3000 \dfrac{\text{lbf}}{\text{in}^2} + 0\right)\right)}{1.45 \times 10^7 \dfrac{\text{lbf}}{\text{in}^2}}$$

$$= 1.202 \times 10^{-3} \text{ in}$$

For the steel shaft,

$$r_o = \frac{6 \text{ in}}{2} = 3 \text{ in}$$

$$r_i = 0$$

From Table 51.2, the circumferential stress at the outside radius due to external pressure, p, is

$$\sigma_{co} = \frac{-(r_o^2 + r_i^2)p}{r_o^2 - r_i^2}$$

$$= \frac{\left((3 \text{ in})^2 + (0)^2\right)\left(3000 \dfrac{\text{lbf}}{\text{in}^2}\right)}{(3 \text{ in})^2 - (0)^2}$$

$$= -3000 \text{ lbf/in}^2$$

The radial stress is

$$\sigma_{ro} = -p = -3000 \text{ lbf/in}^2$$

From Eq. 51.18, the radial deflection at the outside of the steel shaft is

$$\Delta r_{\text{outer}} = \frac{r\left(\sigma_c - \nu(\sigma_r + \sigma_l)\right)}{E}$$

$$= \frac{(3 \text{ in})\left(-3000 \dfrac{\text{lbf}}{\text{in}^2} - (0.30)\left(-3000 \dfrac{\text{lbf}}{\text{in}^2} + 0\right)\right)}{2.9 \times 10^7 \dfrac{\text{lbf}}{\text{in}^2}}$$

$$= -2.172 \times 10^{-4} \text{ in}$$

From Eq. 51.20, the radial interference is

$$I_{\text{radial}} = |\Delta r_{\text{outer}}| + |\Delta r_{\text{inner}}|$$

$$= |-2.172 \times 10^{-4} \text{ in}| + |1.202 \times 10^{-3} \text{ in}|$$

$$= \boxed{1.419 \times 10^{-3} \text{ in}}$$

The answer is (A).

SI Solution

For the cast-iron hub,

$$r_o = \frac{300 \text{ mm}}{2} = 150 \text{ mm}$$

$$r_i = \frac{150 \text{ mm}}{2} = 75 \text{ mm}$$

From Table 51.2, the circumferential stress at the inside radius due to internal pressure, p, is

$$\sigma_{ci} = \frac{(r_o^2 + r_i^2)\, p}{r_o^2 - r_i^2}$$

The ultimate strength of cast iron is $S_{ut} = 210 \text{ MPa}$.

For cast iron, nonlinearity begins at approximately $^1/_6 S_{ut}$.

$$\sigma_{ci} = \frac{S_{ut}}{\text{FS}} = \frac{210 \text{ MPa}}{6} = 35 \text{ MPa}$$

The σ_{ci} relationship is

$$35 \text{ MPa} = \frac{\left((150 \text{ mm})^2 + (75 \text{ mm})^2\right) p}{(150 \text{ mm})^2 - (75 \text{ mm})^2}$$

$$p = 21 \text{ MPa}$$

The radial stress is

$$\sigma_{ri} = -p = -21 \text{ MPa}$$

From Eq. 51.18, the radial deflection at the inside of the cast iron hub is

$$\Delta r_{\text{inner}} = \frac{r\left(\sigma_c - \nu(\sigma_r + \sigma_l)\right)}{E}$$

$$= \frac{(75 \text{ mm})\left(35 \text{ MPa} - (0.27)(-21 \text{ MPa} + 0)\right)}{(100 \text{ GPa})\left(1000 \dfrac{\text{MPa}}{\text{GPa}}\right)}$$

$$= 0.03050 \text{ mm}$$

For the steel shaft,

$$r_o = \frac{150 \text{ mm}}{2} = 75 \text{ mm}$$

$$r_i = 0$$

From Table 51.2, the circumferential stress at the outside radius due to external pressure, p, is

$$\sigma_{co} = \frac{-(r_o^2 + r_i^2)p}{r_o^2 - r_i^2}$$

$$= \frac{-\left((75 \text{ mm})^2 + (0)^2\right)(21 \text{ MPa})}{(75 \text{ mm})^2 - (0)^2}$$

$$= -21 \text{ MPa}$$

The radial stress is

$$\sigma_{ro} = -p = -21 \text{ MPa}$$

From Eq. 51.18, the radial deflection at the outside of the steel shaft is

$$\Delta r_{\text{outer}} = \frac{r\left(\sigma_c - \nu(\sigma_r + \sigma_l)\right)}{E}$$

$$= \frac{(75 \text{ mm})\left(-21 \text{ MPa} - (0.30)(-21 \text{ MPa} + 0)\right)}{(200 \text{ GPa})\left(1000 \dfrac{\text{MPa}}{\text{GPa}}\right)}$$

$$= -0.00551 \text{ mm}$$

From Eq. 51.20, the radial interference is

$$I_{\text{radial}} = |\Delta r_{\text{outer}}| + |\Delta r_{\text{inner}}|$$

$$= |-0.00551 \text{ mm}| + |0.03050 \text{ mm}|$$

$$= \boxed{0.03601 \text{ mm}}$$

The answer is (A).

12. *Customary U.S. Solution*

First, find the properties of the bolt area. For an equilateral triangular layout, the distance from the centroid of the bolt group to each bolt is

$$l = \left(\tfrac{2}{3}\right)(9 \text{ in}) = 6 \text{ in}$$

The area of each bolt is

$$A = \frac{\pi d^2}{4} = \frac{\pi (0.75 \text{ in})^2}{4} = 0.442 \text{ in}^2$$

The polar moment of inertia of a bolt about the centroid is

$$J = Al^2 = (0.442 \text{ in}^2)(6 \text{ in})^2 = 15.91 \text{ in}^4$$

The vertical shear load at each bolt is

$$F_v = \frac{F}{3} = 0.333F \quad [\text{in lbf}]$$

The moment applied to each bolt is

$$M = \frac{(20 \text{ in})F}{3} = 6.67F \quad [\text{in in-lbf}]$$

The vertical shear stress in each bolt is

$$\tau_v = \frac{F_v}{A} = \frac{0.333F}{0.442 \text{ in}^2} = 0.753F \quad [\text{in lbf/in}^2]$$

The torsional shear stress in each bolt is

$$\tau = \frac{Ml}{J} = \frac{(6.67F)(6 \text{ in})}{15.91 \text{ in}^4} = 2.52F \quad [\text{in lbf/in}^2]$$

The most highly stressed bolt is the rightmost bolt. The shear stress configuration is

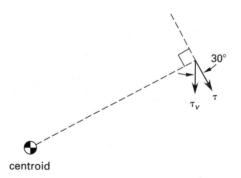

centroid

The stresses are combined to find the maximum stress.

$$\tau_{\text{max}} = \sqrt{(\tau \sin 30°)^2 + (\tau_v + \tau \cos 30°)^2}$$

$$15{,}000 \; \frac{\text{lbf}}{\text{in}^2} = \sqrt{\begin{array}{c}\left((2.52F)(\sin 30°)\right)^2 \\ + \left(0.753F + (2.52F)(\cos 30°)\right)^2\end{array}}$$

$$= 3.19F$$

$$F = \boxed{4702 \text{ lbf}}$$

The answer is (B).

SI Solution

First, find the properties of the bolt area. For an equilateral triangular layout, the distance from the centroid of the bolt group to each bolt is

$$l = \left(\tfrac{2}{3}\right)(230 \text{ mm}) = 153 \text{ mm}$$

The area of each bolt is

$$A = \frac{\pi d^2}{4} = \frac{\pi(19 \text{ mm})^2 \left(\dfrac{1 \text{ m}}{10^3 \text{ mm}}\right)^2}{4} = 2.835 \times 10^{-4} \text{ m}^2$$

The polar moment of inertia of a bolt about the centroid is

$$J = Al^2 = (2.835 \times 10^{-4} \text{ m}^2)(0.153 \text{ m})^2$$
$$= 6.636 \times 10^{-6} \text{ m}^4$$

The vertical shear load at each bolt is

$$F_v = \frac{F}{3} = 0.333F \quad [\text{in N}]$$

The moment applied to each bolt is

$$M = \frac{(0.5 \text{ m})F}{3} = 0.167F \text{ N·m}$$

The vertical shear stress in each bolt is

$$\tau_v = \frac{F_v}{A} = \frac{0.333F}{2.835 \times 10^{-4} \text{ m}^2} = 1.1746 \times 10^3 F \quad [\text{in N/m}^2]$$

The torsional shear stress in each bolt is

$$\tau = \frac{Ml}{J} = \frac{(0.167F)(0.15 \text{ m})}{6.636 \times 10^{-6} \text{ m}^4}$$
$$= 3.775 \times 10^3 F \quad [\text{in N/m}^2]$$

The most highly stressed bolt is the rightmost bolt. The shear stress configuration is

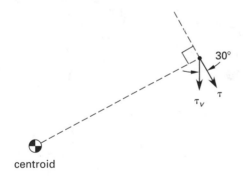

The stresses are combined to find the maximum stress.

$$\tau_{\text{max}} = \sqrt{(\tau \sin 30°)^2 + (\tau_v + \tau \cos 30°)^2}$$

$$\begin{aligned}
(100 \text{ MPa}) \\
\times \left(1000 \, \frac{\text{kPa}}{\text{MPa}}\right)
\end{aligned} = \sqrt{\begin{aligned} &\left((3.775 \times 10^3 F)(\sin 30°)\right)^2 \\ &+ \left(1.1746 \times 10^3 F \right. \\ &\left. + (3.775 \times 10^3 F)(\cos 30°)\right)^2 \end{aligned}}$$

$$= 4.83 \times 10^3 F \quad [\text{in N/m}^2]$$

$$\boxed{F = 20.7 \text{ kN}}$$

The answer is (B).

13. *Customary U.S. Solution*

The fillet weld consists of three basic shapes, each with a throat size of t.

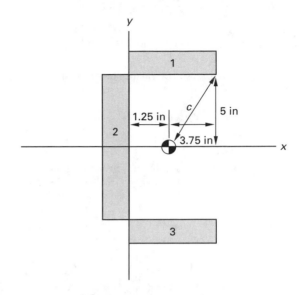

First, find the centroid.

By inspection, $y_c = 0$.

The areas of the basic shapes are

$$A_1 = 5t \quad [\text{in in}^2]$$
$$A_2 = 10t \quad [\text{in in}^2]$$
$$A_3 = 5t \quad [\text{in in}^2]$$

The x-components of the centroids of the basic shapes are

$$x_{c1} = \frac{5 \text{ in}}{2} = 2.5 \text{ in}$$
$$x_{c2} = -\frac{t}{2} = 0 \quad [\text{for small } t]$$
$$x_{c3} = \frac{5 \text{ in}}{2} = 2.5 \text{ in}$$
$$x_c = \frac{\displaystyle\sum_i A_i x_{ci}}{A_i}$$
$$= \frac{(5t)(2.5 \text{ in}) + (10t)(0) + (5t)(2.5 \text{ in})}{5t + 10t + 5t}$$
$$= 1.25 \text{ in}$$

Next, find the centroidal moments of inertia.

The moments of inertia of basic shape 1 about its own centroid are

$$I_{cx1} = \frac{bh^3}{12} = \frac{(5 \text{ in})t^3}{12} = 0.417t^3 \quad [\text{in in}^4]$$
$$I_{cy1} = \frac{bh^3}{12} = \frac{t(5 \text{ in})^3}{12} = 10.417t \quad [\text{in in}^4]$$

The moments of inertia of basic shape 2 about its own centroid are

$$I_{cx2} = \frac{bh^3}{12} = \frac{t(10 \text{ in})^3}{12} = 83.333t \quad [\text{in in}^4]$$

$$I_{cy2} = \frac{bh^3}{12} = \frac{(10 \text{ in})t^3}{12} = 0.833t^3 \quad [\text{in in}^4]$$

The moments of inertia of basic shape 3 about its own centroid are the same as for basic shape 1.

$$I_{cx3} = I_{cx1} = 0.417t^3 \quad [\text{in in}^4]$$

$$I_{cy3} = I_{cy1} = 10.417t \quad [\text{in in}^4]$$

From the parallel axis theorem, the moments of inertia of basic shape 1 about the centroidal axis of the section are

$$I_{x1} = I_{cx1} + A_1 d_1^2 = 0.417t^3 + (5t)(5 \text{ in})^2$$

$$= 0.417t^3 + 125t$$

$$I_{y1} = I_{cy1} + A_1 d_1^2$$

$$= 10.417t + 5t \left(\frac{5 \text{ in}}{2} - 1.25 \text{ in}\right)^2$$

$$= 18.23t \quad [\text{in in}^4]$$

The moments of inertia of basic shape 2 about the centroidal axis of the section are

$$I_{x2} = I_{cx2} + A_2 d_2^2 = 83.333t + (10t)(0)$$

$$= 83.333t \quad [\text{in in}^4]$$

$$I_{y2} = I_{cy2} + A_2 d_2^2 = 0.833t^3 + (10t)(1.25 \text{ in})^2$$

$$= 0.833t^3 + 15.625t$$

The moment of inertia of basic shape 3 about the centroidal axis of the section are the same as for basic shape 1.

$$I_{x3} = I_{x1} = 0.417t^3 + 125t$$

$$I_{y3} = I_{y1} = 18.23t \quad [\text{in in}^4]$$

The total moments of inertia about the centroidal axis of the section are

$$I_x = I_{x1} + I_{x2} + I_{x3}$$

$$= 0.417t^3 + 125t + 83.333t + 0.417t^3 + 125t$$

$$= 0.834t^3 + 333.33t$$

$$I_y = I_{y1} + I_{y2} + I_{y3}$$

$$= 18.23t + 0.833t^3 + 15.625t + 18.23t$$

$$= 0.833t^3 + 52.09t$$

Since t will be small and since the coefficient of the t^3 term is smaller than the coefficient of the t term, the higher-order t^3 term may be neglected.

$$I_x = 333.33t \quad [\text{in in}^4]$$

$$I_y = 52.09t \quad [\text{in in}^4]$$

The polar moment of inertia for the section is

$$J = I_x + I_y = 333.33t + 52.09t$$

$$= 385.42t \quad [\text{in in}^4]$$

The maximum shear stress will occur at the rightmost point of basic shape 1 since this point is farthest from the centroid of the section.

This distance is

$$c = \sqrt{(5 \text{ in} - 1.25 \text{ in})^2 + \left(\frac{10 \text{ in}}{2}\right)^2}$$

$$= 6.25 \text{ in}$$

The applied moment is

$$M = Fd = (10{,}000 \text{ lbf})(12 \text{ in} + 5 \text{ in} - 1.25 \text{ in})$$

$$= 157{,}500 \text{ in-lbf}$$

The torsional shear stress is

$$\tau = \frac{Mc}{J} = \frac{(157{,}500 \text{ in-lbf})(6.25 \text{ in})}{385.42t} = \frac{2554.0}{t} \text{ lbf/in}^2$$

The shear stress is perpendicular to the line from the point to the centroid. The x- and y-components are shown on the figure.

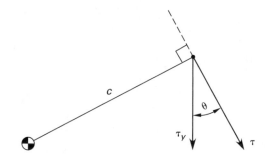

The angle θ is

$$\theta = 90° - \tan^{-1}\left(\frac{5 \text{ in} - 1.25 \text{ in}}{\dfrac{10 \text{ in}}{2}}\right) = 53.1°$$

$$\tau_x = \tau \sin\theta = \left(\frac{2554.0 \dfrac{\text{lbf}}{\text{in}^2}}{t}\right)(\sin 53.1°)$$

$$= \frac{2042.4}{t} \text{ lbf/in}^2$$

$$\tau_y = \tau \cos\theta = \left(\frac{2554.0 \dfrac{\text{lbf}}{\text{in}^2}}{t}\right)(\cos 53.1°)$$

$$= \frac{1553.5}{t} \text{ lbf/in}^2$$

The vertical shear stress due to the load is

$$\tau_{vy} = \frac{F}{\sum A_i} = \frac{10{,}000 \text{ lbf}}{5t + 10t + 5t}$$

$$= \frac{500}{t} \text{ lbf/in}^2$$

The resultant shear stress is

$$\tau = \sqrt{\tau_x^2 + (\tau_y + \tau_{vy})^2}$$

$$= \sqrt{\left(\frac{2042.4}{t}\right)^2 + \left(\frac{1553.5}{t} + \frac{500}{t}\right)^2}$$

$$= \frac{2896.2}{t} \text{ lbf/in}^2$$

For a maximum shear stress of 8000 lbf/in^2,

$$8000 \; \frac{\text{lbf}}{\text{in}^2} = \frac{2896.2}{t} \text{ lbf/in}^2$$

$$t = 0.362 \text{ in}$$

From Eq. 51.42, the required weld size is

$$y = \frac{t}{0.707} = \frac{0.362 \text{ in}}{0.707} = \boxed{0.512 \text{ in}}$$

(App. 51.A can also be used to solve this problem.)

The answer is (C).

SI Solution

The fillet weld consists of three basic shapes, each with a throat size of t.

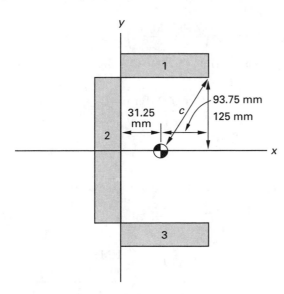

First, find the centroid.

By inspection, $y_c = 0$.

The areas of the basic shapes are

$$A_1 = 125t \quad [\text{in mm}^2]$$
$$A_2 = 250t \quad [\text{in mm}^2]$$
$$A_3 = 125t \quad [\text{in mm}^2]$$

The x-components of the centroids of the basic shapes are

$$x_{c1} = \frac{125 \text{ mm}}{2} = 62.5 \text{ mm}$$

$$x_{c2} = -\frac{t}{2} = 0 \quad [\text{for small } t]$$

$$x_{c3} = \frac{125 \text{ mm}}{2} = 62.5 \text{ mm}$$

$$x_c = \frac{\sum_i A_i x_{ci}}{A_i}$$

$$= \frac{\begin{array}{c}(125t)(62.5 \text{ mm}) \\ +(250t)(0) + (125t)(62.5 \text{ mm})\end{array}}{125t + 250t + 125t}$$

$$= 31.25 \text{ mm}$$

Next, find the centroidal moments of inertia.

The moments of inertia of basic shape 1 about its own centroid are

$$I_{cx1} = \frac{bh^3}{12} = \frac{(125 \text{ mm})t^3}{12} = 10.42t^3 \quad [\text{in mm}^4]$$

$$I_{cy1} = \frac{bh^3}{12} = \frac{t(125 \text{ mm})^3}{12} = 162\,760t \quad [\text{in mm}^4]$$

The moments of inertia of basic shape 2 about its own centroid are

$$I_{cx2} = \frac{bh^3}{12} = \frac{t(250 \text{ mm})^3}{12} = 1\,302\,083t \quad [\text{in mm}^4]$$

$$I_{cy2} = \frac{bh^3}{12} = \frac{(250 \text{ mm})t^3}{12} = 20.83t^3 \quad [\text{in mm}^4]$$

The moments of inertia of basic shape 3 about its own centroid are the same as for basic shape 1.

$$I_{cx3} = I_{cx1} = 10.42t^3 \quad [\text{in mm}^4]$$

$$I_{cy3} = I_{cy1} = 162\,760t \quad [\text{in mm}^4]$$

From the parallel axis theorem, the moments of inertia of basic shape 1 about the centroidal axis of the section are

$$I_{x1} = I_{cx1} + A_1 d_1^2$$

$$= 10.42t^3 + (125t)(125 \text{ mm})^2$$

$$= 10.42t^3 + 1\,953\,125t$$

$$I_{y1} = I_{cy1} + A_1 d_1^2$$

$$= 162\,760t + (125t)\left(\frac{125 \text{ mm}}{2} - 31.25 \text{ mm}\right)^2$$

$$= 284\,830t \quad [\text{in mm}^4]$$

The moments of inertia of basic shape 2 about the centroidal axis of the section are

$$
\begin{aligned}
I_{x2} &= I_{cx2} + A_2 d_2^2 \\
&= 1\,302\,083t + (250t)(0)^2 \\
&= 1\,302\,083t \quad [\text{in mm}^4] \\
I_{y2} &= I_{cy2} + A_2 d_2^2 \\
&= 20.83t^3 + (250t)(31.25 \text{ mm})^2 \\
&= 20.83t^3 + 244\,140t
\end{aligned}
$$

The moments of inertia of basic shape 3 about the centroidal axis of the section are the same as for basic shape 1.

$$
\begin{aligned}
I_{x3} &= I_{x1} = 10.42t^3 + 1\,953\,125t \\
I_{y3} &= I_{y1} = 284\,830t \quad [\text{in mm}^4]
\end{aligned}
$$

The total moments of inertia about the centroidal axis of the section are

$$
\begin{aligned}
I_x &= I_{x1} + I_{x2} + I_{x3} \\
&= 10.42t^3 + 1\,953\,125t + 1\,302\,083t + 10.42t^3 \\
&\quad + 1\,953\,125t \\
&= 20.84t^3 + 5\,208\,333t \\
I_y &= I_{y1} + I_{y2} + I_{y3} \\
&= 284\,830t + 20.83t^3 + 244\,140t \\
&\quad + 284\,830t \\
&= 20.83t^3 + 813\,800t
\end{aligned}
$$

Since t will be small and since the coefficient of the t^3 term is smaller than the coefficient of the t term, the t^3 term may be neglected.

$$
\begin{aligned}
I_x &= 5\,208\,333t \quad [\text{in mm}^4] \\
I_y &= 813\,800t \quad [\text{in mm}^4]
\end{aligned}
$$

The maximum shear stress will occur at the rightmost point of basic shape 1 since this point is farthest from the centroid of the section.

The distance is

$$
\begin{aligned}
c &= \sqrt{(125 \text{ mm} - 31.25 \text{ mm})^2 + \left(\frac{250 \text{ mm}}{2}\right)^2} \\
&= 156.25 \text{ mm}
\end{aligned}
$$

The polar moment of inertia for the section is

$$
\begin{aligned}
J &= I_x + I_y = 5\,208\,333t + 813\,800t \\
&= 6\,022\,133t \quad [\text{in mm}^4]
\end{aligned}
$$

The applied moment is

$$
\begin{aligned}
M &= Fd = (44 \text{ kN})(300 \text{ mm} + 125 \text{ mm} - 31.25 \text{ mm}) \\
&= 17\,325 \text{ kN·mm}
\end{aligned}
$$

The torsional shear stress is

$$
\begin{aligned}
\tau &= \frac{Mc}{J} \\
&= \left(\frac{(17\,325 \text{ kN·mm})(156.25 \text{ mm})}{6\,022\,133t}\right)\left(10^3 \frac{\text{mm}}{\text{m}}\right)^2 \\
&\quad \times \left(\frac{1 \text{ MPa}}{1000 \text{ kPa}}\right) \\
&= \frac{449.5}{t} \text{ MPa}
\end{aligned}
$$

The shear stress is perpendicular to the line from the point to the centroid. The x- and y-components are shown on the figure.

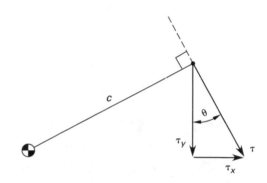

The angle θ is

$$
\theta = 90° - \tan^{-1}\left(\frac{125 \text{ mm} - 31.25 \text{ mm}}{\dfrac{250 \text{ mm}}{2}}\right) = 53.1°
$$

$$
\begin{aligned}
\tau_x &= \tau \sin\theta = \left(\frac{449.5 \text{ MPa}}{t}\right)(\sin 53.1°) \\
&= \frac{359.5 \text{ MPa}}{t} \\
\tau_y &= \tau \cos\theta = \left(\frac{449.5 \text{ MPa}}{t}\right)(\cos 53.1°) \\
&= \frac{269.9 \text{ MPa}}{t}
\end{aligned}
$$

The vertical shear stress due to the load is

$$
\begin{aligned}
\tau_{vy} &= \frac{F}{\sum A_i} \\
&= \frac{(44 \text{ kN})\left(\dfrac{1 \text{ MPa}}{1000 \text{ kPa}}\right)}{(125t + 250t + 125t)\left(\dfrac{1 \text{ m}}{10^3 \text{ mm}}\right)^2} \\
&= \frac{88.0}{t} \text{ MPa}
\end{aligned}
$$

The resultant shear stress is

$$\tau = \sqrt{\tau_x^2 + (\tau_y + \tau_{vy})^2}$$

$$= \sqrt{\left(\frac{359.5}{t}\right)^2 + \left(\frac{269.9}{t} + \frac{88.0}{t}\right)^2}$$

$$= \frac{507.3}{t} \text{ MPa}$$

For a maximum shear stress of 55 MPa,

$$55 \text{ MPa} = \frac{507.3}{t} \text{ MPa}$$

$$t = 9.22 \text{ mm}$$

From Eq. 51.42, the required weld size is

$$y = \frac{t}{0.707} = \frac{9.22 \text{ mm}}{0.707} = \boxed{13.0 \text{ mm}}$$

(App. 51.A can also be used to solve this problem.)

The answer is (C).

14. *Customary U.S. Solution*

The rotational speed is

$$\omega = \frac{\left(3500 \dfrac{\text{rev}}{\text{min}}\right)\left(2\pi \dfrac{\text{rad}}{\text{rev}}\right)}{60 \dfrac{\text{sec}}{\text{min}}} = 366.5 \text{ rad/sec}$$

From Eq. 57.14, the tangential stress at the inside radius of the flywheel is

$$\sigma = \left(\frac{\rho\omega^2}{4g_c}\right)\left((3+\nu)r_o^2 + (1-\nu)r_i^2\right)$$

$$= \left(\frac{\left(0.283 \dfrac{\text{lbm}}{\text{in}^3}\right)\left(366.5 \dfrac{\text{rad}}{\text{sec}}\right)^2}{(4)\left(386 \dfrac{\text{in-lbm}}{\text{lbf-sec}^2}\right)}\right)$$

$$\times \left((3+0.3)(8 \text{ in})^2 + (1-0.3)(1 \text{ in})^2\right)$$

$$= 5217 \text{ lbf/in}^2$$

The strain is

$$\epsilon = \frac{\sigma}{E} = \frac{5217 \dfrac{\text{lbf}}{\text{in}^2}}{2.9 \times 10^7 \dfrac{\text{lbf}}{\text{in}^2}} = 1.799 \times 10^{-4}$$

The change in the inner radius is

$$\Delta r = r\epsilon = (1 \text{ in})(1.799 \times 10^{-4}) = 1.799 \times 10^{-4} \text{ in}$$

Since the shaft and hub are both steel, the radial interference is obtained from Eq. 51.23.

$$I_{\text{radial}} = \frac{I_{\text{diametral}}}{2} = \left(\frac{2pr_{\text{shaft}}}{E}\right)\left(\frac{1}{1 - \left(\dfrac{r_{\text{shaft}}}{r_{o,\text{hub}}}\right)^2}\right)$$

$$= \left(\frac{(2)\left(1250 \dfrac{\text{lbf}}{\text{in}^2}\right)(1 \text{ in})}{2.9 \times 10^7 \dfrac{\text{lbf}}{\text{in}^2}}\right)\left(\frac{1}{1 - \left(\dfrac{1 \text{ in}}{8 \text{ in}}\right)^2}\right)$$

$$= 8.758 \times 10^{-5} \text{ in}$$

The required initial interference is

$$I = \Delta r + I_{\text{radial}}$$

$$= 1.799 \times 10^{-4} \text{ in} + 8.758 \times 10^{-5} \text{ in}$$

$$= \boxed{2.67 \times 10^{-4} \text{ in}}$$

The answer is (D).

SI Solution

The rotational speed is

$$\omega = \frac{\left(3500 \dfrac{\text{rev}}{\text{min}}\right)\left(2\pi \dfrac{\text{rad}}{\text{rev}}\right)}{60 \dfrac{\text{s}}{\text{min}}} = 366.5 \text{ rad/s}$$

From Eq. 57.14, the tangential stress at the inside radius of the flywheel is

$$\sigma = \left(\frac{\rho\omega}{4}\right)\left((3+\nu)r_o^2 + (1-\nu)r_i^2\right)$$

$$= \left(\frac{\left(7830 \dfrac{\text{kg}}{\text{m}^3}\right)\left(366.5 \dfrac{\text{rad}}{\text{s}}\right)^2}{4}\right)$$

$$\times \left((3+0.3)(0.203 \text{ m})^2 + (1-0.3)(0.025 \text{ m})^2\right)$$

$$= 35.87 \times 10^6 \text{ Pa} \quad (35.87 \text{ MPa})$$

The strain is

$$\epsilon = \frac{\sigma}{E} = \frac{35.87 \text{ MPa}}{(200 \text{ GPa})\left(1000 \dfrac{\text{MPa}}{\text{GPa}}\right)} = 1.794 \times 10^{-4}$$

The change in inner radius is

$$\Delta r = r\epsilon = (25 \text{ mm})(1.794 \times 10^{-4})$$

$$= 44.85 \times 10^{-4} \text{ mm}$$

Since the shaft and hub are both steel, the radial interference is obtained from Eq. 51.23.

$$I_{\text{radial}} = \frac{I_{\text{diametral}}}{2} = \left(\frac{2pr_{\text{shaft}}}{E}\right)\left(\frac{1}{1 - \left(\frac{r_{\text{shaft}}}{r_{o,\text{hub}}}\right)^2}\right)$$

$$= \left(\frac{(2)(8.6 \text{ MPa})(25 \text{ mm})}{(200 \text{ GPa})\left(1000 \frac{\text{MPa}}{\text{GPa}}\right)}\right)$$

$$\times \left(\frac{1}{1 - \left(\frac{25 \text{ mm}}{203 \text{ mm}}\right)^2}\right)$$

$$= 21.83 \times 10^{-4} \text{ mm}$$

The required initial interference is

$$I = \Delta r + I_{\text{radial}}$$
$$= 44.85 \times 10^{-4} \text{ mm} + 21.83 \times 10^{-4} \text{ mm}$$
$$= \boxed{6.67 \times 10^{-3} \text{ mm}}$$

The answer is (D).

52 Advanced Machine Design

PRACTICE PROBLEMS

1. A spring with 12 active coils and a spring index of 9 supports a static load of 50 lbf (220 N) with a deflection of 0.5 in (12 mm). The shear modulus of the spring material is 1.2×10^7 lbf/in^2 (83 GPa).

(a) What is the theoretical wire diameter?

- (A) 0.58 in (15 mm)
- (B) 0.63 in (16 mm)
- (C) 0.69 in (18 mm)
- (D) 0.74 in (19 mm)

(b) What is the mean spring diameter?

- (A) 5.2 in (140 mm)
- (B) 6.4 in (160 mm)
- (C) 7.1 in (180 mm)
- (D) 7.9 in (200 mm)

2. A severe service valve spring is to be manufactured from unpeened ASTM A230 steel wire in standard W & M sizes operating continuously between 20 lbf and 30 lbf (100 N and 150 N). The valve lift is 0.3 in (8 mm). The spring index is 10. The factor of safety is 1.5.

(a) What is the wire diameter?

- (A) 0.12 in (3.0 mm)
- (B) 0.15 in (4.1 mm)
- (C) 0.21 in (53 mm)
- (D) 0.28 in (71 mm)

(b) What is the spring constant?

- (A) 10 lbf/in (1.9 kN/m)
- (B) 21 lbf/in (3.9 kN/m)
- (C) 33 lbf/in (6.3 kN/m)
- (D) 48 lbf/in (92 kN/m)

(c) What is the number of active coils if the ends are squared and ground?

- (A) 4.7
- (B) 5.2
- (C) 6.4
- (D) 8.6

(d) What is the total number of coils?

- (A) 6.2
- (B) 6.9
- (C) 7.8
- (D) 8.4

(e) What is the solid height?

- (A) 1.2 in (35 mm)
- (B) 1.7 in (43 mm)
- (C) 2.3 in (58 mm)
- (D) 28 in (71 mm)

(f) What is the spring force at solid height?

- (A) 32 lbf (10 N)
- (B) 46 lbf (250 N)
- (C) 53 lbf (290 N)
- (D) 62 lbf (330 N)

(g) What is the deflection at solid height?

- (A) 1.4 in (39 mm)
- (B) 1.7 in (43 mm)
- (C) 2.1 in (53 mm)
- (D) 2.4 in (61 mm)

(h) What is the minimum free height?

- (A) 2.1 in (53 mm)
- (B) 2.6 in (75 mm)
- (C) 3.2 in (81 mm)
- (D) 4.5 in (110 mm)

3. The material in a spring wire has a shear modulus of 1.2×10^7 lbf/in^2 (83 GPa). The maximum allowable stress under design conditions is 50,000 lbf/in^2 (350 MPa). The spring index is 7. Assume the maximum stress is experienced when a 700 lbm (320 kg) object falls from a height of 46 in (120 cm) above the tip of the spring, impacts squarely on the spring, and deflects the spring 10 in (26 cm).

(a) What is the minimum wire diameter?

- (A) 0.6 in (15 mm)
- (B) 0.9 in (23 mm)
- (C) 1.4 in (36 mm)
- (D) 1.8 in (47 mm)

(b) What is the mean coil diameter?

- (A) 8.6 in (220 mm)
- (B) 10 in (250 mm)
- (C) 13 in (330 mm)
- (D) 16 in (410 mm)

(c) What is the number of active coils?
- (A) 8
- (B) 10
- (C) 12
- (D) 14

4. A 6 in (150 mm) wide, 24 in (610 mm) long cantilever steel spring supports an 800 lbf (3.5 kN) load at its tip. The deflection is to be less than 1 in (25 mm). The bending stress is limited to 50,000 lbf/in² (345 MPa).

(a) What is the minimum thickness as limited by deflection alone?
- (A) 0.37 in (9.4 mm)
- (B) 0.45 in (11 mm)
- (C) 0.63 in (16 mm)
- (D) 0.77 in (20 mm)

(b) What is the minimum thickness as limited by bending stress?
- (A) 0.62 in (15.7 mm)
- (B) 0.66 in (16.7 mm)
- (C) 0.72 in (18.3 mm)
- (D) 0.78 in (19.8 mm)

5. A gear train is to have a speed reduction of 600:1. The gears used can have no fewer than 12 teeth and no more than 96 teeth.

(a) How many stages are needed?

(b) How many teeth should be in each gear in the gear train?

6. A mechanism is driven by a 550 hp (410 kW) motor. The motor turns at 1200 rpm, but a pair of old $14^1/_2°$ gears on 15 in (380 mm) centers is to reduce the speed of the mechanism to 270 rpm. The pinion is to be SAE 1045 steel with an endurance strength of 90,000 lbf/in² (620 MPa). The gear is to be cast steel with an endurance strength of 50,000 lbf/in² (345 MPa). A safety factor of 3 is used in the design. Use the Lewis beam strength theory. Do not check for undercutting.

(a) What are the pitch diameters?

(b) Given a pinion face width of 6 in (150 mm), what is the diametral pitch?

(c) Given a diametral pitch of 1.5 (a module of 17 mm), what should be the face width for the gear?

7. Two 20° involute spur gears are mounted such that their centers are 15 in (380 mm) apart. The pinion is untreated steel with an allowable stress of 30,000 lbf/in² (210 MPa). The gear set reduces the speed of a 250 hp motor (190 kW) with an integral speed reducer from 250 rpm to $83^1/_3$ rpm. The gear is cast steel with a maximum strength of 50,000 lbf/in² (345 MPa). A factor of safety of 3 is used. Use the Lewis beam strength theory.

(a) What are the pitch diameters?

(b) What is the diametral pitch?

(c) Given a diametral pitch of 2 in⁻¹ (0.068 mm⁻¹), how many teeth are on each gear?

(d) Given a diametral pitch of 2 in⁻¹ (0.068 mm⁻¹), what is the minimum face width of the pinion?

(e) Given a diametral pitch of 2 in⁻¹ (0.068 mm⁻¹), what is the minimum face width of the gear?

8. A wet, steel-backed asbestos clutch with hardened steel plates is being designed to transmit 300 in-lbf (33 N·m). The coefficient of friction is 0.12. Slip will occur at 300% of the rated torque. The maximum and minimum friction surface diameters are 4.5 in and 2.5 in (115 mm and 65 mm), respectively. The contact pressure is 100 psi (700 kPa). Uniform wear is expected.

(a) How many plates are needed?
- (A) 2
- (B) 3
- (C) 4
- (D) 5

(b) How many disks are needed?
- (A) 2
- (B) 3
- (C) 4
- (D) 5

9. (*Time limit: one hour*) A 3 in (76 mm) diameter shaft turns at 1200 rpm in a journal bearing. The bearing has an axial length of 3.5 in (90 mm) and a radial clearance ratio (c_r/r) of 0.001. The transverse load allocated to the bearing is 880 lbf (4 kN). The lubricating oil has a temperature of 165°F (75°C) and a viscosity of 1.184×10^{-6} lbf-sec/in² (8.16 cP).

(a) What is the minimum film thickness?

(b) What power is lost to friction?

(c) Is this bearing operating within or outside its optimum capacity?

10. (*Time limit: one hour*) Two concentric springs are constructed with squared-and-ground ends from oil-hardened steel. The ultimate tensile strength for the steel is 204,000 lbf/in² (1.4 GPa). The shear modulus for the steel is 11.5×10^6 lbf/in² (79 GPa). The springs support a static force of 150 lbf (660 N). The spring dimensions and properties are as follows.

inner spring
wire diameter:	0.177 in	(4.5 mm)
mean coil diameter:	1.5 in	(38 mm)
free length:	4.5 in	(115 mm)
total number of coils:	12.75	

outer spring
wire diameter:	0.2253 in	(5.723 mm)
mean coil diameter:	2.0 in	(51 mm)
free length:	3.75 in	(95.3 mm)
total number of coils:	10.25	

(a) What is the deflection of the inner spring?
- (A) 2.0 in
- (B) 2.4 in
- (C) 2.9 in
- (D) 3.1 in

(b) What is the maximum force exerted by the inner spring?
- (A) 60 lbf (240 N)
- (B) 70 lbf (280 N)
- (C) 80 lbf (330 N)
- (D) 90 lbf (360 N)

(c) What is the maximum shear stress in the inner spring?
- (A) 47 ksi (320 MPa)
- (B) 52 ksi (350 MPa)
- (C) 57 ksi (390 MPa)
- (D) 64 ksi (410 MPa)

(d) What is the factor of safety in shear for the inner spring?
- (A) 1.2
- (B) 1.5
- (C) 1.8
- (D) 2.2

(e) Specify the winding helix direction for each spring.
- (A) parallel
- (B) righthand
- (C) clockwise
- (D) reverse

11. (*Time limit: one hour*) The power transmission system shown consists of helical gears and AISI 1045 cold-drawn steel shafting. The gears have a 25° helix angle, a 20° normal pressure angle, and a diametral pitch of 5 (module of 5 mm). The yield strength of the 1045 steel is 69,000 lbf/in^2 (480 MPa). The mesh efficiency of each gear set is 98%. Loading is slow and steady. Use a factor of safety of 2.

(a) What is the speed of the output shaft?
- (A) 7000 rpm
- (B) 9000 rpm
- (C) 11,000 rpm
- (D) 12,000 rpm

(b) What is the torque output?
- (A) 230 in-lbf (26 N·m)
- (B) 280 in-lbf (32 N·m)
- (C) 350 in-lbf (35 N·m)
- (D) 550 in-lbf (61 N·m)

(c) What is the minimum shaft diameter at section A-A assuming static loading?

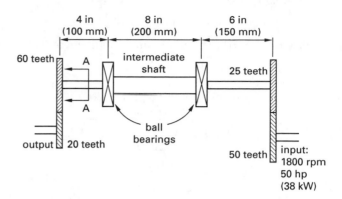

- (A) 0.51 in (13 mm)
- (B) 0.57 in (15 mm)
- (C) 0.61 in (16 mm)
- (D) 0.64 in (19 mm)

12. A piece of equipment is driven by a 15 hp (11.2 kW) motor through a standard B90 V-belt. The motor runs at 1750 rpm. The nominal speed of the equipment must be 800 rpm. The motor's sheave diameter is 10 in (254 mm).

(a) What is the pitch length of the belt?
- (A) 77 in (2000 mm)
- (B) 81 in (2100 mm)
- (C) 86 in (2200 mm)
- (D) 92 in (2300 mm)

(b) What is the pitch diameter of the sheave on the equipment drive?
- (A) 20 in (520 mm)
- (B) 22 in (560 mm)
- (C) 24 in (610 mm)
- (D) 26 in (660 mm)

SOLUTIONS

1. *Customary U.S. Solution*

Select a spring with a spring index of $C = 9$.

(a) From Eq. 52.10, the minimum wire diameter is

$$d = \frac{8FC^3 n_a}{G\delta}$$

$$= \frac{(8)(50 \text{ lbf})(9)^3(12)}{\left(1.2 \times 10^7 \dfrac{\text{lbf}}{\text{in}^2}\right)(0.5 \text{ in})}$$

$$= \boxed{0.583 \text{ in}}$$

The answer is (A).

(b) From Eq. 52.8, the mean spring diameter is

$$D = Cd = (9)(0.583 \text{ in}) = \boxed{5.247 \text{ in}}$$

The answer is (A).

SI Solution

Select a spring with a spring index of $C = 9$.

(a) From Eq. 52.10, the minimum wire diameter is

$$d = \frac{8FC^3 n_a}{G\delta}$$

$$= \left(\frac{(8)(220 \text{ N})(9)^3(12)}{(83 \times 10^9 \text{ Pa})(0.012 \text{ m})}\right)\left(1000 \frac{\text{mm}}{\text{m}}\right)$$

$$= \boxed{15.45 \text{ mm}}$$

The answer is (B).

(b) From Eq. 52.8, the mean spring diameter is

$$D = Cd = (9)(15.45 \text{ mm}) = \boxed{139.1 \text{ mm}}$$

The answer is (A).

2. *Customary U.S. Solution*

From Table 52.2, the ultimate tensile strength of ASTM A230 steel is arbitrarily selected for 0.15 in wire.

$$S_{ut} = 205,000 \text{ lbf/in}^2$$

From Table 52.3, the fatigue loading shear stress factor is 0.30. From Eq. 52.7, the maximum allowable shear stress is

$$\tau_{\max} = \frac{(\text{factor})S_{ut}}{\text{FS}} = \frac{(0.3)\left(205,000 \dfrac{\text{lbf}}{\text{in}^2}\right)}{1.5}$$

$$= 41,000 \text{ lbf/in}^2$$

From Eq. 52.13, the Wahl correction factor is

$$W = \frac{4C - 1}{4C - 4} + \frac{0.615}{C}$$

$$= \frac{(4)(10) - 1}{(4)(10) - 4} + \frac{0.615}{10} = 1.145$$

(a) From Eq. 52.14, the wire diameter is

$$d = \sqrt{\frac{8FCW}{\pi \tau_{\max}}} = \sqrt{\frac{(8)(30 \text{ lbf})(10)(1.145)}{\pi \left(41,000 \dfrac{\text{lbf}}{\text{in}^2}\right)}}$$

$$= 0.146 \text{ in}$$

Use W&M wire #9 with $d = \boxed{0.148 \text{ in.}}$

The answer is (B).

(b) From Eq. 52.2, the spring constant is

$$k = \frac{F_1 - F_2}{\delta_1 - \delta_2} = \frac{30 \text{ lbf} - 20 \text{ lbf}}{0.3 \text{ in}} = \boxed{33.33 \text{ lbf/in}}$$

The answer is (C).

(c) From Table 52.1, the shear modulus of ASTM A230 steel wire is $G = 11.5 \times 10^6 \text{ lbf/in}^2$.

From Eq. 52.12, the number of active coils is

$$n_a = \frac{Gd}{8kC^3} = \frac{\left(11.5 \times 10^6 \dfrac{\text{lbf}}{\text{in}^2}\right)(0.148 \text{ in})}{(8)\left(33.33 \dfrac{\text{lbf}}{\text{in}}\right)(10)^3}$$

$$= \boxed{6.38}$$

The answer is (C).

(d) From Table 52.4, the total number of coils for squared and ground ends is

$$n_t = n_a + 2 = 6.38 + 2 = \boxed{8.38}$$

The answer is (D).

(e) From Table 52.4, the solid height is

$$h_s = dn_t = (0.148 \text{ in})(8.38) = \boxed{1.24 \text{ in}}$$

The answer is (A).

(f) At solid height, the maximum shear stress is

$$\tau_{\max} = (\text{factor})S_{ut} = (0.3)\left(205,000 \frac{\text{lbf}}{\text{in}^2}\right)$$

$$= 61,500 \text{ lbf/in}^2$$

From Eq. 52.14, the force at solid height is

$$F_s = \frac{\tau_{max}\pi d^2}{8CW} = \frac{\left(61{,}500 \ \frac{lbf}{in^2}\right)\pi(0.148 \ in)^2}{(8)(10)(1.145)}$$

$$= \boxed{46.20 \ lbf}$$

The answer is (B).

(g) From Eq. 52.10, the deflection at solid height is

$$\delta_s = \frac{F_s}{k} = \frac{46.20 \ lbf}{33.33 \ \frac{lbf}{in}} = \boxed{1.39 \ in}$$

The answer is (A).

(h) From Eq. 52.15, the minimum free height is

$$h_f = h_s + \delta_s = 1.24 \ in + 1.39 \ in = \boxed{2.63 \ in}$$

The answer is (B).

SI Solution

From Table 52.2, the ultimate tensile strength of ASTM A230 steel wire is selected for 3.8 mm wire.

$$S_{ut} = \left(205 \ \frac{kips}{in^2}\right)\left(6.9 \ \frac{MPa}{\frac{kips}{in^2}}\right) = 1414.5 \ MPa$$

From Table 52.3, the fatigue loading shear stress factor is 0.30. From Eq. 52.7, the maximum allowable shear stress is

$$\tau_{max} = \frac{(factor)S_{ut}}{FS} = \frac{(0.3)(1414.5 \ MPa)}{1.5}$$

$$= 282.9 \ MPa$$

From Eq. 52.13, the Wahl correction factor is

$$W = \frac{4C-1}{4C-4} + \frac{0.615}{C}$$

$$= \frac{(4)(10)-1}{(4)(10)-4} + \frac{0.615}{10} = 1.145$$

(a) From Eq. 52.14, the wire diameter is

$$d = \sqrt{\frac{8FCW}{\pi\tau_{max}}} = \sqrt{\frac{(8)(150 \ N)(10)(1.145)}{\pi(282.9 \times 10^6 \ Pa)\left(\frac{1 \ m}{10^3 \ mm}\right)^2}}$$

$$= 3.93 \ mm$$

Use W&M wire #8 with a diameter of

$$d = (0.162 \ in)\left(25.4 \ \frac{mm}{in}\right) = \boxed{4.115 \ mm}$$

The answer is (B).

(b) From Eq. 52.2, the spring constant is

$$k = \frac{F_1 - F_2}{\delta_1 - \delta_2} = \frac{150 \ N - 100 \ N}{(8 \ mm)\left(\frac{1 \ m}{10^3 \ mm}\right)}$$

$$= \boxed{6.25 \times 10^3 \ N/m}$$

The answer is (C).

(c) From Table 52.1, the shear modulus of ASTM A230 steel wire is

$$G = \left(11.5 \times 10^6 \ \frac{lbf}{in^2}\right)\left(6.89 \times 10^3 \ \frac{Pa}{\frac{lbf}{in^2}}\right)$$

$$= 79.2 \times 10^9 \ Pa$$

From Eq. 52.12, the number of active coils is

$$n_a = \frac{Gd}{8kC^3} = \frac{(79.2 \times 10^9 \ Pa)(4.115 \ mm)\left(\frac{1 \ m}{10^3 \ mm}\right)}{(8)\left(6.25 \times 10^3 \ \frac{N}{m}\right)(10)^3}$$

$$= \boxed{6.52}$$

The answer is (C).

(d) From Table 52.4, the total number of coils for squared and ground ends is

$$n_t = n_a + 2 = 6.52 + 2 = \boxed{8.52}$$

The answer is (D).

(e) From Table 52.4, the solid height is

$$h_s = dn_t = (4.115 \ mm)(8.52) = \boxed{35.06 \ mm}$$

The answer is (A).

(f) At solid height, the maximum shear stress is

$$\tau_{max} = (factor)S_{ut} = (0.3)(1414.5 \ MPa) = 424.4 \ MPa$$

From Eq. 52.14, the force at solid height is

$$F_s = \frac{\tau_{max} \pi d^2}{8CW}$$

$$= \frac{(424.4 \text{ MPa}) \pi (4.115 \text{ mm})^2}{\times \left(\frac{1 \text{ m}}{10^3 \text{ mm}}\right)^2 \left(10^6 \frac{\text{Pa}}{\text{MPa}}\right)}{(8)(10)(1.145)}$$

$$= \boxed{246.5 \text{ N}}$$

The answer is (B).

(g) From Eq. 52.10, the deflection at solid height is

$$\delta_s = \frac{F_s}{k} = \left(\frac{246.5 \text{ N}}{6.25 \times 10^3 \frac{\text{N}}{\text{m}}}\right) \left(1000 \frac{\text{mm}}{\text{m}}\right)$$

$$= \boxed{39.44 \text{ mm}}$$

The answer is (A).

(h) From Eq. 52.15, the minimum free height is

$$h_f = h_s + \delta_s = 35.06 \text{ mm} + 39.44 \text{ mm} = \boxed{74.5 \text{ mm}}$$

The answer is (B).

3. *Customary U.S. Solution*

The potential energy absorbed is

$$\text{PE} = \frac{mg\Delta h}{g_c}$$

$$= \frac{(700 \text{ lbm}) \left(32.2 \frac{\text{ft}}{\text{sec}^2}\right) (46 \text{ in} + 10 \text{ in})}{32.2 \frac{\text{ft-lbm}}{\text{lbf-sec}^2}}$$

$$= 39{,}200 \text{ in-lbf}$$

The work done by the spring is equal to the potential energy.

$$W = \tfrac{1}{2} k \delta^2 = \text{PE}$$
$$k = \frac{2(\text{PE})}{\delta^2} = \frac{(2)(39{,}200 \text{ in-lbf})}{(10 \text{ in})^2} = 784 \text{ lbf/in}$$

The equivalent spring force is

$$F = k\delta = \left(784 \frac{\text{lbf}}{\text{in}}\right)(10 \text{ in}) = 7840 \text{ lbf}$$

From Eq. 52.13, the Wahl correction factor is

$$W = \frac{4C - 1}{4C - 4} + \frac{0.615}{C} = \frac{(4)(7) - 1}{(4)(7) - 4} + \frac{0.615}{7} = 1.213$$

(a) From Eq. 52.14, the wire diameter is

$$d = \sqrt{\frac{8FCW}{\pi \tau_{allowable}}} = \sqrt{\frac{(8)(7840 \text{ lbf})(7)(1.213)}{\pi \left(50{,}000 \frac{\text{lbf}}{\text{in}^2}\right)}}$$

$$= \boxed{1.84 \text{ in}}$$

The answer is (D).

(b) From Eq. 52.8, the mean coil diameter is

$$D = Cd = (7)(1.84 \text{ in}) = \boxed{12.88 \text{ in}}$$

The answer is (C).

(c) From Eq. 52.12, the number of active coils is

$$n_a = \frac{Gd}{8kC^3} = \frac{\left(1.2 \times 10^7 \frac{\text{lbf}}{\text{in}^2}\right)(1.84 \text{ in})}{(8)\left(784 \frac{\text{lbf}}{\text{in}}\right)(7)^3}$$

$$= \boxed{10.3}$$

The answer is (B).

SI Solution

The potential energy absorbed is

$$\text{PE} = mg\Delta h = (320 \text{ kg}) \left(9.81 \frac{\text{m}}{\text{s}^2}\right)(1.2 \text{ m} + 0.26 \text{ m})$$

$$= 4583 \text{ J}$$

The work done by the spring is equal to the potential energy.

$$W_k = \tfrac{1}{2} k \delta^2 = \text{PE}$$
$$k = \frac{2(\text{PE})}{\delta^2} = \frac{(2)(4583 \text{ J})}{(0.26 \text{ m})^2} = 135\,592 \text{ N/m}$$

The equivalent spring force is

$$F = k\delta = \left(135\,592 \frac{\text{N}}{\text{m}}\right)(0.26 \text{ m}) = 35\,254 \text{ N}$$

From Eq. 52.13, the Wahl correction factor is

$$W = \frac{4C - 1}{4C - 4} + \frac{0.615}{C} = \frac{(4)(7) - 1}{(4)(7) - 4} + \frac{0.615}{7} = 1.213$$

(a) From Eq. 52.14, the wire diameter is

$$d = \sqrt{\frac{8FCW}{\pi \tau_{allowable}}} = \sqrt{\frac{(8)(35\,254 \text{ N})(7)(1.213)}{\pi (350 \times 10^6 \text{ Pa}) \left(\frac{1 \text{ m}}{10^3 \text{ mm}}\right)^2}}$$

$$= \boxed{46.7 \text{ mm}}$$

The answer is (D).

(b) From Eq. 52.8, the mean coil diameter is

$$D = Cd = (7)(46.7 \text{ mm}) = \boxed{327 \text{ mm}}$$

The answer is (C).

(c) From Eq. 52.12, the number of active coils is

$$n_a = \frac{Gd}{8kC^3} = \frac{(83 \times 10^9 \text{ Pa})(46.7 \text{ mm}) \left(\dfrac{1 \text{ m}}{10^3 \text{ mm}} \right)}{(8) \left(135\,592 \dfrac{\text{N}}{\text{m}} \right)(7)^3}$$

$$= \boxed{10.4}$$

The answer is (B).

4. *Customary U.S. Solution*

The moment of inertia of the beam cross section is

$$I = \frac{bh^3}{12} = \frac{(6 \text{ in})h^3}{12} = 0.5h^3 \quad [\text{in in}^4]$$

The section modulus is

$$\frac{I}{c} = \frac{0.5h^3}{\dfrac{h}{2}} = h^2 \quad [\text{in in}^3]$$

From Table 52.1, the modulus of elasticity for a steel spring is $E = 30 \times 10^6 \text{ lbf/in}^2$.

(a) The tip deflection is

$$y = \frac{FL^3}{3EI} = \frac{FL^3}{3E(0.5)h^3}$$

$$h = \left(\frac{FL^3}{1.5yE} \right)^{\frac{1}{3}}$$

$$= \left(\frac{(800 \text{ lbf})(24 \text{ in})^3}{(1.5 \text{ in})(1.0 \text{ in}) \left(30 \times 10^6 \dfrac{\text{lbf}}{\text{in}^2} \right)} \right)^{\frac{1}{3}}$$

$$= \boxed{0.627 \text{ in}}$$

The width-thickness ratio is

$$\frac{w}{t} = \frac{6.0 \text{ in}}{0.627 \text{ in}} \approx 9.6$$

Since $w/t < 10$, this is not a wide beam.

The answer is (C).

(b) The moment at the fixed end is

$$M = FL = (800 \text{ lbf})(24 \text{ in}) = 19,200 \text{ in-lbf}$$

The bending stress is

$$\sigma = \frac{M}{\dfrac{I}{c}} = \frac{M}{h^2}$$

$$h = \sqrt{\frac{M}{\sigma}}$$

$$= \sqrt{\frac{19,200 \text{ in-lbf}}{(1.0 \text{ in}) \left(50,000 \dfrac{\text{lbf}}{\text{in}^2} \right)}}$$

$$= \boxed{0.620 \text{ in}}$$

The answer is (A).

SI Solution

The moment of inertia of the beam cross section is

$$I = \frac{bh^3}{12} = \frac{(150 \text{ mm})h^3}{12} = 12.5h^3 \quad [\text{in mm}^4]$$

The section modulus is

$$\frac{I}{c} = \frac{12.5h^3}{\dfrac{h}{2}} = 25h^2 \quad [\text{in mm}^3]$$

From Table 52.1, the modulus of elasticity for a steel spring is

$$E = \left(30 \times 10^6 \frac{\text{lbf}}{\text{in}^2} \right) \left(6.89 \times 10^3 \frac{\text{Pa}}{\dfrac{\text{lbf}}{\text{in}^2}} \right)$$

$$= 206.7 \times 10^9 \text{ Pa}$$

(a) The tip deflection is

$$y = \frac{FL^3}{3EI} = \frac{FL^3}{3E(12.5)h^3}$$

$$h = \left(\frac{FL^3}{37.5yE} \right)^{\frac{1}{3}}$$

$$= \left(\begin{array}{c} \dfrac{(3.5 \text{ kN}) \left(1000 \dfrac{\text{N}}{\text{kN}} \right)(610 \text{ mm})^3}{(37.5)(25 \text{ mm})} \\ \times (206.7 \times 10^9 \text{ Pa}) \left(\dfrac{1 \text{ m}}{10^3 \text{ mm}} \right)^2 \end{array} \right)^{\frac{1}{3}}$$

$$= \boxed{16.0 \text{ mm}}$$

The width-thickness ratio is

$$\frac{w}{t} = \frac{150 \text{ mm}}{16} \approx 9.4$$

Since $w/t < 10$, this is not a wide beam.

The answer is (C).

(b) The moment at the fixed end is

$$M = FL = \frac{(3.5 \text{ kN}) \left(1000 \frac{\text{N}}{\text{kN}}\right)(610 \text{ mm})}{1000 \frac{\text{mm}}{\text{m}}}$$

$$= 2.135 \times 10^3 \text{ N·m}$$

The bending stress is

$$\sigma = \frac{M}{\frac{I}{c}} = \frac{M}{25h^2}$$

$$h = \sqrt{\frac{M}{25\sigma}}$$

$$= \sqrt{\frac{2.135 \times 10^3 \text{ N·m}}{(25 \text{ mm})(345 \text{ MPa}) \left(10^6 \frac{\text{Pa}}{\text{MPa}}\right) \left(\frac{1 \text{ m}}{10^3 \text{ mm}}\right)}}$$

$$= \boxed{15.7 \text{ mm}}$$

The answer is (A).

5. The maximum speed ratio is

$$\frac{N_{\max}}{N_{\min}} = \frac{96 \text{ teeth}}{12 \text{ teeth}} = 8$$

(a) For three stages,

$$(8)^3 = 512$$

Since this is less than 600, $\boxed{\text{four stages are required.}}$

$$\text{check: } (8)^4 = 4096$$

(b) The approximate gear ratio for each stage should be

$$(600)^{\frac{1}{4}} = 4.95$$

Select 12 teeth for the pinion gears for the first three stages. Let the first three stages be

$$\frac{N_1}{12 \text{ teeth}} = 5$$

$$\frac{N_2}{12 \text{ teeth}} = 5$$

$$\frac{N_3}{12 \text{ teeth}} = 5$$

$$N_1 = 60 \text{ teeth}$$
$$N_2 = 60 \text{ teeth}$$
$$N_3 = 60 \text{ teeth}$$

The fourth stage is obtained from

$$\left(\frac{60 \text{ teeth}}{12 \text{ teeth}}\right)\left(\frac{60 \text{ teeth}}{12 \text{ teeth}}\right)\left(\frac{60 \text{ teeth}}{12 \text{ teeth}}\right) R_4 = 600$$

$$R_4 = 4.8$$

Select 15 teeth for the pinion of stage 4.

$$N_4 = R_4 15 = (4.8)(15) = 72 \text{ teeth}$$

The four-stage gear train would have four pairs of gears with the following numbers of teeth.

$$\frac{12}{60}, \frac{12}{60}, \frac{12}{60}, \frac{15}{72}$$

6. *Customary U.S. Solution*

(a) Let gear 1 be the gear and gear 2 be the pinion.

$$\frac{d_1}{2} + \frac{d_2}{2} = \text{center distance} = 15 \text{ in}$$

From Eq. 52.39,

$$\frac{d_1}{d_2} = \frac{n_2}{n_1} = \frac{1200 \text{ rpm}}{270 \text{ rpm}} = 4.44$$

Solving these two equations simultaneously gives

$$d_1 = \boxed{24.49 \text{ in}}$$

$$d_2 = \boxed{5.51 \text{ in}}$$

(b) From Eq. 52.31, the pitch circle velocity is

$$v_t = \pi d_1 n_1$$

$$= \pi (24.49 \text{ in}) \left(\frac{1 \text{ ft}}{12 \text{ in}}\right)(270 \text{ rpm})$$

$$= 1731 \text{ ft/min}$$

From Eq. 52.46, the transmitted load is

$$F_t = \frac{P_{\text{hp}} \left(33,000 \frac{\text{ft-lbf}}{\text{hp-min}}\right)}{v_t}$$

$$= \frac{(550 \text{ hp}) \left(33,000 \frac{\text{ft-lbf}}{\text{hp-min}}\right)}{1731 \frac{\text{ft}}{\text{min}}}$$

$$= 10,485 \text{ lbf}$$

The allowable bending stress for the pinion is

$$\sigma_a = \frac{S_e}{\text{FS}} = \frac{90,000 \frac{\text{lbf}}{\text{in}^2}}{3} = 30,000 \text{ lbf/in}^2$$

From Eq. 52.62, the Barth speed factor is

$$k_d = \frac{a}{a + v_t} = \frac{600 \frac{\text{ft}}{\text{min}}}{600 \frac{\text{ft}}{\text{min}} + 1731 \frac{\text{ft}}{\text{min}}} = 0.257$$

Use a pinion width of $w = 6$ in and an approximate form factor of $Y = 0.3$.

From Eq. 52.61, the diametral pitch is

$$P = \frac{k_d \sigma_a w Y}{F_t}$$

$$= \frac{(0.257)\left(30,000 \frac{\text{lbf}}{\text{in}^2}\right)(6 \text{ in})(0.3)}{10,485 \text{ lbf}}$$

$$= 1.32 \frac{1}{\text{in}}$$

Use a standard diametral pitch of $\boxed{1.25 \ 1/\text{in},}$ rounding down so as not to exceed the allowable stress.

(c) The number of teeth on the pinion is

$$N_2 = Pd_2 = \left(1.50 \frac{1}{\text{in}}\right)(5.51 \text{ in}) = 8.27 \text{ teeth}$$

$$[\text{say 8 teeth}]$$

The number of teeth on the gear is

$$N_1 = Pd_1 = \left(1.50 \frac{1}{\text{in}}\right)(24.49 \text{ in}) = 36.7$$

$$[\text{say 36 teeth}]$$

For a $14^1/_2°$, 36-tooth gear, the form factor is $Y = 0.33$. The allowable bending stress for the gear is

$$\sigma_a = \frac{S_e}{\text{FS}} = \frac{50,000 \frac{\text{lbf}}{\text{in}^2}}{3} = 16,667 \text{ lbf/in}^2$$

From Eq. 52.61, the face width of the gear is

$$w = \frac{PF_t}{k_d \sigma_a Y}$$

$$= \frac{\left(1.5 \frac{1}{\text{in}}\right)(10,485 \text{ lbf})}{(0.257)\left(16,667 \frac{\text{lbf}}{\text{in}^2}\right)(0.33)}$$

$$= \boxed{11.1 \text{ in}}$$

For a $14^1/_2°$, eight-tooth gear, the form factor is $Y \approx 0.2$. The allowable bending stress for the pinion is

$$\sigma_a = \frac{90,000 \frac{\text{lbf}}{\text{in}^2}}{3} = 30,000 \text{ lbf/in}^2$$

From Eq. 52.61, the face width of the pinion is

$$w = \frac{PF_t}{k_d \sigma_a Y} = \frac{(1.5)(10,485 \text{ lbf})}{(0.257)\left(30,000 \frac{\text{lbf}}{\text{in}^2}\right)(0.2)}$$

$$= 10.2 \text{ in}$$

Each gear would be 11.1 in wide or larger. This disregards stress concentration factors.

SI Solution

(a) Let gear 1 be the gear and gear 2 be the pinion.

$$\frac{d_1}{2} + \frac{d_2}{2} = \text{center distance} = 380 \text{ mm}$$

From Eq. 52.39,

$$\frac{d_1}{d_2} = \frac{n_2}{n_1} = \frac{1200 \text{ rpm}}{270 \text{ rpm}} = 4.44$$

Solving these two equations simultaneously gives

$$d_1 = \boxed{620.4 \text{ mm}}$$

$$d_2 = \boxed{139.6 \text{ mm}}$$

(b) From Eq. 52.31, the pitch circle velocity is

$$v_t = \pi d_1 n_1$$

$$= \pi(0.6204 \text{ m})(270 \text{ rpm})\left(\frac{1 \text{ min}}{60 \text{ sec}}\right) = 8.77 \text{ m/s}$$

From Eq. 52.46, the transmitted load is

$$F_t = \frac{P_{\text{kW}}\left(1000 \frac{\text{W}}{\text{kW}}\right)}{v_t} = \frac{(410 \text{ kW})\left(1000 \frac{\text{W}}{\text{kW}}\right)}{8.77 \frac{\text{m}}{\text{s}}}$$

$$= 46\,750 \text{ N} \quad (46.75 \text{ kN})$$

The allowable bending stress for the pinion is

$$\sigma_a = \frac{S_e}{\text{FS}} = \frac{620 \text{ MPa}}{3} = 206.7 \text{ MPa}$$

From Eq. 52.62, the Barth speed factor is

$$k_d = \frac{a}{a + v_t} = \frac{600 \frac{\text{ft}}{\text{min}}}{600 \frac{\text{ft}}{\text{min}} + \left(8.77 \frac{\text{m}}{\text{s}}\right)\left(196.8 \frac{\frac{\text{ft}}{\text{min}}}{\frac{\text{m}}{\text{s}}}\right)}$$

$$= 0.258$$

Use a pinion width of $w = 150$ mm and an approximate form factor of $Y = 0.3$.

From Eq. 52.61, the diametral pitch is

$$P = \frac{k_d \sigma_a w Y}{F_t}$$

$$= \frac{\begin{array}{c}(0.258)(206.7 \text{ MPa})\left(1000 \dfrac{\text{kPa}}{\text{MPa}}\right) \\[2mm] \times (150 \text{ mm})(0.3)\left(\dfrac{1 \text{ m}}{10^3 \text{ mm}}\right)^2\end{array}}{46.75 \text{ kN}}$$

$$= \boxed{0.051 \ \frac{1}{\text{mm}}}$$

(c) The number of teeth on the pinion is

$$N_2 = \frac{d_2}{m} = \frac{139.6 \text{ mm}}{17 \text{ mm}} = 8.21 \text{ teeth}$$

[say 8 teeth]

The number of teeth on the gear is

$$N_1 = \frac{d_1}{m} = \frac{620.4 \text{ mm}}{17 \text{ mm}} = 36.5 \text{ teeth}$$

[say 36 teeth]

For a $14\frac{1}{2}°$, 36-tooth gear, the form factor is $Y = 0.33$. The allowable bending stress for the gear is

$$\sigma_a = \frac{S_e}{\text{FS}} = \frac{345 \text{ MPa}}{3} = 115 \text{ MPa}$$

From Eq. 52.61, the face width of the gear is

$$w = \frac{PF_t}{k_d \sigma_a Y} = \frac{F_t}{m k_d \sigma_a Y}$$

$$= \frac{46.75 \text{ kN}}{\begin{array}{c}(17 \text{ mm})(0.258)(115 \text{ MPa}) \\[2mm] \times \left(1000 \dfrac{\text{kPa}}{\text{MPa}}\right)(0.29)\left(\dfrac{1 \text{ m}}{10^3 \text{ mm}}\right)^2\end{array}}$$

$$= 319.6 \text{ mm}$$

For a $14\frac{1}{2}°$, eight-tooth gear, the form factor is $Y \approx 0.2$. The allowable bending stress for the pinion is

$$\sigma_a = \frac{620 \text{ MPa}}{3} = 206.7 \text{ MPa}$$

From Eq. 52.61, the face width of the pinion is

$$w = \frac{PF_t}{k_d \sigma_a Y} = \frac{F_t}{m k_d \sigma_a Y}$$

$$= \frac{46.75 \text{ kN}}{\begin{array}{c}(17 \text{ mm})(0.258)(206.7 \text{ MPa}) \\[2mm] \times \left(1000 \dfrac{\text{kPa}}{\text{MPa}}\right)(0.20)\left(\dfrac{1 \text{ m}}{1000 \text{ mm}}\right)^2\end{array}}$$

$$= \boxed{257.8 \text{ mm}}$$

Each gear would be 320 mm wide or larger. This disregards stress concentration factors.

7. *Customary U.S. Solution*

(a) Let gear 1 be the gear and gear 2 be the pinion. The center distance is

$$\frac{d_1}{2} + \frac{d_2}{2} = 15 \text{ in}$$

From Eq. 52.39,

$$\frac{d_1}{d_2} = \frac{n_2}{n_1} = \frac{250 \text{ rpm}}{83.33 \text{ rpm}} = 3.0$$

$$d_1 = \boxed{22.5 \text{ in}}$$

$$d_2 = \boxed{7.5 \text{ in}}$$

(b) From Eq. 52.31, the pitch circle velocity is

$$v_t = \pi d_1 n_1 = \pi (22.5 \text{ in})(83.33 \text{ rpm})\left(\frac{1 \text{ ft}}{12 \text{ in}}\right)$$

$$= 490.9 \text{ ft/min}$$

From Eq. 52.46, the transmitted load is

$$F_t = \frac{P_{\text{hp}}\left(33{,}000 \dfrac{\text{ft-lbf}}{\text{hp-min}}\right)}{v_t}$$

$$= \frac{(250 \text{ hp})\left(33{,}000 \dfrac{\text{ft-lbf}}{\text{hp-min}}\right)}{490.9 \dfrac{\text{ft}}{\text{min}}}$$

$$= 16{,}806 \text{ lbf}$$

From Eq. 52.62, the Barth speed factor is

$$k_d = \frac{a}{a + v_t} = \frac{600 \dfrac{\text{ft}}{\text{min}}}{600 \dfrac{\text{ft}}{\text{min}} + 490.9 \dfrac{\text{ft}}{\text{min}}} = 0.55$$

Select a pinion width of $w = 6$ in and a form factor of $Y = 0.3$.

The allowable stress is given so the factor of safety is not used.

From Eq. 52.61, the diametral pitch is

$$P = \frac{k_d \sigma_a w Y}{F_t} = \frac{(0.55)\left(30{,}000 \dfrac{\text{lbf}}{\text{in}^2}\right)(6 \text{ in})(0.3)}{16{,}806 \text{ lbf}}$$

$$= 1.77 \ 1/\text{in}$$

Select $P = 1.75 \ 1/\text{in}$ as a standard size.

(c) The number of teeth on the pinion is

$$N_{\text{pinion}} = Pd_2 = \left(1.75 \, \frac{1}{\text{in}}\right)(7.5 \text{ in})$$

$$= 13.1 \text{ teeth} \quad \boxed{[\text{say 13 teeth}]}$$

The number of teeth on the gear is

$$N_{\text{gear}} = Pd_1 = \left(1.75 \, \frac{1}{\text{in}}\right)(22.5 \text{ in})$$

$$= 39.4 \text{ teeth} \quad \boxed{[\text{say 39 teeth}]}$$

(d) For a 20°, 13-tooth pinion, the form factor is $Y = 0.26$.

From Eq. 52.61, the minimum face width of the pinion is

$$w = \frac{PF_t}{k_d \sigma_a Y} = \frac{\left(1.75 \, \dfrac{1}{\text{in}}\right)(16{,}806 \text{ lbf})}{(0.55)\left(30{,}000 \, \dfrac{\text{lbf}}{\text{in}^2}\right)(0.26)}$$

$$= 6.86 \text{ in} \quad \boxed{[\text{say 7.0 in}]}$$

(e) For a 20°, 39-tooth gear, the form factor is $Y = 0.38$.

The allowable bending stress for the gear is

$$\sigma_a = \frac{\sigma_{\text{max}}}{\text{FS}} = \frac{50{,}000 \, \dfrac{\text{lbf}}{\text{in}^2}}{3} = 16{,}667 \text{ lbf/in}^2$$

From Eq. 52.61, the minimum face width of the gear is

$$w = \frac{PF_t}{k_d \sigma_a Y} = \frac{\left(1.75 \, \dfrac{1}{\text{in}}\right)(16{,}806 \text{ lbf})}{(0.55)\left(16{,}667 \, \dfrac{\text{lbf}}{\text{in}^2}\right)(0.38)}$$

$$= 8.44 \text{ in} \quad \boxed{[\text{say 9.0 in}]}$$

SI Solution

(a) Let gear 1 be the gear and gear 2 be the pinion. The center distance is

$$\frac{d_1}{2} + \frac{d_2}{2} = 380 \text{ mm}$$

From Eq. 52.39,

$$\frac{d_1}{d_2} = \frac{n_2}{n_1} = \frac{250 \text{ rpm}}{83.33 \text{ rpm}} = 3.0$$

$$d_1 = \boxed{570 \text{ mm}}$$

$$d_2 = \boxed{190 \text{ mm}}$$

(b) From Eq. 52.31, the pitch circle velocity is

$$v_t = \pi d_1 n_1 = \pi (0.57 \text{ m})(83.33 \text{ rpm})\left(\frac{1 \text{ min}}{60 \text{ sec}}\right)$$

$$= 2.49 \text{ m/s}$$

From Eq. 52.46, the transmitted load is

$$F_t = \frac{P_{\text{kW}}\left(1000 \, \dfrac{\text{W}}{\text{kW}}\right)}{v_t} = \frac{(190 \text{ kW})\left(1000 \, \dfrac{\text{W}}{\text{kW}}\right)}{2.49 \, \dfrac{\text{m}}{\text{s}}}$$

$$= 76\,305 \text{ W} \quad (76.3 \text{ kW})$$

From Eq. 52.62, the Barth speed factor is

$$k_d = \frac{a}{a + v_t} = \frac{600 \, \dfrac{\text{ft}}{\text{min}}}{600 \, \dfrac{\text{ft}}{\text{min}} + \left(2.49 \, \dfrac{\text{m}}{\text{s}}\right)\left(196.8 \, \dfrac{\dfrac{\text{ft}}{\text{min}}}{\dfrac{\text{m}}{\text{s}}}\right)}$$

$$= 0.55$$

Select a pinion width of $w = 150$ mm and a form factor of $Y = 0.3$.

The allowable stress is given so the factor of safety is not used.

From Eq. 52.61, the diametral pitch is

$$P = \frac{k_d \sigma_a w Y}{F_t}$$

$$= \frac{(0.55)(210 \text{ MPa})\left(1000 \, \dfrac{\text{kPa}}{\text{MPa}}\right)}{76.3 \text{ kN}}$$
$$\times (150 \text{ mm})(0.3)\left(\dfrac{1 \text{ m}}{10^3 \text{ mm}}\right)^2$$

$$= \boxed{0.068 \text{ 1/mm}}$$

(c) The number of teeth on the pinion is

$$N_{\text{pinion}} = Pd_2 = \left(0.068 \, \frac{1}{\text{mm}}\right)(190 \text{ mm}) = 12.9 \text{ teeth}$$

$$\boxed{[\text{say 13 teeth}]}$$

The number of teeth on the gear is

$$N_{\text{gear}} = Pd_1 = \left(0.068 \, \frac{1}{\text{mm}}\right)(570 \text{ mm}) = 38.8 \text{ teeth}$$

$$\boxed{[\text{say 39 teeth}]}$$

(d) For a 20°, 13-tooth pinion, the form factor is $Y = 0.26$.

From Eq. 52.61, the minimum face width of the pinion is

$$w = \frac{PF_t}{k_d \sigma_a Y}$$

$$= \frac{\left(0.068 \ \frac{1}{\text{mm}}\right)(76.3 \text{ kN})}{(0.55)(210 \text{ MPa})\left(1000 \ \frac{\text{kPa}}{\text{MPa}}\right)(0.26)\left(\frac{1 \text{ m}}{10^3 \text{ mm}}\right)^2}$$

$$= \boxed{173 \text{ mm}}$$

(e) For a 20°, 39-tooth gear, the form factor is $Y = 0.38$.

The allowable bending stress for the gear is

$$\sigma_a = \frac{\sigma_{\text{max}}}{\text{FS}} = \frac{345 \text{ MPa}}{3} = 115 \text{ MPa}$$

From Eq. 52.61, the minimum face width of the gear is

$$w = \frac{PF_t}{k_d \sigma_a Y}$$

$$= \frac{\left(0.068 \ \frac{1}{\text{mm}}\right)(76.3 \text{ kN})}{(0.55)(115 \text{ MPa})\left(1000 \ \frac{\text{kPa}}{\text{MPa}}\right)(0.38)\left(\frac{1 \text{ m}}{10^3 \text{ mm}}\right)^2}$$

$$= \boxed{216 \text{ mm}}$$

8. *Customary U.S. Solution*

From Eq. 52.94, the torque per contact surface is

$$T = \pi f p_{\text{max}} r_i (r_o^2 - r_i^2)$$

$$= (\pi)(0.12)\left(100 \ \frac{\text{lbf}}{\text{in}^2}\right)\left(\frac{2.5 \text{ in}}{2}\right)$$

$$\times \left(\left(\frac{4.5 \text{ in}}{2}\right)^2 - \left(\frac{2.5 \text{ in}}{2}\right)^2\right)$$

$$= 164.9 \text{ in-lbf}$$

The slipping torque is

$$T_{\text{slip}} = 3T_{\text{rated}} = (3)(300 \text{ in-lbf}) = 900 \text{ in-lbf}$$

The slipping torque is equal to the torque per contact surface.

$$T_{\text{slip}} = T$$
$$900 \text{ in-lbf} = 164.9M \quad \text{[in in-lbf]}$$
$$M = 5.5$$

Use six contact surfaces. The arrangement is

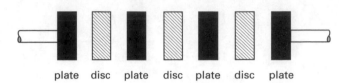

plate   disc   plate   disc   plate   disc   plate

(a) | Use four plates. |

The answer is (C).

(b) | Use three discs. |

The answer is (B).

SI Solution

Assume uniform wear. From Eq. 52.94, the torque per contact surface is

$$T = \pi f p_{\text{max}} r_i (r_o^2 - r_i^2)^2$$

$$= (\pi)(0.12)(700 \text{ kPa})\left(1000 \ \frac{\text{Pa}}{\text{kPa}}\right)\left(\frac{65 \text{ mm}}{2}\right)$$

$$\times \left(\left(\frac{115 \text{ mm}}{2}\right)^2 - \left(\frac{65 \text{ mm}}{2}\right)^2\right)$$

$$\times \left(\frac{1 \text{ m}}{1000 \text{ mmm}}\right)^3 = 19.29 \text{ N·m}$$

The slipping torque is

$$T_{\text{slip}} = 3T_{\text{rated}} = (3)(33 \text{ N·m}) = 99 \text{ N·m}$$

The slipping torque is equal to the torque per contact surface.

$$T_{\text{slip}} = T$$
$$99 \text{ N·m} = 19.30M \quad \text{[in N·m]}$$
$$M = 5.1$$

Use six contact surfaces. The arrangement is

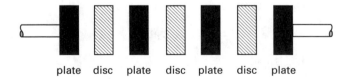

plate   disc   plate   disc   plate   disc   plate

(Eq. 52.96 could also have been used.)

(a) | Use four plates. |

The answer is (C).

(b) Use three discs.

The answer is (B).

9. *Customary U.S. Solution*

From Eq. 52.111, the bearing pressure is

$$p = \frac{\text{lateral shaft load}}{Ld} = \frac{880 \text{ lbf}}{(3.5 \text{ in})(3 \text{ in})} = 83.81 \text{ lbf/in}^2$$

From Eq. 52.110, the ratio c_d/d is

$$\frac{c_d}{d} = \frac{c_r}{r} = 0.001$$

This is the radial clearance ratio.

From Eq. 52.117, the bearing characteristic number is

$$S = \left(\frac{d}{c_d}\right)^2 \left(\frac{\mu n_{\text{rps}}}{p}\right)$$

$$= \left(\frac{r}{c_r}\right)^2 \left(\frac{\mu n_{\text{rps}}}{p}\right)$$

$$= \left(\frac{1}{0.001}\right)^2 \left(\frac{\left(1.184 \times 10^{-6} \, \frac{\text{lbf-sec}}{\text{in}^2}\right) \times \left(1200 \, \frac{\text{rev}}{\text{min}}\right)\left(\frac{1 \text{ min}}{60 \text{ sec}}\right)}{83.81 \, \frac{\text{lbf}}{\text{in}^2}} \right)$$

$$= 0.283$$

The axial length-to-diameter ratio is

$$\frac{L}{d} = \frac{3.5 \text{ in}}{3 \text{ in}} = 1.17$$

(a) From App. 52.B, the minimum film thickness variable is approximately 0.64. From Eq. 52.119, the minimum film thickness is

$$h_o = r\left(\frac{c_d}{d}\right) (\text{minimum film thickness variable})$$

$$= \left(\frac{3 \text{ in}}{2}\right)(0.001)(0.64) = \boxed{0.00096 \text{ in}}$$

(b) From App. 52.B, the coefficient of friction variable is approximately 6.

From Eq. 52.118, the coefficient of friction is

$$f = \left(\frac{c_d}{d}\right) (\text{coefficient of friction variable})$$

$$= (0.001)(6) = 0.006$$

The friction torque is

$$T = fF_r = (0.006)(880 \text{ lbf})\left(\frac{3 \text{ in}}{2}\right)$$

$$= 7.92 \text{ in-lbf}$$

From Eq. 52.32, the power lost to friction is

$$P_{\text{hp}} = \frac{T_{\text{in-lbf}} n_{\text{rpm}}}{63,025} = \frac{(7.92 \text{ in-lbf})\left(1200 \, \frac{\text{rev}}{\text{min}}\right)}{63,025}$$

$$= \boxed{0.151 \text{ hp}}$$

(c) Since the minimum film thickness variable is out of the shaded area of the figure in App. 52.B and to the right of the maximum load-carrying ability curve, the bearing is operating below its optimum capacity.

SI Solution

From Eq. 52.111, the bearing pressure is

$$p = \frac{\text{lateral shaft load}}{Ld}$$

$$= \frac{4 \text{ kN}}{(90 \text{ mm})(76 \text{ mm})\left(\frac{1 \text{ m}}{10^3 \text{ mm}}\right)^2}$$

$$= 0.585 \text{ MPa}$$

From Eq. 52.110, the ratio c_d/d is

$$\frac{c_d}{d} = \frac{c_r}{r} = 0.001$$

This is the radial clearance ratio.

From Eq. 52.117, the bearing characteristic number is

$$S = \left(\frac{d}{c_d}\right)^2 \left(\frac{\mu n_{\text{rps}}}{p}\right)$$

$$= \left(\frac{r}{c_r}\right)^2 \left(\frac{\mu n_{\text{rps}}}{p}\right)$$

$$= \left(\frac{1}{0.001}\right)^2$$

$$\times \left(\frac{(8.16 \text{ cP})\left(1 \, \frac{\frac{\text{N}\cdot\text{s}}{\text{m}^2}}{10^3 \text{ cP}}\right)\left(1200 \, \frac{\text{rev}}{\text{min}}\right)\left(\frac{1 \text{ min}}{60 \text{ s}}\right)}{(0.585 \text{ MPa})\left(10^6 \, \frac{\text{Pa}}{\text{MPa}}\right)} \right)$$

$$= 0.279$$

The axial length-to-diameter ratio is

$$\frac{L}{d} = \frac{90 \text{ mm}}{76 \text{ mm}} = 1.18$$

(a) From App. 52.B, the minimum film thickness variable is approximately 0.64.

From Eq. 52.119, the minimum film thickness is

$$h_o = r\left(\frac{c_d}{d}\right) \text{(minimum film thickness variable)}$$
$$= \left(\frac{76 \text{ mm}}{2}\right)(0.001)(0.64) = \boxed{0.0243 \text{ mm}}$$

(b) From App. 52.B, the coefficient of friction variable is approximately 6. From Eq. 52.118, the coefficient of friction is

$$f = \left(\frac{c_d}{d}\right) \text{(coefficient of friction variable)}$$
$$= (0.001)(6) = 0.006$$

The friction torque is

$$T = fF_r$$
$$= (0.006)(4 \text{ kN})\left(1000 \frac{\text{N}}{\text{kN}}\right)\left(\frac{76 \text{ mm}}{2}\right)\left(\frac{1 \text{ m}}{10^3 \text{ mm}}\right)$$
$$= 0.912 \text{ N·m}$$

From Eq. 52.32(a), the power lost to friction is

$$P_{kW} = \frac{T_{N \cdot m} n_{rpm}}{9549} = \frac{(0.912 \text{ N·m})\left(1200 \frac{\text{rev}}{\text{min}}\right)}{9549}$$
$$= \boxed{0.115 \text{ kW}}$$

(c) Since the minimum film thickness variable is out of the shaded area of the figure in App. 52.B and to the right of the maximum load-carrying ability curve, the bearing is operating below its optimum capacity.

10. *Customary U.S. Solution*

From Table 52.4, the number of active coils are as follows.

For the inner spring,

$$n_a = n_t - 2 = 12.75 - 2 = 10.75$$

For the outer spring,

$$n_a = n_t - 2 = 10.25 - 2 = 8.25$$

From Eq. 52.8, the spring indices are as follows.

For the inner spring,

$$C = \frac{D}{d} = \frac{1.5 \text{ in}}{0.177 \text{ in}} = 8.47$$

For the outer spring,

$$C = \frac{D}{d} = \frac{2.0 \text{ in}}{0.2253 \text{ in}} = 8.88$$

From Eq. 52.12, the spring constants are as follows.

For the inner spring,

$$k_i = \frac{Gd}{8C^3 n_a} = \frac{\left(11.5 \times 10^6 \frac{\text{lbf}}{\text{in}^2}\right)(0.177 \text{ in})}{(8)(8.47)^3(10.75)}$$
$$= 38.95 \text{ lbf/in}$$

For the outer spring,

$$k_o = \frac{Gd}{8C^3 n_a} = \frac{\left(11.5 \times 10^6 \frac{\text{lbf}}{\text{in}^2}\right)(0.2253 \text{ in})}{(8)(8.88)^3(8.25)}$$
$$= 56.06 \text{ lbf/in}$$

Since the inner spring is longer by $\delta_i = 4.5 \text{ in} - 3.75 \text{ in} = 0.75 \text{ in}$, the force absorbed by the inner spring over this distance is

$$F_i = k_i \delta_i = \left(38.95 \frac{\text{lbf}}{\text{in}}\right)(0.75 \text{ in}) = 29.2 \text{ lbf}$$

The force shared by both springs is

$$F = F_s - F_i = 150 \text{ lbf} - 29.2 \text{ lbf} = 120.8 \text{ lbf}$$

The composite spring constant for both springs is

$$k = k_i + k_o = 38.95 \frac{\text{lbf}}{\text{in}} + 56.06 \frac{\text{lbf}}{\text{in}} = 95.0 \text{ lbf/in}$$

The deflection of the composite spring is

$$\delta_c = \frac{F}{k} = \frac{120.8 \text{ lbf}}{95.0 \frac{\text{lbf}}{\text{in}}} = 1.27 \text{ in}$$

(a) The total deflection of the inner spring is

$$\delta_{total} = \delta_i + \delta_c = 0.75 \text{ in} + 1.27 \text{ in} = \boxed{2.02 \text{ in}}$$

The answer is (A).

(b) The maximum force exerted by the inner spring is

$$F = k_i \delta_{total} = \left(38.95 \frac{\text{lbf}}{\text{in}}\right)(2.02 \text{ in}) = \boxed{78.7 \text{ lbf}}$$

The answer is (C).

(c) From Eq. 52.13, the Wahl correction factor for the inner spring is

$$W = \frac{4C - 1}{4C - 4} + \frac{0.615}{C}$$
$$= \frac{(4)(8.47) - 1}{(4)(8.47) - 4} + \frac{0.615}{8.47}$$
$$= 1.173$$

From Eq. 52.14, the maximum shear stress in the inner spring is

$$\tau_{\max} = \frac{8FCW}{\pi d^2}$$

$$= \frac{(8)(78.7 \text{ lbf})(8.47)(1.173)}{\pi (0.177 \text{ in})^2}$$

$$= \boxed{63{,}555 \text{ lbf/in}^2}$$

The answer is (D).

(d) The yield strength is 75% of the ultimate strength.

$$S_{yt} = 0.75 S_{ut}$$

$$= (0.75)\left(204{,}000 \ \frac{\text{lbf}}{\text{in}^2}\right)$$

$$= 153{,}000 \text{ lbf/in}^2$$

According to the maximum shear stress failure theory, the maximum allowable shear stress is

$$\tau_{\max,a} = 0.5 S_{yt}$$

$$= (0.5)\left(153{,}000 \ \frac{\text{lbf}}{\text{in}^2}\right)$$

$$= 76{,}500 \text{ lbf/in}^2$$

The factor of safety in shear for the inner spring is

$$\text{FS} = \frac{\tau_{\max,a}}{\tau_{\max}}$$

$$= \frac{76{,}500 \ \dfrac{\text{lbf}}{\text{in}^2}}{63{,}555 \ \dfrac{\text{lbf}}{\text{in}^2}}$$

$$= \boxed{1.20}$$

The answer is (A).

(e) The inner and outer springs should be wound with opposite direction helixes. This configuration will minimize resonance and prevent coils from one spring from entering the other spring's gaps.

The answer is (D).

SI Solution

From Table 52.4, the number of active coils are as follows.

For the inner spring,

$$n_a = n_t - 2 = 12.75 - 2 = 10.75$$

For the outer spring,

$$n_1 = n_t - 2 = 10.25 - 2 = 8.25$$

From Eq. 52.8, the spring indices are as follows. For the inner spring,

$$C' = \frac{D}{d} = \frac{38 \text{ mm}}{4.5 \text{ mm}} = 8.44$$

For the outer spring,

$$C = \frac{D}{d} = \frac{51 \text{ mm}}{5.723 \text{ mm}} = 8.91$$

From Eq. 52.12, the spring constants are as follows. For the inner spring,

$$k_i = \frac{Gd}{8C^3 n_a} = \frac{(79 \times 10^9 \text{ Pa})(4.5 \text{ mm})\left(\dfrac{1 \text{ m}}{10^3 \text{ mm}}\right)}{(8)(8.44)^3(10.75)}$$

$$= 6876 \text{ N/m}$$

For the outer spring,

$$k_o = \frac{Gd}{8C^3 n_a} = \frac{(79 \times 10^9 \text{ Pa})(5.723 \text{ mm})\left(\dfrac{1 \text{ m}}{10^3 \text{ mm}}\right)}{(8)(8.91)^3(8.25)}$$

$$= 9684 \text{ N/m}$$

Since the inner spring is longer by $\delta_i = 115 \text{ mm} - 95.3 \text{ mm} = 19.7 \text{ mm}$, the force absorbed by the inner spring over this distance is

$$F_i = k_i \delta_i = \left(6876 \ \frac{\text{N}}{\text{m}}\right)(19.7 \text{ mm})\left(\frac{1 \text{ m}}{10^3 \text{ m}}\right)$$

$$= 135.5 \text{ N}$$

The force shared by both springs is

$$F = F_s - F_i$$

$$= 600 \text{ N} - 135.5 \text{ N} = 464.5 \text{ N}$$

The composite spring constant for both springs is

$$k = k_i + k_o = 6876 \ \frac{\text{N}}{\text{m}} + 9684 \ \frac{\text{N}}{\text{m}}$$

$$= 16\,560 \text{ N/m}$$

The deflection of the composite spring is

$$\delta_c = \frac{F}{k} = \frac{464.5 \text{ N}}{\left(16\,560 \ \dfrac{\text{N}}{\text{m}}\right)\left(\dfrac{1 \text{ m}}{10^3 \text{ mm}}\right)} = 28.0 \text{ mm}$$

(a) The total deflection of the inner spring is

$$\delta_{total} = \delta_i + \delta_c = 19.7 \text{ mm} + 28.0 \text{ mm}$$

$$= \boxed{47.7 \text{ mm}}$$

The answer is (A).

(b) The maximum force exerted by the inner spring is

$$F = k_i \delta_{total} = \left(6876 \, \frac{N}{m} \right) (47.7 \text{ mm}) \left(\frac{1 \text{ m}}{10^3 \text{ mm}} \right)$$

$$= \boxed{328.0 \text{ N}}$$

The answer is (C).

(c) From Eq. 52.13, the Wahl correction factor for the inner spring is

$$W = \frac{4C - 1}{4C - 4} + \frac{0.615}{C} = \frac{(4)(8.44) - 1}{(4)(8.44) - 4} + \frac{0.615}{8.44}$$

$$= 1.174$$

From Eq. 52.14, the maximum shear stress in the inner spring is

$$\tau_{max} = \frac{8FCW}{\pi d^2} = \frac{(8)(328.0 \text{ N})(8.44)(1.174)}{\pi (4.5 \text{ mm})^2 \left(\frac{1 \text{ m}}{10^3 \text{ mm}} \right)^2}$$

$$= 4.087 \times 10^6 \text{ Pa} \quad \boxed{(408.7 \text{ MPa})}$$

The answer is (D).

(d) The yield strength is 75% of the ultimate strength.

$$S_{yt} = 0.75 S_{ut}$$
$$= (0.75)(1.4 \times 10^9 \text{ Pa}) = 1.05 \times 10^9 \text{ Pa}$$

According to the maximum shear stress failure theory, the maximum allowable shear stress is

$$\tau_{max,a} = 0.5 S_{yt} = (0.5)(1.05 \times 10^9 \text{ Pa})$$
$$= 5.25 \times 10^8 \text{ Pa} \quad (525 \text{ MPa})$$

The factor of safety in shear for the inner spring is

$$FS = \frac{\tau_{max,a}}{\tau_{max}} = \frac{525 \text{ MPa}}{408.7 \text{ MPa}} = \boxed{1.28}$$

The answer is (A).

(e) The inner and outer springs should be wound with opposite direction helixes. This configuration will minimize resonance and prevent coils from the spring entering from the other spring's gaps.

The answer is (D).

11. *Customary U.S. Solution*

(a) From Eq. 52.39, the speed of the intermediate shaft is

$$n_{int} = n_{input} \left(\frac{N_{input}}{N_{int \text{ at input}}} \right)$$

$$= \left(1800 \, \frac{rev}{min} \right) \left(\frac{50 \text{ teeth}}{25 \text{ teeth}} \right) = 3600 \text{ rev/min}$$

The speed of the output shaft is

$$n_o = n_{int} \left(\frac{N_{int \text{ at output}}}{N_o} \right)$$

$$= \left(3600 \, \frac{rev}{min} \right) \left(\frac{60 \text{ teeth}}{20 \text{ teeth}} \right) = \boxed{10,800 \text{ rev/min}}$$

The answer is (C).

(b) From Eq. 52.32, with two gear sets, the torque output is

$$T_o = \frac{P_o(63,025)}{n_o} = \frac{(\eta_{mesh})^2 P_i (63,025)}{n_o}$$

$$= \frac{(0.98)^2 (50 \text{ hp})(63,025)}{10,800 \, \frac{rev}{min}} = \boxed{280.2 \text{ in-lbf}}$$

The answer is (B).

(c) For the gear at section A-A, the pitch diameter is

$$d = \frac{N}{P} = \frac{60 \text{ teeth}}{5 \, \frac{teeth}{in}} = 12 \text{ in}$$

From Eq. 52.31, the pitch circle velocity is

$$v_t = \pi d n_{rpm} = \pi (12 \text{ in}) \left(\frac{1 \text{ ft}}{12 \text{ in}} \right) \left(3600 \, \frac{rev}{min} \right)$$

$$= 11,310 \text{ ft/min}$$

From Eq. 52.43, the horsepower transmitted is

$$P = \eta_{mesh} P_i = (0.98)(50 \text{ hp}) = 49 \text{ hp}$$

From Eq. 52.46, the transmitted load is

$$F_t = \frac{P_{hp} \left(33,000 \, \frac{ft\text{-}lbf}{hp\text{-}min} \right)}{v_t}$$

$$= \frac{(49 \text{ hp}) \left(33,000 \, \frac{ft\text{-}lbf}{hp\text{-}min} \right)}{11,310 \, \frac{ft}{min}}$$

$$= 143.0 \text{ lbf}$$

From Eq. 52.49, the tangential pressure angle is

$$\phi_t = \arctan\left(\frac{\tan\phi_n}{\cos\psi}\right)$$

$$= \arctan\left(\frac{\tan 20°}{\cos 25°}\right) = 21.9°$$

From Eq. 52.53, the radial force is

$$F_r = F_t \tan\phi_t = (143.0 \text{ lbf})(\tan 21.9°) = 57.5 \text{ lbf}$$

The bending moment on the shaft is

$$M = F_r L = (57.5 \text{ lbf})(4 \text{ in}) = 230 \text{ in-lbf}$$

The torsional moment on the shaft is

$$T = F_t\left(\frac{d}{2}\right) = (143.0 \text{ lbf})\left(\frac{12 \text{ in}}{2}\right) = 858 \text{ in-lbf}$$

According to the maximum shear stress failure theory, the maximum shear stress is

$$\tau_{\max} = 0.5 S_{yt} = (0.5)\left(69{,}000 \frac{\text{lbf}}{\text{in}^2}\right) = 34{,}500 \text{ lbf/in}^2$$

With a factor of safety of 2, the allowable shear stress is

$$\tau_a = \frac{\tau_{\max}}{2} = \frac{34{,}500 \frac{\text{lbf}}{\text{in}^2}}{2}$$

$$= 17{,}250 \text{ lbf/in}^2$$

Disregarding the axial loading, from Eq. 51.52, the minimum shaft diameter is

$$d_s = \left(\frac{16\sqrt{M^2 + T^2}}{\pi\tau_{\max}}\right)^{\frac{1}{3}}$$

$$= \left(\frac{16\sqrt{(230 \text{ in-lbf})^2 + (858 \text{ in-lbf})^2}}{\pi\left(17{,}250 \frac{\text{lbf}}{\text{in}^2}\right)}\right)^{\frac{1}{3}}$$

$$= 0.64 \text{ in}$$

Check the normal stress criterion. The maximum allowable normal stress is

$$\sigma_{\max} = \frac{S_{yt}}{\text{FS}}$$

$$= \frac{69{,}000 \frac{\text{lbf}}{\text{in}^2}}{2} = 34{,}500 \text{ lbf/in}^2$$

From Eq. 51.53, the minimum diameter is

$$d = \left(\left(\frac{16}{\pi\sigma_{\max}}\right)\sqrt{4M^2 + 3T^2}\right)^{\frac{1}{3}}$$

$$= \left(\left(\frac{16}{\pi\left(34{,}500 \frac{\text{lbf}}{\text{in}^2}\right)}\right)\sqrt{\begin{array}{c}(4)(230 \text{ in-lbf})^2 \\ +(3)(858 \text{ in-lbf})^2\end{array}}\right)^{\frac{1}{3}}$$

$$= 0.612 \text{ in}$$

This is smaller than the diameter calculated from the shear stress criterion, so $d = \boxed{0.64 \text{ in.}}$

Check the axial stress.

From Eq. 52.54, the axial force is

$$F_a = F_t \tan\psi = (143.0 \text{ lbf})(\tan 25°) = 66.7 \text{ lbf}$$

The cross-sectional area of the shaft is

$$A = \frac{\pi}{4}d_s^2 = \left(\frac{\pi}{4}\right)(0.64 \text{ in})^2 = 0.322 \text{ in}^2$$

The axial stress is

$$\sigma_{\text{axial}} = \frac{F_a}{A} = \frac{66.7 \text{ lbf}}{0.322 \text{ in}^2} = 207 \text{ lbf/in}^2$$

Since σ_{axial} is much less than the yield strength, it is insignificant in this problem.

The answer is (D).

SI Solution

(a) From Eq. 52.39, the speed of the intermediate shaft is

$$n_{\text{int}} = n_{\text{input}}\left(\frac{N_{\text{input}}}{N_{\text{int at input}}}\right)$$

$$= \left(1800 \frac{\text{rev}}{\text{min}}\right)\left(\frac{50 \text{ teeth}}{25 \text{ teeth}}\right) = 3600 \text{ rev/min}$$

The speed of the output shaft is

$$n_o = n_{\text{int}}\left(\frac{N_{\text{int at output}}}{N_o}\right)$$

$$= \left(3600 \frac{\text{rev}}{\text{min}}\right)\left(\frac{60 \text{ teeth}}{20 \text{ teeth}}\right) = \boxed{10{,}800 \text{ rev/min}}$$

The answer is (C).

(b) From Eq. 52.32, with two gear sets, the torque output is

$$T_o = \frac{P_o(9549)}{n_o} = \frac{(\eta_{\text{mesh}})^2 P_i(9549)}{n_o}$$

$$= \frac{(0.98)^2(38 \text{ kW})(9549)}{10{,}800 \frac{\text{rev}}{\text{min}}} = \boxed{32.27 \text{ N·m}}$$

The answer is (B).

(c) For the gear at section A-A, the pitch diameter is

$$d = Nm = (60 \text{ teeth})\left(5 \frac{\text{mm}}{\text{tooth}}\right) = 300 \text{ mm}$$

From Eq. 52.31, the pitch circle velocity is

$$
\begin{aligned}
v_t &= \pi d n_{\mathrm{rpm}} = \pi (300 \text{ mm}) \left(\frac{1 \text{ m}}{1000 \text{ mm}} \right) \\
&\quad \times \left(3600 \, \frac{\text{rev}}{\text{min}} \right) \left(\frac{1 \text{ min}}{60 \text{ sec}} \right) \\
&= 56.55 \text{ m/s}
\end{aligned}
$$

From Eq. 52.43, the power transmitted is

$$ P = \eta_{\mathrm{mesh}} P_i = (0.98)(38 \text{ kW}) = 37.24 \text{ kW} $$

From Eq. 52.46, the transmitted load is

$$
\begin{aligned}
F_t &= \frac{P_{\mathrm{kW}} \left(1000 \, \frac{\text{W}}{\text{kW}} \right)}{v_t} \\
&= \frac{(37.24 \text{ kW}) \left(1000 \, \frac{\text{W}}{\text{kW}} \right)}{56.55 \, \frac{\text{m}}{\text{s}}} \\
&= 658.5 \text{ N}
\end{aligned}
$$

From Eq. 52.49, the tangential pressure angle is

$$
\begin{aligned}
\phi_t &= \arctan \left(\frac{\tan \phi_n}{\cos \psi} \right) \\
&= \arctan \left(\frac{\tan 20°}{\cos 25°} \right) = 21.9°
\end{aligned}
$$

From Eq. 52.53, the radial force is

$$ F_r = F_t \tan \phi_t = (658.5 \text{ N})(\tan 21.9°) = 264.7 \text{ N} $$

The bending moment on the shaft is

$$
\begin{aligned}
M &= F_r L = (264.7 \text{ N})(100 \text{ mm}) \left(\frac{1 \text{ m}}{1000 \text{ mm}} \right) \\
&= 26.47 \text{ N·m}
\end{aligned}
$$

The torsional moment on the shaft is

$$
\begin{aligned}
T &= F_t \left(\frac{d}{2} \right) = (658.5 \text{ N}) \left(\frac{300 \text{ mm}}{2} \right) \left(\frac{1 \text{ m}}{1000 \text{ mm}} \right) \\
&= 98.78 \text{ N·m}
\end{aligned}
$$

According to the maximum shear stress failure theory, the maximum shear stress is

$$ \tau_{\max} = 0.5 S_{yt} = (0.5)(480 \text{ MPa}) = 240 \text{ MPa} $$

With a factor of safety of 2, the allowable shear stress is

$$ \tau_a = \frac{\tau_{\max}}{2} = \frac{240 \text{ MPa}}{2} = 120 \text{ MPa} $$

Disregarding the axial loading, from Eq. 51.52, the minimum shaft diameter is

$$
\begin{aligned}
d_s &= \left(\frac{16 \sqrt{M^2 + T^2}}{\pi \tau_{\max}} \right)^{\frac{1}{3}} \\
&= \left(\frac{16 \sqrt{(26.47 \text{ N·m})^2 + (98.78 \text{ N·m})^2}}{\pi (120 \times 10^6 \text{ Pa})} \right)^{\frac{1}{3}} \\
&= 0.0163 \text{ m}
\end{aligned}
$$

Check the normal stress criterion. The maximum allowable normal stress is

$$
\begin{aligned}
\sigma_{\max} &= \frac{S_{yt}}{\mathrm{FS}} \\
&= \frac{480 \text{ MPa}}{2} = 240 \text{ MPa}
\end{aligned}
$$

From Eq. 51.53, the minimum diameter is

$$
\begin{aligned}
d &= \left(\left(\frac{16}{\pi \tau_{\max}} \right) \sqrt{4M^2 + 3T^2} \right)^{\frac{1}{3}} \\
&= \left(\left(\frac{16}{\pi (240 \times 10^6 \text{ Pa})} \right) \sqrt{\begin{aligned} &(4)(26.47 \text{ N·m})^2 \\ &+(3)(197.6 \text{ N·m})^2 \end{aligned}} \right)^{\frac{1}{3}} \\
&= 0.0194 \text{ m}
\end{aligned}
$$

This is larger than the diameter calculated from the shear stress criterion, so $d = \boxed{0.0194 \text{ m.}}$

Check the axial stress.

From Eq. 52.54, the axial force is

$$ F_a = F_t \tan \psi = (658.5 \text{ N})(\tan 25°) = 307.1 \text{ N} $$

The cross-sectional area of the shaft is

$$ A = \frac{\pi}{4} d_s^2 = \left(\frac{\pi}{4} \right)(0.0194 \text{ m})^2 = 2.96 \times 10^{-4} \text{ m} $$

The axial stress is

$$ \sigma_{\mathrm{axial}} = \frac{F_a}{A} = \frac{307.1 \text{ N}}{2.96 \times 10^{-4} \text{ m}} = 1.04 \times 10^6 \text{ Pa} $$

Since σ_{axial} is much less than the yield strength, it is insignificant in this problem.

The answer is (D).

12. *Customary U.S. Solution*

(a) For a B-type v-belt, the belt length correction is 1.8 in. The pitch length is

$$
\begin{aligned}
L_p &= L_{\mathrm{inside}} + \text{correction} \\
&= 90 \text{ in} + 1.8 \text{ in} \\
&= \boxed{91.8 \text{ in}}
\end{aligned}
$$

The answer is (D).

(b) The ratio of speeds determines the sheave sizes.

$$d_{equipment} = d_{motor} \left(\frac{n_{motor}}{n_{equipment}} \right)$$

$$= (10 \text{ in}) \left(\frac{1750 \frac{\text{rev}}{\text{min}}}{800 \frac{\text{rev}}{\text{min}}} \right)$$

$$= \boxed{21.875 \text{ in} \quad (21.9 \text{ in})}$$

The answer is (B).

SI Solution

(a) For a B-type v-belt, the belt length correction is 1.8 in. The pitch length is

$$L_p = L_{inside} + \text{correction}$$

$$= (90 \text{ in} + 1.8 \text{ in}) \left(25.4 \frac{\text{mm}}{\text{in}} \right)$$

$$= \boxed{2332 \text{ mm}}$$

The answer is (D).

(b) The ratio of speeds determines the sheave sizes.

$$d_{equipment} = d_{motor} \left(\frac{n_{motor}}{n_{equipment}} \right)$$

$$= (254 \text{ mm}) \left(\frac{1750 \frac{\text{rev}}{\text{min}}}{800 \frac{\text{rev}}{\text{min}}} \right)$$

$$= \boxed{555.625 \text{ mm} \quad (556 \text{ mm})}$$

The answer is (B).

53 Pressure Vessels

PRACTICE PROBLEMS

1. A seamless pressure vessel has an inside diameter of 60 in. The vessel is made from 0.75 in SA-515, grade 60 plate and is designed for service at 200 psig at 750°F. A seamless 10 in diameter nozzle made from 0.5 inch SA-105 is attached. The nozzle is flush with the vessel walls. The corrosion allowance is $1/16$ inch. Full radiography is used.

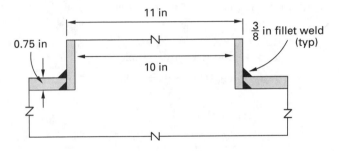

(a) Are the specified thicknesses adequate?

(b) Is a reinforcing pad necessary?

2. The pressure vessel shown is intended for general service. No radiography is performed. The corrosion allowance is 0.125 in. The vessel is seamless, operates at 350 psig and 500°F, and is made from SA-106 steel. The shell has a 48 in outside diameter and a nominal thickness of 0.75 in. The sump is a seamless pipe with a 16 in outside diameter and 0.625 in thickness.

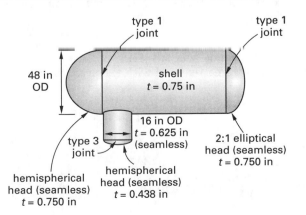

(a) Are the specified thicknesses for the head acceptable?

(b) Are the specified thicknesses for the shell acceptable?

(c) Are the specified thicknesses for the sump acceptable?

3. The pressure vessel shown weighs 21,000 lbf empty. The contents weigh 40,000 lbf. The allowable tensile and compressive stresses in the supporting skirt are 14,000 psi and 17,000 psi, respectively. If the vessel is subjected to the loads shown, is a skirt thickness of 0.25 in adequate?

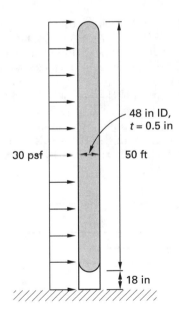

4. A pressure vessel previously used in a refinery has been out of service for a number of years. The vessel is known to be constructed from 0.625 in SA-516, grade 65 plate, but no other documentation is available. The longitudinal seam joint in the vessel is type 1, and the heads are attached with type 2 joints. The vessel is spot radiographed and operates at 850°F. The corrosion allowance is $1/16$ in.

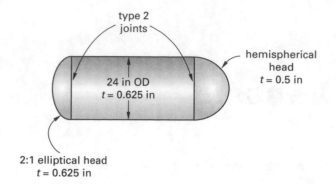

type 2 joints

hemispherical head
$t = 0.5$ in

24 in OD
$t = 0.625$ in

2:1 elliptical head
$t = 0.625$ in

(a) What is the maximum design pressure?
- (A) 300 psig
- (B) 350 psig
- (C) 430 psig
- (D) 530 psig

(b) What is the hydrostatic test pressure most nearly if the vessel is tested at room temperature?
- (A) 600 psig
- (B) 800 psig
- (C) 900 psig
- (D) 1000 psig

5. A pressure vessel with a 48 in inside diameter is fabricated from $^3/_4$ in thick SA-105 plate. The vessel is constructed with a type 1 longitudinal seam weld. A 12 in inside diameter, $^1/_2$ in thick, seamless nozzle made from SA-515, grade 70 is attached. The nozzle does not pass through the seam weld. The operating temperature of the vessel is 750°F at 400 psig. The vessel is fully radiographed. An annular reinforcing pad 0.5 in thick, made from SA-105 steel, abuts the nozzle neck and is attached to the vessel as shown. There is no corrosion allowance.

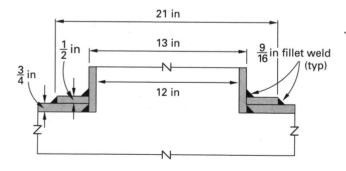

21 in

13 in

$\frac{1}{2}$ in

$\frac{9}{16}$ in fillet weld (typ)

$\frac{3}{4}$ in

12 in

(a) Does the vessel meet Section VIII, Division 1 requirements for shell and nozzle wall thicknesses?

(b) Does the vessel meet Section VIII, Division 1 requirements for weld size?

(c) Does the vessel meet Section VIII, Division 1 requirements for required area?

SOLUTIONS

1. (a) To determine if the specified thicknesses are adequate, compute the minimum required wall thicknesses for the shell and nozzle.

Shell:

$$R = 30 \text{ in} + 0.0625 \text{ in} = 30.0625 \text{ in}$$
$$[\text{correction for corrosion allowance}]$$

$$S = 13{,}000 \text{ psi} \quad [\text{from Table 53.3}]$$
$$E = 1.0 \quad [\text{seamless vessel}]$$
$$t_r = \frac{pR}{SE - 0.6p}$$
$$= \frac{(200 \text{ psig})(30.0625 \text{ in})}{(13{,}000 \text{ psi})(1.0) - (0.6)(200 \text{ psig})}$$
$$= 0.47 \text{ in}$$

Since 0.75 in > 0.47 in, the specified shell thickness is adequate even with a corrosion allowance.

Nozzle:

$$S = 14{,}800 \text{ psi} \quad [\text{from Table 53.3}]$$
$$R = 5 \text{ in} + 0.0625 \text{ in} = 5.0625 \text{ in}$$
$$t_{r,n} = \frac{pR_n}{SE - 0.6p}$$
$$= \frac{(200 \text{ psig})(5.0625 \text{ in})}{(14{,}800 \text{ psi})(1.0) - (0.6)(200 \text{ psig})}$$
$$= 0.069 \text{ in}$$

Since 0.5 in > 0.069 in, the specified thickness is adequate for the working pressure, even with a corrosion allowance.

However, the nozzle thickness also cannot be less than the required shell thickness with a corrosion allowance.

$$t_{\text{min}} = 0.47 \text{ in} + 0.0625 \text{ in}$$
$$= 0.5325 \text{ in}$$

Since 0.5 in < 0.5325 in, the nozzle thickness is inadequate.

(b) To determine whether reinforcement is required, compare the area required with the available area.

$$f_{r1} = f_{r2} = \frac{S_n}{S_v} = \frac{14{,}800 \text{ psi}}{13{,}000 \text{ psi}} > 1.0$$

Therefore,

$$f_{r1} = f_{r2} = 1.0$$

$$D = 10 \text{ in} + 0.125 \text{ in} = 10.125 \text{ in} \begin{bmatrix} \text{correction for} \\ \text{corrosion allowance} \end{bmatrix}$$

The required area is

$$A_r = Dt_r + 2t_n t_r (1 - f_{r1}) = (10.125 \text{ in})(0.47 \text{ in}) + 0$$
$$= 4.76 \text{ in}^2$$

The area available is $A = A_1 + A_2 + A_{41}$.

Calculation of A_1 (area available in shell):

$$t = 0.75 \text{ in} - 0.0625 \text{ in} = 0.6875 \text{ in}$$
$$t_n = 0.5 \text{ in} - 0.0625 \text{ in} = 0.4375 \text{ in}$$

A_1 is the larger of

$$D(Et - t_r) - 2t_n(Et - t_r)(1 - f_{r1})$$
$$= (10.125 \text{ in})\big((1.0)(0.6875 \text{ in}) - 0.47 \text{ in}\big) - 0$$
$$= 2.20 \text{ in}^2$$

and

$$2(t + t_n)(Et - t_r) - 2t_n(Et - t_r)(1 - f_{r1})$$
$$= 2(0.6875 \text{ in} + 0.4375 \text{ in})$$
$$\times \big((1)(0.6875 \text{ in}) - 0.47 \text{ in}\big) - 0$$
$$= 0.49 \text{ in}^2$$

$$A_1 = 2.20 \text{ in}^2$$

Calculation of A_2 (area available from nozzle):

A_2 is the smaller of

$$5(t_n - tr_n)f_{r2}t$$
$$= (5)(0.4375 \text{ in} - 0.069 \text{ in})(1.0)(0.6875 \text{ in})$$
$$= 1.27 \text{ in}^2$$

and

$$5(t_n - t_{rn})f_{r2}t_n$$
$$= (5)(0.4375 \text{ in} - 0.069 \text{ in})(1.0)(0.4375 \text{ in})$$
$$= 0.81 \text{ in}^2$$

$$A_2 = 0.81 \text{ in}^2$$

Calculation of A_{41} (area available from welds):

$$f_{r3} = \frac{S_n}{S_v} = \frac{14{,}800 \text{ psi}}{13{,}000 \text{ psi}} > 1.0, \text{ so use } f_{r3} = 1.0.$$

$$A_{41} = (\text{leg})^2 f_{r3}$$
$$= (0.375 \text{ in})^2 (1.0)$$
$$= 0.14 \text{ in}^2$$

$$\text{total available area} = A_1 + A_2 + A_{41}$$
$$= 2.20 \text{ in}^2 + 0.81 \text{ in}^2 + 0.14 \text{ in}^2$$
$$= 3.15 \text{ in}^2$$

Since $4.76 \text{ in}^2 > 3.15 \text{ in}^2$ (i.e., $A_{\text{required}} > A_{\text{available}}$), additional reinforcement is required. This can be accomplished by adding an annular reinforcing pad around the nozzle neck.

$$f_{r4} = \frac{\text{allowable stress of pad}}{\text{allowable stress of shell}} = 1.0$$

For a 15 in OD pad that is 0.5 in thick, the additional area added is

$$A_5 = (D_p - D - 2t_n)t_p f_{r4}$$
$$= \big((15 \text{ in} - 10 \text{ in} - (2)(0.5 \text{ in})\big)(0.5 \text{ in})$$
$$= 2.0 \text{ in}^2$$

This will satisfy the area requirements.

2. At 500°F, the allowable stress for SA-106 is 17,100 psi (Table 53.3).

(b) Shell:

Circumferential stress (seamless):

$$E = 0.85 \quad [\text{seamless shell, no radiography}]$$
$$t_{\text{corroded}} = 0.750 \text{ in} - 0.125 \text{ in}$$
$$= 0.625 \text{ in}$$

Check this design value against the minimum required thickness to determine whether it is adequate.

$$R = 24.000 \text{ in} - 0.625 \text{ in} = 23.375 \text{ in}$$
$$t = \frac{pR}{SE - 0.6p}$$
$$= \frac{(350 \text{ psig})(23.375 \text{ in})}{(17{,}100 \text{ psig})(0.85) - (0.6)(350 \text{ psig})}$$
$$= 0.571 \text{ in}$$

Longitudinal stress:

$$E = 0.70 \quad [\text{no radiography, type 1 joint}]$$
$$t = \frac{pR}{2SE + 0.4p}$$
$$= \frac{(350 \text{ psig})(23.375 \text{ in})}{(2)(17{,}100 \text{ psi})(0.70) + (0.4)(350 \text{ psi})}$$
$$= 0.340 \text{ in}$$

Since 0.625 in > 0.571 in and 0.340 in, the specified shell thickness is adequate.

(a) Elliptical head (seamless):

$E = 0.85$ [seamless head, no radiography]

$t_{corroded} = 0.750 \text{ in} - 0.125 \text{ in} = 0.625 \text{ in}$

$D = 48 \text{ in} - (2)(0.625 \text{ in}) = 46.75 \text{ in}$

$$K = \left(\frac{1}{6}\right)\left(2 + \left(\frac{D}{2h}\right)^2\right) = \left(\frac{1}{6}\right)\left(2 + (2)^2\right)$$

$$= 1$$

$$t = \frac{pDK}{2SE - 0.2p}$$

$$= \frac{(350 \text{ psig})(46.75 \text{ in})(1)}{(2)(17,100 \text{ psi})(0.85) - (0.2)(350 \text{ psig})}$$

$$= 0.564 \text{ in}$$

Since 0.625 in > 0.564 in, the specified head thickness is adequate.

Hemispherical head (seamless):

$E = 0.70$ $\begin{bmatrix}\text{seamless hemispherical head,} \\ \text{type 1 joint, no radiography}\end{bmatrix}$

$R = 24 \text{ in} - 0.625 \text{ in} = 23.375 \text{ in}$

$$t = \frac{pR}{2SE - 0.2p}$$

$$= \frac{(350 \text{ psig})(23.375 \text{ in})}{(2)(17,100 \text{ psi})(0.70) - (0.2)(350 \text{ psig})}$$

$$= 0.343 \text{ in}$$

Since 0.625 in > 0.343 in, the specified head thickness is acceptable.

(c) Sump shell (seamless, no radiography):

Circumferential stress:

$E = 0.85$ [seamless shell, no radiography]

$t_{corroded} = 0.625 \text{ in} - 0.125 \text{ in} = 0.500 \text{ in}$

$R = 8 \text{ in} - 0.500 \text{ in} = 7.50 \text{ in}$

$$t = \frac{pR}{SE - 0.6p}$$

$$= \frac{(350 \text{ psig})(7.50 \text{ in})}{(17,100 \text{ psi})(0.85) - (0.6)(350 \text{ psig})}$$

$$= 0.183 \text{ in}$$

Longitudinal stress:

$E = 0.60$ [type 3 joint]

$$t = \frac{pR}{2SE + 0.4p}$$

$$= \frac{(350 \text{ psig})(7.50 \text{ in})}{(2)(17,100 \text{ psi})(0.60) + (0.4)(350 \text{ psig})}$$

$$= 0.127 \text{ in}$$

Since 0.500 in > 0.183 in and 0.127 in, the specified sump thickness is adequate.

Hemispherical sump head (type 2 joint):

$E = 0.65$

$t_{corroded} = 0.438 \text{ in} - 0.125 \text{ in} = 0.313 \text{ in}$

$R = 8.000 \text{ in} - 0.313 \text{ in} = 7.687 \text{ in}$

$$t = \frac{pR}{2SE - 0.2p}$$

$$= \frac{(350 \text{ psig})(7.687 \text{ in})}{(2)(17,100 \text{ psi})(0.65) - (0.2)(350 \text{ psig})}$$

$$= 0.121 \text{ in}$$

Since 0.313 in > 0.121 in, the specified sump thickness is adequate.

3. Calculate the stress due to the weight of the vessel and its contents and compare with the allowable stress in the skirt.

Stress on skirt

= axial stress due to vessel and content weight

+ bending stress due to wind load

Axial stress due to the weight of the vessel and its contents:

Assume the skirt is just wide enough to support the vessel.

$$D_o = 48 \text{ in} + (2)(0.5 \text{ in}) = 49 \text{ in}$$

$$D_i = 49 \text{ in} - (2)(0.25 \text{ in}) = 48.5 \text{ in}$$

$$\text{force} = 21,000 \text{ lbf} + 40,000 \text{ lbf} = 61,000 \text{ lbf}$$

$$\text{skirt area} = \left(\frac{\pi}{4}\right)(D_o^2 - D_i^2)$$

$$= \left(\frac{\pi}{4}\right)\left((49 \text{ in})^2 - (48.5 \text{ in})^2\right)$$

$$= 38.29 \text{ in}^2$$

$$\text{axial stress} = \frac{\text{force}}{\text{skirt area}}$$

$$= \frac{61,000 \text{ lbf}}{38.29 \text{ in}^2}$$

$$= 1593 \text{ psi}$$

Bending stress due to wind load:

For thin, circular cross sections, the moment at bottom of the skirt due to a uniform wind load is

$$M = Fy = pAy$$

$$= (30 \text{ psf})\left(\frac{48 \text{ in} + (2)(0.5 \text{ in})}{12 \frac{\text{in}}{\text{ft}}}\right)(51.5 \text{ ft})\left(\frac{51.5 \text{ ft}}{2}\right)$$

$$= 162,450 \text{ ft-lbf}$$

$$R_o = \frac{48 \text{ in} + (2)(0.5 \text{ in})}{2} = 24.5 \text{ in}$$

$$R_i = \frac{48 \text{ in}}{2} = 24 \text{ in}$$

$$I = I_{\text{circle},R_o} - I_{\text{circle},R_i}$$

$$= \frac{\pi}{4}\left(R_o^4 - R_i^4\right)$$

$$= \frac{\pi}{4}\left((24.5 \text{ in})^4 - (24 \text{ in})^4\right)$$

$$= 22{,}403 \text{ in}^4$$

The bending stress is

$$\sigma = \frac{M_c}{I} = \frac{(159{,}135 \text{ ft-lbf})\left(12 \dfrac{\text{in}}{\text{ft}}\right)(24 \text{ in})}{22{,}403 \text{ in}^4}$$

$$= 2046 \text{ lbf/in}^2$$

The maximum tensile stress is

$$2046 \text{ psi} - 1593 \text{ psi} = 453 \text{ psi}$$

The maximum compressive stress is

$$2046 \text{ psi} + 1593 \text{ psi} = 3639 \text{ psi}$$

The compressive and tensile stresses are below the allowable stresses, so a thickness of 0.25 in is adequate.

4. (a) Check the maximum allowable pressure for the shell and heads.

Shell:

Longitudinal stress (circumferential joint):

$$S = 8700 \text{ psi} \quad \text{[from Table 53.3]}$$

$$E = 0.80 \quad \text{[spot radiography, type 2 joint]}$$

$$t_{\text{corroded}} = 0.625 \text{ in} - 0.0625 \text{ in} = 0.563 \text{ in}$$

$$R = 12.000 \text{ in} - 0.563 \text{ in} = 11.437 \text{ in}$$

$$p = \frac{2SEt}{R - 0.4t} = \frac{(2)(8700 \text{ psi})(0.80)(0.563 \text{ in})}{11.437 \text{ in} - (0.4)(0.563 \text{ in})}$$

$$= 699 \text{ psi}$$

Longitudinal joint:

$$E = 0.85 \quad \text{[spot radiography, type 1 joint]}$$

$$p = \frac{SEt}{R + 0.6t} = \frac{(8700 \text{ psi})(0.85)(0.563 \text{ in})}{11.437 \text{ in} + (0.6)(0.563 \text{ in})}$$

$$= 354 \text{ psi}$$

Hemispherical head:

$$E = 0.8 \quad \text{[spot radiography, type 2 joint]}$$

$$t = 0.500 \text{ in} - 0.0625 \text{ in} = 0.438 \text{ in}$$

$$R = 12.000 \text{ in} - 0.438 \text{ in} = 11.562 \text{ in}$$

$$p = \frac{2SEt}{R + 0.2t} = \frac{(2)(8700 \text{ psi})(0.8)(0.438 \text{ in})}{11.562 \text{ in} + (0.2)(0.438 \text{ in})}$$

$$= 523 \text{ psi}$$

Elliptical head:

$$E = 1.0$$

$$K = 1.0 \quad \text{[for 2:1 elliptical heads]}$$

$$t = 0.625 \text{ in} - 0.0625 \text{ in} = 0.563 \text{ in}$$

$$D = 24.000 \text{ in} - (2)(0.563 \text{ in}) = 22.874 \text{ in}$$

$$p = \frac{2SEt}{KD + 0.2t}$$

$$= \frac{(2)(8700 \text{ psi})(1.0)(0.563 \text{ in})}{(1)(22.874 \text{ in}) + (0.2)(0.563 \text{ in})}$$

$$= 426 \text{ psi}$$

The design pressure is the smallest of the calculated pressures = 354 psig, say 350 psig. This will be the MAWP for the vessel.

The answer is (B).

(b)

$$S_{\text{allowable at room temp}} = 18{,}600 \text{ psi}$$

$$S_{\text{allowable at operating temp}} = 8700 \text{ psi}$$

The hydrostatic pressure at room temperature is

$$\frac{1.3 P_{\max} S_{\text{allowable at room temp}}}{S_{\text{allowable at operating temp}}}$$

$$= \frac{(1.3)(350 \text{ psi})(18{,}600 \text{ psi})}{8700 \text{ psi}}$$

$$= 973 \text{ psi}$$

The answer is (D).

5. (a) Calculate shell and nozzle thicknesses.

Determine the allowable stresses in the shell and nozzle from Table 53.3.

$$S_{\text{shell}} = S_{\text{nozzle}} = 14{,}800 \text{ psi}$$

The circumferential stress will determine shell and nozzle thicknesses.

Shell:

$$p = 400 \text{ psig}$$

$$R = 24 \text{ in}$$

$$E = 1.0 \quad \text{[type 1 joint, full radiography]}$$

$$t_r = \frac{pR}{SE - 0.6p}$$

$$= \frac{(400 \text{ psig})(24 \text{ in})}{(14{,}800)(1.0) - (0.6)(400 \text{ psig})}$$

$$= 0.659 \text{ in}$$

Since 0.75 in > 0.659 in, the specified shell thickness is adequate.

Nozzle:

$$R = 6 \text{ in}$$

$$E = 1.0 \quad [\text{seamless pipe, full radiography}]$$

$$t_{r,n} = \frac{pR}{SE - 0.6p}$$

$$= \frac{(400 \text{ psig})(6 \text{ in})}{(14,800 \text{ psi})(1.0) - (0.6)(400 \text{ psig})}$$

$$= 0.165 \text{ in}$$

Since 0.5 in > 0.165 in, the specified thickness is adequate for the working pressure.

However, even without considering an allowance for corrosion, the nozzle thickness is less than the required shell thickness, 0.5 in < 0.659 in.

The nozzle thickness is inadequate.

(b) Calculate the size of the fillet welds.

Inner weld:

The minimum allowable weld throat is $t_w = 0.7t_{\min}$, where $t_{\min} = 0.75$ in or the thickness of the thinner of the parts joined by the weld, whichever is smaller.

The nozzle is 0.50 in thick, so $t_{\min} = 0.5$ in.

$$t_w = (0.7)(0.5 \text{ in}) = 0.35 \text{ in}$$

The actual weld size is

$$t_w = 0.7(\text{weld size}) = (0.7)(0.563 \text{ in}) = 0.394 \text{ in}$$

The inner weld size is acceptable.

Outer weld:

The minimum allowable weld size is

$$0.5t_{\min} = (0.5)(0.5 \text{ in}) = 0.25 \text{ in}$$

The actual weld size is

$$0.5(\text{weld size}) = (0.5)(0.563 \text{ in}) = 0.281 \text{ in}$$

The outer weld size is acceptable.

(c) To determine whether reinforcement is adequate, compare the area required by the Code with the available area.

$$f_{r1} = f_{r2} = \frac{S_n}{S_v} = \frac{14,800 \text{ psi}}{14,800 \text{ psi}} = 1.0$$

The area required is

$$A_r = Dt_r + 2t_n t_r (1 - f_{r1}) = (12 \text{ in})(0.659 \text{ in})$$

$$= 7.91 \text{ in}^2$$

Calculation of A_1 (area available in shell):

A_1 = the larger of

$$D(Et - t_r) - 2t_n(Et - t_r)(1 - f_{r1})$$

$$= (12 \text{ in})\big((1)(0.75 \text{ in}) - 0.659 \text{ in}\big) - 0$$

$$= 1.09 \text{ in}^2$$

and

$$2(t + t_n)(Et - t_r) - 2t_n(Et - t_r)(1 - f_{r1})$$

$$= (2)(0.75 \text{ in} + 0.5 \text{ in})$$

$$\times \big((1)(0.75 \text{ in}) - 0.659 \text{ in}\big) - 0$$

$$= 0.23 \text{ in}^2$$

$$A_1 = 1.09 \text{ in}^2$$

Calculation of A_2 (area available from nozzle):

The equations for A_2 with a reinforcing pad are used.

A_2 = the smaller of

$$5(t_n - t_{rn})t f_{r2}$$

$$= (5)(0.5 \text{ in} - 0.164 \text{ in})(0.75 \text{ in})(1.0)$$

$$= 1.26 \text{ in}^2$$

and

$$2(t_n - t_{rn})(2.5t_n + t_p)f_{r2}$$

$$= (2)(0.5 \text{ in} - 0.164 \text{ in})((2.5)(0.5 \text{ in})$$

$$+ 0.5 \text{ in})(1.0)$$

$$= 1.18 \text{ in}^2$$

$$A_2 = 1.18 \text{ in}^2$$

Calculation of A_4 (area available from welds):

$$f_{r3} = f_{r4} = 1.0$$

$$A_4 = A_{41} + A_{42} = (\text{leg})^2 f_{r3} + (\text{leg})^2 f_{r4}$$

$$= (0.563 \text{ in})^2(1.0) + (0.563 \text{ in})^2(1.0)$$

$$= 0.634 \text{ in}^2$$

Calculation of A_5 (area available from pad):

$$A_5 = (D_p - D - 2t_n)t_p f_{r4}$$

$$= (21 \text{ in} - 12 \text{ in} - (2)(0.5 \text{ in}))(0.5 \text{ in})(1.0)$$

$$= 4.0 \text{ in}^2$$

The available area is

$$A = A_1 + A_2 + A_4 + A_5$$

$$= 1.09 \text{ in}^2 + 1.18 \text{ in}^2 + 0.634 \text{ in}^2 + 4.0 \text{ in}^2$$

$$= 6.90 \text{ in}^2$$

Machine Design

Since 6.90 in^2 < 7.91 in^2, the pad is not acceptable as designed. An additional 1.01 in^2 of reinforcement is required. An acceptable design can be achieved by increasing the outside diameter of the pad by 3.5 in.

$$A_5 = (D_p - D - 2t_n)t_p f_{r4}$$
$$= \big(24.5 \text{ in} - 12 \text{ in} - (2)(0.5 \text{ in})\big)(0.5 \text{ in})(1.0)$$
$$= 5.75 \text{ in}^2$$

The available area is

$$A = A_1 + A_2 + A_4 + A_5$$
$$= 1.09 \text{ in}^2 + 1.18 \text{ in}^2 + 0.634 \text{ in}^2 + 5.75 \text{ in}^2$$
$$= 8.65 \text{ in}^2$$

54 Properties of Solid Bodies

PRACTICE PROBLEMS

Center of Gravity

1. Locate the center of gravity if the weights of the sphere, rod, and cylinder are 64.4, 32.2, and 64.4 pounds (29.2, 14.6, and 29.2 kg), respectively.

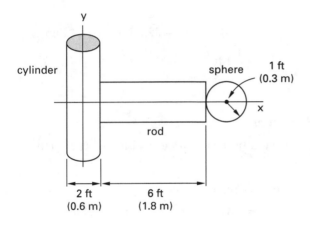

2. Find the center of gravity of the object shown.

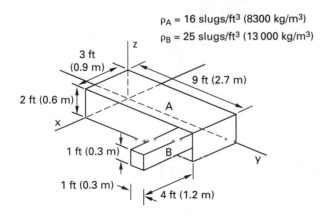

3. Find the center of gravity with respect to the x-axis of the object shown.

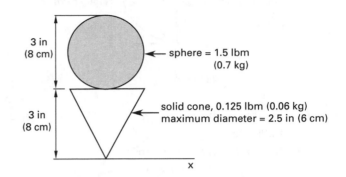

Mass Moments of Inertia

4. What is the mass moment of inertia about the x-axis of the object shown in problem 2?

5. What is the mass moment of inertia about the x-axis for the object shown in problem 3?

6. Find the moment of inertia about the BB axis if I_{AA} is 90 slug-ft^2 (120 kg·m^2) and the mass is 64.4 pounds (29.2 kg).

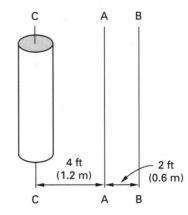

7. A spoked flywheel has an outside diameter of 60 in (1500 mm). The rim thickness is 6 in (150 mm), and the width is 12 in (300 mm). The cylindrical hub has an outside diameter of 12 in (300 mm), a thickness of 3 in (75 mm), and a width of 12 in (300 mm). The rim and hub are connected by six equally spaced cylindrical radial spokes, each with a diameter of 4.25 in (110 mm). All parts of the flywheel are ductile cast iron with a density of 0.256 lbm/in^3 (7080 kg/m^3). What is the rotational mass moment of inertia of the entire flywheel?

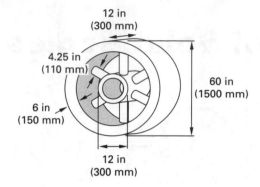

12 in
(300 mm)

4.25 in
(110 mm)

6 in
(150 mm)

60 in
(1500 mm)

12 in
(300 mm)

(A)   14,000 lbm-ft^2 (530 kg·m^2)
(B)   15,000 lbm-ft^2 (570 kg·m^2)
(C)   16,000 lbm-ft^2 (600 kg·m^2)
(D)   17,000 lbm-ft^2 (650 kg·m^2)

8. The rim of a spoked cast-iron flywheel in a punch press has a radius of gyration of approximately 15 in (38.1 cm) and a face width of 12 in (30.5 cm). The flywheel rotates at 200 rpm. It supplies 1500 ft-lbf (2 kJ) of energy and slows to 175 rpm each time a punch press is cycled. The density of the cast iron is 0.26 lbm/in^3 (7200 kg/m^3). The hub and arms increase the rotational moment of inertia by 10%. What is the approximate rim thickness?

(A)   1.9 in (4.7 cm)
(B)   2.3 in (5.8 cm)
(C)   2.9 in (7.4 cm)
(D)   3.5 in (8.9 cm)

SOLUTIONS

1. *Customary U.S. Solution*

Since the object is symmetrical about the x-axis, $y_c = 0$ ft.

$$m_S = 64.4 \, \text{lbm}$$
$$x_{cS} = 1 \, \text{ft} + 6 \, \text{ft} + 1 \, \text{ft} = 8 \, \text{ft}$$
$$m_R = 32.2 \, \text{lbm}$$
$$x_{cR} = 1 \, \text{ft} + 3 \, \text{ft} = 4 \, \text{ft}$$
$$m_C = 64.4 \, \text{lbm}$$
$$x_{cC} = 0 \, \text{ft}$$
$$x_c = \frac{\Sigma m_i x_{ci}}{\Sigma m_i}$$

$$= \frac{\begin{array}{c}(64.4 \, \text{lbm})(8 \, \text{ft}) + (32.2 \, \text{lbm})(4 \, \text{ft}) \\ + (64.4 \, \text{lbm})(0 \, \text{ft})\end{array}}{(64.4 \, \text{lbm} + 32.2 \, \text{lbm} + 64.4 \, \text{lbm})}$$

$$= \boxed{4 \, \text{ft}}$$

SI Solution

Since the object is symmetrical about the x-axis, $y_c = 0$ m.

$$m_S = 29.2 \, \text{kg}$$
$$x_{cS} = 0.3 \, \text{m} + 1.8 \, \text{m} + 0.3 \, \text{m} = 2.4 \, \text{m}$$
$$m_R = 14.6 \, \text{kg}$$
$$x_{cR} = 0.3 \, \text{m} + 0.9 \, \text{m} = 1.2 \, \text{m}$$
$$m_C = 29.2 \, \text{kg}$$
$$x_{cC} = 0 \, \text{m}$$
$$x_c = \frac{\Sigma m_i x_{ci}}{\Sigma m_i}$$

$$= \frac{\begin{array}{c}(29.2 \, \text{kg})(2.4 \, \text{m}) + (14.6 \, \text{kg})(1.2 \, \text{m}) \\ + (29.2 \, \text{kg})(0 \, \text{m})\end{array}}{(29.2 \, \text{kg} + 14.6 \, \text{kg} + 29.2 \, \text{kg})}$$

$$= \boxed{1.20 \, \text{m}}$$

2. *Customary U.S. Solution*

$$m_A = (2 \, \text{ft})\,(3 \, \text{ft})\,(9 \, \text{ft}) \left(16 \, \frac{\text{slug}}{\text{ft}^3}\right)$$
$$= 864 \, \text{slugs}$$
$$m_B = (1 \, \text{ft})\,(1 \, \text{ft})\,(4 \, \text{ft}) \left(25 \, \frac{\text{slug}}{\text{ft}^3}\right)$$
$$= 100 \, \text{slugs}$$

$x_{cA} = 1.5 \text{ ft}$

$x_{cB} = 3 \text{ ft} + 2 \text{ ft} = 5 \text{ ft}$

$$x_c = \frac{\Sigma m_i x_{ci}}{\Sigma m_i} = \frac{(864 \text{ slug})(1.5 \text{ ft}) + (100 \text{ slug})(5 \text{ ft})}{864 \text{ slug} + 100 \text{ slug}}$$

$$= \boxed{1.863 \text{ ft}}$$

$y_{cA} = 4.5 \text{ ft}$

$y_{cB} = 8 \text{ ft} + 0.5 \text{ ft} = 8.5 \text{ ft}$

$$y_c = \frac{(864 \text{ slug})(4.5 \text{ ft}) + (100 \text{ slug})(8.5 \text{ ft})}{864 \text{ slug} + 100 \text{ slug}}$$

$$= \boxed{4.915 \text{ ft}}$$

$z_{cA} = 1 \text{ ft}$

$z_{cB} = 1 \text{ ft} + 0.5 \text{ ft} = 1.5 \text{ ft}$

$$z_c = \frac{(864 \text{ slug})(1 \text{ ft}) + (100 \text{ slug})(1.5 \text{ ft})}{864 \text{ slug} + 100 \text{ slug}}$$

$$= \boxed{1.052 \text{ ft}}$$

SI Solution

$$m_A = (0.6 \text{ m})(0.9 \text{ m})(2.7 \text{ m})\left(8300 \, \frac{\text{kg}}{\text{m}^3}\right)$$

$$= 12\,101 \text{ kg}$$

$$m_B = (0.3 \text{ m})(0.3 \text{ m})(1.2 \text{ m})\left(13\,000 \, \frac{\text{kg}}{\text{m}^3}\right)$$

$$= 1404 \text{ kg}$$

$x_{cA} = 0.45 \text{ m}$

$x_{cB} = 0.9 \text{ m} + 0.6 \text{ m} = 1.5 \text{ m}$

$$x_c = \frac{\Sigma m_i x_{ci}}{\Sigma m_i}$$

$$= \frac{(12\,101 \text{ kg})(0.45 \text{ m}) + (1404 \text{ kg})(1.5 \text{ m})}{12\,101 \text{ kg} + 1404 \text{ kg}}$$

$$= \boxed{0.5592 \text{ m}}$$

$y_{cA} = 1.35 \text{ m}$

$y_{cB} = 2.4 \text{ m} + 0.15 \text{ m} = 2.55 \text{ m}$

$$y_c = \frac{(12\,101 \text{ kg})(1.35 \text{ m}) + (1404 \text{ kg})(2.55 \text{ m})}{12\,101 \text{ kg} + 1404 \text{ kg}}$$

$$= \boxed{1.475 \text{ m}}$$

$z_{cA} = 0.3 \text{ m}$

$z_{cB} = 0.3 \text{ m} + 0.15 \text{ m} = 0.45 \text{ m}$

$$z_c = \frac{(12\,101 \text{ kg})(0.3 \text{ m}) + (1404 \text{ kg})(0.45 \text{ m})}{12\,101 \text{ kg} + 1404 \text{ kg}}$$

$$= \boxed{0.3156 \text{ m}}$$

3. *Customary U.S. Solution*

$m_S = 1.5 \text{ lbm}$

$$y_{cS} = 3 \text{ in} + \frac{3 \text{ in}}{2} = 4.5 \text{ in}$$

$m_C = 0.125 \text{ lbm}$

$$y_{cC} = \frac{3h}{4} = \frac{(3)(3 \text{ in})}{4} = 2.25 \text{ in}$$

$$y_c = \frac{\Sigma m_i y_{ci}}{\Sigma m_i}$$

$$= \frac{(1.5 \text{ lbm})(4.5 \text{ in}) + (0.125 \text{ lbm})(2.25 \text{ in})}{1.5 \text{ lbm} + 0.125 \text{ lbm}}$$

$$= \boxed{4.327 \text{ in from the } x\text{-axis}}$$

SI Solution

$m_S = 0.70 \text{ kg}$

$$y_{cS} = 8 \text{ cm} + \frac{8 \text{ cm}}{2} = 12 \text{ cm}$$

$m_C = 0.06 \text{ kg}$

$$y_{cC} = \frac{3h}{4} = \frac{(3)(8 \text{ cm})}{4} = 6.0 \text{ cm}$$

$$y_c = \frac{\Sigma m_i y_{ci}}{\Sigma m_i}$$

$$= \frac{(0.70 \text{ kg})(12 \text{ cm}) + (0.06 \text{ kg})(6.0 \text{ cm})}{0.70 \text{ kg} + 0.06 \text{ kg}}$$

$$= \boxed{11.53 \text{ cm from the } x\text{-axis}}$$

4. *Customary U.S. Solution*

Since A is a rectangular parallelepiped,

$$I_{cA} = \frac{1}{12} m_A \left(a^2 + b^2\right)$$

$$= \left(\frac{1}{12}\right)(864 \text{ slug})\left((2 \text{ ft})^2 + (9 \text{ ft})^2\right)$$

$$= 6120 \text{ slug-ft}^2$$

The distance, d, from the x-centroidal axis to the x-axis is found by

$$d = \sqrt{(y_c)^2 + (z_c)^2} = \sqrt{(4.5 \text{ ft})^2 + (1 \text{ ft})^2} = 4.61 \text{ ft}$$

$$I_{xA} = I_{cA} + m_A d^2$$

$$= 6120 \text{ slug-ft}^2 + (864 \text{ slug})(4.61 \text{ ft})^2$$

$$= 24{,}482 \text{ slug-ft}^2$$

For B,

$$I_{cB} = \frac{1}{12} m_B \left(a^2 + b^2\right)$$

$$= \left(\frac{1}{12}\right)(100 \text{ slug})\left((1 \text{ ft})^2 + (1 \text{ ft})^2\right)$$

$$= 16.67 \text{ slug-ft}^2$$

Dynamics and Vibrations

$$d = \sqrt{(8.5 \text{ ft})^2 + (1.5 \text{ ft})^2} = 8.63 \text{ ft}$$

$$\begin{aligned} I_{xB} &= I_{cB} + m_B d^2 \\ &= 16.67 \text{ slug-ft}^2 + (100 \text{ slug})(8.63 \text{ ft})^2 \\ &= 7464 \text{ slug-ft}^2 \end{aligned}$$

$$\begin{aligned} I_x &= I_{xA} + I_{xB} \\ &= 24{,}482 \text{ slug-ft}^2 + 7464 \text{ slug-ft}^2 \\ &= \boxed{31{,}946 \text{ slug-ft}^2} \end{aligned}$$

SI Solution

Since A is a rectangular parallelepiped,

$$\begin{aligned} I_{cA} &= \frac{1}{12} m_A \left(a^2 + b^2\right) \\ &= \left(\frac{1}{12}\right)(12\,101 \text{ kg})\left((0.6 \text{ m})^2 + (2.7 \text{ m})^2\right) \\ &= 7714 \text{ kg·m}^2 \end{aligned}$$

The distance, d, from the x-centroidal axis to the x-axis is found by

$$d = \sqrt{(y_c)^2 + (z_c)^2} = \sqrt{(1.35 \text{ m})^2 + (0.3 \text{ m})^2} = 1.38 \text{ m}$$

$$\begin{aligned} I_{xA} &= I_{cA} + m_A d^2 \\ &= 7714 \text{ kg·m}^2 + (12\,101 \text{ kg})(1.38 \text{ m})^2 \\ &= 30\,759 \text{ kg·m}^2 \end{aligned}$$

For B,

$$\begin{aligned} I_{cB} &= \frac{1}{12} m_B \left(a^2 + b^2\right) \\ &= \left(\frac{1}{12}\right)(1404 \text{ kg})\left((0.3 \text{ m})^2 + (0.3 \text{ m})^2\right) \\ &= 21.06 \text{ kg·m}^2 \end{aligned}$$

$$d = \sqrt{(2.55 \text{ m})^2 + (0.45 \text{ m})^2} = 2.59 \text{ m}$$

$$\begin{aligned} I_{xB} &= I_{cB} + m_B d^2 \\ &= 21.06 \text{ kg·m}^2 + (1404 \text{ kg})(2.59 \text{ m})^2 \\ &= 9439 \text{ kg·m}^2 \end{aligned}$$

$$\begin{aligned} I_x &= I_{xA} + I_{xB} \\ &= 30\,759 \text{ kg·m}^2 + 9439 \text{ kg·m}^2 \\ &= \boxed{40\,198 \text{ kg·m}^2} \end{aligned}$$

5. *Customary U.S. Solution*

For the cone,

$$\begin{aligned} I_{xC} &= \frac{3}{5} m_C \left(\frac{r^2}{4} + h^2\right) \\ &= \left(\frac{3}{5}\right)(0.125 \text{ lbm})\left(\frac{(1.25 \text{ in})^2}{4} + (3 \text{ in})^2\right) \\ &= 0.704 \text{ lbm-in}^2 \end{aligned}$$

For the sphere,

$$\begin{aligned} I_{cS} &= \frac{2}{5} m_S r^2 = \left(\frac{2}{5}\right)(1.5 \text{ lbm})(1.5 \text{ in})^2 \\ &= 1.35 \text{ lbm-in}^2 \end{aligned}$$

$$\begin{aligned} I_{xS} &= I_{cS} + m_S d^2 \\ &= 1.35 \text{ lbm-in}^2 + (1.5 \text{ lbm})(4.5 \text{ in})^2 \\ &= 31.725 \text{ lbm-in}^2 \end{aligned}$$

$$I_x = I_{xC} + I_{xS} = 0.704 \text{ lbm-in}^2 + 31.725 \text{ lbm-in}^2$$

$$= \boxed{32.429 \text{ lbm-in}^2}$$

SI Solution

For the cone,

$$\begin{aligned} I_{xC} &= \frac{3}{5} m_C \left(\frac{r^2}{4} + h^2\right) \\ &= \left(\frac{3}{5}\right)(0.06 \text{ kg})\left(\frac{(0.03 \text{ m})^2}{4} + (0.08 \text{ m})^2\right) \\ &= 0.000239 \text{ kg·m}^2 \end{aligned}$$

For the sphere,

$$\begin{aligned} I_{cS} &= \frac{2}{5} m_S r^2 = \left(\frac{2}{5}\right)(0.70 \text{ kg})(0.04 \text{ m})^2 \\ &= 0.000448 \text{ kg·m}^2 \end{aligned}$$

$$\begin{aligned} I_{xS} &= I_{cS} + m_S d^2 \\ &= 0.000448 \text{ kg·m}^2 + (0.7 \text{ kg})(0.12 \text{ m})^2 \\ &= 0.010528 \text{ kg·m}^2 \end{aligned}$$

$$I_x = I_{xC} + I_{xS} = 0.000239 \text{ kg·m}^2 + 0.010528 \text{ kg·m}^2$$

$$= \boxed{0.010767 \text{ kg·m}^2}$$

6. *Customary U.S. Solution*

$$m = (64.4 \text{ lbm})\left(\frac{\text{slug}}{32.2 \text{ lbm}}\right) = 2 \text{ slug}$$

$$\begin{aligned} I_{CC} &= I_{AA} - m d^2 = 90 \text{ slug-ft}^2 - (2 \text{ slug})(4 \text{ ft})^2 \\ &= 58 \text{ slug-ft}^2 \end{aligned}$$

$$I_{BB} = I_{CC} + md'^2 = 58 \text{ slug-ft}^2 + (2 \text{ slug})(6 \text{ ft})^2$$

$$\boxed{= 130 \text{ slug-ft}^2}$$

SI Solution

$$I_{CC} = I_{AA} - md^2$$

$$= 120 \text{ kg·m}^2 - (29.2 \text{ kg})(1.2 \text{ m})^2$$

$$= 77.95 \text{ kg·m}^2$$

$$I_{BB} = I_{CC} + md'^2$$

$$= 77.95 \text{ kg·m}^2 + (29.2 \text{ kg})(1.8 \text{ m})^2$$

$$\boxed{= 172.56 \text{ kg·m}^2}$$

7. *Customary U.S. Solution*

From the hollow circular cylinder in App. 54.A, the mass moment of inertia of the rim is

$$I_{x,1} = \left(\frac{\pi \rho L}{2}\right)(r_o^4 - r_i^4)$$

$$= \left(\frac{\pi}{2}\right)\left(0.256 \frac{\text{lbm}}{\text{in}^3}\right)(12 \text{ in})$$

$$\times \left(\left(\frac{60 \text{ in}}{2}\right)^4 - \left(\frac{60 \text{ in} - 12 \text{ in}}{2}\right)^4\right)\left(\frac{1 \text{ ft}^2}{144 \text{ in}^2}\right)$$

$$= 16{,}025 \text{ lbm-ft}^2$$

From the hollow circular cylinder in App. 54.A, the mass moment of inertia of the hub is

$$I_{x,2} = \left(\frac{\pi \rho L}{2}\right)(r_o^4 - r_i^4)$$

$$= \left(\frac{\pi}{2}\right)\left(0.256 \frac{\text{lbm}}{\text{in}^3}\right)(12 \text{ in})$$

$$\times \left(\left(\frac{12 \text{ in}}{2}\right)^4 - \left(\frac{12 \text{ in} - 6 \text{ in}}{2}\right)^4\right)\left(\frac{1 \text{ ft}^2}{144 \text{ in}^2}\right)$$

$$= 40.72 \text{ lbm-ft}^2$$

The length of a cylindrical spoke is

$$L = 24 \text{ in} - 6 \text{ in} = 18 \text{ in}$$

The mass of a spoke is

$$m = \rho A L = \rho \pi r^2 L$$

$$= \left(0.256 \frac{\text{lbm}}{\text{in}^3}\right) \pi \left(\frac{4.25 \text{ in}}{2}\right)^2 (18 \text{ in})$$

$$= 65.37 \text{ lbm}$$

From the solid circular cylinder in App. 54.A, the mass moment of inertia of a spoke about its own centroidal axis is

$$I_z = \frac{m(3r^2 + L^2)}{12}$$

$$= \frac{(65.37 \text{ lbm})\left((3)\left(\frac{4.25 \text{ in}}{2}\right)^2 + (18 \text{ in})^2\right)\left(\frac{1 \text{ ft}^2}{144 \text{ in}^2}\right)}{12}$$

$$= 13 \text{ lbm-ft}^2$$

Use the parallel axis theorem, Eq. 54.12, to find the mass moment of inertia of a spoke about the axis of the flywheel.

$$d = \frac{12 \text{ in}}{2} + \frac{18 \text{ in}}{2} = 15 \text{ in}$$

$$I_{x,3} \text{ per spoke} = I_z + md^2$$

$$= 13 \text{ lbm-ft}^2 + (65.37 \text{ lbm})\left(\frac{15 \text{ in}}{12 \frac{\text{in}}{\text{ft}}}\right)^2$$

$$= 115 \text{ lbm-ft}^2$$

The total for six spokes is

$$I_{x,3} = 6 I_{x,3} \text{ per spoke}$$

$$= (6)(115 \text{ lbm-ft}^2) = 690 \text{ lbm-ft}^2$$

Finally, the total rotational mass moment of inertia of the flywheel is

$$I = I_{x,1} + I_{x,2} + I_{x,3}$$

$$= 16{,}025 \text{ lbm-ft}^2 + 40.72 \text{ lbm-ft}^2 + 690 \text{ lbm-ft}^2$$

$$\boxed{= 16{,}756 \text{ lbm-ft}^2}$$

The answer is (D).

SI Solution

From the hollow circular cylinder in App. 54.A, the mass moment of inertia of the rim is

$$I_{x,1} = \left(\frac{\pi \rho L}{2}\right)(r_o^4 - r_i^4)$$

$$= \left(\frac{\pi \left(7080 \frac{\text{kg}}{\text{m}^3}\right)(0.3 \text{ m})}{2}\right)$$

$$\times \left(\left(\frac{1.5 \text{ m}}{2}\right)^4 - \left(\frac{1.5 \text{ m} - 0.30 \text{ m}}{2}\right)^4\right)$$

$$= 623.3 \text{ kg·m}^2$$

From the hollow circular cylinder in App. 54.A, the mass moment of inertia of the hub is

$$I_{x,2} = \left(\frac{\pi \rho L}{2}\right)(r_o^4 - r_i^4)$$

$$= \left(\frac{\pi \left(7080 \ \frac{kg}{m^3}\right)(0.3 \ m)}{2}\right)$$

$$\times \left(\left(\frac{0.3 \ m}{2}\right)^4 - \left(\frac{0.3 \ m - 0.15 \ m}{2}\right)^4\right)$$

$$= 1.6 \ kg \cdot m^2$$

The length of a cylindrical spoke is

$$L = 0.6 \ m - 0.15 \ m = 0.45 \ m$$

The mass of a spoke is

$$m = \rho A L = \rho \pi r^2 L$$

$$= \left(7080 \ \frac{kg}{m^3}\right)\pi \left(\frac{0.11 \ m}{2}\right)^2 (0.45 \ m)$$

$$= 30.3 \ kg$$

From the solid circular cylinder in App. 54.A, the mass moment of inertia of a spoke about its own centroidal axis is

$$I_z = \frac{m(3r^2 + L^2)}{12}$$

$$= \frac{(30.3 \ kg)\left((3)\left(\frac{0.11 \ m}{2}\right)^2 + (0.45 \ m)^2\right)}{12}$$

$$= 0.53 \ kg \cdot m^2$$

Use the parallel axis theorem, Eq. 54.12, to find the mass moment of inertia of a spoke about the axis of the flywheel.

$$d = \frac{0.30 \ m}{2} + \frac{0.45 \ m}{2} = 0.375 \ m$$

$$I_{x,3} \text{ per spoke} = I_z + md^2$$

$$= 0.53 \ kg \cdot m^2 + (30.3 \ kg)(0.375 \ m)^2$$

$$= 4.8 \ kg \cdot m^2$$

The total for six spokes is

$$I_{x,3} = 6 I_{x3} \text{ per spoke}$$

$$= (6)(4.8 \ kg \cdot m^2) = 28.8 \ kg \cdot m^2$$

Finally, the total rotational mass moment of inertia of the flywheel is

$$I = I_{x,1} + I_{x,2} + I_{x,3}$$

$$= 623.3 \ kg \cdot m^2 + 1.6 \ kg \cdot m^2 + 28.8 \ kg \cdot m^2$$

$$= \boxed{653.7 \ kg \cdot m^2}$$

The answer is (D).

8. *Customary U.S. Solution*

The angular velocity of the flywheel at the start of the cycle is

$$\omega_1 = \frac{\left(2\pi \ \frac{rad}{rev}\right)\left(200 \ \frac{rev}{min}\right)}{60 \ \frac{sec}{min}} = 20.94 \ rad/sec$$

The angular velocity of the flywheel at the end of the cycle is

$$\omega_2 = \frac{\left(2\pi \ \frac{rad}{rev}\right)\left(175 \ \frac{rev}{min}\right)}{60 \ \frac{sec}{min}} = 18.33 \ rad/sec$$

By the work-energy principle, the decrease in kinetic energy is equal to the work supplied by the flywheel.

$$\Delta KE = W$$

$$\left(\tfrac{1}{2}\right)\left(\frac{I}{g_c}\right)\omega_1^2 - \left(\tfrac{1}{2}\right)\left(\frac{I}{g_c}\right)\omega_2^2 = W$$

Solve for the mass moment of inertia, I.

$$I = \frac{2 g_c W}{\omega_1^2 - \omega_2^2}$$

$$= \frac{(2)\left(32.2 \ \frac{ft\text{-}lbm}{lbf\text{-}sec^2}\right)(1500 \ ft\text{-}lbf)}{\left(20.94 \ \frac{rad}{sec}\right)^2 - \left(18.33 \ \frac{rad}{sec}\right)^2}$$

$$= 942.5 \ lbm\text{-}ft^2$$

Assume all of the rim mass is concentrated at the radius of gyration. The mean circumference is

$$L_{mean} = 2\pi r_{mean}$$

$$= 2\pi (15 \ in) = 94.25 \ in$$

The mass moment of inertia of the rim is

$$I_{rim} = mr_{mean}^2 = \rho tw L_{mean} r_{mean}^2$$

$$= \left(0.26 \ \frac{lbm}{in^3}\right)t(12 \ in)(94.25 \ in)$$

$$\times (15 \ in)^2 \left(\frac{1 \ ft^2}{144 \ in^2}\right)$$

$$= 459.5t$$

Adding 10% for the hub and arms,

$$I_{total} = 1.10 I_{rim} = (1.10)(459.5t)$$

$$= 505.5t$$

Solve for t.

$$t = \frac{I}{\frac{I_{\text{total}}}{t}} = \frac{942.5 \text{ lbm-ft}^2}{505.5 \frac{\text{lbm-ft}^2}{\text{in}}}$$

$$= \boxed{1.86 \text{ in}}$$

The answer is (A).

SI Solution

The angular velocity of the flywheel at the start of the cycle is

$$\omega_1 = \frac{\left(2\pi \frac{\text{rad}}{\text{rev}}\right)\left(200 \frac{\text{rev}}{\text{min}}\right)}{60 \frac{\text{sec}}{\text{min}}} = 20.94 \text{ rad/sec}$$

The angular velocity of the flywheel at the end of the cycle is

$$\omega_2 = \frac{\left(2\pi \frac{\text{rad}}{\text{rev}}\right)\left(175 \frac{\text{rev}}{\text{min}}\right)}{60 \frac{\text{sec}}{\text{min}}} = 18.33 \text{ rad/sec}$$

By the work-energy principle, the decrease in kinetic energy is equal to the work supplied by the flywheel.

$$\Delta \text{KE} = W$$
$$\tfrac{1}{2}I\omega_1^2 - \tfrac{1}{2}I\omega_2^2 = W$$

Solve for the mass moment of inertia, I.

$$I = \frac{2W}{\omega_1^2 - \omega_2^2}$$

$$= \frac{(2)(2 \text{ kJ})\left(1000 \frac{\text{J}}{\text{kJ}}\right)}{\left(20.94 \frac{\text{rad}}{\text{sec}}\right)^2 - \left(18.33 \frac{\text{rad}}{\text{sec}}\right)^2}$$

$$= 39.03 \text{ kg·m}^2$$

Assume all of the rim mass is concentrated at the mean radius. The mean circumference is

$$L_{\text{mean}} = 2\pi r_{\text{mean}}$$
$$= 2\pi \left(\frac{0.762 \text{ m}}{2}\right) = 2.394 \text{ m}$$

The mass moment of inertia of the rim is

$$I_{\text{rim}} = mr_{\text{mean}}^2 = \rho t w L_{\text{mean}} r_{\text{mean}}^2$$
$$= \left(7200 \frac{\text{kg}}{\text{m}^3}\right) t(0.305 \text{ m})(2.394 \text{ m})\left(\frac{0.762 \text{ m}}{2}\right)^2$$
$$= 763.1t$$

Adding 10% for the hub and arms,

$$I_{\text{total}} = 1.10 I_{\text{rim}} = (1.10)(763.1t)$$
$$= 839.4t$$

Solve for t.

$$t = \frac{I}{\frac{I_{\text{total}}}{t}} = \frac{39.03 \text{ kg·m}^2}{839.4 \frac{\text{kg·m}^2}{\text{m}}}$$

$$= \boxed{0.0465 \text{ m}}$$

The answer is (A).

Dynamics and Vibrations

55 Kinematics

PRACTICE PROBLEMS

Linear Particle Motion

1. A particle moves horizontally according to $s = 2t^2 - 8t + 3$.

(a) When $t = 2$, what are the position, velocity, and acceleration?

(b) What are the linear displacement and total distance traveled between $t = 1$ and $t = 3$?

Uniform Acceleration

2. A jet plane acquires a speed of 180 mph (290 km/h) in 60 seconds. What is its acceleration?
- (A) 4.4 ft/sec^2 (1.3 m/s^2)
- (B) 6.3 ft/sec^2 (1.9 m/s^2)
- (C) 8.9 ft/sec^2 (2.7 m/s^2)
- (D) 12 ft/sec^2 (3.6 m/s^2)

3. What is the acceleration of a train that increases its speed from 5 ft/sec to 20 ft/sec (1.5 m/s to 6 m/s) in 2 minutes?
- (A) 0.13 ft/sec^2 (0.038 m/s^2)
- (B) 0.39 ft/sec^2 (0.12 m/s^2)
- (C) 0.82 ft/sec^2 (0.25 m/s^2)
- (D) 1.3 ft/sec^2 (0.39 m/s^2)

4. A car traveling at 60 mph (100 km/h) applies its brakes and stops in 5 seconds. What is its acceleration and distance traveled before stopping?

Projectile Motion

5. A projectile is fired at 45° from the horizontal with an initial velocity of 2700 ft/sec (820 m/s). Find the maximum altitude and range neglecting air friction.

6. A projectile is launched with an initial velocity of 900 ft/sec (270 m/s). The target is 12,000 ft (3600 m) away and 2000 ft (600 m) higher than the launch point. Air friction is to be neglected. At what angle should the projectile be launched?

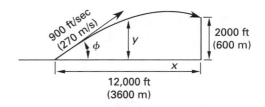

7. A baseball is hit at 60 ft/sec (20 m/s) and 36.87° from the horizontal. It strikes a fence 72 feet (22 m) away. Find the velocity components and the elevation above the origin at impact.

8. A bomb is dropped from a plane that is climbing at 30° and 600 ft/sec (180 m/s) while traveling at 12,000 feet (3600 m) altitude.

(a) What is the bomb's maximum altitude?
- (A) 12,500 ft (3700 m)
- (B) 13,000 ft (3900 m)
- (C) 13,400 ft (4000 m)
- (D) 14,200 ft (4300 m)

(b) How long will it take for the bomb to reach the ground from the release point?
- (A) 24 sec
- (B) 38 sec
- (C) 43 sec
- (D) 55 sec

Rotational Particle Motion

9. A point travels in a circle according to $\omega = 6t^2 - 10t$. At $t = 2$, the direction of motion is clockwise.

(a) What is the angular velocity at $t = 2$?
- (A) 4
- (B) 6
- (C) 8
- (D) 12

(b) What is the displacement between $t = 1$ and $t = 3$?
- (A) 6 rad
- (B) 10 rad
- (C) 12 rad
- (D) 15 rad

(c) What is the total angle turned through between $t = 1$ and $t = 3$? Assume $\theta(0) = 0$.

    (A) 6 rad

    (B) 8 rad

    (C) 12 rad

    (D) 15 rad

10. What is the linear speed of a point on the edge of a 14-inch-diameter (36-cm-diameter) disk turning at 40 rpm?

    (A) 2.4 ft/sec (0.75 m/s)

    (B) 4.5 ft/sec (1.4 m/s)

    (C) 7.8 ft/sec (2.3 m/s)

    (D) 9.2 ft/sec (2.8 m/s)

11. What angular acceleration is required to increase an electric motor's speed from 1200 rpm to 3000 rpm in 10 seconds?

    (A) 10 rad/sec^2

    (B) 14 rad/sec^2

    (C) 18 rad/sec^2

    (D) 25 rad/sec^2

12. An apparatus for determining the speed of a bullet consists of 2 paper disks mounted 5 feet (1.5 m) apart on a single horizontal shaft which is turning at 1750 rpm. A bullet pierces both disks at radius 6 inches (15 cm) and an angle of 18° exists between each hole. What is the bullet velocity?

    (A) 1700 ft/sec (510 m/s)

    (B) 2100 ft/sec (630 m/s)

    (C) 2400 ft/sec (720 m/s)

    (D) 2900 ft/sec (880 m/s)

13. Disks B and C are in contact and rotate without slipping. A and B are splined together and rotate counterclockwise. Angular velocity and acceleration of disk C are 2 rad/sec (2 rad/s) and 6 rad/sec^2 (6 rad/s^2), respectively. What is the velocity and acceleration of point D?

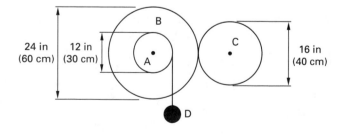

Relative Motion

14. The center of a wheel with an outer diameter of 24 in (610 mm) is moving at 28 mi/hr (12.5 m/s). There is no slippage between the wheel and surface. A valve

stem is mounted 6 in (150 mm) from the center. What are the velocity and direction of the valve stem when it is 45° from the horizontal?

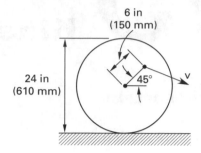

15. A balloon is 200 feet (60 m) above the ground and is rising at a constant 15 ft/sec (4.5 m/s). An automobile passes under it traveling along a straight and level road at 45 mph (72 km/h). How fast is the distance between them changing one second later?

    (A) 13 ft/sec (4.0 m/s)

    (B) 34 ft/sec (10 m/s)

    (C) 37 ft/sec (11 m/s)

    (D) 51 ft/sec (15 m/s)

Rotation About a Fixed Axis

16. Find the velocities of points A and B with respect to point O if the wheel rolls without slipping. The axle to which the wheel is attached moves at 10 ft/sec (3 m/s) to the right.

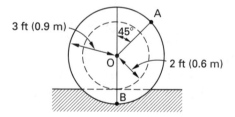

17. What is the velocity of point B with respect to point A in problem 16?

SOLUTIONS

1. *Customary U.S. Solution*

(a)
$$x(t) = 2t^2 - 8t + 3$$
$$v(t) = \frac{dx}{dt} = 4t - 8$$
$$a(t) = \frac{dv}{dt} = 4$$

At $t = 2$:

$$x = (2)(2)^2 - (8)(2) + 3 = \boxed{-5}$$

$$v = (4)(2) - 8 = \boxed{0}$$

$$a = \boxed{4}$$

(b) At $t = 1$:

$$x = (2)(1)^2 - (8)(1) + 3 = \boxed{-3}$$

At $t = 3$:

$$x = (2)(3)^2 - (8)(3) + 3 = \boxed{-3}$$

displacement $= x_3 - x_1 = -3 - (-3) = 0$

total distance traveled from $t = 1$ to $t = 3$

$$= \int_1^3 |v(t)| dt = \int_1^3 |4t - 8| dt$$
$$= \int_1^2 (8 - 4t) dt + \int_2^3 (4t - 8) dt$$
$$= 2 + 2 = \boxed{4}$$

SI Solution

(a)
$$x(t) = 2t^2 - 8t + 3$$
$$v(t) = \frac{dx}{dt} = 4t - 8$$
$$a(t) = \frac{dv}{dt} = 4$$

At $t = 2$:

$$x = (2)(2)^2 - (8)(2) + 3 = \boxed{-5}$$

$$v = (4)(2) - 8 = \boxed{0}$$

$$a = \boxed{4}$$

(b) At $t = 1$:

$$x = (2)(1)^2 - (8)(1) + 3 = \boxed{-3}$$

At $t = 3$:

$$x = (2)(3)^2 - (8)(3) + 3 = \boxed{-3}$$

displacement $= x_3 - x_1 = -3 - (-3) = 0$

total distance traveled from $t = 1$ to $t = 3$

$$= \int_1^3 |v(t)| dt = \int_1^3 |4t - 8| dt$$
$$= \int_1^2 (8 - 4t) dt + \int_2^3 (4t - 8) dt$$
$$= 2 + 2 = \boxed{4}$$

2. *Customary U.S. Solution*

$$a = \frac{\Delta v}{\Delta t} = \frac{\left(180 \dfrac{\text{mile}}{\text{hr}}\right)\left(\dfrac{5280 \text{ ft}}{\text{mile}}\right)\left(\dfrac{\text{hr}}{3600 \text{ sec}}\right)}{60 \text{ sec}}$$
$$= \boxed{4.4 \text{ ft/sec}^2}$$

The answer is (A).

SI Solution

$$a = \frac{\Delta v}{\Delta t} = \frac{\left(290 \dfrac{\text{km}}{\text{h}}\right)\left(\dfrac{1000 \text{ m}}{\text{km}}\right)\left(\dfrac{\text{h}}{3600 \text{ s}}\right)}{60 \text{ s}}$$
$$= \boxed{1.34 \text{ m/s}^2}$$

The answer is (A).

3. *Customary U.S. Solution*

$$a = \frac{v - v_0}{t} = \frac{20 \dfrac{\text{ft}}{\text{sec}} - 5 \dfrac{\text{ft}}{\text{sec}}}{(2 \text{ min})\left(\dfrac{60 \text{ sec}}{\text{min}}\right)}$$
$$= \boxed{0.125 \text{ ft/sec}^2}$$

The answer is (A).

SI Solution

$$a = \frac{v - v_0}{t} = \frac{6\,\frac{m}{s} - 1.5\,\frac{m}{s}}{(2\,\text{min})\left(\frac{60\,s}{\text{min}}\right)}$$

$$= \boxed{0.0375\ \text{m/s}^2}$$

The answer is (A).

4. *Customary U.S. Solution*

Assume uniform deceleration.

$$v_0 = \left(60\,\frac{\text{mile}}{\text{hr}}\right)\left(\frac{5280\ \text{ft}}{\text{mile}}\right)\left(\frac{\text{hr}}{3600\ \text{sec}}\right) = 88\ \text{ft/sec}$$

$$a = \frac{v - v_0}{t} = \frac{0 - 88\,\frac{\text{ft}}{\text{sec}}}{5\ \text{sec}} = \boxed{-17.6\ \text{ft/sec}^2}$$

$$s = \frac{1}{2}t(v + v_0) = \frac{1}{2}tv_0$$

$$= \left(\frac{1}{2}\right)(5\ \text{sec})\left(88\,\frac{\text{ft}}{\text{sec}}\right) = \boxed{220\ \text{ft}}$$

SI Solution

Assume uniform deceleration.

$$v_0 = \left(100\,\frac{\text{km}}{\text{h}}\right)\left(\frac{1000\ \text{m}}{\text{km}}\right)\left(\frac{\text{h}}{3600\ \text{s}}\right) = 27.78\ \text{m/s}$$

$$a = \frac{v - v_0}{t} = \frac{0 - 27.78\,\frac{m}{s}}{5\ \text{s}} = \boxed{-5.56\ \text{m/s}^2}$$

$$s = \frac{1}{2}t(v + v_0) = \frac{1}{2}tv_0$$

$$= \left(\frac{1}{2}\right)(5\ \text{s})\left(27.78\,\frac{m}{s}\right) = \boxed{69.5\ \text{m}}$$

5. *Customary U.S. Solution*

$$H = \frac{v_0^2 \sin^2 \phi}{2g}$$

$$= \frac{\left(2700\,\frac{\text{ft}}{\text{sec}}\right)^2 (\sin^2 45°)}{(2)\left(32.2\,\frac{\text{ft}}{\text{sec}^2}\right)} = \boxed{56{,}599\ \text{ft}}$$

$$R = \frac{v_0^2 \sin 2\phi}{g}$$

$$= \frac{\left(2700\,\frac{\text{ft}}{\text{sec}}\right)^2 (\sin 90°)}{32.2\,\frac{\text{ft}}{\text{sec}^2}} = \boxed{226{,}398\ \text{ft}}$$

SI Solution

$$H = \frac{v_0^2 \sin^2 \phi}{2g}$$

$$= \frac{\left(820\,\frac{m}{s}\right)^2 (\sin^2 45°)}{(2)\left(9.81\,\frac{m}{s^2}\right)} = \boxed{17\,136\ \text{m}}$$

$$R = \frac{v_0^2 \sin 2\phi}{g}$$

$$= \frac{\left(820\,\frac{m}{s}\right)^2 (\sin 90°)}{9.81\,\frac{m}{s^2}} = \boxed{68\,542\ \text{m}}$$

6. *Customary U.S. Solution*

From Table 55.2, the x-distance is

$$x = (v_0 \cos \phi)t$$

Solve for t.

$$t = \frac{x}{v_0 \cos \phi} = \frac{12{,}000\ \text{ft}}{\left(900\,\frac{\text{ft}}{\text{sec}}\right)(\cos \phi)} = \frac{13.33}{\cos \phi}$$

From Table 55.2, the y-distance is

$$y = (v_0 \sin \phi)t - \tfrac{1}{2}gt^2$$

Substitute t and the value for y.

$$2000\ \text{ft} = \left(900\,\frac{\text{ft}}{\text{sec}}\right)(\sin \phi)\left(\frac{13.33}{\cos \phi}\right)$$

$$- \left(\tfrac{1}{2}\right)\left(32.2\,\frac{\text{ft}}{\text{sec}^2}\right)\left(\frac{13.33}{\cos \phi}\right)^2$$

Simplify.

$$1 = 6.0 \tan \phi - \frac{1.43}{\cos^2 \phi}$$

Use the identity.

$$\frac{1}{\cos^2 \phi} = 1 + \tan^2 \phi$$

$$1 = 6.0 \tan \phi - 1.43 - 1.43 \tan^2 \phi$$

Simplify.

$$\tan^2 \phi - 4.20 \tan \phi + 1.70 = 0$$

Use the quadratic formula.

$$\tan \phi = \frac{4.20 \pm \sqrt{(4.20)^2 - (4)(1)(1.70)}}{(2)(1)}$$

$$= 3.75,\ 0.454 \quad [\text{radians}]$$

$$\phi = \tan^{-1}(3.75),\ \tan^{-1}(0.454)$$

$$= \boxed{75.1°,\ 24.4°}$$

SI Solution

From Table 55.2, the x-distance is

$$x = (v_0 \cos \phi)t$$

Solve for t.

$$t = \frac{x}{v_0 \cos \phi} = \frac{3600 \text{ m}}{\left(270 \frac{\text{m}}{\text{s}}\right)(\cos \phi)} - \frac{13.33}{\cos \phi}$$

From Table 55.2, the y-distance is

$$y = (v_0 \sin \phi)t - \frac{1}{2}gt^2$$

Substitute t and the value of y.

$$600 \text{ m} = \left(270 \frac{\text{m}}{\text{s}}\right)(\sin \phi)\left(\frac{13.33}{\cos \phi}\right)$$
$$- \left(\frac{1}{2}\right)\left(9.81 \frac{\text{m}}{\text{s}^2}\right)\left(\frac{13.33}{\cos \phi}\right)^2$$

Simplify.

$$1 = 6.0 \tan \phi - \frac{1.45}{\cos^2 \phi}$$

Use the identity.

$$\frac{1}{\cos^2 \phi} = 1 + \tan^2 \phi$$
$$1 = 6.0 \tan \phi - 1.45 - 1.45 \tan^2 \phi$$

Simplify.

$$\tan^2 \phi - 4.14 \tan \phi + 1.69 = 0$$

Use the quadratic formula.

$$\tan \phi = \frac{4.14 \pm \sqrt{(4.14)^2 - (4)(1)(1.69)}}{(2)(1)}$$
$$= 3.68, \ 0.459 \quad [\text{radians}]$$
$$\phi = \tan^{-1}(3.68), \ \tan^{-1}(0.459)$$
$$= \boxed{74.8°, \ 24.7°}$$

7. *Customary U.S. Solution*

Neglect air friction.

$$v_x = v_0 \cos \phi = \left(60 \frac{\text{ft}}{\text{sec}}\right)(\cos 36.87°)$$
$$= \boxed{48 \text{ ft/sec}}$$

$$t = \frac{s}{v_x} = \frac{72 \text{ ft}}{48 \frac{\text{ft}}{\text{sec}}} = 1.5 \text{ sec}$$

$$v_y = v_0 \sin \phi - gt$$
$$= \left(60 \frac{\text{ft}}{\text{sec}}\right)(\sin 36.87°) - \left(32.2 \frac{\text{ft}}{\text{sec}^2}\right)(1.5 \text{ sec})$$
$$= \boxed{-12.3 \text{ ft/sec}}$$

$$y = v_0 \sin \phi \, t - \frac{1}{2}gt^2$$
$$= \left(60 \frac{\text{ft}}{\text{sec}}\right)(\sin 36.87°)(1.5 \text{ sec})$$
$$- \left(\frac{1}{2}\right)\left(32.2 \frac{\text{ft}}{\text{sec}^2}\right)(1.5 \text{ sec})^2$$
$$= \boxed{17.78 \text{ ft}}$$

SI Solution

Neglect air friction.

$$v_x = v_0 \cos \phi = \left(20 \frac{\text{m}}{\text{s}}\right)(\cos 36.87°)$$
$$= \boxed{16 \text{ m/s}}$$

$$t = \frac{s}{v_x} = \frac{22 \text{ m}}{16 \frac{\text{m}}{\text{s}}} = 1.375 \text{ s}$$

$$v_y = v_0 \sin \phi - gt$$
$$= \left(20 \frac{\text{m}}{\text{s}}\right)(\sin 36.87°) - \left(9.81 \frac{\text{m}}{\text{s}^2}\right)(1.375 \text{ s})$$
$$= \boxed{-1.49 \text{ m/s}}$$

$$y = v_0 \sin \phi t - \frac{1}{2}gt^2$$
$$= \left(20 \frac{\text{m}}{\text{s}}\right)(\sin 36.87°)(1.375 \text{ s})$$
$$- \left(\frac{1}{2}\right)\left(9.81 \frac{\text{m}}{\text{s}^2}\right)(1.375 \text{ s})^2$$
$$= \boxed{7.227 \text{ m}}$$

8. *Customary U.S. Solution*

(a) $$H = z + \frac{v_0^2 \sin^2 \phi}{2g}$$
$$= 12{,}000 \text{ ft} + \frac{\left(600 \frac{\text{ft}}{\text{sec}}\right)^2 (\sin^2 30°)}{(2)\left(32.2 \frac{\text{ft}}{\text{sec}^2}\right)}$$
$$= \boxed{13{,}398 \text{ ft}}$$

The answer is (C).

(b) Let t_1 be the time the bomb takes to reach the maximum altitude.

$$t_1 = \frac{1}{2}\left(\frac{2v_0\sin\phi}{g}\right) = \frac{\left(600\ \frac{\text{ft}}{\text{sec}}\right)(\sin 30°)}{32.2\ \frac{\text{ft}}{\text{sec}^2}} = 9.32\ \text{sec}$$

Let t_2 be the time the bomb takes to fall from H.

$$t_2 = \sqrt{\frac{2H}{g}} = \sqrt{\frac{(2)(13{,}398\ \text{ft})}{32.2\ \frac{\text{ft}}{\text{sec}^2}}} = 28.85\ \text{sec}$$

$$t = t_1 + t_2 = 9.32\ \text{sec} + 28.85\ \text{sec}$$

$$= \boxed{38.17\ \text{sec}}$$

The answer is (B).

SI Solution

(a) $\qquad H = z + \dfrac{v_0^2\sin^2\phi}{2g}$

$$= 3600\ \text{m} + \frac{\left(180\frac{\text{m}}{\text{s}}\right)^2(\sin^2 30°)}{(2)\left(9.81\frac{\text{m}}{\text{s}^2}\right)}$$

$$= \boxed{4012.8\ \text{m}}$$

The answer is (C).

(b) Let t_1 be the time the bomb takes to reach the maximum altitude.

$$t_1 = \frac{1}{2}\left(\frac{2\text{ft}_0\sin\phi}{g}\right) = \frac{\left(180\frac{\text{m}}{\text{s}}\right)(\sin 30°)}{9.81\frac{\text{m}}{\text{s}^2}} = 9.17\ \text{s}$$

Let t_2 be the time the bomb takes to fall from H.

$$t_2 = \sqrt{\frac{2H}{g}} = \sqrt{\frac{(2)(4012.8\ \text{m})}{9.81\frac{\text{m}}{\text{s}^2}}} = 28.60\ \text{s}$$

$$t = t_1 + t_2 = 9.17\ \text{s} + 28.60\ \text{s}$$

$$= \boxed{37.77\ \text{s}}$$

The answer is (B).

9. *Customary U.S. Solution*

(a) $\omega(2) = (6)(2)^2 - (10)(2) = \boxed{4\ \text{clockwise}}$

(b) $\qquad \theta = \theta(3) - \theta(1) = \displaystyle\int_1^3 \omega(t)\,dt$

$$= \int_1^3 (6t^2 - 10t)\,dt = 2t^3 - 5t^2 + C\Big|_1^3$$

$$[C = 0\ \text{since}\ \theta(0) = 0]$$

$$= (2)(3)^3 - (5)(3)^2 - (2)(1)^3 + (5)(1)^2$$

$$= \boxed{12}$$

To find the total distance traveled, check for sign reversals in $\omega(t)$ over the interval $t = 1$ to $t = 3$.

$$6t^2 - 10t = 0$$

$$\text{sign reversal at}\ t = \frac{5}{3}$$

(c) The total angle turned is

$$\int_1^3 |\omega(t)|\,dt = \int_1^{\frac{5}{3}} (10t - 6t^2)\,dt + \int_{\frac{5}{3}}^3 (6t^2 - 10t)\,dt$$

$$= \boxed{15.26\ \text{rad}}$$

The answer is (D).

SI Solution

(a) $\qquad \omega(2) = (6)(2)^2 - (10)(2) = \boxed{4\ \text{clockwise}}$

(b) $\qquad \theta = \theta(3) - \theta(1) = \displaystyle\int_1^3 \omega(t)\,dt$

$$= \int_1^3 (6t^2 - 10t)\,dt = \boxed{12}$$

To find the total distance traveled, check for sign reversals in $\omega(t)$ over the interval $t = 1$ to $t = 3$.

$$6t^2 - 10t = 0$$

$$\text{sign reversal at}\ t = \frac{5}{3}$$

(c) The total angle turned is

$$\int_1^3 |\omega(t)|\,dt = \int_1^{\frac{5}{3}} (10t - 6t^2)\,dt + \int_{\frac{5}{3}}^3 (6t^2 - 10t)\,dt$$

$$= \boxed{15.26\ \text{rad}}$$

The answer is (D).

10. *Customary U.S. Solution*

$$v = r\omega = r2\pi f = d\pi f$$
$$= (14 \text{ in})\left(\frac{\text{ft}}{12 \text{ in}}\right)(\pi)\left(40 \frac{\text{rev}}{\text{min}}\right)\left(\frac{\text{min}}{60 \text{ sec}}\right)$$
$$= \boxed{2.44 \text{ ft/sec}}$$

The answer is (A).

SI Solution

$$v = r\omega = r2\pi f = d\pi f$$
$$= (0.36 \text{ m})(\pi)\left(40 \frac{\text{rev}}{\text{min}}\right)\left(\frac{\text{min}}{60 \text{ s}}\right)$$
$$= \boxed{0.754 \text{ m/s}}$$

The answer is (A).

11. *Customary U.S. Solution*

$$\omega_1 = 2\pi f_1 = \left(2\pi \frac{\text{rad}}{\text{rev}}\right)\left(1200 \frac{\text{rev}}{\text{min}}\right)\left(\frac{\text{min}}{60 \text{ sec}}\right)$$
$$= 125.66 \text{ rad/sec}$$

$$\omega_2 = \left(2\pi \frac{\text{rad}}{\text{rev}}\right)\left(3000 \frac{\text{rev}}{\text{min}}\right)\left(\frac{\text{min}}{60 \text{ sec}}\right)$$
$$= 314.16 \text{ rad/sec}$$

$$\alpha = \frac{\omega_2 - \omega_1}{\Delta t} = \frac{314.16 \frac{\text{rad}}{\text{sec}} - 125.66 \frac{\text{rad}}{\text{sec}}}{10 \text{ sec}}$$
$$= \boxed{18.85 \text{ rad/sec}^2}$$

The answer is (C).

SI Solution

$$\omega_1 = 2\pi f_1 = \left(2\pi \frac{\text{rad}}{\text{rev}}\right)\left(1200 \frac{\text{rev}}{\text{min}}\right)\left(\frac{\text{min}}{60 \text{ s}}\right)$$
$$- 125.66 \text{ rad/s}$$

$$\omega_2 = \left(2\pi \frac{\text{rad}}{\text{rev}}\right)\left(3000 \frac{\text{rev}}{\text{min}}\right)\left(\frac{\text{min}}{60 \text{ s}}\right)$$
$$= 314.16 \text{ rad/s}$$

$$\alpha = \frac{\omega_2 - \omega_1}{\Delta t} = \frac{314.16 \frac{\text{rad}}{\text{s}} - 125.66 \frac{\text{rad}}{\text{s}}}{10 \text{ s}}$$
$$= \boxed{18.85 \text{ rad/s}^2}$$

The answer is (C).

12. *Customary U.S. Solution*

$$f = \left(1750 \frac{\text{rev}}{\text{min}}\right)\left(\frac{\text{min}}{60 \text{ sec}}\right) = 29.167 \text{ rev/sec}$$

$$\theta = (18°)\left(\frac{\text{rev}}{360°}\right) = 0.05 \text{ rev}$$

$$t = \frac{\theta}{f} = \frac{0.05 \text{ rev}}{29.167 \frac{\text{rev}}{\text{sec}}} = 0.001714 \text{ sec}$$

$$v = \frac{s}{t} = \frac{5 \text{ ft}}{0.001714 \text{ sec}} = \boxed{2916.7 \text{ ft/sec (fps)}}$$

The answer is (D).

SI Solution

$$\omega = \left(1750 \frac{\text{rev}}{\text{min}}\right)\left(\frac{\text{min}}{60 \text{ s}}\right)\left(\frac{2\pi \text{ rad}}{\text{rev}}\right) = 183.26 \text{ rad/s}$$

$$\theta = (18°)\left(\frac{2\pi \text{ rad}}{360°}\right) = 0.314 \text{ rad}$$

$$t = \frac{\theta}{\omega} = \frac{0.314 \text{ rad}}{183.26 \frac{\text{rad}}{\text{s}}} = 0.001713 \text{ s}$$

$$v = \frac{s}{t} = \frac{1.5 \text{ m}}{0.001713 \text{ s}} = \boxed{875.66 \text{ m/s}}$$

The answer is (D).

13. *Customary U.S. Solution*

$$\omega_B = \left(\frac{16}{24}\right)\omega_C = \left(\frac{16}{24}\right)\left(2 \frac{\text{rad}}{\text{sec}}\right) = 1.333 \text{ rad/sec}$$

$$\alpha_B = \left(\frac{16}{24}\right)\alpha_C = \left(\frac{16}{24}\right)\left(6 \frac{\text{rad}}{\text{sec}^2}\right) = 4 \text{ rad/sec}^2$$

Since A and B are splined together,

$$\omega_A = \omega_B$$
$$\alpha_A = \alpha_B$$

$$v_D = r_A\omega_A = (0.5 \text{ ft})\left(1.333 \frac{\text{rad}}{\text{sec}}\right)$$
$$= \boxed{0.6665 \text{ ft/sec (fps)}}$$

$$a_D = r_A\alpha_A = (0.5 \text{ ft})\left(4 \frac{\text{rad}}{\text{sec}^2}\right)$$
$$= \boxed{2 \text{ ft/sec}^2}$$

SI Solution

$$\omega_B = \left(\frac{40}{60}\right)\omega_C = \left(\frac{40}{60}\right)\left(2\,\frac{\text{rad}}{\text{s}}\right) = 1.333 \text{ rad/s}$$

$$\alpha_B = \left(\frac{40}{60}\right)\alpha_C = \left(\frac{40}{60}\right)\left(6\,\frac{\text{rad}}{\text{s}^2}\right) = 4 \text{ rad/s}^2$$

Since A and B are splined together,

$$\omega_A = \omega_B$$

$$\alpha_A = \alpha_B$$

$$v_D = r_A\omega_A = (0.15 \text{ m})\left(1.333\,\frac{\text{rad}}{\text{s}}\right)$$

$$= \boxed{0.2 \text{ m/s}}$$

$$a_D = r_A\alpha_A = (0.15 \text{ m})\left(4\,\frac{\text{rad}}{\text{s}^2}\right)$$

$$= \boxed{0.6 \text{ m/s}^2}$$

14. *Customary U.S. Solution*

The angular velocity of the wheel is

$$\omega = \frac{v_c}{r}$$

$$= \frac{\left(28\,\frac{\text{mi}}{\text{hr}}\right)\left(5280\,\frac{\text{ft}}{\text{mi}}\right)\left(12\,\frac{\text{in}}{\text{ft}}\right)\left(\frac{1 \text{ hr}}{3600 \text{ sec}}\right)}{12 \text{ in}}$$

$$= 41.07 \text{ rad/sec}$$

The distance from the valve stem to the instant center of the point of contact of the wheel and the surface is determined from the law of cosines for the triangle defined by the valve stem, instant center, and center of wheel.

$$l^2 = (12 \text{ in})^2 + (6 \text{ in})^2 - (2)(12 \text{ in})(6 \text{ in})(\cos 135°)$$

$$l = 16.79 \text{ in}$$

Use Eq. 55.52.

$$v = l\omega$$

$$= (16.79 \text{ in})\left(\frac{1 \text{ ft}}{12 \text{ in}}\right)\left(41.07\,\frac{\text{rad}}{\text{sec}}\right)$$

$$= \boxed{57.46 \text{ ft/sec}}$$

The direction is perpendicular to the direction of l.

SI Solution

The angular velocity of the wheel is

$$\omega = \frac{v_c}{r} = \frac{12.5\,\frac{\text{m}}{\text{s}}}{0.305 \text{ m}} = 40.98 \text{ rad/s}$$

The distance from the valve stem to the instant center at the point of contact of the wheel and the surface is determined from the law of cosines for the triangle defined by the valve stem, instant center, and center of wheel.

$$l^2 = (0.305 \text{ m})^2 + (0.150 \text{ m})^2$$
$$- (2)(0.305 \text{ m})(0.150 \text{ m})(\cos 135°)$$

$$l = 0.425 \text{ m}$$

Use Eq. 55.52.

$$v = l\omega$$

$$= (0.425 \text{ m})\left(40.98\,\frac{\text{rad}}{\text{s}}\right)$$

$$= \boxed{17.4 \text{ m/s}}$$

The direction is perpendicular to the direction of l.

15. *Customary U.S. Solution*

$$v_{\text{car}} = \left(45\,\frac{\text{miles}}{\text{hr}}\right)\left(\frac{5280 \text{ ft}}{\text{mile}}\right)\left(\frac{\text{hr}}{3600 \text{ sec}}\right)$$

$$= 66 \text{ ft/sec}$$

The separation distance after 1 second is

$$s(1) = \sqrt{\left(200 \text{ ft} + \left(15\,\frac{\text{ft}}{\text{sec}}\right)(1 \text{ sec})\right)^2 + \left(\left(66\,\frac{\text{ft}}{\text{sec}}\right)(1 \text{ sec})\right)^2}$$

$$= 224.9 \text{ ft}$$

The separation velocity is the difference in components of the car's and balloon's velocities along a mutually parallel line. Use the separation vector as this line.

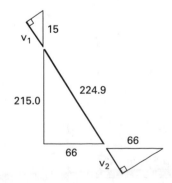

$$v_1 = \left(15 \frac{\text{ft}}{\text{sec}}\right)\left(\frac{215}{224.9}\right) = 14.34 \frac{\text{ft}}{\text{sec}}$$

$$v_2 = \left(66 \frac{\text{ft}}{\text{sec}}\right)\left(\frac{66}{224.9}\right) = 19.37 \frac{\text{ft}}{\text{sec}}$$

$$\Delta v = v_1 + v_2 = 14.34 \frac{\text{ft}}{\text{sec}} + 19.37 \frac{\text{ft}}{\text{sec}}$$

$$= \boxed{33.71 \text{ ft/sec}}$$

The answer is (B).

SI Solution

$$v_{\text{car}} = \left(72 \frac{\text{km}}{\text{h}}\right)\left(\frac{1000 \text{ m}}{\text{km}}\right)\left(\frac{\text{h}}{3600 \text{ s}}\right)$$

$$= 20 \text{ m/s}$$

The separation distance after 1 second is

$$s(1) = \sqrt{\left(60 \text{ m} + \left(4.5 \frac{\text{m}}{\text{s}}\right)(1 \text{ s})\right)^2 + \left(\left(20 \frac{\text{m}}{\text{s}}\right)(1 \text{ s})\right)^2}$$

$$= 67.53 \text{ m}$$

The separation velocity is the difference in components of the car's and balloon's velocities along a mutually parallel line. Use the separation vector as this line.

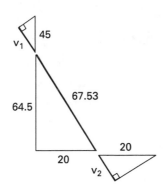

$$v_1 = \left(4.5 \frac{\text{m}}{\text{s}}\right)\left(\frac{64.5}{67.53}\right) = 4.298 \frac{\text{m}}{\text{s}}$$

$$v_2 = \left(20 \frac{\text{m}}{\text{s}}\right)\left(\frac{20}{67.53}\right) = 5.923 \frac{\text{m}}{\text{s}}$$

$$\Delta v = v_1 + v_2 = 4.298 \frac{\text{m}}{\text{s}} + 5.923 \frac{\text{m}}{\text{s}}$$

$$= \boxed{10.22 \text{ m/s}}$$

The answer is (B).

16. *Customary U.S. Solution*

$$\omega = \frac{v_0}{r_{\text{inner}}} = \frac{10 \frac{\text{ft}}{\text{sec}}}{2 \frac{\text{ft}}{\text{rad}}} = 5 \text{ rad/sec}$$

$$v_{\text{A/O}} = r_{\text{outer}}\omega = \left(3 \frac{\text{ft}}{\text{rad}}\right)\left(5 \frac{\text{rad}}{\text{sec}}\right)$$

$$= \boxed{\begin{array}{l} 15 \text{ ft/sec}, 45° \text{ below the horizontal,} \\ \text{to the right} \end{array}}$$

$$v_{\text{B/O}} = r_{\text{outer}}\omega = \boxed{15 \text{ ft/sec, horizontal, to the left}}$$

SI Solution

$$\omega = \frac{v_0}{r_{\text{inner}}} = \frac{3 \frac{\text{m}}{\text{s}}}{0.6 \frac{\text{m}}{\text{rad}}} = 5 \text{ rad/s}$$

$$v_{\text{A/O}} = r_{\text{outer}}\omega = \left(0.9 \frac{\text{m}}{\text{rad}}\right)\left(5 \frac{\text{rad}}{\text{s}}\right)$$

$$= \boxed{\begin{array}{l} 4.5 \text{ m/s}, 45° \text{ below the horizontal,} \\ \text{to the right} \end{array}}$$

$$v_{\text{B/O}} = r_{\text{outer}}\omega = \boxed{4.5 \text{ m/s, horizontal, to the left}}$$

17. *Customary U.S. Solution*

$$|AB|^2 = (3 \text{ ft})^2 + (3 \text{ ft})^2 - (2)(3 \text{ ft})^2(\cos 135°)$$

$$= 30.728 \text{ ft}^2$$

$$|AB| = 5.543 \text{ ft}$$

$$v_{\text{B/A}} = |AB|\omega = (5.543 \text{ ft})\left(5 \frac{\text{rad}}{\text{sec}}\right)$$

$$= \boxed{\begin{array}{l} 27.72 \text{ ft/sec}, 22.5° \text{ above the horizontal,} \\ \text{to the left} \end{array}}$$

SI Solution

$$|AB|^2 = (0.9 \text{ m})^2 + (0.9 \text{ m})^2 - (2)(0.9 \text{ m})^2(\cos 135°)$$

$$= 2.7655 \text{ m}^2$$

$$|AB| = 1.663 \text{ m}$$

$$v_{\text{B/A}} = |AB|\omega = (1.663 \text{ m})\left(5 \frac{\text{rad}}{\text{s}}\right)$$

$$= \boxed{\begin{array}{l} 8.315 \text{ m/s}, 22.5° \text{ above the horizontal,} \\ \text{to the left} \end{array}}$$

Dynamics and Vibrations

56 Kinetics

PRACTICE PROBLEMS

Centripetal and Centrifugal Forces

1. Calculate the superelevation in percent necessary on a curve with a 6000-foot (1800-m) radius so that at 60 mph (100 km/h) cars will not have to rely on friction to stay on the roadway.

- (A) 4%
- (B) 7%
- (C) 11%
- (D) 14%

2. What is the angle between the pole and the wire if the radius of the path is 4 feet (1.2 m)? The 8.05-pound (3.65-kg) object is rotating at 20 ft/sec (6 m/s).

- (A) 46°
- (B) 58°
- (C) 72°
- (D) 79°

3. A 10-pound (5-kg) mass is tied to a 2-ft (50-cm) string and whirled at 5 revolutions per second horizontally to the ground.

(a) Find the centripetal acceleration.
- (A) 1400 ft/sec² (350 m/s²)
- (B) 1600 ft/sec² (400 m/s²)
- (C) 1800 ft/sec² (450 m/s²)
- (D) 2000 ft/sec² (490 m/s²)

(b) Find the centrifugal force.
- (A) 490 lbf (1900 N)
- (B) 610 lbf (2500 N)
- (C) 750 lbf (2900 N)
- (D) 820 lbf (3100 N)

(c) Find the centripetal force.
- (A) 490 lbf (1900 N)
- (B) 610 lbf (2500 N)
- (C) 750 lbf (2900 N)
- (D) 820 lbf (3100 N)

(d) Find the angular momentum.
- (A) 39 ft-lbf-sec (39 J·s)
- (B) 44 ft-lbf-sec (44 J·s)
- (C) 53 ft-lbf-sec (53 J·s)
- (D) 71 ft-lbf-sec (71 J·s)

Friction

4. A 100-pound (50-kg) body has an initial velocity of 12.88 ft/sec (3.93 m/s) while moving on a plane with coefficient of friction of 0.2. What distance will it travel before coming to rest?

- (A) 13 ft (3.9 m)
- (B) 15 ft (4.5 m)
- (C) 18 ft (5.4 m)
- (D) 24 ft (7.2 m)

5. A box is dropped onto a conveyor belt moving at 10 ft/sec (3 m/s). The coefficient of friction between the box and the belt is 0.333. How long will it take before the box stops slipping on the belt?

- (A) 0.48 sec
- (B) 0.67 sec
- (C) 0.93 sec
- (D) 1.2 sec

6. A motorcycle and rider weigh 400 pounds (200 kg). They travel horizontally around the inside of a hollow, right-angle cylinder of 100 feet (30 m) inside diameter. What most nearly is the coefficient of friction that will allow a speed of 40 mph (60 km/h)?

- (A) 0.3
- (B) 0.4
- (C) 0.5
- (D) 0.6

7. When a force acts on a 10-pound (5-kg) body initially at rest, a speed of 12 ft/sec (3.6 m/s) is attained in 36 feet (11 m). What is the force if the coefficient of friction is 0.25 and the acceleration is constant?

   (A) 1.8 lbf (8.6 N)

   (B) 2.4 lbf (13 N)

   (C) 3.1 lbf (15 N)

   (D) 4.6 lbf (22 N)

8. A 130-pound (60-kg) block slides up a 22.62° incline with coefficient of friction of 0.1 and initial velocity of 30 ft/sec (9 m/s). How far will the block slide up the incline before coming to rest?

   (A) 18 ft (5.4 m)

   (B) 29 ft (8.7 m)

   (C) 42 ft (13 m)

   (D) 57 ft (17 m)

9. A 100-pound (50-kg) block initially at rest is acted upon by a 100-pound (400-N) force while resting on a horizontal surface with coefficient of friction of 0.2. A 50-pound (25-kg) block sits on top of the 100-pound (50-kg) block. What is the minimum coefficient of friction between the 50- and 100-pound (25- and 50-kg) blocks for there to be no slipping?

Rigid Body Motion

10. A constant force of 20 pounds (90 N) is applied to a 100-pound (50-kg) door supported on rollers at A and B.

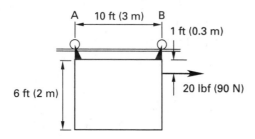

(a) What is the acceleration of the door?

(b) What are the reactions at A and B?

(c) Where would the 20 lbf (90-N) force have to be applied if the reactions at A and B were to be equal?

Constrained Motion

11. A solid sphere rolls without slipping down a 30° incline, starting from rest. What is its speed after two seconds?

   (A) 18 ft/sec (5.5 m/s)

   (B) 23 ft/sec (7.0 m/s)

   (C) 34 ft/sec (10 m/s)

   (D) 86 ft/sec (26 m/s)

12. Object A weighs 10 pounds (5 kg) and rests on a frictionless plane with a 36.87° slope. Object B weighs 20 pounds (10 kg). What is the velocity of B three seconds after release?

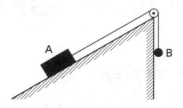

   (A) 14 ft/sec (4.2 m/s)

   (B) 22 ft/sec (6.6 m/s)

   (C) 38 ft/sec (11 m/s)

   (D) 45 ft/sec (14 m/s)

Impulse and Momentum

13. Sand drops at the rate of 560 lbm/min (250 kg/min) onto a conveyor belt moving with a velocity of 3.2 ft/sec (0.98 m/s). What force is required to keep the belt moving?

   (A) 0.93 lbf (4.1 N)

   (B) 1.2 lbf (5.3 N)

   (C) 2.3 lbf (10 N)

   (D) 4.8 lbf (14 N)

14. What is the impulse imparted to a 0.4-pound (0.2-kg) baseball that approaches the batter at 90 ft/sec (30 m/s) and leaves at 130 ft/sec (40 m/s)?

   (A) 1.8 lbf-sec (9.4 N·s)

   (B) 2.7 lbf-sec (14 N·s)

   (C) 4.1 lbf-sec (21 N·s)

   (D) 8.9 lbf-sec (46 N·s)

15. At what velocity will a 1000-lbm (500-kg) gun mounted on wheels recoil if a 2.6-lbm (1.2-kg) projectile is propelled to 2100 ft/sec (650 m/s)?

   (A) 3.7 ft/sec (1.1 m/s)

   (B) 4.8 ft/sec (1.4 m/s)

   (C) 5.5 ft/sec (1.6 m/s)

   (D) 15 ft/sec (4.5 m/s)

16. A 0.15 lbm (60-g) bullet traveling 2300 ft/sec (700 m/s) embeds itself in a 9-lbm (4.5-kg) wooden block. What will be the block's initial velocity?

   (A) 24 ft/sec (7.2 m/s)

   (B) 38 ft/sec (9.2 m/s)

   (C) 66 ft/sec (200 m/s)

   (D) 110 ft/sec (330 m/s)

17. A nozzle discharges 40 gal/min (0.15 m³/min) of water at 60 ft/sec (20 m/s). Find the total force required to hold a flat plate in front of the nozzle. Assume that there is no splashback.

    (A)   10 lbf (50 N)

    (B)   20 lbf (100 N)

    (C)   40 lbf (200 N)

    (D)   80 lbf (400 N)

18. In a water turbine, 100 gallons (0.4 m³) per second impinge on a stationary blade. The water is turned through an angle of 160° and exits at 57 ft/sec (17 m/s). The water impinges at 60 ft/sec (20 m/s).

(a) Find the force exerted by the blade on the stream.

(b) What is the angle from the horizontal of the force?

Impacts

19. An electron collides with a hydrogen atom initially at rest. The electron's initial velocity is 1500 ft/sec (500 m/s). Its final velocity is 65 ft/sec (20.2 m/s) with a path 30° from its original path. Find the velocity of the hydrogen atom in the x-direction if it recoils 1.2° from the original path of the electron.

    (A)   0.043 ft/sec (0.013 m/s)

    (B)   0.79 ft/sec (0.26 m/s)

    (C)   4.6 ft/sec (1.4 m/s)

    (D)   53 ft/sec (16 m/s)

20. Two freight cars weighing 5 tons (5000 kg) each roll toward each other and couple. The left car has a velocity of 5 ft/sec (1.5 m/s) and the right car has velocity of 4 ft/sec (1.2 m/s) prior to the impact. What is the velocity of the two cars coupled together after the impact?

    (A)   0

    (B)   0.5 ft/sec (0.15 m/s)

    (C)   1.0 ft/sec (0.30 m/s)

    (D)   2.0 ft/sec (0.60 m/s)

21. Two billiard balls collide as shown below. What are the velocities after impact? Use $e = 0.8$.

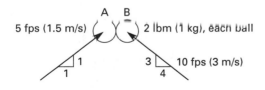

22. A 10 lb (5 kg) pendulum is released from rest and strikes a 50 lb (25 kg) block with $e = 0.7$. The block slides on a frictionless surface.

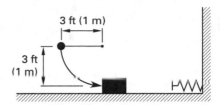

(a) Find the velocity of the pendulum at impact.

    (A)   14 ft/sec (4.4 m/s)

    (B)   26 ft/sec (7.8 m/s)

    (C)   38 ft/sec (11 m/s)

    (D)   51 ft/sec (15 m/s)

(b) Find the tension in the cord at impact.

    (A)   10 lbf (50 N)

    (B)   20 lbf (100 N)

    (C)   30 lbf (150 N)

    (D)   40 lbf (200 N)

(c) Find the block's velocity immediately after impact.

    (A)   2.2 ft/sec (0.66 m/s)

    (B)   3.9 ft/sec (1.3 m/s)

    (C)   4.6 ft/sec (1.4 m/s)

    (D)   5.8 ft/sec (1.7 m/s)

(d) Find the required spring constant to stop the block with less than 6 in (15 cm) deflection.

    (A)   54 lbf/ft (1200 N/m)

    (B)   62 lbf/ft (1400 N/m)

    (C)   75 lbf/ft (1600 N/m)

    (D)   96 lbf/ft (1800 N/m)

SOLUTIONS

1. *Customary U.S. Solution*

$$v_t = \left(60 \frac{\text{miles}}{\text{hr}}\right)\left(\frac{5280 \text{ ft}}{\text{mile}}\right)\left(\frac{\text{hr}}{3600 \text{ sec}}\right) = 88 \text{ ft/sec}$$

$$\tan\phi = \frac{v_t^2}{gr} = \frac{\left(88 \frac{\text{ft}}{\text{sec}}\right)^2}{\left(32.2 \frac{\text{ft}}{\text{sec}^2}\right)(6000 \text{ ft})} = \boxed{0.04 \ (4\%)}$$

The answer is (A).

SI Solution

$$v_t = \left(100 \frac{\text{km}}{\text{h}}\right)\left(\frac{1000 \text{ m}}{\text{km}}\right)\left(\frac{\text{h}}{3600 \text{ s}}\right) = 27.78 \text{ m/s}$$

$$\tan\phi = \frac{v_t^2}{gr} = \frac{\left(27.78 \frac{\text{m}}{\text{s}}\right)^2}{\left(9.81 \frac{\text{m}}{\text{s}^2}\right)(1800 \text{ m})} = \boxed{0.0437 \ (4.37\%)}$$

The answer is (A).

2. *Customary U.S. Solution*

The net acceleration vector is directed along the string.

$$\phi = \arctan\left(\frac{a_c}{a_n}\right) = \arctan\left(\frac{v_t^2}{gr}\right)$$

$$= \arctan\left(\frac{\left(20 \frac{\text{ft}}{\text{sec}}\right)^2}{\left(32.2 \frac{\text{ft}}{\text{sec}^2}\right)(4 \text{ ft})}\right)$$

$$= \boxed{72.15° \text{ (independent of object's mass)}}$$

The answer is (C).

SI Solution

The net acceleration vector is directed radially along the string.

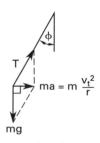

$$\frac{mv_t^2}{r}\cos\phi = mg\sin\phi$$

$$\tan\phi = \frac{v_t^2}{rg}$$

$$\phi = \arctan\left(\frac{v_t^2}{rg}\right)$$

$$= \arctan\left(\frac{\left(6 \frac{\text{m}}{\text{s}}\right)^2}{\left(9.81 \frac{\text{m}}{\text{s}^2}\right)(1.2 \text{ m})}\right)$$

$$= \boxed{71.89° \text{ (independent of object's mass)}}$$

The answer is (C).

3. *Customary U.S. Solution*

(a) $$v_t = \omega r = 2\pi f r$$

$$= \left(2\pi \frac{\text{rad}}{\text{rev}}\right)\left(5 \frac{\text{rev}}{\text{sec}}\right)(2 \text{ ft}) = 62.83 \text{ ft/sec}$$

$$a_n = \frac{v_t^2}{r} = \frac{\left(62.83 \frac{\text{ft}}{\text{sec}}\right)^2}{2 \text{ ft}} = \boxed{1974 \text{ ft/sec}^2}$$

The answer is (D).

(b) $$F_{\text{centrifugal}} = \frac{ma_n}{g_c} = \frac{(10 \text{ lbm})\left(1974 \frac{\text{ft}}{\text{sec}^2}\right)}{32.2 \frac{\text{lbm-ft}}{\text{sec}^2\text{-lbf}}}$$

$$= \boxed{613.04 \text{ lbf (directed outward)}}$$

The answer is (B).

(c) $$|\mathbf{F}_{\text{centripetal}}| = |\mathbf{F}_{\text{centrifugal}}|$$

$$F_{\text{centripetal}} = \boxed{613.04 \text{ lbf (directed inward)}}$$

The answer is (B).

(d) $$\mathbf{H} = \frac{\mathbf{r} \times m\mathbf{v}_t}{g_c}$$

$$H = \frac{rm v_t}{g_c} \text{ since } \mathbf{r} \perp \mathbf{v}_t$$

$$= \frac{(2 \text{ ft})(10 \text{ lbm})\left(62.83 \frac{\text{ft}}{\text{sec}}\right)}{32.2 \frac{\text{lbm-ft}}{\text{sec}^2\text{-lbf}}}$$

$$= \boxed{39.02 \text{ ft-lbf-sec}}$$

The answer is (A).

SI Solution

(a) $\quad v_t = \omega r = 2\pi f r$

$$= \left(2\pi \frac{\text{rad}}{\text{rev}}\right)\left(5 \frac{\text{rev}}{\text{s}}\right)(0.5 \text{ m}) = 15.708 \text{ m/s}$$

$$a_n = \frac{v_t^2}{r} = \frac{\left(15.708 \frac{\text{m}}{\text{s}}\right)^2}{0.5 \text{ m}} = \boxed{493.5 \text{ m/s}^2}$$

The answer is (D).

(b) $\quad F_{\text{centrifugal}} = ma_n = (5 \text{ kg})\left(493.5 \frac{\text{m}}{\text{s}^2}\right)$

$$= \boxed{2467.5 \text{ N (directed outward)}}$$

The answer is (B).

(c) $\quad |\mathbf{F}_{\text{centripetal}}| = |\mathbf{F}_{\text{centrifugal}}|$

$$F_{\text{centripetal}} = \boxed{2467.5 \text{ N (directed inward)}}$$

The answer is (B).

(d) $\quad \mathbf{H} = \mathbf{r} \times m\mathbf{v}_t$

$\quad H = rmv_t \quad$ since $\mathbf{r} \perp \mathbf{v}_t$

$$= (0.5 \text{ m})(5 \text{ kg})\left(15.708 \frac{\text{m}}{\text{s}}\right) = \boxed{39.27 \text{ J·s}}$$

The answer is (A).

4. *Customary U.S. Solution*

$$a = fg = (0.2)\left(32.2 \frac{\text{ft}}{\text{sec}^2}\right) = 6.44 \text{ ft/sec}^2$$

$$s_{\text{skidding}} = \frac{v^2}{2a} = \frac{\left(12.88 \frac{\text{ft}}{\text{sec}}\right)^2}{(2)\left(6.44 \frac{\text{ft}}{\text{sec}^2}\right)} = \boxed{12.88 \text{ ft}}$$

The answer is (A).

SI Solution

$$a = fg = (0.2)\left(9.81 \frac{\text{m}}{\text{s}^2}\right) = 1.962 \text{ m/s}^2$$

$$s_{\text{skidding}} = \frac{v^2}{2a} = \frac{\left(3.93 \frac{\text{m}}{\text{s}}\right)^2}{(2)\left(1.962 \frac{\text{m}}{\text{s}^2}\right)} = \boxed{3.936 \text{ m}}$$

The answer is (A).

5. *Customary U.S. Solution*

$$a = fg = (0.333)\left(32.2 \frac{\text{ft}}{\text{sec}^2}\right) = 10.72 \text{ ft/sec}^2$$

$$\Delta t = \frac{v}{a} = \frac{10 \frac{\text{ft}}{\text{sec}}}{10.72 \frac{\text{ft}}{\text{sec}^2}} = \boxed{0.933 \text{ sec}}$$

The answer is (C).

SI Solution

$$a = fg = (0.333)\left(9.81 \frac{\text{m}}{\text{s}^2}\right) = 3.267 \text{ m/s}^2$$

$$\Delta t = \frac{v}{a} = \frac{3 \frac{\text{m}}{\text{s}}}{3.267 \frac{\text{m}}{\text{s}^2}} = \boxed{0.918 \text{ s}}$$

The answer is (C).

6. *Customary U.S. Solution*

$$v_t = \left(40 \frac{\text{mile}}{\text{hr}}\right)\left(\frac{5280 \text{ ft}}{\text{mile}}\right)\left(\frac{\text{hr}}{3600 \text{ sec}}\right)$$

$$= 58.67 \text{ ft/sec}$$

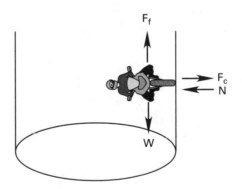

$$N = F_c = \frac{ma_n}{g_c} = \left(\frac{m}{g_c}\right)\left(\frac{v_t^2}{r}\right)$$

$$W = F_f = fN$$

$$f = \frac{W}{N} = \frac{\frac{mg}{g_c}}{\frac{mv_t^2}{g_c r}} = \frac{gr}{v_t^2} = \frac{\left(32.2 \frac{\text{ft}}{\text{sec}^2}\right)(50 \text{ ft})}{\left(58.67 \frac{\text{ft}}{\text{sec}}\right)^2}$$

$$= \boxed{0.468}$$

The answer is (C).

SI Solution

$$v_t = \left(60 \, \frac{km}{h}\right) \left(\frac{1000 \, m}{km}\right) \left(\frac{h}{3600 \, s}\right)$$

$$= 16.67 \, m/s$$

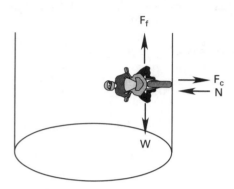

$$N = F_c = ma_n = m\left(\frac{v_t^2}{r}\right)$$

$$W = F_f = fN$$

$$f = \frac{W}{N} = \frac{mg}{\frac{mv_t^2}{r}} = \frac{gr}{v_t^2} = \frac{\left(9.81 \, \frac{m}{s^2}\right)(15 \, m)}{\left(16.67 \, \frac{m}{s}\right)^2}$$

$$= \boxed{0.530}$$

The answer is (C).

7. *Customary U.S. Solution*

$$a = \frac{v^2}{2s} = \frac{\left(12 \, \frac{ft}{sec}\right)^2}{(2)(36 \, ft)} = 2 \, ft/sec^2$$

$$F_f = fN = fW = (0.25)(10 \, lbf) = 2.5 \, lbf$$

$$P = F_f + \frac{ma}{g_c} = 2.5 \, lbf + \frac{(10 \, lbm)\left(2 \, \frac{ft}{sec^2}\right)}{32.2 \, \frac{lbm\text{-}ft}{lbf\text{-}sec^2}}$$

$$= \boxed{3.12 \, lbf}$$

The answer is (C).

SI Solution

$$a = \frac{v^2}{2s} = \frac{\left(3.6 \, \frac{m}{s}\right)^2}{(2)(11 \, m)} = 0.589 \, m/s^2$$

$$F_f = fN = fW = (0.25)(5 \, kg)\left(9.81 \, \frac{m}{s^2}\right) = 12.26 \, N$$

$$P = F_f + ma = 12.26 \, N + (5 \, kg)\left(0.589 \, \frac{m}{s^2}\right)$$

$$= \boxed{15.21 \, N}$$

The answer is (C).

8. *Customary U.S. Solution*

$$a = g\sin\theta + fg\cos\theta$$

$$= \left(32.2 \, \frac{ft}{sec^2}\right)\left(\sin(22.62°) + (0.1)(\cos 22.62°)\right)$$

$$= 15.36 \, ft/sec^2$$

$$s = \frac{v_0^2}{2a} = \frac{\left(30 \, \frac{ft}{sec}\right)^2}{(2)\left(15.36 \, \frac{ft}{sec^2}\right)} = \boxed{29.30 \, ft}$$

The answer is (B).

SI Solution

$$a = g\sin\theta + fg\cos\theta$$

$$= \left(9.81 \, \frac{m}{s^2}\right)\left(\sin 22.62° + (0.1)(\cos 22.62°)\right)$$

$$= 4.679 \, m/s^2$$

$$s = \frac{v_0^2}{2a} = \frac{\left(9 \, \frac{m}{s}\right)^2}{(2)\left(4.679 \, \frac{m}{s^2}\right)} = \boxed{8.656 \, m}$$

The answer is (B).

9. *Customary U.S. Solution*

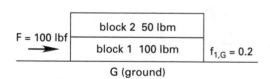

$$F_{f(1,G)} = (W_1 + W_2)f_{1,G} = (100 \, lbf + 50 \, lbf)(0.2)$$

$$= 30 \, lbf \quad \left[\begin{array}{l} < F; \text{ blocks move together at accel-} \\ \text{eration} = a, \text{ assuming no slipping} \end{array}\right]$$

$$a = \frac{(F - F_{f(1,G)})g_c}{m_1 + m_2}$$

$$= \frac{(100 \, lbf - 30 \, lbf)\left(32.2 \, \frac{lbm\text{-}ft}{lbf\text{-}sec^2}\right)}{100 \, lbm + 50 \, lbm} = 15.03 \, ft/sec^2$$

If block 2 is not slipping,

$$F_{f(1,2)} = f_{1,2}W_2 = \frac{m_2 a}{g_c}$$

$$f_{1,2} = \frac{a}{g} = \frac{15.03 \, \frac{ft}{sec^2}}{32.2 \, \frac{ft}{sec^2}} = \boxed{0.467}$$

SI Solution

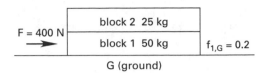

$$F_{f(1,G)} = (W_1 + W_2)f_{1,G}$$

$$= (50 \text{ kg} + 25 \text{ kg})\left(9.81 \frac{\text{m}}{\text{s}^2}\right)(0.2)$$

$$= 147.2 \text{ N} \begin{bmatrix} < F; \text{ blocks move together at accel-} \\ \text{eration } a, \text{ assuming no slipping} \end{bmatrix}$$

$$a = \frac{F - F_{f(1,G)}}{m_1 + m_2}$$

$$= \frac{400 \text{ N} - 147.2 \text{ N}}{50 \text{ kg} + 25 \text{ kg}} = 3.371 \text{ m/s}^2$$

If block 2 is not slipping,

$$F_{f(1,2)} = f_{1,2}W_2 = m_2 a$$

$$f_{1,2} = \frac{a}{g} = \frac{3.371 \frac{\text{m}}{\text{s}^2}}{9.81 \frac{\text{m}}{\text{s}^2}} = \boxed{0.344}$$

10. *Customary U.S. Solution*

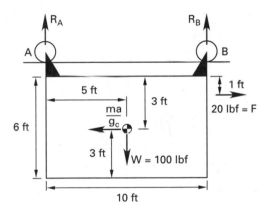

(a) $$a_x = \frac{F_x g_c}{m} = \frac{(20 \text{ lbf})\left(32.2 \frac{\text{lbm-ft}}{\text{lbf-sec}^2}\right)}{100 \text{ lbm}}$$

$$= \boxed{6.44 \text{ ft/sec}^2}$$

(b) The inertial force (ma/g_c) is also 20 lbf.

$$\sum M_A = -(5 \text{ ft})(100 \text{ lbf}) + (10 \text{ ft})R_B$$

$$+ (1 \text{ ft})(20 \text{ lbf}) - (3 \text{ ft})(20 \text{ lbf}) = 0$$

$$R_B = \boxed{54 \text{ lbf (upward)}}$$

$$\sum F_y = R_A + 54 \text{ lbf} - 100 \text{ lbf} = 0$$

$$R_A = \boxed{46 \text{ lbf (upward)}}$$

(c) Let y be the distance (positive upwards) from the center of gravity to F's line of action for $R_A = R_B = R$.

$$\sum M_{CG} = W(0) - (5 \text{ ft})R + (5 \text{ ft})R + y(20 \text{ lbf}) = 0$$

$$y = 0$$

Apply the force in line with the center of gravity.

SI Solution

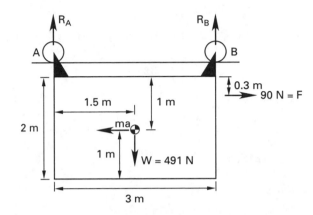

(a) $$a_x = \frac{F_x}{m} = \frac{90 \text{ N}}{50 \text{ kg}} = \boxed{1.8 \text{ m/s}^2}$$

(b) The inertial force is also 90 N.

$$\sum M_A = -(1.5 \text{ m})(491 \text{ N}) + (3 \text{ m})R_B$$

$$+ (0.3 \text{ m})(90 \text{ N}) - (1 \text{ m})(90 \text{ N}) = 0$$

$$R_B = \boxed{266.5 \text{ N (upward)}}$$

$$\sum F_y = R_A + 266.5 \text{ N} - 491 \text{ N} = 0$$

$$R_A = \boxed{224.5 \text{ N (upward)}}$$

(c) Let y be the distance (positive upward) from the center of gravity to F's line of action for $R_A = R_B = R$.

$$\sum M_{CG} = W(0) - (1.5 \text{ m})R + (1.5 \text{ m})R + y(90 \text{ N}) = 0$$

$$y = 0$$

Apply the force in line with the center of gravity.

11. *Customary U.S. Solution*

$$F_f = m\frac{g}{g_c}\sin\phi - \frac{ma_o}{g_c} \qquad \text{[Eq. 1]}$$

a_o is the acceleration at the centroid.

Dynamics and Vibrations

For constrained motion, with a_t being the tangential acceleration,

$$F_f r = \frac{I_o}{g_c}\alpha = \left(\frac{2}{5}mr^2\right)\left(\frac{a_t}{r}\right)\left(\frac{1}{g_c}\right)$$

$$= \frac{2}{5}\frac{ma_t r}{g_c}$$

$$F_f = \frac{2}{5}\frac{ma_t}{g_c} \qquad \text{[Eq. 2]}$$

From Eq. 55.53,

$$a_t = 2a_o \qquad \text{[Eq. 3]}$$

From Eq. 1 and Eq. 2, $mg\sin\phi - ma_o = \frac{2}{5}ma_t = \frac{4}{5}ma_o$.

$$a_o = \left(\frac{5}{9}\right)g\sin\phi = \left(\frac{5}{9}\right)\left(32.2\,\frac{\text{ft}}{\text{sec}^2}\right)(\sin 30°)$$

$$= 8.94\text{ ft/sec}^2$$

$$v_o = a_o t = \left(8.94\,\frac{\text{ft}}{\text{sec}^2}\right)(2\text{ sec}) = \boxed{17.9\text{ ft/sec (fps)}}$$

The answer is (A).

SI Solution

$$F_f = mg\sin\phi - ma_o \qquad \text{[Eq. 1]}$$

a_o is the acceleration at the centroid. For constrained motion, with a_t being the tangential acceleration,

$$F_f r = I_o\alpha = \left(\frac{2}{5}mr^2\right)\left(\frac{a_t}{r}\right) = \frac{2}{5}mar$$

$$F_f = \frac{2}{5}ma_t \qquad \text{[Eq. 2]}$$

From Eq. 55.53,

$$a_t = 2a_o \qquad \text{[Eq. 3]}$$

From Eq. 1 and Eq. 2, $mg\sin\phi - ma_o = \frac{2}{5}ma_t = \frac{4}{5}ma_o$.

$$a_o = \left(\frac{5}{9}\right)g\sin\phi = \left(\frac{5}{9}\right)\left(9.81\,\frac{\text{m}}{\text{s}^2}\right)(\sin 30°)$$

$$= 2.73\text{ m/s}^2$$

$$v_o = a_o t = \left(2.73\,\frac{\text{m}}{\text{s}^2}\right)(2\text{ s}) = \boxed{5.46\text{ m/s}}$$

The answer is (A).

12. *Customary U.S. Solution*

Let B's acceleration and velocity be a and v, positive upwards. Then A's acceleration and velocity are $-a$ and $-v$, respectively.

$$\text{For B:}\quad T - \frac{m_B g}{g_c} = \frac{m_B a}{g_c}$$

$$T = \frac{m_B a + m_B g}{g_c}$$

$$\text{For A:}\quad T - \frac{m_A g\sin\phi}{g_c} = \frac{m_A(-a)}{g_c}$$

$$T = \frac{m_A g\sin\phi - m_A a}{g_c}$$

$$a = \frac{m_A g\sin\phi - m_B g}{m_B + m_A}$$

$$= \frac{\begin{array}{c}(10\text{ lbm})\left(32.2\,\frac{\text{ft}}{\text{sec}^2}\right)(\sin(36.87°))\\[1mm] - (20\text{ lbm})\left(32.2\,\frac{\text{ft}}{\text{sec}^2}\right)\end{array}}{20\text{ lbm} + 10\text{ lbm}}$$

$$= -15.03\text{ ft/sec}^2$$

$$v = at = \left(-15.03\,\frac{\text{ft}}{\text{sec}^2}\right)(3\text{ sec})$$

$$= -45.1\text{ ft/sec} = \boxed{45.1\text{ ft/sec (downward)}}$$

The answer is (D).

SI Solution

Let B's acceleration and velocity be a and v, positive upwards. Then A's acceleration and velocity are $-a$ and $-v$, respectively.

$$\text{For B:}\quad T - m_B g = m_B a$$

$$T = m_B a + m_B g$$

$$\text{For A:}\quad T - m_A g\sin\phi = m_A(-a)$$

$$T = m_A g\sin\phi - m_A a$$

$$a = \frac{m_A g\sin\phi - m_B g}{m_A + m_B}$$

$$= \frac{(5\text{ kg})\left(9.81\,\frac{\text{m}}{\text{s}^2}\right)(\sin 36.87°) - (10\text{ kg})\left(9.81\,\frac{\text{m}}{\text{s}^2}\right)}{5\text{ kg} + 10\text{ kg}}$$

$$= -4.578\text{ m/s}^2$$

$$v = at = \left(-4.578\,\frac{\text{m}}{\text{s}^2}\right)(3\text{ s})$$

$$= -13.7\text{ m/s} = \boxed{13.7\text{ m/s (downward)}}$$

The answer is (D).

13. *Customary U.S. Solution*

$$F = \frac{\dot{m}\Delta v}{g_c} = \frac{\left(560\,\frac{\text{lbm}}{\text{min}}\right)\left(\frac{\text{min}}{60\text{ sec}}\right)\left(3.2\,\frac{\text{ft}}{\text{sec}}\right)}{32.2\,\frac{\text{lbm-ft}}{\text{lbf-sec}^2}}$$

$$= \boxed{0.9275\text{ lbf}}$$

The answer is (A).

SI Solution

$$F = \dot{m}\Delta v = \left(250\,\frac{\text{kg}}{\text{min}}\right)\left(\frac{\text{min}}{60\text{ s}}\right)\left(0.98\,\frac{\text{m}}{\text{s}}\right)$$

$$= \boxed{4.083\text{ N}}$$

The answer is (A).

14. *Customary U.S. Solution*

$$|\mathbf{Imp}| = \frac{|\Delta \mathbf{p}|}{g_c} = \frac{|\mathbf{p}_2 - \mathbf{p}_1|}{g_c} = \frac{mv_2 - (-mv_1)}{g_c}$$

$$= \frac{(0.4 \text{ lbm}) \left(90 \dfrac{\text{ft}}{\text{sec}} + 130 \dfrac{\text{ft}}{\text{sec}} \right)}{32.2 \dfrac{\text{lbm-ft}}{\text{lbf-sec}^2}}$$

$$= \boxed{2.73 \text{ lbf-sec}}$$

The answer is (B).

SI Solution

$$|\mathbf{Imp}| = |\Delta \mathbf{p}| = |\mathbf{p}_2 - \mathbf{p}_1| = mv_2 - (-mv_1)$$

$$= (0.2 \text{ kg}) \left(40 \dfrac{\text{m}}{\text{s}} + 30 \dfrac{\text{m}}{\text{s}} \right) = \boxed{14 \text{ N·s}}$$

The answer is (B).

15. *Customary U.S. Solution*

$$\Delta p_{\text{gun}} = -\Delta p_{\text{projectile}}$$

$$\frac{m_{\text{gun}} v_{\text{gun}}}{g_c} = \frac{-m_{\text{projectile}} v_{\text{projectile}}}{g_c}$$

$$v_{\text{gun}} = -\frac{m_{\text{projectile}} v_{\text{projectile}}}{m_{\text{gun}}} = \frac{-(2.6 \text{ lbm}) \left(2100 \dfrac{\text{ft}}{\text{sec}} \right)}{1000 \text{ lbm}}$$

$$= \boxed{-5.46 \text{ ft/sec (opposite projectile)}}$$

The answer is (C).

SI Solution

$$\Delta p_{\text{gun}} = -\Delta p_{\text{projectile}}$$

$$m_{\text{gun}} v_{\text{gun}} = -m_{\text{projectile}} v_{\text{projectile}}$$

$$v_{\text{gun}} = -\frac{m_{\text{projectile}} v_{\text{projectile}}}{m_{\text{gun}}} = \frac{-(1.2 \text{ kg}) \left(650 \dfrac{\text{m}}{\text{s}} \right)}{500 \text{ kg}}$$

$$= \boxed{-1.56 \text{ m/s (opposite projectile)}}$$

The answer is (C).

16. *Customary U.S. Solution*

In absence of external forces, $\Delta \mathbf{p} = 0$.

$$\frac{m_{\text{bullet}} v_{\text{bullet}}}{g_c} = \frac{m_{(\text{bullet+block})} v_{(\text{bullet+block})}}{g_c}$$

$$v_{(\text{bullet+block})} = \frac{m_{\text{bullet}} v_{\text{bullet}}}{m_{(\text{bullet+block})}}$$

$$= \frac{(0.15 \text{ lbm}) \left(2300 \dfrac{\text{ft}}{\text{sec}} \right)}{0.15 \text{ lbm} + 9 \text{ lbm}}$$

$$= \boxed{37.7 \text{ ft/sec}}$$

The answer is (B).

SI Solution

In absence of external forces, $\Delta \mathbf{p} = 0$.

$$m_{\text{bullet}} v_{\text{bullet}} = m_{(\text{bullet+block})} v_{(\text{bullet+block})}$$

$$v_{(\text{bullet+block})} = \frac{m_{\text{bullet}} v_{\text{bullet}}}{m_{(\text{bullet+block})}} = \frac{(0.06 \text{ kg}) \left(700 \dfrac{\text{m}}{\text{s}} \right)}{4.5 \text{ kg} + 0.06 \text{ kg}}$$

$$= \boxed{9.21 \text{ m/s}}$$

The answer is (B).

17. *Customary U.S. Solution*

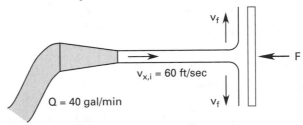

Neglecting gravity, water is turned through an angle of 90° in equal portions in all directions.

$$\dot{m} = \rho Q$$

$$= \left(62.4 \dfrac{\text{lbm}}{\text{ft}^3} \right) \left(40 \dfrac{\text{gal}}{\text{min}} \right) \left(\dfrac{\text{min}}{60 \text{ sec}} \right) \left(\dfrac{0.13368 \text{ ft}^3}{\text{gal}} \right)$$

$$= 5.56 \text{ lbm/sec}$$

For every direction other than along the x-axis, equal amounts of water are directed in opposite senses; no net force is applied in these directions.

$$F_x = \frac{\dot{m} \Delta v_x}{g_c} = \frac{-\dot{m} v_{x,i}}{g_c} = \frac{-\left(5.56 \dfrac{\text{lbm}}{\text{sec}} \right) \left(60 \dfrac{\text{ft}}{\text{sec}} \right)}{32.2 \dfrac{\text{lbm-ft}}{\text{lbf-sec}^2}}$$

$$= \boxed{10.36 \text{ lbf (opposite water direction)}}$$

The answer is (A).

SI Solution

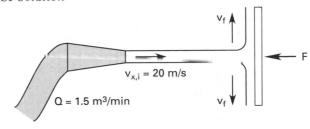

The mass flow rate is

$$\dot{m} = \rho Q$$

$$= \left(1000 \dfrac{\text{kg}}{\text{m}^3} \right) \left(0.15 \dfrac{\text{m}^3}{\text{min}} \right) \left(\dfrac{\text{min}}{60 \text{ s}} \right) = 2.5 \text{ kg/s}$$

Neglecting gravity, water is turned through an angle of $90°$ in equal portions in all directions. For every direction other than along the x-axis, equal amounts of water are directed in opposite senses; no net force is applied in these directions.

$$F_x = \dot{m}\Delta v_x = -\dot{m}v_{x,i} = -\left(2.5\,\frac{\text{kg}}{\text{s}}\right)\left(20\,\frac{\text{m}}{\text{s}}\right)$$

$$= \boxed{50\text{ N (opposite water direction)}}$$

The answer is (A).

18. *Customary U.S. Solution*

(a)
$$\dot{m} = \rho Q = \left(62.4\,\frac{\text{lbm}}{\text{ft}^3}\right)\left(100\,\frac{\text{gal}}{\text{sec}}\right)\left(\frac{0.13368\text{ ft}^3}{\text{gal}}\right)$$

$$= 834.16\text{ lbm/sec}$$

Let inward directions be positive.

$$\dot{p}_{\text{out},x} = \dot{m}v_{\text{out}}\cos(160°)$$

$$= \left(834.16\,\frac{\text{lbm}}{\text{sec}}\right)\left(57\,\frac{\text{ft}}{\text{sec}}\right)(\cos 160°)$$

$$= -44{,}680\,\frac{\text{lbm-ft}}{\text{sec}^2}$$

$$\dot{p}_{\text{out},y} = \dot{m}v_{\text{out}}\sin(160°)$$

$$= \left(834.16\,\frac{\text{lbm}}{\text{sec}}\right)\left(57\,\frac{\text{ft}}{\text{sec}}\right)(\sin 160°)$$

$$= 16{,}262\text{ lbm-ft/sec}^2$$

$$\dot{p}_{\text{in},x} = \dot{m}v_{\text{in}} = \left(834.16\,\frac{\text{lbm}}{\text{sec}}\right)\left(60\,\frac{\text{ft}}{\text{sec}}\right)$$

$$= 50{,}050\text{ lbm-ft/sec}^2$$

$$F_x = \frac{\Delta\dot{p}_x}{g_c} = \frac{\dot{p}_{\text{out},x} - \dot{p}_{\text{in},x}}{g_c}$$

$$= \frac{-44{,}680\,\dfrac{\text{lbm-ft}}{\text{sec}^2} - 50{,}050\,\dfrac{\text{lbm-ft}}{\text{sec}^2}}{32.2\,\dfrac{\text{lbm-ft}}{\text{lbf-sec}^2}}$$

$$= -2942\text{ lbf}$$

$$F_y = \frac{\Delta\dot{p}_y}{g_c} = \frac{16{,}262\,\dfrac{\text{lbm-ft}}{\text{sec}^2} - 0}{32.2\,\dfrac{\text{lbm-ft}}{\text{lbf-sec}^2}} = 505.0\text{ lbf}$$

$$F = \sqrt{F_x{}^2 + F_y{}^2} = \sqrt{(-2942\text{ lbf})^2 + (505\text{ lbf})^2}$$

$$= \boxed{2985\text{ lbf}}$$

(b) $\phi = \arctan\left(\dfrac{F_y}{F_x}\right)$

$$= \arctan\left(\frac{505\text{ lbf}}{-2942\text{ lbf}}\right)$$

$$= \boxed{170.26°\text{ from the horizontal, counterclockwise}}$$

SI Solution

(a)
$$\dot{m} = \rho Q = \left(1000\,\frac{\text{kg}}{\text{m}^3}\right)\left(0.4\,\frac{\text{m}^3}{\text{s}}\right)$$

$$= 400\text{ kg/s}$$

Let inward directions be positive.

$$\dot{p}_{\text{out},x} = \dot{m}v_{\text{out}}\cos 160°$$

$$= \left(400\,\frac{\text{kg}}{\text{s}}\right)\left(17\,\frac{\text{m}}{\text{s}}\right)(\cos 160°) = -6390\text{ N}$$

$$\dot{p}_{\text{out},y} = \dot{m}v_{\text{out}}\sin 160°$$

$$= \left(400\,\frac{\text{kg}}{\text{s}}\right)\left(17\,\frac{\text{m}}{\text{s}}\right)(\sin 160°) = 2326\text{ N}$$

$$\dot{p}_{\text{in},x} = \dot{m}v_{\text{in}} = \left(400\,\frac{\text{kg}}{\text{s}}\right)\left(20\,\frac{\text{m}}{\text{s}}\right) = 8000\text{ N}$$

$$F_x = \Delta\dot{p}_x = \dot{p}_{\text{out},x} - \dot{p}_{\text{in},x}$$

$$= -6390\text{ N} - 8000\text{ N} = -14{,}390\text{ N}$$

$$F_y = \Delta\dot{p}_y = 2326\text{ N} - 0 = 2326\text{ N}$$

$$F = \sqrt{F_x^2 + F_y^2} = \sqrt{(-14{,}390\text{ N})^2 + (2326\text{ N})^2}$$

$$= \boxed{14{,}576\text{ N}}$$

(b) $\phi = \arctan\left(\dfrac{F_y}{F_x}\right)$

$$= \arctan\left(\frac{2326\text{ N}}{-14{,}390\text{ N}}\right)$$

$$= \boxed{170.82°\text{ from the horizontal, counterclockwise}}$$

19. *Customary U.S. Solution*

Since the electron is deflected from its original path, the collision is oblique. The ratio of the masses is

$$\frac{m_H}{m_e} = \frac{1.007277u}{0.0005486u} = 1836$$

Kinetic energy may or may not be conserved in this collision; there is insufficient information to make the determination. Momentum is always conserved, regardless of the axis along which it is evaluated. Consider the original path of the electron to be parallel to the x-axis. Then, by conservation of momentum in the x-direction,

$$m_e v_{e,x} + m_H v_{H,x} = m_e v'_{e,x} + m_H v'_{H,x}$$

Recognizing that $v_{H,x} = 0$ and substituting the ratio of masses,

$$v_{e,x} = v'_e\cos\theta_e + 1836\,v'_{H,x}$$

$$1500\,\frac{\text{ft}}{\text{sec}} = \left(65\,\frac{\text{ft}}{\text{sec}}\right)\cos 30° + (1836)(v'_{H,x})$$

$$v'_{H,x} = \boxed{0.786\text{ ft/sec}}$$

Notice that the x-component of v'_H was requested. Thus, the 1.2° recoil angle was not used. However, since $\cos 1.2° \approx 1.0$, the resultant and x-component velocities are nearly identical.

The answer is (B).

SI Solution

Since the electron is deflected from its original path, the collision is oblique. The ratio of the masses is

$$\frac{m_H}{m_e} = \frac{1.673 \times 10^{-27} \text{ kg}}{9.11 \times 10^{-31} \text{ kg}} = 1836$$

Kinetic energy may or may not be conserved in this collision; there is insufficient information to make the determination. Momentum is always conserved, regardless of the axis along which it is evaluated. Consider the original path of the electron to be parallel to the x-axis. Then, by conservation of momentum in the x-direction,

$$m_e v_{e,x} + m_H v_{H,x} = m_e v'_{e,x} + m_H v'_{H,x}$$

Recognizing that $v_{H,x} = 0$ and substituting the ratio of masses,

$$v_{e,x} = v'_e \cos \theta_e + 1836 \, v'_{H,x}$$

$$500 \, \frac{\text{m}}{\text{s}} = \left(20.2 \, \frac{\text{m}}{\text{s}}\right)(\cos 30°) + (1836)(v'_{H,x})$$

$$v'_{H,x} = \boxed{0.263 \text{ m/s}}$$

Notice that the x-component of v'_H was requested. Thus, the 1.2° recoil angle was not used. However, since $\cos 1.2° \approx 1.0$, the resultant and x-component velocities are nearly identical.

The answer is (B).

20. *Customary U.S. Solution*

$$m_\text{left} v_\text{left} + m_\text{right} v_\text{right} = m_\text{couple} v_\text{couple}$$

$$v_\text{couple} = \frac{(5 \text{ ton})\left(5 \, \frac{\text{ft}}{\text{sec}}\right) + (5 \text{ ton})\left(-4 \, \frac{\text{ft}}{\text{sec}}\right)}{10 \text{ ton}}$$

$$= \boxed{\frac{1}{2} \text{ ft/sec (to the right)}}$$

The answer is (D).

SI Solution

$$m_\text{left} v_\text{left} + m_\text{right} v_\text{right} = m_\text{couple} v_\text{couple}$$

$$v_\text{couple} = \frac{(5000 \text{ kg})\left(1.5 \, \frac{\text{m}}{\text{s}}\right) + (5000 \text{ kg})\left(-1.2 \, \frac{\text{m}}{\text{s}}\right)}{10\,000 \text{ kg}}$$

$$= \boxed{0.15 \text{ m/s (to the right)}}$$

The answer is (B).

21. *Customary U.S. Solution*

$$v_{Ay} = 5 \sin 45° = 3.536 \text{ ft/sec}$$

$$v_{Ax} = 5 \cos 45° = 3.536 \text{ ft/sec}$$

$$v_{By} = \left(10 \, \frac{\text{ft}}{\text{sec}}\right)\left(\frac{3}{5}\right) = 6 \text{ ft/sec}$$

$$v_{Bx} = \left(10 \, \frac{\text{ft}}{\text{sec}}\right)\left(-\frac{4}{5}\right) = -8 \text{ ft/sec}$$

The force of impact is in the x-direction only.

$$v'_{Ay} = v_{Ay}$$

$$v'_{By} = v_{By}$$

In the x-direction,

$$e = \frac{v'_{Ax} - v'_{Bx}}{v_{Bx} - v_{Ax}} = \frac{v'_{Ax} - v'_{Bx}}{-8 \, \frac{\text{ft}}{\text{sec}} - 3.536 \, \frac{\text{ft}}{\text{sec}}}$$

$$v'_{Ax} - v'_{Bx} = (0.8)\left(-8 \, \frac{\text{ft}}{\text{sec}} - 3.536 \, \frac{\text{ft}}{\text{sec}}\right)$$

$$= -9.229 \text{ ft/sec} \qquad \text{[Eq. 1]}$$

$$m_A v_{Ax} + m_B v_{Bx} = m_A v'_{Ax} + m_B v'_{Bx}$$

Since $m_A = m_B$,

$$v'_{Ax} + v'_{Bx} = 3.536 \, \frac{\text{ft}}{\text{sec}} - 8 \, \frac{\text{ft}}{\text{sec}} = -4.464 \text{ ft/sec} \quad \text{[Eq. 2]}$$

Solving Eq. 1 and Eq. 2 simultaneously,

$$v'_{Ax} = -6.846 \text{ ft/sec}$$

$$v'_{Bx} = 2.382 \text{ ft/sec}$$

$$v'_A = \sqrt{(v'_{Ax})^2 + (v'_{Ay})^2}$$

$$= \sqrt{\left(-6.846 \, \frac{\text{ft}}{\text{sec}}\right)^2 + \left(3.536 \, \frac{\text{ft}}{\text{sec}}\right)^2} = \boxed{7.705 \text{ ft/sec}}$$

$$v'_B = \sqrt{\left(6 \, \frac{\text{ft}}{\text{sec}}\right)^2 + \left(2.382 \, \frac{\text{ft}}{\text{sec}}\right)^2} = \boxed{6.456 \text{ ft/sec}}$$

$$\phi_A = \arctan\left(\frac{3.536}{-6.846}\right) = \boxed{152.7°}$$

$$\phi_B = \arctan\left(\frac{6}{2.382}\right) = \boxed{68.3°}$$

SI Solution

$$v_{Ay} = 1.5 \sin 45° = 1.061 \text{ m/s}$$

$$v_{Ax} = 1.5 \cos 45° = 1.061 \text{ m/s}$$

$$v_{By} = \left(3 \, \frac{\text{m}}{\text{s}}\right)\left(\frac{3}{5}\right) = 1.8 \text{ m/s}$$

$$v_{Bx} = \left(3 \, \frac{\text{m}}{\text{s}}\right)\left(-\frac{4}{5}\right) = -2.4 \text{ m/s}$$

The force of impact is in the x-direction only.

$$v'_{Ay} = v_{Ay}$$
$$v'_{By} = v_{By}$$

In the x-direction,

$$e = \frac{v'_{Ax} - v'_{Bx}}{v_{Bx} - v_{Ax}} = \frac{v'_{Ax} - v'_{Bx}}{-2.4\,\frac{m}{s} - 1.061\,\frac{m}{s}}$$

$$v'_{Ax} - v'_{Bx} = (0.8)\left(-2.4\,\frac{m}{s} - 1.061\,\frac{m}{s}\right)$$
$$= -2.769 \text{ m/s} \qquad [\text{Eq. 1}]$$

$$m_A v_{Ax} + m_B v_{Bx} = m_A v'_{Ax} + m_B v'_{Bx}$$

Since $m_A = m_B$,

$$v'_{Ax} + v'_{Bx} = 1.061\,\frac{m}{s} - 2.4\,\frac{m}{s} = -1.339 \text{ m/s} \quad [\text{Eq. 2}]$$

Solving Eq. 1 and Eq. 2 simultaneously,

$$v'_{Ax} = -2.054 \text{ m/s}$$
$$v'_{Bx} = 0.715 \text{ m/s}$$
$$v'_A = \sqrt{(v'_{Ax})^2 + (v'_{Ay})^2}$$
$$= \sqrt{\left(-2.054\,\frac{m}{s}\right)^2 + \left(1.061\,\frac{m}{s}\right)^2} = \boxed{2.312 \text{ m/s}}$$

$$v'_B = \sqrt{\left(0.715\,\frac{m}{s}\right)^2 + \left(1.8\,\frac{m}{s}\right)^2} = \boxed{1.937 \text{ m/s}}$$

$$\phi_A = \arctan\left(\frac{1.061}{-2.054}\right) = \boxed{152.7°}$$

$$\phi_B = \arctan\left(\frac{1.8}{0.715}\right) = \boxed{68.34°}$$

22. *Customary U.S. Solution*

(a)
$$mgh = \frac{1}{2}mv^2$$

The answer is (A).

$$v = \sqrt{2gh} = \sqrt{(2)\left(32.2\,\frac{ft}{sec^2}\right)(3 \text{ ft})} = \boxed{13.9 \text{ ft/sec}}$$

(b)
$$F_c = \frac{mv^2}{g_c r} = \frac{(10 \text{ lbm})\left(13.9\,\frac{ft}{sec}\right)^2}{\left(32.2\,\frac{lbm\text{-}ft}{lbf-sec^2}\right)(3 \text{ ft})} = 20 \text{ lbf}$$

$$T = W + F_c = 10 \text{ lbf} + 20 \text{ lbf} = \boxed{30 \text{ lbf}}$$

The answer is (C).

(c)
$$e = \frac{v_1' - v_2'}{v_2 - v_1} = \frac{v_1' - v_2'}{0 - 13.9\,\frac{ft}{sec}}$$

$$v_1' - v_2' = (0.7)\left(-13.9\,\frac{ft}{sec}\right) = -9.73 \text{ ft/sec} \quad [\text{Eq. 1}]$$

$$m_1 v_1 + m_2 v_2 = m_1 v_1' + m_2 v_2'$$
$$(10 \text{ lbm})v_1' + (50 \text{ lbm})v_2' = 139 \text{ ft/sec} \qquad [\text{Eq. 2}]$$

Solving Eq. 1 and Eq. 2 simultaneously,

$$v_1' = -5.79 \text{ ft/sec}$$
$$v_2' = \boxed{3.94 \text{ ft/sec}}$$

The answer is (B).

(d)
$$\left(\frac{1}{2}\right)\left(\frac{m}{g_c}\right)v^2 = \frac{1}{2}kx^2$$

$$k = \frac{mv^2}{g_c x^2} = \frac{(50 \text{ lbm})\left(3.94\,\frac{ft}{sec}\right)^2}{\left(32.2\,\frac{lbm\text{-}ft}{lbf-sec^2}\right)(0.5 \text{ ft})^2}$$

$$= \boxed{96.4 \text{ lbf/ft}}$$

The answer is (D).

SI Solution

(a) $mgh = \dfrac{1}{2}mv^2$

$$v = \sqrt{2gh} = \sqrt{(2)\left(9.81\,\dfrac{m}{s^2}\right)(1\,m)} = \boxed{4.43\,m/s}$$

The answer is (A).

(b) $\quad F_c = \dfrac{mv^2}{r} = \dfrac{(5\,kg)\left(4.43\,\dfrac{m}{s}\right)^2}{1\,m} = 98.1\,N$

$$T = W + F_c = (5\,kg)\left(9.81\,\dfrac{m}{s^2}\right) + 98.1\,N$$

$$= \boxed{147\,N}$$

The answer is (C).

(c) $\qquad e = \dfrac{v_1{}' - v_2{}'}{v_2 - v_1} = \dfrac{v_1{}' - v_2{}'}{0 - 4.43\,\dfrac{m}{s}}$

$$v_1{}' - v_2{}' = (0.7)\left(-4.43\,\dfrac{m}{s}\right) = -3.1\,m/s \qquad [Eq.\ 1]$$

$$m_1 v_1 + m_2 v_2 = m_1 v_1{}' + m_2 v_2{}'$$

$$(5\,kg)v_1{}' + (25\,kg)v_2{}' = 22.15\,m/s \qquad [Eq.\ 2]$$

Solving Eq. 1 and Eq. 2 simultaneously,

$$v_1{}' = -1.845\,m/s$$

$$v_2{}' = \boxed{1.255\,m/s}$$

The answer is (B).

(d) $\qquad \dfrac{1}{2}mv^2 = \dfrac{1}{2}kx^2$

$$k = \dfrac{mv^2}{x^2} = \dfrac{(25\,kg)\left(1.255\,\dfrac{m}{s}\right)^2}{(0.15\,m)^2}$$

$$= \boxed{1750\,N/m}$$

The answer is (D).

Dynamics and Vibrations

57 Mechanisms and Power Transmission Systems

PRACTICE PROBLEMS

1. A simple epicyclic gearbox with one planet has gears with 24, 40, and 104 teeth on the sun, planet, and internal ring gears, respectively. The sun rotates clockwise at 50 rpm. The ring gear is fixed. What is the rotational velocity of the planet carrier?

(A) 9.4 rpm

(B) 12 rpm

(C) 17 rpm

(D) 23 rpm

2. Refer to the epicyclic gear set illustrated. Gear A rotates counterclockwise on a fixed center at 100 rpm. The ring gear rotates. The planet carrier rotates clockwise at 60 rpm. Each gear has the number of teeth indicated. What is the rotational speed of the sun gear?

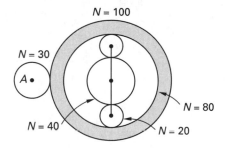

(A) 30 rpm

(B) 60 rpm

(C) 90 rpm

(D) 120 rm

3. An epicyclic gear train consists of a ring gear, three planets, and a fixed sun gear. 15 hp (11 kW) are transmitted through the input ring gear, which turns clockwise at 1500 rpm. The diametral pitch is 10 per inch with a 20° pressure angle. The pitch diameters are at 5 in, $2^1/_2$ in, and 10 in (127 mm, 63.5 mm, 254 mm) for the sun, planets, and ring gear, respectively.

(a) How many teeth are on each gear?

(b) What are the speeds of the sun, ring, and carrier gears?

(c) In what directions do the sun, ring, and carrier gears turn?

(d) What torques are on the input and output shafts?

4. A flywheel is designed as a solid disk, 2 in (51 mm) thick, 20 in (510 mm) in diameter, and with a concentric 4 in (102 mm) diameter mounting hole. It is manufactured from cast iron with an ultimate tensile strength of 30,000 lbf/in² (207 MPa) and a Poisson's ratio of 0.27. The density of the cast iron is 0.26 lbm/in³ (7200 kg/m³). Using a factor of safety of 10, what is the maximum safe speed for the flywheel?

(A) 1700 rpm

(B) 2200 rpm

(C) 3500 rpm

(D) 8700 rpm

5. *(Time limit: one hour)* An automobile differential gear set is arranged as shown. Gear C is turned by the automobile driveshaft. The axle shafts are driven by gears A and B.

(a) What is the speed of rotation for gear B?

(b) What is the direction of rotation for gear B?

gear A: 30 teeth, 50 rpm counterclockwise as viewed from the gear looking up the shaft

gear B: 30 teeth

gear C: 18 teeth, 600 rpm counterclockwise as viewed from the gear looking up the shaft

gear D: 54 teeth

gear E: 15 teeth

gear F: 15 teeth

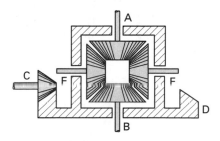

6. *(Time limit: one hour)* The epicyclic gear set shown has an overall speed reduction of 3:1. The planet carrier is the driven element. The driving gear is the sun gear, which turns clockwise at 1000 rpm. The planet has 20 teeth and a diametral pitch of 10.

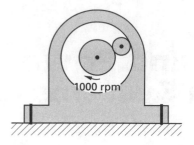

(a) What is the ratio of the numbers of teeth on the ring and sun gears?

(A) 0.5

(B) 2.0

(C) 3.3

(D) 4.0

(b) What is the angular velocity (in rpm) of the planet with respect to the carrier's axis of rotation?

(A) 333 rpm

(B) 667 rpm

(C) 1000 rpm

(D) 1200 rpm

(c) What is the angular velocity (in rpm) of the planet with respect to its own axis of rotation?

(A) 333 rpm

(B) 500 rpm

(C) 1000 rpm

(D) 1200 rpm

7. *(Time limit: one hour)* A gear train is constructed as shown. The output turns at 250 rpm counterclockwise. How many teeth does gear C have?

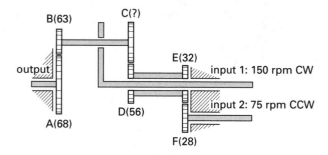

(A) 11

(B) 14

(C) 21

(D) 38

8. *(Time limit: one hour)* A cam is turning at a constant speed of 120 rpm. A radial follower starting from rest rises 0.5 in (12 mm) with constant acceleration as the cam turns 60°. The follower returns to rest with constant acceleration during the next 90° of cam movement.

(a) How far does the follower move during the 150° of rotation?

(A) 0.5 in (12 mm)

(B) 0.75 in (19 mm)

(C) 1.0 in (25 mm)

(D) 1.3 in (30 mm)

(b) What is the magnitude of the acceleration during the first 60°?

(A) 60 in/sec^2 (1.4 m/s^2)

(B) 92 in/sec^2 (2.2 m/s^2)

(C) 120 in/sec^2 (2.9 m/s^2)

(D) 144 in/sec^2 (3.5 m/s^2)

(c) What is the magnitude of the deceleration during the last 90°?

(A) 80 in/sec^2 (1.9 m/s^2)

(B) 100 in/sec^2 (2.4 m/s^2)

(C) 120 in/sec^2 (2.9 m/s^2)

(D) 150 in/sec^2 (3.6 m/s^2)

SOLUTIONS

1. If the arm was locked and the ring was free to rotate, the sun and ring gears would rotate in different directions. Therefore, the train value is negative.

From Eq. 57.29,

$$\text{TV} = -\frac{N_{\text{ring}}}{N_{\text{sun}}} = -\frac{104 \text{ teeth}}{24 \text{ teeth}} = -4.333$$

From Eq. 57.30, the rotational velocity of the sun is

$$\omega_{\text{sun}} = (\text{TV})\omega_{\text{ring}} + (1 - \text{TV})\omega_{\text{carrier}}$$

Since the ring gear is fixed, $\omega_{\text{ring}} = 0$ and the rotational velocity of the carrier are

$$\omega_{\text{carrier}} = \frac{\omega_{\text{sun}}}{1 - \text{TV}} = \frac{50 \dfrac{\text{rev}}{\text{min}}}{1 - (-4.333)}$$

$$= \boxed{9.38 \text{ rev/min}}$$

The answer is (A).

2. Since the ring gear rotates in a different direction from gear A, the rotational velocity of the ring gear is

$$\omega_{\text{ring}} = \omega_A \left(-\frac{N_A}{N_{\text{ring}}}\right)$$

$$= \left(-100 \frac{\text{rev}}{\text{min}}\right)\left(-\frac{30 \text{ teeth}}{100 \text{ teeth}}\right)$$

$$= 30 \text{ rev/min} \quad [\text{clockwise}]$$

Since the ring and sun gears rotate in different directions, the train value is negative. From Eq. 57.29,

$$\text{TV} = -\frac{N_{\text{ring}}}{N_{\text{sun}}} = -\frac{80 \text{ teeth}}{40 \text{ teeth}} = -2$$

From Eq. 57.30, the rotational velocity of the sun gear is

$$\omega_{\text{sun}} = (\text{TV})\omega_{\text{ring}} + (1 - \text{TV})\omega_{\text{carrier}}$$

$$= (-2)\left(30 \frac{\text{rev}}{\text{min}}\right) + (1 - (-2))\left(60 \frac{\text{rev}}{\text{min}}\right)$$

$$= \boxed{120 \text{ rev/min}} \quad [\text{clockwise}]$$

The answer is (D).

3. *Customary U.S. Solution*

(a) From Eq. 52.34, the number of teeth are as follows.

For the sun gear,

$$N_{\text{sun}} = Pd = \left(10 \frac{\text{teeth}}{\text{in}}\right)(5 \text{ in}) = \boxed{50 \text{ teeth}}$$

For the planet gears,

$$N_{\text{planet}} = Pd = \left(10 \frac{\text{teeth}}{\text{in}}\right)(2.5 \text{ in}) = \boxed{25 \text{ teeth}}$$

For the ring gear,

$$N_{\text{ring}} = Pd = \left(10 \frac{\text{teeth}}{\text{in}}\right)(10 \text{ in}) = \boxed{100 \text{ teeth}}$$

(b) Since the sun gear is fixed, its rotational speed is zero ($\omega_{\text{sun}} = 0$).

The rotational speed of the ring gear is $\omega_{\text{ring}} = 1500$ rev/min.

Since the ring and sun gears rotate in different directions, the train value is negative. From Eq. 57.29,

$$\text{TV} = -\frac{N_{\text{ring}}}{N_{\text{sun}}} = -\frac{100 \text{ teeth}}{50 \text{ teeth}} = -2$$

From Eq. 57.30, the rotational speed of the sun gear is

$$\omega_{\text{sun}} = (\text{TV})\omega_{\text{ring}} + (1 - \text{TV})\omega_{\text{carrier}}$$

$$0 \frac{\text{rev}}{\text{min}} = (-2)\left(1500 \frac{\text{rev}}{\text{min}}\right) + (1 - (-2))\omega_{\text{carrier}}$$

Solve for the rotational speed of the carrier gear.

$$\omega_{\text{carrier}} = \frac{(2)\left(1500 \dfrac{\text{rev}}{\text{min}}\right)}{3} = \boxed{1000 \text{ rev/min}}$$

(c) From part (b), the direction of rotation of the ring and carrier gears is clockwise. The sun gear is fixed.

(d) From Eq. 52.32, the torque on the input shaft is

$$T_{\text{in-lbf}} = \frac{P_{\text{hp}}(63,025)}{n_{\text{rpm}}}$$

$$= \frac{P_{\text{hp}}(63,025)}{n_{\text{ring}}} = \left(\frac{(15 \text{ hp})(63,025)}{1500 \dfrac{\text{rev}}{\text{min}}}\right)\left(\frac{1 \text{ ft}}{12 \text{ in}}\right)$$

$$= \boxed{52.52 \text{ ft-lbf}}$$

The torque on the output shaft is

$$T_{\text{in-lbf}} = \frac{P_{\text{hp}}(63,025)}{n_{\text{carrier}}} = \left(\frac{(15 \text{ hp})(63,025)}{1000 \dfrac{\text{rev}}{\text{min}}}\right)\left(\frac{1 \text{ ft}}{12 \text{ in}}\right)$$

$$= \boxed{78.78 \text{ ft-lbf}}$$

Dynamics and Vibrations

SI Solution

(a) From Eq. 52.34, the number of teeth are as follows. For the sun gear,

$$N_{sun} = Pd = \left(10 \; \frac{\text{teeth}}{\text{in}}\right)\left(\frac{1 \text{ in}}{25.4 \text{ mm}}\right)(127 \text{ mm})$$

$$= \boxed{50 \text{ teeth}}$$

For the planet gears,

$$N_{planet} = Pd = \left(10 \; \frac{\text{teeth}}{\text{in}}\right)\left(\frac{1 \text{ in}}{25.4 \text{ mm}}\right)(63.5 \text{ mm})$$

$$= \boxed{25 \text{ teeth}}$$

For the ring gear,

$$N_{ring} = Pd = \left(10 \; \frac{\text{teeth}}{\text{in}}\right)\left(\frac{1 \text{ in}}{25.4 \text{ mm}}\right)(254 \text{ mm})$$

$$= \boxed{100 \text{ teeth}}$$

(b) Since the sun gear is fixed, its rotational speed is zero ($\omega_{sun} = 0$).

The rotational speed of the ring gear is $\omega_{ring} = 1500$ rev/min.

Since the ring and sun gears rotate in different directions, the train value is negative.

From Eq. 57.29,

$$TV = -\frac{N_{ring}}{N_{sun}} = -\frac{100 \text{ teeth}}{50 \text{ teeth}} = -2$$

From Eq. 57.30, the rotational speed of the sun gear is

$$\omega_{sun} = (TV)\omega_{ring} + (1 - TV)\omega_{carrier}$$

$$0 \; \frac{\text{rev}}{\text{min}} = (-2)\left(1500 \; \frac{\text{rev}}{\text{min}}\right) + (1 - (-2))\omega_{carrier}$$

Solve for the rotational speed of the carrier gear.

$$\omega_{carrier} = \frac{(2)\left(1500 \; \frac{\text{rev}}{\text{min}}\right)}{3} = \boxed{1000 \text{ rev/min}}$$

(c) From part (b), the direction of rotation of the ring and carrier gears is clockwise. The sun gear moves in the opposite direction of the ring gear, so the sun gear rotates counterclockwise.

(d) From Eq. 52.32, the torque on the input shaft is

$$T_{N \cdot m} = \frac{P_{kW}(9549)}{n_{ring}}$$

$$= \frac{(11 \text{ kW})(9549)}{1500 \; \frac{\text{rev}}{\text{min}}} = \boxed{70.03 \text{ N·m}}$$

The torque on the output shaft is

$$T_{N \cdot m} = \frac{P_{kW}(9549)}{n_{carrier}} = \frac{(11 \text{ kW})(9549)}{1000 \; \frac{\text{rev}}{\text{min}}} = \boxed{105.0 \text{ N·m}}$$

4. *Customary U.S. Solution*

Treat the flywheel as a rotating hub. First, consider the maximum tangential stress.

For a safety factor of 10,

$$\sigma_{t,max} = \frac{30,000 \; \frac{\text{lbf}}{\text{in}^2}}{10} = 3000 \text{ lbf/in}^2$$

Use Eq. 57.14 to solve for ω.

$$\omega = \sqrt{\frac{4g_c\sigma_{t,max}}{\rho\left((3+\nu)r_o^2 + (1-\nu)r_i^2\right)}}$$

$$= \sqrt{\frac{(4)\left(32.2 \; \frac{\text{ft-lbm}}{\text{lbf-sec}^2}\right)\left(12 \; \frac{\text{in}}{\text{ft}}\right)\left(3000 \; \frac{\text{lbf}}{\text{in}^2}\right)}{\left(0.26 \; \frac{\text{lbm}}{\text{in}^3}\right)\left(3+0.27\left(\frac{20 \text{ in}}{2}\right)^2\right.}}$$
$$\left. \overline{+ (1-0.27)\left(\frac{4 \text{ in}}{2}\right)^2\right)}$$

$$= 232.5 \text{ rad/sec}$$

$$n = \frac{\omega}{2\pi} = \left(232.5 \; \frac{\text{rad}}{\text{sec}}\right)\left(\frac{1 \text{ rev}}{2\pi \text{ rad}}\right)\left(60 \; \frac{\text{sec}}{\text{min}}\right)$$

$$= 2220 \text{ rev/min}$$

Next, consider the maximum radial stress. Use Eq. 57.15 to solve for ω.

$$\omega = \sqrt{\frac{8g_c\sigma_{r,max}}{\rho(3+\nu)(r_o - r_i)^2}}$$

$$= \sqrt{\frac{(8)\left(32.2 \; \frac{\text{ft-lbm}}{\text{lbf-sec}^2}\right)\left(12 \; \frac{\text{in}}{\text{ft}}\right)\left(3000 \; \frac{\text{lbf}}{\text{in}^2}\right)}{\left(0.26 \; \frac{\text{lbm}}{\text{in}^3}\right)(3+0.27)\left(\frac{20 \text{ in}}{2} - \frac{4 \text{ in}}{2}\right)^2}}$$

$$= 412.8 \text{ rad/sec}$$

$$n = \frac{\omega}{2\pi} = \left(412.8 \; \frac{\text{rad}}{\text{sec}}\right)\left(\frac{1 \text{ rev}}{2\pi \text{ rad}}\right)\left(60 \; \frac{\text{sec}}{\text{min}}\right)$$

$$= 3942 \text{ rev/min}$$

The maximum safe speed for the flywheel is

$$\boxed{2220 \text{ rev/min.}}$$

The answer is (B).

SI Solution

Treat the flywheel as a rotating hub. First, consider the maximum tangential stress.

For a safety factor of 10,

$$\sigma_{t,\text{max}} = \frac{207 \times 10^6 \text{ Pa}}{10} = 20.7 \times 10^6 \text{ Pa}$$

Use Eq. 57.14 to solve for ω.

$$\omega = \sqrt{\frac{4\sigma_{t,\text{max}}}{\rho((3+\nu)r_o^2 + (1-\nu)r_i^2)}}$$

$$= \sqrt{\frac{(4)\left(20.7 \times 10^6 \text{ Pa}\right)}{\left(7200 \dfrac{\text{kg}}{\text{m}^3}\right)\left((3+0.27)\left(\dfrac{0.51 \text{ m}}{2}\right)^2 + (1-0.27)\left(\dfrac{0.102 \text{ m}}{2}\right)^2\right)}}$$

$$= 231.5 \text{ rad/s}$$

$$n = \frac{\omega}{2\pi} = \left(231.5 \frac{\text{rad}}{\text{s}}\right)\left(\frac{1 \text{ rev}}{2\pi \text{ rad}}\right)\left(60 \frac{\text{s}}{\text{min}}\right)$$

$$= 2211 \text{ rev/min}$$

Next, consider the maximum radial stress as the worst case. Use Eq. 57.15 to solve for ω.

For a safety factor of 10,

$$\sigma_{r,\text{max}} = \frac{207 \times 10^6 \text{ Pa}}{10} = 20.7 \times 10^6 \text{ Pa}$$

$$\omega = \sqrt{\frac{8\sigma_{r,\text{max}}}{\rho(3+\nu)(r_o - r_i)^2}}$$

$$= \sqrt{\frac{(8)\left(20.7 \times 10^6 \text{ Pa}\right)}{\left(7200 \dfrac{\text{kg}}{\text{m}^3}\right)(3+0.27)\left(\dfrac{0.51 \text{ m}}{2} - \dfrac{0.102 \text{ m}}{2}\right)^2}}$$

$$= 411.1 \text{ rad/s}$$

$$n = \frac{\omega}{2\pi} = \left(411.1 \frac{\text{rad}}{\text{s}}\right)\left(\frac{1 \text{ rev}}{2\pi \text{ rad}}\right)\left(60 \frac{\text{s}}{\text{min}}\right)$$

$$= 3926 \text{ rev/min}$$

The maximum safe speed for the flywheel is

$$\boxed{2211 \text{ rev/min.}}$$

The answer is (B).

5. (This problem can be solved quite quickly with Eq. 57.33.) The automobile differential is equivalent to the following imaginary gear set.

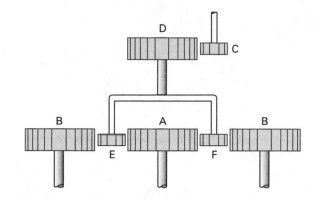

This is the same as a conventional epicyclic gear train except that the ring gear is replaced with two external gears B.

(a) Start with $\omega_C = +600$ rev/min.

Gear D rotates in the opposite direction of gear C.

$$\omega_D = -\left(\frac{N_C}{N_D}\right)\omega_C = -\left(\frac{18 \text{ teeth}}{54 \text{ teeth}}\right)\left(600 \frac{\text{rev}}{\text{min}}\right)$$

$$= -200 \text{ rev/min}$$

Gear A rotates in the same direction as gear D.

$$\omega_A = -50 \text{ rev/min}$$

From the actual gear set, turn gear A. Then gear B moves in the opposite direction of gear A. Thus, the train value is negative. From Eq. 57.29,

$$\text{TV} = -\frac{N_B}{N_A} = -\frac{30 \text{ teeth}}{30 \text{ teeth}} = -1.0$$

From Eq. 57.30,

$$\omega_A = (\text{TV})\omega_B + (1 - \text{TV})\omega_D$$

$$\omega_B = \frac{\omega_A - (1 - \text{TV})\omega_D}{\text{TV}}$$

$$= \frac{-50 \dfrac{\text{rev}}{\text{min}} - (1 - (-1.0))\left(-200 \dfrac{\text{rev}}{\text{min}}\right)}{-1.0}$$

$$= \boxed{-350 \text{ rev/min}}$$

(b) Gear B rotates in a direction opposite to gear C; that is, it rotates clockwise.

6. From the overall speed reduction of 3:1, the speed of the carrier is

$$\omega_{\text{carrier}} = \frac{\omega_{\text{sun}}}{3} = \frac{1000 \dfrac{\text{rev}}{\text{min}}}{3}$$

$$= 333.3 \text{ rev/min}$$

The speed of the ring gear is zero. From Eq. 57.29,

$$TV = \frac{\omega_{\text{sun}} - \omega_{\text{carrier}}}{\omega_{\text{ring}} - \omega_{\text{carrier}}} = \frac{1000 \dfrac{\text{rev}}{\text{min}} - 333.3 \dfrac{\text{rev}}{\text{min}}}{0 - 333.3 \dfrac{\text{rev}}{\text{min}}}$$

$$= -2$$

(a) From Eq. 57.29, the ratio of the number of teeth on the ring and sun gears is

$$\frac{N_{\text{ring}}}{N_{\text{sun}}} = TV = \boxed{2} \quad \text{[negative sign is not required]}$$

The answer is (B).

(b) Assume $N_{\text{sun}} = 40$ teeth and $N_{\text{ring}} = 80$ teeth to satisfy part (a).

Then, from Eq. 52.34,

$$d_{\text{sun}} = \frac{N_{\text{sun}}}{P} = \frac{40 \text{ teeth}}{10 \dfrac{\text{teeth}}{\text{in}}} = 4 \text{ in}$$

$$d_{\text{ring}} = \frac{N_{\text{ring}}}{P} = \frac{80 \text{ teeth}}{10 \dfrac{\text{teeth}}{\text{in}}} = 8 \text{ in}$$

$$d_{\text{planet}} = \frac{N_{\text{planet}}}{P} = \frac{20 \text{ teeth}}{10 \dfrac{\text{teeth}}{\text{in}}} = 2 \text{ in}$$

Check the sum of the gear diameters.

$$d_{\text{ring}} = d_{\text{sun}} + 2d_{\text{planet}}$$
$$8 \text{ in} = 4 \text{ in} + (2)(2 \text{ in}) = 8 \text{ in}$$

Since the gears fit, $N_{\text{sun}} = 40$ teeth and $N_{\text{ring}} = 80$ teeth is a valid solution. Use Eq. 57.31 to solve for ω_{planet}.

$$\omega_{\text{planet}} = \omega_{\text{carrier}} - \left(\frac{N_{\text{sun}}}{N_{\text{planet}}}\right)(\omega_{\text{sun}} - \omega_{\text{carrier}})$$

$$= 333.3 \frac{\text{rev}}{\text{min}}$$

$$- \left(\frac{40 \text{ teeth}}{20 \text{ teeth}}\right)\left(1000 \frac{\text{rev}}{\text{min}} - 333.3 \frac{\text{rev}}{\text{min}}\right)$$

$$= \boxed{-1000 \text{ rev/min}}$$

The answer is (C).

(c) The planet's rotational speed with respect to its own axis can found by holding the arm (carrier) from moving. (The ring must be imagined to rotate.) For this reason, the planet's rotational speed will be with respect to the planet's shaft only. The planet will be driven by the sun and will act as an intermediate (idler)

gear between the sun and ring. Solve Eq. 57.31 with $\omega_{\text{carrier}} = 0$.

$$\omega_{\text{planet}} = \omega_{\text{carrier}} - \left(\frac{N_{\text{sun}}}{N_{\text{planet}}}\right)(\omega_{\text{sun}} - \omega_{\text{carrier}})$$

$$= 0 - \left(\frac{40 \text{ teeth}}{20 \text{ teeth}}\right)\left(1000 \frac{\text{rev}}{\text{min}} - 0\right)$$

$$= \boxed{-2000 \text{ rev/min}}$$

The negative sign indicates that the planet rotates in an opposite direction to the sun.

The answer is (D).

7. First, simplify the problem. Input #2 causes gear D–E to rotate at

$$(-75 \text{ rpm})\left(-\frac{N_F}{N_E}\right) = (-75 \text{ rpm})\left(-\frac{28 \text{ teeth}}{32 \text{ teeth}}\right)$$

$$= 65.625 \text{ rpm}$$

Thus, $\omega_D = 65.625$ rpm.

Use the eight-step procedure from Sec. 57-14.

step 1: Gears A and D have the same center of rotation as the arm.

	A		D	
row 1				
row 2				
row 3				

step 2: Write ω_{carrier} in the first row.

	A		D	
row 1	ω_{carrier}		ω_{carrier}	
row 2				
row 3				

step 3: Arbitrarily select gear D as the gear with unknown speed.

	A		D	
row 1	ω_{carrier}		ω_{carrier}	
row 2				
row 3			ω_D	

step 4:

	A		D	
row 1	ω_{carrier}		ω_{carrier}	
row 2			$\omega_D - \omega_{\text{carrier}}$	
row 3			ω_D	

step 5: The translational path D to A is a compound
mesh.

From Fig. 57.8,

$$\omega_A = \omega_D \left(\frac{N_B N_D}{N_A N_C}\right)$$

$$= \omega_D \left(\frac{(63 \text{ teeth})(56 \text{ teeth})}{(68 \text{ teeth})N_C}\right)$$

$$= 51.8823\omega_D/N_C$$

step 6:

	A		D
row 1	ω_{carrier}		ω_{carrier}
row 2	$\left(\frac{51.8823}{N_C}\right)(\omega_D - \omega_{\text{carrier}})$		$\omega_D - \omega_{\text{carrier}}$
row 3			ω_D

step 7: Insert the known values for ω_{carrier} and ω_D.

	A		D
row 1	150		150
row 2	$\left(\frac{51.8823}{N_C}\right)(65.625-150)$		$65.625-150$
row 3	-250		65.625

step 8: From Eq. 57.32, the characteristic equation
for column 1 is

row 1 + row 2 = row 3

$$150 \text{ rpm} + \left(\frac{51.8823}{N_C}\right)(65.625 \text{ rpm} - 150 \text{ rpm})$$

$$= -250 \text{ rpm}$$

$$N_C = \boxed{10.94 \quad (11 \text{ teeth})}$$

The answer is (A).

8. *Customary U.S. Solution*

The angular velocity is

$$\omega = 2\pi n_{\text{rps}} = \frac{(2\pi)\left(120 \ \dfrac{\text{rev}}{\text{min}}\right)}{60 \ \dfrac{\text{sec}}{\text{min}}} = 12.57 \text{ rad/sec}$$

(b) The time to turn 60° is

$$t = \frac{\theta}{\omega} = \frac{(60°)\left(\dfrac{\pi \text{ rad}}{180°}\right)}{12.57 \ \dfrac{\text{rad}}{\text{sec}}} = 0.08331 \text{ sec}$$

The constant acceleration during the first 60° is

$$a = \frac{2x}{t^2} = \frac{(2)(0.5 \text{ in})}{(0.08331 \text{ sec})^2} = \boxed{144.08 \text{ in/sec}^2}$$

The answer is (D).

(c) The velocity at the end of the first 60° of rotation is

$$v = at = \left(144.08 \ \frac{\text{in}}{\text{sec}^2}\right)(0.08331 \text{ sec})$$

$$= 12.0 \text{ in/sec}$$

The time between $\theta = 60°$ and $\theta = 150°$ is

$$\Delta t = \frac{\Delta\theta}{\omega} = \frac{(150° - 60°)\left(\dfrac{\pi \text{ rad}}{180°}\right)}{12.57 \ \dfrac{\text{rad}}{\text{sec}}} = 0.1250 \text{ sec}$$

Since the cam follower returns to rest at $\theta = 150°$, its
velocity is zero there. The constant deceleration during
the last 90° is

$$a = \frac{\Delta v}{\Delta t} = \frac{0 \ \dfrac{\text{in}}{\text{sec}} - 12.0 \ \dfrac{\text{in}}{\text{sec}}}{0.1250 \text{ sec}} = \boxed{-96 \text{ in/sec}^2}$$

The answer is (B).

(a) The distance moved by the cam follower from $\theta = 60°$ to $\theta = 150°$ is

$$x = \tfrac{1}{2}at^2 = \left(\tfrac{1}{2}\right)\left(95.92 \ \frac{\text{in}}{\text{sec}^2}\right)(0.1250 \text{ sec})^2$$

$$= 0.75 \text{ in}$$

The total follower movement for 150° of rotation is

$$x_{\text{total}} = 0.50 \text{ in} + 0.75 \text{ in} = \boxed{1.25 \text{ in}}$$

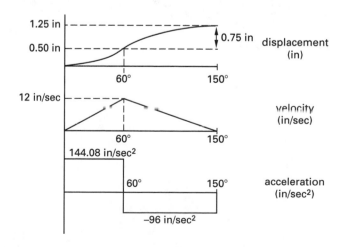

The answer is (D).

SI Solution

The angular velocity is

$$\omega = 2\pi n_{\text{rps}} = \frac{(2\pi)\left(120\,\dfrac{\text{rev}}{\text{min}}\right)}{60\,\dfrac{\text{s}}{\text{min}}} = 12.57\text{ rad/s}$$

(b) The time it takes to turn 60° is

$$t = \frac{\theta}{\omega} = \frac{(60°)\left(\dfrac{\pi\text{ rad}}{180°}\right)}{12.57\,\dfrac{\text{rad}}{\text{s}}} = 0.08331\text{ s}$$

The constant acceleration during the first 60° is

$$a = \frac{2x}{t^2} = \frac{(2)(0.012\text{ m})}{(0.08331\text{ s})^2} = \boxed{3.458\text{ m/s}^2}$$

The answer is (D).

(c) The velocity during the first 60° is

$$\text{v} = at = \left(3.458\,\frac{\text{m}}{\text{s}^2}\right)(0.08331\text{ s})$$
$$= 0.288\text{ m/s}$$

The time between $\theta = 60°$ and $\theta = 150°$ is

$$\Delta t = \frac{\Delta\theta}{\omega} = \frac{(150° - 60°)\left(\dfrac{\pi\text{ rad}}{180°}\right)}{12.57\,\dfrac{\text{rad}}{\text{s}}} = 0.1250\text{ s}$$

Since the cam follower returns to rest at $\theta = 150°$, its velocity is zero there. The constant deceleration during the last 90° is

$$a = \frac{\Delta\text{v}}{\Delta t} = \frac{0\,\dfrac{\text{m}}{\text{s}} - 0.288\,\dfrac{\text{m}}{\text{s}}}{0.1250\text{ s}} = \boxed{-2.30\text{ m/s}^2}$$

The answer is (B).

(a) The distance moved by the cam follower from $\theta = 60°$ to $\theta = 150°$ is

$$x = \tfrac{1}{2}at^2 = \left(\tfrac{1}{2}\right)\left(2.30\,\frac{\text{m}}{\text{s}^2}\right)(0.1250\text{ s})^2\left(1000\,\frac{\text{mm}}{\text{m}}\right)$$
$$= 18.0\text{ mm}$$

The total follower movement for 150° of rotation is

$$x_{\text{total}} = 12.0\text{ mm} + 18.0\text{ mm} = \boxed{30.0\text{ mm}}$$

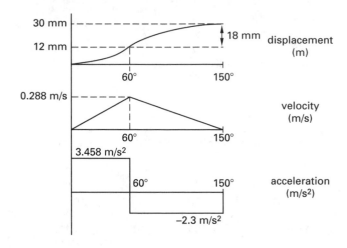

The answer is (D).

58 Vibrating Systems

PRACTICE PROBLEMS

1. A 2 in (50 mm) steel shaft 40 in (1020 mm) long is supported on frictionless bearings at its two ends. The shaft carries a 100 lbm disk (45 kg) 15 in (380 mm) from the left bearing, and a 75 lbm (34 kg) disk 25 in (640 mm) from the left bearing. The shaft weight is negligible. There is no damping. What is the critical speed of the shaft?

(A) 1500 rpm

(B) 1700 rpm

(C) 2000 rpm

(D) 2400 rpm

2. A 300 lbm (140 kg) electromagnet at the end of a cable holds 200 lbm (90 kg) of scrap metal. The total equivalent stiffness of the cable and crane boom is 1000 lbf/in (175 kN/m). The current to the electromagnet is cut off suddenly, and the scrap falls away. Neglect damping.

(a) What is the frequency of oscillation of the electromagnet?

(A) 5.7 Hz

(B) 10 Hz

(C) 23 Hz

(D) 84 Hz

(b) What will be the minimum cable tension?

(A) 100 lbf (490 N)

(B) 200 lbf (980 N)

(C) 300 lbf (1500 N)

(D) 400 lbf (2000 N)

3. An 800 lbm (360 kg), single-cylinder vertical compressor operates at 1200 rpm. The compressor is mounted to the floor on four identical, equally loaded springs at its corners. It is desired to reduce the maximum transmitted force from 25 lbf to 3 lbf (110 N to 13 N). Damping is negligible. What will be the new maximum oscillation?

(A) 0.49×10^{-4} in (1.2×10^{-5} m)

(B) 0.63×10^{-4} in (1.6×10^{-5} m)

(C) 0.72×10^{-4} in (1.8×10^{-5} m)

(D) 8.6×10^{-4} in (2.2×10^{-5} m)

4. A uniform bar with a mass of 5 lbm (2.3 kg) carries a concentrated mass of 3 lbm (1.4 kg) at its free end. The bar is hinged at one end and supported by an outboard spring as shown. The deflection of the spring from its unstressed position is 0.55 in (1.4 cm). There is no damping. What is the natural frequency of vibration?

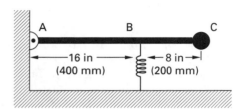

(A) 4 Hz

(B) 7 Hz

(C) 16 Hz

(D) 50 Hz

5. When an 8 lbm (3.6 kg) mass is attached on the end of a spring, the spring stretches 5.9 in (150 mm). A dashpot with a damping coefficient of 0.50 lbf-sec/ft (7.3 N·s/m) opposes movement of the mass. A forcing function of $4 \cos 2t$ lbf ($18 \cos 2t$ N) is applied to the mass. The mass is initially at rest.

(a) What is the natural frequency?

(b) What is the damping ratio?

(c) What is the maximum excursion?

(d) What is the response of the system?

6. A 50 lbm (23 kg) motor is supported by four identical, equally loaded springs, each with a spring constant of 1000 lbf/in (175 kN/m). When the motor is turning at 800 rpm, the rotor imbalance is equivalent to a 1 oz (30 g) mass located 5 in (130 mm) from the shaft's longitudinal axis. The damping factor is $^1/_8$ of critical damping. What is the maximum vertical vibration amplitude?

(A) 0.0012 in (3.0×10^{-5} m)

(B) 0.0018 in (5.0×10^{-5} m)

(C) 0.0046 in (1.2×10^{-4} m)

(D) 0.0099 in (2.5×10^{-4} m)

7. *(Time limit: one hour)* A 175 lbm (80 kg), single-cylinder air compressor is mounted on four identical, equally loaded corner springs. The motor turns at 1200 rpm. During each cycle, a disturbing force is generated by a 3.6 lbm (1.6 kg) imbalance acting at a radius of 3 in (75 mm). Damping is insignificant.

(a) What individual spring stiffness is required so that only 5% of the dynamic force is transmitted to the base?

   (A) 44 lbf/in (7.5 kN/m)

   (B) 68 lbf/in (12 kN/m)

   (C) 85 lbf/in (15 kN/m)

   (D) 170 lbf/in (29 kN/m)

(b) What is the amplitude of vibration?

   (A) 0.018 in (0.00045 m)

   (B) 0.033 in (0.00083 m)

   (C) 0.065 in (0.0016 m)

   (D) 0.097 in (0.0024 m)

8. *(Time limit: one hour)* Eight horizontal high-strength steel plates are supported on rigid rollers and support a 20,000 lbm (9100 kg) load, as shown. Each of the plates is 40 in × 30 in × $\frac{1}{2}$ in (1020 mm × 760 mm × 12 mm). The modulus of elasticity of the steel is 2.9×10^7 lbf/in^2 (200 GPa) and Poisson's ratio of 0.3. The masses of the plates and rollers are insignificant compared to the supported load. The yield point of the steel is not exceeded.

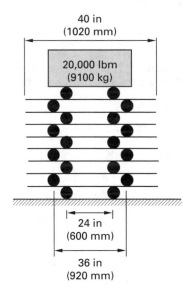

40 in
(1020 mm)

20,000 lbm
(9100 kg)

24 in
(600 mm)

36 in
(920 mm)

(a) What is the total vertical static deflection of the load?

   (A) 4.0 in (0.13 m)

   (B) 4.4 in (0.15 m)

   (C) 6.2 in (0.16 m)

   (D) 6.9 in (0.17 m)

(b) What is the maximum stress in the plates?

   (A) 48 ksi (390 MPa)

   (B) 60 ksi (480 MPa)

   (C) 96 ksi (780 MPa)

   (D) 120 ksi (970 MPa)

(c) What is the natural frequency of oscillation in the vertical direction?

   (A) 1.9 Hz

   (B) 2.7 Hz

   (C) 4.2 Hz

   (D) 7.6 Hz

9. A centrifugal fan has 8 driving blades and 64 fan blades. The fan turns at 600 rpm and is driven by a 1725 rpm, 60 Hz, 4 pole motor. The fan and motor pulley are 11$\frac{1}{2}$ in (290 mm) and 4 in (100 mm) in diameter, respectively. The drive belt has a total length of 72 in (1.83 m). What frequencies of sound and vibration are produced?

10. When not running, a hydraulic oil pump compresses a cork mounting pad 0.02 in (0.5 mm). The pump is turned at 1725 rpm. What is the transmissibility of the pad?

   (A) 25% increase

   (B) 50% increase

   (C) 75% increase

   (D) 100% increase

SOLUTIONS

1. *Customary U.S. Solution*

The moment of inertia of the circular cross section is

$$I = \frac{\pi r^4}{4} = \frac{\pi \left(\frac{2.0 \text{ in}}{2}\right)^4}{4} = 0.7854 \text{ in}^4$$

Use App. 49.A. The deflection due to the 100 lbm disk at the 100 lbm disk is

$$\delta = \frac{Fa^2b^2}{3EIL}$$

$$= \frac{(100 \text{ lbm}) \left(\dfrac{32.2 \, \dfrac{\text{ft}}{\text{sec}^2}}{32.2 \, \dfrac{\text{ft-lbm}}{\text{lbf-sec}^2}}\right) (15 \text{ in})^2 (25 \text{ in})^2}{(3) \left(30 \times 10^6 \, \dfrac{\text{lbf}}{\text{in}^2}\right) (0.7854 \text{ in}^4)(40 \text{ in})}$$

$$= 0.00497 \text{ in}$$

This installation is symmetrical with respect to load positioning.

The deflection due to the 100 lbm disk at the 75 lbm disk is

$$\delta = \left(\frac{Fbx}{6EIL}\right)(L^2 - b^2 - x^2)$$

$$\begin{bmatrix} \text{look at the shaft from the} \\ \text{back to justify this formula} \end{bmatrix}$$

$$= \frac{(100 \text{ lbm}) \left(\dfrac{32.2 \, \dfrac{\text{ft}}{\text{sec}^2}}{32.2 \, \dfrac{\text{ft-lbm}}{\text{lbf-sec}^2}}\right) (15 \text{ in})(15 \text{ in})}{}$$

$$= \frac{\times \left((40 \text{ in})^2 - (15 \text{ in})^2 - (15 \text{ in})^2\right)}{(6) \left(30 \times 10^6 \, \dfrac{\text{lbf}}{\text{in}^2}\right) (0.7854 \text{ in}^4)(40 \text{ in})}$$

$$= 0.00458 \text{ in}$$

The deflection due to the 75 lbm disk at the 75 lbm disk is

$$\delta = \frac{Fa^2b^2}{3EIL}$$

$$= \frac{(75 \text{ lbm}) \left(\dfrac{32.2 \, \dfrac{\text{ft}}{\text{sec}^2}}{32.2 \, \dfrac{\text{ft-lbm}}{\text{lbf-sec}^2}}\right) (25 \text{ in})^2 (15 \text{ in})^2}{(3) \left(30 \times 10^6 \, \dfrac{\text{lbf}}{\text{in}^2}\right) (0.7854 \text{ in}^4)(40 \text{ in})}$$

$$= 0.00373 \text{ in}$$

The deflection due to the 75 lbm disk at the 100 lbm disk is

$$\delta = \left(\frac{Fbx}{6EIL}\right)(L^2 - b^2 - x^2)$$

$$= \frac{(75 \text{ lbm}) \left(\dfrac{32.2 \, \dfrac{\text{ft}}{\text{sec}^2}}{32.2 \, \dfrac{\text{ft-lbm}}{\text{lbf-sec}^2}}\right) (15 \text{ in})(15 \text{ in})}{}$$

$$= \frac{\times \left((40 \text{ in})^2 - (15 \text{ in})^2 - (15 \text{ in})^2\right)}{(6) \left(30 \times 10^6 \, \dfrac{\text{lbf}}{\text{in}^2}\right) (0.7854 \text{ in}^4)(40 \text{ in})}$$

$$= 0.00343 \text{ in}$$

The total deflection at the 100 lbm disk is

$$\delta_{\text{st},1} = 0.00497 \text{ in} + 0.00343 \text{ in} = 0.00840 \text{ in}$$

The total deflection at the 75 lbm disk is

$$\delta_{\text{st},2} = 0.00458 \text{ in} + 0.00373 \text{ in} = 0.00831 \text{ in}$$

Use Eq. 58.60 to find the natural frequency.

$$f = \left(\frac{1}{2\pi}\right) \sqrt{\frac{g \sum m_i \delta_{\text{st},i}}{\sum m_i \delta_{\text{st},i}^2}}$$

$$= \left(\frac{1}{2\pi}\right) \sqrt{\frac{\left(386.4 \, \dfrac{\text{in}}{\text{sec}^2}\right) \begin{pmatrix} (100 \text{ lbm})(0.00840 \text{ in}) \\ + (75 \text{ lbm})(0.00831 \text{ in}) \end{pmatrix}}{\begin{matrix} (100 \text{ lbm})(0.00840 \text{ in})^2 \\ + (75 \text{ lbm})(0.00831 \text{ in})^2 \end{matrix}}}$$

$$= 34.21 \text{ Hz}$$

The critical speed of the shaft is

$$n = (34.21 \text{ Hz}) \left(60 \, \frac{\text{sec}}{\text{min}}\right)$$

$$= \boxed{2053 \text{ rpm}}$$

The answer is (C).

SI Solution

The moment of inertia of the circular cross section is

$$I = \frac{\pi r^4}{4} = \frac{\pi \left(\frac{0.05 \text{ m}}{2}\right)^4}{4} = 3.068 \times 10^{-7} \text{ m}^4$$

The deflection due to the 45 kg disk at the 45 kg disk is

$$\delta = \frac{Fa^2b^2}{3EIL}$$

$$= \frac{(45 \text{ kg}) \left(9.81 \, \dfrac{\text{m}}{\text{s}^2}\right) (0.38 \text{ m})^2 (0.64 \text{ m})^2}{(3) \left(200 \times 10^9 \text{ Pa}\right) (3.068 \times 10^{-7} \text{ m}^4)(1.02 \text{ m})}$$

$$= 0.000139 \text{ m}$$

The deflection due to the 45 kg disk at the 34 kg disk is

$$\delta = \left(\frac{Fbx}{6EIL}\right)(L^2 - b^2 - x^2)$$

$$= \frac{\begin{array}{c}(45 \text{ kg})\left(9.81 \dfrac{\text{m}}{\text{s}^2}\right)(0.38 \text{ m})(0.38 \text{ m}) \\ \times ((1.02 \text{ m})^2 - (0.38 \text{ m})^2 - (0.38 \text{ m})^2)\end{array}}{(6)(200 \times 10^9 \text{ Pa})(3.068 \times 10^{-7} \text{ m}^4)(1.02 \text{ m})}$$

$$= 0.000128 \text{ m}$$

The deflection due to the 34 kg disk at the 34 kg disk is

$$\delta = \frac{Fa^2b^2}{3EIL}$$

$$= \frac{(34 \text{ kg})\left(9.81 \dfrac{\text{m}}{\text{s}^2}\right)(0.64 \text{ m})^2(0.38 \text{ m})^2}{(3)(200 \times 10^9 \text{ Pa})(3.068 \times 10^{-7} \text{ m}^4)(1.02 \text{ m})}$$

$$= 0.000105 \text{ m}$$

The deflection due to the 34 kg disk at the 45 kg disk is

$$\delta = \left(\frac{Fbx}{6EIL}\right)(L^2 - b^2 - x^2)$$

$$= \frac{\begin{array}{c}(34 \text{ kg})\left(9.81 \dfrac{\text{m}}{\text{s}^2}\right)(0.38 \text{ m})(0.38 \text{ m}) \\ \times ((1.02 \text{ m})^2 - (0.38 \text{ m})^2 - (0.38 \text{ m})^2)\end{array}}{(6)(200 \times 10^9 \text{ Pa})(3.068 \times 10^{-7} \text{ m}^4)(1.02 \text{ m})}$$

$$= 0.000096 \text{ m}$$

The total deflection at the 45 kg disk is

$$\delta_{\text{st},1} = 0.000139 \text{ m} + 0.000096 \text{ m} = 0.000235 \text{ m}$$

The total deflection at the 34 kg disk is

$$\delta_{\text{st},2} = 0.000128 \text{ m} + 0.000105 \text{ m} = 0.000233 \text{ m}$$

Use Eq. 58.60 to find the natural frequency.

$$f = \left(\frac{1}{2\pi}\right)\sqrt{\frac{g\sum m_i\delta_{\text{st},i}}{\sum m_i\delta_{\text{st},i}^2}}$$

$$= \left(\frac{1}{2\pi}\right)\sqrt{\frac{\left(9.81 \dfrac{\text{m}}{\text{s}^2}\right)\begin{pmatrix}(45 \text{ kg})(0.000235 \text{ m}) \\ + (34 \text{ kg})(0.000233 \text{ m})\end{pmatrix}}{\begin{array}{c}(45 \text{ kg})(0.000235 \text{ m})^2 \\ + (34 \text{ kg})(0.000233 \text{ m})^2\end{array}}}$$

$$= 32.6 \text{ Hz}$$

The critical speed of the shaft is

$$n = (32.6 \text{ Hz})\left(60 \frac{\text{sec}}{\text{min}}\right)$$

$$= \boxed{1956 \text{ rpm}}$$

The answer is (C).

2. *Customary U.S. Solution*

The static deflection caused by the electromagnet is

$$\delta_{\text{st}} = \frac{\text{weight}}{k} = \frac{m\left(\dfrac{g}{g_c}\right)}{k}$$

$$= \left(\frac{300 \text{ lbm}}{1000 \dfrac{\text{lbf}}{\text{in}}}\right)\left(\frac{32.2 \dfrac{\text{ft}}{\text{sec}^2}}{32.2 \dfrac{\text{ft-lbm}}{\text{lbf-sec}^2}}\right)$$

$$= 0.3 \text{ in}$$

(a) From Eqs. 58.4 and 58.7, the natural frequency is

$$f = \frac{\omega}{2\pi} = \left(\frac{1}{2\pi}\right)\sqrt{\frac{g}{\delta_{\text{st}}}} = \left(\frac{1}{2\pi}\right)\sqrt{\frac{386.4 \dfrac{\text{in}}{\text{sec}^2}}{0.3 \text{ in}}}$$

$$= \boxed{5.71 \text{ Hz}}$$

The answer is (A).

(b) The minimum tension occurs at the upper limit of travel. The decrease in tension at that point is the same as the increase in tension at the lower limit caused by the scrap.

$$F_{\min} = (300 \text{ lbm} - 200 \text{ lbm})\left(\frac{32.2 \dfrac{\text{ft}}{\text{sec}^2}}{32.2 \dfrac{\text{ft-lbm}}{\text{lbf-sec}^2}}\right)$$

$$= \boxed{100 \text{ lbf}}$$

The answer is (A).

SI Solution

The static deflection caused by the electromagnet is

$$\delta_{\text{st}} = \frac{mg}{k} = \frac{(140 \text{ kg})\left(9.81 \dfrac{\text{m}}{\text{s}^2}\right)}{\left(175 \dfrac{\text{kN}}{\text{m}}\right)\left(1000 \dfrac{\text{N}}{\text{kN}}\right)}$$

$$= 0.00785 \text{ m}$$

(a) From Eqs. 58.4 and 58.7, the natural frequency is

$$f = \frac{\omega}{2\pi} = \left(\frac{1}{2\pi}\right)\sqrt{\frac{g}{\delta_{st}}} = \left(\frac{1}{2\pi}\right)\sqrt{\frac{9.81 \frac{m}{s^2}}{0.00785 \text{ m}}}$$

$$= \boxed{5.63 \text{ Hz}}$$

The answer is (A).

(b) The minimum tension occurs at the upper limit of travel. The decrease in tension at that point is the same as the increase in tension at the lower limit caused by the scrap.

$$F_{min} = (140 \text{ kg} - 90 \text{ kg})\left(9.81 \frac{m}{s^2}\right) = \boxed{490.5 \text{ N}}$$

The answer is (A).

3. *Customary U.S. Solution*

The transmissibility is

$$\text{TR} = \frac{|F_{transmitted}|}{F_{applied}} = \frac{3 \text{ lbf}}{25 \text{ lbf}} = 0.12$$

The forcing frequency is

$$\omega_f = \frac{\left(1200 \frac{rev}{min}\right)\left(2\pi \frac{rad}{rev}\right)}{60 \frac{sec}{min}} = 125.7 \text{ rad/sec}$$

For negligible damping, Eqs. 58.55 and 58.56 can be simplified.

$$\text{TR} = \frac{1}{|1-r^2|} = \frac{1}{\left|1-\left(\frac{\omega_f}{\omega}\right)^2\right|}$$

$$= \frac{1}{\left(\frac{\omega_f}{\omega}\right)^2 - 1} \quad [\text{for } \omega_f > \omega]$$

$$\frac{\omega_f}{\omega} = \sqrt{\frac{1}{\text{TR}} + 1} = \sqrt{\frac{1}{0.12} + 1} = 3.055$$

The required natural frequency is

$$\omega = \frac{\omega_f}{3.055} = \frac{125.7 \frac{rad}{sec}}{3.055} = 41.15 \text{ rad/sec}$$

From Eq. 58.3, the equivalent stiffness of the springs is

$$k = \frac{m\omega^2}{g_c} = \frac{(800 \text{ lbm})\left(41.15 \frac{rad}{sec}\right)^2}{32.2 \frac{ft\text{-}lbm}{lbf\text{-}sec^2}}$$

$$= 42{,}070 \text{ lbf/ft}$$

The reduced pseudo-static deflection is

$$\frac{F_0}{k} = \frac{25 \text{ lbf}}{42{,}070 \frac{lbf}{ft}} = 5.94 \times 10^{-4} \text{ ft}$$

From Eq. 58.50, the magnification factor is

$$\beta = \frac{D}{\frac{F_0}{k}} = \left|\frac{1}{1-\left(\frac{\omega_f}{\omega}\right)^2}\right| = \text{TR} = 0.12$$

The new maximum oscillation is

$$D = \left(\frac{F_0}{k}\right)\beta = (5.94 \times 10^{-4} \text{ ft})\left(12 \frac{in}{ft}\right)(0.12)$$

$$= \boxed{8.56 \times 10^{-4} \text{ in}}$$

The answer is (D).

SI Solution

The transmissibility is

$$\text{TR} = \frac{|F_{transmitted}|}{F_{applied}} = \frac{13 \text{ N}}{110 \text{ N}} = 0.118$$

The forcing frequency is

$$\omega_f = \frac{\left(1200 \frac{rev}{min}\right)\left(2\pi \frac{rad}{rev}\right)}{60 \frac{s}{min}} = 125.7 \text{ rad/s}$$

For negligible damping, Eqs. 58.55 and 58.56 can be simplified.

$$\text{TR} = \frac{1}{|1-r^2|} = \frac{1}{\left|1-\left(\frac{\omega_f}{\omega}\right)^2\right|}$$

$$= \frac{1}{\left(\frac{\omega_f}{\omega}\right)^2 - 1} \quad [\text{for } \omega_f > \omega]$$

$$\frac{\omega_f}{\omega} = \sqrt{\frac{1}{\text{TR}} + 1} = \sqrt{\frac{1}{0.118} + 1} = 3.078$$

The required natural frequency is

$$\omega = \frac{\omega_f}{2.73} = \frac{125.7 \frac{rad}{s}}{3.078} = 40.84 \text{ rad/s}$$

From Eq. 58.3, the equivalent stiffness of the springs is

$$k = m\omega^2 = (360 \text{ kg})\left(40.84 \frac{rad}{s}\right)^2$$

$$= 600\,446 \text{ N/m}$$

Dynamics and Vibrations

The pseudo-static deflection is

$$\frac{F_0}{k} = \frac{110 \text{ N}}{600\,446 \, \frac{\text{N}}{\text{m}}} = 1.83 \times 10^{-4} \text{ m}$$

From Eq. 58.50, the magnification factor is

$$\beta = \frac{D}{\frac{F_0}{k}} = \left| \frac{1}{1 - \left(\frac{\omega_f}{\omega}\right)^2} \right| = \text{TR} = 0.118$$

The new maximum oscillation is

$$D = \left(\frac{F_0}{k}\right)\beta = (1.83 \times 10^{-4} \text{ m})(0.118)$$

$$= \boxed{2.16 \times 10^{-5} \text{ m}}$$

The answer is (D).

4. *Customary U.S. Solution*

First, consider static equilibrium. The bar mass is considered concentrated at 12 in from the hinge.

$$\sum M_{\text{A}} = 0$$

$$(3 \text{ lbm}) \left(\frac{32.2 \, \frac{\text{ft}}{\text{sec}^2}}{32.2 \, \frac{\text{ft-lbm}}{\text{lbf-sec}^2}} \right) (24 \text{ in})$$

$$+ (5 \text{ lbm}) \left(\frac{32.2 \, \frac{\text{ft}}{\text{sec}^2}}{32.2 \, \frac{\text{ft-lbm}}{\text{lbf-sec}^2}} \right) (12 \text{ in}) - M_{\text{spring}}$$

$$= 0$$

$$M_{\text{spring}} = 132 \text{ in-lbf}$$

The angle of rotation is

$$\theta = \frac{\delta}{L} = \frac{0.55 \text{ in}}{16 \text{ in}} = 0.0344 \text{ rad}$$

The equivalent torsional spring constant is

$$k_r = \frac{M_{\text{spring}}}{\theta} = \frac{132 \text{ in-lbf}}{0.0344 \text{ rad}} = 3837 \text{ in-lbf}$$

From App. 54.A, the mass moment of inertia of the bar rotating about its end is

$$I_{\text{bar}} = \tfrac{1}{3}mL^2$$
$$= \left(\tfrac{1}{3}\right)(5 \text{ lbm})(24 \text{ in})^2$$
$$= 960 \text{ lbm-in}^2$$

The mass moment of inertia of the concentrated mass is

$$I_{\text{mass}} = mL^2 = (3 \text{ lbm})(24 \text{ in})^2 = 1728 \text{ lbm-in}^2$$

The total mass moment of inertia of the system is

$$I = I_{\text{bar}} + I_{\text{mass}} = 960 \text{ lbm-in}^2 + 1728 \text{ lbm-in}^2$$
$$= 2688 \text{ lbm-in}^2$$

From Eq. 58.22, the natural frequency is

$$\omega = \sqrt{\frac{k_r g_c}{I}} = \sqrt{\frac{(3837 \text{ in-lbf})\left(386.4 \, \frac{\text{lbm-in}}{\text{lbf-sec}^2}\right)}{2688 \text{ lbm-in}^2}}$$
$$= 23.49 \text{ rad/sec}$$

From Eq. 58.4,

$$f = \frac{\omega}{2\pi} = \frac{23.49 \, \frac{\text{rad}}{\text{sec}}}{2\pi \, \frac{\text{rad}}{\text{rev}}} = \boxed{3.74 \text{ Hz}} \begin{bmatrix} \text{Table 58.1 could also} \\ \text{have been used.} \end{bmatrix}$$

The answer is (A).

SI Solution

First, consider static equilibrium. The bar mass is considered concentrated at 0.30 m from the hinge.

$$\sum M_{\text{A}} = 0$$

$$(1.4 \text{ kg}) \left(9.81 \, \frac{\text{m}}{\text{s}^2}\right) (0.6 \text{ m})$$

$$+ (2.3 \text{ kg}) \left(9.81 \, \frac{\text{m}}{\text{s}^2}\right) (0.3 \text{ m}) - M_{\text{spring}} = 0$$

$$M_{\text{spring}} = 15.01 \text{ N·m}$$

The angle of rotation is

$$\theta = \frac{\delta}{L} = \frac{0.014 \text{ m}}{0.40 \text{ m}} = 0.035 \text{ rad}$$

The equivalent torsional spring constant is

$$k_r = \frac{M_{\text{spring}}}{\theta} = \frac{15.01 \text{ N·m}}{0.035 \text{ rad}} = 428.9 \text{ N·m}$$

From App. 54.A, the mass moment of inertia of the bar rotating about its end is

$$I_{\text{bar}} = \tfrac{1}{3}mL^2$$
$$= \left(\tfrac{1}{3}\right)(2.3 \text{ kg})(0.60 \text{ m})^2$$
$$= 0.276 \text{ kg·m}^2$$

The mass moment of inertia of the concentrated mass is

$$I_{\text{mass}} = mL^2 = (1.4 \text{ kg})(0.60 \text{ m})^2 = 0.504 \text{ kg·m}^2$$

The total mass moment of inertia of the system is

$$I = I_{\text{bar}} + I_{\text{mass}} = 0.276 \text{ kg·m}^2 + 0.504 \text{ kg·m}^2$$
$$= 0.78 \text{ kg·m}^2$$

From Eq. 58.22, the natural frequency is

$$\omega = \sqrt{\frac{k_r}{I}} = \sqrt{\frac{428.9 \text{ N·m}}{0.78 \text{ kg·m}^2}} = 23.45 \text{ rad/s}$$

From Eq. 58.4,

$$f = \frac{\omega}{2\pi} = \frac{23.45 \ \dfrac{\text{rad}}{\text{s}}}{2\pi \ \dfrac{\text{rad}}{\text{rev}}} = \boxed{3.73 \text{ Hz}} \quad \begin{bmatrix} \text{Table 58.1 could also} \\ \text{have been used.} \end{bmatrix}$$

The answer is (A).

5. *Customary U.S. Solution*

The static deflection is

$$\delta_{\text{st}} = \frac{\text{weight}}{k} = \frac{m\left(\dfrac{g}{g_c}\right)}{k}$$

$$k = \frac{m\left(\dfrac{g}{g_c}\right)}{\delta_{\text{st}}}$$

$$= \left(\frac{8 \text{ lbm}}{5.9 \text{ in}}\right)\left(\frac{32.2 \ \dfrac{\text{ft}}{\text{sec}^2}}{32.2 \ \dfrac{\text{ft-lbm}}{\text{lbf-sec}^2}}\right)$$

$$= 1.356 \text{ lbf/in}$$

(a) From Eqs. 58.3 and 58.7, the natural frequency is

$$\omega = \sqrt{\frac{kg_c}{m}} = \sqrt{\frac{g}{\delta_{\text{st}}}} = \sqrt{\frac{386.4 \ \dfrac{\text{in}}{\text{sec}^2}}{5.9 \text{ in}}}$$

$$= \boxed{8.09 \text{ rad/sec}}$$

(b) The damping ratio is

$$\zeta = \frac{C}{2\sqrt{\dfrac{km}{g_c}}}$$

$$= \frac{0.50 \ \dfrac{\text{lbf-sec}}{\text{ft}}}{2\sqrt{\left(1.356 \ \dfrac{\text{lbf}}{\text{in}}\right)\left(12 \ \dfrac{\text{in}}{\text{ft}}\right) \times (8 \text{ lbm})\left(\dfrac{1}{32.2 \ \dfrac{\text{ft-lbm}}{\text{lbf-sec}^2}}\right)}}$$

$$= \boxed{0.124}$$

(c) The forcing frequency is $\omega_f = 2 \text{ rad/sec}$.

The pseudo-static deflection is

$$\frac{F_0}{k} = \frac{4 \text{ lbf}}{1.356 \ \dfrac{\text{lbf}}{\text{in}}}$$

$$= 2.95 \text{ in}$$

The ratio of frequencies is

$$r = \frac{\omega_f}{\omega}$$

$$= \frac{2 \ \dfrac{\text{rad}}{\text{sec}}}{8.09 \ \dfrac{\text{rad}}{\text{sec}}}$$

$$= 0.247$$

From Eq. 58.53, the magnification factor is

$$\beta = \frac{D}{\dfrac{F_0}{k}}$$

$$= \left| \frac{1}{\sqrt{(1 - r^2)^2 + (2\zeta r)^2}} \right|$$

$$= \left| \frac{1}{\sqrt{\left(1 - (0.247)^2\right)^2 + \left((2)(0.124)(0.247)\right)^2}} \right|$$

$$= 1.063$$

The maximum excursion of the system is

$$D = \beta\left(\frac{F_0}{k}\right)$$

$$= (1.063)(2.95 \text{ in})$$

$$= \boxed{3.14 \text{ in}}$$

(d) From Eq. 58.52, the differential equation of motion is

$$\left(\frac{m}{g_c}\right)\left(\frac{d^2x}{dt^2}\right) = -kx - C\left(\frac{dx}{dt}\right) + F(t)$$

$$\left(\frac{8 \text{ lbm}}{32.2 \ \dfrac{\text{ft-lbm}}{\text{lbf-sec}^2}}\right)x'' = \frac{-(8 \text{ lbm})\left(\dfrac{32.2 \ \dfrac{\text{ft}}{\text{sec}^2}}{32.2 \ \dfrac{\text{ft-lbm}}{\text{lbf-sec}^2}}\right) \times \left(12 \ \dfrac{\text{in}}{\text{ft}}\right)x}{5.9 \text{ in}}$$

$$- 0.50 \ \frac{\text{lbf-sec}}{\text{ft}} x' + 4\cos 2t$$

$$0.25x'' + 0.50x' + 16.27x = 4\cos 2t$$

$$x'' + 2x' + 65x = 16\cos 2t$$

[coefficients rounded for convenience]

Initial conditions are

$$x_0 = 0$$
$$x_0' = 0$$

There are a variety of methods to solve this differential equation. Use Laplace transforms.

Taking the Laplace transform of both sides,

$$\mathcal{L}(x'') + \mathcal{L}(2x') + \mathcal{L}(65x) = \mathcal{L}(16\cos 2t)$$

$$s^2\mathcal{L}(x) - sx_0 - x_0' + 2s\mathcal{L}(x)$$

$$- 2x_0 + 65\mathcal{L}(x) = (16)\left(\frac{s}{s^2+4}\right)$$

$$\mathcal{L}(x) = \frac{16s}{(s^2+4)(s^2+2s+65)}$$

Use partial fractions.

$$\mathcal{L}(x) = \frac{16s}{(s^2+4)(s^2+2s+65)}$$

$$= \frac{As+B}{s^2+4} + \frac{Cs+D}{s^2+2s+65}$$

$$16s = As^2 + Bs^2 + 2As^2 + 2Bs + 65As + 65B$$
$$+ Cs^3 + Ds^2 + 4Cs + 4D$$

Then,

$$A + C = 0 \qquad C = -A = -\frac{61}{8}B$$

$$B + 2A + D = 0 \quad B + 2A - \frac{65}{4}B = 0 \rightarrow A = \frac{61}{8}B$$

$$65B + 4D = 0 \qquad D = -\frac{65}{4}B$$

$$2B + 65A + 4C = 16$$

$$2B + (65)\left(\frac{61}{8}B\right) + (4)\left(-\frac{61}{8}B\right) = 16$$

$$B = 0.0342521$$
$$A = 0.2611721$$
$$C = -0.2611721$$
$$D = -0.5565962$$

$$\mathcal{L}(x) = \frac{0.2611721s + 0.0342521}{s^2 + 4}$$
$$- \frac{0.2611721s + 0.5566}{(s+1)^2 + (8)^2}$$

$$= (0.26)\left(\frac{s}{s^2+(2)^2}\right) + (0.017)\left(\frac{2}{s^2+(2)^2}\right)$$

$$- (0.26)\left(\frac{\dfrac{s-(-1)}{(s-(-1))^2 + (8)^2}}{+ \dfrac{1.1311472}{(s-(-1))^2 + (8)^2}}\right)$$

Take the inverse transform. The response is

$$\boxed{\begin{aligned} x(t) &= 0.26\cos 2t + 0.017\sin 2t \\ &\quad - (0.26)(e^{-t}\cos 8t + 0.14e^{-t}\sin 8t) \end{aligned}}$$

Compare the values here with those obtained previously.

$$D = (0.26 \text{ ft})\left(12\ \frac{\text{in}}{\text{ft}}\right) = 3.12 \text{ in}$$

This checks with part (c).

The natural frequency is 8 rad/sec, which corresponds to ω from part (a). (The coefficient of 0.25 in the differential equation was rounded from 0.248, which accounts for the difference.)

SI Solution

The static deflection is

$$\delta_{\text{st}} = \frac{\text{weight}}{k} = \frac{mg}{k}$$

$$k = \frac{mg}{\delta_{\text{st}}} = \frac{(3.6 \text{ kg})\left(9.81\ \dfrac{\text{m}}{\text{s}^2}\right)}{0.15 \text{ m}} = 235.4 \text{ N/m}$$

(a) From Eqs. 58.3 and 58.7, the natural frequency is

$$\omega = \sqrt{\frac{k}{m}} = \sqrt{\frac{g}{\delta_{\text{st}}}} = \sqrt{\frac{9.81\ \dfrac{\text{m}}{\text{s}^2}}{0.15 \text{ m}}}$$

$$= \boxed{8.09 \text{ rad/s}}$$

(b) The damping ratio is

$$\zeta = \frac{C}{2\sqrt{km}} = \frac{7.3\ \dfrac{\text{N·s}}{\text{m}}}{2\sqrt{\left(235.4\ \dfrac{\text{N}}{\text{m}}\right)(3.6 \text{ kg})}} = \boxed{0.125}$$

(c) The forcing frequency is $\omega_f = 2$ rad/s.

The pseudo-static deflection is

$$\frac{F_0}{k} = \frac{18 \text{ N}}{235.4 \frac{\text{N}}{\text{m}}} = 0.0765 \text{ m}$$

The ratio of frequencies is

$$r = \frac{\omega_f}{\omega} = \frac{2 \frac{\text{rad}}{\text{s}}}{8.09 \frac{\text{rad}}{\text{s}}} = 0.247$$

From Eq. 58.53, the magnification factor is

$$\beta = \frac{D}{\frac{F_0}{k}} = \left| \frac{1}{\sqrt{(1-r^2)^2 + (2\zeta r)^2}} \right|$$

$$= \left| \frac{1}{\sqrt{\left(1 - (0.247)^2\right)^2 + \left((2)(0.125)(0.247)\right)^2}} \right|$$

$$= 1.063$$

The response of the system is

$$D = \beta \left(\frac{F_0}{k} \right) = (1.063)(0.0765 \text{ m}) = \boxed{0.0813 \text{ m}}$$

(d) From Eq. 58.52, the differential equation of motion is

$$m \left(\frac{d^2 x}{dt^2} \right) = -kx - C \left(\frac{dx}{dt} \right) + F(t)$$

$$(3.6 \text{ kg}) x'' = \left(\frac{(-3.6 \text{ kg}) \left(9.81 \frac{\text{m}}{\text{s}^2} \right)}{0.15 \text{ m}} \right) x$$

$$- \left(7.3 \frac{\text{N·m}}{\text{s}} \right) x' + 18 \cos 2t$$

$$3.6 x'' + 7.3 x' + 235.4 x = 18 \cos 2t$$
$$x'' + 2x' + 65x = 5 \cos 2t$$

[coefficients rounded for convenience]

Initial conditions are

$$x_0 = 0$$
$$x_0' = 0$$

There are a variety of methods to solve this differential equation. Use Laplace transforms.

Take the Laplace transform of both sides.

$$\mathcal{L}(x'') + \mathcal{L}(2x') + \mathcal{L}(65x) = \mathcal{L}(5 \cos 2t)$$

$$s^2 \mathcal{L}(x) - sx_0 - x_0' + 2s\mathcal{L}(x) - 2x_0 + 65\mathcal{L}(x)$$

$$= (5) \left(\frac{s}{s^2 + 4} \right)$$

$$\mathcal{L}(x) = \frac{5s}{(s^2 + 4)(s^2 + 2s + 65)}$$

Use partial fractions.

$$\mathcal{L}(x) = \frac{5s}{(s^2 + 4)(s^2 + 2s + 65)}$$

$$= \frac{As + B}{s^2 + 4} + \frac{Cs + D}{s^2 + 2s + 65}$$

$$5s = As^2 + Bs^2 + 2As^2 + 2Bs + 65As + 65B$$
$$+ Cs^3 + Ds^2 + 4Cs + 4D$$

Then,

$$A + C = 0 \qquad C = -A = -\frac{61}{8}B$$

$$B + 2A + D = 0 \quad B + 2A - \frac{65}{4}B = 0 \rightarrow A = \frac{61}{8}B$$

$$65B + 4D = 0 \qquad D = -\frac{65}{4}B$$

$$2B + 65A + 4C = 5$$

$$2B + (65) \left(\frac{61}{8}B \right) + (4) \left(-\frac{61}{8}B \right) = 5$$

$$B = 0.0107$$
$$A = 0.0816$$
$$C = -0.0816$$
$$D = -0.1739$$

$$\mathcal{L}(x) = \frac{0.0816s + 0.0107}{s^2 + 4} - \frac{0.0816 + 0.1739}{(s+1)^2 + (8)^2}$$

$$= (0.0816) \left(\frac{s}{s^2 + (2)^2} \right)$$

$$+ (0.00535) \left(\frac{2}{s^2 + (2)^2} \right)$$

$$- (0.0816) \left(\frac{s - (-1)}{\left(s - (-1)\right)^2 + (8)^2} + \frac{1.1311}{\left(s - (-1)\right)^2 + (8)^2} \right)$$

Take the inverse transform. The response is

$$\boxed{\begin{aligned} x(t) &= 0.0816 \cos 2t + 0.00535 \sin 2t \\ &\quad - (0.0816)(e^{-t} \cos 8t + 0.14 e^{-t} \sin 8t) \end{aligned}}$$

Compare the values here with those obtained previously.

$$D = 0.0816 \text{ m}$$

This checks with part (c).

The natural frequency is 8 rad/sec, which corresponds to ω from part (a).

6. *Customary U.S. Solution*

The equivalent spring constant is

$$k_{eq} = (4) \left(1000 \; \frac{\text{lbf}}{\text{in}} \right) = 4000 \; \text{lbf/in}$$

From Eq. 58.3, the natural frequency of the system is

$$\omega = \sqrt{\frac{k_{eq} g_c}{m}} = \sqrt{\frac{\left(4000 \; \frac{\text{lbf}}{\text{in}} \right) \left(386.4 \; \frac{\text{in-lbm}}{\text{lbf-sec}^2} \right)}{50 \; \text{lbm}}}$$

$$= 175.8 \; \text{rad/sec}$$

The forcing frequency is

$$\omega_f = \frac{\left(800 \; \frac{\text{rev}}{\text{min}} \right) \left(2\pi \; \frac{\text{rad}}{\text{rev}} \right)}{60 \; \frac{\text{sec}}{\text{min}}} = 83.77 \; \text{rad/sec}$$

The out-of-balance force caused by the rotating eccentric mass is

$$F_0 = \frac{m_0 \omega^2 e}{g_c} = \frac{\left(\frac{1 \; \text{oz}}{16 \; \frac{\text{oz}}{\text{lbm}}} \right) \left(83.77 \; \frac{\text{rad}}{\text{sec}} \right)^2 (5 \; \text{in})}{386.4 \; \frac{\text{in-lbm}}{\text{lbf-sec}^2}}$$

$$= 5.675 \; \text{lbf}$$

The pseudo-static deflection is

$$\frac{F_0}{k_{eq}} = \frac{5.675 \; \text{lbf}}{4000 \; \frac{\text{lbf}}{\text{in}}} = 0.00142 \; \text{in}$$

The ratio of frequencies is

$$r = \frac{\omega_f}{\omega} = \frac{83.77 \; \frac{\text{rad}}{\text{sec}}}{175.8 \; \frac{\text{rad}}{\text{sec}}} = 0.477$$

From Eq. 58.53, the magnification factor is

$$\beta = \frac{D}{\frac{F_0}{k}} = \left| \frac{1}{\sqrt{(1 - r^2)^2 + (2\zeta r)^2}} \right|$$

$$= \left| \frac{1}{\sqrt{\left(1 - (0.477)^2 \right)^2 + \left((2)(0.125)(0.477) \right)^2}} \right|$$

$$= 1.28$$

The maximum vertical displacement is

$$D = \beta \left(\frac{F_0}{k_{eq}} \right) = (1.28)(0.00142 \; \text{in}) = \boxed{0.00182 \; \text{in}}$$

The answer is (B).

SI Solution

The equivalent spring constant is

$$k_{eq} = (4) \left(175 \; \frac{\text{kN}}{\text{m}} \right) \left(1000 \; \frac{\text{N}}{\text{kN}} \right) = 700\,000 \; \text{N/m}$$

From Eq. 58.3, the natural frequency of the system is

$$\omega = \sqrt{\frac{k_{eq}}{m}} = \sqrt{\frac{700\,000 \; \frac{\text{N}}{\text{m}}}{23 \; \text{kg}}} = 174.5 \; \text{rad/s}$$

The forcing frequency is

$$\omega_f = \frac{\left(800 \; \frac{\text{rev}}{\text{min}} \right) \left(2\pi \; \frac{\text{rad}}{\text{rev}} \right)}{60 \; \frac{\text{s}}{\text{min}}} = 83.77 \; \text{rad/s}$$

The out-of-balance force caused by the rotating eccentric mass is

$$F_0 = m_0 \omega^2 e$$

$$= (30 \; \text{g}) \left(\frac{1 \; \text{kg}}{1000 \; \text{g}} \right) \left(83.77 \; \frac{\text{rad}}{\text{s}} \right)^2 (0.130 \; \text{m})$$

$$= 27.37 \; \text{N}$$

The pseudo-static deflection is

$$\frac{F_0}{k_{eq}} = \frac{27.37 \; \text{N}}{700\,000 \; \frac{\text{N}}{\text{m}}} = 3.91 \times 10^{-5} \; \text{m}$$

The ratio of frequencies is

$$r = \frac{\omega_f}{\omega} = \frac{83.77 \; \frac{\text{rad}}{\text{s}}}{174.5 \; \frac{\text{rad}}{\text{s}}} = 0.48$$

From Eq. 58.53, the magnification factor is

$$\beta = \frac{D}{\frac{F_0}{k}} = \left| \frac{1}{\sqrt{(1 - r^2)^2 + (2\zeta r)^2}} \right|$$

$$= \left| \frac{1}{\sqrt{\left(1 - (0.48)^2 \right)^2 + \left((2)(0.125)(0.48) \right)^2}} \right|$$

$$= 1.28$$

The maximum vertical displacement is

$$D = \beta \left(\frac{F_0}{k_{eq}} \right) = (1.28)(3.91 \times 10^{-5} \text{ m})$$

$$= \boxed{5.0 \times 10^{-5} \text{ m}}$$

The answer is (B).

7. *Customary U.S. Solution*

The forcing frequency is

$$\omega_f = \frac{\left(1200 \frac{\text{rev}}{\text{min}}\right)\left(2\pi \frac{\text{rad}}{\text{rev}}\right)}{60 \frac{\text{sec}}{\text{min}}} = 125.7 \text{ rad/sec}$$

The out-of-balance force caused by the rotating imbalance is

$$F_0 = \frac{m_0 \omega^2 e}{g_c} = \frac{(3.6 \text{ lbm})\left(125.7 \frac{\text{rad}}{\text{sec}}\right)^2 (3 \text{ in})}{386.4 \frac{\text{in-lbm}}{\text{lbf-sec}^2}}$$

$$= 441.6 \text{ lbf}$$

The transmissibility is

$$\text{TR} = \frac{|F_{\text{transmitted}}|}{F_{\text{applied}}} = 0.05$$

From Eqs. 58.55 and 58.56, the transmissibility for negligible damping and a value of TR < 1 is

$$\text{TR} = \frac{1}{\left(\frac{\omega_f}{\omega}\right)^2 - 1}$$

$$\frac{\omega_f}{\omega} = \sqrt{\frac{1}{\text{TR}} + 1} = \sqrt{\frac{1}{0.05} + 1} = 4.5826$$

The required natural frequency is

$$\omega = \frac{\omega_f}{4.5826} = \frac{125.7 \frac{\text{rad}}{\text{sec}}}{4.5826} = 27.43 \text{ rad/sec}$$

(a) From Eq. 58.3, the required stiffness of the system is

$$k_{eq} = \frac{m\omega^2}{g_c} = \frac{(175 \text{ lbm})\left(27.43 \frac{\text{rad}}{\text{sec}}\right)^2}{386.4 \frac{\text{in-lbm}}{\text{lbf-sec}^2}} = 340.8 \text{ lbf/in}$$

For four identical springs in parallel, the required stiffness for an individual spring is

$$k = \frac{k_{eq}}{4} = \frac{340.8 \frac{\text{lbf}}{\text{in}}}{4} = \boxed{85.2 \text{ lbf/in}}$$

The answer is (C).

(b) The pseudo-static deflection is

$$\frac{F_0}{k_{eq}} = \frac{441.6 \text{ lbf}}{340.8 \frac{\text{lbf}}{\text{in}}} = 1.30 \text{ in}$$

From Eq. 58.50, the amplitude of vibration is

$$D = \left(\frac{F_0}{k_{eq}}\right) \left| \frac{1}{1 - \left(\frac{\omega_f}{\omega}\right)^2} \right| = \left(\frac{F_0}{k_{eq}}\right)(\text{TR})$$

$$= (1.30 \text{ in})(0.05)$$

$$= \boxed{0.065 \text{ in}}$$

The answer is (C).

SI Solution

The forcing frequency is

$$\omega_f = \frac{\left(1200 \frac{\text{rev}}{\text{min}}\right)\left(2\pi \frac{\text{rad}}{\text{rev}}\right)}{60 \frac{\text{s}}{\text{min}}} = 125.7 \text{ rad/s}$$

The out-of-balance force caused by the rotating imbalance is

$$F_0 = m_0 \omega^2 e = (1.6 \text{ kg})\left(125.7 \frac{\text{rad}}{\text{s}}\right)^2 (0.075 \text{ m})$$

$$= 1896 \text{ N}$$

The transmissibility is

$$\text{TR} = \frac{|F_{\text{transmitted}}|}{F_{\text{applied}}} = 0.05$$

From Eqs. 58.55 and 58.56, the transmissibility for negligible damping and a value of TR < 1 is

$$\text{TR} = \frac{1}{\left(\frac{\omega_f}{\omega}\right)^2 + 1}$$

$$\frac{\omega_f}{\omega} = \sqrt{\frac{1}{\text{TR}} + 1} = \sqrt{\frac{1}{0.05} + 1} = 4.5826$$

The required natural frequency is

$$\omega = \frac{\omega_f}{4.5826} = \frac{125.7 \ \dfrac{\text{rad}}{\text{s}}}{4.5826} = 27.43 \text{ rad/s}$$

(a) From Eq. 58.3, the required stiffness of the system is

$$k_{eq} = m\omega^2 = (80 \text{ kg}) \left(27.43 \ \frac{\text{rad}}{\text{s}}\right)^2 = 60\,192 \text{ N/m}$$

For four identical springs in parallel, the required stiffness for an individual spring is

$$k = \frac{k_{eq}}{4} = \frac{60\,192 \ \dfrac{\text{N}}{\text{m}}}{4} = \boxed{15\,048 \text{ N/m}}$$

The answer is (C).

(b) The pseudo-static deflection is

$$\frac{F_0}{k_{eq}} = \frac{1896 \text{ N}}{60\,192 \ \dfrac{\text{N}}{\text{m}}} = 0.0315 \text{ m}$$

From Eq. 58.50, the amplitude of vibration is

$$D = \left(\frac{F_0}{k_{eq}}\right) \left|\frac{1}{1 - \left(\frac{\omega_f}{\omega}\right)^2}\right| = \left(\frac{F_0}{k_{eq}}\right)(\text{TR})$$

$$= (0.0315 \text{ m})(0.05)$$

$$= \boxed{0.00157 \text{ m}}$$

The answer is (C).

8.   *Customary U.S. Solution*

The first plate under the load is a simple beam with two concentrated forces and may be modeled as in case 8 of App. 49.A.

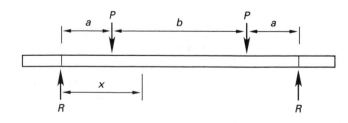

The load P is

$$P = \frac{20,000 \text{ lbf}}{2} = 10,000 \text{ lbf}$$

The locations of P are

$$a = \frac{36 \text{ in} - 24 \text{ in}}{2} = 6 \text{ in}$$

$$b = 24 \text{ in} \quad [\text{given}]$$

The moment of inertia of the cross section of the plate is

$$I = \frac{bh^3}{12} = \frac{(30 \text{ in})(0.5 \text{ in})^3}{12} = 0.3125 \text{ in}^4$$

The overhangs do not contribute to the rigidity of the plate. The length of the beam model is

$$L = 2a + b = (2)(6 \text{ in}) + 24 \text{ in} = 36 \text{ in}$$

(a) The deflection at the load $P(x = a)$ is

$$y = \left(\frac{P}{6EI}\right)(3Lax - 3a^2x - x^3)$$

$$= \left(\frac{Px}{6EI}\right)\left((3a)(L - a) - x^2\right)$$

$$= \frac{(10,000 \text{ lbf})(6 \text{ in})\left((3)(6 \text{ in})(36 \text{ in} - 6 \text{ in}) - (6 \text{ in})^2\right)}{(6)\left(2.9 \times 10^7 \ \dfrac{\text{lbf}}{\text{in}^2}\right)(0.3125 \text{ in}^4)}$$

$$= 0.556 \text{ in}$$

The wide-beam correction is needed here. The deflection is multiplied by

$$1 - \nu^2 = 1 - (0.3)^2 = 0.91$$

$$y = (0.91)(0.556 \text{ in}) = 0.506 \text{ in}$$

The second plate is loaded exactly the same as the first plate, only upside down. Therefore, its deflection is also 0.556 in at the load. Since plates 3, 5, and 7 are loaded the same as plate 1 and since plates 4, 6, and 8 are loaded the same as plate 2, the total static deflection for the eight plates is

$$y_{\text{total}} = (4)(0.506 \text{ in}) + (4)(0.506 \text{ in})$$

$$= \boxed{4.048 \text{ in}}$$

The answer is (A).

(b) The maximum moment in the plates is at the load P and is

$$M_{\text{max}} = Pa = (10,000 \text{ lbf})(6 \text{ in}) = 60,000 \text{ in-lbf}$$

The maximum stress in the plates is at the extreme fiber and is

$$\sigma_{\text{max}} = \frac{M_{\text{max}}c}{I}$$

$$= \frac{(60,000 \text{ in-lbf})\left(\dfrac{0.50 \text{ in}}{2}\right)}{0.3125 \text{ in}^4}$$

$$= \boxed{48,000 \text{ lbf/in}^2}$$

The answer is (A).

(c) From Eqs. 58.4 and 58.7, the natural frequency of oscillation is

$$f = \left(\frac{1}{2\pi}\right)\sqrt{\frac{g}{\delta_{\text{st}}}}$$

$$= \left(\frac{1}{2\pi}\right)\sqrt{\frac{g}{y_{\text{total}}}}$$

$$= \left(\frac{1}{2\pi}\right)\sqrt{\frac{386.4\ \dfrac{\text{in}}{\text{sec}^2}}{4.048\ \text{in}}}$$

$$= \boxed{1.55\ \text{Hz}}$$

The answer is (A).

SI Solution

The first plate under the load is a simple beam with two concentrated forces and may be modeled as in case 8 of App. 49.A.

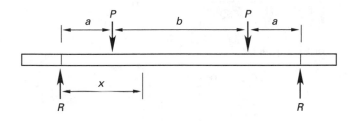

The load P is

$$P = \frac{(9100\ \text{kg})\left(9.81\ \dfrac{\text{m}}{\text{s}^2}\right)}{2} = 44\,636\ \text{N}$$

The locations of P are

$$a = \frac{0.92\ \text{m} - 0.60\ \text{m}}{2} = 0.16\ \text{m}$$

$$b = 0.60\ \text{m}\quad[\text{given}]$$

The moment of inertia of the cross section of the plate is

$$I = \frac{bh^3}{12} = \frac{(0.76\ \text{m})(0.012\ \text{m})^3}{12} = 1.094 \times 10^{-7}\ \text{m}^4$$

The overhangs do not contribute to the rigidity of the plate. The length of the beam model is

$$L = 2a + b = (2)(0.16\ \text{m}) + 0.60\ \text{m} = 0.92\ \text{m}$$

(a) The deflection at the load $P(x = a)$ is

$$y = \left(\frac{P}{6EI}\right)(3Lax - 3a^2x - x^3)$$

$$= \left(\frac{Px}{6EI}\right)\left((3a)(L - a) - x^2\right)$$

$$= \frac{\begin{array}{c}(44\,636\ \text{N})(0.16\ \text{m})((3)(0.16\ \text{m}) \\ \times\,(0.92\ \text{m} - 0.16\ \text{m}) - (0.16\ \text{m})^2)\end{array}}{(6)\,(200 \times 10^9\ \text{Pa})\,(1.094 \times 10^{-7}\ \text{m}^4)}$$

$$= 0.01845\ \text{m}$$

The wide-beam correction is needed here. The deflection is multiplied by

$$1 - \nu^2 = 1 - (0.3)^2 = 0.91$$

$$y = (0.91)(0.01845\ \text{m})$$

$$= 0.01679\ \text{m}$$

The second plate is loaded exactly the same as the first plate, only upside down. Therefore, its deflection is also 0.01845 m at the load. Since plates 3, 5, and 7 are loaded the same as plate 1 and since plates 4, 6, and 8 are loaded the same as plate 2, the total static deflection for the eight plates is

$$y_{\text{total}} = (4)(0.01679\ \text{m}) + (4)(0.01679\ \text{m}) = \boxed{0.1343\ \text{m}}$$

The answer is (A).

(b) The maximum moment in the plates is at the load P and is

$$M_{\text{max}} = Pa = (44\,636\ \text{N})(0.16\ \text{m}) = 7142\ \text{N·m}$$

The maximum stress in the plates is at the extreme fiber and is

$$\sigma_{\text{max}} = \frac{M_{\text{max}}c}{I} = \frac{(7142\ \text{N·m})\left(\dfrac{0.012\ \text{m}}{2}\right)}{1.094 \times 10^{-7}\ \text{m}^4}$$

$$= 3.92 \times 10^8\ \text{Pa}\quad\boxed{(392\ \text{MPa})}$$

The answer is (A).

(c) From Eqs. 58.4 and 58.7, the natural frequency of oscillation is

$$f = \left(\frac{1}{2\pi}\right)\sqrt{\frac{g}{\delta_{\text{st}}}} = \left(\frac{1}{2\pi}\right)\sqrt{\frac{g}{y_{\text{total}}}}$$

$$= \left(\frac{1}{2\pi}\right)\sqrt{\frac{9.81\ \dfrac{\text{m}}{\text{s}^2}}{0.1343\ \text{m}}}$$

$$= \boxed{1.36\ \text{Hz}}$$

The answer is (A).

9. *Customary U.S. Solution*

From the fan rotation,

$$\frac{600 \frac{\text{rev}}{\text{min}}}{60 \frac{\text{sec}}{\text{min}}} = \boxed{10 \text{ Hz}}$$

From the driving blades,

$$\frac{\left(600 \frac{\text{rev}}{\text{min}}\right)(8)}{60 \frac{\text{sec}}{\text{min}}} = \boxed{80 \text{ Hz}}$$

From the fan blades,

$$\frac{\left(600 \frac{\text{rev}}{\text{min}}\right)(64)}{60 \frac{\text{sec}}{\text{min}}} = \boxed{640 \text{ Hz}}$$

From the motor rotation,

$$\frac{1725 \frac{\text{rev}}{\text{min}}}{60 \frac{\text{sec}}{\text{min}}} \approx \boxed{29 \text{ Hz}}$$

From the poles,

$$\frac{\left(1725 \frac{\text{rev}}{\text{min}}\right)(4)}{60 \frac{\text{sec}}{\text{min}}} = \boxed{115 \text{ Hz}}$$

The electrical hum is $\boxed{60 \text{ Hz.}}$

The pulleys are

$$\text{motor pulley (same as motor)} = \boxed{29 \text{ Hz}}$$

$$\text{fan pulley (same as fan)} = \boxed{10 \text{ Hz}}$$

For the belt,

$$\text{belt speed} = \pi D n = \pi(4 \text{ in})\left(\frac{1725 \frac{\text{rev}}{\text{min}}}{60 \frac{\text{sec}}{\text{min}}}\right)$$

$$= 361.3 \text{ in/sec}$$

The frequency is

$$f = \frac{361.3 \frac{\text{in}}{\text{sec}}}{72 \text{ in}} \approx \boxed{5 \text{ Hz}}$$

SI Solution

From the fan rotation,

$$\frac{600 \frac{\text{rev}}{\text{min}}}{60 \frac{\text{sec}}{\text{min}}} = \boxed{10 \text{ Hz}}$$

From the driving blades,

$$\frac{\left(600 \frac{\text{rev}}{\text{min}}\right)(8)}{60 \frac{\text{sec}}{\text{min}}} = \boxed{80 \text{ Hz}}$$

From the fan blades,

$$\frac{\left(600 \frac{\text{rev}}{\text{min}}\right)(64)}{60 \frac{\text{sec}}{\text{min}}} = \boxed{640 \text{ Hz}}$$

From the motor rotation,

$$\frac{1725 \frac{\text{rev}}{\text{min}}}{60 \frac{\text{sec}}{\text{min}}} \approx \boxed{29 \text{ Hz}}$$

From the poles,

$$\frac{\left(1725 \frac{\text{rev}}{\text{min}}\right)(4)}{60 \frac{\text{sec}}{\text{min}}} = \boxed{115 \text{ Hz}}$$

The electrical hum is $\boxed{60 \text{ Hz.}}$

The pulleys are

$$\text{motor pulley (same as motor)} = \boxed{29 \text{ Hz}}$$

$$\text{fan pulley (same as fan)} = \boxed{10 \text{ Hz}}$$

For the belt,

$$\text{belt speed} = \frac{(\pi)(100 \text{ mm})\left(\frac{1725 \frac{\text{rev}}{\text{min}}}{60 \frac{\text{s}}{\text{min}}}\right)}{1000 \frac{\text{mm}}{\text{m}}}$$

$$= 9.032 \text{ m/s}$$

The frequency is

$$f = \frac{9.032 \frac{\text{m}}{\text{s}}}{1.83 \text{ m}} = \boxed{4.94 \text{ Hz}}$$

10. *Customary U.S. Solution*

The forcing frequency is

$$f_f = \frac{1725 \frac{\text{rev}}{\text{min}}}{60 \frac{\text{sec}}{\text{min}}} = 28.75 \text{ Hz}$$

The natural frequency is

$$f = \left(\frac{1}{2\pi}\right)\sqrt{\frac{g}{\delta_{\text{st}}}} = \left(\frac{1}{2\pi}\right)\sqrt{\frac{386.4 \frac{\text{in}}{\text{sec}^2}}{0.02 \text{ in}}}$$
$$= 22.12 \text{ Hz}$$

From Eq. 58.56 with negligible damping,

$$\text{TR} = \frac{1}{\sqrt{(1-r^2)^2}} = \frac{1}{\sqrt{\left(1-\left(\frac{f_f}{f}\right)^2\right)^2}}$$
$$= \frac{1}{\sqrt{\left(1-\left(\frac{28.75 \text{ Hz}}{22.12 \text{ Hz}}\right)^2\right)^2}}$$
$$= 1.451$$

This is a 45.1% increase in force.

The answer is (B).

SI Solution

The forcing frequency is

$$f_f = \frac{1725 \frac{\text{rev}}{\text{min}}}{60 \frac{\text{sec}}{\text{min}}} = 28.75 \text{ Hz}$$

The natural frequency is

$$f = \left(\frac{1}{2\pi}\right)\sqrt{\frac{g}{\delta_{\text{st}}}} = \left(\frac{1}{2\pi}\right)\sqrt{\frac{9.81 \frac{\text{m}}{\text{s}^2}}{(0.5 \text{ mm})\left(\frac{1 \text{ m}}{1000 \text{ mm}}\right)}}$$
$$= 22.29 \text{ Hz}$$

From Eq. 58.56 with negligible damping,

$$\text{TR} = \frac{1}{\sqrt{(1-r^2)^2}} = \frac{1}{\sqrt{\left(1-\left(\frac{f_f}{f}\right)^2\right)^2}}$$
$$= \frac{1}{\sqrt{\left(1-\left(\frac{28.75 \text{ Hz}}{22.29 \text{ Hz}}\right)^2\right)^2}}$$
$$= 1.507$$

This is a 50.7% increase in force.

The answer is (B).

59 Modeling of Engineering Systems

PRACTICE PROBLEMS

For each of the systems of ideal elements shown, (a) draw the system diagram and (b) write the differential equations.

1.

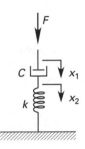

2.

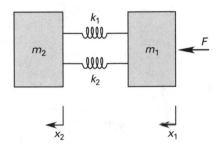

3.

uniform beam
mass m
infinite stiffness
frictionless pivot

4.

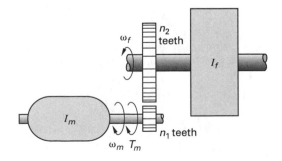

5.

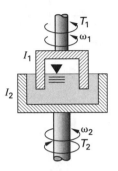

6. The coupling of a railroad car is modeled as the mechanical system shown. Assume all elements are linear. What are the system equations that describe the positions x_1 and x_2 as functions of time?

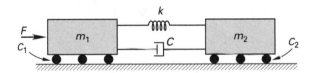

7. Water is discharged freely at a constant rate into an open tank. Water flows out of the tank through a drain with a resistance to flow. (a) Draw the system diagram using idealized elements, and (b) write the differential equations that describe the response of the system.

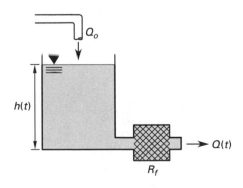

8. Water is pumped into the bottom of an open tank. (a) Draw the system diagram using idealized elements, and (b) write the differential equations that describe the response of the system.

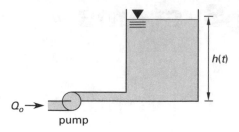

SOLUTIONS

1. (a) The velocity of the plunger is v_1. The velocity of the body of the damper is the same as the upper part of the spring, v_2. By Rule 59.2, the other end of the force and the spring is attached to the stationary wall at $v = 0$. The system diagram is

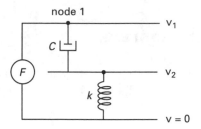

(b) By Rule 59.3, the force from the source is the same force experienced by the dashpot. One of the system equations is based on node 1. Using Rule 59.4 and expanding with Eq. 59.5,

$$F = F_C = C(v_1 - v_2) = C(x_1' - x_2')$$

By Rule 59.3, the force from the source is the same force experienced by the spring. A second system equation is based on node 2. Using Rule 59.4 and expanding with Eq. 59.4,

$$F = F_k = k(x_2 - 0) = kx_2$$

2. (a) The velocity of the ends of the springs connected to m_i is v_1. The velocity of the ends of the springs connected to m_2 is v_2. By Rule 59.1, the other end of each mass connects to $v = 0$. The system diagram is

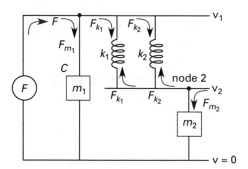

(b) The force leaving the source splits: some of it goes through m_1, some of it goes through k_1, and some of it goes through k_2. One of the system equations is based on node 1.

$$F = F_{m_1} + F_{k_1} + F_{k_2}$$

Using Rule 59.4 and expanding with Eqs. 59.3 and 59.4,

$$F = m_1 a_1 + k_1(x_1 - x_2) + k_2(x_1 - x_2)$$
$$= m_1 x_1'' + (k_1 + k_2)(x_1 - x_2)$$

A second system equation is based on node 2. The conservation law is written to conserve force in the v_2 line.

$$0 = F_{m_2} + F_{k_2} + F_{k_1}$$

Using Rule 59.4 and expanding with Eqs. 59.3 and 59.4,

$$0 = m_2 a_2 + k_2(x_2 - x_1) + k_1(x_2 - x_1)$$
$$= m_2 x_2'' + (k_1 + k_2)(x_2 - x_1)$$

3. (a) Treat this as a rotational system. The applied rotational torque is

$$T = FL$$

The equivalent torsional spring constant is

$$k_r = \frac{M_{\text{resisting}}}{\theta} = \frac{F_k l}{\theta} = \frac{kx_2 l}{\theta}$$

However, $x_2 = l \sin \theta$ and $\theta \approx \sin \theta$ for small angles.

$$k_r = kl^2$$

The moment of inertia of the beam about the hinge point is

$$I = \tfrac{1}{3}mL^2$$

The equivalent rotational system is

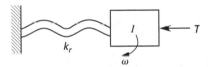

The angular velocity of the end of the spring connected to the inertial element is ω. By Rule 59.2, the other end of the spring is attached to the stationary wall at $\omega = 0$. The system diagram is

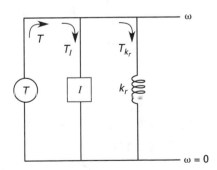

(b) The torque leaving the source splits: some of it goes through I and some of it goes through k_r. The conservation law is written to conserve torque in the ω line.

$$T = T_I + T_{k_r}$$

Using Rule 59.4 and expanding with Eqs. 59.6 and 59.7,

$$T = I\alpha + k_r(\theta - 0)$$
$$FL = \left(\tfrac{1}{3}mL^2\right)\theta'' + kl^2\theta$$

4. (a) The angular velocity of the small gear is ω_m, and the angular velocity of the large gear is ω_f. By Rule 59.2, the other end of each inertia connects to $\omega = 0$. The gearing transforms the torque and angular displacement from gear 1 to gear 2. The system diagram is

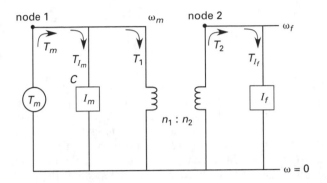

(b) The conservation law based on node 1 is written to conserve torque in the ω_m line.

$$T_m = T_{I_m} + T_1 = I_m \alpha_m + T_1 = I_m \theta_m'' + T_1$$

The same conservation principle based on node 2 is used to conserve torque in the ω_f line.

$$T_2 = T_{I_f} = I_f \alpha_f = I_f \theta_f''$$

The transformer equations are

$$T_2 = \left(\frac{n_2}{n_1}\right) T_1$$

$$\theta_m = \left(\frac{n_2}{n_1}\right) \theta_f$$

5. (a) Consider the fluid to act as a damper with coefficient C_r. The plunger is connected to velocity ω_1, and the body is connected to velocity ω_2. By Rule 59.2, the ends of the inertia elements are connected to $\omega = 0$. The system diagram is

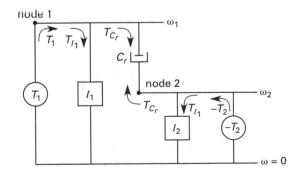

(b) One of the system equations is based on conservation of torque at node 1.

$$T_1 = T_{I_1} + T_{C_r}$$

Using Rule 59.4 and expanding with Eqs. 59.6 and 59.8,

$$T_1 = I_1\alpha_1 + C_r(\omega_1 - \omega_2) = I_1\theta_1'' + C_r(\theta_1' - \theta_2')$$

The second system equation is based on conservation of torque at node 2.

$$-T_2 = T_{I_2} + T_{C_r} = I_2\alpha_2 + C_r(\omega_2 - \omega_1)$$
$$= I_2\theta_2'' + C_r(\theta_2' - \theta_1')$$

6. (a) The velocity of the end of the spring connected to m_1 is v_1. This is also the velocity of the plunger and the velocity of the viscous damper, C_1. The velocity of the end of the spring connected to m_2 is v_2. This is also the velocity of the body of the damper and the velocity of the viscous damper C_2. By Rule 59.2, the other end of each mass connects to $v = 0$. The system diagram is

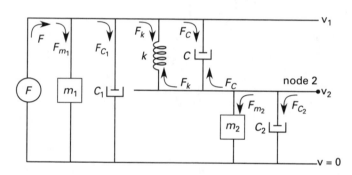

(b) The force leaving the source splits: some of it goes through m_1, C_1, k, and C. The conservation law is written to conserve force in the v_1 line. The equation is based on node 1.

$$F = F_{m_1} + F_{C_1} + F_k + F_C$$

Using Rule 59.4 and expanding with Eqs. 59.3, 59.4, and 59.5,

$$F = m_1a_1 + C_1(v_1 - 0) + C(v_1 - v_2) + k(x_1 - x_2)$$
$$= m_1x_1'' + C_1x_1' + C(x_1' + x_2') + k(x_1 - x_2)$$

The same conservation principle based on node 2 is used to conserve force in the v_2 line. Using Rule 59.4 and expanding with Eqs. 59.3, 59.4, and 59.5,

$$0 = F_{C_2} + F_{m_2} + F_C + F_k$$
$$= C_2(v_2 - 0) + m_2a_2 + C(v_2 - v_1) + k(x_2 - x_1)$$
$$= C_2x_2' + m_2x_2'' + C(x_2' - x_1') + k(x_2 - x_1)$$

7. (a) The fluid capacitance of the water in the tank is C_f. From Rule 59.2, one end of each of the two energy sources Q_1 and Q_2 connects to $p = 0$. The system diagram is

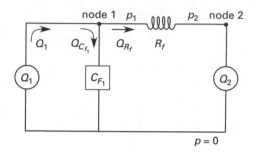

(b) From Eq. 59.10, the flow through the capacitor is

$$Q_{C_{f_1}} = C_{f_1}\left(\frac{dp_1}{dt}\right)$$

From Eq. 59.12, the flow through the resistor is

$$Q_{R_f} = \frac{p_1 - p_2}{R_f}$$

One of the system equations is based on conservation of flow at node 1.

$$Q_1 = Q_{f_1} + Q_{R_f} = C_{f_1}\left(\frac{dp_1}{dt}\right) + \left(\frac{1}{R_f}\right)(p_1 - p_2)$$

The second system equation is based on conservation of flow at node 2.

$$Q_2 = \left(\frac{1}{R_f}\right)(p_1 - p_2)$$

8. (a) The fluid resistance in the entrance pipe is R_f. The pressure at the entrance is p_1, and the pressure in the tank is p_2. The fluid capacitance of the water is C_f. The pressure at the open top of the tank is $p = 0$. The system diagram is

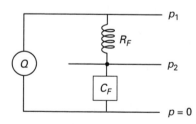

(b) By Rule 59.3, the source flow Q is the same flow through the resistor and the capacitor. Use Eq. 59.12 for the resistor.

$$Q = \frac{p_1 - p_2}{R_f}$$

Use Eq. 59.10 for the capacitor.

$$Q = C_f\left(\frac{dp_2}{dt}\right)$$

60 Analysis of Engineering Systems

PRACTICE PROBLEMS

1. Simplify the following block diagrams and determine the overall system gain.

(a)

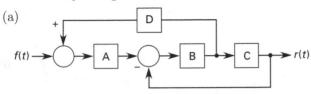

(b)

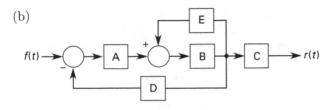

2. *(Time limit: one hour)* A mass of 100 lbm (45 kg) is supported uniformly by a spring system. The spring system has a combined stiffness of 1200 lbf/ft (17.5 kN/m). A dashpot with a damping coefficient of 60 lbf-sec/ft (880 N·s/m) has been installed.

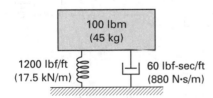

(a) What is the undamped natural frequency?

(b) What is the damping ratio?

(c) Sketch the magnitude and phase characteristics of the frequency response.

(d) Sketch the response to a unit step input.

3. *(Time limit: one hour)* A constant-speed motor/magnetic clutch drive train is monitored and controlled by a speed-sensing tachometer. The entire system is modeled as a control system block diagram, as shown on the following page. (The lowercase letters represent small-signal increments from the reference values.) When the control system is operating, the desired motor speed, n (in rpm), is set with a speed-setting potentiometer. The setting is compared to the tachometer output. The comparator output error (in volts), controls the clutch. A current, i (in amps), passes through the clutch coil. The external load torque, $t_L m$ (in in-lbf), is seen by the clutch and is countered by the clutch output torque, t (in in-lbf).

(a) Plot the open-loop frequency response.

(b) What is the open-loop steady-state gain?

(c) Plot the unity feedback closed-loop frequency response.

(d) What is the closed-loop steady-state gain?

(e) Plot the system sensitivity.

(f) Describe the closed-loop response to a step change in the desired output angular velocity. Is it damped or oscillatory? Is there a steady-state error? Why or why not?

(g) Describe the closed-loop response to a step change in the load torque. Is the response damped or oscillatory? Is there a steady-state error? Why or why not?

(h) Assume that you have to select the comparator gain and that it doesn't have to be 0.1. Using the root-locus method or either the Routh or Nyquist stability criterion, find the limits of the comparator gain that cause the closed-loop system to be unstable.

(i) How can you improve the steady-state response of the closed-loop system to constant disturbances in the load torque, t_L?

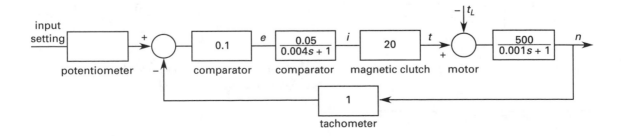

SOLUTIONS

1. (a) Draw the first block diagram.

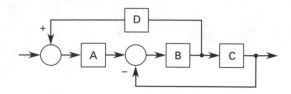

From the rules of simplifying block diagrams, use case 7 to move the extreme right pick-off point to the left of C.

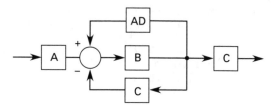

Use case 6 to combine the two summing points on the left.

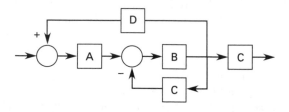

Use case 1 to combine boxes in series in the upper feedback loop.

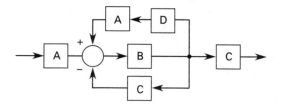

Use case 2 to combine the two feedback loops.

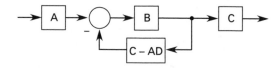

Use case 3 to simplify the remaining feedback loop.

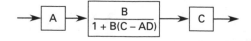

Use case 1 to combine boxes in series to determine the system gain.

$$G_{\text{loop}} = \frac{ABC}{1 + BC - ABD}$$

(b) Draw the second block diagram.

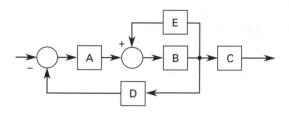

From the rules of simplifying block diagrams, use case 6 to combine the two summing points on the left.

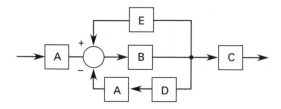

Use case 1 to combine boxes in series in the lower feedback loop.

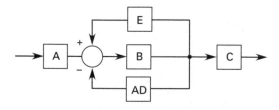

Use case 2 to combine the two feedback loops.

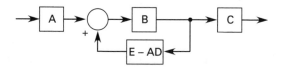

Use case 3 to simplify the remaining feedback loop.

Use case 1 to combine boxes in series to determine the system gain.

$$G_{\text{loop}} = \frac{ABC}{1 - BE + ABD}$$

2. *Customary U.S. Solution*

The system differential equation for a force input f is

$$mx'' + Bx' + kx = f$$

Divide by m to write the equation in terms of natural frequency and damping factor.

$$x'' + \left(\frac{B}{m}\right)x' + \left(\frac{k}{m}\right)x = \frac{f}{m}$$

$$x'' + 2\zeta\omega_n x' + \omega_n^2 x = \frac{f}{\frac{k}{\omega_n^2}} = \omega_n^2\left(\frac{f}{k}\right)$$

Define the forcing function as $h = f/k$.

The equation is the same as Eq. 60.3.

$$x'' + 2\zeta\omega_n x' + \omega_n^2 x = \omega_n^2 h$$

(a) The undamped natural frequency is

$$\omega_n = \sqrt{\frac{kg_c}{m}} = \sqrt{\frac{\left(1200\ \frac{\text{lbf}}{\text{ft}}\right)\left(32.2\ \frac{\text{lbm-ft}}{\text{lbf-sec}^2}\right)}{100\ \text{lbm}}}$$

$$= \boxed{19.66\ \text{rad/sec}}$$

(b) The damping ratio is

$$\zeta = \frac{\frac{B}{m}}{2\omega_n} = \frac{B}{2\omega_n m} = \frac{\left(60\ \frac{\text{lbf-sec}}{\text{ft}}\right)\left(32.2\ \frac{\text{lbm-ft}}{\text{lbf-sec}^2}\right)}{(2)\left(19.66\ \frac{\text{rad}}{\text{sec}}\right)(100\ \text{lbm})}$$

$$= \boxed{0.491}$$

(c) Take the Laplace transform of Eq. 60.3. Consider zero initial conditions.

$$s^2 x(s) + 2\zeta\omega_n s x(s) + \omega_n^2 x(s) = \omega_n^2 H(s)$$

Determine the transfer function.

$$T(s) = \frac{x(s)}{H(s)} = \frac{\omega_n^2}{s^2 + 2\zeta\omega_n s + \omega_n^2}$$

The frequency response is obtained by letting $s = j\omega$.

$$T(j\omega) = \frac{\omega_n^2}{(\omega_n^2 - \omega^2) + j2\zeta\omega_n\omega}$$

Write the equation in polar form.

$$T(j\omega) = |T(j\omega)|e^{j\phi(\omega)}$$

The magnitude is

$$|T(j\omega)| = \frac{\omega_n^2}{\sqrt{(\omega_n^2 - \omega^2)^2 + (2\zeta\omega_n\omega)^2}}$$

At $\omega = 0$, $|T(j\omega)| = 1$.

At $\omega = \omega_n$,

$$|T(j\omega)| = \frac{1}{2\zeta} = \frac{1}{(2)(0.491)} \approx 1.0$$

At $\omega \to \infty$, $|T(j\omega)| \to 0$.

A peak occurs near $\omega = \omega_n$. To obtain the location ω_p, set the derivative of $|T(j\omega)|$ equal to zero.

$$\frac{d|T(j\omega)|}{d\omega} = -\left(\frac{1}{2}\right)\omega_n^2\left((\omega_n^2 - \omega^2) + (2\zeta\omega_n\omega)^2\right)^{-\frac{3}{2}}$$
$$\times \left((2)(\omega_n^2 - \omega^2)(-2\omega)\right.$$
$$\left. + (2)(2\zeta\omega_n\omega)(2\zeta\omega_n)\right)$$
$$= 0$$

$$\omega_p = \omega_n\sqrt{1 - 2\zeta^2}$$
$$= \left(19.66\ \frac{\text{rad}}{\text{sec}}\right)\sqrt{1 - (2)(0.491)^2}$$
$$= 14.15\ \text{rad/sec}$$

At $\omega = \omega_p$,

$$|T(j\omega)| = \frac{\omega_n^2}{\sqrt{(\omega_n^2 - \omega_p^2)^2 + (2\zeta\omega_n\omega_p)^2}}$$

$$= \frac{\left(19.66\ \frac{\text{rad}}{\text{sec}}\right)^2}{\sqrt{\left(\left(19.66\ \frac{\text{rad}}{\text{sec}}\right)^2 - \left(14.15\ \frac{\text{rad}}{\text{sec}}\right)^2\right)^2 + \left((2)(0.491)\left(19.66\ \frac{\text{rad}}{\text{sec}}\right) \times \left(14.15\ \frac{\text{rad}}{\text{sec}}\right)\right)^2}}$$

$$= 1.17$$

A sketch of the frequency response magnitude is

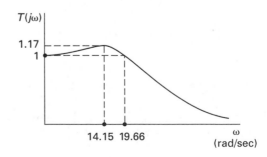

The phase is

$$\phi(\omega) = \tan^{-1}\left(\frac{-2\zeta\omega_n\omega}{\omega_n^2 - \omega^2}\right)$$

At $\omega = 0$, $\phi(\omega) = 0$.

At $\omega = \omega_n$, $\phi(\omega) = -\pi/2$.

At $\omega \to \infty$, $\phi(\omega) = -\pi$.

A sketch of the phase is

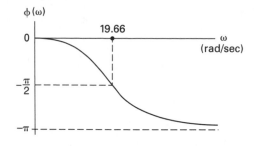

(d) The final value of the step response is obtained from the final value theorem.

$$x_{\text{final}} = \lim_{s\to 0} sx(s) = \lim_{s\to 0} sT(s)H(s)$$

$$= \lim_{s\to 0} sT(s)\left(\frac{1 \text{ ft}}{s}\right) = \left(\frac{\omega_n^2}{\omega_n^2}\right)(1 \text{ ft})$$

$$= 1.0 \text{ ft}$$

From Eq. 60.7, the fraction of overshoot is

$$M_p = \exp\left(\frac{-\pi\zeta}{\sqrt{1-\zeta^2}}\right) = \exp\left(\frac{-\pi(0.491)}{\sqrt{1-(0.491)^2}}\right)$$

$$= 0.17$$

The peak is $x_{\max} = 1.17$ ft.

The damped natural frequency is

$$\omega_d = \omega_n\sqrt{1-\zeta^2} = \left(19.66 \frac{\text{rad}}{\text{sec}}\right)\sqrt{1-(0.491)^2}$$

$$= 17.1 \text{ rad/sec}$$

The peak time from Eq. 60.6 is

$$t_p = \frac{\pi}{\omega_d} = \frac{\pi}{17.1 \dfrac{\text{rad}}{\text{sec}}} = 0.18 \text{ sec}$$

The 5% criterion settling time from Eq. 60.9 is

$$t_s = \frac{3.00}{\zeta\omega_n} = \frac{3.00}{(0.491)\left(19.66 \dfrac{\text{rad}}{\text{sec}}\right)} = 0.31 \text{ sec}$$

The sketch of the response to a unit step input is obtained from Fig. 60.2 and the preceding calculations.

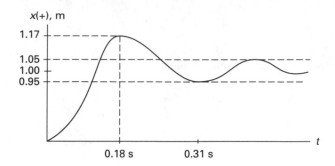

SI Solution

The system differential equation for a force input f is

$$mx'' + Bx' + kx = f$$

Divide by m to write the equation in terms of natural frequency and damping factor.

$$x'' + \left(\frac{B}{m}\right)x' + \left(\frac{k}{m}\right)x = \frac{f}{m}$$

$$x'' + 2\zeta\omega_n x' + \omega_n^2 x = \frac{f}{\dfrac{k}{\omega_n^2}} = \omega_n^2\left(\frac{f}{k}\right)$$

Define the forcing function as $h = f/k$.

The equation is the same as Eq. 60.3.

$$x'' + 2\zeta\omega_n x' + \omega_n^2 x = \omega_n^2 h$$

(a) The undamped natural frequency is

$$\omega_n = \sqrt{\frac{k}{m}} = \sqrt{\frac{17\,500 \dfrac{\text{N}}{\text{m}}}{45 \text{ kg}}} = \boxed{19.7 \text{ rad/s}}$$

(b) The damping ratio is

$$\zeta = \frac{\dfrac{B}{m}}{2\omega_n} = \frac{\dfrac{880 \dfrac{\text{N·s}}{\text{m}}}{45 \text{ kg}}}{(2)\left(19.7 \dfrac{\text{rad}}{\text{s}}\right)} = \boxed{0.50}$$

(c) Take the Laplace transform of Eq. 60.3. Consider zero initial conditions.

$$s^2 x(s) + 2\zeta\omega_n sx(s) + \omega_n^2 x(s) = \omega_n^2 H(s)$$

Determine the transfer function.

$$T(s) = \frac{x(s)}{H(s)} = \frac{\omega_n^2}{s^2 + 2\zeta\omega_n s + \omega_n^2}$$

The frequency response is obtained by letting $s = j\omega$.

$$T(j\omega) = \frac{\omega_n^2}{(\omega_n^2 - \omega^2) + j2\zeta\omega_n\omega}$$

Write the equation in polar form.

$$T(j\omega) = |T(j\omega)|e^{j\phi(\omega)}$$

The magnitude is

$$|T(j\omega)| = \frac{\omega_n^2}{\sqrt{(\omega_n^2 - \omega^2)^2 + (2\zeta\omega_n\omega)^2}}$$

At $\omega = 0$, $|T(j\omega)| = 1$.

At $\omega = \omega_n$,

$$|T(j\omega)| = \frac{1}{2s} = \frac{1}{(2)(0.50)} = 1.0$$

At $\omega \to \infty$, $|T(j\omega)| \to 0$.

A peak occurs near $\omega = \omega_n$. To obtain the location ω_p, set the derivative of $|T(j\omega)|$ equal to zero.

$$\frac{d|T(j\omega)|}{d\omega} = -\left(\frac{3}{2}\right)\omega_n^2\left((\omega_n^2 - \omega^2) + (2\zeta\omega_n\omega)^2\right)^{-\frac{3}{2}}$$

$$\times \left((2)(\omega_n^2 - \omega^2)(-2\omega) + (2)(2\zeta\omega_n\omega)2\zeta\omega_n\right)$$

$$= 0$$

$$\omega_p = \omega_n\sqrt{1 - 2\zeta^2}$$

$$= \left(19.7\,\frac{\text{rad}}{\text{s}}\right)\sqrt{1 - (2)(0.50)^2}$$

$$= 13.9\text{ rad/s}$$

At $\omega = \omega_p$,

$$|T(j\omega)| = \frac{\omega_n^2}{\sqrt{(\omega_n^2 - \omega_p^2)^2 + (2\zeta\omega_n\omega_p)^2}}$$

$$= \frac{\left(10.7\,\dfrac{\text{rad}}{\text{s}}\right)^2}{\sqrt{\begin{array}{l}\left(\left(19.7\,\dfrac{\text{rad}}{\text{s}}\right)^2 - \left(13.9\,\dfrac{\text{rad}}{\text{s}}\right)^2\right)^2 \\ + \left(\begin{array}{c}(2)(0.50)\left(19.7\,\dfrac{\text{rad}}{\text{s}}\right) \\ \times \left(13.9\,\dfrac{\text{rad}}{\text{s}}\right)\end{array}\right)^2\end{array}}}$$

$$= 1.16$$

A sketch of the frequency response magnitude is

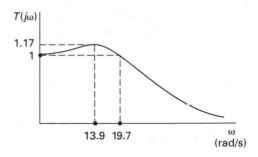

The phase is

$$\phi(\omega) = \tan^{-1}\left(\frac{-2\zeta\omega_n\omega}{\omega_n^2 - \omega^2}\right)$$

At $\omega = 0$, $\phi(\omega) = 0$.

At $\omega = \omega_n$, $\phi(\omega) = -\pi/2$.

At $\omega \to \infty$, $\phi(\omega) = -\pi$.

A sketch of the phase is

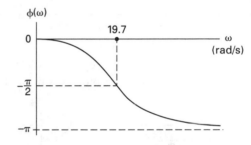

(d) The final value of the step response is obtained from the final value theorem.

$$x_{\text{final}} = \lim_{s \to 0} sx(s) = \lim_{s \to 0} sT(s)H(s)$$

$$= \lim_{s \to 0} sT(s)\left(\frac{1\text{ m}}{s}\right) = \left(\frac{\omega_n^2}{\omega_n^2}\right)(1\text{ m})$$

$$= 1.0\text{ m}$$

From Eq. 60.7, the fraction of overshoot is

$$M_p = \exp\left(\frac{-\pi\zeta}{\sqrt{1 - \zeta^2}}\right) = \exp\left(\frac{-\pi(0.50)}{\sqrt{1 - (0.50)^2}}\right)$$

$$= 0.16$$

The peak is $x_{\text{max}} = 1.16$ m.

The damped natural frequency is

$$\omega_d = \omega_n\sqrt{1 - \zeta^2} = \left(19.7\,\frac{\text{rad}}{\text{s}}\right)\sqrt{1 - (0.50)^2}$$

$$= 17.1\text{ rad/s}$$

The peak time from Eq. 60.6 is

$$t_p = \frac{\pi}{\omega_d} = \frac{\pi}{17.1 \ \frac{\text{rad}}{\text{s}}} = 0.18 \text{ s}$$

The 5% criterion settling time from Eq. 60.9 is

$$t_s = \frac{3.00}{\zeta \omega_n} = \frac{3.00}{(0.50)\left(19.7 \ \frac{\text{rad}}{\text{s}}\right)} = 0.30 \text{ s}$$

The sketch of the response to a unit step input is obtained from Fig. 60.2 and the preceding calculations.

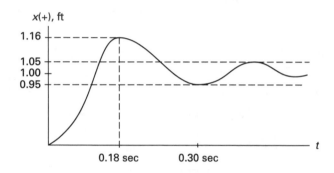

3. First, redraw the system in more traditional form.

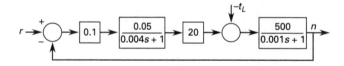

From Fig. 60.5, use case 1 to combine boxes in series.

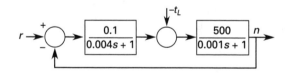

Use case 5 to combine the two summing points.

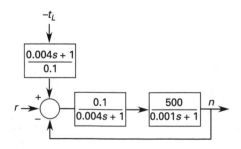

Use case 1 to combine boxes in series.

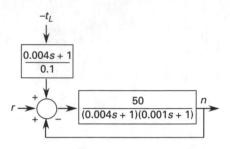

(a) Ignoring the feedback loop, the open-loop transfer function from r to n is

$$T_1(s) = \frac{N(s)}{R(s)}$$

$$= \frac{50}{(0.004s + 1)(0.001s + 1)}$$

The open-loop transfer function from $-t_L$ to n is

$$T_2(s) = \frac{N(s)}{-T_L(s)}$$

$$= \left(\frac{0.004s + 1}{0.1}\right)\left(\frac{50}{(0.004s + 1)(0.001s + 1)}\right)$$

$$= \frac{500}{0.001s + 1}$$

The open-loop frequency response for $T_1(s)$ is

$$T_1(j\omega) = |T_1(j\omega)|e^{j\phi_1(\omega)}$$

From Eq. 60.34, the gain is $20 \log |T_1(j\omega)|$ (in dB).

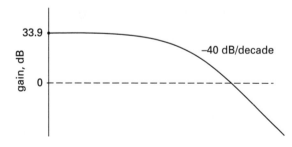

The phase is $\phi_1(\omega)$.

The open-loop frequency response for $T_2(s)$ is

$$T_2(j\omega) = |T_2(j\omega)|e^{j\phi_2(\omega)}$$

From Eq. 63.34, the gain is 20 log $|T_2(j\omega)|$ (in dB).

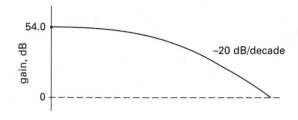

The phase is $\phi_2(\omega)$.

(b) The open-loop steady-state gain for $T_1(s)$ is

$$T_1(0) = \frac{50}{\big((0.004)(0)+1\big)\big((0.001)(0)+1\big)} = \boxed{50}$$

The open-loop steady-state gain for $T_2(s)$ is

$$T_2(0) = \frac{500}{(0.001)(0)+1} = \boxed{500}$$

(c) From Fig. 60.5, use case 3 to simplify the feedback loop.

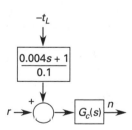

$$
\begin{aligned}
G_c(s) &= \frac{\dfrac{50}{(0.004s+1)(0.001s+1)}}{1 + \dfrac{50}{(0.004s+1)(0.001s+1)}} \\[2mm]
&= \frac{50}{(0.004s+1)(0.001s+1)+50} \\[2mm]
&= \frac{50}{(4 \times 10^{-6})s^2 + 0.005s + 51}
\end{aligned}
$$

The closed-loop transfer function from r to n is

$$
\begin{aligned}
T_1(s) &= \frac{N(s)}{R(s)} = G_c(s) \\[2mm]
&= \frac{50}{(4 \times 10^{-6})s^2 + 0.005s + 51}
\end{aligned}
$$

The closed-loop transfer function from $-t_L$ to n is

$$
\begin{aligned}
T_2(s) &= \frac{N(s)}{-T_L(s)} = \left(\frac{0.004s+1}{0.1}\right) G_c(s) \\[2mm]
&= \frac{(500)(0.004s+1)}{(4 \times 10^{-6})s^2 + 0.005s + 51}
\end{aligned}
$$

The gain for the closed-loop frequency response of $T_1(s)$ is

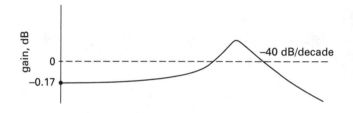

The phase for the closed-loop frequency response of $T_1(s)$ is

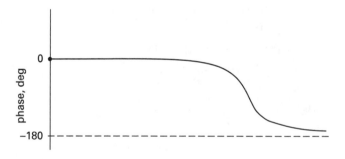

The gain for the closed-loop frequency response of $T_2(s)$ is

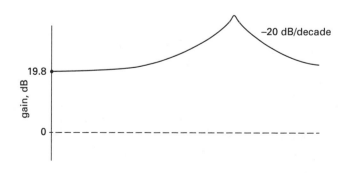

The phase for the closed-loop frequency response of $T_2(s)$ is

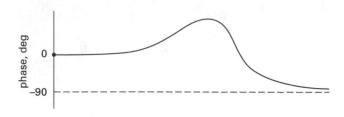

(d) The closed-loop steady-state gain for $T_1(s)$ is

$$T_1(0) = \frac{50}{(4 \times 10^{-6})(0)^2 + (0.005)(0) + 51} = \boxed{0.98}$$

The closed-loop steady-state gain for $T_2(s)$ is

$$T_2(0) = \frac{(500)\big((0.004)(0) + 1\big)}{(4 \times 10^{-6})(0)^2 + (0.005)(0) + 51} = \boxed{9.80}$$

(e) From Fig. 60.4, use Eq. 60.22 to find the system sensitivity.

$$S = \frac{1}{1 + GH} = \frac{1}{1 + \left(\dfrac{50}{(0.004s + 1)(0.001s + 1)}\right)} \quad (1)$$

$$= \frac{(0.004s + 1)(0.001s + 1)}{50 + (0.004s + 1)(0.001s + 1)}$$

$$= \frac{(0.004s + 1)(0.001s + 1)}{(4 \times 10^{-6})s^2 + 0.005s + 51}$$

The frequency response for S is

$$S(j\omega) = |S(j\omega)|e^{j\phi(\omega)}$$

From Eq. 60.34, the gain is $20 \log |S(j\omega)|$ (in dB).

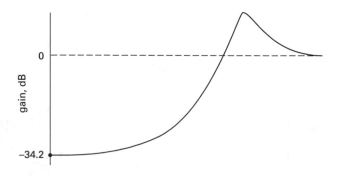

The phase is $\phi(\omega)$.

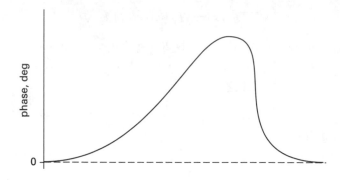

(f) From part (c), the response to a step change in input r is

$$N(s) = R(s)G_c(s)$$

$$= \left(\frac{R}{s}\right)\left(\frac{50}{4 \times 10^{-6}s^2 + 0.005s + 51}\right)$$

$$= \frac{(1.25 \times 10^7)\left(\dfrac{R}{s}\right)}{s^2 + 1250s + 1.275 \times 10^7}$$

Equate the denominator of $N(s)$ to the standard second-order form $s^2 + 2\zeta\omega_n s + \omega_n^2$.

$$\omega_n = \sqrt{1.275 \times 10^7 \ \frac{\text{rad}^2}{\text{sec}^2}} = 3570.7 \ \text{rad/sec}$$

Set $2\zeta\omega_n$ equal to 1250 and solve for ζ.

$$\zeta = \frac{1250 \ \dfrac{\text{rad}}{\text{sec}}}{2\omega_n} = \frac{1250 \ \dfrac{\text{rad}}{\text{sec}}}{(2)\left(3570.7 \ \dfrac{\text{rad}}{\text{sec}}\right)} = 0.175$$

Since ζ is less than one, the response is oscillatory.

Use Eq. 60.7 to find the fraction overshoot.

$$M_p = \exp\left(\frac{-\pi\zeta}{\sqrt{1 - \zeta^2}}\right) = \exp\left(\frac{-\pi(0.175)}{\sqrt{1 - (0.175)^2}}\right)$$

$$= 0.57$$

Use Eq. 60.5 to find the 90% rise time.

$$t_r = \frac{\pi - \arccos\zeta}{\omega_d} = \frac{\pi - \arccos\zeta}{\omega_n\sqrt{1 - \zeta^2}}$$

$$= \frac{\pi - \arccos(0.175)}{\left(3570.7 \ \dfrac{\text{rad}}{\text{sec}}\right)\sqrt{1 - (0.175)^2}}$$

$$= 4.97 \times 10^{-4} \ \text{sec}$$

The response is very fast but highly oscillatory.

Use the final value theorem, Eq. 60.27, to find the steady-state value.

$$n = \lim_{s \to 0} sN(s) = \lim_{s \to 0} \left(\frac{s(1.25 \times 10^7)\left(\dfrac{R}{s}\right)}{s^2 + 1250s + 1.275 \times 10^7} \right)$$

$$= 0.98R$$

The steady-state error is

$$e = R - n = R - 0.98R = 0.02R$$

There is a steady-state error proportional to the desired output angular velocity R.

The plot for $n(t)$ is

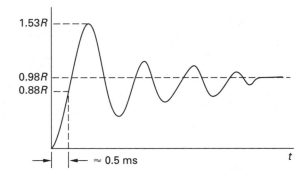

(g) From part (c), the response to a step change in input t_L is

$$N(s) = -T_L(s)\left(\frac{0.004s + 1}{0.1}\right) G_c(s)$$

$$= \left(\frac{-T_L}{s}\right)\left(\frac{(500)(0.004s + 1)}{(4 \times 10^{-6})s^2 + 0.005s + 51}\right)$$

$$= \frac{-(1.25 \times 10^8)\left(\dfrac{T_L}{s}\right)(0.004s + 1)}{s^2 + 1250s + 1.275 \times 10^7}$$

Since the denominator is the same as that for $N(s)$ in Sol. 3.6, the response to a step change in t_L will be similar to a step change in r. However, the numerator, $0.004s + 1$, will cause the response to deviate from second order. The response will still be oscillatory and very fast.

Use the final value theorem, Eq. 60.27, to find the steady-state error.

$$n = \lim_{s \to 0} sN(s)$$

$$= \lim_{s \to 0} \left(\frac{-s(1.25 \times 10^8)\left(\dfrac{T_L}{s}\right)(0.004s + 1)}{s^2 + 1250s + 1.27 \times 10^7} \right)$$

$$= -9.84T_L$$

The new steady-state error for both r and t_L inputs is

$$e = R - n = R - (0.98R - 9.84T_L) = 0.02R + 9.84T_L$$

Thus the load torque, $-t_L$, contributes to the steady-state error.

The response is oscillatory. The plot of the response due to a step change in t_L is

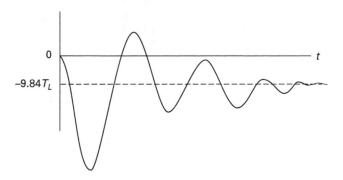

(h) Replace the comparator gain of 0.1 with K. The reduced system is

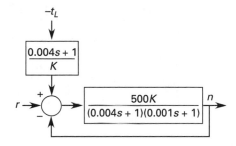

The closed-loop transfer function from r to n is

$$T_1(s) = \frac{N(s)}{R(s)} = \frac{\dfrac{500K}{(0.004s + 1)(0.001s + 1)}}{1 + \dfrac{500K}{(0.004s + 1)(0.001s + 1)}}$$

$$= \frac{500K}{(4 \times 10^{-6})s^2 + 0.005s + 1 + 500K}$$

The Routh-Hurwitz table is

$$\begin{bmatrix} a_0 & a_2 \\ a_1 & a_3 \\ b_1 & b_2 \end{bmatrix} = \begin{bmatrix} 4 \times 10^{-6} & 1 + 500K \\ 0.005 & 0 \\ b_1 & b_2 \end{bmatrix}$$

Use Eq. 60.40.

$$b_1 = \frac{a_1 a_2 - a_0 a_3}{a_1}$$

$$= \frac{(0.005)(1 + 500K) - (4 \times 10^{-6})(0)}{0.005}$$

$$= 1 + 500K$$

For a stable system, there can be no sign changes in the first column of the table.

Thus,

$$1 + 500K > 0$$

$$K > -0.002$$

(i) The closed-loop system steady-state response can be improved by adding integral control. This will effectively compensate for any steady-state disturbances due to t_L and will provide a zero steady-state error for a step input for r. This addition has a side effect of reducing the stability margin of the system. However, if properly designed, the system will still be stable.

61 Management Science

PRACTICE PROBLEMS

1. *(Time limit: one hour)* Printed circuit boards are manufactured in four consecutive departmental operations. Each operation occurs on a different machine. Employees in all departments work from 8:00 a.m. to 5:00 p.m. and have one hour total for lunch and personal breaks. The target rate is 900,000 completed units per year. Units found to be defective are discarded. No units are produced during set-up, downtime, maintenance, or record-keeping periods.

	department			
	1	2	3	4
production time (sec/unit)	6	10	11	45
set-up time (min/day)	16	8	20	5
downtime (min/day)	12	10	15	0
maintenance time (min/day)	8	12	8	0
record-keeping (min/day)	6	6	6	30
percentage defects	4%	6%	3%	2%

(a) What is the maximum number of completed circuit boards that can be produced in one year if there is one machine per department?

(b) If additional machines can be added for any or all operations, what is the most efficient method of meeting the target production rate? (There are no changes in the defect rates.)

(c) What is the efficiency of each department if the capacity is increased per part (b)?

2. *(Time limit: one hour)* Four workers perform operations 1, 2, 3, and 4 in sequence on a manual assembly line. Each station performs its operation only once on the product before sending the product on to the next operation. Operation times at the stations are as given. (Travel times are included in the operation times.)

station	time (min)
1	0.6
2	0.6
3	0.9
4	0.8

A fifth "floating" station has the ability to assist any of the four stations. The fifth station works with the same efficiencies and times as the four stations. There is no fixed assignment for this fifth station. The fifth station is allowed to help any station that needs it.

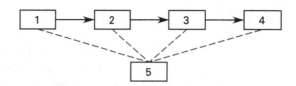

The operators of all five stations are permitted a 10 minute break each hour.

(a) What is the maximum number of products that can be produced assuming that the fifth station is assigned to work optimally? Neglect the initial (transient) performance.

(b) What is the fraction of station 5's time allocated to each operation?

3. The activities that constitute a project are listed below. The project starts at $t = 0$.

activity	predecessors	successors	duration
start	–	A	0
A	start	B,C,D	7
B	A	G	6
C	A	E,F	5
D	A	G	2
E	C	H	13
F	C	H,I	4
G	D,B	I	18
H	E,F	finish	7
I	F,G	finish	5
finish	H,I		0

(a) Draw the critical path network.

(b) Indicate the critical path.

(c) What is the earliest finish?

(d) What is the latest finish?

(e) What is the slack along the critical path?

(f) What is the float along the critical path?

4. PERT activities constituting a short project are listed with their characteristic completion times. If the project starts at $t = 15$, what is the probability that the project will be completed on or before $t = 42$?

activity	predecessors	successors	t_{min}	t_{likely}	t_{max}
start	–	A	0	0	0
A	start	B,D	1	2	5
B	A	C	7	9	20
C	B	D	5	12	18
D	A,C	finish	2	4	7
finish	D	–	0	0	0

    (A) 18%

    (B) 29%

    (C) 56%

    (D) 71%

5. *(Time limit: one hour)* Your manufacturing facility produces two models of municipal transit buses, designated B-1 and B-2. You can sell as many of either model as you produce. The per-bus profits are $800,000 for model B-1 and $650,000 for model B-2. You would like to maximize your company's profit by determining the number of each model buses to produce. However, it is not a a simple matter of producing only B-1 models because certain common parts are in limited supply.

part	number available
gage sending units	2000
wheel housing flares	1800
intake grilles	3600

Due to differences in design, the quantity of each common part varies between the two models.

part	number in model B-1	number in model B-2
gage sending units	8	10
wheel housing flares	6	4
intake grilles	3	2

6. A small company makes two chemicals, designated C-1 and C-2. The process for both includes fermentation and purification. Labor limits all fermentation operations to 300 hr per month, and purification is limited to 120 hr per month. Each unit of product C-1 requires 10 hr for fermentation and 8 hr for purification. Each unit of product C-2 requires 20 hr for fermentation and 3 hr for purification. The profit per unit of product C-1 is $3000; the profit per unit of product C-2 is $5000.

(a) How much should the company make of each chemical per month if partial units are permitted?

(b) How much should the company make of each chemical per month if only whole units are permitted?

7. A linear wage incentive program provides for 50% participation and a bonus that begins at a productivity level of 66.7% of standard. A worker produces 1900 units in 8 hr. The standard time for each unit is 0.004 hr. What is the worker's relative earnings?

    (A) 14%

    (B) 21%

    (C) 120%

    (D) 160%

SOLUTIONS

1. Assume the following.

$$5 \; \frac{\text{work days}}{\text{week}}; \; 52 \; \frac{\text{weeks}}{\text{year}}$$

$$\left(5 \; \frac{\text{work days}}{\text{week}}\right)\left(52 \; \frac{\text{weeks}}{\text{year}}\right)$$

$$= 260 \text{ work days/year}$$

Let t equal production time, in minutes/unit, for 1 unit.

Let m equal the minutes/day available for production.

Note:

$$\left(8 \; \frac{\text{hours}}{\text{day}}\right)\left(60 \; \frac{\text{minutes}}{\text{hour}}\right) = 480 \text{ minutes/day}$$

$$m = 480 \; \frac{\text{minutes}}{\text{day}} - \left(\frac{\text{setup time}}{\text{day}} + \frac{\text{downtime}}{\text{day}}\right.$$
$$\left. + \frac{\text{maintenance time}}{\text{day}} + \frac{\text{recordkeeping time}}{\text{day}}\right)$$

$$\left(900{,}000 \; \frac{\text{units}}{\text{year}}\right)\left(\frac{1 \text{ year}}{260 \text{ work days}}\right)$$

$$= 3462 \text{ units/work day}$$

$n =$ number of machines needed to reach a production rate of 3462 units/day

$e =$ efficiency of each department

$$= \frac{\begin{array}{c}\text{number of units} \\ \text{produced in each department}\end{array}}{\begin{array}{c}\text{maximum possible number of units} \\ \text{produced in each department}\end{array}}$$

$p =$ percentage of defective units

$\left(\dfrac{m}{t}\right)(1 - p) =$ maximum production in each department using one machine

Based on the preceding definitions,

$$n = \frac{3462}{\left(\dfrac{m}{t}\right)(1-p)} + 1 \quad \text{[greatest integer function]}$$

$$e = \frac{3462}{n\left(\dfrac{m}{t}\right)(1-p)}$$

dept	t	m	$\left(\dfrac{m}{t}\right)(1-p)$	n	e
1	6/60	438	4204	1	82.4%
2	10/60	444	2504	2	69.1%
3	11/60	431	2280	2	75.9%
4	45/60	445	581	6	99.3%

(a) Since each circuit board must go through all four departments, maximum production is 581 units/day.

$$\left(581 \; \frac{\text{units}}{\text{day}}\right)\left(260 \; \frac{\text{work days}}{\text{year}}\right)$$

$$= \boxed{151{,}060 \text{ units/year}}$$

(b) The values of n are computed in the chart.

(c) The values of e are computed in the chart.

2. (b) For the time being, disregard the 10 min per hour shift break since this break reduces the capacity of all stations by the same percentage.

Determine the maximum output per hour for each station.

station	output
1	$\dfrac{60 \; \frac{\text{min}}{\text{hr}}}{0.6 \; \frac{\text{min}}{\text{unit}}} = 100 \text{ units/hr}$
2	$\dfrac{60 \; \frac{\text{min}}{\text{hr}}}{0.6 \; \frac{\text{min}}{\text{unit}}} = 100 \text{ units/hr}$
3	$\dfrac{60 \; \frac{\text{min}}{\text{hr}}}{0.9 \; \frac{\text{min}}{\text{unit}}} = 66.67 \text{ units/hr}$
4	$\dfrac{60 \; \frac{\text{min}}{\text{hr}}}{0.8 \; \frac{\text{min}}{\text{unit}}} = 75 \text{ units/hr}$

Stations 3 and 4 are the bottleneck operations. Intuitively, we would want to help operation 3 the most, followed by helping operation 4.

Start by allocating station 5 capacity to the slowest operation, operation 3. Try to bring station 3 up to the same capacity as stations 1 and 2. To do so requires station 5 to produce $100 - 66.67 = 33.33$ units per hour. Since station 5 works at the same speed as station 3, the fraction of time station 5 needs to assist station 3 is $33.33/66.67 = 0.5$ (50%). This leaves 50% of station 5's time available to assist other stations.

Next, allocate the remaining station 5 time to station 4. To bring station 4 up to 100 units per hour will require station 5 to produce $100 - 75 = 25$ units per hour. Since station 5 works at the same rate as station 4, the fraction of time station 5 needs to assist station 4 is $25/75 = 0.3333$ (33.33%).

So, we have brought all of the stations up to 100 units per hour. Station 5 still has some remaining capacity: $100\% - 50\% - 33.33\% = 16.67\%$. This remaining capacity needs to be allocated to all of the remaining stations to bring them all up to the same output rate.

Suppose we want to raise the output of the assembly line by 1 unit (i.e., from 100 to 101 units/hr). How much time would this take? It would take 0.6 min for operation 1, 0.6 min for operation 2, 0.9 min for operation 3, and 0.8 min for operation 4. Suppose we want to raise the output by 2 units/hr. That would take 1.2 min for operation 1, 1.2 min for operation 2, 1.8 min for operation 3, and 1.6 min for operation 4. Notice that the ratios of times between stations remain the same. The additional time for operation 2, for example, is always the same as for operation 1.

All of the extra time must come from the remaining capacity of station 5, since all other stations are working at their individual capacities. Station 5 has 17% of its time left, and it must allocate its time in the same fraction (ratio) as the assembly times.

operation	time	fraction of total	ratio × 17%
1	0.6	0.2069	3.52%
2	0.6	0.2069	3.52%
3	0.9	0.3103	5.27%
4	0.8	0.2759	4.69%
totals	2.9	1.0000	17.00%

Therefore, 3.52% of station 5's time will be given to operations 1 and 2. Operation 3 will receive $50\% + 5.27\% = 55.27\%$ of station 5's time. Operation 4 will receive $33.33\% + 4.69\% = 38.02\%$.

(a) The production rates of each of the operations are as follows.

operations 1 and 2:
$$\frac{(1.0352)\left(60\ \dfrac{\text{min}}{\text{hr}}\right)}{0.6\ \dfrac{\text{min}}{\text{unit}}} = 103.5\ \text{units/hr}$$

operation 3:
$$\frac{(1.5527)\left(60\ \dfrac{\text{min}}{\text{hr}}\right)}{0.9\ \dfrac{\text{min}}{\text{unit}}} = 103.5\ \text{units/hr}$$

operation 4:
$$\frac{(1.3802)\left(60\ \dfrac{\text{min}}{\text{hr}}\right)}{0.8\ \dfrac{\text{min}}{\text{unit}}} = 103.5\ \text{units/hr}$$

So, the capacity of the assembly line is 103 units per hour.

(Check: The total time required per product is 2.9 minutes. With all five 5 stations working optimally, the total available time per hour is (5 stations)(60 min/station) = 300 min. The optimal production rate would be (300 min/hr)/(2.9 min/unit) = 103.5 units/hr. This checks.)

However, everybody takes a 10 min break per hour, and the stations only work 50 min per hour. So, the overall capacity is reduced proportionally.

$$\text{capacity} = \left(103.5\ \frac{\text{units}}{\text{hr}}\right)\left(\frac{50\ \text{min}}{60\ \text{min}}\right)$$

$$= \boxed{86.3\ \text{units/hr}}$$

3. (a) The critical path network diagram is as follows.

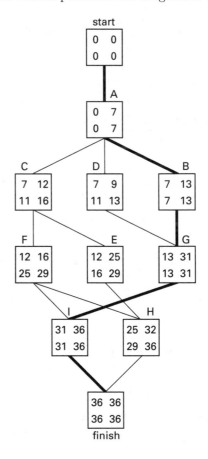

ES (earliest start) Rule: The earliest start time for an activity leaving a particular node is equal to the largest of the earliest finish times for all activities entering the node.

LF (latest finish) Rule: The latest finish time for an activity entering a particular node is equal to the smallest of the latest start times for all activities leaving the node.

The activity is critical if the earliest start equals the latest start.

(b) The critical path is $\boxed{\text{A-B-G-I.}}$

(c) The earliest finish is $\boxed{36.}$

(d) The latest finish is $\boxed{36.}$

(e) The slack along the critical path is $\boxed{0.}$

(f) The float along the critical path is $\boxed{0.}$

4. From Eq. 61.1,

$$\mu = \frac{t_{\text{minimum}} + 4t_{\text{likely}} + t_{\text{maximum}}}{6}$$

From Eq. 61.2,

$$\sigma^2 = \left(\frac{t_{\text{maximum}} - t_{\text{minimum}}}{6} \right)^2$$

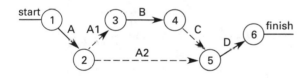

The critical path is A-A1-B-C-D.

The following probability calculations assume that all activities are independent. Use the following theorems for the sum of independent random variables and use the normal distribution for T (project time).

$$\mu_{\text{total}} = t_A + t_B + t_C + t_D$$
$$\sigma^2_{\text{total}} = \sigma^2_A + \sigma^2_B + \sigma^2_C + \sigma^2_D$$

The variance is 10.52778 and the standard deviation is 3.244654.

$$\mu_{\text{total}} = 43.83333$$
$$\sigma^2_{\text{total}} = 10.52778$$
$$\sigma_{\text{total}} = 3.244654$$
$$z = \frac{t - \mu_{\text{total}}}{\sigma} = \left| \frac{42 - 43.83333}{3.244654} \right| = 0.565$$

The probability of finishing for $T \leq 42$ is 0.286037 $\boxed{(28.6\%).}$

The answer is (B).

5. This is a two-dimensional linear programming problem.

$$x_1 = \text{no. of B-1 buses produced}$$
$$x_2 = \text{no. of B-2 buses produced}$$
$$Z = \text{total profit}$$

PERT Analysis for Prob. 4

no.	name	activity exp. time	variance	earliest start	latest start	earliest finish	latest finish	slack LS-ES
1	A	+2.33333	+0.44444	15	15	+17.3333	+17.3333	0
2	A1	0	0	+17.3333	+17.3333	+17.3333	+17.3333	0
3	A2	0	0	+17.3333	+39.6667	+17.3333	+39.6667	+22.3333
4	B	+10.5000	+4.69444	+17.3333	+17.3333	+27.8333	+27.8333	0
5	C	+11.8333	+4.69444	+27.8333	+27.8333	+39.6667	+39.6667	0
6	D	+4.16667	+0.69444	+39.6667	+39.6667	+43.8333	+43.8333	0

expected completion time = 43.83333

The objective function is

$$\text{maximize } Z = 800{,}000x_1 + 650{,}000x_2$$

The constraints are

$$8x_1 + 10x_2 \leq 2000$$
$$6x_1 + 4x_2 \leq 1800$$
$$3x_1 + 2x_2 \leq 3600$$
$$x_1 \geq 0, x_2 \geq 0$$

The simplex theory states that the optimal solution will be a feasible corner point.

feasible corner points (x_1, x_2)	Z
$(0, 0)$	0
$(0, 200)$	$130{,}000{,}000$
optimal solution $\rightarrow (250, 0)$	$200{,}000{,}000$

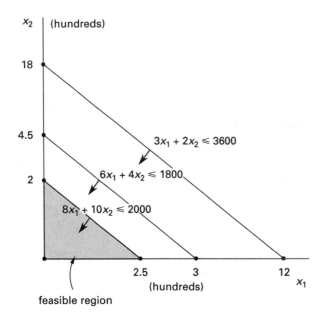

feasible region

6. (a) Solve as a linear programming problem, since partial units are permitted.

$$x_1 = \text{no. of units of C-1 produced/month}$$
$$x_2 = \text{no. of units of C-2 produced/month}$$
$$Z = \text{total profit}$$

The objective function is

$$\text{maximize } Z = 3000x_1 + 5000x_2$$

The constraints are

$$10x_1 + 20x_2 \leq 300$$
$$8x_1 + 3x_2 \leq 120$$
$$x_1 \geq 0, x_2 \geq 0$$

The simplex theory states that the optimal solution will be a feasible corner point.

feasible corner points (x_1, x_2)	Z
$(0, 0)$	0
$(0, 15)$	$75{,}000$
optimal solution $\rightarrow \left(\dfrac{150}{13}, \dfrac{120}{13}\right)$	$80{,}769.23$
$(15, 0)$	$45{,}000$

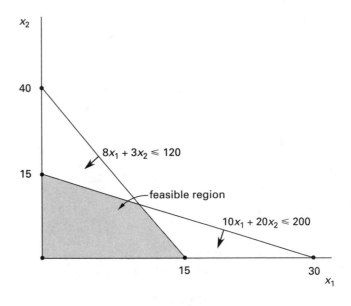

At the intersection point of the two constraints,

$$65x_2 = 600$$
$$x_2 = \frac{600}{65} = 120/13$$

Substituting x_2 into the first constraint yields

$$10x_1 + (20)\left(\frac{120}{13}\right) \leq 300$$
$$x_1 = 150/13$$
$$(x_1, x_2) = \left(\frac{150}{13}, \frac{120}{13}\right)$$

(b) Integer programming methods are required when variables are constrained to integer values. Merely deleting the fractional parts in the solution doesn't always work. In this case, deleting the fractional parts yields

$$x_1 = \text{INT}\left(\frac{150}{13}\right) = 11$$
$$x_2 = \text{INT}\left(\frac{120}{13}\right) = 9$$
$$Z = 3000x_1 + 5000x_2 = 78{,}000$$

However, $x_1 = x_2 = 10$ also satisfies the constraints and yields the optimum $Z = 80,000$. In the absence of integer programming tools, trial and error in the vicinity of the feasible corners is required.

7.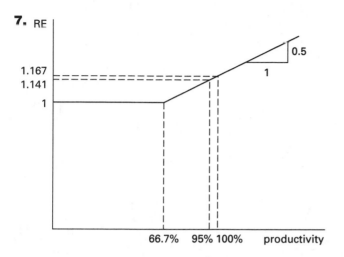

This is a linear gain sharing plan because the participation is 50% (i.e., less than 100%).

$$\text{participation} = 0.5$$

The productivity is the number of units produced in an hour divided by the number of standard units.

$$\text{productivity} = \frac{\text{actual production}}{\text{ideal production}} = \frac{\dfrac{1900 \text{ units}}{8 \text{ hr}}}{0.004 \dfrac{\text{units}}{\text{hr}}}$$

$$= 0.95$$

In this plan, relative earnings are 1.0 until productivity is 0.667 (i.e., 66.7%). Since the productivity of 95% is greater than 66.7%, the employee will be paid a bonus.

The bonus factor is determined from Eq. 61.75.

$$\text{bonus factor} = \frac{1}{\text{minimum bonus productivity}} - 1$$
$$= \frac{1}{0.667} - 1$$
$$= 0.50$$

The equation that describes the relative earnings, RE, is

$$\text{RE} = 1 - \text{participation} + (\text{productivity})(\text{participation})$$
$$\times (1 + \text{bonus factor})$$
$$= 1 - 0.5 + (0.95)(0.50)(1 + 0.50)$$
$$= 1.2125$$

The answer is (C).

62 Instrumentation and Measurements

PRACTICE PROBLEMS

Chapter 62 of the *Mechanical Engineering Reference Manual* contains no practice problems.

63 Manufacturing Processes

PRACTICE PROBLEMS

1. A 10 mil (250 μm) adhesive with a shear strength of 1500 lbf/in^2 (10 MPa) is used in a lap joint between two 0.20 in (5 mm) aluminum sheets. The aluminum has a yield strength of 15,000 lbf/in^2 (100 MPa). Assume the adhesive is loaded in pure shear, but use a stress concentration factor of 2. If the joint is to be as strong as the aluminum, what width (overlap) of adhesive joint is required?

(A) 2.3 in (0.058 m)

(B) 4.0 in (0.10 m)

(C) 5.3 in (0.13 m)

(D) 6.9 in (0.17 m)

SOLUTIONS

1.

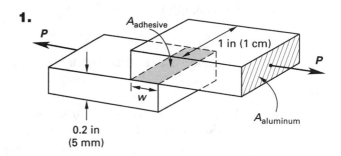

Customary U.S. Solution

step 1: Assume the unit length ($L = 1$ in) for the joint.

$$A_{\text{adhesive}} = wL = w_{\text{in}}(1 \text{ in}) = w$$
$$A_{\text{aluminum}} = Lt = (1 \text{ in})(0.2 \text{ in}) = 0.2 \text{ in}^2$$

step 2: Find the tensile force P as follows.

The yield strength is

$$P = S_y A_{\text{aluminum}}$$
$$= \left(15{,}000 \ \frac{\text{lbf}}{\text{in}^2}\right)(0.2 \text{ in}^2)$$
$$= 3000 \text{ lbf}$$

step 3: Find the required width, w, with a stress concentration factor of 2.

The shear stress is

$$\tau = \frac{2P}{A_{\text{adhesive}}}$$
$$1500 \ \frac{\text{lbf}}{\text{in}^2} = \frac{(2)(3{,}000 \text{ lbf})}{w}$$
$$w = \boxed{4 \text{ in}}$$

The answer is (B).

SI Solution

step 1: Assume the unit length ($L = 1$ cm) for the joint.

$$A_{\text{adhesive}} = wL = w_{\text{in}}(0.01 \text{ m}) = 0.01w$$
$$A_{\text{aluminum}} = Lt = (0.01 \text{ m})(0.005 \text{ m})$$
$$= 0.00005 \text{ m}^2$$

step 2: Find the tensile force P as follows.

The yield strength is

$$P = S_y A_{\text{aluminum}}$$
$$= (100 \times 10^6 \text{ Pa})(0.00005 \text{ m}^2)$$
$$= 5000 \text{ N}$$

step 3: Find the required width, w, with a stress concentration factor of 2.

The shear stress is

$$\tau = \frac{2P}{A_{\text{adhesive}}}$$
$$10 \times 10^6 \text{ Pa} = \frac{(2)(5000 \text{ N})}{0.01w}$$
$$w = \boxed{0.1 \text{ m}}$$

The answer is (B).

64 Materials Handling and Processing

PRACTICE PROBLEMS

Chapter 64 of the *Mechanical Engineering Reference Manual* contains no practice problems.

65 Fire Protection Systems

PRACTICE PROBLEMS

Chapter 65 of the *Mechanical Engineering Reference Manual* contains no practice problems.

66

Environmental Engineering

Chapter 66 of the *Mechanical Engineering Reference Manual* contains no practice problems.

67 Electricity and Electrical Equipment

PRACTICE PROBLEMS

Induction Motors

1. Calculate the full-load phase current drawn by a 440 V (rms) 20-hp (per phase) induction motor having a full-load efficiency of 86% and a full-load power factor of 76%.

2. A 200-hp, three-phase, four-pole, 60 Hz, 440 V (rms) squirrel-cage induction motor operates at full load with an efficiency of 85%, power factor of 91%, and 3% slip.

(a) Find the speed in rpm.

(b) Find the torque developed.

(c) Find the line current.

3. A factory's induction motor load draws 550 kW at 82% power factor. What size synchronous motor is required to carry 250 hp and raise the power factor to 95%? The line voltage is 220 V (rms).

4. The nameplate of an induction motor lists 960 rpm as the full-load speed. For what frequency was the motor designed?

SOLUTIONS

1. $I_{\text{phase}} = \dfrac{P_{\text{phase}}}{\eta V \cos \phi}$

$= \dfrac{(20 \text{ hp}) \left(0.7457 \times 10^3 \, \dfrac{\text{W}}{\text{hp}} \right)}{(0.86)(440 \text{ V})(0.76)} = \boxed{51.86 \text{ A}}$

2. (a) $n_r = \left(\dfrac{(2) \left(60 \, \dfrac{\text{sec}}{\text{min}} \right)}{p} \right) (1 - s)$

$= \left(\dfrac{(2) \left(60 \, \dfrac{\text{sec}}{\text{min}} \right) (60 \text{ Hz})}{4} \right) (1 - 0.03)$

$= \boxed{1746 \text{ rpm}}$

(b) $T = \dfrac{P}{\omega} = \dfrac{(200 \text{ hp}) \left(550 \, \dfrac{\text{ft-lbf}}{\text{hp-sec}} \right)}{2\pi \left(1746 \, \dfrac{\text{rev}}{\text{min}} \right) \left(\dfrac{\text{min}}{60 \text{ sec}} \right)}$

$= \boxed{602 \text{ ft-lbf}}$

(c) $I_l = \left(\dfrac{1}{\sqrt{3}} \right) \left(\dfrac{P}{\eta V_l \cos \phi} \right)$

$= \left(\dfrac{1}{\sqrt{3}} \right) \left(\dfrac{(200 \text{ hp}) \left(0.7457 \times 10^3 \, \dfrac{\text{W}}{\text{hp}} \right)}{(0.85)(440 \text{ V})(0.91)} \right)$

$= \boxed{253 \text{ A}}$

3.

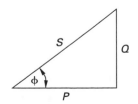

The original power angle is

$\phi_i = \arccos(0.82) = 34.92°$

$P_1 = 550 \text{ kW}$

$Q_i = P_1 \tan \phi_i = (550 \text{ kW})(\tan 34.92°) = 384.0 \text{ kVAR}$

The new conditions are

$$P_2 = (250 \text{ hp})\left(0.7457 \frac{\text{kW}}{\text{hp}}\right) = 186.4 \text{ kW}$$

$$\phi_f = \arccos(0.95) = 18.19°$$

Since both motors perform real work,

$P_f = P_1 + P_2 = 550 \text{ kW} + 186.4 \text{ kW} = 736.4 \text{ kW}$

The new reactive power is

$$Q_f = P_f \tan \phi_f = (736.4 \text{ kW})(\tan 18.19°)$$
$$= 242.0 \text{ kVAR}$$

The change in reactive power is

$$\Delta Q = 384.0 \text{ kVAR} - 242.0 \text{ kVAR} = 142 \text{ kVAR}$$

Synchronous motors used for power factor correction are rated by apparent power.

$S = \sqrt{(\Delta P)^2 + (\Delta Q)^2}$

$= \sqrt{(186.4 \text{ kW})^2 + (142 \text{ kVAR})^2} = \boxed{234.3 \text{ kVA}}$

4. $f = \dfrac{pn_s}{(20)\left(60 \frac{\text{sec}}{\text{min}}\right)} = \dfrac{pn}{(2)\left(60 \frac{\text{sec}}{\text{min}}\right)(1-s)}$

The slip and number of poles are unknown. Assume $s = 0$ and $p = 4$.

$$f = \frac{(4)\left(960 \frac{\text{rev}}{\text{min}}\right)}{(2)\left(60 \frac{\text{sec}}{\text{min}}\right)} = 32 \text{ Hz}$$

This is not close to anything in commercial use. Try $p = 6$.

$$f = \frac{(6)\left(960 \frac{\text{rev}}{\text{min}}\right)}{(2)\left(60 \frac{\text{sec}}{\text{min}}\right)} = 48 \text{ Hz}$$

With a 4% slip, $f = 50 \text{ Hz}$.

$\boxed{50 \text{ Hz (European)}}$

68 Illumination and Sound

PRACTICE PROBLEMS

1. What is the total sound pressure level when a 40 dB source is placed adjacent to a 35 dB source?

(A) 17 dB
(B) 28 dB
(C) 37 dB
(D) 41 dB

2. With no machinery operating, the background noise in a room has a sound pressure level of 43 dB. With the machinery operating, the sound pressure level is 45 dB. What is the sound pressure level due to the machinery alone?

(A) 31 dB
(B) 41 dB
(C) 47 dB
(D) 49 dB

3. An unenclosed source produces a sound pressure level of 100 dB. An enclosure is constructed from a material having a transmission loss of 30 dB. The sound pressure level inside the enclosure from the enclosed source increases to 110 dB. What is the reduction in sound pressure level outside the enclosure?

(A) 20 dB
(B) 30 dB
(C) 80 dB
(D) 90 dB

4. 4 ft (1.2 m) from an isotropic sound source, the sound pressure level is 92 dB. What is the sound pressure level 12 ft (3.6 m) from the source?

(A) 62 dB
(B) 73 dB
(C) 83 dB
(D) 87 dB

5. Octave band measurements of a noise source were made. The measurements were 85, 90, 92, 87, 82, 78, 65, and 54 dB at frequencies of 63, 125, 250, 500, 1000, 2000, 4000, and 8000 Hz, respectively. What is the overall A-weighted sound pressure?

(A) 78 dB
(B) 81 dB
(C) 85 dB
(D) 89 dB

6. What is the maximum possible reduction in sound pressure level if the number of sabins is 50% of the total room area?

(A) 1.2 dB decrease
(B) 3.0 dB decrease
(C) 4.6 dB decrease
(D) 6.1 dB decrease

7. A storage room has dimensions of 100 ft × 400 ft × 20 ft (30 m × 120 m × 6 m). All surfaces are plain concrete. 40% of the walls are treated acoustically with a material having a sound absorption coefficient of 0.8. What is the reduction in sound pressure level?

(A) 1.2 dB decrease
(B) 3.0 dB decrease
(C) 4.6 dB decrease
(D) 6.1 dB decrease

8. A meeting room has dimensions of 20 ft × 50 ft × 10 ft (6 m × 15 m × 3 m). The floor is covered with roll vinyl. The walls and ceiling are sheetrock. 20% of the walls are glass windows. There are 15 seats, lightly upholstered, with 15 occupants, and 5 miscellaneous sabins. After complaints from the occupants, the ceiling is treated with a sound absorbing material with a sound absorption coefficient of 0.7. What is the reduction in sound level?

(A) 3.5 dB decrease
(B) 5.7 dB decrease
(C) 6.3 dB decrease
(D) 12 dB decrease

9. A room has dimensions of 15 ft × 20 ft × 10 ft (4.5 m × 6 m × 3 m). The sound absorption coefficients are 0.03, 0.5, and 0.06 for the floor, ceiling, and walls, respectively. A machine with a sound power level of 65 dB is located at the intersection of the floor and wall, 7.5 ft (2.25 m) from the perpendicular walls. The ambient sound pressure level is 50 dB everywhere in the room. What is the sound pressure level 5 ft (1.5 m) from the machine?

(A) 45 dB
(B) 61 dB
(C) 73 dB
(D) 81 dB

SOLUTIONS

1. Use Eq. 68.16.

$$L = 10 \log \sum 10^{\frac{W_i}{10}} = 10 \log \left(10^{\frac{40}{10}} + 10^{\frac{35}{10}} \right)$$

$$= \boxed{41.2 \text{ dB}}$$

The answer is (D).

2. Use Eq. 68.16.

$$L = 10 \log \sum 10^{\frac{W_i}{10}} = 10 \log \left(10^{\frac{43}{10}} + 10^{\frac{W}{10}} \right) = 45 \text{ dB}$$

Solve for the unknown machinery sound pressure level, W.

$$10^{\frac{W}{10}} = 10^{\frac{45}{10}} - 10^{\frac{43}{10}}$$

$$W = 10 \log \left(10^{\frac{45}{10}} - 10^{\frac{43}{10}} \right) = \boxed{40.7 \text{ dB}}$$

The answer is (B).

3. Define $L_{W,1}$ as the sound pressure level inside the enclosure, and define $L_{W,2}$ as the sound pressure level outside the enclosure. Use Eq. 68.23.

$$L_{W,2} = L_{W,1} - \text{TL} = 110 \text{ dB} - 30 \text{ dB} = 80 \text{ dB}$$

Define $L_{W,1}$ as the sound pressure level for the unenclosed source, and define $L_{W,2}$ as the sound pressure level for the enclosed source.

Use Eq. 68.22 to solve for the insertion loss.

$$\text{IL} = 100 \text{ dB} - 80 \text{ dB} = \boxed{20 \text{ dB}}$$

The answer is (A).

4. The free-field sound pressure is inversely proportional to the square of the distance from the source.

$$\frac{p_2}{p_1} = \left(\frac{r_1}{r_2} \right)^2$$

From Eq. 68.13,

$$L_{p,2} = L_{p,1} + 10 \log \left(\frac{p_1}{p_2} \right)^2$$

$$= L_{p,1} + 20 \log \left(\frac{p_1}{p_2} \right) = L_{p,1} + 20 \log \left(\frac{r_1}{r_2} \right)^2$$

Customary U.S. Solution

$$L_{p,2} = 92 \text{ dB} + 20 \log \left(\frac{4 \text{ ft}}{12 \text{ ft}} \right)^2$$

$$= 92 \text{ dB} - 19.1 \text{ dB}$$

$$= \boxed{72.9 \text{ dB}}$$

The answer is (B).

SI Solution

$$L_{p,2} = 92 \text{ dB} + 20 \log \left(\frac{1.2 \text{ m}}{3.6 \text{ m}} \right)^2$$

$$= 92 \text{ dB} - 19.1 \text{ dB}$$

$$= \boxed{72.9 \text{ dB}}$$

The answer is (B).

5. Add the corrections from Table 68.5 to the measurements.

frequency	measurement	correction	corrected value
63	85	−26.2	58.8
125	90	−16.1	73.9
250	92	−8.6	83.4
500	87	−3.2	83.8
1000	82	0	82.0
2000	78	+1.2	79.2
4000	65	+1.0	66.0
8000	54	−1.1	52.9

Use Eq. 68.16.

$$L = 10 \log \sum 10^{\frac{W_i}{10}}$$

$$= 10 \log \left(10^{5.88} + 10^{7.39} + 10^{8.34} + 10^{8.38} + 10^{8.2} \right.$$
$$\left. + 10^{7.92} + 10^{6.6} + 10^{5.29} \right)$$

$$= \boxed{88.6 \text{ dBA}}$$

The answer is (D).

6. Define A as the total room area.

$$\sum S_1 = 0.50A$$

The maximum number of sabins is equal to the room area.

$$\sum S_2 = A$$

Use Eq. 68.21.

$$\text{NR} = 10\log\left(\frac{\sum S_1}{\sum S_2}\right) = 10\log\left(\frac{0.50A}{A}\right)$$
$$= 10\log(0.50)$$
$$= \boxed{-3.0 \text{ dB} \quad [\text{decrease}]}$$

The answer is (B).

7. *Customary U.S. Solution*

The surface area of the room walls is

$$A_1 = ((2)(100\text{ ft}) + (2)(400\text{ ft}))(20\text{ ft}) = 20{,}000\text{ ft}^2$$

The surface area of the room floor and ceiling is

$$A_2 = (2)(100\text{ ft})(400\text{ ft}) = 80{,}000\text{ ft}^2$$

The sound absorption coefficient of concrete is the NRC value of 0.02 from App. 68.A.

Define the sound absorption coefficient of precast concrete as α_{concrete}.

The sabin area of the room with all precast concrete is

$$\sum S_2 = \alpha_{\text{concrete}}(A_1 + A_2)$$
$$= (0.02)(20{,}000\text{ ft}^2 + 80{,}000\text{ ft}^2)$$
$$= 2000\text{ ft}^2$$

Define the sound absorption coefficient of the wall acoustical treatment as α_{wall}.

The sabin area of the room with 40% of the walls treated with $\alpha_{\text{wall}} = 0.8$ is

$$\sum S_1 = \alpha_{\text{concrete}}A_2 + \alpha_{\text{concrete}}(0.6A_1) + \alpha_{\text{wall}}(0.4A_1)$$
$$= (0.02)(80{,}000\text{ ft}^2) + (0.02)(0.6)(20{,}000\text{ ft}^2)$$
$$+ (0.8)(0.4)(20{,}000\text{ ft}^2)$$
$$= 8240\text{ ft}^2$$

Use Eq. 68.21.

$$\text{NR} = 10\log\left(\frac{\sum S_1}{\sum S_2}\right)$$
$$= 10\log\left(\frac{8240\text{ ft}^2}{2000\text{ ft}^2}\right) = \boxed{6.1\text{ dB}}$$

The answer is (D).

SI Solution

The surface area of the room walls is

$$A_1 = ((2)(30\text{ m}) + (2)(120\text{ m}))(6\text{ m}) = 1800\text{ m}^2$$

The surface area of the room floor and ceiling is

$$A_2 = (2)(30\text{ m})(120\text{ m}) = 7200\text{ m}^2$$

The sound absorption coefficient of precast concrete is the NRC value of 0.02 from App. 68.A.

Define the sound absorption coefficient of precast concrete as α_{concrete}.

The sabin area of the room with all precast concrete is

$$\sum S_2 = \alpha_{\text{concrete}}(A_1 + A_2)$$
$$= (0.02)(1800\text{ m}^2 + 7200\text{ m}^2)$$
$$= 180\text{ m}^2$$

Define the sound absorption coefficient of the wall acoustical treatment as α_{wall}.

The sabin area of the room with 40% of the walls treated with $\alpha_{\text{wall}} = 0.8$ is

$$\sum S_1 = \alpha_{\text{concrete}}A_2 + \alpha_{\text{concrete}}(0.6A_1) + \alpha_{\text{wall}}(0.4A_1)$$
$$= (0.02)(7200\text{ m}^2) + (0.02)(0.6)(1800\text{ m}^2)$$
$$+ (0.8)(0.4)(1800\text{ m}^2)$$
$$= 741.6\text{ m}^2$$

Use Eq. 68.21.

$$\text{NR} = 10\log\left(\frac{\sum S_1}{\sum S_2}\right)$$
$$= 10\log\left(\frac{741.6\text{ m}^2}{180\text{ m}^2}\right) = \boxed{6.1\text{ dB}}$$

The answer is (D).

8. *Customary U.S. Solution*

The gross area of the walls is

$$A_1 = ((2)(20\text{ ft}) + (2)(50\text{ ft}))(10\text{ ft}) = 1400\text{ ft}^2$$
$$A_{\text{walls}} = (1 - 0.20)(1400\text{ ft}^2) = 1120\text{ ft}^2$$
$$A_{\text{glass}} = (0.20)(1400\text{ ft}^2) = 280\text{ ft}^2$$

From App. 68.A, the sound absorption coefficient of glass is $\alpha_1 = 0.03$.

The area of the floor is

$$A_2 = (20\text{ ft})(50\text{ ft}) = 1000\text{ ft}^2$$

From App. 68.A, the sound absorption coefficient of roll vinyl is $\alpha_2 = 0.03$.

The area of the ceiling is

$$A_3 = (20\text{ ft})(50\text{ ft}) = 1000\text{ ft}^2$$

The sound absorption coefficient of the sheetrock is $\alpha_3 = 0.05$.

From App. 68.A, the seats have approximately 1.5 sabins each and the occupants have approximately 5.0 sabins each.

The total sabin area of the untreated room is

$$\sum S_2 = \alpha_{glass}S_{glass} + \alpha_{walls}S_{walls} + \alpha_{floor}S_{floor}$$
$$+ \alpha_{ceiling}S_{ceiling} + seats + occupants$$
$$+ miscellaneous$$
$$= (0.03)(280 \text{ ft}^2) + (0.05)(1120 \text{ ft}^2)$$
$$+ (0.03)(1000 \text{ ft}^2) + (0.05)(1000 \text{ ft}^2)$$
$$+ (15)(1.5 \text{ ft}^2) + (15)(5.0 \text{ ft}^2) + 5.0 \text{ ft}^2$$
$$= 246.9 \text{ ft}^2$$

The total sabin area excluding the ceiling is

$$246.9 \text{ ft}^2 - \alpha_3 S_3 = 246.9 \text{ ft}^2 - (0.03)(1000 \text{ ft}^2)$$
$$= 216.9 \text{ ft}^2$$

The total sabin area of the room with the ceiling treated with $\alpha_3 = 0.7$ sound absorption is

$$\sum S_1 = 216.9 \text{ ft}^2 + \alpha_3 S_3$$
$$= 216.9 \text{ ft}^2 + (0.7)(1000 \text{ ft}^2)$$
$$= 916.9 \text{ ft}^2$$

Use Eq. 68.21.

$$NR = 10 \log \left(\frac{\sum S_1}{\sum S_2} \right)$$
$$= 10 \log \left(\frac{916.9 \text{ ft}^2}{246.9 \text{ ft}^2} \right) = \boxed{5.7 \text{ dB}}$$

The answer is (B).

SI Solution

The gross area of the walls is

$$A_1 = \big((2)(6 \text{ m}) + (2)(15 \text{ m})\big)(3 \text{ m}) = 126 \text{ m}^2$$
$$A_{walls} = (1 - 0.20)(126 \text{ m}^2) = 100.8 \text{ m}^2$$
$$A_{glass} = (0.20)(126 \text{ m}^2) = 25.2 \text{ m}^2$$

From App. 68.A, the sound absorption coefficient of glass is $\alpha_1 = 0.03$.

The area of the floor is

$$A_2 = (6 \text{ m})(15 \text{ m}) = 90 \text{ m}^2$$

From App. 68.A, the sound absorption coefficient of roll vinyl is $\alpha_2 = 0.03$.

The area of the ceiling is

$$A_3 = (6 \text{ m})(15 \text{ m}) = 90 \text{ m}^2$$

The sound absorption coefficient of the sheetrock is $\alpha_3 = 0.05$.

From App. 68.A, the seats have approximately 1.5 sabins each and the occupants have approximately 5.0 sabins each.

The total sabin area (m^2) of the untreated room is

$$\sum S_2 = \alpha_{glass}S_{glass} + \alpha_{walls}S_{walls} + \alpha_{floor}S_{floor}$$
$$+ \alpha_{ceiling}S_{ceiling} + seats + occupants$$
$$+ miscellaneous$$
$$= (0.03)(25.2 \text{ m}^2) + (0.05)(100.8 \text{ m}^2)$$
$$+ (0.03)(90 \text{ m}^2) + (0.05)(90 \text{ m}^2) + (15)(1.5 \text{ ft}^2)$$
$$\times \left(\frac{1 \text{ m}}{3.28 \text{ ft}} \right)^2 + (15)(5.0 \text{ ft}^2) \left(\frac{1 \text{ m}}{3.28 \text{ ft}} \right)^2$$
$$+ (5.0 \text{ ft}^2) \left(\frac{1 \text{ m}}{3.28 \text{ ft}} \right)^2$$
$$= 22.5 \text{ m}^2$$

The total sabin area excluding the ceiling is

$$18.7 \text{ m}^2 - \alpha_3 S_3 = 22.5 \text{ m}^2 - (0.03)(90 \text{ m}^2)$$
$$= 19.8 \text{ m}^2$$

The total sabin area of the room with the ceiling treated with $\alpha_3 = 0.7$ sound absorption is

$$\sum S_1 = 19.8 \text{ m}^2 + \alpha_3 S_3$$
$$= 19.8 \text{ m}^2 + (0.7)(90 \text{ m}^2)$$
$$= 82.8 \text{ m}^2$$

Use Eq. 68.21.

$$NR = 10 \log \left(\frac{\sum S_1}{\sum S_2} \right) = 10 \log \left(\frac{82.8 \text{ m}^2}{22.5 \text{ m}^2} \right)$$
$$= \boxed{5.7 \text{ dB}}$$

The answer is (B).

9. *Customary U.S. Solution*

Define the sound absorption coefficients of the floor, ceiling, and walls as α_1, α_2, and α_3, respectively.

The floor area is

$$A_1 = (15 \text{ ft})(20 \text{ ft}) = 300 \text{ ft}^2$$

The ceiling area is

$$A_2 = (15 \text{ ft})(20 \text{ ft}) = 300 \text{ ft}^2$$

The area of the walls is

$$A_3 = \big((2)(15 \text{ ft}) + (2)(20 \text{ ft})\big)(10 \text{ ft}) = 700 \text{ ft}^2$$

The total surface area of the room is

$$A = \sum A_i = A_1 + A_2 + A_3$$
$$= 300 \text{ ft}^2 + 300 \text{ ft}^2 + 700 \text{ ft}^2$$
$$= 1300 \text{ ft}^2$$

From Eq. 68.18, the average sound absorption coefficient of the room is

$$\overline{\alpha} = \frac{\sum S_i}{\sum A_i} = \frac{\alpha_1 A_1 + \alpha_2 A_2 + \alpha_3 A_3}{A}$$
$$= \frac{(0.03)(300 \text{ ft}^2) + (0.5)(300 \text{ ft}^2) + (0.06)(700 \text{ ft}^2)}{1300 \text{ ft}^2}$$
$$= 0.155$$

From Eq. 68.19, the room constant is

$$R = \frac{\overline{\alpha} A}{1 - \overline{\alpha}} = \frac{(0.155)(1300 \text{ ft}^2)}{1 - 0.155} = 238.5 \text{ ft}^2$$

From Eq. 68.15, the sound pressure level due to the machine is

$$L_{p,\text{dB}} - 10.5 + L_W + 10 \log \left(\frac{Q}{4\pi r^2} + \frac{4}{R} \right)$$
$$= 10.5 + 65 + 10 \log \left(\frac{4}{4\pi(5 \text{ ft})^2} + \frac{4}{238.5 \text{ ft}^2} \right)$$
$$= 60.2 \text{ dB}$$

Use Eq. 68.16 to combine the machine sound pressure level with the ambient sound pressure level.

$$L = 10 \log \sum 10^{\frac{W_i}{10}} = 10 \log \left(10^{\frac{60.2}{10}} + 10^{\frac{50}{10}} \right)$$
$$= \boxed{60.6 \text{ dB}}$$

The answer is (B).

SI Solution

Define the sound absorption coefficients of the floor, ceiling, and walls as α_1, α_2, and α_3, respectively.

The floor area is

$$A_1 = (4.5 \text{ m})(6 \text{ m}) = 27 \text{ m}^2$$

The ceiling area is

$$A_2 = (4.5 \text{ m})(6 \text{ m}) = 27 \text{ m}^2$$

The area of the walls is

$$A_3 = \big((2)(4.5 \text{ m}) + (2)(6 \text{ m})\big)(3 \text{ m}) = 63 \text{ m}^2$$

The total surface area of the room is

$$A = \sum A_i = A_1 + A_2 + A_3$$
$$= 27 \text{ m}^2 + 27 \text{ m}^2 + 63 \text{ m}^2$$
$$= 117 \text{ m}^2$$

From Eq. 68.18, the average sound absorption coefficient of the room is

$$\overline{\alpha} = \frac{\sum S_i}{\sum A_i} = \frac{\alpha_1 A_1 + \alpha_2 A_2 + \alpha_3 A_3}{A}$$
$$= \frac{(0.03)(27 \text{ m}^2) + (0.5)(27 \text{ m}^2) + (0.06)(63 \text{ m}^2)}{117 \text{ m}^2}$$
$$= 0.155$$

From Eq. 68.19, the room constant is

$$R = \frac{\overline{\alpha} A}{1 - \overline{\alpha}} = \frac{(0.155)(117 \text{ m}^2)}{1 - 0.155} = 21.5 \text{ m}^2$$

From Eq. 68.15, the sound pressure level due to the machine is

$$L_{p,\text{dB}} = L_W + 10 \log \left(\frac{Q}{4\pi r^2} + \frac{4}{R} \right)$$
$$= 65 + 10 \log \left(\frac{4}{4\pi(1.5 \text{ m})^2} + \frac{4}{21.5 \text{ m}^2} \right)$$
$$= 60.2 \text{ dB}$$

Use Eq. 68.16 to combine the machine sound pressure level with the ambient sound pressure level.

$$L = 10 \log \sum \frac{\frac{W_i}{10}}{10} = 10 \log \left(10^{\frac{60.2}{10}} + 10^{\frac{50}{10}} \right) = \boxed{60.6 \text{ dB}}$$

The answer is (B).

69 Engineering Economic Analysis

PRACTICE PROBLEMS

1. At 6% effective annual interest, how much will be accumulated if $1000 is invested for ten years?

2. At 6% effective annual interest, what is the present worth of $2000 that becomes available in four years?

3. At 6% effective annual interest, how much should be invested to accumulate $2000 in 20 years?

4. At 6% effective annual interest, what year-end annual amount deposited over seven years is equivalent to $500 invested now?

5. At 6% effective annual interest, what will be the accumulated amount at the end of ten years if $50 is invested at the end of each year for ten years?

6. At 6% effective annual interest, how much should be deposited at the start of each year for ten years (a total of ten deposits) in order to empty the fund by drawing out $200 at the end of each year for ten years (a total of ten withdrawals)?

7. At 6% effective annual interest, how much should be deposited at the start of each year for five years to accumulate $2000 on the date of the last deposit?

8. At 6% effective annual interest, how much will be accumulated in ten years if three payments of $100 are deposited every other year, with the first payment occurring at $t = 0$?

9. $500 is compounded monthly at a 6% nominal annual interest rate. How much will have accumulated in five years?

10. What is the effective annual rate of return on an $80 investment that pays back $120 in seven years?

11. A new machine will cost $17,000 and will have a resale value of $14,000 after five years. Special tooling will cost $5000. The tooling will have a resale value of $2500 after five years. Maintenance will be $2000 per year. The effective annual interest rate is 6%. What will be the average annual cost of ownership during the next five years?

12. An old covered wooden bridge can be strengthened at a cost of $9000, or it can be replaced for $40,000. The present salvage value of the old bridge is $13,000. It is estimated that the reinforced bridge will last for 20 years, will have an annual cost of $500, and will have a salvage value of $10,000 at the end of 20 years. The estimated salvage value of the new bridge after 25 years is $15,000. Maintenance for the new bridge would cost $100 annually. The effective annual interest rate is 8%. Which is the best alternative?

13. A firm expects to receive $32,000 each year for 15 years from sales of a product. An initial investment of $150,000 will be required to manufacture the product. Expenses will run $7530 per year. Salvage value is zero, and straight-line depreciation is used. The income tax rate is 48%. What is the after-tax rate of return?

14. A highway public works project has initial costs of $1,000,000, benefits of $1,500,000, and disbenefits of $300,000.

(a) What is the Winfrey benefit/cost ratio?

(b) What is the excess of benefits over costs?

15. A speculator in land pays $14,000 for property that he expects to hold for ten years. $1000 is spent in renovation, and a monthly rent of $75 is collected from the tenants. (Use the year-end convention.) Taxes are $150 per year, and maintenance costs are $250 per year. What must be the sale price in ten years to realize a 10% rate of return?

16. What is the effective annual interest rate for a payment plan of 30 equal payments of $89.30 per month when a lump sum payment of $2000 would have been an outright purchase?

17. A depreciable item is purchased for $500,000. The salvage value at the end of 25 years is estimated at $100,000.

(a) What is the depreciation in each of the first three years using the straight line method?

(b) What is the depreciation in each of the first three years using the sum-of-the-years' digits method?

(c) What is the depreciation in each of the first three years using the double declining balance method?

18. Equipment that is purchased for $12,000 now is expected to be sold after ten years for $2000. The estimated maintenance is $1000 for the first year, but it is expected to increase $200 each year thereafter. The effective annual interest rate is 10%.

(a) What is the present worth?

(b) What is the annual cost?

19. A new grain combine with a twenty year life can remove seven pounds of rocks from its harvest per hour. Any rocks left in its output hopper will cause $25,000 damage in subsequent processes. Several investments are available to increase the rock-removal capacity, as listed in the table. The effective annual interest rate is 10%. What should be done?

rock removal rate	probability of exceeding rock removal rate	required investment to achieve removal rate
7	0.15	0
8	0.10	$15,000
9	0.07	$20,000
10	0.03	$30,000

20. (*Time limit: one hour*) A mechanism that costs $10,000 has operating costs and salvage values as given. An effective annual interest rate of 20% is to be used.

year	operating cost	salvage value
1	$2000	$8000
2	$3000	$7000
3	$4000	$6000
4	$5000	$5000
5	$6000	$4000

(a) What is the economic life of the mechanism?

(b) Assuming that the mechanism has been owned and operated for four years already, what is the cost of owning and operating the mechanism for one more year?

21. (*Time limit: one hour*) A salesperson intends to purchase a car for $50,000 for personal use, driving 15,000 miles per year. Insurance for personal use costs $2000 per year, and maintenance costs $1500 per year. The car gets 15 miles per gallon, and gasoline costs $1.50 per gallon. The resale value after five years will be $10,000. The salesperson's employer has asked that the car be used for business driving of 50,000 miles per year and has offered a reimbursement of $0.30 per mile. Using the car for business would increase the insurance cost to $3000 per year and maintenance to $2000 per year. The salvage value after five years would be reduced to $5000. If the employer purchased a car for the salesperson to use, the initial cost would be the same, but insurance, maintenance, and salvage would be $2500, $2000, and $8000, respectively. The salesperson's effective annual interest rate is 10%.

(a) Is the reimbursement offer adequate?

(b) With a reimbursement of $0.30 per mile, how many miles must the car be driven per year to justify the employer buying the car for the salesperson to use?

22. (*Time limit: one hour*) Alternatives A and B are being evaluated. The effective annual interest rate is 10%. What alternative is economically superior?

	alternative A	alternative B
first cost	$80,000	$35,000
life	20 years	10 years
salvage value	$7000	0
annual costs		
years 1–5	$1000	$3000
years 6–10	$1500	$4000
years 11–20	$2000	0
additional cost		
year 10	$5000	0

23. (*Time limit: one hour*) A car is needed for three years. Plans A and B for acquiring the car are being evaluated. An effective annual interest rate of 10% is to be used. Which plan is economically superior?

plan A: lease the car for $0.25/mile (all inclusive)

plan B: purchase the car for $30,000
keep the car for three years
sell the car after three years for $7200
pay $0.14 per mile for oil and gas
pay other costs of $500 per year

24. (*Time limit: one hour*) Two methods are being considered to meet strict air pollution control requirements over the next ten years. Method A uses equipment with a life of ten years. Method B uses equipment with a life of five years that will be replaced with new equipment with an additional life of five years. Capacities of the two methods are different, but operating costs do not depend on the throughput. Operation is 24 hours per day, 365 days per year. The effective annual interest rate for this evaluation is 7%.

	method A	method B	
	years 1–10	years 1–5	years 6–10
installation cost	$13,000	$6000	$7000
equipment cost	$10,000	$2000	$2200
operating cost			
per hour	$10.50	$8.00	$8.00
salvage value	$5000	$2000	$2000
capacity (tons/yr)	50	20	20
life	10 years	5 years	5 years

(a) What is the uniform annual cost per ton for each method?

(b) Over what range of throughput (in tons/yr) does each method have the minimum cost?

25. (*Time limit: one hour*) A transit district has asked for your assistance in determining the proper fare for its bus system. An effective annual interest rate of 7% is to be used. The following additional information was compiled for your study.

cost per bus	$60,000
bus life	20 years
salvage value	$10,000
miles driven per year	37,440
number of passengers per year	80,000
operating cost	$1.00 per mile in the first year, increasing $0.10 per mile each year thereafter

(a) If the fare is to remain constant for the next 20 years, what is the break-even fare per passenger?

(b) If the transit district decides to set the per-passenger fare at $0.35 for the first year, by what amount should the per-passenger fare go up each year thereafter such that the district can break even in 20 years?

(c) If the transit district decides to set the per-passenger fare at $0.35 for the first year and the per-passenger fare goes up $0.05 each year thereafter, what additional governmental subsidy (per passenger) is needed for the district to break even in 20 years?

26. Make a recommendation to your client to accept one of the following alternatives. Use the present worth comparison method. (Initial costs are the same.)

A. a 25 year annuity paying $4800 at the end of each year, where the interest rate is a nominal 12% per annum

B. a 25 year annuity paying $1200 every quarter at 12% nominal annual interest

27. A firm has two alternatives for improvement of its existing production line. The data are as follows.

	alternative A	alternative B
initial installment cost	$1500	$2500
annual operating cost	$800	$650
service life	5 years	8 years
salvage value	0	0

Determine the best alternative using an interest rate of 15%.

28. Two mutually exclusive alternatives requiring different investments are being considered. The life of both alternatives is estimated at 20 years with no salvage values. The minimum rate of return that is considered acceptable is 4%. Which alternative is best?

	alternative A	alternative B
investment required	$70,000	$40,000
net income per year	$5620	$4075
rate of return on total investment	5%	8%

29. Compare the costs of two plant renovation schemes, A and B. Assume equal lives of 25 years, no salvage values, and interest at 25%.

	alternative A	alternative B
first cost	$20,000	$25,000
annual expenditure	$3000	$2500

(a) Make the comparison on the basis of present worth. Which alternative is best?

(b) Make the comparison on the basis of capitalized cost. Which alternative is best?

(c) Make the comparison on the basis of annual cost. Which alternative is best?

30. With interest at 8%, obtain the solutions to the following to the nearest dollar.

(a) A machine costs $18,000 and has a salvage value of $2000. It has a useful life of eight years. What is its book value at the end of five years using straight line depreciation?

(b) Using data from part (a), find the depreciation in the first three years using the sinking fund method.

(c) Repeat part (a) using double declining balance depreciation to find the first five years' depreciation.

31. A chemical pump motor unit is purchased for $14,000. The estimated life is eight years, after which it will be sold for $1800. (a) Find the depreciation in the first two years by the sum-of-the-years' digits method. (b) Calculate the after-tax depreciation recovery over the entire lifetime using 15% interest with 52% income tax.

32. A soda ash plant has the water effluent from processing equipment treated in a large settling basin. The settling basin eventually discharges into a river that runs alongside the basin. Recently enacted environmental regulations require all rainfall on the plant to be diverted and treated in the settling basin. A heavy rainfall will cause the entire basin to overflow. An uncontrolled overflow will cause environmental damage and heavy fines. The construction of additional height on the existing basic walls is under consideration.

Data on the costs of construction and expected costs for environmental cleanup and fines are shown. Data on 50 typical winters have been collected. The soda ash plant management considers 12% to be their minimum rate of return, and it is felt that after 15 years the plant will be closed. The company wants to select the alternative that minimizes its total expected costs.

additional basin height (ft)	number of winters with basin overflow	expense for environmental clean up per year	construction cost
0	24	$550,000	0
5	14	$600,000	$600,000
10	8	$650,000	$710,000
15	3	$700,000	$900,000
20	1	$800,000	$1,000,000
	50		

33. A wood processing plant installed a waste gas scrubber at a cost of $30,000 to remove pollutants from the exhaust discharged into the atmosphere. The scrubber has no salvage value and will cost $18,700 to operate next year, with operating costs expected to increase at the rate of $1200 per year thereafter. When should the company consider replacing the scrubber? Money can be borrowed at 12%.

34. Two alternative piping schemes are being considered by a water treatment facility. On the basis of a ten year life and an interest rate of 12%, determine the number of hours of operation for which the two installations will be equivalent.

	alternative A	alternative B
pipe diameter	4 in	6 in
head loss for required flow	48 ft	26 ft
size motor required	20 hp	7 hp
energy cost per hour of operation	$ 0.30	$ 0.10
cost of motor installed	$3600	$2800
cost of pipes and fittings	$3050	$5010
salvage value at end of 10 years	$200	$280

35. An 88% learning curve is used with an item whose first production time was six weeks.

(a) How long will it take to produce the fourth item?

(b) How long will it take to produce the 6th through 14th items?

36. (*Time limit: one hour*) A company is considering two alternatives, only one of which can be selected.

alternative	initial investment	salvage value	annual net profit	life
A	$120,000	$15,000	$57,000	5 yr
B	$170,000	$20,000	$67,000	5 yr

The net profit is after operating and maintenance costs, but before taxes. The company pays 45% of its year-end profit as income taxes. Use straight line depreciation. Do not use investment tax credit. Find the best alternative if the company's minimum attractive rate of return is 15%.

37. (*Time limit: one hour*) A company is considering the purchase of equipment to expand its capacity. The equipment cost is $300,000. The equipment is needed for five years, after which it will be sold for $50,000. The company's before-tax cash flow will be improved $90,000 annually by the purchase of the asset.

The corporate tax rate is 48%, and straight line depreciation will be used. The company will take an investment tax credit of 6.67%. What is the after-tax rate of return associated with this equipment purchase?

38. (*Time limit: one hour*) A 120-room hotel is purchased for $2,500,000. A 25 year loan is available for 12%. A study was conducted to determine the various occupancy rates.

occupancy	probability
65% full	0.40
70%	0.30
75%	0.20
80%	0.10

The operating costs of the hotel are as follows.

taxes and insurance	$20,000 annually
maintenance	$50,000 annually
operating	$200,000 annually

The life of the hotel is figured to be 25 years when operating 365 days per year. The salvage value after 25 years is $500,000.

Neglect tax credit and income taxes. Determine the average rate that should be charged per room per night to return 15% of the initial cost each year.

39. (*Time limit: one hour*) A company is insured for $3,500,000 against fire. The insurance rate is $0.69/1000. The insurance company will decrease the rate to $0.47/1000 if fire sprinklers are installed. The initial cost of the sprinklers is $7500. Annual costs are $200; additional taxes are $100 annually. The system life is 25 years. What is the rate of return on this investment?

40. (*Time limit: one hour*) Heat losses through the walls in an existing building cost a company $1,300,000 per year. This amount is considered excessive, and two alternatives are being evaluated. Neither of the alternatives will increase the life of the existing building beyond the current expected life of six years, and neither of the alternatives will produce a salvage value.

Alternative A: Do nothing. Continue with current losses.

Alternative B: Spend $2,000,000 immediately to upgrade the building and reduce the loss by 80%. This alternative will require annual maintenance of $150,000.

Alternative C: Spend $1,200,000 immediately. Repeat the $1,200,000 expenditure three years from now. Heat loss the first year will be reduced 80%. Due to deterioration, the reduction will be 55% and 20% in the second and third years. (The pattern is repeated starting after the second expenditure.) There are no maintenance costs.

All energy and maintenance costs are considered expenses for tax purposes. The company's tax rate is 48%, and straight line depreciation is used. 15% is regarded as the effective annual interest rate. Evaluate each alternative on an after-tax basis, and recommend the best alternative.

41. *(Time limit: one hour)* You have been asked to determine if a seven year old machine should be replaced. Give a full explanation for your recommendation. Base your decision on a before-tax interest rate of 15%.

The existing machine is presumed to have a ten year life. It has been depreciated on a straight line basis from its original value of $1,250,000 to a current book value of $620,000. Its ultimate salvage value was assumed to be $350,000 for purposes of depreciation. Its present salvage value is estimated at $400,000, and this is not expected to change over the next three years. The current operating costs are not expected to change from $200,000 per year.

A new machine costs $800,000, with operating costs of $40,000 the first year, and increasing by $30,000 each year thereafter. The new machine has an expected life of ten years. The salvage value depends on the year the new machine is retired.

year retired	salvage
1	$600,000
2	$500,000
3	$450,000
4	$400,000
5	$350,000
6	$300,000
7	$250,000
8	$200,000
9	$150,000
10	$100,000

SOLUTIONS

1.

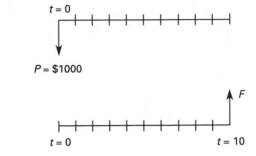

$i = 6\%$ a year

By the formula from Table 69.1,

$$F = P(1 + i)^n = (\$1000)(1 + 0.06)^{10} = \boxed{\$1790.85}$$

By the factor converting P to F, $(F/P, i, n) = 1.7908$ for $i = 6\%$ a year and $n = 10$ years.

$$F = P(F/P, 6\%, 10) = (\$1000)(1.7908) = \boxed{\$1790.80}$$

2.

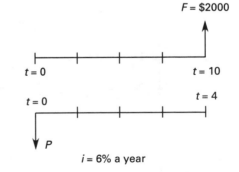

$i = 6\%$ a year

By the formula from Table 69.1,

$$P = \frac{F}{(1+i)^n} = \frac{\$2000}{(1+0.06)^4} = \boxed{\$1584.19}$$

From the factor converting F to P, $(P/F, i, n) = 0.7921$ for $i = 6\%$ a year and $n = 4$ years.

$$P = F(P/F, 6\%, 4) = (\$2000)(0.7921) = \boxed{\$1584.20}$$

Economics

3.

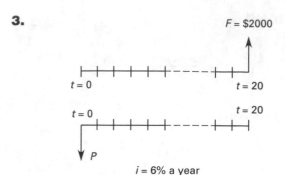

$i = 6\%$ a year

By the formula from Table 69.1,

$$P = \frac{F}{(1+i)^n} = \frac{\$2000}{(1+0.06)^{20}} = \boxed{\$623.61}$$

From the factor converting F to P, $(P/F, i, n) = 0.3118$ for $i = 6\%$ a year and $n = 20$ years.

$$P = F(P/F, 6\%, 20) = (\$2000)(0.3118) = \boxed{\$623.60}$$

4.

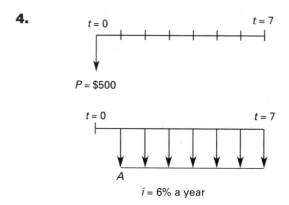

$i = 6\%$ a year

By the formula from Table 69.1,

$$A = P\left(\frac{i(1+i)^n}{(1+i)^n - 1}\right) = (\$500)\left(\frac{(0.06)(1+0.06)^7}{(1+0.06)^7 - 1}\right)$$

$$= \boxed{\$89.57}$$

By the factor converting P to A, $(A/P, i, n) = 0.17914$ for $i = 6\%$ a year and $n = 7$ years.

$$A = P(A/P, 6\%, 7) = (\$500)(0.17914) = \boxed{\$89.57}$$

5.

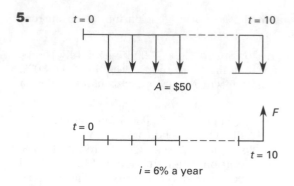

$i = 6\%$ a year

By the formula from Table 69.1,

$$F = A\left(\frac{(1+i)^n - 1}{i}\right) = (\$50)\left(\frac{(1+0.06)^{10} - 1}{0.06}\right)$$

$$= \boxed{\$659.04}$$

By the factor converting A to F, $(F/A, i, n) = 13.181$ for $i = 6\%$ a year and $n = 10$ years.

$$F = A(F/A, 6\%, 10) = (\$50)(13.181) = \boxed{\$659.05}$$

6.

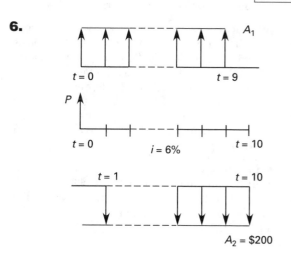

By the formula from Table 69.1, for each cash flow diagram,

$$P = A_1 + A_1\left(\frac{(1+0.06)^9 - 1}{(0.06)(1+0.06)^9}\right)$$

$$= A_2\left(\frac{(1+0.06)^{10} - 1}{(0.06)(1+0.06)^{10}}\right)$$

Therefore for $A_2 = \$200$,

$$A_1 + A_1\left(\frac{(1+0.06)^9 - 1}{(0.06)(1+0.06)^9}\right)$$

$$= (\$200)\left(\frac{(1+0.06)^{10} - 1}{(0.06)(1+0.06)^{10}}\right)$$

$$7.80A_1 = \$1472.02$$

$$A_1 = \boxed{\$188.72}$$

By the factor converting A to P,

$$(P/A, 6\%, 9) = 6.8017$$
$$(P/A, 6\%, 10) = 7.3601$$
$$A_1 + A_1(6.8017) = (\$200)(7.3601)$$
$$7.8017A_1 = \$1472.02$$
$$A_1 = \frac{\$1472.02}{7.8017} = \boxed{\$188.08}$$

7.

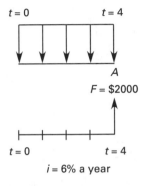

$$i = 6\% \text{ a year}$$

Find and equate the present worth of each cash flow.

$$(\$2000)(P/F, 6\%, 4) = A\big(1 + (P/A, 6\%, 4)\big)$$
$$(\$2000)(0.7921) = A(1 + 3.4651)$$
$$A = \boxed{\$354.79}$$

8.

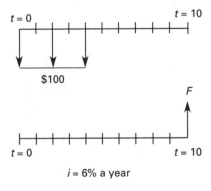

$$i = 6\% \text{ a year}$$

By the formula from Table 69.1, $F = P(1+i)^n$. If each deposit is considered as P, each will accumulate interest for periods of 10, 8, and 6 years.

Therefore,

$$F = (\$100)(1 + 0.06)^{10} + (\$100)(1 + 0.06)^8$$
$$+ (\$100)(1 + 0.06)^6$$
$$= (\$100)(1.7908 + 1.5938 + 1.4185)$$
$$= (\$100)(4.8031)$$
$$= \boxed{\$480.31}$$

By the factor converting P to F,

$$(F/P, i, n) = 1.7908 \text{ for } i = 6\% \text{ and } n = 10$$
$$= 1.5938 \text{ for } i = 6\% \text{ and } n = 8$$
$$= 1.4185 \text{ for } i = 6\% \text{ and } n = 6$$

By summation,

$$F = (\$100)(1.7908 + 1.5938 + 1.4185)$$
$$= (\$100)(4.8031)$$
$$= \boxed{\$480.31}$$

9.

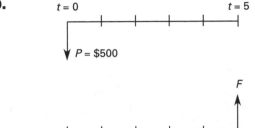

$$r = 6\% \text{ a year}$$

Since the deposit is compounded monthly, the effective interest rate should be calculated as shown by Eq. 69.54.

$$i = \left(1 + \frac{r}{k}\right)^k - 1 = \left(1 + \frac{0.06}{12}\right)^{12} - 1$$
$$= \boxed{0.061677 \quad (6.1677\%)}$$

By the formula from Table 69.1,

$$F = P(1+i)^n = (\$500)(1 + 0.061677)^5 = \$674.42$$

To use a table of factors, interpolation is required.

$i\%$	factor F/P
6	1.3382
6.1677	desired
7	1.4026

$$i = \left(\frac{6.1677 - 6}{7 - 6}\right)(1.4026 - 1.3382)$$
$$= 0.0108$$

Therefore,

$$F/P = 1.3382 + 0.0108 = 1.3490$$
$$F = P(F/P, 6.1677\%, 5) = (\$500)(1.3490)$$
$$= \boxed{\$674.50}$$

10.

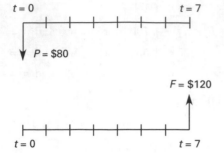

By the formula from Table 69.1,

$$F = P(1+i)^n$$

Therefore,

$$(1+i)^n = F/P$$

$$i = (F/P)^{\frac{1}{n}} - 1 = \left(\frac{\$120}{\$80}\right)^{\frac{1}{7}} - 1$$

$$= 0.059 \approx \boxed{6\%}$$

By the factor coverting P to F,

$$F = P(F/P, i\%, 7)$$

$$(F/P, i\%, 7) = F/P = \frac{\$120}{\$80} = 1.5$$

By checking the interest tables,

$$(F/P, i\%, 7) = 1.4071 \text{ for } i = 5\%$$
$$= 1.5036 \text{ for } i = 6\%$$
$$= 1.6058 \text{ for } i = 7\%$$

Therefore, $i = \boxed{6\%}$.

11.

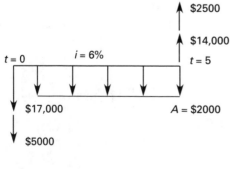

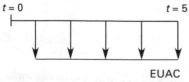

Annual cost of ownership, EUAC, can be obtained by the factors converting P to A and F to A.

$$P = \$17,000 + \$5000$$

$$= \$22,000$$

$$F = \$14,000 + \$2500$$

$$= \$16,500$$

$$\text{EUAC} = A + P(A/P, 6\%, 5) - F(A/F, 6\%, 5)$$

$$(A/P, 6\%, 5) = 0.23740$$

$$(A/F, 6\%, 5) = 0.17740$$

$$\text{EUAC} = \$2000 + (\$22,000)(0.23740)$$

$$- (\$16,500)(0.17740)$$

$$= \boxed{\$4295.70}$$

12.

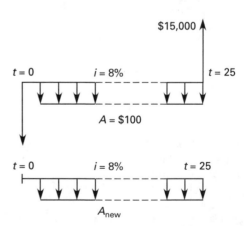

Consider the salvage value as a benefit lost (cost).

$$\text{EUAC}_{old} = \$500 + (\$22,000)(A/P, 8\%, 20)$$

$$- (\$10,000)(A/F, 8\%, 20)$$

$$(A/P, 8\%, 20) = 0.10185$$

$$(A/F, 8\%, 20) = 0.02185$$

$$\begin{aligned} \text{EUAC}_{\text{old}} &= \$500 + (\$22{,}000)(0.10185) \\ &\quad - (\$10{,}000)(0.02185) \\ &= \$2522.20 \end{aligned}$$

Similarly,

$$\begin{aligned} \text{EUAC}_{\text{new}} &= \$100 + (\$40{,}000)(A/P, 8\%, 25) \\ &\quad - (\$15{,}000)(A/F, 8\%, 25) \end{aligned}$$

$$(A/P, 8\%, 25) = 0.09368$$

$$(A/F, 8\%, 25) = 0.01368$$

$$\begin{aligned} \text{EUAC}_{\text{new}} &= \$100 + (\$40{,}000)(0.09368) \\ &\quad - (\$15{,}000)(0.01368) \\ &= \$3642 \end{aligned}$$

Therefore, the new bridge is going to be more costly.

The best alternative is to strengthen the old bridge.

13.

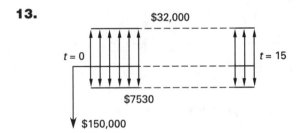

The annual depreciation is

$$\begin{aligned} D &= \frac{C - S_n}{n} = \frac{\$150{,}000}{15} \\ &= \$10{,}000/\text{year} \end{aligned}$$

The taxable income is

$$\$32{,}000 - \$7530 - \$10{,}000 = \$14{,}470/\text{year}$$

Taxes paid are

$$(\$14{,}470)(0.48) = \$6945.60/\text{year}$$

The after-tax cash flow is

$$\$24{,}470 - \$6945.60 = \$17{,}524.40$$

The present worth of the alternate is zero when evaluated at its ROR.

$$0 = -\$150{,}000 + (\$17{,}524.40)(P/A, i\%, 15)$$

Therefore,

$$(P/A, i\%, 15) = \frac{\$150{,}000}{\$17{,}524.40} = 8.55949$$

By checking the tables, this factor matches $i = 8\%$.

$$\boxed{\text{ROR} = 8\%}$$

14. The Winfrey benefit/cost ratio is

$$B/C = \frac{B - D}{C}$$

(a) The benefit/cost ratio will be

$$B/C = \frac{\$1{,}500{,}000 - \$300{,}000}{\$1{,}000{,}000} = \boxed{1.2}$$

(b) The excess of benefits over cost are $\boxed{\$200{,}000.}$

15.

The annual rent is

$$(\$75)\left(12 \, \frac{\text{months}}{\text{year}}\right) = \$900$$

$$P = P_1 + P_2 = \$15{,}000$$

$$A_1 = -\$900$$

$$A_2 = \$250 + \$150 = \$400$$

By the factors converting P to F and A to F,

$$\begin{aligned} F &= (\$15{,}000)(F/P, 10\%, 10) \\ &\quad + (\$400)(F/A, 10\%, 10) \\ &\quad - (\$900)(F/A, 10\%, 10) \end{aligned}$$

$$(F/P, 10\%, 10) = 2.5937$$

$$(F/A, 10\%, 10) = 15.937$$

$$\begin{aligned} F &= (\$15{,}000)(2.5937) + (\$400)(15.937) \\ &\quad - (\$900)(15.937) \\ &= \boxed{\$30{,}937} \end{aligned}$$

16.

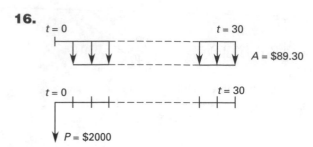

By the formula relating P to A,

$$P = A\left(\frac{(1+i)^n - 1}{i(1+i)^n}\right)$$

$$\frac{(1+i)^{30} - 1}{i(1+i)^{30}} = \frac{\$2000}{\$89.30} = 22.30$$

By trial and error,

i	$(1+i)^{30}$	$\dfrac{(1+i)^{30} - 1}{i(1+i)^{30}}$
10	17.45	9.42
6	5.74	13.76
4	3.24	17.28
2	1.81	22.37

2% per month is close.

$$i = (1 + 0.02)^{12} - 1 = \boxed{0.2682 \quad (26.82\%)}$$

17. (a) Use the straight line method, Eq. 69.25.

$$D = \frac{C - S_n}{n}$$

Each year depreciation will remain the same.

$$D = \frac{\$500,000 - \$100,000}{25} = \boxed{\$16,000}$$

(b) Sum-of-the years digits (SOYD) can be calculated as shown by Eq. 69.28,

$$D_j = \frac{(C - S_n)(n - j + 1)}{T}$$

Use Eq. 69.27.

$$T = \tfrac{1}{2}n(n+1) = \left(\tfrac{1}{2}\right)(25)(25+1) = 325$$

$$D_1 = \frac{(\$500,000 - \$100,000)(25 - 1 + 1)}{325}$$

$$= \boxed{\$30,769}$$

$$D_2 = \frac{(\$500,000 - \$100,000)(25 - 2 + 1)}{325}$$

$$= \boxed{\$29,538}$$

$$D_3 = \frac{(\$500,000 - \$100,000)(25 - 3 + 1)}{325}$$

$$= \boxed{\$28,308}$$

(c) The double-declining balance (DDB) method can be used. By Eq. 69.32,

$$D_j = dC(1 - d)^{j-1}$$

Use Eq. 69.31.

$$d = \frac{2}{n}$$

$$= \frac{2}{25}$$

$$D_1 = \left(\frac{2}{25}\right)(\$500,000)\left(1 - \frac{2}{25}\right)^0 = \boxed{\$40,000}$$

$$D_2 = \left(\frac{2}{25}\right)(\$500,000)\left(1 - \frac{2}{25}\right)^1 = \boxed{\$36,800}$$

$$D_3 = \left(\frac{2}{25}\right)(\$500,000)\left(1 - \frac{2}{25}\right)^2 = \boxed{\$33,856}$$

18.

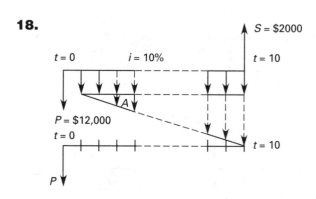

(a) $A = \$1000$ and $G = \$200$ for $t = n - 1 = 9$ years.

$$F = S = \$2000$$

$$P = \$12,000 + A(P/A, 10\%, 10) + G(P/G, 10\%, 10)$$
$$\quad - F(P/F, 10\%, 10)$$
$$= \$12,000 + (\$1000)(6.1446) + (\$200)(22.8913)$$
$$\quad - (\$2000)(0.3855)$$
$$= \boxed{\$21,952}$$

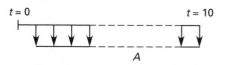

(b) $A = (\$12,000)(A/P, 10\%, 10) + \$1000$
$$\quad + (\$200)(A/G, 10\%, 10)$$
$$\quad - (\$2000)(A/F, 10\%, 10)$$
$$= (\$12,000)(0.16275) + \$1000 + (\$200)(3.7255)$$
$$\quad - (\$2000)(0.06275)$$
$$= \boxed{\$3572.60}$$

19. An increase in rock removal capacity can be achieved by a 20-year loan (investment). Different cases available can be compared by equivalent uniform annual cost (EUAC).

$$\text{EUAC} = \text{annual loan cost}$$
$$+ \text{expected annual damage}$$
$$= \text{cost } (A/P, 10\%, 20)$$
$$+ (\$25,000)(\text{probability})$$

$$(A/P, 10\%, 20) = 0.11746$$

A table can be prepared for different cases. Notice that the exceedance probability is given in the problem statement and does not need to be calculated by accumulation.

rock removal rate	cost ($)	annual loan cost ($)	proba-bility	expected annual damage ($)	EUAC ($)
7	0	0	0.15	3750	3750.00
8	15,000	1761.90	0.10	2500	4261.90
9	20,000	2349.20	0.07	1750	4099.20
10	30,000	3523.80	0.03	750	4273.80

It is cheapest to do nothing.

20. Calculate the cost of owning and operating for each year.

$$A_1 = (\$10,000)(A/P, 20\%, 1) + \$2000$$
$$- (\$8000)(A/F, 20\%, 1)$$

$$(A/P, 20\%, 1) = 1.2$$
$$(A/F, 20\%, 1) = 1.0$$

$$A_1 = (\$10,000)(1.2) + \$2000 - (\$8000)(1.0)$$
$$= \$6000$$

$$A_2 = (\$10,000)(A/P, 20\%, 2) + \$2000$$
$$+ (\$1000)(A/G, 20\%, 2)$$
$$(\$7000)(A/F, 20\%, 2)$$

$$(A/P, 20\%, 2) = 0.6545$$
$$(A/G, 20\%, 2) = 0.4545$$
$$(A/F, 20\%, 2) = 0.4545$$

$$A_2 = (\$10,000)(0.6545) + \$2000$$
$$+ (\$1000)(0.4545) - (\$7000)(0.4545)$$
$$= \$5818$$

$$A_3 = (\$10,000)(A/P, 20\%, 3) + \$2000$$
$$+ (\$1000)(A/G, 20\%, 3)$$
$$- (\$6000)(A/F, 20\%, 3)$$

$$(A/P, 20\%, 3) = 0.4747$$
$$(A/G, 20\%, 3) = 0.8791$$
$$(A/F, 20\%, 3) = 0.2747$$

$$A_3 = (\$10,000)(0.4747) + \$2000$$
$$+ (\$1000)(0.8791)$$
$$- (\$6000)(0.2747)$$
$$= \$5977.90$$

$$A_4 = (\$10,000)(A/P, 20\%, 4)$$
$$+ \$2000 + (\$1000)(A/G, 20\%, 4)$$
$$- (\$5000)(A/F, 20\%, 4)$$

$$(A/P, 20\%, 4) = 0.3863$$
$$(A/G, 20\%, 4) = 1.2762$$
$$(A/F, 20\%, 4) = 0.1863$$

$$A_4 = (\$10,000)(0.3863) + \$2000$$
$$+ (\$1000)(1.2762) - (\$5000)(0.1863)$$
$$= \$6207.70$$

$$A_5 = (\$10,000)(A/P, 20\%, 5) + \$2000$$
$$+ (\$1000)(A/G, 20\%, 5)$$
$$- (\$4000)(A/F, 20\%, 5)$$

$$(A/P, 20\%, 5) = 0.3344$$
$$(A/G, 20\%, 5) = 1.6405$$
$$(A/F, 20\%, 5) = 0.1344$$

$$A_5 = (\$10,000)(0.3344) + \$2000$$
$$+ (\$1000)(1.6405) - (\$4000)(0.1344)$$
$$= \$6446.90$$

(a) Since the annual owning and operating cost is smallest after two years of operation, it is advantageous to sell the mechanism after the second year.

The economic life is two years.

(b) After four years of operation, the owning and operating cost of the mechanism for one more year will be

$$A = \$6000 + (\$5000)(1 + i) - \$4000$$
$$i = 0.2 \quad (20\%)$$
$$A = \$6000 + (\$5000)(1.2) - \$4000$$
$$= \$8000$$

Economics

21. (a) To find out if the reimbursement is adequate, calculate the business-related expense.

Charge the company for business travel.

$$\text{insurance: } \$3000 - \$2000 = \$1000$$
$$\text{maintenance: } \$2000 - \$1500 = \$500$$
$$\text{drop in salvage value: } \$10{,}000 - \$5000 = \$5000$$

The annual portion of the drop in salvage value is

$$A = (\$5000)(A/F, 10\%, 5)$$
$$(A/F, 10\%, 5) = 0.1638$$
$$A = (\$5000)(0.1638) = \$819/\text{year}$$

The cost of gas is

$$\left(\frac{50{,}000 \text{ mi}}{15 \frac{\text{mi}}{\text{gal}}}\right)\left(\frac{\$1.50}{\text{gal}}\right) = \$5000/\text{yr}$$

$$\text{EUAC per mile} = \frac{\$1000 + \$500 + \$819 + \$5000}{50{,}000 \text{ mi}}$$
$$= \boxed{\$0.14638/\text{mi}}$$

Since the reimbursement per mile was \$0.30 and since \$0.30 > \$0.14638, the reimbursement is adequate.

(b) Next, determine (with reimbursement) how many miles must the car be driven to break even.

If the car is driven M miles per year,

$$\left(\frac{\$0.30}{1 \text{ mi}}\right)M = (\$50{,}000)(A/P, 10\%, 5) + \$2500$$
$$+ \$2000 - (\$8000)(A/F, 10\%, 5)$$
$$+ \left(\frac{M}{15 \frac{\text{mi}}{\text{gal}}}\right)(\$1.50)$$

$$(A/P, 10\%, 5) = 0.2638$$
$$(A/F, 10\%, 5) = 0.1638$$
$$0.3M = (\$50{,}000)(0.2638) + \$2500 + \$2000$$
$$- (\$8000)(0.1638) + 0.1M$$
$$0.2M = \$16{,}379.60$$
$$M = \frac{\$16{,}379.60}{0.2 \frac{\$}{\text{mi}}} = \boxed{81{,}898 \text{ mi}}$$

22.

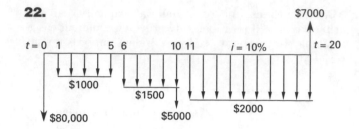

$$P_{\text{A}} = \$80{,}000 + (\$1000)(P/A, 10\%, 5)$$
$$+ (\$1500)(P/A, 10\%, 5)(P/F, 10\%, 5)$$
$$+ (\$2000)(P/A, 10\%, 10)(P/F, 10\%, 10)$$
$$+ (\$5000)(P/F, 10\%, 10)$$
$$- (\$7000)(P/F, 10\%, 20)$$

$$(P/A, 10\%, 5) = 3.7908$$
$$(P/F, 10\%, 5) = 0.6209$$
$$(P/A, 10\%, 10) = 6.1446$$
$$(P/F, 10\%, 10) = 0.3855$$
$$(P/F, 10\%, 10) = 0.3855$$
$$(P/F, 10\%, 20) = 0.1486$$

$$P_{\text{A}} = \$80{,}000 + (\$1000)(3.7908)$$
$$+ (\$1500)(3.7908)(0.6209)$$
$$+ (\$2000)(6.1446)(0.3855)$$
$$+ (\$5000)(0.3855) - (\$7000)(0.1486)$$
$$= \$92{,}946.15$$

Since the lives are different, compare by EUAC.

$$\text{EUAC}(A) = (\$92{,}946.14)(A/P, 10\%, 20)$$
$$= (\$92{,}946.14)(0.1175) = \$10{,}921$$

Similarly, evaluate alternative B.

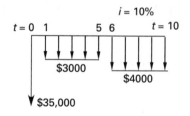

$$P_{\text{B}} = \$35{,}000 + (\$3000)(P/A, 10\%, 5)$$
$$+ (\$4000)(P/A, 10\%, 5)(P/F, 10\%, 5)$$

$$(P/A, 10\%, 5) = 3.7908$$
$$(P/F, 10\%, 5) = 0.6209$$

$$P_{\text{B}} = \$35{,}000 + (\$3000)(3.7908)$$
$$+ (\$4000)(3.7908)(0.6209)$$
$$= \$55{,}787.23$$

$$\text{EUAC}(B) = (\$55{,}787.23)(A/P, 10\%, 10)$$
$$= (\$55{,}787.23)(0.1627) = \$9077$$

Since $\text{EUAC}(B) < \text{EUAC}(A)$,

> Alternative B is economically superior.

23. For both cases, if the annual cost is compared with a total annual mileage of M,

$$A_\text{A} = \$0.25M$$
$$A_\text{B} = (\$30,000)(A/P, 10\%, 3) + \$0.14M$$
$$+ \$500 - (\$7200)(A/F, 10\%, 3)$$
$$(A/P, 10\%, 3) = 0.40211$$
$$(A/F, 10\%, 3) = 0.30211$$
$$A_\text{B} = (\$30,000)(0.40211) + \$0.14M + \$500$$
$$- (\$7200)(0.30211)$$
$$= \$12,063.30 + \$0.14M$$
$$+ \$500 - \$2175.19$$
$$= \$10,388.11 + \$0.14M$$

For an equal annual cost $A_\text{A} = A_\text{B}$,

$$\$0.25M = \$10,388.11 + \$0.14M$$

An annual mileage would be $M = 94,437$ mi.

For an annual mileage less than that, $A_\text{A} < A_\text{B}$.

> Plan A is economically superior until 94,437 mi annually is exceeded.

24. ((a) and (b))

Method A:

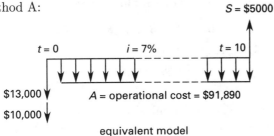

$$24 \text{ hours/day}$$
$$365 \text{ days/year}$$
$$\text{total of } (24)(365) = 8760 \text{ hours/year}$$
$$\$10.50 \text{ operational cost/hour}$$
$$\text{total of } (8760)(\$10.50) = \$91,890 \text{ operational cost/year}$$

$$A = \$91,980 + (\$23,000)(A/P, 7\%, 10)$$
$$- (\$5000)(A/F, 7\%, 10)$$
$$(A/P, 7\%, 10) = 0.14238$$
$$(A/F, 7\%, 10) = 0.07238$$
$$A = \$91,980 + (\$23,000)(0.14238)$$
$$- (\$5000)(0.07238)$$
$$= \$94,892.84/\text{yr}$$

Therefore, the uniform annual cost per ton each year will be

$$\frac{\$94,892.84}{50 \text{ ton}} = \boxed{\$1897.86}$$

Method B:

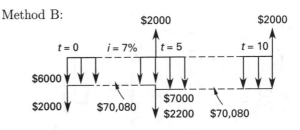

$$8760 \text{ hours/year}$$
$$\$8 \text{ operational cost/hour}$$
$$\text{total of } \$70,080 \text{ operational cost/year}$$

$$A = \$70,080 + (\$6000 + \$2000)(A/P, 7\%, 10)$$
$$+ (\$7000 + \$2200 - \$2000)(P/F, 7\%, 5)$$
$$\times (A/P, 7\%, 10)$$
$$- (\$2000)(A/F, 7\%, 10)$$
$$(A/P, 7\%, 10) = 0.1424$$
$$(A/F, 7\%, 10) = 0.07238$$
$$(P/F, 7\%, 5) = 0.7130$$
$$A = \$70,080 + (\$8000)(0.1424)$$
$$+ (\$7200)(0.7130)(0.1424)$$
$$- (\$2000)(0.07238)$$
$$= \$71,805.46/\text{yr}$$

Therefore, the uniform annual cost per ton each year will be

$$\frac{\$71,805.46}{20 \text{ ton}} = \boxed{\$3590.27}$$

tons/yr	cost of using A		cost of using B		cheapest
0–20	$94,893	(1x)	$71,805	(1x)	B
20–40	$94,893	(1x)	$143,610	(2x)	A
40–50	$94,893	(1x)	$215,415	(3x)	A
50–60	$189,786	(2x)	$215,415	(3x)	A
60–80	$189,786	(2x)	$287,220	(4x)	A

25.

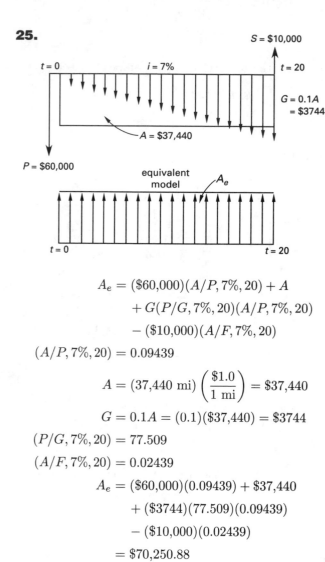

$$A_e = (\$60,000)(A/P, 7\%, 20) + A$$
$$+ G(P/G, 7\%, 20)(A/P, 7\%, 20)$$
$$- (\$10,000)(A/F, 7\%, 20)$$
$$(A/P, 7\%, 20) = 0.09439$$

$$A = (37,440 \text{ mi})\left(\frac{\$1.0}{1 \text{ mi}}\right) = \$37,440$$

$$G = 0.1A = (0.1)(\$37,440) = \$3744$$

$$(P/G, 7\%, 20) = 77.509$$
$$(A/F, 7\%, 20) = 0.02439$$
$$A_e = (\$60,000)(0.09439) + \$37,440$$
$$+ (\$3744)(77.509)(0.09439)$$
$$- (\$10,000)(0.02439)$$
$$= \$70,250.88$$

(a) With 80,000 passengers a year, the break-even fare per passenger would be

$$\text{fare} = \frac{A_e}{80,000} = \frac{\$70,250.88}{80,000} = \boxed{\$0.878/\text{passenger}}$$

(b)
$$\$0.878 = \$0.35 + G(A/G, 7\%, 20)$$
$$G = \frac{\$0.878 - \$0.35}{7.3163}$$
$$= \boxed{\$0.072 \text{ increase per year}}$$

(c) As in part (b), the subsidy should be

subsidy = cost − revenue
$$P = \$0.878 - (\$0.35 + (\$0.05)(A/G, 7\%, 20))$$
$$= \$0.878 - (\$0.35 + (\$0.05)(7.3163))$$
$$= \boxed{\$0.162}$$

26.
$$P(A) = (\$4800)(P/A, 12\%, 25)$$
$$= (\$4800)(7.8431)$$
$$= \$37,646.88$$

(4 quarters)(25 years) = 100 compounding periods
$$P(B) = (\$1200)(P/A, 3\%, 100)$$
$$= (\$1200)(31.5989)$$
$$= \$37,918.68$$

Alternative B is economically superior.

27.
$$\text{EUAC(A)} = (\$1500)(A/P, 15\%, 5) + \$800$$
$$= (\$1500)(0.2983) + \$800$$
$$= \$1247.45$$
$$\text{EUAC(B)} = (\$2500)(A/P, 15\%, 8) + \$650$$
$$= (\$2500)(0.2229) + \$650$$
$$= \$1207.25$$

Alternative B is economically superior.

28. The data given imply that both investments return 4% or more. However, the increased investment of $30,000 may not be cost effective. Do an incremental analysis.

incremental cost = $70,000 − $40,000 = $30,000
$$\text{incremental income} = \$5620 - \$4075 = \$1545$$
$$0 = -\$30,000 + (\$1545)(P/A, i\%, 20)$$
$$(P/A, i\%, 20) = 19.417$$
$$i \approx 0.25\% < 4\%$$

Alternative B is economically superior.

(The same conclusion could be reached by taking the present worths of both alternatives at 4%.)

29. (a) $P(A) = (-\$3000)(P/A, 25\%, 25) - \$20,000$
$$= (-\$3000)(3.9849) - \$20,000$$
$$= -\$31,954.70$$
$$P(B) = (-\$2500)(3.9849) - \$25,000$$
$$= -\$34,962.25$$

A is better.

(b) $CC(A) = \$20,000 + \dfrac{\$3000}{0.25} = \$32,000$

$CC(B) = \$25,000 + \dfrac{\$2500}{0.25} = \$35,000$

A is better.

(c) $EUAC(A) = (\$20,000)(A/P, 25\%, 25) + \$3000$
$= (\$20,000)(0.2509) + \$3000$
$= \$8018.00$
$EUAC(B) = (\$25,000)(0.2509) + \$2500$
$= \$8772.50$

A is better.

30. (a) $BV = \$18,000 - (5)\left(\dfrac{\$18,000 - \$2000}{8}\right)$

$= \boxed{\$8000}$

(b) With the sinking fund method, the basis is
$(\$18,000 - \$2000)(A/F, 8\%, 8)$
$= (\$18,000 - \$2000)(0.0940)$
$= \$1504$

$D_1 = (\$1504)(1.000) = \boxed{\$1504}$

$D_2 = (\$1504)(1.0800) = \boxed{\$1624}$

$D_3 = (\$1504)(1.0800)^2 = (\$1504)(1.1664)$

$= \boxed{\$1754}$

(c) $D_1 = \left(\tfrac{2}{8}\right)(\$18,000) = \boxed{\$4500}$

$D_2 = \left(\tfrac{2}{8}\right)(\$18,000 - \$4500) = \boxed{\$3375}$

$D_3 = \left(\tfrac{2}{8}\right)(\$18,000 - \$4500 - \$3375) = \boxed{\$2531}$

$D_4 = \left(\tfrac{2}{8}\right)(\$18,000 - \$4500 - \$3375 - \$2531)$

$- \boxed{\$1898}$

$D_5 = \left(\tfrac{2}{8}\right)(\$18,000 - \$4500 - \$3375$
$- \$2531 - \$1898)$

$= \boxed{\$1424}$

$BV = \$18,000 - \$4500 - \$3375 - \$2531$
$- \$1898 - \$1424$

$= \boxed{\$4272}$

31. (a) $T = \left(\tfrac{1}{2}\right)(8)(9) = 36$

$D_1 = \left(\tfrac{8}{36}\right)(\$14,000 - \$1800) = \boxed{\$2711}$

$\Delta D = \left(\tfrac{1}{36}\right)(\$14,000 - \$1800) = \$339$

$D_2 = \$2711 - \$339 = \boxed{\$2372}$

(b) $DR = (0.52)(\$2711)(P/A, 15\%, 8)$
$- (0.52)(\$339)(P/G, 15\%, 8)$
$= (0.52)(\$2711)(4.4873)$
$- (0.52)(\$339)(12.4807)$

$= \boxed{\$4125.74}$

32. The annual cost of doing nothing, $EUAC_{0\ ft}$, is in excess of \$550,000.

$EUAC_{5\ ft} = (\$600,000)(A/P, 12\%, 15)$
$+ \left(\tfrac{14}{50}\right)(\$600,000) + \left(\tfrac{8}{50}\right)(\$650,000)$
$+ \left(\tfrac{3}{50}\right)(\$700,000) + \left(\tfrac{1}{50}\right)(\$800,000)$
$= \$418,080$

$EUAC_{10\ ft} = (\$710,000)(A/P, 12\%, 15)$
$+ \left(\tfrac{8}{50}\right)(\$650,000) + \left(\tfrac{3}{50}\right)(\$700,000)$
$+ \left(\tfrac{1}{50}\right)(\$800,000)$
$= \$266,228$

$EUAC_{15\ ft} = (\$900,000)(A/P, 12\%, 15)$
$+ \left(\tfrac{3}{50}\right)(\$700,000) + \left(\tfrac{1}{50}\right)(\$800,000)$
$= \$190,120$

$EUAC_{20\ ft} = (\$1,000,000)(A/P, 12\%, 15)$
$+ \left(\tfrac{1}{50}\right)(\$800,000)$
$= \$162,800$

Build to 20 ft.

33. Assume replacement after 1 year.
$EUAC(1) = (\$30,000)(A/P, 12\%, 1) + \$18,700$
$= (\$30,000)(1.12) + \$18,700 = \$52,300$

Assume replacement after 2 years.
$EUAC(2) = (\$30,000)(A/P, 12\%, 2)$
$+ \$18,700 + (\$1200)(A/G, 12\%, 2)$
$= (\$30,000)(0.5917) + \$18,700$
$+ (\$1200)(0.4717) = \$37,017$

Assume replacement after 3 years.
$EUAC(3) = (\$30,000)(A/P, 12\%, 3)$
$+ \$18,700 + (\$1200)(A/G, 12\%, 3)$
$= (\$30,000)(0.4163) + \$18,700$
$+ (\$1200)(0.9246) = \$32,299$

Economics

Similarly, calculate to obtain the numbers in the following table.

years in service	EUAC
1	$52,300
2	$37,017
3	$32,299
4	$30,207
5	$29,152
6	$28,602
7	$28,335
8	$28,234
9	$28,240
10	$28,312

Replace after 8 yr.

34. Assume the head and horsepower data are already reflected in the hourly operating costs.

Let N = no. of hours operated each year.

$$\text{EUAC(A)} = (\$3600 + \$3050)(A/P, 12\%, 10)$$
$$- (\$200)(A/F, 12\%, 10) + 0.30N$$
$$= (\$6650)(0.1770)$$
$$- (\$200)(0.0570) + 0.30N$$
$$= 1165.65 + 0.30N$$
$$\text{EUAC(B)} = (\$2800 + \$5010)(A/P, 12\%, 10)$$
$$- (\$280)(A/F, 12\%, 10) + 0.10N$$
$$= (\$7810)(0.1770)$$
$$- (\$280)(0.0570) + 0.10N$$
$$= 1366.41 + 0.10N$$
$$\text{EUAC(A)} = \text{EUAC(B)}$$
$$1165.65 + 0.30N = 1366.41 + 0.10N$$
$$N = \boxed{1003.8 \text{ hr}}$$

35. (a) From Eq. 69.78,

$$\frac{T_2}{T_1} = 0.88 = 2^{-b}$$
$$\log 0.88 = -b \log 2$$
$$-0.0555 = -(0.3010)b$$
$$b = 0.1843$$
$$T_4 = (6)(4)^{-0.1843} = \boxed{4.65 \text{ wk}}$$

(b) From Eq. 69.79,

$$T_{6-14} = \left(\frac{6}{1 - 0.1843} \right)$$
$$\times \left(\left(14 + \tfrac{1}{2}\right)^{1-0.1843} - \left(6 - \tfrac{1}{2}\right)^{1-0.1843} \right)$$
$$= \left(\frac{6}{0.8157} \right) (8.857 - 4.017)$$
$$= \boxed{35.6 \text{ wk}}$$

36. First check that both alternatives have an ROR greater than the MARR. Work in thousands of dollars.

Evaluate alternative A.

$$P(A) = -\$120 + (\$15)(P/F, i\%, 5)$$
$$+ (\$57)(P/A, i\%, 5)(1 - 0.45)$$
$$+ \left(\frac{\$120 - \$15}{5} \right) (P/A, i\%, 5)(0.45)$$
$$= -\$120 + (\$15)(P/F, i\%, 5)$$
$$+ (\$40.8)(P/A, i\%, 5)$$

Try 15%.

$$P(A) = -\$120 + (\$15)(0.4972) + (\$40.8)(3.3522)$$
$$= \$24.23$$

Try 25%.

$$P(A) = -\$120 + (\$15)(0.3277) + (\$40.8)(2.6893)$$
$$= -\$5.36$$

Since $P(A)$ goes through 0,

$$(\text{ROR})_A > \text{MARR} = 15\%$$

Next, evaluate alternative B.

$$P(B) = -\$170 + (\$20)(P/F, i\%, 5)$$
$$+ (\$67)(P/A, i\%, 5)(1 - 0.45)$$
$$+ \left(\frac{\$170 - \$20}{5} \right) (P/A, i\%, 5)(0.45)$$
$$= -\$170 + (\$20)(P/F, i\%, 5)$$
$$+ (\$50.35)(P/A, i\%, 5)$$

Try 15%.

$$P(B) = -\$170 + (\$20)(0.4972) + (\$50.35)(3.352)$$
$$= \$8.72$$

Since $P(B) > 0$ and will decrease as i increases,

$$(ROR)_B > 15\%$$

ROR > MARR for both alternatives.

Do an incremental analysis to see if it is worthwhile to invest the extra $170 − $120 = $50.

$$P(B - A) = -\$50 + (\$20 - \$15)(P/F, i\%, 5)$$
$$+ (\$50.35 - \$40.8)(P/A, i\%, 5)$$

Try 15%.

$$P(B - A) = -\$50 + (\$5)(0.4972)$$
$$+ (\$9.55)(3.3522)$$
$$= -\$15.50$$

Since $P(B-A) < 0$ and would become more negative as i increases, the ROR of the added investment is $< 15\%$.

Alternative A is superior.

37. Use the year-end convention with the tax credit. The purchase is made at $t = 0$. However, the tax credit is received at $t = 1$ and must be multiplied by $(P/F, i\%, 1)$.

(Note that 0.0667 is actually ²/₃ of 10%.)

$$P = -\$300,000 + (0.0667)(\$300,000)(P/F, i\%, 1)$$
$$+ (\$90,000)(P/A, i\%, 5)(1 - 0.48)$$
$$+ \left(\frac{\$300,000 - \$50,000}{5}\right)(P/A, i\%, 5)(0.48)$$
$$+ (\$50,000)(P/F, i\%, 5)$$
$$= -\$300,000 + (\$20,000)(P/F, i\%, 1)$$
$$+ (\$46,800)(P/A, i\%, 5)$$
$$+ (\$24,000)(P/A, i\%, 5)$$
$$+ (\$50,000)(P/F, i\%, 5)$$

By trial and error,

i	P
10%	$17,616
15%	−$20,412
12%	$1448
13%	−$6142
$12\frac{1}{4}\%$	−$479

i is between 12% and $12^1/_4$%.

38. Assume loan payments are made at the end of each year. Find the annual payment.

$$\text{payment} = (\$2,500,000)(A/P, 12\%, 25)$$
$$= (\$2,500,000)(0.1275)$$
$$= \$318,750$$
$$\text{distributed profit} = (0.15)(\$2,500,000)$$
$$= \$375,000$$

After paying all expenses and distributing the 15% profit, the remainder should be 0.

$$0 = EUAC = \$20,000 + \$50,000 + \$200,000$$
$$+ \$375,000 + \$318,750 - \text{annual receipts}$$
$$- (\$500,000)(A/F, 15\%, 25)$$
$$= \$963,750 - \text{annual receipts}$$
$$- (\$500,000)(0.0047)$$

This calculation assumes $i = 15\%$, which equals the desired return. However, this assumption only affects the salvage calculation, and since the number is so small, the analysis is not sensitive to the assumption.

$$\text{annual receipts} = \$961,400$$

The average daily receipts are

$$\frac{\$961,400}{365} = \$2634$$

Use the expected value approach. The average occupancy is

$$(0.40)(0.05) + (0.30)(0.70) + (0.20)(0.75)$$
$$+ (0.10)(0.80) = 0.70$$

The average number of rooms occupied each night is

$$(0.70)(120 \text{ rooms}) = 84 \text{ rooms}$$

The minimum required average daily rate per room is

$$\frac{\$2634}{84} = \boxed{\$31.36}$$

39. $\quad \dfrac{\text{annual}}{\text{savings}} = \left(\dfrac{0.69 - 0.47}{1000}\right)(\$3,500,000) = \$770$

$$P = -\$7500 + (\$770 - \$200 - \$100)$$
$$\times (P/A, i\%, 25) = 0$$
$$(P/A, i\%, 25) = 15.957$$

Searching the tables and interpolating,

$$i \approx \boxed{3.75\%}$$

Economics

40. Work in millions of dollars.

$$P(\text{A}) = -(\$1.3)(1 - 0.48)(P/A, 15\%, 6)$$
$$= (\$1.3)(0.52)(3.7845)$$
$$= -\$2.56 \quad [\text{millions}]$$

Since this is an after-tax analysis and since the salvage value was mentioned, assume that the improvements can be depreciated.

Use straight line depreciation.

Evaluate alternative B.

$$D_j = \frac{2}{6} = 0.333$$
$$P(\text{B}) = -\$2 - (1 - 0.80)(\$1.3)(1 - 0.48)(P/A, 15\%, 6)$$
$$\quad - (\$0.15)(1 - 0.48)(P/A, 15\%, 6)$$
$$\quad + (\$0.333)(0.48)(P/A, 15\%, 6)$$
$$= -\$2 - (0.20)(\$1.3)(0.52)(3.7845)$$
$$\quad - (\$0.15)(0.52)(3.7845)$$
$$\quad + (\$0.333)(0.48)(3.7845)$$
$$= -\$2.206 \quad [\text{millions}]$$

Next, evaluate alternative C.

$$D_j = \frac{1.2}{3} = 0.4$$
$$P(\text{C}) = -(\$1.2)\big(1 + (P/F, 15\%, 3)\big)$$
$$\quad - (1 - 0.80)(\$1.3)(1 - 0.48)$$
$$\quad \times \big((P/F, 15\%, 1) + (P/F, 15\%, 4)\big)$$
$$\quad - (0.45)(\$1.3)(1 - 0.48)$$
$$\quad \times \big((P/F, 15\%, 2) + (P/F, 15\%, 5)\big)$$
$$\quad - (0.80)(\$1.3)(1 - 0.48)$$
$$\quad \times \big((P/F, 15\%, 3) + (P/F, 15\%, 6)\big)$$
$$\quad + (0.4)(\$0.48)(P/A, 15\%, 6)$$
$$= -(\$1.2)(1.6575)$$
$$\quad - (0.20)(\$1.3)(0.52)(0.8696 + 0.5718)$$
$$\quad - (0.45)(\$1.3)(0.52)(0.7561 + 0.4972)$$
$$\quad - (0.80)(\$1.3)(0.52)(0.6575 + 0.4323)$$
$$\quad + (0.4)(\$0.48)(3.7845)$$
$$= -\$2.436 \quad [\text{millions}]$$

> Alternative B is superior.

41. This is a replacement study. Since production capacity and efficiency are not a problem with the defender, the only question is when to bring in the challenger.

Since this is a before-tax problem, depreciation is not a factor, nor is book value.

The cost of keeping the defender one more year is

$$\text{EUAC(defender)} = \$200{,}000 + (0.15)(\$400{,}000)$$
$$= \$260{,}000$$

For the challenger,

$$\text{EUAC(challenger)}$$
$$= (\$800{,}000)(A/P, 15\%, 10) + \$40{,}000$$
$$\quad + (\$30{,}000)(A/G, 15\%, 10)$$
$$\quad - (\$100{,}000)(A/F, 15\%, 10)$$
$$= (\$800{,}000)(0.1993) + \$40{,}000$$
$$\quad + (\$30{,}000)(3.3832)$$
$$\quad - (\$100{,}000)(0.0493)$$
$$= \$296{,}006$$

Since the defender is cheaper, keep it. The same analysis next year will give identical answers. Therefore, keep the defender for the next 3 years, at which time the decision to buy the challenger will be automatic.

Having determined that it is less expensive to keep the defender than to maintain the challenger for 10 years, determine whether the challenger is less expensive if retired before 10 years.

If retired in 9 years,

$$\text{EUAC(challenger)} = (\$800{,}000)(A/P, 15\%, 9) + \$40{,}000$$
$$\quad + (\$30{,}000)(A/G, 15\%, 9)$$
$$\quad - (\$150{,}000)(A/F, 15\%, 9)$$
$$= (\$800{,}000)(0.2096)$$
$$\quad + \$40{,}000 + (\$30{,}000)(3.0922)$$
$$\quad - (\$150{,}000)(0.0596)$$
$$= \$291{,}506$$

Similar calculations yield the following results for all the retirement dates.

n	EUAC
10	$296,000
9	$291,506
8	$287,179
7	$283,214
6	$280,016
5	$278,419
4	$279,909
3	$288,013
2	$313,483
1	$360,000

Since none of these equivalent uniform annual costs is less than that of the defender, it is not economical to buy and keep the challenger for any length of time.

> Keep the defender.

70 Engineering Law

PRACTICE PROBLEMS

Company Ownership

1. List the different forms of company ownership.What are the advantages and disadvantages of each?

General Contracts

2. Define the requirements for a contract to be enforceable.

3. What standard features should a written contract include?

Consulting Fee Structure

4. Describe the ways a consulting fee can be structured.

5. What is a retainer fee?

SOLUTIONS

1. The three different forms of company ownership are the (1) sole proprietorship, (2) partnership, and (3) corporation.

A *sole proprietor* is his or her own boss. This satisfies the proprietor's ego and facilitates quick decisions, but unless the proprietor is trained in business, the company will usually operate without the benefit of expert or mitigating advice. The sole proprietor also personally assumes all the debts and liabilities of the company. A sole proprietorship is terminated upon the death of the proprietor.

A *partnership* increases the capitalization and the knowledge base beyond that of a proprietorship, but offers little else in the way of improvement. In fact, the partnership creates an additional disadvantage of one partner's possible irresponsible actions creating debts and liabilities for the remaining partners.

A *corporation* has sizable capitalization (provided by the stockholders) and a vast knowledge base (provided by the board of directors). It keeps the company and owner liability separate. It also survives the death of any employee, officer, or director. Its major disadvantage is the administrative work required to establish and maintain the corporate structure.

2. To be legal, a contract must contain an *offer*, some form of *consideration* (which does not have to be equitable), and an *acceptance* by both parties. To be enforceable, the contract must be voluntarily entered into, both parties must be competent and of legal age, and the contract cannot be for illegal activities.

3. A written contract will identify both parties, state the purpose of the contract and the obligations of the parties, give specific details of the obligations (including relevant dates and deadlines), specify the consideration, state the boilerplate clauses to clarify the contract terms, and leave places for signatures.

4. A consultant will either charge a fixed fee, a variable fee, or some combination of the two. A one-time fixed fee is known as a *lump-sum fee*. In a *cost plus fixed fee* contract, the consultant will also pass on certain costs to the client. Some charges to the client may depend on other factors, such as the salary of the consultant's staff, the number of days the consultant works, or the eventual cost or value of an item being designed by the consultant.

5. A *retainer* is a (usually) nonreturnable advance paid by the client to the consultant. While the retainer may be intended to cover the consultant's initial expenses until the first big billing is sent out, there does not need to be any rational basis for the retainer. Often, a small retainer is used by the consultant to qualify the client (i.e., to make sure the client is not just shopping around and getting free initial consultations) and as a security deposit (to make sure the client does not change consultants after work begins).

71 Engineering Ethics

PRACTICE PROBLEMS

(Note: Each problem has two parts. Determine whether the situation is (or can be) permitted legally. Then, determine whether the situation is permitted ethically.)

1. (a) Was it legal and/or ethical for an engineer to sign and seal plans that were not prepared by him or prepared under his responsible direction, supervision, or control?

(b) Was it legal and/or ethical for an engineer to sign and seal plans that were not prepared by him but were prepared under his responsible direction, supervision, and control?

2. Under what conditions would it be legal and/or ethical for an engineer to rely on the information (e.g., elevations and amounts of cuts and fills) furnished by a grading contractor?

3. Was it legal and/or ethical for an engineer to alter the soils report prepared by another engineer for his client?

4. Under what conditions would it be legal and/or ethical for an engineer to assign work called for in his contract to another engineer?

5. A licensed professional engineer was convicted of a felony totally unrelated to his consulting engineering practice.

(a) What actions would you recommend be taken by the state registration board?

(b) What actions would you recommend be taken by the professional or technical society (e.g., ASCE, ASME, IEEE, NSPE, etc.)?

6. An engineer came across some work of a predecessor. After verifying the validity and correctness of all assumptions and calculations, the engineer used the work. Under what conditions would such use be legal and/or ethical?

7. A building contractor made it a policy to provide cellular car telephones to the engineers of projects he was working on. Under what conditions could the engineers accept the telephones?

8. An engineer designed a tilt-up slab building for a client. The design engineer sent the design out to another engineer for checking. The checking engineer himself sent the plans to a concrete contractor for review. The concrete contractor made suggestions that were incorporated into the recommendations of the checking contractor. These recommendations were subsequently incorporated into the plans by the original design engineer. What steps must be taken to keep the design process legal and/or ethical?

9. A consulting engineer registered his corporation as "John Williams, P.E. and Associates, Inc." even though he had no associates. Under what conditions would this name be legal and/or ethical?

10. When it became known that a chemical plant was planning on producing a toxic product, an engineer wrote to the local newspaper condemning the action. Under what conditions would this action be legal and/or ethical?

11. An engineer signed a contract with a client. The fee the client agreed to pay was based on the engineer's estimate of time required. The engineer was able to complete the contract satisfactorily in half the time he expected. Under what conditions would it be legal and/or ethical for the engineer to keep the full fee?

12. After working on a project for a client, the engineer was asked by a competitor of the client to perform design services. Under what conditions would it be legal and/or ethical for the engineer to work for the competitor?

13. Two engineers submitted bids to a prospective client for a design project. The client told engineer A how much engineer B had bid and invited engineer A to beat the amount. Under what conditions could engineer A legally/ethically submit a lower bid?

14. A registered civil engineer specializing in well-drilling, irrigation pipelines, and farmhouse sanitary systems took a booth at a county fair located in a farming town. By a random drawing, the engineer's booth was located next to a hog-breeder's booth, complete with live (prize) hogs. The engineer gave away helium balloons with his name and phone number to all visitors to the booth. Did the engineer violate any laws/ethical guidelines?

15. While in a developing country supervising construction of a project he designed, an engineer discovered his client's project manager was treating local workers in an unsafe and inhuman (but for that country, legal) manner. When he objected, the client told the engineer to mind his own business. Later, the local workers asked the engineer to participate in a walkout and strike with them.

(a) What legal/ethical positions should the engineer take?

(b) Should it have made any difference if the engineer had or had not yet accepted any money from the client?

16. While working for a client, an engineer learns confidential knowledge of a proprietary production process being used by the client's chemical plant. The process is clearly destructive to the environment, but the client will not listen to the objections of the engineer. To inform the proper authorities will require the engineer to release information that he gained in confidence. Is it legal and/or ethical for the engineer to expose the client?

17. While working for an engineering design firm, an engineer was moonlighting as a soils engineer. At night, the engineer used the facilities of his employer to analyze and plot the results of soils tests. He then used his employer's computers and word processors to write his reports. The equipment, computers, and word processors would otherwise be unused. Under what conditions could the engineer's actions be considered legal and/or ethical?

SOLUTIONS

Introduction to the Answers

Case studies in law and ethics can be interpreted in many ways. The problems presented are simple thumbnail outlines. In most real cases, there will be more facts to influence a determination than are presented in the case scenarios. In some cases, a state may have specific laws affecting the determination; in other cases, prior case law will have been established.

The determination of whether an action is legal can be made in two ways. The obvious interpretation of an illegal action is one that violates a specific law or statute. An action can also be *found to be illegal* if it is judged in court to be a breach of a written, verbal, or implied contract. Both of these approaches are used in the following solutions.

These answers have been developed to teach legal and ethical principals. While being realistic, they are not necessarily based on actual incidents or prior case law.

1. (a) Stamping plans for someone else is illegal. The registration laws of all states permit a registered engineer to stamp/sign/seal only plans that were prepared by him personally or were prepared under his direction, supervision, or control. This is sometimes called being in *responsible charge*. The stamping/signing/sealing, for a fee or gratis, of plans produced by another person, whether that person is registered or not and whether that person is an engineer or not, is illegal.

(b) The act is illegal. An illegal act, being a concealed act, is intrinsically unethical. In addition, stamping/signing/sealing plans that have not been checked violates the rule contained in all ethical codes that requires an engineer to protect the public.

2. Unless the engineer and contractor worked together such that the engineer had personal knowledge that the information was correct, accepting the contractor's information is illegal. Not only would using unverified data violate the state's registration law (for the same reason that stamping/signing/sealing unverified plans in problem 1 was illegal), but the engineer's contract clause dealing with assignment of work to others would probably be violated.

The act is unethical. An illegal act, being a concealed act, is intrinsically unethical. In addition, using unverified data violates the rule contained in all ethical codes that requires an engineer to protect the client.

3. It is illegal to alter a report to bring it "more into line" with what the client wants unless the alterations represent actual, verified changed conditions. Even when the alterations are warranted, however, use of the unverified remainder of the report is a violation of the state registration law requiring an engineer only to stamp/sign/seal plans developed by or under him. Furthermore, this would be a case of fraudulent misrepresentation unless the originating engineer's name was removed from the report.

Unless the engineer who wrote the original report has given permission for the modification, altering the report would be unethical.

4. Assignment of engineering work is legal (1) if the engineer's contract permitted assignment, (2) all prerequisites (i.e., notifying the client) were met, and (3) the work was performed under the direction of another licensed engineer.

Assignment of work is ethical (1) if it is not illegal, (2) if it is done with the awareness of the client, and (3) if the assignor has determined that the assignee is competent in the area of the assignment.

5. (a) The registration laws of many states require a hearing to be held when a licensee is found guilty of unrelated, but nevertheless unforgivable, felonies (e.g., moral turpitude). The specific action (e.g., suspension, revocation of license, public censure, etc.) taken depends on the customs of the state's registration board.

(b) By convention, it is not the responsibility of technical and professional organizations to monitor or judge the personal actions of their members. Such organizations do not have the authority to discipline members (other than to revoke membership), nor are they immune from possible retaliatory libel/slander lawsuits if they publicly censure a member.

6. The action is legal because, by verifying all the assumptions and checking all the calculations, the engineer effectively does the work himself. Very few engineering procedures are truly original; the fact that someone else's effort guided the analysis does not make the action illegal.

The action is probably ethical, particularly if the client and the predecessor are aware of what has happened (although it is not necessary for the predecessor to be told). It is unclear to what extent (if at all) the predecessor should be credited. There could be other extenuating circumstances that would make referring to the original work unethical.

7. Gifts, per se, are not illegal. Unless accepting the phones violates some public policy or other law, or is in some way an illegal bribe to induce the engineer to favor the contractor, it is probably legal to accept the phones.

Ethical acceptance of the phones requires (among other considerations) that (1) the phones be required for the job, (2) the phones be used for business only, (3) the phones are returned to the contractor at the end of the job, and (4) the contractor's and engineer's clients know and approve of the transaction.

8. There are two issues here: (1) the assignment and (2) the incorporation of work done by another. To avoid a breach, the contracts of both the design and checking engineers must permit the assignments. To avoid a violation of the state registration law requiring engineers to be in responsible charge of the work they stamp/sign/seal, both the design and checking engineers must verify the validity of the changes.

To be ethical, the actions must be legal and all parties (including the design engineer's client) must be aware that the assignments have occurred and that the changes have been made.

9. The name is probably legal. If the name was accepted by the state's corporation registrar, it is a legally formatted name. However, some states have engineering registration laws that restrict what an engineering corporation may be named. For example, all individuals listed in the name (e.g., "Cooper, Williams, and Somerset—Consulting Engineers") may need to be registered. Whether having "Associates" in the name is legal depends on the state.

Using the name is unethical. It misleads the public and represents unfair competition with other engineers running one-person offices.

10. Unless the engineeer's accusation is known to be false or exaggerated, or the engineer has signed an agreement (e.g., confidentiality, non-disclosure, etc.) with his employer forbidding the disclosure, the letter to the newspaper is probably not illegal.

The action is probably unethical. (If the letter to the newspaper is unsigned it is a concealed action and is definitely unethical.) While whistle-blowing to protect the public is implicitly an ethical procedure, unless the engineer is reasonably certain that manufacture of the toxic product represents a hazard to the public, he has a responsibility to the employer. Even then, the engineer should exhaust all possible remedies to render the manufacture nonhazardous before blowing the whistle. Of course, the engineer may quit working for the chemical plant and be as critical as the law allows without violating engineer-employer ethical considerations.

11. Unless the engineer's payment was explicitly linked in the contract to the amount of time spent on the job, taking the full fee would not be illegal or a breach of the contract.

An engineer has an obligation to be fair in estimates of cost, particularly when the engineer knows no one else is providing a competitive bid. Taking the full fee would be ethical if the original estimate was arrived at logically and was not meant to deceive or take advantage of the client. An engineer is permitted to take advantage of economies of scale, state-of-the-art techniques, and break-through methods. (Similarly, when a job costs more than the estimate, the engineer may be ethically bound to stick with the original estimate.)

12. In the absence of a nondisclosure or noncompetition agreement or similar contract clause, working for the competitor is probably legal.

Working for both clients is unethical. Even if both clients know and approve, it is difficult for the engineer not to "cross-pollinate" his work and improve one client's position with knowledge and insights gained at the expense of the other client. Furthermore, the mere appearance of a conflict of interest of this type is a violation of most ethical codes.

13. In the absence of a sealed-bid provision mandated by a public agency and requiring all bids to be opened at once (and the award going to the lowest bidder), the action is probably legal.

It is unethical for an engineer to undercut the price of another engineer. Not only does this violate a standard of behavior expected of professionals, it unfairly benefits one engineer because a similar chance is not given to the other engineer. Even if both engineers are bidding openly against each other (in an auction format), the client must understand that a lower price means reduced service. Each reduction in price is an incentive to the engineer to reduce the quality or quantity of service.

14. It is generally legal for an engineer to advertise his services. Unless the state has relevant laws, the engineer probably did not engage in illegal actions.

Most ethical codes prohibit unprofessional advertising. The unfortunate location due to a random drawing might be excusable, but the engineer should probably refuse to participate. In any case, the balloons are a form of unprofessional advertising, and as such, are unethical.

15. (a) As stated in the scenario statement, the client's actions are legal for that country. The fact that the actions might be illegal in another country is irrelevant. Whether or not the strike is legal depends on the industry and the laws of the land. Some or all occupations (e.g., police and medical personnel) may be forbidden to strike. Assuming the engineer's contract does not prohibit his own participation, the engineer should determine the legality of the strike before making a decision to participate.

If the client's actions are inhuman, the engineer has an ethical obligation to withdraw from the project. Not doing so associates the profession of engineering with human misery.

(b) The engineer has a contract to complete the project for the client. (It is assumed that the contract between the engineer and client was negotiated in good faith, that the engineer had no knowledge of the work conditions prior to signing, and that the client did not falsely induce the engineer to sign.) Regardless of the reason for withdrawing, the engineer is breaching his contract. In the absence of proof of illegal actions by the client, withdrawal by the engineer requires a return of all fees received. Even if no fees have been received, withdrawal exposes the engineer to other delay-related claims by the client.

16. A contract for an illegal action cannot be enforced. Therefore, any confidentiality or nondisclosure agreement that the engineer has signed is unenforceable if the production process is illegal, uses illegal chemicals, or violates laws protecting the environment. If the production process is not illegal, it is not legal for the engineer to expose the client.

Society and the public are at the top of the hierarchy of an engineer's responsibilities. Obligations to the public take precedence over the client. If the production process is illegal, it would be ethical to expose the client.

17. It is probably legal for the engineer to use the facilities, particularly if the employer is aware of the use. (The question of whether the engineer is trespassing or violating a company policy cannot be answered without additional information.)

Moonlighting, in general, is not ethical. Most ethical codes prohibit running an engineering consulting business while receiving a salary from another employer. The rationale is that the moonlighting engineer is able to offer services at a much lower price, placing other consulting engineers at a competitive disadvantage. The use of someone else's equipment only compounds the problem. Since the engineer does not have to pay for using the equipment, he does not have to charge his clients for it. This places him at an unfair competitive advantage compared to other consultants who have invested heavily in equipment.

72 Engineering Registration in the United States

PRACTICE PROBLEMS

Chapter 72 of the *Mechanical Engineering Reference Manual* contains no practice problems.